Energy Conversion Factors

To convert from:	To:	Multiply by:
calories	ergs	4.184×10^7
calories	joules	4.184
calories	kilowatt-hours	1.162×10^{-6}
electron volts	kcal mol^{-1}	23.058
electron volts	ergs	1.602×10^{-12}
electron volts	joules	1.602×10^{-19}
electron volts	kilojoules mol^{-1}	96.474
ergs	calories	2.390×10^{-8}
ergs	joules	10^{-7}
ergs molecule^{-1}	joules mol^{-1}	6.022×10^{16}
ergs molecule^{-1}	kcal mol^{-1}	1.439×10^{13}
joules	calories	0.2390
joules	ergs	10^7
joules	kilowatt-hours	2.778×10^{-7}
joules mol^{-1}	ergs molecule^{-1}	1.661×10^{-17}
joules mol^{-1}	kcal mol^{-1}	2.390×10^{-4}
kilowatt-hours	calories	8.606×10^5
kilowatt-hours	joules	3.600×10^6
kcal mol^{-1}	electron volts	0.04337
kcal mol^{-1}	ergs molecule^{-1}	6.949×10^{-14}
kcal mol^{-1}	joules mol^{-1}	4.184×10^3
kilojoules mol^{-1}	electron volts	0.01036
kilojoules mol^{-1}	ergs molecule^{-1}	1.661×10^{-14}
kilojoules mol^{-1}	kcal mol^{-1}	0.2390

Miscellaneous Conversions and Abbreviations

Length

1 ångstrom $= 10^{-10}$ m $= 10^{-8}$ cm $= 0.1$ nm
1 inch $= 2.54$ cm
1 foot $= 30.48$ cm
1 mile $= 5280$ ft $= 1609$ m

Mass

1 pound $= 453.6$ g
1 kg $= 2.205$ pounds

Energy

1 erg $= 1$ g cm^2 s^{-2}
1 joule $= 1$ kg m^2 s^{-2}

Force

1 newton $= 1$ kg m s^{-2} $= 10^5$ dyn
1 dyne $= 1$ g cm s^{-2} $= 10^{-5}$ N

Pressure

1 atmosphere $= 760$ mmHg (Torr.) $= 14.70$ lb in^{-2}
  $= 1.013 \times 10^6$ dyn cm^{-2}
  $= 1.013 \times 10^5$ newton m^{-2}
  $= 1.033 \times 10^4$ kg-force m^{-2}
  $= 1.013 \times 10^5$ pascals
1 newton m^{-2} $= 1$ pascal

Power

1 watt $= 1$ J s^{-1} $= 1$ VA

Volume

1 gallon $= 3.785$ L

Viscosity

1 poise (P) $= 1$ g cm^{-1} s^{-1} $= 10^{-1}$ Pa s
1 pascal sec (Pa s) $= 1$ kg m^{-1} s^{-1} $= 10$ P
1 mPa s $= 1$ cP

Å =	ångstrom	P =	poise
A =	ampere	rad =	radian
cal =	calorie	V =	volt
hertz (Hz) =	cycles per second	W =	watt
hr =	hour	y =	year

Physical Chemistry
Principles and Applications in Biological Sciences

FOURTH EDITION

Physical Chemistry
Principles and Applications in Biological Sciences

Ignacio Tinoco, Jr.
University of California, Berkeley

Kenneth Sauer
University of California, Berkeley

James C. Wang
Harvard University

Joseph D. Puglisi
Stanford University

Prentice Hall
Upper Saddle River, New Jersey, 07458

Editor in Chief: *John Challice*
Senior Marketing Manager: *Steve Sartori*
Executive Managing Editor: *Kathleen Schiaparelli*
Assistant Managing Editor: *Beth Sturla*
Production Supervision/Composition: *G&S Typesetters, Inc.*
Assistant Editor: *Kristen Kaiser*
Art Director: *Jayne Conte*
Cover Designer: *Bruce Kenselaar*
Manufacturing Manager: *Trudy Pisciotti*
Assistant Manufacturing Manager: *Michael Bell*
Managing Editor, Audio/Video Assets: *Grace Hazeldine*
Art Editor: *Shannon Sims*
Art Studio: *Prepare, Inc.*
Editorial Assistant: *Eliana Ortiz*
Vice President of Production and Manufacturing: *David W. Riccardi*

© 2002 by Prentice-Hall, Inc.
Upper Saddle River, New Jersey 07458

Printed in the United States of America
10 9 8 7 6 5 4 3

Reprinted with corrections July, 2003.

ISBN 0-13-095943-X

About the Cover: The cover shows the structure of the dimeric protein BmrR from *Bacillus subtilis* (purple and red) bound to its target DNA sequence (gold) and a drug called TPSb (bronze). Such protein-DNA interactions are common and in fact form the basis for regulation of transcription in both prokaryotes and eukaryotes. In the presence of the drug, the protein dimer distorts the DNA by unwinding the central bases of the helix and inducing an approximately 50° bend. This in turn allows this segment of DNA to be transcribed. Physical biochemists use many of the techniques described in this book to analyze and understand such intermolecular interactions and the structural features of these biological systems. Zheleznova-Heldwein, E. E., Brennan, R. G. (2001) "Crystal Structure of the Transcription Activator BmrR Bound to DNA and a Drug" Nature 409:378-382. Structure rendered by Jonathan Parrish, University of Alberta.

Pearson Education Ltd., *London*
Pearson Education Australia Pty., Limited, *Sydney*
Pearson Education Singapore, Pte. Ltd.
Pearson Education North Asia Ltd., *Hong Kong*
Pearson Education Canada, Ltd., *Toronto*
Pearson Educación de Mexico, S.A. de C.V.
Pearson Education—Japan, *Tokyo*
Pearson Education Malaysia, Pte. Ltd.

CONTENTS

CHAPTER 1

Introduction

CHAPTER 2

The First Law: Energy Is Conserved

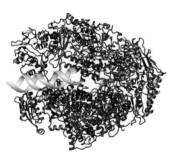

CHAPTER 3

The Second Law: The Entropy of the Universe Increases

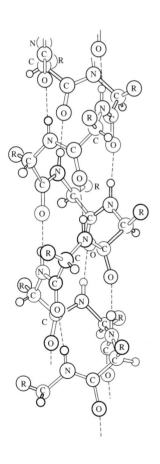

CHAPTER 4

Free Energy and Chemical Equilibria

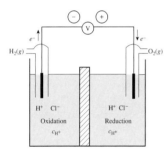

CHAPTER 5

Free Energy and Physical Equilibria

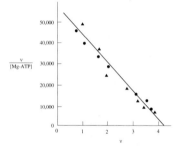

C H A P T E R 6

Molecular Motion and Transport Properties

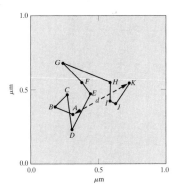

C H A P T E R **7**

Kinetics: Rates of Chemical Reactions

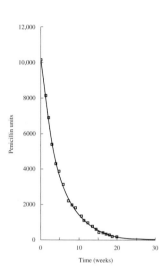

C H A P T E R 8

Enzyme Kinetics

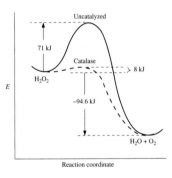

C H A P T E R 9

Molecular Structures and Interactions: Theory

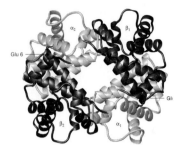

C H A P T E R 10

Molecular Structures and Interactions: Spectroscopy

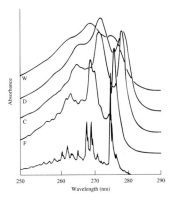

CHAPTER 11

Molecular Distributions and Statistical Thermodynamics

CHAPTER 12

Macromolecular
Structure and
X-Ray Diffraction

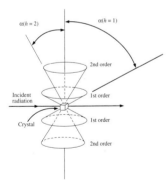

PREFACE

There is a deep sense of pleasure to be experienced when the patterns and symmetry of nature are revealed. Physical chemistry provides the methods to discover and understand these patterns. We think that not only is it important to learn and apply physical chemistry to biological problems, it may even be fun. In this book, we have tried to capture some of the excitement of making new discoveries and finding answers to fundamental questions.

This is not an encyclopedia of physical chemistry. Rather, we have written this text specifically with the life-science student in mind. We present a streamlined treatment that covers the core aspects of biophysical chemistry (thermodynamics and kinetics as well as quantum mechanics, spectroscopy, and X-ray diffraction), which are of great importance to students of biology and biochemistry. Essentially all applications of the concepts are to systems of interest to life-science students; nearly all the problems apply to life-science examples.

For this fourth edition we are joined by Joseph Puglisi, a new, young author who strengthens the structural biology content of the book. We have also tried to make the book more reader-friendly. In particular, we omit fewer steps in the explanations to make the material more understandable, and we have followed the many helpful and specific recommendations of our reviewers to improve the writing throughout. Important new topics, such as single-molecule thermodynamics, kinetics, and spectroscopy, are introduced. Subjects that have become less pertinent to current biophysical chemistry have been deleted or de-emphasized. Reference lists for each chapter have been updated. However, the format and organization of the book is essentially unchanged.

Chapter 1 introduces representative areas of active current research in biophysical chemistry and molecular biology: the human genome, the transfer of genetic information from DNA to RNA to protein, ion channels, and cell-to-cell communication. We encourage students to read the current literature to see how the vocabulary and concepts of physical chemistry are used in solving biological problems.

Chapters 2 through 5 cover the laws of thermodynamics and their applications to chemical reactions and physical processes. Essentially all of the examples and problems deal with biochemical and biological systems. For example, after defining work as a force multiplied by the distance moved (the displacement), we discuss the experimental measurement of the work necessary to stretch a single DNA molecule from its random-coiled form to an extended rod. Molecular interpretations of energies and entropies are emphasized in each of the chapters. Chapter 4, "Free Energy and Chemical Equilibria," now starts with the application of the chemical potential to chemical reactions. We think that this will make it easier to understand the logic relating activities and equilibrium constants to free energy. Binding of ligands and equilibria between phases are described in chapter 5, "Free Energy and Physical Equilibria." We discuss in detail the allosteric effect and the cooperative binding of oxygen by hemoglobin. We also describe the formation of lipid monolayers, lipid bilayers, and micelles, and their structures are compared to biological membranes.

Chapters 6 through 8 cover molecular motion and chemical kinetics. Chapter 6, "Molecular Motion and Transport Properties," starts with the Brownian motion on an aqueous surface of a single lipid molecule labeled with a fluorescent dye. The random motion of the molecule can be followed to test Einstein's equation relating average distance traveled by a single molecule to a bulk diffusion coefficient. Following this direct experimental demonstration of thermal motion of a molecule, we introduce the kinetic theory of gases and discuss transport properties (diffusion, sedimentation, and electrophoresis) of macromolecules. The next two chapters deal with general chemical kinetics and enzyme kinetics. New topics include Marcus's theory of charge-transfer reactions, allosteric effects in enzyme kinetics, and single-molecule enzyme kinetics.

Chapter 9, "Molecular Structures and Interactions: Theory," has been rearranged to begin with the origins of the quantum theory, continue through quantum mechanics of simple models, and, finally, discuss the semi-empirical methods applied to macromolecules. This logical progression should make it easier for students to understand and appreciate the applications of quantum mechanics to macromolecular structure. In chapter 10 we emphasize absorption, fluorescence, and nuclear magnetic resonance—the spectroscopic methods most used in structural biology.

In chapter 11, "Molecular Distributions and Statistical Thermodynamics," we present a detailed discussion of the effect of cooperativity or anticooperativity on the binding of successive ligands.

Chapter 12 discusses X-ray diffraction, electron microscopy, and scanning microscopies (such as atomic force microscopy), and emphasizes how structures are determined experimentally. We describe the many methods used to solve the phase problem in X-ray diffraction—including MAD, multiwavelength anomalous diffraction.

The problems have been revised and checked for clarity and the answers in the back of the book and in the solutions manual have been checked for accuracy. We thank Christopher Ackerson, Ruben Gonzalez, Michael Sykes, and Anne Roberts for checking the problems.

We are gratified by the number of faculty who have elected to use this book over the many years since it was first published. We are also grateful for the many students and faculty who have given us their thoughts and impressions. Such feedback has helped improve the book from edition to edition. We are particularly grateful to those of our colleagues who commented on the third edition, reviewed the manuscript for this edition, and checked our manuscript for accuracy: Fritz Allen, University of Minnesota; Carey Bagdassarian, College of William and Mary; Wallace Brey, University of Florida; Mark Britt, Baylor University; Kuang Yu Chen, Rutgers University; Gerald S. Harbison, University of Nebraska–Lincoln; Roger Koeppel, University of Arkansas; Philip Reiger, Brown University; Gianluigi Veglia, University of Minnesota; and Danny Yeager, Texas A&M University.

We welcome your comments.

Ignacio Tinoco, Jr.
INTinoco@lbl.gov
Kenneth Sauer
James C. Wang
Joseph D. Puglisi

About the Authors

Ignacio Tinoco was an undergraduate at the University of New Mexico, a graduate student at the University of Wisconsin, and a postdoctoral fellow at Yale. He then went to the University of California, Berkeley, where he has remained. His research interest has been on the structures of nucleic acids, particularly RNA. He was chairman of the Department of Energy committee that recommended in 1987 a major initiative to sequence the human genome. His present research is on unfolding single RNA molecules by force.

Kenneth Sauer grew up in Cleveland, Ohio, and received his A.B. in chemistry from Oberlin College. Following his Ph.D. studies in gas-phase physical chemistry at Harvard, he spent three years teaching at the American University of Beirut, Lebanon. A postdoctoral opportunity to learn from Melvin Calvin about photosynthesis in plants led him to the University of California, Berkeley, where he has been since 1960. Teaching general chemistry and biophysical chemistry in the Chemistry Department has complemented research in the Physical Biosciences Division of the Lawrence Berkeley National Lab involving spectroscopic studies of photosynthetic light reactions and their role in water oxidation. His other activities include reading, renaissance and baroque choral music, canoeing, and exploring the Sierra Nevada with his family and friends.

James C. Wang was on the faculty of the University of California, Berkeley, from 1966 to 1977. He then joined the faculty of Harvard University, where he is presently Mallinckrodt Professor of Biochemistry and Molecular Biology. His research focuses on DNA and enzymes that act on DNA, especially a class of enzymes known as DNA topoisomerases. He has taught courses in biophysical chemistry and molecular biology and has published over 200 research articles. He is a member of Academia Sinica, the American Academy of Arts and Sciences, and the U.S. National Academy of Sciences.

Joseph Puglisi was born and raised in New Jersey. He received his B.A. in chemistry from The Johns Hopkins University in 1984 and his Ph.D. from the University of California, Berkeley, in 1989. He has studied and taught in Strasbourg, Boston, and Santa Cruz, and is currently professor of structural biology at Stanford University. His research interests are in the structure and mechanism of the ribosome and the use of NMR spectroscopy to study RNA structure. He has been a Dreyfus Scholar, Sloan Scholar, and Packard Fellow.

Physical Chemistry
Principles and Applications in Biological Sciences

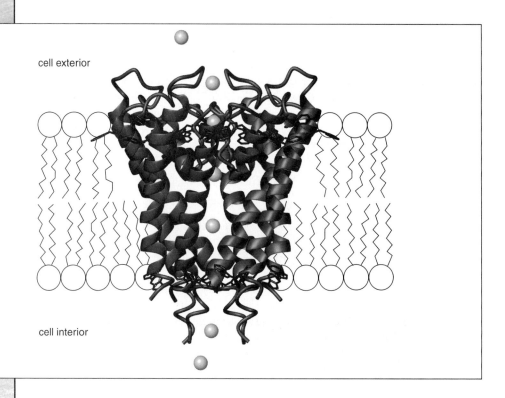

cell exterior

cell interior

Introduction

Physical chemistry is everywhere. Physical chemical principles are basic to the methods used to determine the sequence of the human genome, obtain atomic resolution structures of proteins and nucleic acids, and learn how biochemicals react and interact to make a cell function.

Physical chemistry is a set of principles and experimental methods for exploring chemical systems. The power of physical chemistry lies in its generality. The principles described in this book can be applied to systems as large as the cosmos and as small as an individual atom. Physical chemistry has been especially powerful in understanding fundamental biological processes. In the following chapters, we will present the principles of thermodynamics, transport properties, kinetics, quantum mechanics and molecular interactions, spectroscopy, and scattering and diffraction. We will also discuss various experimentally measurable properties such as enthalpy, electrophoretic mobility, light absorption, and X-ray diffraction. All these experimental and theoretical methods give useful information about the part of the universe in which you are interested. We emphasize the molecular interpretation of these methods and stress biochemical and biological applications. By learning the principles behind the methods, you will be able to judge the conclusions obtained from them. This is the first step in inventing new methods or discovering new concepts.

First, a quick tour of the book. Chapters 2 through 5 cover the fundamentals of thermodynamics and their applications to chemical reactions and physical processes. Because these chapters review material usually covered in beginning chemistry courses, we emphasize the applications to biological macromolecules. Chapter 6 covers transport properties and describes the effect of sizes and shapes of molecules on their motions in gases, liquids, and gels. The driving forces for molecular motion are either random thermal forces that cause diffusion or the directed forces in sedimentation, flow, and electrophoresis. Chapter 7 describes general kinetics, and chapter 8 concentrates on the kinetics of enzyme-catalyzed reactions. Chapter 9, which deals with molecular structures and intermolecular interactions, introduces the quantum mechanical principles necessary for understanding bonding and spectroscopy, and describes calculations of protein and nucleic acid conformations using classical force fields (Coulomb's law, van der Waals' potential). Chapter 10 includes the main spectroscopic methods used for studying molecules in solution: ultraviolet, visible, and infrared absorption; fluorescence emission; circular dichroism and optical rotatory dispersion; and nuclear magnetic resonance. Chapter 11 introduces the principles of statistical thermodynamics and describes their biochemical applications. Topics such as helix-coil transitions in polypeptides (proteins) and polynucleotides (nucleic acids) and the binding of small molecules to macromolecules are emphasized. Chapter 12 starts with the scattering of electromagnetic radiation from one electron and proceeds through the diffraction of X rays by crystals. Scanning microscope methods are introduced. The appendix contains numerical data used throughout the book, unit conversion tables, and the structures of many of the biological molecules mentioned in the text.

We encourage you to consult other books for background information and further depth of coverage. Standard physical chemistry texts offer applications of physical chemistry to other areas. Biochemistry and molecular biology texts can provide specific information about such areas as protein and nucleic acid structures, enzyme mechanisms, and metabolic pathways. Finally, a good physics textbook is useful for learning or reviewing the fundamentals of forces, charges, electromagnetic fields, and energy. A list of such books is given at the end of this chapter.

In the following sections, we highlight several important biological problems that physical chemistry can address. These examples are meant to give you an overview of how physical chemistry is applied in the biological sciences. Read them for pleasure, without trying to memorize them. Our aim is simply to illustrate some current research from the scientific literature and to point out the principles and methods that are used. We hope to motivate you to learn the material discussed in the following chapters. Many articles in journals such as *Nature* or *Science* apply the methods and concepts described in this book. Read such an article to learn how the book will improve your understanding.

The Human Genome and Beyond

The *Human Genome Project* has nearly determined the complete sequence of all 3 billion (3×10^9) base pairs that make up the genetic information of humans—the human genome (see *Nature* 2001 409:860-921). Genes are sequences of base pairs in double-stranded DNA. In human sperm the DNA is packaged in 23 chromosomes: 22 autosomes plus a male Y chromosome or a female X chromosome. In human eggs the DNA is packaged in 22 autosomes plus a female X chromosome. Thus, each of us acquires 23 pairs of chromosomes; the XX pair makes us female, the XY male. Surprisingly, there are at most 40,000 genes, only twice that of fruit flies. Less than 5% of the human genome consists of genes. The remaining DNA may be structural regions that fold the DNA into a required three-dimensional shape or may be sequences that have lost their function during evolution. Although long regions of highly repetitive sequences are useful for distinguishing human DNA from that of other species, they have no known biological functions. Determining the precise sequence of 3 billion nucleotides is a heroic task, which has been made possible by the use of fluorescently labeled DNA analogs. These analogs allow the sequence to be read using automated equipment. Also, DNA sequences have been placed on silicon chips and glass slides. Using the principles of Watson–Crick base pairing, scientists can rapidly identify changes in DNA sequences. These "genes on a chip" are revolutionizing the way that genes and gene expression are analyzed. Many times in science, fundamental advances are allowed by improvements in instrumentation to measure physical properties.

Once the sequence of an organism's genome has been determined, a difficult task begins. What does the string of four different letters of the DNA alphabet (A, C, G, T) mean? The DNA sequence is first transcribed into the RNA alphabet (A, C, G, U). The RNA is then translated into a protein sequence of 20 amino acids in a three-letter code.* As there are 4^3 (64) three-letter words with an alphabet of four letters, the code must be redundant. In the genetic dictionary, most amino acids are coded by two or four different words. Three amino acids have six words each (arginine, leucine, and serine), and two have only one word (tryptophan, methionine). Three of the words do not code for amino acids but instead signal for protein synthesis to stop: UAA, UAG, UGA. One word, AUG, codes for the start of protein synthesis (it

---

*The structures and names of the nucleic acid bases and the protein amino acids are given in table A.9 in the appendix.

also codes for methionine). Sequences before the starting AUG and after the terminating UAA, UAG, or UGA control and regulate the synthesis of the protein. Using the known genetic code, scientists can predict the sequence for a protein that is coded by a given gene. What does this protein do? What does it look like?

Physical chemistry provides the principles that allow bioinformatics scientists to make sense of the vast DNA sequence data of a genome. The protein sequence predicted from a gene is first compared to known protein sequences. If the protein is an essential part of some biochemical process that is common to many or all organisms, related proteins have likely been studied. Computer sequence comparisons establish the relationship between a novel protein and known proteins. The sequences of two proteins of similar biological function from different organisms are almost never identical. However, different protein sequences can adopt similar three-dimensional structures to perform similar functions. Different amino acids can have similar physical properties. For example, both isoleucine and valine have greasy aliphatic side chains and can often be exchanged for each other in a protein with little effect on its activity. Likewise, negatively charged amino acids (aspartic acid or glutamic acid) can often be swapped, and so on. Using this type of logic, computer programs can sometimes predict the function of the unknown gene by its relation to a known protein.

The weakness of this approach is obvious. You require the sequence of a known protein with which to compare the new gene. In addition, what truly determines the function of a protein is not exactly the sequence of amino acids but rather how these amino acids fold into a three-dimensional structure that can perform a specific function—for example, catalysis of an enzymatic reaction. Biophysical chemists can determine the three-dimensional structures of biological macromolecules, using methods described in this book. Unfortunately, the rate at which structures can be determined lags behind that of sequencing a gene. Nonetheless, comparisons of protein structures often reveal similarities that simple protein-sequence comparisons miss. Triosephosphate isomerase is a protein involved in metabolism, and it has a barrel-like three-dimensional structure. This structure is a rather common motif in proteins, but sequence comparisons by computer can rarely identify its presence.

Determining the three-dimensional structure of a protein would be easy if it could be predicted from its sequence. A protein's amino acid sequence contains the physical characteristics that determine the most stable three-dimensional fold. Biophysical chemists have shown that proteins almost always adopt the most stable three-dimensional structure as determined by the principles of thermodynamics. Thus, physical chemistry provides the framework to predict protein structure. However, predicting the most stable three-dimensional structure of a protein is a very difficult task because a large number of relatively weak interactions stabilize its structure (chapter 9). Precise treatment of these interactions is impossible, so biophysicists and computation biologists use a number of approximations to calculate a protein structure from its sequence (for an example, see Xia et al. 2000). This is a valid approach to many complex biological problems. How can a scientist know whether a computer program is actually working? Well, she could try it on a sequence of a protein of known structure. But this is of course biased, for our scientist already knows the answer. Scientists in this field in fact resort to friendly competitions. They are asked to predict the structures of proteins whose structures are not known at the beginning of the competition but will be revealed by the end. This provides an unbiased test of various algorithms. This example shows a glimpse of the human side of the scientific process. Although current algorithms cannot predict the structures of protein to the same precision as experimental methods, they are improving. Computer prediction of protein folding and RNA folding is now a highly active area of biophysical research.

Transcription and Translation

Genetic information must be faithfully transmitted from DNA to messenger RNA to protein. Copying DNA to RNA is called *transcription;* reading RNA to protein is called *translation.* Two central macromolecular machines are responsible for these processes: RNA polymerases transcribe RNA, and the ribosome translates proteins. In both systems a series of repetitive tasks must be performed with *high fidelity.* These machines must be directional, because they copy information in only one direction. The machines are processive, in that once they start the process of transcription or translation, they continue through hundreds or even thousands of steps of the process. Finally, these biological processes are highly regulated. Associated factors determine when, where, and how rapidly these processes begin and end. Physical methods have provided great insights into how transcription and translation occur.

The process of transcription was first investigated in simple organisms such as bacteria. The protein that catalyzes transcription consists of only one or a few polypeptide chains. In contrast, in eukaryotic organisms, the RNA polymerase enzyme consists of ten or more polypeptide chains, reflecting the higher degree of regulation in higher organisms. Transcription begins at specific signals in the DNA called *promoters.* These DNA sequences bind specific protein *transcription factors* that enhance or prevent transcription. This is an essential feature in the regulation of gene expression. The activity of these transcription factors can be affected by attaching a phosphate group to the protein or by binding of a small molecule cofactor. The classic example is a protein that binds both to small sugars and to DNA, like the lac repressor (Lewin 1999). These *DNA-binding proteins* recognize specific promoter sequences, which control the expression of genes for sugar metabolism enzymes. When the lactose concentration reaches a certain level, the sugar binds to specific sites on the protein and changes its conformation, such that it binds much more tightly to its DNA site, thus turning off transcription of genes that would produce more sugars. This is an example of *feedback inhibition.* This example of biological regulation can be explained by the laws of chemical equilibrium and thermodynamics, discussed in chapters 2 through 5.

The high fidelity of transcription is ensured by an elegant kinetic mechanism, determined using the methods of enzyme kinetics described in chapter 8. During a round of polymerization, a nucleoside triphosphate enters the active site of RNA polymerase and pairs with the single-stranded DNA, which has been opened from its double helical form (figure 1.1). The three-dimensional structure of this essential enzyme from eukaryotic organisms has been solved (Cramer et al. 2000). The shape of the active site is such that only the correct Watson–Crick base pair is tolerated; the wrong nucleoside triphosphate does not make a good fit into the active site and is more rapidly

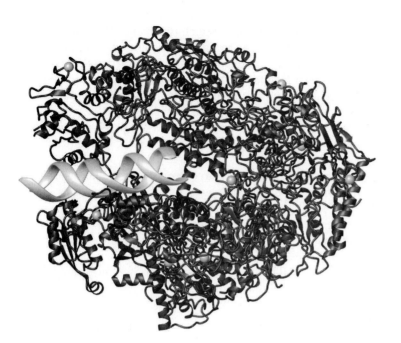

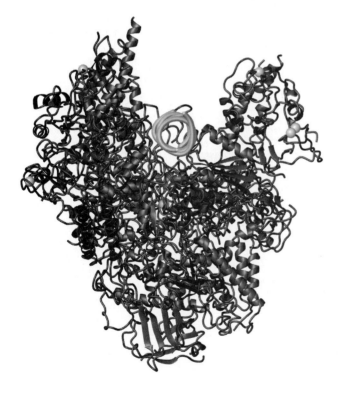

◀ FIGURE 1.1

Three-dimensional structure of the core of yeast RNA polymerase II, the enzyme responsible for transcribing all messenger RNAs in yeast. The core of the polymerase contains the essential functions of the enzyme and consists of ten individual polypeptide chains, shown in black. Two views of the enzyme, rotated by 180°, are shown. The ribbons represent the polypeptide backbone of the proteins. A DNA double helix (in white), which represents the substrate for transcription, is modeled into the active site. Note that the helix cannot continue through the protein; one function of the polymerase is to dissociate the two strands of DNA. (Courtesy of P. Cramer and R. Kornberg)

ejected. For DNA polymerase, the enzyme that copies DNA during cell division, the push for fidelity is so strong that the enzyme contains an editing function. If a wrong nucleotide is incorporated into the DNA, it is snipped out, and the correct nucleotide is incorporated. This drive for fidelity is understandable, consisidering the drastic effects mutations can have on protein function. On the other hand, the polymerases have to perform their functions rapidly, so they have evolved a trade-off between perfect fidelity and reasonable rates of polymerization. Such trade-offs are a hallmark of biological chemistry.

The regulation of transcription is a central process in biology; the requirement for a complex macromolecular assembly to perform RNA transcription in higher organisms (see figure 1.1) derives from the need for regulation. Cells must sense outside stimuli and respond, usually by rapidly synthesizing or degrading a protein or proteins. Recent biochemical experiments have revealed elaborate *signal transduction pathways*. A protein on the surface of a cell, called a *receptor*, will bind to an external signal, which may be a specific hormone or other extracellular-signaling molecules. The receptor molecule spans the cell membrane, and the binding of the hormone causes a change in its three-dimensional structure, activitating an enzymatic activity (a *kinase*) that adds a phosphate group to a protein. When a phosphate group modifies a protein, the protein's shape and activity can change. Often, a cascade of kinase events occurs, where protein 1 phosphorylates protein 2, which in turn phosphorylates protein 3, and so on. The final targets of these cascades are often transcription factors, which can turn transcription on or off depending on the desired result of a signaling event. Certain human cancers occur when these signaling pathways—and thus the ability of a cell to respond to external stimulus—are disrupted. Signaling pathways are very complex, and biologists are still identifying their many components. Physical methods and reasoning, however, will be required to unravel the mechanisms of these signaling pathways.

The *ribosome* (figure 1.2), where translation occurs, is more complex than RNA polymerases. It consists of two subunits, which in bacteria weigh 0.80×10^6 and 1.4×10^6 daltons. These enormous subunits each consist of at least one RNA chain and 20 to 30 proteins. The adaptors between the genetic code of RNA and the protein amino acid, first proposed by Crick, are called *transfer RNAs* (tRNAs). A single loop of the tRNA contains three nucleotides —the *anticodon*—that can form Watson–Crick base pairs with a given codon; the amino acid that corresponds to that codon is attached at the 3'-end of the tRNA. The three-dimensional structure of tRNA shows that these two parts are located 75 angstroms (Å) apart. The ribosome is able to select the correct tRNA that binds to the appropriate codon. The messenger RNA (mRNA) runs through a cleft between the subunits, and the anticodon portion of the tRNA interacts with the smaller (30S) subunit. Once the correct tRNA is selected, the 3'-end of the tRNA sits within the larger subunit, where peptide bond formation is catalyzed between the amino acid and a peptide-chain containing tRNA (which is bound at the adjacent codon). The ribosome then must shuttle down three nucleotides of the mRNA to the next codon; this precise directional movement is called *translocation*.

The basic mechanism of translation was delineated over 30 years ago, but molecular details of how the ribosome performs the task of protein syn-

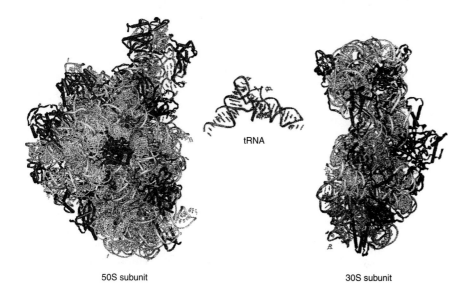

50S subunit   30S subunit

▲ FIGURE 1.2

The architecture of the large (left) and small (right) ribosomal subunits solved by
X-ray crystallographic methods (see chapter 11). Each subunit is made up of at least
one long RNA chain and multiple proteins held together by noncovalent interactions. For reference, a transfer RNA (a substrate for the ribosome) is shown. The two
extended ends of the L-shaped tRNA fit into clefts formed by the ribosome. (From
Puglisi et al. 2000.)

thesis have been revealed only recently (reviewed in Puglisi, Blanchard, and
Green 2000). First, it is apparent that the biological functions of the ribosome
are dominated by the RNA components; RNA catalyzes the formation of the
peptide bond, making it an RNA enzyme. Kinetic studies, similar to those
performed on polymerase enzymes, have revealed the origins of translational fidelity. The strategy used by the ribosome is somewhat similar to that used
by polymerases. In the case of the ribosome, the base pairing between codon
and anticodon occurs about 75 Å from the site of peptide bond formation;
the ribosome couples this base pairing to another chemical reaction: hydrolysis of guanosine triphosphate, GTP, which is bound to a protein factor that
escorts the tRNA to the ribosome. Rate contstants for tRNA dissociation and
ribosomal conformational changes are modulated by whether the correct or
the incorrect tRNA is present. Structural biologists have obtained the first
detailed views of the ribosomal particles. The two subunits of the ribosome
interact through an interface that is entirely RNA. Adjustments of this interface allow the ribosome to translocate down the mRNA. As biochemical
experiments have predicted, the structures show that RNA forms the critical active sites for tRNA binding and peptidyl transfer. The RNA folds into a
complex three-dimensional structure, which the protein components of the
ribosome (many of which bind to ribosomal RNA) stabilize. The molecular
rationale for how the ribosome performs translation will only be revealed by
physical chemical investigations.

Ion Channels

Cells perform spectacular feats of chemistry. *Ion channels* are proteins that span the lipid membrane of a cell and specifically allow one ion type to traverse the channel. Ion channels are critical for many biological processes, including signaling by neurons. Ion channels can be remarkably selective. Potassium ion (K^+) channels are about 10,000-fold more selective for K^+ than for Na^+ (sodium) even though their ionic radii are 1.33 and 0.95 Å, respectively. Also, the ion channels must allow a large number of ions to pass across a membrane in a directional manner in a short time period. Finally, many ion channels are controlled by external conditions. They are opened or closed to ion passage by factors such as the voltage difference across the membrane. The methods of physical chemistry have been invaluable in determining how ion channels work (Doyle et al. 1998). When ion channels do not function properly, the results can be disastrous. Many human diseases are linked to impairment in these molecular highways. For example, cystic fibrosis, one of the most common genetic diseases, is caused by mutations in a Cl^- (chloride ion) channel. The disrupted function for this channel leads to a buildup of thick, fibrous mucus in the lungs, which impairs breathing.

Determining the three-dimensional structure of a K^+ channel to atomic resolution was a significant breakthrough in understanding how ion channels work. The protein is a tetramer of identical subunits. Long rods of alpha helix span the membrane. The protein is not merely a tube through which potassium flows. The overall shape of the protein is like a flower, with the petals opening toward the outside of the membrane and narrowing at the inside of the membrane (figure 1.3). The ions pass through a channel in the center of the tetrameric protein. How are potassium ions specifically selected and transported? The top of the flower-shaped tetramer is a selectivity filter.

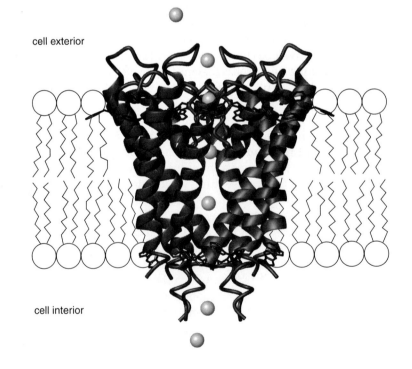

cell exterior

cell interior

▶ FIGURE 1.3

The three-dimensional structure of a K^+ channel, showing schematically the position of the channel within a cellular membrane. Intracellular (in the cell interior) and intercellular (on the cell surface) domains are indicated. Potassium ions are shown as spheres and are transported directionally from the exterior to the interior of the cell. (Courtesy of R. Mackinnon)

This region of the protein is rich in negatively charged amino acids, which specifically select for cation binding. The shape and chemical properties of this filter region are such that potassium ions bind most favorably, while smaller or larger cations are excluded. The presence of a second potassium ion helps nudge the first further down the channel. Beyond this selectivity filter, the channel widens and is lined by mainly hydrophobic side chains. This may seem surprising, as one might expect the whole pore to be lined by negatively charged amino acids. However, the function of the pore is to transport a large number of cations through the channel. If the pore was too negatively charged, the laws of electrostatics (chapter 9) predict that the cation would stick within the channel. The three-dimensional structure—combined with the principles of physical chemistry—takes the mystery out of ion channels.

What turns an ion channel on or off? Voltage-gated ion channels are required for nerve signaling and control the flow of ions in response to transmembrane voltages. Biophysicists have studied voltage-gated potassium channels that have sequence similarity to the ion channel whose structure the preceding paragraph describes (Glauner et al. 1999; Cha et al. 1999). Based on this similarity, a model for what the related protein might look like in three dimensions was created. This low-resolution model allows the design of experiments to test how the structure of the protein responds to voltage gating. We will see in chapter 10 that spectroscopy allows the investigation of protein structure as well as its time-dependent changes. With the technique of fluorescence energy transfer, in which fluorescent dyes are attached to the ion channel at defined places, biophysicists use spectroscopy to measure the distances between the dyes and to determine changes in distance caused by conformational alterations. For instance, an alpha helix rich in positively charged amino acids senses the change in voltage and twists and changes its position. This twisting movement changes the shape of the pore and allows ions to pass. This example shows the synergy of various physical measurements. Structural studies lead to models of activity that spectroscopic and kinetic measurements can test. Physical studies can transform what was originally just a gene sequence for the ion channel into a drama of molecular forms and movements.

References

The following textbooks can be useful for the entire course.

Physical Chemistry

Alberty, R. A., and R. J. Silbey. 1997. *Physical Chemistry*, 2d ed. New York: Wiley.

Atkins, P. W. 1998. *Physical Chemistry*. 6th ed. New York: Freeman.

Barrow, G. M. 1988. *Physical Chemistry*. 5th ed. New York: McGraw-Hill.

Levine, I. N. 1995. *Physical Chemistry*. 4th ed. New York: McGraw-Hill.

Moore, W. J. 1983. *Basic Physical Chemistry*. Englewood Cliffs, NJ: Prentice Hall.

Biophysical Chemistry

Cantor, C. R., and P. R. Schimmel. 1980. *Biophysical Chemistry*. Pts. 1–3. San Francisco: Freeman.

Van Holde, K. E., W. C. Johnson, and P. S. Ho. 1998. *Physical Biochemistry*. Upper Saddle River, NJ: Prentice Hall.

Biochemistry

Mathews, C. K., K. E. van Holde, K.G. Ahern. 2000. *Biochemistry*, 3rd ed. Reading, MA: Addison-Wesley.

Stryer, L. 1995. *Biochemistry*. 4th ed. San Francisco: Freeman.

Voet, D., J. G. Voet, and C. W. Pratt. 1999. *Fundamentals of Biochemistry*. New York: Wiley.

Molecular Biology

Alberts, B., D. Bray, J. Lewis, M. Raff, K. Roberts, and J. D. Watson. 1994. *Molecular Biology of the Cell.* 3d ed. New York: Garland.

Fasman, G. D., ed. 1976. *Handbook of Biochemistry and Molecular Biology.* 3d ed. Cleveland: CRC Press. This handbook is useful for compilations of data.

Lewin, B. 1999. *Genes 7.* New York: Oxford University Press.

Watson, J. D., N. H. Hopkins, J. W. Roberts, J. A. Steitz, and A. M. Weiner. 1994. *Molecular Biology of the Gene.* 4th ed. Vol. 1 and 2. Redwood City, CA: Benjamin/Cummings.

Physics

Giancoli, D., 2001. *Physics for Scientists and Engineers.* Upper Saddle River, NJ: Prentice Hall.

Halliday, D., R. Resnick, and J. Walker. 2000. *Fundamentals of Physics.* 6th ed. New York: Wiley.

Suggested Reading

The human genome sequence is described in two special issues of *Nature* and *Science.* International Human Genome Sequencing Consortium. 2001. Initial Sequencing and Analysis of the Human Genome. *Nature,* 409:860–921. The Human Genome. 2001. *Science,* 291.

Ban, N., P. Nissen, J. Hansen, P. B. Moore, and T. A. Steitz. 2000. The Complete Atomic Structure of the Large Ribosomal Subunit at 2.4Å Resolution. *Science* 289:905–920.

Berg, P., and M. Singer. 1992. *Dealing with Genes: The Language of Heredity.* Mill Valley, CA: University Science Books.

Cha, A., G. E. Snyder, P. R. Selvin, and F. Bezanilla. 1999. Atomic Scale Movement of the Voltage-Sensing Region in a Potassium Channel Measured via Spectroscopy. *Nature* 402:809–813.

Cramer, P., D. A. Bushnell, J. Fu, A. L. Gnatt, B. Maier-Davis, N. E. Thompson, R. R. Burgess, A. M. Edwards, P. R. David, and R. D. Kornberg. 2000. Structure of RNA Polymerase II and Implications for the Transcription Mechanism. *Science* 288:640–649.

Doyle, D. A., C. J. Morais, R. A. Pfuetzner, A. Kuo, J. M. Gulbis, S. L. Cohen, B. T. Chait, and R. MacKinnon. 1998. The Structure of the Potassium Channel: Molecular Basis of K^+ Conduction and Selectivity. *Science* 280:69–77.

Glauner, K. S., L. M. Mannuzzu, C. S. Gandhi, and E. Y. Isacoff. 1999. Spectroscopic Mapping of Voltage Sensor Movement in the *Shaker* Potassium Channel. *Nature* 402:813–817.

Johnson, A. D., A. R. Poleete, G. Lauer, R. T. Sauer, G. K. Ackers, and M. Ptashne. 1981. λ Repressor and *cro*-Components of an Efficient Molecular Switch. *Nature* 294:217–223.

Nissen, P., J. Hansen, N. Ban, P. B. Moore, and T. A. Steitz. 2000. The Structural Basis of Ribosome Activity in Peptide Bond Synthesis. *Science* 289:920–930.

Ptashne, M., A. D. Johnson, and C. O. Pabo. 1982. A Genetic Switch in a Bacterial Virus, *Sci. Am.* 247 (November):128–140.

Puglisi, J. D, S. C. Blanchard, and R. Green. 2000. Approaching the Ribosome at Atomic Resolution. *Nature Structural Biol.* 7:855–861.

Xia, Y., E. S. Huang, M. Levitt, and R. Samudrala. 2000. Ab Initio Construction of Protein Tertiary Structures Using a Hierarchical Approach. *J. Mol. Biol.* 300:171–185.

Problem

1. Read a paper in the scientific literature that sounds interesting to you.

   (a) Record the complete reference to it: authors, title, journal, volume, first and last pages, and year.

   (b) Summarize the purpose of the paper and why it is worthwhile.

   (c) List the methods used and state how the measurements are related to the results.

   (d) What further experiments could be done to learn more about the problem being studied?

CHAPTER 2

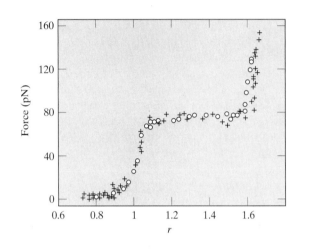

The First Law: Energy Is Conserved

The three laws of thermodynamics have been summarized as: You can't win, you can't break even, and you can't get out of the game. The next three chapters give more precise and quantitative definitions.

Concepts

A *scientific law* is an attempt to describe, concisely, one aspect of nature. Therefore, scientific laws are usually limited in their applicability: They may be incomplete or approximate, and they may be modified or rejected based on new experimental findings.

The branch of science known as *thermodynamics* deals with interchanges among different forms of energy. The laws of thermodynamics are excellent examples of both the generality and the limitations of scientific laws. The *first law of thermodynamics* states that energy is conserved: Different forms of energy can interconvert, but their sum remains unchanged. The law was originally based on experiments carried out in the early nineteenth century. In one of these experiments, a falling weight tied to a string was used to turn some paddles in a bucket of water, and the temperature of the water was found to increase. This experiment showed that the potential energy of the weight was converted to the internal energy of the water. Other forms of energy were later recognized and included in the first law.

As late as 1923, however, 18 years after Einstein postulated his famous equation that $E = mc^2$ (energy equals mass times the square of the speed of light), thermodynamicists were still debating whether the first law of thermodynamics applied to radioactive materials. Now we know that the laws of conservation of mass and conservation of energy are each incomplete, but that the law of conservation of mass-energy is correct. In a radioactive decay process, there is a very small but appreciable decrease in the total mass, and a large amount of energy is released according to Einstein's equation. Therefore, to make the first law of thermodynamics applicable to radioactive materials, mass itself must be considered as a form of energy. In processes that do not involve nuclear reactions or radioactive decay, however, the change in total mass does not contribute significantly to the change in energy, and we need not consider mass changes in the application of the first law. Thus, the first law of thermodynamics evolved from a simple description of a few experiments to a general statement about *all* forms of energy. Any new forms of energy that may be discovered can presumably be incorporated into the first law.

The *second law of thermodynamics* has had a different history. It also started in the early nineteenth century with considerations of the flow of heat from a warmer to a colder reservoir and the efficiencies of heat engines that converted some of this heat into mechanical work. These considerations led to a surprising finding: No engine, no matter how well it was designed, can possibly convert all heat taken up from the warmer reservoir to work. Some of this heat must be discharged to the colder reservoir, and the maximum efficiency achievable for any heat engine is determined by the temperatures of the warmer and colder reservoirs. The second law was later generalized to provide a criterion for *spontaneous processes*. All people, especially scientists, like to predict whether various processes will occur and in which direction; the second law is therefore enormously useful. As we will see in chapter 3, the second law also introduced the word *entropy* into the general lexicon. It was eventually realized, however, that the second law does not always apply to *very* small systems that contain too few molecules for precise statistical prediction. If we flip a coin 1,000 times, we can be reasonably sure of obtaining around 500 heads and 500 tails, but we cannot predict with confidence whether one particular flip will give a head or tail. For very small systems, the second law requires amendment.

The *third law of thermodynamics* is the most recent addition to the principles of thermodynamics and was clearly stated in the 1920s. It is concerned with the thermodynamic properties of matter when the temperature approaches 0 K (zero Kelvin or absolute zero). Apparent discrepancies between experimental measurements and predictions of the third law were subsequently noted, but in each case satisfactory explanations were found, and the third law has not been significantly modified since its original formulation. Because bioscientists rarely deal with very low temperatures, in this book the third law will not receive as much attention as the first and the second laws.

The additions, exceptions, and corrections to the thermodynamic laws —and to all other scientific laws—provide some of the reasons for our continuing study of science. We assume that new ideas will lead to new experiments and eventually to new or improved laws and a clearer understanding of nature. The laws of thermodynamics, as we understand them now, represent a concise summary of a very large body of experimental and theoretical studies. They are among most scientists' short list of fundamental laws of nature, and their wide applicability gives them a prominent place in diverse disciplines ranging from astrophysics to engineering.

Applications

Thermodynamics applies to everything from black holes so massive that even light cannot escape from them, to massless neutrino particles. Thermodynamics can answer questions like: How do you calculate the work done when a muscle contracts or stretches? How can chemical reactions be used to do work or to produce heat? How much heat can be generated by burning 1 gram (g) of sugar or from eating and digesting 1 g of sugar? Many examples of the applications of thermodynamics will be illustrated in this and the next three chapters.

Energy Conversion and Conservation

A large number of experiments, done over a period of many years, have shown that *energy can be converted from one form to another but that the total amount of energy remains constant.* We will eventually discuss this quantitatively, but a couple of examples will make the idea clearer.

Consider a brick on the window ledge of the fifth floor of an apartment building. Owing to its height above the sidewalk, it possesses gravitational potential energy. If the brick falls, most of the potential energy will first become kinetic energy of motion, and a small amount is converted to heat, because of friction, as the brick moves through air. What happens when the brick hits bottom? The kinetic energy is converted into many new forms of energy. Much heat will be produced. If the brick hits the sidewalk, there might be some light energy in the form of sparks. Some energy is used to break and make chemical bonds in the brick and sidewalk fragments. Some sound energy is produced. Although complicated changes in different forms of energy are involved, the first law tells us that the total energy will remain constant.

A question of more interest is about the total amount of energy arriving from the Sun and into what forms of energy it is transformed. Sunlight hitting a desert or a solar collector is mainly transformed into heat. However, some of the light energy can be converted into electrical energy by the use of solar energy cells, and sunlight striking a green leaf is partly transformed into useful chemical energy through photosynthesis. It is vitally important for us to know and understand the various kinds of energy that are available and the limits, if any, of their interconversion.

Systems and Surroundings

For a quantitative treatment of the interconversion among different forms of energy, it is necessary to clearly define some terms. The term *system* is defined as the specific part of the universe on which we choose to focus. It might be the Sun, Earth, a person, a liver, a single living cell, or a mole of liquid water. Everything else we call the *surroundings*, and what separates the system from the surroundings is termed the *boundaries* (figure 2.1).

The simplest system in thermodynamics is one that has no exchange of any kind with the surroundings; such a system is called an *isolated system*. There is neither matter nor energy that passes through the boundaries between an isolated system and the surroundings. It is difficult to actually construct an isolated system, but it is useful to think of one. The contents of a sealed and thermally insulated flask comes very close to one, especially over a short period of time with negligible heat flow in and out. Thermodynamicists, and scientists in general, often define situations that can be obtained only approximately. Such idealized situations are more easily analyzed, and their choice helps illuminate the basic principles involved. A *closed system* is defined as one that does not exchange matter with the surroundings, but that does allow energy exchanges across the boundaries. A closed system can be simply one with a physical enclosure. Many chemical reactions are performed in a closed system, a sealed flask being an example. The chemicals stay inside the flask, but heat can come in or out. The most difficult type of system to consider (and also the most common and useful) is the open system. An *open system* can exchange both matter and energy with the surroundings. The example illustrated in figure 2.1 is an open system. A fertilized egg being hatched by a hen is another example: Oxygen goes in and carbon dioxide comes out of the egg; heat is also exchanged between the egg and its surroundings.

We should emphasize that it is entirely up to us to specify our system and to define real or imaginery boundaries that separate it from its surroundings. If we specify as our system 10 mol of H_2O (180 g) that was poured into a beaker and left to evaporate, then the liquid water that remains in the beaker [$(180 - x)$ g], plus the water that has evaporated from it (x g), constitutes a closed system. As another example, if a solution containing the enzyme catalase is added to an open beaker containing a hydrogen peroxide solution, the enzyme will accelerate the reaction $H_2O_2 \rightarrow H_2O + \frac{1}{2}O_2$, and oxygen gas will come out of the beaker. If we choose the liquid content in the beaker as our system, we have an open system. But if we choose the liquid content plus

▲ FIGURE 2.1

In thermodynamics, the *system* is what we focus our attention on. The *surroundings* is everything else in the universe, but we need to consider only the part that interacts with the system. Here, a beakerful of water is sketched. If we pick the liquid water inside the beaker as our system, then the water–glass and water–air interfaces form the boundaries. In this example, the system and surroundings may exchange both material and energy through the boundaries: Water may evaporate into the surrounding air, and air may dissolve into water; heat may flow from the water to the surroundings if the water is warmer than its surroundings, or flow the other way if the water is cooler than the surroundings.

the oxygen evolved as our system, we have a closed system. We can choose various systems based on our specific interest, objectives, and which system is most convenient for a particular problem; we must define the chosen system clearly, however, to avoid confusions.

Energy Exchanges

Having defined a system, we can then focus on how its energy can be changed. In thermodynamics, it is the *change* of energy in which we are interested, and we are less concerned about the absolute value of energy for the system. We can add energy to a system in various ways. Adding matter to an open system, for example, increases the chemical energy of the system because the matter can undergo various physical and/or chemical reactions. But we do not have to think about the large amount of energy potentially available from nuclear reactions unless such reactions are actually occuring in the system; that is, usually we do not have to include the $E = mc^2$ energy term, because this term does not change significantly in an ordinary reaction.

It is convenient to divide energy exchange between system and surroundings into different types. Two of the most common types of energy exchanges are work and heat.

Work

A system can do work on the surroundings, or the surroundings can do work on the system. In classical mechanics, *work* is defined as the product of a *force* times a *distance*. To calculate the work done by the system or on the system, multiply the *external force on the system* by the distance moved—the *displacement*:

$$\text{work} \equiv \text{external force} \cdot \text{displacement} \tag{2.1}$$

We must be consistent about the sign of the work to keep proper accounting of the energy exchanges between the system and the surroundings. In this book, we follow the convention that *the work is positive if the surroundings are doing work on the system, and negative if the system is doing work on the surroundings.* In the use of Eq. (2.1) for the various situations illustrated below, we keep a watchful eye to make sure that the sign of work, which depends on the proper choice of signs for the force and the displacement, is always consistent with this convention. Some other books have work done *by* the system as positive. Sign conventions can be a pain, but as long we specify these clearly the pain can be kept to a minimum.

Work of extending a spring.   Suppose we have a spring with a length x_0 when it is not subject to an external force, and an external force f_{ex} is applied to extend the spring from x_0 to a longer length x. According to *Hooke's law,* the spring force f_{sp} that counters f_{ex} is directly proportional to the change in the length of the spring. To calculate the work of extending the spring, we choose an x-axis with one end of the spring fixed at $x = 0$. The other end is free to move along this axis; its position when there are no forces acting on it is x_0; in general its position is at x.

According to Hooke's law,

$$f_{sp} = \text{spring force} = -k(x - x_0) \tag{2.2}$$

where k is a constant for a given spring (if we do not extend it beyond its elastic limits). The magnitude of k will be different for different springs. The negative sign states that the direction of the spring force is opposite to the direction of displacement from the position at x_0. The external force is the negative of the spring force, as the two are equal in magnitude but opposite in direction:

$$f_{ex} = \text{external force} = -f_{sp} = k(x - x_0) \tag{2.3}$$

Because the force depends on the displacement, we must integrate the product of force times distance to calculate the work w:

$$w = \int f_{ex}\, dx$$

Note that in the preceding equation w is positive if the external force f_{ex} and the displacement dx are of the same sign, and negative if they are of opposite signs. This is consistent with our sign convention for w: If the direction of the external force is the same as the direction of the displacement, the external force is doing work on the system (the spring), and w is positive; if the direction of the external force is opposite to the direction of the displacement, the spring is doing work on the surroundings, and w is negative.

The work done on the spring when its length is changed from x_1 to x_2 is

$$w = \int_{x_1}^{x_2} f_{ex}\, d(x - x_0)$$

$$= \int_{x_1}^{x_2} k(x - x_0)\, d(x - x_0)$$

$$= \frac{1}{2}k[(x_2 - x_0)^2 - (x_1 - x_0)^2]$$

$$= k(x_2 - x_1)\left(\frac{x_2 + x_1}{2} - x_0\right) \tag{2.4}$$

We have thus quantitatively calculated the work done on the system by extending the spring. If the spring was originally extended, the system would do work on the surroundings when the spring returned to its equilibrium position. A muscle fiber does work in a similar fashion. It can do work by stretching or contracting against an external force.

In *Standard International (SI) units*, work is expressed in joules (J) if k is in newtons meter^{-1} (N m^{-1}) and x is in meters (m): 1 N m = 1 J. One newton is the magnitude of the force that will cause an acceleration of 1 m s^{-2} when applied to a mass of 1 kilogram (kg): 1 N = 1 kg m s^{-2}. Other units for work are the erg: 1 erg = 10^{-7} J; and the calorie: 1 cal = 4.184 J. Additional energy conversion tables can be found in table A.3 in the appendix. The trend now is to use SI units, which we will be using most often in this book. The basic units of length, mass, and time in the SI system are meter (m), kilogram (kg), and second (s). Units such as erg, cal, and kcal (kilocalorie) are still widely used by biochemists, however.

EXAMPLE 2.1

Measurements with single DNA molecules in aqueous solution gave the data shown in the figure below.

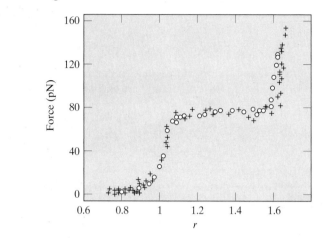

In this experiment, the DNA used was that of a bacterial virus called λ (P. Cluzel, A. Lebrun, C. Heller, R. Lavery, J.-L. Viovy, D. Chatenay, and F. Caron 1996, DNA: An Extensible Molecule, *Science* 271:792–794). One end of a double-stranded λ DNA, which has about 50,000 base pairs, was attached to an optical fiber, which would bend when pulled and served as a force sensor; the other end of the DNA was attached to a device termed a *piezo translator,* which would move the attached DNA end by a distance in proportion to the voltage applied to the device. The measured force, in pN (1 pN, or piconewton, is 10^{-12} N), is shown in the figure as a function of r, the distance between the ends of a DNA molecule divided by the contour length of the DNA. A ratio r of 1.00 corresponds to an actual end-to-end separation of about 15 μm, the contour length l_0 of an unstretched λ DNA (1 μm, or micrometer, is 10^{-6} m).

When the force is zero, the separation between the ends is less than l_0, as the unperturbed DNA molecule is coiled up rather than extended. The DNA extends readily to its contour length ($r = 1$) by the application of a small force, and thereafter the force rises more rapidly until a plateau is reached around 80 pN. The presence of this plateau suggests a tension-induced structural transition of the DNA, the length of which can be increased by at least 60% from that of the well-known Watson–Crick B-form of DNA. Stretching beyond an r of 1.7 leads to irreversible changes in the DNA.

Calculate the total work done on (a) a single λ DNA molecule when it is stretched from $r = 1$ to $r = 1.6$ and (b) 1 mole (mol) of λ DNA molecules.

Solution

a. Clearly, Hooke's law does not apply to stretching a DNA molecule; here the force is a complex function of the displacement. The work is

$$w = \int f_{ex}\,dl = \int f_{ex}\,d(l_0 r) = l_0 \int_{1.0}^{1.6} f_{ex}\,dr$$

The integral $\int_{1.0}^{1.6} f_{ex}\,dr$ is equal to the shaded area in the sketch (a) below and can be evaluated by a number of numerical methods (the use of a computer program is probably the most convenient). As an approximation, we visually estimate that the desired area in sketch (a) can be approximated by the combined areas of the two shaded rectangles in sketch (b).

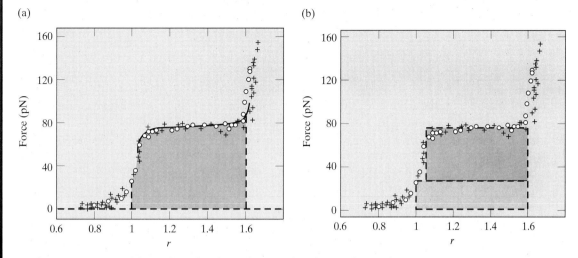

(a)

(b)

These areas are readily evaluated to be $26\,l_0$ pN and $15\,l_0$ pN. The work is therefore

$$w = l_0 \int_{1.0}^{1.6} f_{ex}\,dr = 15 \times 10^{-6}\,\text{m} \times (26 + 15)10^{-12}\,\text{N} = 6.15 \times 10^{-16}\,\text{N m}$$

$$= 6.15 \times 10^{-16}\,\text{J}$$

b. For 1 mol of the DNA, the total work becomes 6.15×10^{-16} J $\times 6.023 \times 10^{23}$ (6.023×10^{23} is the number of molecules per mole, *Avogadro's number*), or 3.7×10^8 J or 3.7×10^5 kJ. This seems to be a huge amount of work! We note, however, that 1 mol of a large molecule like λ DNA is a lot of material: The total mass in grams is $50,000 \times 660$ (the average molecular mass of a base pair), which is 3.3×10^7 g, or 33 metric tons!

Work of increasing or decreasing a volume.   Consider a system such as a gas or liquid enclosed in a container with a movable wall or piston (figure 2.2). The system can expand and do work on the surroundings if the external pressure P_{ex} is less than the pressure of the system P, or the surroundings can do work on the system by compressing it if P_{ex} is greater than P.

This example is very similar to the spring example (2.1). By definition, pressure is just force per unit area, and the external pressure P_{ex} is related to the external force f_{ex} by $P_{ex} = f_{ex}/A$, where A is the area of the moving boundary surface between the system and surroundings (in the example illustrated in figure 2.2, A is the cross-sectional area of the piston). The product $f_{ex}\,dx$ is

$$f_{ex}\,dx = \left(\frac{f_{ex}}{A}\right)A\,dx = P_{ex} \cdot A\,dx$$

The quantity $A\,dx$ is equal to dV, dV being the change in volume of the gas container. Thus, from the equation above, force times displacement is equivalent to pressure times change of volume. The work done on the system is

$$w = -\int_{V_1}^{V_2} P_{ex}\,dV \qquad \textbf{(2.5)}$$

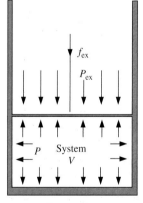

Surroundings

▲ FIGURE 2.2

Work done on a system by an external force f_{ex}, acting against an opposing force provided by the internal pressure P of the system. The work is positive when the volume V decreases and negative when V increases.

The negative sign is needed because of the sign conventions for w, P_{ex}, and dV. Compression, a decrease in volume or a negative dV, means the surroundings have done work on the system, and w is thus positive. For a compression, P_{ex} is replaced by the pressure of the gas P_{gas} in Eq. (2.5). Expansion, an increase in volume or a positive dV, means the system does work on the surroundings, and w is thus negative. For brevity, in arriving at Eq. (2.5), we specifically address the sign only in the final equation and not during its derivation. The sign conventions can be bewildering sometimes; both P_{ex} and P, for example, are commonly taken as positive quantities, even though they oppose each other.

Work associated with a change in the volume of the system is often termed *pressure–volume work*, or *PV* work. To calculate work in joules, we must use pressure in N m^{-2} and volume in m^3. If the pressure is in atmospheres* and the volume in liters (L), the result in L atm must be converted using 1 L atm = 101.325 J.

If the opposing pressure is kept constant, Eq. (2.5) can be easily integrated. For *constant pressure*,

$$w = -P_{ex}(V_2 - V_1) \tag{2.6}$$

EXAMPLE 2.2

Calculate the *PV* work done when a system containing a gas expands from 1.0 L to 2.0 L against a constant external pressure of 10 atm.

Solution

$$w = -P_{ex}(V_2 - V_1) \qquad \text{(expansion)}$$

$$= -(10 \text{ atm})(2 \text{ L} - 1 \text{ L})$$

$$= -10 \text{ L atm}$$

$$= -(10 \text{ L atm})\left(101.3 \frac{\text{J}}{\text{L atm}}\right)$$

$$= -1013 \text{ J}$$

The system does work; the sign of w is negative.

EXAMPLE 2.3

Calculate the *PV* work done in joules when a sphere of water 1.00 μm (1 μm = 10^{-6} m) in diameter freezes to ice at 0°C and 1 atm pressure.

Solution

We first find the volume of the sphere of water and then use the density of ice and liquid water at 0°C to calculate the volume of the frozen sphere.

---

*Atmospheric pressure is the force per unit area exerted by the mass of air above the surface of Earth. The *standard atmosphere* is defined as 1.01325 $\times$ 10^5 Pa (pascals) = 1.01325 $\times$ 10^5 N m^{-2}.

The volume of the liquid water is

$$V_l = \frac{1}{6}\pi D^3 = \left(\frac{3.142}{6}\right)(10^{-6}\,\text{m})^3$$

$$= 5.237 \times 10^{-19}\,\text{m}^3$$

The density of liquid water at 0°C is 1.000 g cm^{-3}. The density of ice at 0°C is 0.915 g cm^{-3}. The volume of the solid water is

$$V_s = (5.237 \times 10^{-19}\,\text{m}^3)\left(\frac{1.000}{0.915}\right)$$

$$= 5.723 \times 10^{-19}\,\text{m}^3$$

The work is

$$w = -P_{ex}(V_2 - V_1)$$

$$= -(1\,\text{atm})(5.723 \times 10^{-19}\,\text{m}^3 - 5.237 \times 10^{-19}\,\text{m}^3)$$

$$= -0.486 \times 10^{-19}\,\text{m}^3\,\text{atm}$$

$$= (-0.486 \times 10^{-19}\,\text{m}^3\,\text{atm})(1.013 \times 10^5\,\text{J m}^{-3}\,\text{atm}^{-1})$$

$$= -4.92 \times 10^{-15}\,\text{J}$$

The system expands; therefore, work is done by the system, and the sign of w is negative.

In general, the external pressure P_{ex} and the pressure of the system P are not equal when the system is undergoing compression or expansion. If P_{ex} is kept nearly equal to P during the process (that is, they differ only infinitesimally), the process is said to be *reversible*. A very small change in pressure will reverse the process from an expansion to a compression and vice versa. The sign of the work will change from positive to negative, depending on an infinitesimal change in the pressure. If P_{ex} and P differ significantly, *irreversible* expansion or compression will occur, depending on whether P_{ex} is smaller or greater than P. In the special case when a system is expanding irreversibly against a vacuum, P_{ex} is zero and no work is done. Reversible heating or cooling is similar to reversible expansion or compression: Here, the temperature difference between the system and the surroundings must be very small—infinitesimal. Heat will flow in or out, depending on very small changes in temperature. If the temperatures of two bodies differ significantly, irreversible heat exchange occurs when the two come into contact.

Expansion work done by a system containing only gases has been important in thermodynamic analysis of heat engines. Engineers (and others) need to know how much work can be done by the expanding gas inside the cylinder of a car, for example. Biologists may be interested in the work done by the lungs on the air we breathe or the work done in systems containing liquids. The work done when liquid water expands upon freezing is important because of the large forces that may be generated during the process. Freezing of water in a crevice, for example, can cause severe structural damage. Preventing damage caused by ice formation is also of key importance in preserving biological materials at low temperatures.

Friction. The force of friction causes an energy change whenever two surfaces in contact move relative to each other. The frictional force is opposite in direction to the force causing the motion. We do not discuss frictional effects in this chapter, but they may be important in considering interconversions of different forms of energy in real systems. Engineers, of course, try to maximize expansion work and minimize friction in engines. Frictional forces are also important in biological engines. As arteries become rougher and narrower with aging, the work needed to circulate blood increases. The blood pressure must then also increase, and the heart must work harder. Friction is not all bad, however. Driving on icy roads is treacherous enough; think of what would happen if there were no friction at all!

Work in a gravitational field. All processes that occur on Earth are affected by Earth's gravitational field (and to a much lesser extent by the Moon's and other astronomical bodies' gravitational fields). This means that all systems on Earth can do gravitational work. If an object of mass m (the system) is lowered at a constant velocity from a height h_1 to a height h_2 above Earth's surface, the force is mg with g the acceleration of gravity. The work done on the system is

$$w \text{ (gravitational)} = mg(h_2 - h_1) \tag{2.7}$$

The work done on the system by the external force is negative ($h_2 - h_1$ is negative); in other words, the system is doing work on the surroundings.

For mass m in kg, height h in meters, and g in m s^{-2} (the standard acceleration of gravity = 9.807 m s^{-2}), the work is obtained in joules.

Work in an electric field. If a system contains electrical charges, an electric field will produce a force on the charges that will cause them to move, and thus a current will flow. The work done on the system can be shown to be

$$w \text{ (electrical)} = -EIt \tag{2.8}$$

where E is the voltage = potential difference = electromotive force, I is the current, and t is the time. For E in volts, I in amperes, and t in seconds, the units of work obtained are joules. The cost of electricity is usually calculated in kilowatt hours (kW h). Since a watt second (W s) is a joule, 1 kW h is 3.6×10^6 J.

EXAMPLE 2.4

Calculate the electrical work that can be done, in joules and calories, by a 12.0-V storage battery that discharges 0.1 ampere (A) for 1.00 h.

Solution
The electrical work is

$$w \text{ (electrical)} = -EIt \tag{2.8}$$

$$= -(12 \text{ V})(0.1 \text{ A})(1 \text{ h})$$

$$= -1.2 \text{ V A h} = -1.2 \text{ W h}$$

$$= -(1.2 \text{ W h})(3600 \text{ s h}^{-1})$$

$$= -4.32 \times 10^3 \text{ W s}$$

$$= -4.32 \times 10^3 \text{ J}$$

$$= (-4.32 \times 10^3 \text{ J})(0.2389 \text{ cal J}^{-1})$$

$$= -1.03 \times 10^3 \text{ cal}$$

The minus sign tells us that the battery is doing work on the surroundings. Because some of the electrical energy is converted to heat (see the next section) when the battery discharges, the calculated electrical work is a theoretical maximum.

Heat

When two bodies are in contact with each other, their temperatures tend to become equal. Energy is being exchanged. The hot body will lose energy and cool down; the cold body will gain energy and warm up. This energy exchange is said to occur by *heat transfer,* and the energy that passes through the system-surroundings boundaries because of a temperature difference between the two is called *heat.* The sign convention for heat is similar to that for work: *Heat is positive if it flows into the system and negative if it flows out of the system.*

For a closed system, the transfer of a small quantity of heat dq will result in a change dT in its temperature. The ratio dq/dT is called the *heat capacity* of the system and is given the symbol C:

$$\frac{dq}{dT} = C \tag{2.9}$$

In general, C is different for different materials, and for a given material, C will vary with temperature. Thus, heat q gained when the temperature of the system changes from T_1 to T_2 should be evaluated from the integral

$$q = \int_{T_1}^{T_2} C \, dT \tag{2.9a}$$

If C is constant in the temperature range T_1 to T_2, then Eq. (2.9a) reduces to

$$q = C(T_2 - T_1) \tag{2.9b}$$

For a hot body in contact with a cold body, the heat capacity of each body must be known, and Eqs. (2.9–2.9b) must be applied to each body separately. The heat capacity C increases with the amount of material in the body. For a pure chemical substance, $C = n\overline{C}$, where n is the number of moles and $\overline{C}$ is the *molar heat capacity,* or heat capacity per mole. Heat capacity for every material is a quantity that characterizes how much heat is necessary to raise its temperature by 1°C or 1 K. It is usually determined by measuring the changes in temperature when known amounts of electrical energy are dissipated in this material. The units are often in $J \text{ K}^{-1} \text{ mol}^{-1}$ or $\text{cal K}^{-1} \text{ mol}^{-1}$. It is easy to remember that the heat capacity of liquid water is about $1 \text{ cal K}^{-1} \text{g}^{-1}$. The heat capacity for a substance in a given phase (gas, liquid, or solid) will in general

TABLE 2.1   Heat Capacities at Constant Pressure, 1 atm, of Various Substances Near 25°C

Substance	$\overline{C}_P$	Substance	$\overline{C}_P$	Substance	$C_P{}^*$
Gases	Molar heat capacities ($J\ K^{-1}\ mol^{-1}$)	Liquids	Molar heat capacities ($J\ K^{-1}\ mol^{-1}$)	Solids	Specific heat capacities ($J\ K^{-1}\ kg^{-1}$)
He	20.8	Hg	28.0	Au	129
H_2	28.8	H_2O	75.2	Fe	452
O_2	29.4	Ethanol	111.4	C (diamond)	510
N_2	29.1	Benzene	136.1	Glass (Pyrex)	840
H_2O	33.6	n-Hexane	195.0	Brick	~800
CH_4	35.8			Al	902
CO_2	37.1			Glucose	1250
				Urea	2199
				H_2O (0°C)	2100
				Wood	~2000

increase with increasing temperature. It will also depend on whether P or V is held constant during the heating. The symbol C_P means heat capacity at constant pressure, and C_V is heat capacity at constant volume. As we will soon see, C_P is larger than C_V, although for solids and liquids they are nearly equal; for gases, the difference between C_P and C_V is significant. Molar heat capacities at constant pressure for several substances are tabulated in the first two columns of table 2.1. In the third column, values of *specific heat capacity*, $C_P{}^*$, or heat capacity per kilogram, are listed.

EXAMPLE 2.5

Calculate the heat, in joules, necessary to change the temperature of 100.0 g of liquid water by 50°C at constant pressure. The heat capacity of liquid water at constant pressure is 1.00 cal g^{-1} deg^{-1} and is nearly independent of temperature.

Solution

The heat absorbed by the system is

$$q = \int_{T_1}^{T_2} C\ dT = C \int_{T_1}^{T_2} dT$$

$$= C(T_2 - T_1)$$

$$= (100.0\ \text{g})\left(1.00\ \frac{\text{cal}}{\text{g deg}}\right)(50\ \text{deg})\left(4.184\ \frac{\text{J}}{\text{cal}}\right)$$

$$= 20.9\ \text{kJ}$$

There are many practical applications of Eq. (2.9). We often want to know how much heat can be transferred from one system to another and what the final temperature will be. One of the main problems in solar heating is how to store the energy for use at night. It is simple to raise the temperature of a storage system such as water or rocks during the day and to transfer the heat at night to the cold air in a house. The amount of heat transferred depends on the temperature difference and the heat capacity of the storage system and of the house.

EXAMPLE 2.6

Cold air at 0°C is passed through 100 kg of hot crushed rock that has been heated to 110°C. The air is heated to 20°C by the time it leaves the rock and is admitted into a house for heating. Estimate the maximal volume of 20°C air that can be obtained by this process. The heat capacity at constant pressure of the rock is 800 J K^{-1} kg^{-1} and of the air is 1000 J K^{-1} kg^{-1}. The density of air at 1 atm and 20°C is 1.20 × 10^{-3} g cm^{-3}; the density of the rock is 2.5 g cm^{-3}.

Solution

As cold air is passed through the hot crushed rock, the temperature of the rock will drop. The maximal amount of cold air that can be warmed by the hot rock to 20°C can be estimated by assuming that this volume of 0°C air and 100 kg of 110°C crushed rock would come to thermal equilibrium at a temperature of 20°C. The total heat loss by the rock when its temperature drops from 110 to 20°C is

$$q = C(T_2 - T_1)$$
$$= (100 \text{ kg})(800 \text{ J K}^{-1} \text{ kg}^{-1})(90 \text{ K})$$
$$= 7.2 \times 10^6 \text{ J}$$

The weight of the air heated to 20°C by this amount of heat is

$$\text{air wt} = \frac{7.2 \times 10^6 \text{ J}}{(1000 \text{ J K}^{-1} \text{ kg}^{-1})(20 \text{ K})}$$
$$= 360 \text{ kg}$$

The volume of the air heated to 20°C is

$$\text{air vol} = \frac{3.6 \times 10^5 \text{ g}}{1.2 \times 10^{-3} \text{ g cm}^{-3}}$$
$$= 3 \times 10^8 \text{ cm}^3$$
$$= 3 \times 10^2 \text{ m}^3$$

This corresponds to the volume of a medium-sized room. The volume of crushed rock necessary is only 100 × 10^3 g/2.5 g cm^{-3}, or 4 × 10^4 cm^3 = 0.04 m^3.

Radiation

A very important form of energy exchange is that of radiation. Most of the energy coming into Earth everyday is from the Sun's radiation. The energy present in a given number (N) of photons of frequency ν, given in s^{-1}, can be calculated from

$$\text{radiation energy} = Nh\nu \qquad (2.10)$$

where h = Planck's constant = 6.6261 × 10^{-34} J s.

Another useful equation, the Stefan–Boltzmann equation, describes how much energy is radiated by a body as a function of its temperature. The equation is for an ideal body that radiates and absorbs all wavelengths, a *blackbody*:

$$\text{radiation energy m}^{-2} \text{ s}^{-1} = \sigma T^4 \qquad (2.11)$$

where $\sigma \equiv$ Stefan–Boltzmann constant = 5.67 × 10^{-8} J m^{-2} s^{-1} K^{-4}

T = absolute temperature

EXAMPLE 2.7

The average surface temperature of the Sun is about 6000 K; its diameter is about 1.4×10^9 m. Estimate the total energy radiated by the Sun in J s^{-1}.

Solution

The Stefan–Boltzmann equation provides the radiation rate per m^2. From the diameter of the Sun, we can obtain the surface area and thus the total energy radiated per second:

$$\text{radiation} = \sigma T^4 \tag{2.11}$$

$$= (5.67 \times 10^{-8} \text{ J m}^{-2} \text{ s}^{-1} \text{ K}^{-4})(6000 \text{ K})^4$$

$$= 7.35 \times 10^7 \text{ J m}^{-2} \text{ s}^{-1}$$

$$\text{area} = \pi D^2$$

$$= (\pi)(1.4 \times 10^9 \text{ m})^2$$

$$= 6.2 \times 10^{18} \text{ m}^2$$

$$\text{energy radiation per second} \equiv \text{luminosity}$$

$$= (7.35 \times 10^7 \text{ J m}^{-2} \text{ s}^{-1})(6.2 \times 10^{18} \text{ m}^2)$$

$$\text{luminosity} = 4.6 \times 10^{26} \text{ J s}^{-1}$$

This number agrees well with the measured luminosity of the Sun. By comparison, the intensity of bright summer sunlight at the surface of Earth is about 1000 J m^{-2} s^{-1}.

EXAMPLE 2.8

Radiation can cause chemical reactions to occur. For some reactions, each photon produces one molecule of product. How many photons are there per joule of red light? The wavelength is 700 nanometers (nm); the frequency is 4.28×10^{14} s^{-1}.

Solution

$$N = \frac{\text{energy}}{h\nu} \tag{2.10}$$

$$= \frac{1 \text{ J}}{(6.63 \times 10^{-34} \text{ J s})(4.28 \times 10^{14} \text{ s}^{-1})}$$

$$= 3.5 \times 10^{18} \text{ photons}$$

First Law of Thermodynamics

The first law of thermodynamics states that energy is conserved: It can be transferred between the system and surroundings, but the total energy of the system plus surroundings is constant. If we focus on the system, then the first law requires that the change in energy or ΔE of a system is equal to the amount of energy that entered it (from the surroundings) minus the energy that went out of it (into the surroundings). These statements might sound obvious and even trivial, but they represent an important description of a fundamental law of nature. Nineteenth-century thermodynamicists who thought about the Sun were worried. As example 2.7 indicates, the Sun radiates an enormous amount of energy. Because not much energy goes into the Sun, the first law says that the Sun must be losing energy at a high rate. What might be producing this energy in the Sun, and how long would this supply last? The mass of the Sun and its composition of roughly half hydrogen and half helium have long been known. What kinds of reactions involving hydrogen and helium could be the source of solar energy? If no known chemical reaction could account for even a tiny fraction of the energy the sun radiates, either the sun should have burnt out long ago, or thermodynamics does not work. It came as a great relief to learn about the discovery of nuclear energy as the source of solar energy—which saved thermodynamics, as well as all lifeforms on Earth, from an early demise. The Sun will eventually use up all its energy source, but we don't have to worry for a long time.

For a closed system, if heat and work are the only forms of energy that the system exchanges with the surroundings, the first law of thermodynamics requires that

$$\Delta E = q + w \qquad\qquad \textbf{(2.12)}$$

where ΔE = change in energy of the system
$\qquad q$ = net heat transferred to the system (heat *in*)
$\qquad w$ = net work done on the system (work *in*)

Additional terms can be added to Eq. (2.12) if there are energy exchanges other than heat and work between the system and surroundings or if the system is an open one. In the special case of an isolated system that exchanges neither matter nor energy with the surroundings, the first law requires that its energy be constant:

$$\Delta E = 0 \qquad \text{(for an isolated system)} \qquad \textbf{(2.13)}$$

In the next section, we see that the first law also makes the energy E a property of a system: *If a system changes from an initial state to a final state, the change in the energy of the system, ΔE, is dependent only on the initial and final state, not on the process by which the change occurred.*

Describing the State of a System
Variables of State

We have been using terms like *pressure* and *volume* without defining them further because they are familiar to us. We must pay attention to units, however. For pressure, the various units we will use are atm, torr (= mm Hg) and pascals (Pa = N m^{-2}).

$$1 \text{ atm} = 760 \text{ torr} = 1.01325 \times 10^5 \text{ Pa}$$

For volume, we use liter, or sometimes cm^3 = milliliter. For the temperature T, we use Kelvin, K. In the SI system, the unit of pressure is the pascal. Whenever T occurs in thermodynamic equations, Kelvin is indicated. To convert from other temperature scales, use

$$K = {}^\circ C + 273.15$$

$$= \frac{{}^\circ F - 32}{1.8} + 273.15$$

where $^\circ C$ = degrees Celsius, or centigrade
$\qquad ^\circ F$ = degrees Fahrenheit

We now want to describe in more detail the variables that help specify the state of a system. Consider a closed system in the absence of all external fields. This statement describes a system that we can only approximate on Earth. We cannot turn off gravity, for example. However, for many practical purposes we can ignore the effects of gravity.

If the system consists of a pure liquid, specifying the pressure P, volume V, and temperature T of the liquid is sufficient to specify many other properties of the liquid, such as its density, surface tension, refractive index, and so forth. These other properties of the system, and P, V, and T, are called *variables of state. The variables of state depend only on the state of the system, not on how the system arrived at that state.* The useful characteristic of variables of state is that when a few (P, V, T, chemical composition, etc.) are used to specify a system, all other variables of state are determined implicitly.

Early thermodynamic studies indicated that if work was done on a closed system with thermally insulated or *adiabatic* walls (so that there was negligible heat exchange between the system and the surroundings), the change of the state of the system was dependent only on the amount of the work done on the system, not on the particular type of work or how it was done. Similarly, if exchanges of both work and heat were involved, the change of the state of the system was found to depend only on the sum of the work w and the heat q. Thus, it appeared that there was a certain function of the state that changed according to w if $q = 0$ and to the sum of q and w if both were involved. From our discussion of the first law of thermodynamics, it is clear that this function is the energy E. Similar to P, V, or T, E is *a state function or a variable of state.* Because E is a property of the system, it is sometimes called the *internal energy of the system.* In contrast, it will become clear to us that heat and work are *not* state functions; they depend specifically on the way one state is changed to another. In other words, they depend on the *path* between states.

Any property of the system that depends only on the variables of the state must itself be a variable of the state. Sometimes it is convenient to define new variables of state by combining previously defined ones. The *enthalpy H,* for example, is such a variable. It is defined as

$$H \equiv E + PV \qquad \qquad \textbf{(2.14)}$$

We will see later in this chapter why H is a very useful parameter. The units of H are the same as E, usually in J or kJ or cal or kcal. In combining E and PV to give H, the units of PV must be the same as that used for E.

The variables of state are particularly convenient to use. The energy or enthalpy change when 1 teaspoon of sugar is converted to CO_2 and H_2O is the same whether sugar is burned in a reaction vessel or metabolized in the human body, as long as the initial and final states for the reaction are the same in the two cases. It does not matter whether the conversion involves a direct reaction with O_2 (combustion) or multiple enzyme-catalyzed steps inside a human being: *As long as a system goes from the same initial state to the same final state, the same value of ΔE or ΔH is obtained independent of the actual process that changes the system from its initial state to its final state.* A corollary of this statement is that *if a system undergoes a complex cycle of steps and returns to its initial state, ΔE, ΔH, or the change in any other variable of the state, is zero for the system.* In other words, *the change in any state variable of a system is zero for a cyclic process.*

Variables of state are divided into two classes: extensive and intensive. An *intensive variable of state* is one that remains unchanged when a system is subdivided. If 100 g of water at a uniform temperature is divided into two portions, the temperature of each part is unchanged. Thus, temperature is an intensive property. Similarly, the pressure of a system is also an intensive property. The value of an *extensive variable of state*, on the other hand, is changed by such a subdivision. The volume of a system, for example, is an extensive property. If the mass of a system of a fixed chemical composition is changed, the magnitude of any extensive variable is changed proportionally. Of the variables of state that we have considered so far, the extensive variables other than V are E, H, and the heat capacity C. However, we can always change an extensive variable to an intensive variable by expressing it in per-unit amount of material. Energy is extensive, but energy mol^{-1} or energy g^{-1} is intensive; mass is extensive, but density or mass volume^{-1} is intensive; heat capacity is extensive, but molar heat capacity is intensive.

Equations of State

A few variables of state are usually sufficient to specify all others. This means that equations exist that can relate the variables of state. Such equations are termed *equations of state*. The simplest and most frequently used equations of state link P, V, and T.

Solids or liquids

The volume of a solid or a liquid does not change very much with either pressure or temperature. For the time being, we will not consider changes from solid to liquid or any other change except P, V, and T. Therefore, a first approximation for the *equation of state of a solid or liquid is $V \cong$ constant*. This means that calculating the volume of a solid or liquid simply requires finding the density or specific volume at one temperature and using that value for any temperature and pressure. The volume is related to the density by

$$V = \frac{\text{mass}}{\text{density}}$$

Actually, the volume of a solid or liquid does change somewhat with T and P. Experimental data for 1 mol of liquid water are shown in figure 2.3.

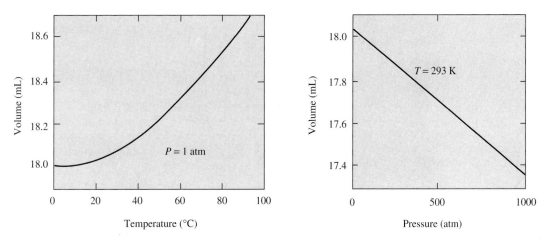

▲ FIGURE 2.3

Volume of 1 mol of liquid water as a function of temperature and pressure. Note that the volume changes by less than 5% within the temperature and pressure ranges shown.

The molar volume is plotted versus temperature at constant pressure in the left panel, and the molar volume is plotted versus pressure at constant temperature in the right panel. Equations can be obtained for V as a function of P and T for liquid water from these data.

For example, at 293 K the linear empirical equation

$$\overline{V} = 18.07(1 - 45.9 \times 10^{-6}\, P)\ \text{mL mol}^{-1} \qquad (P \text{ in atm})$$

closely represents the plot shown in the right panel of figure 2.3. The number $45.9 \times 10^{-6}\ \text{atm}^{-1}$ is the *isothermal compressibility* of liquid water at 293 K. Isothermal compressibilities, which represent the fractional decrease in volume per atmosphere increase in pressure at a particular temperature, are tabulated in handbooks of chemical and physical data. The dependence of the volume of liquid water on temperature (left panel of figure 2.3) is more complicated. Not only is the plot *not* a straight line, but also the molar volume of water actually has a minimum value at 277 K (4°C). For most of our applications, we use the first approximation, that V is independent of T and P for a solid or liquid.

Gases

For gases, the volume varies greatly with T and P, but the variation is nearly independent of the type of gas. Thus, there is a simple approximate equation of state for gases. The *ideal gas equation* is

$$PV = nRT \qquad\qquad\qquad \textbf{(2.15)}$$

where n = number of moles
$\quad R \equiv$ universal gas constant = 0.08206 L atm deg^{-1} mol^{-1}
$\quad\quad\ = 8.314\ \text{J K}^{-1}\ \text{mol}^{-1}$

A plot of how the volume of 1 mol of an ideal gas depends on pressure is shown in figure 2.4. Note that, although 1000 atm is necessary to change the volume of liquid water by 3%, a change from 1 atm to 1000 atm for an ideal gas will cause a change in volume by a factor of 1000. The ideal gas equation

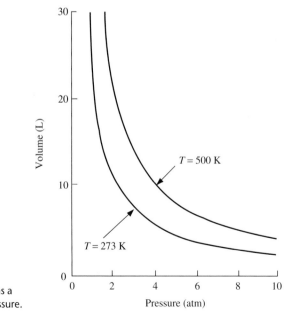

▶ FIGURE 2.4

Volume of 1 mol of an ideal gas as a function of temperature and pressure.

has the great advantage that it contains no constants applying to individual gases; it applies to all gases if the pressure is low enough. It is an exact limiting equation for all gases as P approaches zero. For higher pressures, it is an approximation for real gases. The answers obtained using Eq. (2.15) are usually accurate within ±10% for most gases near room temperature and atmospheric pressure. Of course, if the pressure causes the gas to liquify, the ideal gas equation cannot be used to calculate the volume of the liquid.

Several more accurate equations of state for gases have been developed. They are often corrections to the ideal gas equation that contain parameters relating to individual gases. One example is the *van der Waals gas equation:*

$$\left(P + \frac{n^2a}{V^2}\right)(V - nb) = nRT \qquad \text{(2.16)}$$

where a and b are constants that are different for each gaseous substance. The van der Waals a constant is a measure of the attractive forces between molecules, and the b constant is a measure of the intrinsic volume of the gas molecules themselves. When both of these are zero, Eq. (2.16) simplifies to the ideal gas equation.

The equations of state that we have been discussing have all been applied to systems containing only one component. For mixtures, the number of grams or moles of each component must be specified, and the equation of state will depend on the amounts of the components. For gases that can be approximated by the ideal gas equation, the results are particularly simple. The ideal gas equation can be applied to each gas in the mixture as if the others were not there. The partial pressure of each gas can be calculated as follows:

$$P_i = \frac{n_iRT}{V} \qquad \text{(2.17)}$$

The total pressure is just the sum of the partial pressures:

$$P_{\text{tot}} = \sum_i P_i \qquad \text{(2.18)}$$

$$= \frac{n_{\text{tot}}RT}{V} \qquad \text{(2.19)}$$

The partial pressures can also be obtained from the total pressure and the mole fractions X_i of each component:

$$P_i = X_iP_{\text{tot}}$$

$$X_i = \frac{n_i}{\Sigma_i\, n_i} \qquad \text{(2.20)}$$

The equations for partial pressures are useful because often we are interested in only one of the components of a gas mixture. For example, in the air we breathe, the partial pressure of oxygen or carbon dioxide is much more important than the total pressure.

Paths Connecting Different States

There is an infinite number of paths to get from one state to another, but some are more convenient than others, and they have received special attention. A path of much conceptual importance is the *reversible path.* We have already

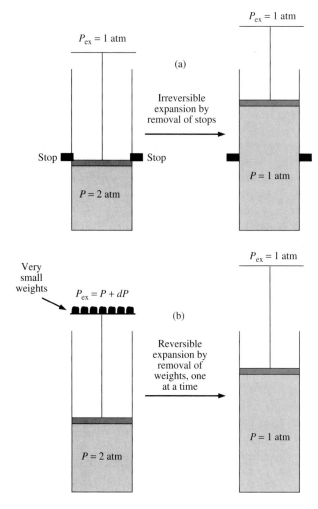

▶ FIGURE 2.5

Comparison of (a) an irreversible expansion and (b) a reversible expansion. In a reversible expansion, the internal pressure is always nearly equal to the external pressure.

mentioned reversible expansion or compression of a gas and reversible heating or cooling of a system (p. 23). In a reversible path, the system remains very near equilibrium at all points along the path. To examine such a path more closely, consider the expansion of an ideal gas in a cylinder with a frictionless and weightless piston, as illustrated in figure 2.5(a). The cylinder is assumed to conduct heat well, so the temperature of the system (the gas in the cylinder) is always the same as the temperature of the surroundings, which we keep constant. Suppose that the pressure outside the cylinder is always at 1 atm and the pressure inside the cylinder is initially at 2 atm. If we remove the stops that hold the piston in position, the gas will expand irreversibly until a final state is reached at which its pressure becomes 1 atm. During the course of expansion, the pressure of the system is always significantly greater than that of the surroundings (that is why the expansion is called *irreversible*). The two become the same only at the end of the expansion.

We can carry out the expansion in a different way. Instead of holding the piston in position with stops at the beginning, we put many small weights on top of the piston to make up for the pressure difference [figure 2.5(b)]. The expansion is then carried out in a stepwise manner by removing one weight at a time. If the number of weights is very large (and the weight of each very small), the pressure of the system is always almost the same as that of the surroundings during the course of expansion. Furthermore, if we add rather than remove a weight, the process will be reversed.

A reversible path is one in which the process can be reversed at any instant by an infinitesimal change of the variable that controls the process. The path illustrated in figure 2.5(a) is not a reversible one; the path illustrated in figure 2.5(b) becomes a reversible one when the weights are very very small and their number approaches infinity.

As a second example, let's consider the transition from liquid water to gaseous water (steam). The transition is reversible at 100°C and 1 atm. If we maintain the temperature at 100°C, liquid water will evaporate to steam if the pressure is lowered to slightly below 1 atm, and steam will condense to liquid water if the pressure is increased to slightly above 1 atm. Alternatively, if the pressure is maintained exactly at 1 atm, the direction of the change can be reversed by causing the temperature of the surroundings to change from slightly above to slightly below 100°C or vice versa.

There are many possible reversible paths for a particular process. Suppose, for example, we want to use a reversible path to calculate the energy needed to evaporate 1 mol of liquid water at 25°C and 1 atm. We can choose either of the reversible paths shown in figure 2.6. Each breaks the overall process into three

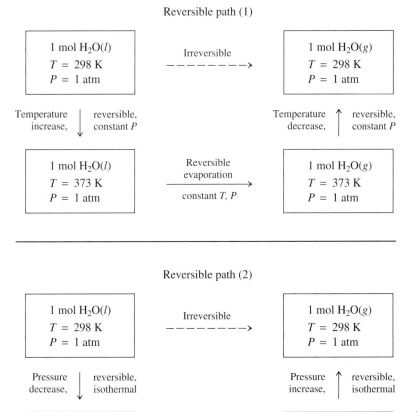

◀ FIGURE 2.6

Two possible reversible paths that can be taken between the same initial and final states. The letters in parentheses (*l*) and (*g*) denote, respectively, liquid and gas.

successive steps. Reversible path (1) consists of raising the temperature of liquid water reversibly to 100°C, the normal boiling point, then evaporating the liquid reversibly at 1 atm and finally cooling the vapor reversibly back to 25°C. In reversible path (2), the pressure on the liquid water is decreased reversibly until it will evaporate reversibly at 25°C; the final step then is to bring the water vapor pressure reversibly back to 1 atm. Reversible path (1) is a constant-pressure path; it is said to be *isobaric*. Reversible path (2) is at constant temperature; it is said to be *isothermal*. Several other named paths are listed below:

Restricted thermodynamic paths		
Constant pressure	Isobaric	$\Delta P_{sys} = 0$
Constant temperature	Isothermal	$\Delta T_{sys} = 0$
Constant volume	Isochoric	$\Delta V = 0$
No heat transferred	Adiabatic	$q = 0$
Final state = initial state	Cyclic	No change in any property of state of system

Whichever path we choose, the same value for $\Delta E = E_2 - E_1$, where E_1 is the energy of the system at its initial state and E_2 is that of the system at its final state, would be obtained. As we have mentioned before, *the change in any variable of state of the system during a process depends only on the initial and final states of the system and not on the path along which the process occurs.* This statement also holds for the direct, irreversible path shown in figure 2.6. Reversible paths are stressed because it is often easier to calculate q and w along reversible paths, from which ΔE can be then calculated from Eq. (2.12). As we will see later in this chapter and again in the next chapter, the work a system does on the surroundings for a given change in energy is also a maximum along a reversible path.

Experimentally, supercooled gaseous water with a temperature of 25°C and a pressure of 1 atm has never been obtained. The thermodynamic properties of water vapor under these conditions can be readily calculated, however, from the choice of any convenient path. This particular example serves to illustrate the power of thermodynamics.

Dependence of the Energy and Enthalpy of a Pure Substance on *P, V,* and *T*

It is often important to know how temperature and pressure affect phase changes (the melting of a solid, the condensation of a gas, etc.), chemical reactions, or other chemical and physical processes. We first consider the effects of T and P on the energy and enthalpy of pure solids, liquids, and gases. The results can then be applied directly to the effects of T and P on the enthalpy and energy changes of phase transitions and of chemical reactions. The methods we use in the next several sections are similar to those necessary to calculate the effects of T and P on other thermodynamic properties (see later chapters).

Liquids or solids

Consider a one-component system that undergoes a change in *P, V,* and *T* only. External fields, friction, and surface effects are ignored. Only two of the three variables P, V, and T need be specified because the equation of state allows calculation of the third. The number of moles or grams of the component does have to be specified (V is an extensive property). Our goal here is to calculate the energy and enthalpy changes for a variety of processes. One can get from the initial state to the final state by many paths. It is usually easiest to use a series of reversible steps, where either P, V, or T is held constant in each one; the energy or enthalpy contributions from the individual steps are then calculated and summed up to give the overall energy or enthalpy change. For illustration we use one of the most important biological molecules there is, water. Some properties of water are given in table 2.2.

TABLE 2.2   Physical Properties of Water, H_2O (mol wt = 18.016), at 1 atm*

Solid H_2O = ice
(at 0°C)

Density = 0.915 g cm^{-3}; specific volume = 1.093 cm^3 g^{-1}
Vapor pressure = 4.579 torr
Heat of melting = 333.4 kJ kg^{-1} = 6.007 kJ mol^{-1}
Absolute molar entropy = 41.0 J K^{-1} mol^{-1}
Specific heat capacity = 2.113 kJ K^{-1} kg^{-1}
Molar heat capacity = 38.07 J K^{-1} mol^{-1}

Liquid H_2O

Temperature (°C)	Density (g cm^{-3})	Surface tension (mN m^{-1})	Vapor pressure (Torr)	Heat of vaporization (kJ kg^{-1})	Viscosity (mPa s)
0	0.9999	75.64	4.579	2493	1.7921
20	0.9982	72.75	17.535	2447	1.0050
40	0.9922	69.56	55.324	2402	0.6560
60	0.9832	66.18	149.38	2356	0.4688
80	0.9718	62.61	355.1	2307	0.3565
100	0.9584	58.85	760.00	2257	0.2838

Absolute molar entropy = 63.2 J K^{-1} mol^{-1} at 0°C
= 87.0 J K^{-1} mol^{-1} at 100°C
Specific heat capacity = 4.18 kJ K^{-1} kg^{-1} between 0 and 100°C
Molar heat capacity = 75.4 J K^{-1} mol^{-1}
Heat of freezing = -333.4 kJ kg^{-1} at 0°C

Gaseous H_2O = steam
(at 100°C)

Density = 5.880 × 10^{-4} g cm^{-3}; specific volume = 1701 cm^3 g^{-1}
Absolute molar entropy = 196.2 J K^{-1} mol^{-1}
Specific heat capacity at constant pressure = 1.874 kJ K^{-1} kg^{-1}
Molar heat capacity at constant pressure = 33.76 J K^{-1} mol^{-1}
Heat of condensation = -2257 kJ kg^{-1} = -40.66 kJ mol^{-1}

*Some of the properties listed will be defined and discussed in later chapters.

First, let's consider heating or cooling some liquid water in an open flask at a constant pressure provided by the atmosphere. The reaction is

$$n \text{ mol } H_2O(l) \text{ at } T_1, P_1, V_1 \longrightarrow n \text{ mol } H_2O(l) \text{ at } T_2, P_1, V_2$$

The heat is calculated from Eq. (2.9a); at constant P:

$$q_P = \int_{T_1}^{T_2} C_P \, dT \tag{2.21}$$

The subscript P on the heat q and the heat capacity C reminds us that the heat effects depend on the path. C_P is the heat capacity at constant pressure. In general, C_P will depend on the pressure and the temperature. However, we can usually neglect the effects of P and often neglect the effect of temperature, that is, C_P for 1 mol of liquid water is close to 75.4 J K^{-1} from 0 to 100°C and for pressures less than a few hundred atmospheres. Therefore, for a temperature change at constant P, if C_P is independent of T,

$$q_P = C_P(T_2 - T_1) \tag{2.22}$$

Because values of heat capacity are often given per mole, we can write

$$q_P = n\overline{C}_P(T_2 - T_1) \tag{2.22a}$$

If T_2 is greater than T_1, we know that heat is absorbed, which is consistent with the positive sign of q_P.

The work is calculated from Eq. (2.6). At constant P,

$$w_P = -P_1(V_2 - V_1) \tag{2.6}$$

But if we assume that the volume change of the liquid water is negligible, then $V_2 - V_1 \cong 0$ and $w_P \cong 0$.

If the water is heated in a closed and very strong container that keeps the volume constant ($T_1, P_1, V_1 \rightarrow T_2, P_2, V_1$), the heat and work are, at constant V,

$$q_V = n\overline{C}_V(T_2 - T_1) \tag{2.23}$$

$$w_V = 0 \tag{2.24}$$

C_V is the heat capacity at constant volume; for a solid or a liquid, it is not very different in magnitude from C_P. The PV work for a constant volume process is obviously zero.

If the water is kept at constant temperature while the pressure is changed, there is only a negligible volume change, and there is no appreciable work done or heat transferred. Thus, for an isothermal process involving liquid or solid

$$q_T \cong 0 \tag{2.25}$$

$$w_T \cong 0 \tag{2.26}$$

To obtain ΔE for the changes discussed above, we just sum up q and w:

$$\Delta E = q_P + w_P \tag{2.12a}$$

$$= n\overline{C}_P(T_2 - T_1) - P(V_2 - V_1) \tag{2.12b}$$

where $P = P_1 = P_2$ because the pressure remains unchanged for an isobaric (constant-pressure) process. ΔE can also be expressed if an isochoric (constant-volume) process is used:

$$\Delta E = q_V + w_V = q_V + 0 = n\overline{C}_V(T_2 - T_1) \qquad \textbf{(2.12c)}$$

To obtain ΔH we use the definition of enthalpy:

$$\Delta H = \Delta(E + PV) = \Delta E + \Delta(PV) \qquad \textbf{(2.14a)}$$

Substituting $\Delta E = n\overline{C}_P(T_2 - T_1) - P(V_2 - V_1)$ for the constant pressure process into the above equation, we see that the $P\Delta V$ terms cancel to give

$$\Delta H = n\overline{C}_P(T_2 - T_1) \qquad \textbf{(2.27)}$$

Because the volume of a solid or a liquid does not change much with temperature or pressure, the term $-P(V_2 - V_1)$ in Eq. (2.12b) is small. Thus, for temperature or pressure changes of a solid or a liquid, to a good approximation

$$\Delta E = \Delta H$$

$$C_P = C_V$$

The exact relations between C_P and C_V or ΔE and ΔH depend on the equation of state.

To summarize, for any change of P_1, V_1, T_1 to P_2, V_2, T_2 we can obtain ΔE and ΔH by choosing a convenient path and then combining the q's and w's. For example, we could use an isothermal plus an isobaric path or an isothermal plus an isochoric path.

EXAMPLE 2.9

Calculate ΔE and ΔH in joules for heating 1 mol of liquid water from 0°C and 1 atm to 100°C and 10 atm. The volume per gram of the water is essentially independent of pressure; it can be calculated from the average density of water, 0.98 g cm^{-3}, given in table 2.2.

Solution

Choose a path such as an isothermal path plus an isobaric path. First, the pressure is raised from 1 atm to 10 atm at 0°C. Then, the temperature is raised from 0°C to 100°C at 10 atm. The overall energy change ΔE is the sum of energy changes for the two steps, ΔE (isothermal) for the constant-temperature path and ΔE (isobaric) for the constant-pressure path:

$$\Delta E = \Delta E \text{ (isothermal)} + \Delta E \text{ (isobaric)}$$

$$= q_T + w_T + q_P + w_P$$

For a liquid, q_T and $w_T \cong 0$ [Eqs. (2.25), (2.26)] and $w_P \cong 0$ because the volume change is small. Thus,

$$\Delta E \cong q_P$$

$$= n\overline{C}_P(T_2 - T_1)$$

$$= (1 \text{ mol})(75.4 \text{ J mol}^{-1} \text{ deg}^{-1})(100 \text{ deg})$$

$$= 7540 \text{ J}$$

To calculate ΔH, we use Eq. (2.14a):

$$\Delta H = \Delta E + (P_2V_2 - P_1V_1)$$

Because the volume of a liquid is essentially independent of pressure, $V_2 \cong V_1$ and $P_2V_2 \cong P_2V_1$. Thus,

$$
\begin{aligned}
\Delta H &= \Delta E + (P_2V_2 - P_1V_1) \\
&= \Delta E + (P_2V_1 - P_1V_1) \\
&= \Delta E + (P_2 - P_1)V_1 \\
&= 7540\ \text{J} + (10\ \text{atm} - 1\ \text{atm})\left(\frac{18.0\ \text{g}}{\text{mol}}\right)\left(\frac{\text{cm}^3}{0.98\ \text{g}}\right)(1\ \text{mol}) \\
&= 7540\ \text{J} + (165\ \text{cm}^3\ \text{atm})\left(0.1013\ \frac{\text{J}}{\text{cm}^3\ \text{atm}}\right) \\
&= 7540\ \text{J} + 16.7\ \text{J} \\
&= 7557\ \text{J}
\end{aligned}
$$

Gases

We will now calculate ΔE and ΔH when gaseous H_2O changes from P_1, V_1, T_1 to P_2, V_2, T_2. The heat transferred at constant P or V has the same form as for solid or liquid water:

$$q_P = n\overline{C}_P(T_2 - T_1) \tag{2.22a}$$

$$q_V = n\overline{C}_V(T_2 - T_1) \tag{2.23}$$

Of course, the heat capacities for gaseous H_2O must be used here.

The constant-pressure work deserves more discussion. In general, for solids or liquids, we can ignore the expansion work, but for gases it is significant. The equation for calculating constant-pressure work is the usual one:

$$w_P = -P_{ex}(V_2 - V_1) \qquad \text{(expansion, constant } P_{ex})$$

Figure 2.7 illustrates the process. Note that for an expansion the external pressure P_{ex} can be any pressure smaller than P. The volumes V_1 and V_2 are fixed by stops that hold the cylinder at these chosen values. The temperatures T_1 and T_2 are controlled by thermostats, and the pressure of the gas P depends on the number of moles of the gas and the equation of state. For the ideal gas $PV_1 = nRT_1$ and $PV_2 = nRT_2$. The maximum constant-pressure work would be done by the gas if the gas could be heated reversibly, keeping the external pressure just slightly less than the gas pressure P at all points during the expansion. For such a reversible expansion or compression,

$$P_{ex} \cong P \qquad \text{(reversible)}$$

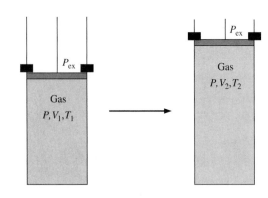

▶ FIGURE 2.7

Expansion of a gas at constant pressure, P_{ex}. For a reversible expansion, $P_{ex} = P$; for an irreversible expansion, $P_{ex} < P$.

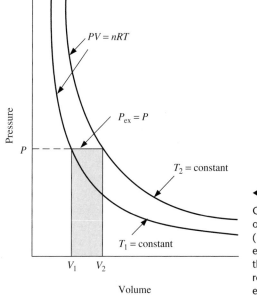

Pressure

$PV = nRT$

$P_{ex} = P$

P

$T_2 = \text{constant}$

$T_1 = \text{constant}$

V_1 V_2

Volume

◀ FIGURE 2.8

Constant-pressure work is the area of rectangle of height P_{ex} and width $(V_2 - V_1)$ on a P versus V plot. The shaded area represents the work done by the system when $P_{ex} = P$. The curves represent paths of isothermal expansion.

Combining Eq. (2.6) and the ideal gas equation, we obtain

$$w_P = -P(V_2 - V_1) = -nR(T_2 - T_1) \qquad \textbf{(2.28)}$$

Clearly, if the pressure is constant and the volume changes, the temperature must also change as required by the ideal gas law. The difference in volume $(V_2 - V_1)$ and the pressure determine the difference in temperature.

Figure 2.8 illustrates that constant-pressure work can be thought of as the area of a rectangle of height P_{ex} and width $(V_2 - V_1)$ in a P versus V plot. If P_{ex} equals P for the gas, the constant-pressure work done by the gas is a maximum for a given change of volume, $V_2 - V_1$. If P_{ex} equals zero, the work is zero. For any P_{ex} between these extremes, the work done is intermediate. In practical engines, the external pressure must be significantly less than the gas pressure inside the piston, however.

If a gas is expanded *reversibly* and *isothermally* from an initial volume V_1 to a final volume V_2, the pressure cannot be kept constant during the expansion. In this case, we use the general expression for the work and write P as a function of volume:

$$w = -\int_{V_1}^{V_2} P_{ex}\, dV$$

$$P_{ex} \cong P \qquad \text{(reversible)}$$

$$w = -\int_{V_1}^{V_2} P\, dV$$

$$= -\int_{V_1}^{V_2} \frac{nRT}{V}\, dV \qquad \text{(ideal gas)}$$

$$= -nRT \int_{V_1}^{V_2} \frac{dV}{V} \qquad \begin{array}{l} (T = \text{constant for an isothermal process;} \\ \quad n = \text{constant for a fixed number of moles of a gas)} \end{array}$$

$$= -nRT \ln \frac{V_2}{V_1} \qquad \begin{array}{l} \text{(reversible, isothermal expansion} \\ \text{or compression of ideal gas)} \end{array}$$

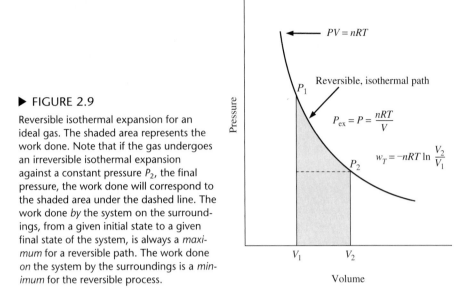

▶ **FIGURE 2.9**

Reversible isothermal expansion for an ideal gas. The shaded area represents the work done. Note that if the gas undergoes an irreversible isothermal expansion against a constant pressure P_2, the final pressure, the work done will correspond to the shaded area under the dashed line. The work done *by* the system on the surroundings, from a given initial state to a given final state of the system, is always a *maximum* for a reversible path. The work done *on* the system by the surroundings is a *minimum* for the reversible process.

This work is represented by the crosshatched area in figure 2.9. For an isothermal, reversible, ideal gas expansion or compression,

$$w_T = -nRT \ln \frac{V_2}{V_1} \tag{2.29a}$$

Because volume is inversely proportional to pressure for an ideal gas at constant temperature,

$$w_T = -nRT \ln \frac{P_1}{P_2} \tag{2.29b}$$

For an ideal gas, we will show later that *its energy E at a constant temperature is independent of the volume* (see "Relations Among Partial Derivatives" in chapter 3). A way of rationalizing this conclusion, from a molecular point of view, is that the ideal gas molecules do not interact with each other, and thus the energy of the gas will not depend on whether the molecules are close together (small volume) or far apart (large volume). If ΔE is zero for isothermal expansion of an ideal gas, then the first law requires that

$$q_T = -w_T$$

Thus for an isothermal, reversible expansion or compression of an ideal gas,

$$q_T = nRT \ln \frac{V_2}{V_1} = nRT \ln \frac{P_1}{P_2} \tag{2.30}$$

To calculate ΔE and ΔH for an ideal gas changing from an initial state specified by P_1, V_1, T_1 to a final state specified by P_2, V_2, T_2, we can use any convenient path. Moreover, we need to specify only two of the three variables P, V, and T because $PV = nRT$. For example, we can choose an isothermal path followed by an isobaric path $(T_1, P_1) \rightarrow (T_1, P_2) \rightarrow (T_2, P_2)$, as we did in example 2.9 for liquid water. We can choose an isothermal path followed by an isochoric path $(T_1, V_1) \rightarrow (T_1, V_2) \rightarrow (T_2, V_2)$. In either choice we need to specify only two of the three parameters T, P, and V at each step, because the third can be calculated from the other two from the equation of state $PV = nRT$.

To calculate the energy change and the enthalpy change for the isothermal plus isobaric path, we use

$$\Delta E = \Delta E \text{ (isothermal)} + \Delta E \text{ (isobaric)}$$
$$= 0 + \Delta E \text{ (isobaric)} \qquad [\Delta E \text{ (isothermal)} = 0 \text{ for ideal gas}]$$
$$= q_P + w_P$$
$$= n\overline{C}_P(T_2 - T_1) + [-nR(T_2 - T_1)] \qquad [\text{Eqs. (2.22a), (2.28)}]$$
$$= n(\overline{C}_P - R)(T_2 - T_1)$$
$$\Delta H = \Delta E + \Delta(PV)$$
$$= n(\overline{C}_P - R)(T_2 - T_1) + nR(T_2 - T_1)$$
$$= n\overline{C}_P(T_2 - T_1)$$

For an isothermal path plus an isochoric path, $(T_1, V_1) \rightarrow (T_1, V_2) \rightarrow (T_2, V_2)$

$$\Delta E = \Delta E \text{ (isothermal)} + \Delta E \text{ (isochoric)}$$
$$= 0 + q_V + w_V$$
$$= 0 + n\overline{C}_V(T_2 - T_1) + 0$$
$$= n\overline{C}_V(T_2 - T_1)$$
$$\Delta H = \Delta E + \Delta(PV)$$
$$= n\overline{C}_V(T_2 - T_1) + nR(T_2 - T_1)$$
$$= n(\overline{C}_V + R)(T_2 - T_1)$$

Because ΔH and ΔE must be independent of path, a comparison of the results obtained for the isothermal plus isobaric path and those for the isothermal plus isochoric path shows that for an ideal gas

$$\overline{C}_P = \overline{C}_V + R \qquad \text{(2.31a)}$$

or

$$C_P = C_V + nR \qquad \text{(2.31b)}$$

For *any* change of P, V, and T for an ideal gas,

$$\Delta E = n\overline{C}_V(T_2 - T_1) \qquad \text{(2.32)}$$
$$\Delta H = n\overline{C}_P(T_2 - T_1) \qquad \text{(2.33)}$$

We have assumed for simplicity that C_P and C_V are independent of temperature in Eqs. (2.32) and (2.33). For real gases, the heat capacity will be a function of temperature. This is often expressed in terms of a power series

$$\overline{C} = a + bT + cT^2 + \cdots$$

where values of the coefficients a, b, and c are determined empirically and tabulated for different gases. Using such data, which are often valid over a temperature range from 300 to 1500 K, energy and enthalpy changes can be calculated by integration:

$$\Delta E = n \int_{T_1}^{T_2} \overline{C}_V \, dT \tag{2.34}$$

$$\Delta H = n \int_{T_1}^{T_2} \overline{C}_P \, dT \tag{2.35}$$

Relations Between Heat Exchanges and ΔE and ΔH

Although the heat exchange q between the system and the surroundings is dependent on path, for particular paths q can be readily calculated from ΔE or ΔH, which are state functions independent of path. *If we consider processes that involve no work other than PV work, then for a closed system undergoing an isochoric process the work w_V is zero and the heat absorbed q_V is equal to the energy change:*

$$q_V = \Delta E \tag{2.12c}$$

Similarly, *for a closed system undergoing an isobaric process, the heat absorbed is equal to the enthalpy change if the process involves no work other than the PV-type work:*

$$
\begin{aligned}
\Delta H &= \Delta E + \Delta(PV) & &\text{(definition of } H) \\
&= \Delta E + P\Delta V & &\text{(constant pressure)} \\
&= q_P + w_P + P\Delta V & &\text{(first law)} \\
&= q_P - P\Delta V + P\Delta V & &\text{(pressure–volume work only)} \\
&= q_P
\end{aligned}
$$

Thus, at constant pressure,

$$q_P = \Delta H \tag{2.36}$$

The enthalpy H is also known as the *heat content,* because the enthalpy change ΔH at constant pressure is equal to the heat absorbed. Equations (2.35) and (2.36) are generally applicable whether the system is a gas, a liquid, or a solid; the system may undergo heating or cooling, or it may undergo a phase transition or a chemical reaction (see later sections). We emphasize, however, that *the derivation of Eqs. (2.35) and (2.36) requires that only pressure–volume work occurs. If the system does other types of work (electrical, for example) during a process, these equations do not apply.* For heating or cooling of a system in the absence of a phase transition or chemical reaction,

$$\frac{dq}{dT} = C \tag{2.9}$$

If the process is isochoric, combining Eqs. (2.12c) and (2.34) gives

$$\Delta E = q_V = \int_{T_1}^{T_2} C_V \, dT \tag{2.37}$$

Similarly, for an isobaric process,

$$\Delta H = q_P = \int_{T_1}^{T_2} C_P \, dT \tag{2.38}$$

Note that Eqs. (2.37) and (2.38) lead readily to Eqs. (2.32) and (2.33), which we derived earlier for the specific case of an ideal gas.

Phase Changes

The preceding section has dealt with changes of P, V, and T only. Now we want to consider changes in phase. The names of some of the phase changes are as follows:

Phase change	Name
Gas → liquid or solid	Condensation
Solid → liquid	Fusion, melting
Liquid → solid	Freezing
Liquid → gas	Vaporization
Solid → gas	Sublimation

There can also be phase changes between different solid phases and between different liquid phases. We are particularly interested in the thermodynamics of reversible changes that occur at constant T and P. If water in a beaker is heated on a hot plate at a pressure of 1 atm, the temperature of the water increases until a temperature of 100°C is reached. It then undergoes a phase transition (vaporization): The temperature stays at 100°C until all the water is boiled off. A phase change is usually a good way to store energy. It takes about 75 J to heat 1 mol of water from 99°C to 100°C, but it takes 40,660 J to change 1 mol of water from liquid to gas at 100°C.

Consider a reversible phase change from phase a to phase b at constant T and P. The work done on the system is

$$w_P = -P\Delta V$$

where ΔV = volume change = V(phase b) − V(phase a). The heat absorbed by the system at constant P is q_P. It is equal to $\Delta H = H$(phase b) − H(phase a), according to Eq. (2.36).

Values of ΔH have been tabulated for various reversible phase changes. For any process at constant pressure, the heat and enthalpy change are equal if only PV-type work is involved. Thus, one can speak about a heat of vaporization, or a heat of fusion, or an enthalpy of vaporization or fusion. The energy change of a phase change at constant P is

$$\Delta E = \Delta H - P\Delta V \qquad \qquad \textbf{(2.39)}$$

Often a ΔH or a ΔE value is known at one set of T and P but is needed at another. Suppose we want to calculate the amount of heat removed when liquid water evaporates at human skin temperature (around 35°C). We want to know how effectively vaporization of sweat can cool us. If we know ΔH for vaporization of water only at 100°C, we can use an indirect path to calculate the ΔH for the desired reaction at 35°C:

$$H_2O(l),\ T_2 = 35°C,\ P = 1\ \text{atm} \xrightarrow{\ \Delta H(35°C)\ } H_2O(g),\ T_2 = 35°C,\ P = 1\ \text{atm}$$

constant P ↓   ↗ constant P

$$H_2O(l),\ T_1 = 100°C,\ P = 1\ \text{atm} \xrightarrow{\ \Delta H(100°C)\ } H_2O(g),\ T_1 = 100°C,\ P = 1\ \text{atm}$$

We can calculate ΔH for each step in the chosen path and add the ΔH's to obtain the overall ΔH:

$$\Delta H(35°C) = C_P(l)(373 - 308) + \Delta H(100°C) + C_P(g)(308 - 373)$$
$$= \Delta H(100°C) + n[\overline{C}_P(g) - \overline{C}_P(l)](-65)$$

The generalization for the temperature dependence of enthalpy of a phase change is

$$\Delta\overline{H}(T_2) = \Delta\overline{H}(T_1) + \Delta\overline{C}_P(T_2 - T_1) \tag{2.40}$$

where $\Delta\overline{H} = \overline{H}(\text{phase b}) - \overline{H}(\text{phase a})$
$\Delta\overline{C}_P = \overline{C}_P(\text{phase b}) - \overline{C}_P(\text{phase a})$

EXAMPLE 2.10

Check the value for ΔH of vaporization of water at 20°C given in table 2.2 by using Eq. (2.40) and ΔH (100°C) from the table.

Solution

$$\Delta\overline{H}(20°C) = \Delta\overline{H}(100°C) + \Delta\overline{C}_P(T_2 - T_1) \tag{2.40}$$

Equation (2.40) is written for 1 mol of substance, but it is easily adapted to 1 kg of substance:

$$\Delta H(20°C, 1\text{ kg } H_2O) = 2257\text{ kJ kg}^{-1} + (1.874\text{ kJ K}^{-1}\text{ kg}^{-1}$$
$$- 4.18\text{ kJ K}^{-1}\text{ kg}^{-1})(-80\text{ K})$$
$$= 2257\text{ kJ kg}^{-1} + 184.5\text{ kJ kg}^{-1}$$
$$= 2442\text{ kJ kg}^{-1}$$

The table gives 2447 kJ kg^{-1}. The temperature dependence of the heat capacities accounts for the small discrepancy.

The energy of a phase change at constant P can be obtained from Eq. (2.39). If one of the phases in the phase change is a gas (condensation, vaporization, or sublimation), the volume of the solid or liquid is so much smaller than the volume of the gas that it can be ignored. Furthermore, the gas phase can be approximated as an ideal gas. As an example, for vaporization

$$\Delta\overline{E} = \Delta\overline{H} - P[\overline{V}(g) - \overline{V}(l)]$$
$$\cong \Delta\overline{H} - P\overline{V}(g)$$
$$\cong \Delta\overline{H} - RT$$

EXAMPLE 2.11

Calculate (a) the change of energy on freezing 1.00 kg of liquid water at 0°C and 1 atm, and (b) the change of energy on vaporizing 1.00 kg of liquid water at 0°C and 1 atm.

Solution

a. $\Delta E = \Delta H - P\Delta V$ **(2.39)**

   $\qquad = -333.4 \text{ kJ kg}^{-1}$

   $\qquad -(1 \text{ atm})(1.093 \text{ cm}^3 \text{ g}^{-1} - 1.000 \text{ cm}^3 \text{ g}^{-1})(1000 \text{ g kg}^{-1})$

   $\qquad = -333.4 \text{ kJ kg}^{-1} - 93 \text{ cm}^3 \text{ atm kg}^{-1}$

   $\qquad = -333.4 \text{ kJ kg}^{-1} - (93 \text{ cm}^3 \text{ atm kg}^{-1})\left(1.013 \times 10^{-4}\dfrac{\text{kJ}}{\text{cm}^3 \text{ atm}}\right)$

   $\qquad = -333.4 \text{ kJ kg}^{-1} - 0.009 \text{ kJ kg}^{-1}$

   $\qquad = -333.4 \text{ kJ kg}^{-1}$

b. $\Delta E = \Delta H - nRT$

   $\qquad = 2493 \text{ kJ kg}^{-1}$

   $\qquad -(8.314 \text{ J K}^{-1} \text{ mol}^{-1})(273.1 \text{ K})\left(\dfrac{1 \text{ mol}}{0.018016 \text{ kg}}\right)\left(\dfrac{1 \text{ kJ}}{1000 \text{ J}}\right)$

   $\qquad = 2493 \text{ kJ kg}^{-1} - 126.0 \text{ kJ kg}^{-1}$

   $\qquad = 2367 \text{ kJ kg}^{-1}$

Note that there is a significant difference between ΔH and ΔE when one of the phases is a gas [part (b)], but the difference is insignificant when neither phase is a gas [part (a)].

Chemical Reactions

We come next to the most important way the energy and enthalpy of a system can be changed: A chemical reaction can occur. The chemical change in a system can be represented by the general reaction

$$n_{\text{A}}\text{A} + n_{\text{B}}\text{B} \longrightarrow n_{\text{C}}\text{C} + n_{\text{D}}\text{D}$$

That is, n_{A} mol of A react with n_{B} mol of B to give n_{C} mol of C and n_{D} mol of D. More terms can be included if there are more reactants and/or products. The conditions of P, V, and T must be specified for both the products and the reactants. The change in a property of the state for the chemical reaction, such as ΔH, can be expressed in terms of the algebraic sum of this property for the reactants and products:

$$\Delta H = H(\text{products}) - H(\text{reactants})$$
$$\Delta H = n_{\text{C}}\overline{H}_{\text{C}} + n_{\text{D}}\overline{H}_{\text{D}} - n_{\text{B}}\overline{H}_{\text{B}} - n_{\text{A}}\overline{H}_{\text{A}} \qquad \textbf{(2.41)}$$

where $\overline{H}$ = enthalpy mol^{-1}. Similar equations can be written for other variables of state. To emphasize, *for any change in a variable of state, it does not matter how the reaction actually takes place; only the initial and final states are important.*

Heat Effects of Chemical Reactions

We have already shown that when there is no work other than the PV type,

$$\Delta E = q_V \qquad \textbf{(2.12c)}$$

$$\Delta H = q_P \qquad \textbf{(2.36)}$$

By measurement of the heat effects of chemical reactions at constant V or P, values of ΔE and ΔH for the reactions are obtained. The heat effect is usually measured by surrounding the reaction vessel with a known amount of a liquid (water, for example) and measuring the temperature rise of the liquid. (The heat capacity of the surrounding liquid can be calibrated by measuring the temperature rise when a given amount of electrical energy is passed through a heating coil immersed in it.) If the reaction is carried out in a vessel with strong walls that keep the volume constant (a "bomb calorimeter"), the measured heat of a reaction gives ΔE; if the reaction is carried out at constant pressure, the measured heat of reaction gives ΔH. Note that if either ΔE or ΔH is measured, the other can be obtained from the relation

$$\Delta H \equiv \Delta E + \Delta(PV) \qquad \textbf{(2.14a)}$$

In the equation above, $\Delta(PV)$ is equal to PV of the products minus PV of the reactants. If gases are involved in the reaction, we can ignore the volumes of the solids or liquids and use the ideal gas equation for the gases.

Reactions that give off heat (q negative, ΔE negative at constant volume, and ΔH negative at constant pressure) are said to be *exothermic*, and reactions that absorb heat (q positive, ΔE positive at constant volume, and ΔH positive at constant pressure) are termed *endothermic*. The amount of heat that can be obtained from a chemical reaction is of great practical importance.

The biochemical reactions necessary to sustain life in an adult person produce about 6000 kJ day^{-1} of heat, which must be replenished in the form of chemical energy from food. This basal metabolic rate is about the same as the power requirement of a 70-W lightbulb (1 W = 1 J s^{-1}). If one does more than resting in bed, more than this basal level of energy is needed, typically in the range of 8000–12,000 kJ day^{-1} per person. Each gram of protein or carbohydrate provides about 15 kJ, and each gram of fat provides about 35 kJ. (The energy unit used by nutritionists is Cal, 1 Cal = 1 kcal = 4.184 kJ).

The heat at constant pressure, and thus the enthalpy changes for many reactions, have been measured. These reactions and their enthalpies can be combined to calculate the enthalpies for many other reactions. For example, suppose we know ΔH_1 for the oxidation of solid glycine at 25°C to form CO_2, ammonia, and liquid water:

(1) $3O_2(g, 1 \text{ atm}) + 2NH_2CH_2COOH(s) \longrightarrow$
                          glycine

     $4CO_2(g, 1 \text{ atm}) + 2H_2O(l) + 2NH_3(g, 1 \text{ atm})$ $\Delta H_1 = -1163.5 \text{ kJ mol}^{-1}$

The ΔH_2 for the hydrolysis of solid urea is also known:

(2) $H_2O(l) + H_2NCONH_2(s) \longrightarrow CO_2(g, 1 \text{ atm}) + 2NH_3(g, 1 \text{ atm})$
                urea

$$\Delta H_2 = 133.3 \text{ kJ mol}^{-1}$$

If we subtract these two reactions, treating the chemicals and their ΔH's as algebraic quantities, we get

$3O_2(g, 1 \text{ atm}) + 2 \text{ glycine}(s) - \text{urea}(s) - H_2O(l) \longrightarrow$

            $4CO_2(g, 1 \text{ atm}) + 2H_2O(l) + 2NH_3(g, 1 \text{ atm})$

                        $- CO_2(g, 1 \text{ atm}) - 2NH_3(g, 1 \text{ atm})$

Rearranging and canceling, we get

(3) $3O_2(g, 1 \text{ atm}) + 2 \text{ glycine}(s) \longrightarrow$

$$1 \text{ urea}(s) + 3CO_2(g, 1 \text{ atm}) + 3H_2O(l)$$

$$\Delta H_3 = \Delta H_1 - \Delta H_2$$
$$= -1296.8 \text{ kJ mol}^{-1}$$

This equation is of more biochemical interest because urea, rather than ammonia, is the main oxidative metabolic product of amino acids.* However, the biological reaction does not involve solid glycine and solid urea but rather aqueous solutions. Therefore, we use the reactions and heats for the dissolution of 1 mol of urea and of glycine:

(4) $\text{glycine}(s) + \infty H_2O(l) \longrightarrow \text{glycine}(aq)$

$$\Delta H_4 = 15.69 \text{ kJ (mol glycine)}^{-1}$$

(5) $\text{urea}(s) + \infty H_2O(l) \longrightarrow \text{urea}(aq) \quad \Delta H_5 = 13.93 \text{ kJ (mol urea)}^{-1}$

The enthalpies of solution will depend on concentration: Here we will use enthalpies for very dilute solutions (designated aq) and assume that they do not depend on concentration. That means that we choose the number of moles of $H_2O(l)$ to be a very large number, infinite ($\equiv \infty$), in the preceding equations. From reaction (3) we now subtract two times reaction (4) and add reaction (5):

(6) $3O_2(g, 1 \text{ atm}) + 2 \text{ glycine}(s) - 2 \text{ glycine}(s)$

$$- \infty H_2O(l) + \text{urea}(s) + \infty H_2O(l) \longrightarrow$$

$$\text{urea}(s) - 3CO_2(g, 1 \text{ atm}) + 3H_2O(l) - 2 \text{ glycine}(aq) + \text{urea}(aq)$$

(6) $3O_2(g, \text{atm}) + 2 \text{ glycine}(aq) \longrightarrow \text{urea}(aq) + 3CO_2(g, 1 \text{ atm}) + 3H_2O(l)$

$$\Delta H_6 = \Delta H_3 - 2\Delta H_4 + \Delta H_5$$
$$= -1314.2 \text{ kJ mol}^{-1}$$

We now know the enthalpy change for the reaction of a dilute aqueous solution of glycine with O_2 gas to form a dilute aqueous solution of urea plus CO_2 gas plus 3 mol of liquid H_2O. Note that we included the term $+ \infty H_2O(l)$

---

*Students are often confused by the units of ΔH for an equation like reaction (3). The value of ΔH_3 given is the enthalpy change associated with the formation of 1 mol of urea, but at the same time it forms 3 mol of CO_2 and 3 mol of H_2O. To deal with this problem, we adopt the convention of calculating thermodynamic quantities (ΔE, ΔH, etc.) per *mole of reaction* as it is written. In the present case, a mole of reaction is defined for reaction (3) as the amount of reaction that produces 1 mol of urea, 3 mol of CO_2, and so on. Obviously, it is just as valid to describe the reaction as follows:

(3a) $O_2(g, 1 \text{ atm}) + \frac{2}{3} \text{glycine}(s) \longrightarrow \frac{1}{3} \text{urea}(s) + CO_2(g, 1 \text{ atm}) + H_2O(l)$

For this reaction, $\Delta H = \frac{1}{3}(-1296.8) = -432.3 \text{ kJ mol}^{-1}$.

in reactions (4) and (5) to indicate the presence of a large amount of water as the solvent, but there is no net change in the total number of moles of water in these reactions. In reaction (6), however, there is a net synthesis of 3 mol of water. The enthalpy of formation of 3 mol of water makes a very significant contribution to ΔH_6, even though reaction (6) is carried out in a dilute solution containing lots of water $[+ \infty H_2O(l)]$.

The enthalpy for reaction (6) will be quite close to that of the naturally occurring reaction in the human body. However, it should be clear that, by adding or subtracting the enthalpies of other reactions, we can calculate ΔH for *any* reaction we like. That is, if we think it is important, we can find the heat of solution of glycine in a defined buffer solution instead of pure water. We can specify that instead of $CO_2(g, 1 \text{ atm})$ as a product, we have a carbonic acid solution of a certain pH. It may be very difficult to measure directly the ΔH of the reaction we want, but we can always find the ΔH by using a convenient alternative path. In other words, if we want the ΔH for reaction $A \rightarrow B$, we may use a path that is a sum of many other reactions:

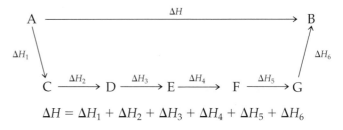

$$\Delta H = \Delta H_1 + \Delta H_2 + \Delta H_3 + \Delta H_4 + \Delta H_5 + \Delta H_6$$

It is convenient to remember that if

$$A \longrightarrow C \quad \text{has} \quad \Delta H_1$$

$$\text{then} \quad C \longrightarrow A \quad \text{has} \quad -\Delta H_1$$

$$\text{and} \quad nA \longrightarrow nC \quad \text{has} \quad n\Delta H_1$$

Temperature Dependence of ΔH

By the same reasoning used for phase changes, if ΔH is known at one temperature, it can be calculated at other temperatures:

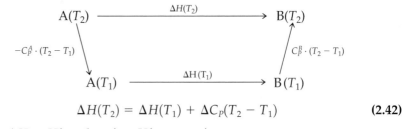

$$\Delta H(T_2) = \Delta H(T_1) + \Delta C_P(T_2 - T_1) \tag{2.42}$$

where $\Delta H = H(\text{products}) - H(\text{reactants})$
$\Delta C_P = C_P(\text{products}) - C_P(\text{reactants})$

and we have incorporated the assumption that C_P is not a function of temperature.

The same equation can also be derived by a different approach. We will take the reaction

$$n_A A + n_B B \longrightarrow n_C C + n_D D$$

and express ΔH in terms of the molar enthalpies of the reactants and products:

$$\Delta H = n_C \overline{H}_C + n_D \overline{H}_D - n_B \overline{H}_B - n_A \overline{H}_A \tag{2.41}$$

The derivative of ΔH with respect to temperature is

$$\frac{d\Delta H}{dT} = n_C \left(\frac{d\overline{H}_C}{dT} \right) + n_D \left(\frac{d\overline{H}_D}{dT} \right) - n_B \left(\frac{d\overline{H}_B}{dT} \right) - n_A \left(\frac{d\overline{H}_A}{dT} \right)$$

$$= n_C \overline{C}_{P,C} + n_D \overline{C}_{P,D} - n_B \overline{C}_{P,B} - n_A \overline{C}_{P,A} \tag{2.43}$$

$$= C_{P,C} + C_{P,D} - C_{P,B} - C_{P,A} \tag{2.43a}$$

$$= \Delta C_P \tag{2.43b}$$

where $\overline{C}_{P,A}$, $\overline{C}_{P,B}$, etc. are the molar heat capacities of the compounds A, B, and so on at constant pressure. Thus,

$$\int_{\Delta H(T_1)}^{\Delta H(T_2)} d\Delta H = \int_{T_1}^{T_2} \Delta C_P \, dT$$

or

$$\Delta H(T_2) - \Delta H(T_1) = \int_{T_1}^{T_2} \Delta C_P \, dT \tag{2.44}$$

If ΔC_P is independent of temperature,

$$\Delta H(T_2) - \Delta H(T_1) = \Delta C_P (T_2 - T_1) \tag{2.44a}$$

which is the same as Eq. (2.42).

The Energy Change ΔE for a Reaction

For a reaction at constant pressure, ΔE can be calculated from ΔH by the use of Eq. (2.39):

$$\Delta E = \Delta H - P\Delta V \tag{2.39}$$

If gases are involved, we ignore the volumes of solids and liquids and approximate the gases as ideal gases:

$$\Delta E = \Delta H - \Delta n RT \tag{2.45}$$

where Δn is the number of moles of *gaseous* products minus the number of moles of *gaseous* reactants.

Standard Enthalpies (or Heats) of Formation

We have commented earlier that thermodynamicists are interested in the *change* of energy of a system rather than its absolute value. In principle, we can calculate the absolute energy of a system from its mass by the use of the equation $E = mc^2$. In practice, this calculation gives such an enormous number that we cannot measure the masses of reactants and products with sufficient accuracy for the calculation of the energy change of a chemical reaction. The same is true for many other variables of state including H.

In calculating the enthalpy change ΔH for the reaction

$$n_A A + n_B B \longrightarrow n_C C + n_D D$$

$$\Delta H = n_C \overline{H}_C + n_D \overline{H}_D - n_B \overline{H}_B - n_A \overline{H}_A \tag{2.41}$$

we need to know the molar enthalpies of the reactants and products, but fortunately only in relative rather than absolute terms for the calculation of their algebraic sum. We can therefore arbitrarily choose a zero of enthalpy, just as we arbitrarily choose mean sea level as the reference point for measuring elevations on the surface of Earth. For example, in the reaction to form water vapor from its elements

$$H_2(g, 1\text{ atm}) + \frac{1}{2}O_2(g, 1\text{ atm}) \xrightarrow{\Delta H} H_2O(g, 1\text{ atm})$$

$$\Delta H = 1\overline{H}(H_2O, g, 1\text{ atm}) - \frac{1}{2}\overline{H}(O_2, g, 1\text{ atm}) - 1\overline{H}(H_2, g, 1\text{ atm})$$

If we choose $\overline{H}$ for the elements O_2 and H_2 equal to zero, the enthalpy of their compound H_2O is equal to the enthalpy of the formation reaction:

$$\Delta H = 1\overline{H}(H_2O, g, 1\text{ atm}) - \frac{1}{2}\cdot 0 - 1\cdot 0$$

$$\overline{H}(H_2O, g, 1\text{ atm}) = \Delta H$$

Thermodynamicists have adopted the convention of assigning zero enthalpy to all elements in their most stable states at 1 atm pressure. These are called *standard states* and are designated by a superscript zero. *The standard enthalpy mol*$^{-1}$ *of a compound is defined to be equal to the enthalpy of formation of 1 mol of the compound at 1 atm pressure from its elements in their standard states.*

$$\overline{H}^0(\text{compound}) \equiv \Delta\overline{H}_f^0 \tag{2.46}$$

where $\Delta\overline{H}_f^0$ = enthalpy of formation of 1 mol from elements under standard conditions. The superscript zero means 1 atm pressure and the most stable form of the element. It should be clear that the standard enthalpy is a defined quantity that depends on the choice of the standard state. Assigning the elements *in their most stable state* at 1 atm pressure to have zero enthalpy is purely arbitrary.

The standard enthalpies of thousands of substances have been determined, usually by measuring q_P at a particular temperature or by measuring q_V at a particular temperature and then adding on the calculated $\Delta(PV)$ terms. Tables A.5–7 in the appendix gives a few standard enthalpies at 25°C. The enthalpies at 25°C and 1 atm pressure for many reactions can be calculated from such tablated $\Delta\overline{H}_f^0$ data. Note that 1 atm pressure is part of the definition of the standard state, but the temperature is not specified in the definition. However, nearly all the tables available are for 25°C. For our general reaction at 25°C,

$$\Delta H^0(298) = n_C\overline{H}_C^0 + n_D\overline{H}_D^0 - n_A\overline{H}_A^0 - n_B\overline{H}_B^0$$

where $\overline{H}^0$ = standard enthalpy mol^{-1} at 25°C = $\Delta\overline{H}_f^0$

$\Delta\overline{H}_f^0 = 0$ for all elements in their standard states

---

EXAMPLE 2.12

Use the tables A.5–7 in the appendix to calculate the value of ΔH for reacting 1 g of solid glycylglycine with oxygen to form solid urea, CO_2 gas, and liquid H_2O at 25°C, 1 atm.

Solution

It is important to begin with a balanced stoichiometric equation for the reaction:

$$3O_2(g) + C_4H_8N_2O_3(s) \longrightarrow CH_4N_2O(s) + 3CO_2(g) + 2H_2O(l)$$

glycylglycine   urea

$$\Delta H^0 = \overline{H}^0(\text{urea}) + 3\overline{H}^0(CO_2) + 2\overline{H}^0(H_2O, l) - \overline{H}^0(\text{glycylglycine}) - 3\overline{H}^0(O_2)$$

$$= -333.17 + 3(-393.51) + 2(-285.83) - (-745.25) - 3(0)$$

$$= -1340.11 \text{ kJ mol}^{-1}$$

The enthalpy change calculated is for 1 mol of glycylglycine. To find ΔH per gram, we must divide by the molecular mass, 132.12 g mol^{-1}:

$$\Delta H = -10.14 \text{ kJ g}^{-1}$$

The negative sign means that heat is given off in the reaction.

What about other temperatures and pressures? The enthalpy at any temperature can be obtained from Eq. (2.44) or Eq. (2.44a). The temperature dependence of the standard chemical reaction is written as

$$\Delta H^0_P(T) = \Delta H^0(298) + \Delta C^0_P(T - 298) \qquad \textbf{(2.47)}$$

where $\Delta C^0_P = n_C \overline{C}^0_{P,C} + n_D \overline{C}^0_{P,D} - n_A \overline{C}^0_{P,A} - n_B \overline{C}^0_{P,B}$. It is necessary to remember that ΔC^0_P here must include all products and reactants. The value of $\overline{H}^0$ for elements is chosen as zero, but $\overline{C}^0_P$ is not zero. Furthermore, if we use a set of data where $\overline{H}^0$ for the elements is zero at 298 K, then their enthalpies at other temperatures will *not* be equal to zero.

$$\overline{H}^0_T(\text{elements}) = 0 + \overline{C}^0_P(\text{elements})(T - 298)$$

The pressure dependence of ΔH is not large, and we will ignore it. For ideal gases, H is independent of P; for solids or liquids, H is not very dependent on P. For geologic processes or biology in the deep ocean where the pressures may become very large, these approximations are not valid. We will also generally ignore the effect of concentration on ΔH.

It should be clear that a heat of reaction and a ΔH can be measured for any condition of concentration, solvent, pH, pressure, and so on. We make these approximations because they are accurate enough for most purposes and because they simplify the calculations.

Bond Energies

Enthalpies of formation for many compounds have been determined very precisely. These are tabulated in extensive compilations of thermodynamic data. Some that are of particular interest to biologists and biochemists are listed with the references at the end of this chapter. There are many more compounds whose ΔH^0_f are not known. However, we can approximate ΔH^0_f and other heats of reaction by using bond dissociation energies D. This essentially involves finding an alternative path for the reaction where the individual steps involve breaking and making chemical bonds. The *bond dissociation energy* is the enthalpy at 25°C and 1 atm for the reaction

$$A - B(g) \longrightarrow A(g) + B(g)$$

TABLE 2.3	Average Bond Dissociation Energies at 25°C
Bond	D (kJ mol^{-1})
C—C	344
C=C	615
C≡C	812
C—H	415
C—N	292
C—O	350
C=O	725
C—S	259
N—H	391
O—O	143
O—H	463
S—H	339
H$_2$	436.0
N$_2$	945.4
O$_2$	498.3
C (graphite)	716.7

Source: After L. Pauling and P. Pauling, *Chemistry* (San Francisco: Freeman, 1975).

The bond dissociation energy should be called *a bond dissociation enthalpy,* but the tradition for energy is strong. In many cases, the amount of energy necessary to break a particular type of bond in a molecule is not too dependent on the molecule. For example, the average energy necessary to break a C—H bond is 415 kJ mol^{-1} ±3% in a wide range of organic compounds.

Some average bond dissociation energies are given in table 2.3. We expect the value for ΔH estimated from the table to be reasonable. One notable exception is for molecules that are stabilized greatly by electron delocalization. For example, the data that were used to obtain values for C—C and C=C bonds in table 2.3 came from molecules with single bonds and isolated double bonds. Molecules with conjugated double bonds (C=C bonds separated by only one C—C bond) are more stable than those containing the same number of isolated single and double bonds. This difference in energy is called the *resonance energy.*

Resonance energy is important for aromatic molecules such as benzene or phenylalanine, for the carboxyl groups in carboxylic acids, for amino acids, porphyrins, carotenoids, and so on. For such molecules, the difference in heat of formation calculated from bond energies and that experimentally measured is an estimate of the resonance energy. In other cases, ring strain produces energy effects that need to be added to the average bond dissociation energies.

The easiest way to show how table 2.3 can be used is with some examples.

EXAMPLE 2.13

Calculate the heat of formation for gaseous cyclohexane using table 2.3 and compare with the measured value in the appendix (table A.6).

Solution

The reaction for the formation of cyclohexane is

$$6C(\text{graphite}) + 6H_2(g) \longrightarrow C_6H_{12}(g)$$

We can write it as a sum of bond-breaking and bond-forming reactions:

(1) $6C(\text{graphite}) \longrightarrow 6C(g)$

$$\Delta H_1 = 6D(\text{graphite}) = (6)(716.7) = 4300 \text{ kJ}$$

This is the enthalpy required to remove 6 mol of carbon atoms from a crystalline lattice of graphite, which is the standard state for elemental carbon.

(2) $6H_2(g) \longrightarrow 12H(g)$

$$\Delta H_2 = 6D(H_2) = (6)(436.0) = 2616.0 \text{ kJ}$$

(3) $6C(g) + 12H(g) \longrightarrow (6\,C{-}C + 12\,C{-}H) = C_6H_{12}(g)$

$$\Delta H_3 = -6D(C{-}C) - 12D(C{-}H) = -(6)(344) - (12)(415)$$
$$= -7044 \text{ kJ}$$

Note that bond formation energies are just the negative of bond dissociation energies:

$$\Delta \overline{H}_f^0(C_6H_{12}) = \Delta H_1 + \Delta H_2 + \Delta H_3$$
$$= -128 \text{ kJ}$$

The value in table A.6 in the appendix is -123.15 kJ for cyclohexane, which is in good agreement. This is because cyclohexane is a molecule that contains normal bonds; also, there is no significant bond angle strain in the six-membered ring.

EXAMPLE 2.14

Calculate the heat of formation for gaseous benzene using table 2.3 and compare with the measured value in table A.6 in the appendix.

Solution

Because of resonance effects, we expect the measured value for benzene to differ significantly from that calculated for the classical structure of benzene. An explanation of the large thermodynamic stability of benzene was an important goal for chemists interested in chemical bonding. The reaction for benzene is

$$6C(\text{graphite}) + 3H_2(g) \longrightarrow C_6H_6(g)$$

(1) $6C(\text{graphite}) \longrightarrow 6C(g)$

$$\Delta H_1 = (6)(716.7) = 4300 \text{ kJ}$$

(2) $3H_2(g) \longrightarrow 6H(g)$

$$\Delta H_2 = (3)(436.0) = 1308 \text{ kJ}$$

(3) $6C(g) + 6H(g) \longrightarrow (3\,C{-}C + 3\,C{=}C + 6\,C{-}H) = C_6H_6(g)$

$$\Delta H_3 = -(3)(344) - (3)(615) - (6)(415)$$
$$= -5367 \text{ kJ}$$

$$\Delta H_f^0(C_6H_6) = \Delta H_1 + \Delta H_2 + \Delta H_3$$
$$= 241 \text{ kJ}$$

The value in table A.6 in the appendix is 82.93 kJ for $C_6H_6(g)$! Benzene is about 158 kJ lower in enthalpy than would be expected for a molecule made up of three

C—C single bonds, three C=C double bonds, and six C—H single bonds. This energy is what we call the *resonance energy.*

Molecular Interpretations of Energy and Enthalpy

We have considered various ways of changing the energy and enthalpy of a system. They include:

- Heat and work exchanges between the system and surroundings
- Chemical reactions or phase changes within the system

Let's consider what the molecules are doing when the energy and enthalpy change. For example, if the system consists of an ideal gas, we find that its internal energy is a function of temperature but not of pressure or volume. We can increase the internal energy (and the temperature) by adding heat to the system or doing work on it. The individual ideal gas molecules have higher translational, rotational, vibrational, and (at high enough temperatures) electronic energies at the higher temperature. The internal energy of an ideal gas is just the sum of the energies of all the individual gas molecules. There are no interactions between ideal gas molecules. The heat capacity is a measure of how much energy is required to raise the temperature by 1 degree. Its magnitude is a measure of how many ways the molecules have of storing energy. Even from the limited set of data in table 2.1, one can see that gases made up of larger and more complex molecules have larger heat capacities than those consisting of simpler molecules or monatomic gases. For normal temperatures, monatomic gases have only increased translational kinetic energies—increased velocities—to store energy.

For real gases and all liquids and solids, interactions between molecules become significant. For these systems, the energy and enthalpy can change at constant temperature. Energy can be stored in intermolecular interactions. Thus, compressing the system at constant temperature can increase its energy by increasing the repulsion between molecules as the distances between them decrease. The energy of a system of liquids, solids, or real gases is the sum of the energies of all the molecules plus the energy of interactions among all the molecules. Raising the temperature of a liquid, for example, raises the energy of individual molecules and also changes the intermolecular interactions. The heat capacity of a substance in the liquid phase is thus greater than that in the gas phase. Chemical reactions and phase changes cause abrupt changes in energy and enthalpy. For chemical reactions, the changes are a consequence of making and breaking bonds. For phase changes, the changes result from the different intermolecular interactions in solids, liquids, and gases. The energy and enthalpy changes in a chemical reaction can be hundreds of kilojoules per mole; in a phase change, tens of kilojoules can be involved. These are both large compared to the effects of temperature alone; a temperature change of 100 degrees may result in no more than 1 kJ mol^{-1} energy change.

The energy stored in chemical bonds represents one of our greatest energy resources. We run automobiles on the energy derived from the chemical combustion (oxidation) of the hydrocarbons of gasoline. One would not attempt to run a vehicle on the energy released by the cooling of 20 gallons of water from 100°C, even if all that thermal energy could be converted into mechanical work.

It is important to keep the relative magnitudes of these quantities in mind, especially when approximations are made in thermodynamic calculations.

Summary
State Variables

Name	Symbol	Units	Definition
Volume	V	liters (L), mL, cm^3	(length)^3
Pressure	P	atm, torr, mm Hg, dyn cm^{-2}, pascal (Pa), N m^{-2}	force area^{-1}
Temperature	T	K, °C, °F	
Energy	E	J, erg, cal, Cal	
Enthalpy	H	J, erg, cal, Cal	$H \equiv E + PV$

Unit Conversions
Volume

$$1\ \text{L} = 1000\ \text{mL} = 1000\ \text{cm}^3$$

Pressure

$$P = \text{force area}^{-1}:$$

$$1\ \text{atm} = 760\ \text{torr} = 1.01325 \times 10^6\ \text{dyn cm}^{-2} = 1.01325 \times 10^5\ \text{Pa}$$

$$1\ \text{torr} \equiv 1\ \text{mm Hg}$$

$$1\ \text{dyn cm}^{-2} \equiv 1\ \text{g cm}^{-1}\text{s}^{-2}$$

$$1\ \text{pascal} \equiv 1\ \text{N m}^{-2} \equiv 1\ \text{kg m}^{-1}\text{s}^{-2}$$

Temperature

$$\text{K} = \text{°C} + 273.15$$

$$\text{°C} = \frac{\text{°F} - 32}{1.8}$$

Energy and enthalpy

$$1\ \text{cal} = 4.184\ \text{J} = 4.184 \times 10^7\ \text{erg}$$

$$1\ \text{J} \equiv 1\ \text{kg m}^2\text{s}^{-2} = 1 \times 10^7\ \text{erg}$$

$$1\ \text{erg} = 1\ \text{g cm}^2\text{s}^{-2}$$

$$1\ \text{L atm} = 24.22\ \text{cal}$$

General Equations
Energy, E
(In a closed system, heat and work being the only forms of energy the system exchanges with surroundings.)

$$\Delta E = q + w \tag{2.12}$$

Enthalpy

$$H \equiv E + PV \tag{2.14}$$

$$\Delta H = \Delta E + P_2 V_2 - P_1 V_1 \tag{2.14a}$$

Heat, q

Heat absorbed by system is positive:

$$q = \int_{T_1}^{T_2} C \, dT \tag{2.9a}$$

$$C = \text{heat capacity} = \frac{dq}{dT} \tag{2.9}$$

Work, w

Work done on system is positive.
Stretching or compressing a spring

$$w = k(x_2 - x_1)\left(\frac{x_2 + x_1}{2} - x_0\right) \tag{2.4}$$

where k = Hooke's law constant
x_0 = length of spring in the absence of a force
x_1, x_2 = initial and final lengths of the spring, respectively

Expansion or compression of a gas

$$w = -\int_{V_1}^{V_2} P_{ex} \, dV \tag{2.5}$$

P_{ex} = external pressure

$$w_P = -P_{ex}(V_2 - V_1) \qquad \text{(constant external pressure)} \tag{2.6}$$

Electrical work done by a system

$$w = -EIt \tag{2.8}$$

where E = voltage
I = current
t = time

Pressure–Volume Work Only

$$\Delta E = E_2 - E_1 = q_V = \int_{T_1}^{T_2} C_V \, dT \qquad \text{(constant volume)} \tag{2.37}$$

$$\Delta H = H_2 - H_1 = q_P = \int_{T_1}^{T_2} C_P \, dT \qquad \text{(constant pressure)} \tag{2.38}$$

Solids and Liquids

We assume in these equations that for a solid or liquid the volume is independent of T and P, and that $C_P = C_V = C$, independent of T and P.

$$\Delta E = n\overline{C}(T_2 - T_1) \qquad \text{(any change of } P, T)$$

$$\Delta H = n\overline{C}(T_2 - T_1) + (P_2 - P_1)V \qquad \text{(any change of } P, T)$$

where n = number of moles
$\overline{C}$ = heat capacity per mole

Gases

We assume that gas properties can be approximated by the ideal gas equation and that C_P and C_V are independent of T. $PV = nRT$ and $\overline{C}_P = \overline{C}_V + R$.

$$\Delta E = n\overline{C}_V(T_2 - T_1) \qquad \text{(any change of } P, V, T) \qquad \textbf{(2.32)}$$

$$\Delta H = n\overline{C}_P(T_2 - T_1) \qquad \text{(any change of } P, V, T) \qquad \textbf{(2.33)}$$

where n = number of moles
$\overline{C}_P$ = heat capacity per mole at constant P
$\overline{C}_V$ = heat capacity per mole at constant V

$$w_P = -nR(T_2 - T_1) \qquad \text{(reversible, constant } P) \qquad \textbf{(2.28)}$$

$$w_T = -nRT \ln \frac{V_2}{V_1} = -nRT \ln \frac{P_1}{P_2} \qquad \text{(reversible, constant } T) \qquad \textbf{(2.29a)}$$

$$q_T = nRT \ln \frac{V_2}{V_1} = nRT \ln \frac{P_1}{P_2} \qquad \text{(reversible, constant } T) \qquad \textbf{(2.30)}$$

where $\ln x = 2.303 \log x$
$R = 8.314 \text{ J K}^{-1} \text{ mol}^{-1}$

Phase Changes

For a phase change, phase a $\rightarrow$ phase b, which occurs at constant T and P,

$$\Delta H = q_P \qquad \textbf{(2.36)}$$

where $\Delta H = H(\text{phase b}) - H(\text{phase a})$

$$\Delta E = \Delta H - P\Delta V \qquad \textbf{(2.39)}$$

where $\Delta E = E(\text{phase b}) - E(\text{phase a})$
$\Delta V = V(\text{phase b}) - V(\text{phase a})$

$$\Delta H(T_2) = \Delta H(T_1) + n\Delta\overline{C}_P(T_2 - T_1) \qquad \textbf{(2.40)}$$

where n = number of moles
$\Delta\overline{C}_P$ = heat capacity per mole at constant P of phase b
$\quad$ − heat capacity per mole at constant P of phase a

$$\Delta H(P_2) \cong \Delta H(P_1)$$

$$w_P = -P\Delta V$$

Chemical Reactions

For a chemical reaction

$$n_A A + n_B B \longrightarrow n_C C + n_D D$$

that occurs at constant T and P,

$$\Delta H = n_C \overline{H}_C + n_D \overline{H}_D - n_A \overline{H}_A - n_B \overline{H}_B = q_P$$

$$\overline{H}^0_{298}(A) \equiv \Delta \overline{H}^0_{f,\,298}(A) = \text{enthalpy (heat) of formation of A}$$
$$\text{per mole from the elements in their}$$
$$\text{most stable states at standard}$$
$$\text{conditions (1 atm) and } 25°C$$

$$\Delta E = \Delta H - \Delta(PV)$$

where $\Delta(PV) = PV(\text{products}) - PV(\text{reactants})$
    1 L atm = 101.325 J

$$\Delta E = \Delta H - \Delta n R T \tag{2.45}$$

where Δn = number of moles of *gaseous* products $-$ number of
    moles of *gaseous* reactants
    $R = 8.314 \text{ J K}^{-1} \text{mol}^{-1}$

$$\Delta H(T_2) = \Delta H(T_1) + \Delta C_P(T_2 - T_1) \tag{2.44a}$$

$$\Delta C_P = n_C \overline{C}_P(C) + n_D \overline{C}_P(D) - n_A \overline{C}_P(A) - n_B \overline{C}_P(B)$$

$$\Delta H(P_2) \cong \Delta H(P_1)$$

Mathematics Needed for Chapter 2

You should be able to integrate simple powers of x.

Indefinite integral of ax^n

$$\int ax^n \, dx = \frac{ax^{n+1}}{n+1} \qquad (n \neq -1) \tag{1}$$

$$\int ax^{-1} \, dx = a \int \frac{dx}{x} = a \ln x \tag{2}$$

Here a is a constant independent of x.

Definite integral of ax^n

$$\int_{x_1}^{x_2} ax^n \, dx = \frac{a(x_2^{n+1} - x_1^{n+1})}{n+1} \qquad (n \neq -1) \tag{3}$$

$$\int_{x_1}^{x_2} a \frac{dx}{x} = a(\ln x_2 - \ln x_1) = a \ln \frac{x_2}{x_1} \tag{4}$$

Remember that $\ln ab = \ln a + \ln b$; $\ln(a/b) = \ln a - \ln b$; $\ln a = 2.303 \log a$.

EXAMPLE 2.15

$P = aV + bV^2$, with a and b constant. Using Eqs. (1) and (3), we obtain

$$\int P\,dV = \int (aV + bV^2)\,dV = a\frac{V^2}{2} + b\frac{V^3}{3}$$

$$\int_{V_1}^{V_2} P\,dV = \frac{a}{2}(V_2^2 - V_1^2) + \frac{b}{3}(V_2^3 - V_1^3)$$

References

Freshman chemistry texts are good for reviewing basic thermodynamics.

Munowitz, M. 1999. *Principles of Chemistry.* New York: Norton.

Oxtoby, D. W., H. P. Gillis, and N. H. Nachtrieb. 1999. *Principles of Modern Chemistry.* 4th ed. Philadelphia: Saunders.

The following textbooks on thermodynamics can be useful as supplements to chapters 2 through 5.

DeVoe, H. 2000. *Thermodynamics and Chemistry.* Upper Saddle River, NJ: Prentice Hall.

Fenn, J. B. 1982. *Engines, Energy, and Entropy.* New York: Freeman.

Hammes, G. G. 2000. *Thermodynamics and Kinetics for the Biological Sciences.* New York: Wiley.

Klotz, I. M., and R. M. Rosenberg. 2000. *Chemical Thermodynamics.* 6th ed. New York: Wiley.

Smith, E. B. 1990. *Basic Chemical Thermodynamics.* 4th ed. New York: Oxford University Press.

Thermodynamic data for inorganic and organic chemicals can be found on the internet.

The Committee on Data for Science and Technology (CODATA) has a web page (http://www.codata.org/) that has links to thermodynamic databases and publications.

The National Institute of Science and Technology (NIST) web page (http://webbook.nist.gov/chemistry/) has a very extensive database of chemical and thermodynamic properties of organic compounds.

For articles on applications of thermodynamics to biological molecules, see

Jones, M. N., ed. 1988. *Biochemical Thermodynamics.* 2d ed. Amsterdam: Elsevier.

Suggested Reading

Energy for Planet Earth. 1990. *Sci. Am.* 263 (special issue).

Dostrosvsky, I. 1991. Chemical Fuels from the Sun. *Sci. Am.* 265:102–107.

Problems

1. (a) A hiker caught in a rainstorm might absorb 1.00 Liter of water in his clothing. If it is windy so that this amount of water is evaporated quickly at 20°C, how much heat would be required for this process?

(b) If all this heat were removed from the hiker (no significant heat was generated by metabolism during this time), what drop in body temperature would the hiker experience? The clothed hiker weighs 60 kg, and you can approximate the heat capacity of hiker and clothes as equal to that of water. (The conclusion from this calculation: Stay out of the wind if you get your clothes wet.)

(c) How many grams of sucrose would the hiker have to metabolize (quickly) to replace the heat of evaporating 1.00 L of water so that his temperature would not change? You can use the heat of reaction at 25°C; the reaction is

$$C_{12}H_{22}O_{11}(s) + 12O_2(g) \longrightarrow 12CO_2(g) + H_2O(g)$$

2. Photosynthesis by land plants leads to the fixation each year of about 1 kg of carbon on the average for each square meter of an actively growing forest. The atmosphere is approximately 20% O_2 and 80% N_2 but contains 0.046% CO_2 by weight.

(a) What volume of air (25°C, 1 atm) is needed to provide this 1 kg of carbon?

(b) How much carbon is present in the entire atmosphere lying above each square meter of Earth's surface? (*Hint:* Remember that atmospheric pressure is the consequence of the force exerted by all the air above the surface; 1 atm is equivalent to 1.033×10^4 kg m^{-2}.)

(c) At the current rate of utilization, how long would it take to use all the CO_2 in the entire atmosphere directly above the forest? (This assumes that atmospheric circulation and replenishment from the oceans, rocks, combustion of fuels, respiration of animals, and decay of biological materials are cut off.)

3. Calculate the work (in joules) done on the system for each of the following examples (the *system* is given in italics). Specify the sign of the work.

(a) A *box of groceries* weighing 10 kg is carried up three flights of stairs (10 m up altogether).

(b) A 6.0-V *storage battery* is charged by a power supply for 2 h with a current of 5.5 A.

(c) A *muscle* of 1 cm^2 cross section and 10.0 cm length is stretched to 11.0 cm by hanging a mass on it. The muscle behaves like a spring that follows Hooke's law. The Hooke's law constant for the muscle was determined by finding that the muscle exerts a force of 5.00 N when it is stretched from 10.0 cm to 10.5 cm.

(d) The volume of an *ideal gas* changes from 1.00 L to 3.00 L at an initial temperature of 25°C and a constant pressure of 1 atm.

(e) The volume of an *ideal gas* changes from 1.00 L to 3.00 L at an initial temperature of 25°C and a constant pressure of 1.00×10^{-6} atm.

(f) The volume of an *ideal gas* changes from 1.00 L to 3.00 L at a constant temperature of 25°C, and the expansion is done reversibly for an amount of gas corresponding to an initial pressure of 1 atm.

4. (a) As heat is added to a sample of pure ice (solid H_2O) at constant pressure, the temperature rises until the melting point of the ice is reached. The ice then melts at the (constant) melting temperature T_m. After the ice is completely melted, the temperature continues to rise. Complete the plots below for heat added, q, vs. T for the three stages of this process. The slope of q plotted vs. T at constant pressure is the heat capacity, C_P. Plot C_P vs. temperature for this process of melting ice.

(b) As heat is added to a dilute buffer solution, the temperature rises. The change in temperature is approximately linear with added heat, as shown in the sketches.

If a solution of DNA in dilute buffer is heated, the DNA double-stranded helix separates to single strands over a narrow temperature range. Heat is absorbed. The midpoint of the temperature range is called T_m. Sketch a q vs. T curve and a C_P vs. T curve for a DNA solution and describe how the heat necessary to denature the DNA could be obtained from each curve.

5. Recently, biological organisms have been discovered living at great depths at the bottom of the ocean. The properties of common substances are greatly altered there because of enormous hydrostatic pressures caused by the weight of the ocean lying above. The organisms need to equilibrate their internal pressures with this external hydrostatic pressure to prevent being crushed.

(a) Estimate the hydrostatic pressure (in atm) in the ocean at depth 2500 m. Assume the density of sea water is 1.025 kg L^{-1}.

(b) Except near sources of heat, such as vents of hot gases coming from Earth's mantle, the temperature is close to 4°C. What is the percentage increase in density of liquid water under these conditions compared with that at the surface? The isothermal compressibility β of liquid water at 4°C is 49.5×10^{-6} atm^{-1}.

$$\beta = \frac{-\Delta V}{V_0 \Delta P} = \text{the fractional change in volume,}$$
$$\Delta V / V_0, \text{ with change of pressure, } \Delta P$$

(c) Consider a balloon filled with oxygen O_2 at 1 atm pressure and occupying a 10-L volume at 293 K. Assuming ideal gas behavior, what would be its volume at a depth of 2500 m under the ocean surface at 277 K? What is the percentage increase in density of the O_2?

(d) Repeat the calculation of part (c) for O_2 gas, taking into account its deviation from ideal behavior. The van der Waals constants for O_2 are $a = 1.360$ L^2 atm mol^{-2} and $b = 3.183 \times 10^{-2}$ L mol^{-1}. (A cubic equation must be solved.)

6. Calculate the heat (in joules) absorbed by the system for each of the following examples (the system is given in italics). Specify the sign of the heat.

(a) 100 g of liquid *water* is heated from 0°C to 100°C at 1 atm.

(b) 100 g of liquid *water* is frozen to ice at 0°C at 0.01 atm.

(c) 100 g of liquid *water* is evaporated to steam at 100°C at 1 atm.

7. One mole of an ideal gas initially at 27°C and 1 atm pressure is heated and allowed to expand reversibly at constant pressure until the final temperature is 327°C. For this gas, $C_V = 20.8$ J K^{-1} mol^{-1} and is constant over the temperature range.

(a) Calculate the work w done on the gas in this expansion.

(b) What are ΔE and ΔH for the process?

(c) What is the amount of heat q absorbed by the gas?

8. One mole of an ideal monatomic gas initially at 300 K is expanded from an initial pressure of 10 atm to a final pressure of 1 atm. Calculate ΔE, q, w, ΔH, and the final temperature T_2 for this expansion carried out according to each of the following paths. The molar heat capacity at constant volume for a monatomic gas is $C_V = \frac{3}{2}R$.

(a) An isothermal, reversible expansion

(b) An expansion against a constant external pressure of 1 atm in a thermally isolated (adiabatic) system

(c) An expansion against zero external pressure (that is, against a vacuum) in an adiabatic system.

9. For the following processes, calculate the indicated quantities for a system consisting of 1 mol of N$_2$ gas. Assume ideal gas behavior.

(a) The gas initially at 10 atm pressure is expanded tenfold in volume against a constant external pressure of 1 atm, all at 298 K. Calculate q, w, ΔE, and ΔH for the gas.

(b) After the expansion in part (a), the volume is fixed, and heat is added until the temperature reaches 373 K. Calculate q, w, ΔE, and ΔH.

(c) What will be the pressure of the gas at the end of the process in part (b)?

(d) If the gas at 373 K is next allowed to expand against 1 atm pressure under adiabatic conditions (no heat transferred), will the final temperature be higher, remain unchanged, or be lower than 373 K? Explain your answer.

10. For the following processes, state whether each of the thermodynamic quantities q, w, ΔE, and ΔH is greater than, equal to, or less than zero for the *system specified*. Explain your answers briefly.

(a) An *ideal gas* expands adiabatically against an external pressure of 1 atm.

(b) An *ideal gas* expands isothermally against an external pressure of 1 atm.

(c) An *ideal gas* expands adiabatically into a vacuum.

(d) A *liquid* at its boiling point is converted reversibly into its *vapor*, at constant temperature and 1 atm pressure.

(e) H$_2$ *gas* and O$_2$ *gas* are caused to react in a closed bomb at 25°C, and the product water is brought back to 25°C.

11. One mole of liquid water at 100°C is heated until the liquid is converted entirely to vapor at 100°C and 1 atm pressure. Calculate q, w, ΔE, and ΔH for each of the following:

(a) The vaporization is carried out in a cylinder where the external pressure on the piston is maintained at 1 atm throughout.

(b) The cylinder is first expanded against a vacuum ($P_{ex} = 0$) to the same volume as in part (a), and then sufficient heat is added to vaporize the liquid completely to 1 atm pressure.

12. An ice cube weighing 18 g is removed from a freezer, where it has been at -20°C.

(a) How much heat is required to warm it to 0°C without melting it?

(b) How much additional heat is required to melt it to liquid water at 0°C?

(c) Suppose the ice cube was placed initially in a 180-g sample of liquid water at +20°C in an insulated (thermally isolated) container. Describe the final state when the system has reached equilibrium.

13. Describe what happens to the temperature, the volume, and the phase composition (number of moles of liquid or vapor present) following the removal of 400 J of heat from 1 mol of pure water initially in each of the following states. Assume that the pressure is maintained constant at 1 atm throughout and that the gas behaves ideally.

(a) Water vapor initially at 125°C

(b) Liquid water initially at 100°C

(c) Water vapor initially at 100°C

(d) Which of the three processes above involves the largest amount of work done on or by the system? What is the sign of w? Explain your answer.

14. Consider the reaction H$_2$O(l, 1 atm) $\rightarrow$ H$_2$O(g, 1 atm) at 298 K, as illustrated in figure 2.6. Using the data in table 2.2 and stating any important assumptions that you need to make,

(a) Calculate ΔH^0_{298} for the process by reversible path (1).

(b) Calculate ΔH^0_{298} for reversible path (2).

(c) Compare your answers to parts (a) and (b) with the value obtained, using the standard enthalpies of formation in table A.5 of the appendix.

(d) Express the range of values that you obtained in the above calculations in terms of a percentage deviation. Which of the values is the most accurate? Why?

15. For the following processes, state whether each of the four thermodynamic quantities q, w, ΔE, and ΔH is greater than, equal to, or less than zero for the *system specified*. Consider all gases to behave ideally. State explicitly any reasonable assumptions that you may need to make.

    (a) *Two copper bars*, one initially at 80°C and the other initially at 20°C, are brought into contact with each other in a thermally insulated compartment and then allowed to come to equilibrium.

    (b) A *sample of liquid* in a thermally insulated container (a calorimeter) is stirred for 1 h by a mechanical linkage to a motor in the surroundings.

    (c) A *sample of H_2 gas* is mixed with *an equimolar amount of N_2 gas* at the same temperature and pressure under conditions where no chemical reaction occurs between them.

16. Some thermodynamic equations or relations discussed in this chapter are true in general; others are true only under restricted conditions, such as at constant volume, for an adiabatic process only, for an ideal gas, for PV work only, and so on. For each of the following equations, state what are the minimum conditions sufficient to make them true in the framework of chemical thermodynamics. (Some subscripts normally present have been omitted. You may assume that the system is uniform in properties: T, P, E, and so on are everywhere the same.)

    (a) $\Delta E = q + w$

    (b) $q = \Delta H$

    (c) $C_P = \dfrac{\Delta H}{\Delta T}$

    (d) $\Delta H = \Delta E + \Delta(nRT)$

    (e) $\left(P + \dfrac{n^2 a}{V^2}\right)(V - nb) = nRT$

    (f) $w = -P_{ex}\Delta V$

17. If you set out to explore the surface of the Moon, you would want to wear a space suit with thermal insulation. In such activity, you might expect to generate roughly 4 kJ of heat per kilogram of mass per hour. If all this heat is retained by your body, how much would your body temperature increase per hour owing to this rate of heat production? (Assume that your heat capacity is roughly that of water.) What time limit would you recommend for a moon walk under these conditions?

18. If a breath of air, with a volume of 0.5 L, is drawn into the lungs and comes to thermal equilibrium with the body at 37°C while the pressure remains constant, calculate the increase in enthalpy of the air if the initial air temperature is 20°C and the pressure is 1 atm. At a breathing rate of 12 per minute, how much heat is lost in this fashion in 1 day? Compare your answer (and that of problem 17, assuming a body weight of 80 kg) with a typical daily intake of 12,000 kJ of food energy. What problems might you foresee in arctic climates, where the air temperature can reach −40°C and below? The heat capacity of air is about 30 J K^{-1} mol^{-1}.

19. Human beings expend energy during expansion and contraction of their lungs in breathing. Each exhalation from the lungs of an adult involves pushing out about 0.5 L of gas against 1 atm of pressure. This occurs about 15,000 times in a 24-h day.

    (a) Estimate the amount of work in breathing done by each person in the course of 24 h.

    (b) To get a feeling for how much work this represents, imagine using it to raise a mass to the top of a 30-story building (about 100 m). First, make a guess at how large a weight could be raised by this amount of work: 1 kg (2.2 lb), 10 kg, 100 kg, 1000 kg (about 1 ton). Now do the calculation and draw some conclusions about how much your body is working even when you are "resting."

20. One hundred grams of liquid H_2O at 55°C is mixed with 10 g of ice at −10°C. The pressure is kept constant, and no heat is allowed to leave the system. The process is adiabatic. Calculate the final temperature for the system.

21. A reaction that is representative of those in the glycolytic pathway is the catabolism of glucose by complete oxidation to carbon dioxide and water:

$$C_6H_{12}O_6(s) + 6O_2(g) \longrightarrow 6CO_2(g) + 6H_2O(l)$$

    Calculate the ΔH^0_{298} for the glucose oxidation.

22. Alcoholic fermentation by microorganisms involves the breakdown of glucose into ethanol and carbon dioxide by the reaction

$$\text{glucose}(s) \longrightarrow 2\,\text{ethanol}(l) + 2CO_2(g)$$

    (a) Calculate the amount of heat liberated in a yeast brew upon fermentation of 1 mol of glucose at 25°C, 1 atm.

    (b) What fraction is the heat calculated in part (a) of the amount of heat liberated by the complete combustion (reaction with O_2) of glucose to carbon dioxide and liquid water at 298 K, calculated in problem 21?

23. The enzyme catalase catalyzes the decomposition of hydrogen peroxide by the exothermic reaction

$$H_2O_2(aq) \xrightarrow{\text{catalase}} H_2O(l) + \frac{1}{2}O_2(g)$$

    Estimate the minimum detectable concentration of H_2O_2 if a small amount of catalase (solid) is added to a hydrogen peroxide solution in a calorimeter. Assume that a temperature rise of 0.02°C can be distinguished.

You can use a heat capacity of 4.18 kJ K^{-1} kg^{-1} for the hydrogen peroxide solution.

24. Consider the reaction

$$CH_3OH(l) \longrightarrow CH_4(g) + \frac{1}{2}O_2(g)$$

   (a) Calculate ΔH^0_{298}.

   (b) Estimate ΔE^0_{298}.

   (c) Write an equation that would allow you to obtain ΔH at 500°C and 1 atm.

25. Use the standard enthalpies of formation in tables A.5–7 in the appendix to verify the value given in the text of this chapter for the oxidation of glycine to give urea and CO_2.

$$3O_2(g, 1\ atm) + 2NH_2CH_2COOH(s) \longrightarrow$$

$$H_2NCOH_2(s) + 3CO_2(g, 1\ atm) + 3H_2O(l)$$

26. Yeasts and other organisms can convert glucose ($C_6H_{12}O_6$) to ethanol or acetic acid. Calculate the change in enthalpy ΔH when 1 mol of glucose is oxidized to (a) ethanol or (b) acetic acid by the following path at 298 K:

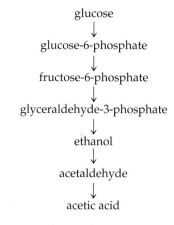

glucose
↓
glucose-6-phosphate
↓
fructose-6-phosphate
↓
glyceraldehyde-3-phosphate
↓
ethanol
↓
acetaldehyde
↓
acetic acid

You can ignore all heats of solution of products or reactants. The overall reactions are

$$C_6H_{12}O_6(s) \longrightarrow 2CH_3CH_2OH(l) + 2CO_2(g)$$

$$2O_2(g) + C_6H_{12}O_6(s) \longrightarrow 2CH_3COOH(l)$$

$$+ 2CO_2(g) + 2H_2O(l)$$

   (c) Calculate the ΔH for the complete combustion of glucose to $CO_2(g)$ and $H_2O(l)$.

27. Estimate the change in ΔH if each reaction in problems 26(a) and 26(b) is carried out by a thermophilic bacterium at 80°C. C_P for ethanol(l) = 111.5 J K^{-1} mol^{-1}, for acetic acid(l) = 123.5 J K^{-1} mol^{-1}, and for glucose(s) ≅ 210 J K^{-1} mol^{-1}.

28. A good yield of photosynthesis for agricultural crops in bright sunlight is 20 kg of carbohydrate (for example, sucrose) per hectare per hour. (1 hectare = 10^4 m^2.) The net reaction for sucrose formation in photosynthesis can be written

$$12CO_2(g) + 11H_2O(l) \xrightarrow[\text{light}]{} C_{12}H_{22}O_{11}(s) + 12O_2(g)$$

   (a) Use standard enthalpies of formation to calculate ΔH^0 for the production of 1 mol of sucrose at 25°C by the reaction above.

   (b) Calculate the energy equivalent of photosynthesis that yields 20 kg of sucrose (hectare h)$^{-1}$.

   (c) Bright sunlight corresponds to radiation incident on the surface of Earth at about 1 kW m^{-2}. What percentage of this energy can be "harvested" as carbohydrate in photosynthesis?

29. (a) Calculate the enthalpy change on burning 1 g of $H_2(g)$ to $H_2O(l)$ at 25°C and 1 atm.

   (b) Calculate the enthalpy change on burning 1 g of n-octane (g) to $CO_2(g)$ and $H_2O(l)$ at 25°C and 1 atm.

   (c) Compare $H_2(g)$ and n-octane(g) in terms of joules of heat available per gram.

30. Consider the reaction in which 1 mol of aspartic acid(s) is converted to alanine(s) and $CO_2(g)$ at 25°C and 1 atm pressure. The balanced reaction is

$$H_2NCH(CH_2COOH)COOH \rightleftharpoons$$

$$H_2NCH(CH_3)COOH + CO_2$$

   (a) How much heat is evolved or absorbed?

   (b) Write a cycle you could use to calculate the heat effect for the reaction at 50°C. State what properties of molecules you would need to know and what equations you would use to calculate the answer.

31. What experiments would you have to do to measure the (a) energy and (b) enthalpy change for the following reaction at 25°C and 1 atm?

$$ATP^{4-} + H_2O \rightleftharpoons ADP^{3-} + HPO_4^{2-} + H^+$$

The reaction takes place in aqueous solution using the sodium salts of each compound. Give as much detail as possible and show the equations you would need to use. ATP^{4-} is adenosine triphosphate; ADP^{3-} is adenosine diphosphate.

32. A household uses 22 kW h day^{-1} of electricity.

   (a) How many kJ day^{-1} are used?

   (b) About 1 kW m^{-2} of energy from the Sun hits the surface of Earth. With a 10% efficient solar battery, what area (in m^2) of solar battery would be needed to provide sufficient solar energy to supply the household? Assume an average of 5 h day^{-1} of sunshine. (A suitable energy-storage device would of course be needed to provide electricity at night.)

33. The enzyme catalase efficiently catalyzes the decomposition of hydrogen peroxide to give water and oxy-

gen. At room temperature, the reaction goes essentially to completion.

(a) Using heats of formation, calculate ΔH^0_{298} for the reaction

$$2H_2O_2(g) \longrightarrow 2H_2O(g) + O_2(g)$$

$\Delta H^0_f(298)$ for gaseous H_2O_2 is -133.18 kJ mol^{-1}.

(b) Calculate the bond dissociation energy for the O—O single bond.

(c) The enzyme normally acts on an aqueous solution of hydrogen peroxide, for which the equation is

$$2H_2O_2(aq) \longrightarrow 2H_2O(l) + O_2(g)$$

What is ΔH^0_{298} for this process?

(d) A solution initially 0.01 M in H_2O_2 and at 25.00°C is treated with a small amount of the enzyme. If all the heat liberated in the reaction is retained by the solution, what would be the final temperature? (Take the heat capacity of the solution to be 4.18 kJ K^{-1} kg^{-1}.)

34. Use bond-energy data to calculate the enthalpy of formation for each of the following compounds at 25°C.

(a) *n*-Octane, $C_8H_{18}(g)$

(b) Naphthalene, $C_{10}H_8(g)$

(c) Formaldehyde, $H_2CO(g)$

(d) Formic acid, $HCOOH(g)$

Compare your answers with enthalpies of formation given in tables A.5–7 in the appendix. (The enthalpy of vaporization of formic acid is 46.15 kJ at 25°C.) Give the most likely reasons for the largest discrepancies between your calculated values and the ones in the tables.

35. The standard enthalpies of formation at 25°C of gaseous *trans*-2-butene and *cis*-2-butene are -11.1 and -7.0 kJ mol^{-1}, respectively. In the *cis* compound,

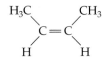

the relatively bulky methyl groups interact repulsively with one another because they are both constrained to remain on the same side of the molecule. (There is no rotation about the C=C bond.) Calculate the thermodynamic enthalpy attributable to this steric repulsion in the *cis* compound relative to the *trans*

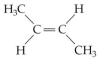

where the methyl groups are on the opposite sides of the molecule. This is an example of how detailed molecular information can be deduced from thermochemical data.

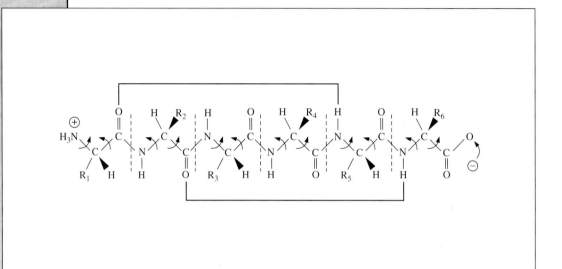

The Second Law: The Entropy of the Universe Increases

Ludwig Boltzmann's tombstone in Vienna has inscribed on it the equation $S = k \ln N$. Entropy is proportional to the logarithm of the number of different ways a system can be rearranged without changing it. Entropy had been defined and used as a thermodynamic variable of state about 100 years earlier, but Boltzmann made the discovery that entropy was a quantitative measure of disorder.

Concepts

Entropy S is a measure of disorder; the greater the disorder, the greater is the entropy. Disorder occurs spontaneously and thus entropy tends to increase. In an isolated system that exchanges neither energy nor matter with the surroundings, every change within the system *increases* its entropy. Fluctuations may, however, decrease the entropy of an isolated system slightly for a short time. If the system is not isolated from the surroundings, then the entropy of the system can decrease. This decrease is accompanied by a larger increase in entropy of the surroundings. Thus, the entropy of the universe—that is, the system plus the surroundings—always increases.

Gibbs free energy G, defined as the enthalpy H minus the product of the absolute temperature T and the entropy S, always *decreases* or does not change, in a process that occurs spontaneouly at constant temperature and pressure. Thus, by calculating a free-energy change for a reaction, we can tell if the reaction can proceed spontaneouly at constant temperature and pressure.

Applications

Chemists are interested in predicting the outcome of chemical reactions. In chapter 2, we learned from the first law of thermodynamics that energy is conserved: One form of energy can convert into another, but their sum remains unchanged. The first law does not tell us, however, whether a particular reaction can spontaneously occur; it just says that whichever way a reaction goes, the energy must be conserved. But is there a criterion, or a set of criteria, that can tell us whether a physical or chemical process can occur? This question is answered by the second law of thermodynamics.

A knowledge of thermodynamics allows us to decide which reactions are impossible under given conditions and how the conditions can be changed to make impossible reactions probable. An inventor might ask, for example, whether diamonds could be made out of graphite, as both are just different forms of the element carbon. Thermodynamics tells us that the answer is yes, if the conversion is carried out at high pressures. Thermodynamics does not tell us, however, how fast a reaction will actually occur or how the rate depends on the reaction conditions. The conversion of graphite to diamond at high pressures, for example, is usually carried out at high temperature to speed the reaction. (The rates of reactions are discussed in the chapters on kinetics.)

The application of thermodynamics is not limited to chemical reactions. The same inventor might try, on a different day, to design a highly efficient heat engine that could convert all the heat generated by burning its fuel to work. The second law of thermodynamics says, however, that this cannot be done: No engine can be constructed, no matter how ingenious its mechanical design, that can convert heat into work without also discharging some heat to the surroundings.

Historical Development of the Second Law: The Carnot Cycle

An important milestone in the development of the second law of thermodynamics was Carnot's analysis in the early nineteenth century of the efficiency of a heat engine. Carnot considered an idealized heat engine that underwent a particular cycle of four steps to return to its original state. Unlike

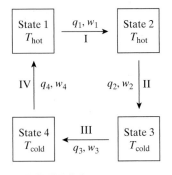

▲ **FIGURE 3.1**

The Carnot cycle of an idealized heat engine. Each step is carried out reversibly. Steps I and III are carried out isothermally (step I at T_{hot} and step III at T_{cold}) and steps II and IV adiabatically (that is, $q_2 = q_4 = 0$).

the operation of a real engine, such as a steam engine or an automobile engine, all steps in this idealized engine are reversible. Carnot's simple yet elegant analysis of the efficiency of such an engine led to the discovery of an important thermodynamic variable of state—entropy.

The four steps of the *Carnot cycle* of an engine with an ideal gas as its working substance are outlined in figures 3.1 and 3.2. In step I, a hot ideal gas at T_{hot} expands isothermally and reversibly in a cylinder. The work and heat terms associated with this step are denoted w_1 and q_1, respectively. The gas then expands adiabatically and reversibly in step II ($q_2 = 0$). In step II, as the engine does work (w_2 is negative) without heat input, its energy drops and the temperature of the gas is reduced to T_{cold}. In step III, the gas is compressed isothermally and reversibly (w_3, q_3). In the last step, the gas is compressed adiabatically and reversibly (w_4 is positive; $q_4 = 0$) to return the gas to its original condition.

For one complete cycle, the total work is $w = (w_1 + w_2 + w_3 + w_4)$, and the heat absorbed is $q = (q_1 + q_2 + q_3 + q_4)$. Because the initial and final states are the same for the cyclic process, we know that there is no change in energy, and thus $\Delta E = q + w = 0$ and $q = -w$.

The q's and w's of each step of the Carnot cycle can be calculated from what we have learned in chapter 2. For the first step (an isothermal reversible expansion),

$$w_1 = -\int_{V_1}^{V_2} P \, dV = -nRT_{hot} \ln \frac{V_2}{V_1}$$

Because

$$E_2 - E_1 = 0 \qquad (E \text{ for an ideal gas is dependent only on temperature, as noted in chapter 2})$$

and

$$q_1 + w_1 = E_2 - E_1 = 0 \qquad \text{(first law)}$$

$$q_1 = -w_1 = nRT_{hot} \ln \frac{V_2}{V_1}$$

For the second step (an adiabatic reversible expansion),

$$q_2 = 0 \qquad \text{(adiabatic)}$$

$$w_2 = E_3 - E_2$$

$$= C_V(T_{cold} - T_{hot}) \qquad \text{(See Eq. 2.23)}$$

We can derive an additional important relation for an adiabatic reversible expansion or compression of an ideal gas. For a small change dV in volume, there will be a corresponding change in temperature dT. The energy change dE is $C_V \, dT$ (Eq. 2.37). Because the process is adiabatic, the energy change is equal to the work, $-P \, dV$:

$$C_V \, dT = -P \, dV$$

$$= -\frac{nRT}{V} dV$$

Dividing both sides by T and integrating, we have for step II,

$$C_V \int_{T_{hot}}^{T_{cold}} \frac{dT}{T} = -nR \int_{V_2}^{V_3} \frac{dV}{V}$$

$$C_V \ln \frac{T_{cold}}{T_{hot}} = -nR \ln \frac{V_3}{V_2} = nR \ln \frac{V_2}{V_3} \qquad \textbf{(3.1a)}$$

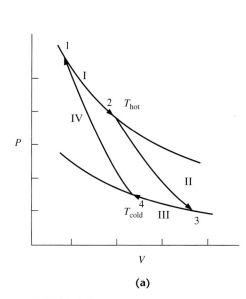

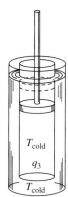

I. Isothermal reversible expansion
$P_1, V_1 \rightarrow P_2, V_2$
T_{hot} is constant
q_1 is positive
w_1 is negative

II. Adiabatic reversible expansion
$P_2, V_2 \rightarrow P_3, V_3$
$T_{hot} \rightarrow T_{cold}$
$q_2 = 0$
w_2 is negative

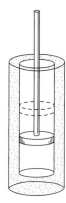

III. Isothermal reversible compression
$P_3, V_3 \rightarrow P_4, V_4$
T_{cold} is constant
q_3 is negative
w_3 is positive

IV. Adiabatic reversible compression
$P_4, V_4 \rightarrow P_1, V_1$
$T_{cold} \rightarrow T_{hot}$
$q_4 = 0$
w_4 is positive

(a)

▲ FIGURE 3.2

(a) A *PV* diagram for a Carnot cycle, illustrating the four reversible steps taken in moving between the states of figure 3.1. There are two reversible isothermal paths (steps I and III) and two reversible adiabatic paths (steps II and IV). (b) A graphical sketch of the four steps of the Carnot-cycle heat engine. The engine is in thermal contact with a hot heat reservoir at T_{hot} (step I), a cold heat reservoir at T_{cold} (step III), or is thermally isolated (steps II and IV). The first two steps (I and II) are expansions in which work is done by the engine. The next two steps (III and IV) are compressions in which work is done on the engine.

(b)

Thus, we have a relation between the temperature and volume changes for an adiabatic reversible expansion (or compression) of an ideal gas. The third and fourth steps are similar to the first and second, respectively:

$$q_3 = -w_3 = nRT_{cold} \ln \frac{V_4}{V_3} \qquad \text{(isothermal)}$$

$$q_4 = 0 \qquad w_4 = C_V(T_{hot} - T_{cold}) \qquad \text{(adiabatic)}$$

$$-C_V \ln \frac{T_{cold}}{T_{hot}} = nR \ln \frac{V_4}{V_1} \qquad \textbf{(3.1b)}$$

The total heat absorbed is

$$q = q_1 + q_2 + q_3 + q_4 = nRT_{hot} \ln \frac{V_2}{V_1} + 0 + nRT_{cold} \ln \frac{V_4}{V_3} + 0$$

The total work done by the engine is

$$-w = -(w_1 + w_2 + w_3 + w_4) = nRT_{hot} \ln \frac{V_2}{V_1} + nRT_{cold} \ln \frac{V_4}{V_3}$$

because w_2 and w_4 cancel. Therefore, the total work done by the engine is just

$$-w = q$$

This is the same conclusion that we reached earlier by using $\Delta E = 0$ for any cyclic process. Note that unlike ΔE neither q nor w is zero for this cyclic path, a reminder that heat and work are not variables of state (chapter 2).

Carnot made an important finding that although the sum of q_i for all the steps of this reversible cycle is not zero, the sum of changes in the quantity q_{rev}/T is zero. We use the subscript "rev" here to emphasize that the heat exchange occurs reversibly. In general, a change in q_{rev}/T can be evaluated by integrating dq_{rev}/T, with dq_{rev} representing an infinitesimal heat change along a reversible path and T being the temperature at which dq_{rev} is exchanged.* For the isothermal steps I and III, T is constant and

$$\int \frac{dq_{rev}}{T} = \frac{1}{T_{hot}} \int dq_{rev} = \frac{q_1}{T_{hot}}$$

$$\text{path I}$$

$$\int \frac{dq_{rev}}{T} = \frac{1}{T_{cold}} \int dq_{rev} = \frac{q_3}{T_{cold}}$$

$$\text{path III}$$

For the reversible steps II and IV, the temperature is not constant, but dq_{rev} is always zero (adiabatic); therefore,

$$\int \frac{dq_{rev}}{T} = 0$$

$$\text{paths II, IV}$$

The sum $\int dq_{rev}/T$ for the cyclic path is therefore

$$\frac{q_1}{T_{hot}} + \frac{q_3}{T_{cold}} = nR \ln \frac{V_2}{V_1} + nR \ln \frac{V_4}{V_3} = nR \ln \frac{V_2 V_4}{V_1 V_3}$$

If we add Eqs. (3.1a) and (3.1b), we see that

$$nR \ln \frac{V_2 V_4}{V_3 V_1} = 0$$

---

*Although we use the symbol dq to express a differential change in heat, we should keep in mind that there is an important difference between dq and the differential of a state function, such as dE. For any cyclic path from state 1 back to state 1, the integral dE is always zero. But dq is dependent on the particular path, and its integral is generally not zero for a cyclic path. Mathematically, we say that dE is an exact differential, but dq (and dw as well) is not.

and therefore,

$$\frac{q_1}{T_{hot}} + \frac{q_3}{T_{cold}} = 0 \qquad (3.2)$$

Equation (3.2) is an important finding. Our system (an ideal gas in a cylinder) has gone through a cycle and returned to its original state. The sum of the quantity $\int dq_{rev}/T$ for this cyclic path is zero. This is what we learned about state functions such as E and H: For a cyclic path, state functions exhibit no change. Carnot recognized that by dividing dq_{rev}, a path-dependent quantity, by T, a path-independent function, a new state function was obtained. But before we get too excited about this realization, we must consider first whether Eq. (3.2) is limited to the special cyclic path we have just considered or to the choice of an ideal gas as the working substance of the engine. Carnot deduced that the conclusion is general.

A New State Function, Entropy

Historically, the argument that led to the realization that dq_{rev}/T is a state function for any system is as follows. The efficiency of a heat engine is expressed as the total work done *by* the engine in one complete cycle $-w$ divided by the heat absorbed at the hot heat reservoir q_{hot}. For the Carnot engine, $q_{hot} = q_1$:

$$efficiency = -\frac{w}{q_{hot}} = -\frac{w}{q_1} \qquad (3.3)$$

The rest of the energy is discharged as heat q_{cold} at the cooler heat reservoir ($q_{cold} = q_3$ for the Carnot engine) and is wasted, as far as the efficiency for converting heat into work is concerned.

With a little algebra, we can solve for efficiency in terms of the temperatures of the heat reservoirs. From Eq. (3.2),

$$\frac{q_1}{T_{hot}} = -\frac{q_3}{T_{cold}} \qquad (3.4)$$

From the first law, the total work done by the engine in one complete cycle is

$$-w = q_1 + q_3$$

Thus,

$$efficiency = -\frac{w}{q_{hot}} = -\frac{w}{q_1} = \frac{q_1 + q_3}{q_1}$$

$$= 1 + \frac{q_3}{q_1} \qquad (3.3a)$$

It follows directly from Eq. (3.4) that

$$\frac{q_3}{q_1} = -\frac{T_{cold}}{T_{hot}}$$

Therefore,

$$\text{efficiency} = 1 + \frac{q_3}{q_1}$$

$$= 1 - \frac{T_{\text{cold}}}{T_{\text{hot}}} \tag{3.5}$$

For any heat engine that operates reversibly at all steps, the engine can also be operated in the reverse direction as a refrigerator or heat pump. In the forward direction as a heat engine, there is a net transfer of heat from the hot reservoir to the cold one, and the system does work on the surroundings. In the reverse direction, as a heat pump, the surroundings perform work on the system, and there is a net flow of heat from the cold reservoir to the hot one. Carnot concluded that all heat engines operating with reversible cycles between T_{hot} and T_{cold} must have the same efficiency.

Suppose T_{hot} and T_{cold} are 1200 and 300 K, respectively. The efficiency of a Carnot engine is therefore $1 - 300/1200$, or 0.75 (75%). If the engine takes up 100 kJ of heat in step I ($q_{\text{hot}} = 100$ kJ), 25 kJ would be discharged in step III ($q_{\text{cold}} = -25$ kJ), and the engine does 75 kJ of work ($w = -75$ kJ) in each cycle. Suppose another engine, like the Carnot engine, also operates on a cycle of reversible steps but has a lower efficiency of 50%. To get 75 kJ of work out of it ($w = -75$ kJ), 150 kJ would have to be absorbed at T_{hot} ($q_{\text{hot}} = 150$ kJ), and 75 kJ would be discharged at T_{cold} ($q_{\text{cold}} = -75$ kJ). Although this second engine is an inferior heat engine compared with the Carnot engine, it becomes a superior heat pump when operated in reverse: An input of 75 kJ of work ($w = 75$ kJ) would lead to the absorption of 75 kJ of heat at the cooler heat reservoir ($q_{\text{cold}} = 75$ kJ) and the delivery of a total of 150 kJ of heat to the hotter heat reservoir ($q_{\text{hot}} = -150$ kJ). If the work output of the Carnot engine is used to drive this heat pump, we can see that the total q_{hot} for the combined engine is 100 kJ $-$ 150 kJ $= -50$ kJ, and the total q_{cold} for the combined engine is -25 kJ $+$ 75 kJ $= 50$ kJ. In other words, the combined effect of the two reversible engines is a net flow of 50 kJ of heat per cycle from the cooler reservoir to the hotter reservoir, with no change in the surroundings. From our experience, however, heat flows spontaneously only from a hot body to a cooler one. If any two heat engines operating with reversible cycles between T_{hot} and T_{cold} differ in their efficiencies, we can always operate the less efficient heat engine in reverse as a heat pump and use the more efficient one as a heat engine, the work output of which is used to drive the heat pump. As illustrated by the specific calculations, the combined actions of these two engines would contradict our experience that heat can flow spontaneously only from a hotter reservoir to a cooler reservoir without changing the surroundings. Therefore, if we do not believe that heat can flow spontaneously from a colder to a hotter body, we must conclude that all reversible engines have the same efficiency.

If all reversible engines have the same efficiency, then Eq. (3.4) and therefore Eq. (3.2) cannot be a consequence of using an ideal gas in our engine or of the particular paths chosen. In other words, we have defined a new state function S, *entropy:*

$$\Delta S = \int dS = \int \frac{dq_{\text{rev}}}{T}$$

$$= \frac{q_{\text{rev}}}{T} \qquad \text{(for isothermal processes only; } T = \text{constant)} \tag{3.6}$$

Entropy is an extensive variable of state; ΔS depends on only the initial and final states of the system. From Eq. (3.6), one sees that the dimensions of entropy are energy/temperature; we will use units of $J\,K^{-1}$. The unit for entropy in the centimeter-gram-second (cgs) system is cal deg^{-1}:

$$4.184\,J\,K^{-1} = 1\,\text{cal deg}^{-1}$$

A calorie per degree is also called an *entropy unit*, eu.

We have "proven" that entropy is a state function only because we accepted as true that heat flows spontaneously from a hot body to a cooler one. We can also consider Eq. (3.6) as *defining* a thermodynamic temperature scale. The temperature scale T is the one that makes S a state function. The fact that Eq. (3.6) can also be derived for an ideal gas, as we have done, means that the thermodynamic temperature scale defined by Eq. (3.6) is identical to the temperature scale defined by the ideal gas law, $PV = nRT$.

The Second Law of Thermodynamics: Entropy Is Not Conserved

When we calculate the entropy changes of both the system and the surroundings for simple processes that can occur spontaneously—such as the flow of heat from a hot body to a cooler one, the expansion of an ideal gas into a vacuum, or the flow of water down a hill—we find that the sum of the entropy changes is not zero. In other words, unlike energy, entropy is not conserved. The generalization of a great deal of experience, which is the second law of thermodynamics, is that the sum of the entropy changes of the system and the surroundings is always positive. Even zero values can be approached only as a limit, and negative values are never found. If there is a decrease in entropy in a system, there must be an equal or larger increase in entropy in the surroundings:

$$\Delta S\,(\text{system}) + \Delta S\,(\text{surroundings}) \geq 0 \qquad \textbf{(3.7)}$$

For an *isolated system,* since there is no energy or material exchange between such a system and the surroundings, there is no change in the surroundings. Therefore,

$$\Delta S\,(\text{isolated system}) \geq 0 \qquad \textbf{(3.8)}$$

The $\geq$ sign means greater than or equal to; the former applies for irreversible processes and the latter for reversible ones.

There are many different ways of stating the second law. That heat spontaneously flows from a hot body to a cold body, for example, can be taken as a statement of this law. We express the second law by Eqs. (3.7) and (3.8) because these choices are more convenient for problems in which chemists and bioscientists are interested.

EXAMPLE 3.1

One mole of an ideal gas initially at $P_1 = 2$ atm, T, and V_1 expands to $P_2 = 1$ atm, T, and $2V_1$. Consider two different paths: (a) the expansion occurs irreversibly into a vacuum, as shown below, and (b) the expansion is reversible. Calculate q_{irrev}, ΔS (system) and ΔS (surroundings) for (a) and q_{rev}, ΔS (system), and ΔS (surroundings) for (b).

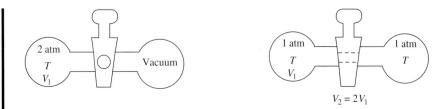

Solution

We recall that ΔS(system) is independent of path and is therefore the same for parts (a) and (b) because the initial and final stages of the system are the same for (a) and (b). However, the effect on the surroundings is quite different for (a) and (b); q depends on the path, and the change in the state of the surroundings is different for (a) and (b). Thus, the values of ΔS(surroundings) will be different.

a.

$$w = 0 \qquad \text{(no work done against surroundings)}$$

$$\Delta E = 0 \qquad (E \text{ for an ideal gas is independent of volume})$$

$$q_{\text{irrev}} = \Delta E - w = 0$$

$$\Delta S \text{ (surroundings)} = 0 \qquad \text{(no work or heat exchange between system and surroundings; surroundings not affected in any way by this process)}$$

To calculate ΔS (system), we can choose any path because ΔS (system) is independent of path. We choose the path of (b), in which the gas expands isothermally and reversibly, because we can readily calculate ΔS (system) for this path.

$$w_{\text{rev}} = -\int_{V_1}^{V_2} P \, dV = -RT \int_{V_1}^{V_2} \frac{dV}{V} = -RT \ln \frac{V_2}{V_1}$$

$$= -RT \ln 2$$

$$\Delta E = 0$$

Therefore,

$$q_{\text{rev}} = \Delta E - w_{\text{rev}} = RT \ln 2$$

and

$$\Delta S \text{ (system)} = \frac{q_{\text{rev}}}{T} = R \ln 2$$

Note that for this spontaneous, irreversible process,

$$\Delta S \text{ (system)} + \Delta S \text{ (surroundings)} = R \ln 2 + 0$$

$$> 0$$

consistent with Eqs. (3.7) and (3.8).

b. We have already calculated that

$$q_{\text{rev}} = RT \ln 2$$

$$\Delta S \text{ (system)} = R \ln 2$$

where C_V = heat capacity at constant V. Because C_P and C_V are always positive, Eqs. (3.15) and (3.17) show that raising the temperature will always increase the entropy; this is expected from the increase in disorder.

EXAMPLE 3.3

Calculate the change in entropy at constant P when 1 mol of liquid water at 100°C is brought in contact with 1 mol of liquid water at 0°C. Assume that the heat capacity of liquid water is independent of temperature and is equal to 75 J mol^{-1} deg^{-1}. No heat is lost to the surroundings.

Solution

As we bring equal amounts of the water in contact, C_P is constant and no heat is lost; the first law tells us that the final temperature of the mixture will be the average of the two temperatures, 50°C. We use Eq. (3.15) to calculate the change in entropy of the hot water and of the cold water and add the results:

Hot water:

$$S_1 = S(50°C) - S(100°C) = \bar{C}_P \ln \frac{323}{373}$$

$$= -10.79 \text{ J K}^{-1}$$

Cold water:

$$S_2 = S(50°C) - S(0°C) = \bar{C}_P \ln \frac{323}{273}$$

$$= 12.61 \text{ J K}^{-1}$$

Sum for $H_2O(100°C) + H_2O(0°C) \longrightarrow 2H_2O(50°C)$:

$$S_1 + S_2 = 1.82 \text{ J K}^{-1}$$

The entropy change is positive, as it must be for a spontaneous process in an isolated system.

Temperature Dependence of the Entropy Change for a Chemical Reaction

To calculate the entropy change for a chemical reaction at 1 atm and some temperature other than 25°C, we can use Eq. (3.15). We consider a cycle in which products and reactants are heated or cooled to the new temperature and then add the entropy change for the heating or cooling to give ΔS^0 (25°C, 1 atm):

$$
\begin{array}{ccc}
A\,(T_2) & \xrightarrow{\;\;\Delta S^0(T_2)\;\;} & B\,(T_2) \\
\Big\downarrow & & \Big\uparrow \\
A\,(25°C) & \xrightarrow{\;\;\Delta S^0(25°C)\;\;} & B\,(25°C)
\end{array}
$$

$$\Delta S^0(T_2) = \Delta S^0(25°C) + \int_{T_2}^{298} C_P(A)\frac{dT}{T} + \int_{298}^{T_2} C_P(B)\frac{dT}{T}$$

We can generalize the result and rewrite the equation in a more compact form by using the identity

$$\int_a^b = -\int_b^a$$

$$\Delta S^0(T_2) = \Delta S^0(T_1) + \int_{T_1}^{T_2} \Delta C_P \frac{dT}{T} \tag{3.18}$$

where $\Delta C_P = C_P \text{(products)} - C_P \text{(reactants)}$.

EXAMPLE 3.4

If a spark is applied to a mixture of $H_2(g)$ and $O_2(g)$, an explosion occurs and water is formed. The gaseous water is cooled to 100°C. Calculate the entropy change when 2 mol of gaseous H_2O is formed at 100°C and 1 atm from $H_2(g)$ and $O_2(g)$ at the same temperature and each at a partial pressure of 1 atm.

Solution

The reaction for 2 mol of H_2O is

$$2H_2(g) + O_2(g) \longrightarrow 2H_2O(g)$$

The entropy change at 25°C, 1 atm, is

$$\Delta S^0(25°C) = 2\bar{S}^0_{H_2O(g)} - \bar{S}^0_{O_2(g)} - 2\bar{S}^0_{H_2(g)}$$

$$= 2(188.72) - 205.04 - 2(130.57)$$

$$= -88.74 \text{ J K}^{-1}$$

To find ΔS (100°C, 1 atm), we need to use Eq. (3.18); therefore, we need to know the heat capacities of $H_2(g)$, $O_2(g)$, and $H_2O(g)$. They can be taken as constants over the temperature range 25°C to 100°C with the values given in table 2.1:

$$\Delta C_P^0 = 2\bar{C}^0_{PH_2O(g)} - \bar{C}^0_{PO_2(g)} - 2\bar{C}^0_{PH_2(g)}$$

$$= 2(33.6) - 29.4 - 2(28.8)$$

$$= -19.8 \text{ J K}^{-1}$$

From Eq. (3.18),

$$\Delta S^0(100°C) = \Delta S^0(25°C) + \int_{298}^{373} \Delta C_P \frac{dT}{T}$$

$$= \Delta S(25°C) + \Delta C_P \ln \frac{373}{298}$$

$$= -88.74 - (19.8)(0.224)$$

$$= -93.18 \text{ J K}^{-1}$$

The entropy of the system decreased; the second law thus requires that there be an increase in the entropy of the surroundings. The reaction is exothermic, and heat is lost to the surroundings at constant temperature.

Entropy Change for a Phase Transition

On heating many compounds from 0 K to room temperature, various phase transitions, such as melting and boiling, may occur. The reversible heat absorbed divided by the *equilibrium* temperature of the transition gives the entropy change for the phase transition. It is important to stress equilibrium transition temperature. Otherwise, the transition is not reversible, and the heat absorbed is not the reversible heat. For a phase transition, at constant P and T,

$$q_P = q_{rev} = \Delta H_{tr} \tag{2.36}$$

$$\Delta S_{tr} = \frac{\Delta H_{tr}}{T_{tr}} \tag{3.19}$$

where ΔS_{tr} = entropy of phase transition and ΔH_{tr} = enthalpy of phase transition at equilibrium temperature T_{tr}.

Third-law entropies or absolute entropies, such as those in tables A.5–7 in the appendix, are obtained by adding together (1) the entropy of the substance at 0 K (which will be zero if it is a perfect crystal), (2) the entropy increase associated with increasing the temperature, and (3) the entropy changes associated with any phase changes. For example, to obtain the third-law entropy of a liquid compound at 25°C and 1 atm, we would use the equation

$$\overline{S}^0(25°C, 1\ atm) = \overline{S}^0(0\ K) + \int_0^{T_m} \overline{C}_P(s) \frac{dT}{T} + \frac{\Delta H_m^0}{T_m} + \int_{T_m}^{298} \overline{C}_P(l) \frac{dT}{T}$$

where T_m is the melting temperature at 1 atm and we have assumed that there are no solid–solid transitions. If there are, we would add $(\Delta H_{tr}^0/T_{tr})$ for each transition and use an appropriate $\overline{C}_P$ for each solid phase.

Pressure Dependence of Entropy

For solids and liquids, we ignore the direct effect of pressure on entropy:

$$\Delta S = S(P_2) - S(P_1) \cong 0 \tag{3.20}$$

For gases, we approximate the effect of pressure at constant temperature by that on an ideal gas. The change in entropy dS for a small change dP in pressure is

$$dS = \frac{dq_{rev}}{T}$$

$$= \frac{dE - dw_{rev}}{T}$$

For an ideal gas that changes its pressure from P_1 to P_2 at constant temperature, dE is zero because E is independent of P at constant temperature. Thus,

$$dS = -\frac{dw_{rev}}{T} = \frac{P\ dV}{T}$$

Because $PV = nRT$ is a constant at constant temperature,

$$d(PV) = P\ dV + V\ dP = 0 \qquad \text{(constant temperature)}$$

and

$$P\ dV = -V\ dP \qquad \text{(constant temperature)}$$

Therefore,

$$dS = \frac{P\,dV}{T}$$

$$= -\frac{V\,dP}{T}$$

$$= -\frac{nR\,dP}{P}$$

and

$$\Delta S = -nR \int_{P_1}^{P_2} \frac{dP}{P} \tag{3.21}$$

or

$$\Delta S = S(P_2) - S(P_1) = -nR \ln \frac{P_2}{P_1} \tag{3.21a}$$

We notice that in Eq. (3.21a) if either P_2 or P_1 is zero, the entropy of n mol of gas becomes infinite. This should not surprise us because zero pressure means infinite volume, and infinite volume implies infinite disorder of the molecules in the volume. The point to remember is that as the pressure of a gas decreases at constant temperature, the entropy will increase. Equations (3.20) and (3.21a) can be used to calculate the entropy change of a substance when the pressure is changed or for a chemical reaction at some pressure other than 1 atm.

A useful application of Eq. (3.21a) is in the calculation of the entropy change of mixing two gases. Two gases, n_A mol of A and n_B mol of B, are initially in two separate flasks connected by a tube with a valve in it, similar to the arrangement illustrated in figure 3.3. We consider the special case with the two gases at the same pressure (that is, $P_A = P_B = P$) and temperature T (the volumes of the flasks are in general not the same; they are the same only if $n_A = n_B$). After opening the valve, the two gases will eventually become completely mixed, as both our experience and the second law of thermodynamics tell us. Because we start with two gases at the same pressure and we also assume that the two gases behave like ideal gases and do not react with each other, there will be no change in the total presure P after mixing. The partial pressures of A and B after mixing, however, become $P_A = P[n_A/(n_A + n_B)] = X_A P$ and $P_B = P[n_B/(n_A + n_B)] = X_B P$, respectively. [$X_A = n_A/(n_A + n_B)$ and $X_B = n_B/(n_A + n_B)$ are the mole fractions of A and B, respectively; see Eq. (2.20).] From Eq. (3.21a), the entropy change ΔS_A accompanying a change of the pressure of A from P to $X_A P$ is, using the ideal gas approximation,

$$\Delta S_A = -n_A R \ln\left(\frac{X_A P}{P}\right) = -n_A R \ln X_A$$

Similarly,

$$\Delta S_B = -n_B R \ln\left(\frac{X_B P}{P}\right) = -n_B R \ln X_B$$

The total entropy change for mixing the two ideal gases is therefore

$$\Delta S_{mix} = \Delta S_A + \Delta S_B = -R[n_A \ln X_A + n_B \ln X_B] \tag{3.22}$$

It is interesting to note that if the two gases are identical, there is no entropy change upon mixing the two sides. The entropy increases when the valve is opened only if the two gases are *distinguishable*. One gas can be an isotope of the other, but as long as we can tell them apart, we calculate the same ΔS_{mix}. If the two are *indistinguishable*, ΔS_{mix} is zero. This conclusion is entirely consistent with the molecular interpretation of entropy. For two indistinguishable gases, swapping molecules between the two flasks does not alter the total number of microscopic states.

Spontaneous Chemical Reactions

We began this chapter by wondering if graphite could be converted into diamonds. One way to answer this question is to learn if the entropy of the universe increases when the reaction occurs. We can easily learn from tables A.5–7 in the appendix whether the entropy of the system increases, but it is not so easy to learn about the change in entropy of the surroundings. For processes that occur in an *isolated system*, however, the surroundings do not change, and we can limit our attention to the system: The sign of the entropy change tells whether the reaction can occur. Very few reactions occur at constant energy and volume and thus do not affect the surroundings. Conversion of an optically active molecule to a racemic mixture is an example. The D form and the L form have identical energies, volumes, and entropies, but the mixture has a larger entropy and nearly the same energy and volume. The conversion reaction is spontaneous. Other examples are the mixing of two ideal gases or the transfer of heat from a hot object to a cold one.

Gibbs Free Energy

For most chemical reactions, there are large changes in energy and enthalpy, and we are interested in whether a reaction will occur at constant T and P. These are the conditions under which most experiments are done in the laboratory: The system is free to exchange heat with the surroundings to remain at room temperature, and it can expand or contract in volume to remain at atmospheric pressure. We need a criterion of spontaneity that applies to the system for these conditions. A new thermodynamic variable of state, the Gibbs free energy, is useful in this situation. The *Gibbs free energy, G*, is defined as a combination of enthalpy, temperature, and entropy:

$$G \equiv H - TS \tag{3.23}$$

Because G is specified by the state variables H, T, and S and because both H and TS are quantities that increase in proportion to the amounts of materials, the Gibbs free energy is an extensive variable of state and has the same units as enthalpy or energy. Its definition in Eq. (3.23) was chosen because it can thus characterize whether a process will occur spontaneously at constant temperature and pressure. We will see in the next two sections that if ΔG for a process at constant T and P is negative, the process can occur spontaneously; if ΔG at constant T and P is positive, the process will not occur spontaneously; if ΔG at constant T and P is zero, the system is at equilibrium.

ΔG and a System's Capacity to Do Nonexpansion Work

To see how the change in Gibbs free energy is related to the spontaneity of a process, we first examine the relation between ΔG and a system's capacity to do work. From the definition of G, for a closed system a small change dG in the Gibbs free energy is related to changes in the other thermodynamic parameters by

$$dG = dH - T\,dS - S\,dT \qquad \text{[definition of } G;\ \text{Eq. (3.23)]}$$

$$= dE + P\,dV + V\,dP - T\,dS - S\,dT \qquad (H = E + PV)$$

$$= dq_{rev} + dw_{rev} + P\,dV + V\,dP - T\,dS - S\,dT \qquad \text{(first law, reversible path)}$$

$$= T\,dS + dw_{rev} + P\,dV + V\,dP - T\,dS - S\,dT \qquad \text{(definition of } S)$$

$$= dw_{rev} + P\,dV + V\,dP - S\,dT \qquad \textbf{(3.24)}$$

We express dw_{rev} as the sum of two types of work, the pressure–volume work, $-P\,dV$, owing to a change dV in the volume of the system, and all other forms of work, dw_{rev}^*:

$$dw_{rev} = -P\,dV + dw_{rev}^* \qquad \textbf{(3.25)}$$

Equation (3.24) then becomes

$$dG = V\,dP - S\,dT + dw_{rev}^* \qquad \textbf{(3.26)}$$

At constant temperature and pressure, $dP = dT = 0$ and

$$dG = dw_{rev}^* \qquad \textbf{(3.27)}$$

or, upon integration,

$$\Delta G = w_{rev}^* \qquad \textbf{(3.28)}$$

Equation (3.28) states that ΔG of a process at constant pressure and temperature is equal to w_{rev}^* or that $-\Delta G$ is equal to $-w_{rev}^*$. In any process, such as in a chemical reaction, a system must do the necessary PV work associated with the change of the system from its initial state to its final state. Therefore, the *useful work* that can be obtained from a reversible process at constant temperature and pressure is $-w_{rev}^*$, the nonexpansion work. If a reversible process does electrical work other than the PV work, then $-w_{rev}^*$ is the electrical work that the process is capable of doing. *Because the work done by a system is maximal when a process is carried out reversibly, $-w_{rev}^* = -\Delta G$ is the maximal amount of nonexpansion work a system can do on the surroundings by a process at constant temperature and pressure.*

Spontaneous Reactions at Constant *T* and *P*

Equation (3.28) provides a criterion for processes that can spontaneously occur at constant pressure and temperature. Our experience tells us that if a system undergoes a spontaneous process, it is capable of performing useful work: water flowing down a fall can be utilized to generate electricity, and heat flowing from a hot to a cold reservoir can be utilized to power a heat engine. Conversely, for a nonspontaneous process, such as water flowing uphill or heat flowing from cold to hot, the surroundings must do work on the system. Thus, $-w_{rev}^*$, the maximal nonexpansion work a system can do on the surroundings, is positive for a spontaneous process. From Eq. (3.28), it follows that $-\Delta G = -w_{rev}^* > 0$ for a spontaneous process at constant temperature and pressure, or

$$\Delta G < 0 \qquad \text{(spontaneous process at constant } T \text{ and } P) \qquad \textbf{(3.29)}$$

If a process does not occur spontaneously, w_{rev}^* must be positive for it to occur and

$$\Delta G > 0 \qquad \text{(nonspontaneous process at constant } T \text{ and } P) \qquad \textbf{(3.30)}$$

If a system is already at equilibrium, it cannot perform useful work. Thus, $w_{rev}^* = 0$ and

$$\Delta G = 0 \qquad \text{(system at equilibrium at constant } T \text{ and } P) \qquad \textbf{(3.31)}$$

The above discussion is also useful in clarifying the meaning of "Conserve energy!"—a message of ever-increasing importance as we become more appreciative of our environment. If the first law of thermodynamics says that energy is conserved, then why worry about conserving energy? The preceding discussion tells us that a system undergoing any spontaneous process has the potential of doing useful work. At constant temperature and pressure, the decrease in Gibbs free energy corresponds to the maximal useful work a system can do. Thus, if a certain amount of gasoline is consumed to power an automobile engine, the maximal amount of mechanical work that can be extracted from this process is characterized by the Gibbs free energy for the combustion of gasoline. The mechanical work from this process can in turn make our life more pleasant. The better we design the engine, the higher the fraction of the maximal useful work we can harness. If we simply burn the gasoline in the air, we get no useful work, and the potential of the process to do useful work is lost (the heat released into the environment by this process cannot be efficiently recovered to do useful work). Therefore, when we say "Conserve energy!" we actually mean "Conserve useful work!" We mean that natural resources that can be utilized to do useful work are limited, and thus we must not waste these resources.

Calculation of Gibbs Free Energy

The Gibbs free-energy change for a reaction at constant temperature can be obtained from the enthalpy and entropy changes. At constant temperature,

$$\Delta G = \Delta H - T\Delta S \qquad \textbf{(3.32)}$$

Tables A.5–7 in the appendix give values of $\overline{H}^0 \equiv \Delta H_f^0$ and $\overline{S}^0$ for various substances at 25°C and 1 atm. These values can be used to calculate ΔH^0 and ΔS^0 and hence ΔG^0, at 25°C and 1 atm.

Values of ΔG_f^0, the standard free energy of formation, of various substances are also given in tables A.5–7 in the appendix. The molar standard free energy of formation is defined as the free energy of formation of 1 mol of any compound at 1 atm pressure from its elements in their standard states at 1 atm. In a manner completely analogous to our discussion on standard enthalpy of formation, we arbitrarily assign the elements in their most stable state at 1 atm to have zero free energy.

EXAMPLE 3.5

Calculate the Gibbs free energy for the following reaction at 25°C and 1 atm. Will the reaction occur spontaneously?

$$H_2O(l) \longrightarrow H_2(g) + \frac{1}{2}O_2(g)$$

Solution

$$\Delta G^0(25°C) = \overline{G}^0_{H_2(g)} + \frac{1}{2}\overline{G}^0_{O_2(g)} - \overline{G}^0_{H_2O(l)}$$

$$= 0 + 0 - (-237.19)$$

$$= 237.19 \text{ kJ mol}^{-1}$$

The Gibbs free-energy change is just the negative of the Gibbs free energy of formation of liquid water.

Alternatively, we can calculate ΔH^0 and ΔS^0 from table A.5 in the appendix:

$$\Delta H^0 = \overline{H}^0_{H_2(g)} + \frac{1}{2}\overline{H}^0_{O_2(g)} - \overline{H}^0_{H_2O(l)}$$

$$= 0 + 0 - (-285.83)$$

$$= 285.83 \text{ kJ mol}^{-1}$$

$$\Delta S^0 = \overline{S}^0_{H_2(g)} + \frac{1}{2}\overline{S}^0_{O_2(g)} - \overline{S}^0_{H_2O(l)}$$

$$= 130.57 + \frac{1}{2}(205.04) - 69.95$$

$$= 163.14 \text{ J K}^{-1} \text{ mol}^{-1}$$

ΔG^0 can then be obtained:

$$\Delta G^0 = \Delta H^0 - T\Delta S^0$$

$$= 285.83 \text{ kJ mol}^{-1} - (298)(163.14 \times 10^{-3}) \text{ kJ mol}^{-1}$$

$$= 237.19 \text{ kJ mol}^{-1}$$

The same answer is obtained. The reaction will *not* occur spontaneously because ΔG is positive.

A proposed method of storing solar energy is to use sunlight to decompose water. The sunlight provides the driving force to overcome the large positive free energy. The hydrogen and oxygen gases produced make an excellent fuel. The light reactions of green plant photosynthesis involve a positive free-energy change that is almost exactly the same as that for the reaction to make oxygen gas by the direct decomposition of water. The difference is that green plants do not make hydrogen gas; instead, they reduce carbon dioxide to carbohydrates and other products. The driving force for this reaction with a large increase in ΔG is the sunlight absorbed by chlorophyll and the other plant pigments.

EXAMPLE 3.6

We wish to know whether proteins in aqueous solution are unstable with respect to their constituent amino acids. As an example, let's calculate the standard free energy of hydrolysis for the dipeptide glycylglycine at 25°C and 1 atm in dilute aqueous solution.

Solution

The reaction is

$$^+H_3NCH_2CONHCH_2COO^-(aq) + H_2O(l) \longrightarrow 2\,^+H_3NCH_2COO^-(aq)$$

glycylglycine   glycine

The standard free-energy change when solid glycine dissolves has been measured and is small. We will assume that this is also true for solid glycylglycine. Therefore, we use the free-energy values from tables A.5–7 in the appendix for solid glycine and glycylglycine in the following calculation:

$$\Delta G^0(25°C) = 2\overline{G}^0(\text{glycine}, s) - \overline{G}^0(\text{glycylglycine}, s) - \overline{G}^0(H_2O, l)$$

$$= 2(-377.69) - (-490.57) - (-237.13)$$

$$= -27.68 \text{ kJ mol}^{-1}$$

The reaction is spontaneous, but it normally occurs slowly. However, appropriate catalysts such as proteolytic enzymes can cause the reaction to occur rapidly. There are many such enzymes in living organisms. Proteolytic enzymes in digestive systems break down proteins in food, and a large number of more specific proteases perform many important tasks in living cells, including the regulation of the cellular functions of various proteins and the programmed death of cells.

Temperature Dependence of Gibbs Free Energy

In the preceding examples, for reactions carried out at 25°C and 1 atm the values of ΔG can often be calculated from tabulations of enthalpy of formation and entropy data. If ΔG of a reaction is known at 25°C, how do we obtain its value at a different temperature T? It is clear from the definition of Gibbs free energy that G (and therefore ΔG of a reaction) depends explicitly on T. For temperatures not too different from 25°C, changes in ΔH and ΔS over a narrow temperature are relatively small, and Eq. (3.32) offers a simple way to calculate ΔG at other temperatures:

$$\Delta G(T) \cong \Delta H(25°C) - T\Delta S(25°C) \tag{3.33}$$

For example, values of ΔH and ΔS at 25°C from tables A.5–7 in the appendix can be used to calculate the approximate free energy of the reaction at a "physiological temperature" of 37°C. Another useful equation that is easily derived from Eq. (3.32) is

$$\Delta G(T) - \Delta G(25°C) \cong -(T - 298)\Delta S(25°C) \tag{3.34}$$

This equation shows that the sign of ΔS for a reaction indicates how ΔG will change with temperature. If ΔS is negative, ΔG increases with increasing temperature.

A general expression of the temperature dependence of ΔG can be obtained by considering a reversible path involving only PV work. For such a path, Eq. (3.26) becomes

$$dG = V\,dP - S\,dT \tag{3.35}$$

Although we derived Eq. (3.35) for a reversible path with only PV work, it is important to note that all variables in this equation are state functions that do not depend on a particular path. Therefore, this equation is generally applicable to all paths and not just to the particular path chosen for its derivation. At constant pressure, $V\,dP$ is zero, and Eq. (3.35) becomes

$$\frac{dG}{dT} = -S \qquad (P = \text{constant}) \tag{3.36}$$

which can be integrated to give ΔG at constant pressure when the temperature is changed from T_1 to T_2:

$$\Delta G = G(T_2) - G(T_1) = -\int_{T_1}^{T_2} S\,dT \qquad (P = \text{constant}) \tag{3.36a}$$

In Eqs. (3.36) and (3.36a), S at constant pressure is a function of T [see Eqs. (3.14)–(3.17)].

Eqs. (3.36) and (3.36a) can also be written in terms of the Gibbs free energy $\overline{G}$ and entropy $\overline{S}$ per mole of a substance:

$$\frac{d\overline{G}}{dT} = -\overline{S} \qquad (P = \text{constant}) \tag{3.37}$$

$$\Delta \overline{G} = \overline{G}(T_2) - \overline{G}(T_1) = -\int_{T_1}^{T_2} \overline{S}\,dT \qquad (P = \text{constant}) \tag{3.37a}$$

To calculate the temperature dependence of the Gibbs free-energy change of a chemical reaction

$$n_A A + n_B B \longrightarrow n_C C + n_D D$$

we use the same approach for the temperature dependence of ΔH [Eq. (2.43)]. The free-energy change of the reaction is

$$\Delta G = n_C \overline{G}_C + n_D \overline{G}_D - n_A \overline{G}_A - n_B \overline{G}_B \tag{3.38}$$

where $\overline{G}_C$, $\overline{G}_D$, and so on are the molar free energies of compounds C, D, and so on. [Here we consider the reactants and products as pure substances that are present as separate phases (solid zinc pellets dropped into a dilute hydrochloric acid solution, steam passed over hot charcoal, etc.). A more precise definition of $\overline{G}_C$, $\overline{G}_D$, etc. is needed when a reaction occurs in a solution. We briefly return to this point later in this chapter and in more detail in chapter 4.]

At constant pressure, taking the derivative of Eq. (3.38) with respect to temperature gives

$$\frac{d\Delta G}{dT} = \frac{n_C d\overline{G}_C}{dT} + \frac{n_D d\overline{G}_D}{dT} - \frac{n_A d\overline{G}_A}{dT} - \frac{n_B d\overline{G}_B}{dT}$$

$$= -n_C \overline{S}_C - n_D \overline{S}_D + n_A \overline{S}_A + n_B \overline{S}_B$$

$$= -\Delta S \tag{3.39}$$

where

$$\Delta S = n_C \overline{S}_C + n_D \overline{S}_D - n_A \overline{S}_A - n_B \overline{S}_B$$

is the entropy change of the reaction. Integrating Eq. (3.39) gives

$$\int_{\Delta G(T_1)}^{\Delta G(T_2)} d\Delta G = -\int_{T_1}^{T_2} \Delta S \, dT$$

$$\Delta G(T_2) - \Delta G(T_1) = -\int_{T_1}^{T_2} \Delta S \, dT \qquad \textbf{(3.39a)}$$

If ΔS of the reaction does not change significantly in the temperature range from T_1 to T_2, Eq. (3.39a) becomes

$$\Delta G(T_2) - \Delta G(T_1) = -\Delta S(T_2 - T_1) \qquad \textbf{(3.39b)}$$

Another useful equation for evaluating the temperature dependence of ΔG at constant pressure is

$$\frac{\Delta G(T_2)}{T_2} - \frac{\Delta G(T_1)}{T_1} = -\int_{T_1}^{T_2} \frac{\Delta H(T)}{T^2} \, dT \qquad \textbf{(3.40)}$$

which is called the *Gibbs–Helmholtz equation*. This equation can be derived by taking the differential of (G/T) with respect to T at constant pressure:

$$\frac{d(G/T)}{dT} = G\left[\frac{d(1/T)}{dT}\right] + \left(\frac{1}{T}\right)\left(\frac{dG}{dT}\right)$$

$$= -\frac{G}{T^2} + \left(\frac{1}{T}\right)\left(\frac{dG}{dT}\right)$$

From Eq. (3.36), at constant pressure $dG/dT = -S$. Thus,

$$\frac{d(G/T)}{dT} = -\frac{G}{T^2} - \frac{S}{T}$$

$$= \frac{-G - TS}{T^2}$$

$$= \frac{-(H - TS) - TS}{T^2}$$

$$= -\frac{H}{T^2} \qquad \textbf{(3.41)}$$

Multiplying both sides of Eq. 3.41 by T^2 and recalling that $dT/T^2 = -d(1/T)$, we obtain

$$\frac{d(G/T)}{d(1/T)} = H \qquad (P = \text{constant}) \qquad \textbf{(3.42)}$$

For a chemical reaction

$$n_A A + n_B B \longrightarrow n_C C + n_D D$$

it follows directly from Eqs. (3.41) and (3.42) that

$$\frac{\Delta G(T_2)}{T_2} - \frac{\Delta G(T_1)}{T_1} = -\int_{T_1}^{T_2} \frac{\Delta H(T)}{T^2} \, dT \qquad \textbf{(3.40)}$$

$$= \int_{1/T_1}^{1/T_2} \Delta H \, d\left(\frac{1}{T}\right) \qquad \textbf{(3.40a)}$$

where ΔH is the enthalpy change of the reaction

$$\Delta H = n_C \overline{H}_C + n_D \overline{H}_D - n_A \overline{H}_A - n_B \overline{H}_B$$

If the change in ΔH is small over the temperature range, Eq. (3.40a) becomes

$$\frac{\Delta G(T_2)}{T_2} - \frac{\Delta G(T_1)}{T_1} = \Delta H\left[\frac{1}{T_2} - \frac{1}{T_1}\right] \qquad \textbf{(3.40b)}$$

EXAMPLE 3.7

What is the free energy of hydrolysis of glycylglycine at 37°C and 1 atm?

Solution

We can use Eq. (3.39b) to find ΔG^0 (37°C) if we assume that ΔS is independent of temperature. Or we can use Eq. (3.40a) if we assume that ΔH is independent of temperature. We will use both equations and thus show that the two equations are indeed equivalent. Because the temperature range is small, the constancy of ΔH and ΔS is a good approximation. To calculate ΔS and ΔH, we use the data for the solid compounds:

$$\Delta S^0(25°C) = 2\bar{S}^0(\text{glycine}) - \bar{S}^0(\text{glycylglycine}) - \bar{S}^0(H_2O, l)$$

$$= 2(103.51) - 190.0 - 69.95$$

$$= -52.9 \text{ J K}^{-1} \text{ mol}^{-1}$$

$$\Delta H^0(25°C) = 2(-537.2) - (-745.25) - (-285.83)$$

$$= -43.32 \text{ kJ mol}^{-1}$$

With Eq. (3.39b),

$$\Delta G^0(37°C) = \Delta G^0(25°C) - 12\Delta S^0(25°C)$$

$$= -27.62 \text{ kJ mol}^{-1} - (12 \text{ K})(-0.0529 \text{ kJ K}^{-1} \text{ mol}^{-1})$$

$$= -26.98 \text{ kJ mol}^{-1}$$

With Eq. (3.40a),

$$\frac{\Delta G^0(37°C)}{310} - \frac{\Delta G^0(25°C)}{298} = \left(\frac{1}{310} - \frac{1}{298}\right)\Delta H^0(25°C)$$

$$\frac{\Delta G^0(37°C)}{310} = \frac{-27.62}{298} + (-1.299 \times 10^{-4})(-43.32)$$

$$= -0.0871$$

$$\Delta G^0(37°C) = -26.99 \text{ kJ mol}^{-1}$$

Both methods lead to a decrease in the magnitude of the free energy of hydrolysis and agree well.

Pressure Dependence of Gibbs Free Energy

From values in tables A.5–7 in the appendix we can calculate the Gibbs free-energy change ΔG^0 at 25°C and at other temperatures by the use of the appropriate equations. However, the tabulated values of ΔG^0 are all defined for 1 atm pressure. To calculate free energy at some other pressure we must know its dependence on pressure. From Eq. (3.35), we see that at constant temperature

$$dG = V \, dP - S \, dT \qquad \textbf{(3.35)}$$

$$= V \, dP \qquad \text{(at constant temperature)} \qquad \textbf{(3.43)}$$

or

$$G(P_2) - G(P_1) = \int_{P_1}^{P_2} V\, dP \tag{3.43a}$$

For a solid or liquid, at moderate pressures the volume is approximately constant independent of pressure, and can be taken out of the integral:

$$G(P_2) - G(P_1) \cong V(P_2 - P_1) \tag{3.44}$$

For a gas, we use the equation of state to write V as a function of P. For an ideal gas, and approximately for any gas, we can substitute the ideal gas equation in Eq. (3.43a):

$$G(P_2) - G(P_1) = \int_{P_1}^{P_2} \frac{nRT}{P}\, dP = nRT \ln \frac{P_2}{P_1} \tag{3.45}$$

To calculate the effect of pressure on the free energy of a chemical reaction, we just apply Eqs. (3.44) and (3.45) to each product and reactant. If all products and reactants are solids or liquids, we use Eq. (3.44):

$$\Delta G(P_2) - \Delta G(P_1) = \Delta V(P_2 - P_1) \tag{3.46}$$

where $\Delta V = V(\text{products}) - V(\text{reactants})$. Equation (3.46) shows that if the volume of products is greater than the volume of reactants (ΔV is positive), increasing the pressure will increase the free energy of the reaction. If at least one gaseous product or reactant is involved, we can ignore the volume of the solids or liquids compared to that of the gases and use Eq. (3.45).

$$\Delta G(P_2) - \Delta G(P_1) = \Delta n\, RT \ln \frac{P_2}{P_1} \tag{3.47}$$

where Δn is the number of moles of *gaseous* products minus the number of moles of *gaseous* reactants.

EXAMPLE 3.8

Can graphite (pencil lead) be converted spontaneously into diamond at 25°C and 1 atm?

Solution

The reaction at 25°C and 1 atm

$$C(s, \text{graphite}) \longrightarrow C(s, \text{diamond})$$

has a Gibbs free energy of +2.90 kJ mol^{-1} from table A.5 in the appendix. Therefore, the answer is no; graphite will not spontaneously convert into diamond at 25°C and 1 atm. This result also means that diamond will spontaneously convert into graphite. For those who treasure their diamonds, it is fortunate that the rate of this conversion is extremely slow under these conditions.

EXAMPLE 3.9

Will increasing the pressure favor the conversion of graphite to diamond? If so, what is the minimum pressure necessary to make this reaction spontaneous at 25°C? Will increasing the temperature favor the reaction?

Solution

We can use Eq. (3.46) to answer the first two questions. We need to know the volume change for the reaction. We will use the densities of graphite and diamond at 25°C to calculate the volumes and assume that the volumes are independent of pressure. Just knowing that diamond is denser than graphite allows us to conclude that ΔV for the reaction is negative (V of diamond is less than V of the same amount of graphite). This means that *increasing the pressure* will decrease the free energy and *favor* the reaction graphite to diamond.

To find the minimum pressure necessary to allow the spontaneous reaction, calculate the pressure that makes $\Delta G = 0$. The densities at 25°C and 1 atm are as follows:

	Density (g cm^{-3})
C (graphite)	2.25
C (diamond)	3.51

The molar volumes at 25°C and 1 atm are obtained by dividing the atomic mass of carbon by the densities:

	$\overline{V}$ (cm^3 mol^{-1})
C (graphite)	5.33
C (diamond)	3.42

For the reaction

$$C\,(\text{graphite}) \longrightarrow C\,(\text{diamond})$$

the Gibbs free energy at 25°C and P atm is, from Eq. (3.46),

$$\Delta G(P) = \Delta G(1 \text{ atm}) + \Delta V(P - 1)$$

$$= 2.84 \text{ kJ mol}^{-1} + (3.42 - 5.33) \text{ cm}^3 \text{ mol}^{-1} \cdot (P - 1) \text{ atm}$$

We must use the same units throughout our equation, so we convert cm^3 atm to kJ by multiplying by a ratio of gas constants, $R/R = 8.314 \times 10^{-3}$ kJ/82.05 cm^3 atm $= 1.0133 \times 10^{-4}$ kJ cm^{-3} atm^{-1}:

$$\Delta G(P) = 2.84 - 1.935 \times 10^{-4}(P - 1) \text{ kJ mol}^{-1}$$

We want to find the pressure that makes $\Delta G(P) = 0$.

$$0 = 2.84 - 1.935 \times 10^{-4}(P - 1)$$

$$P - 1 = \frac{2.84}{1.935 \times 10^{-4}}$$

$$P = 15{,}000 \text{ atm}$$

The assumption that ΔV is constant is *not* valid over this large pressure range, and the actual pressure needed is higher than 15,000 atm. Nevertheless, the effect is real, and diamonds have been made in this way for many years, mostly for industrial uses.

To decide whether increasing temperature will favor the reaction, we have to know the sign of ΔS [see Eqs. (3.39a) and (3.39)]. From table A.5 in the appendix, we find that $\Delta S = 2.38 - 5.74 = -3.36$ J K^{-1} mol^{-1}; therefore, *increasing the temperature* will *not favor* the formation of diamond. Equation (3.39a) rather than Eq. (3.39b) must be used to calculate at what low temperature ΔG becomes zero, because ΔS changes significantly with temperature over the wide temperature change. The industrial process does use high temperatures, but for kinetic reasons. The increase in pressure provides the necessary free-energy change.

Phase Changes

For a phase change that takes place at its equilibrium temperature and pressure, the change in Gibbs free energy is zero:

$$\Delta G = 0 \qquad \text{(at equilibrium)}$$

To calculate the Gibbs free energy at other temperatures and pressures, we can use

$$\Delta G = \Delta H - T\Delta S \tag{3.32}$$

if ΔH and ΔS are known for the phase change at arbitrary T and P. Otherwise, we can calculate the change of ΔG with T and P by the methods described in earlier sections of this chapter, "Temperature Dependence of Gibbs Free Energy" and "Pressure Dependence of Gibbs Free Energy."

Helmholtz Free Energy

For a process that takes place at constant T and V, the sign of the Gibbs free energy is not a criterion for equilibrium. A new thermodynamic variable of state is needed; it is called the *Helmholtz free energy, A*. The Helmholtz free energy is defined as

$$A \equiv E - TS \tag{3.48}$$

and the criterion for equilibrium is, *at constant T and V,*

$$\Delta A < 0 \qquad \text{(spontaneous reaction)} \tag{3.49}$$

$$\Delta A = 0 \qquad \text{(reaction at equilibrium)} \tag{3.50}$$

$$\Delta A > 0 \qquad \text{(not a spontaneous reaction)} \tag{3.51}$$

We will not use the Helmholtz free energy very much in this book, although for reactions at constant volume it is as useful as the Gibbs free energy is for reactions at constant pressure. The Helmholtz free energy is widely used for geochemical problems, where the pressure may vary widely and the constant-volume restriction is more appropriate.

Noncovalent Reactions

The examples that we have given so far of the changes in H, S, and G for chemical reactions involve the breaking and formation of covalent bonds. Many important biochemical processes involve reactions in which no covalent bonds are made or broken; only weaker bonds and interactions are involved. Examples include the reaction of an antigen with its antibody, the binding of many hormones and drugs to nucleic acids and proteins, the reading of the genetic message (codon–anticodon recognition), denaturation of proteins and nucleic acids, and so on. Thermodynamics can help us understand these reactions and decide which proposed mechanisms are

TABLE 3.1 Some Enthalpies of Noncovalent Bonds and Interactions*

Reaction	Characteristic interaction	ΔH^0 (kJ mol^{-1})
$Na^+(g) + Cl^-(g) \rightarrow NaCl(s)$	Ionic	-785
$NaCl(s) + \infty H_2O(l) \rightarrow Na^+(aq) + Cl^-(aq)$	Ionic and ion–dipole	4
$Argon(g) \rightarrow Argon(s)$	London	-8
$n\text{-Butane}(g) \rightarrow n\text{-Butane}(l)$	London–van der Waals	-20
$Acetone(g) \rightarrow Acetone(l)$	London–van der Waals	-30
$2\ [\text{Methanol}](g) \rightarrow [\text{H}_3\text{C}-\text{O}\cdots\text{H}-\text{O}-\text{CH}_3](g)$	Hydrogen bond (g)	-20
$2\ [\text{Ammonia}](g) \rightarrow [\text{H}_2\text{N}\cdots\text{H}-\text{NH}_2](g)$	Hydrogen bond (g)	-15
$2\ [N\text{-Methyl formamide}](\text{benzene}) \rightarrow [\text{dimer}]$	Hydrogen bond (benzene)	-15
$[\text{Urea}\cdots\text{HOH}](aq) + [\text{HOH}\cdots\text{Urea}](aq) \rightarrow [\text{Urea}\cdots\text{Urea}] + [\text{HOH}\cdots\text{HOH}](aq)$	Hydrogen bond (aqueous)	-5
$C_3H_6(l) + \infty H_2O(l) \rightarrow C_3H_6(aq)$	Hydrophobic	-10
$Benzene(l) + \infty H_2O(l) \rightarrow benzene(aq)$	Hydrophobic	0

*The enthalpies were obtained near room temperature except for the vaporization of solid argon; the standard state is the dilute solution extrapolated to 1 *M*. Data are from various sources and are rounded to the nearest integer. For the precise values, see G. C. Pimentel and A. L. McClellan, *The Hydrogen Bond* (San Francisco: Freeman, 1960); W. Kauzmann, *Adv. Protein Chem.* 14(1) (1959); and J. A. A. Ketallar, *Chemical Constitution* (Amsterdam: Elsevier, 1958).

reasonable and which ones are not. Table 3.1 gives some measured enthalpies for simple reactions, which illustrate the forces involved.

Charged species can interact with each other by ionic interactions. In a NaCl crystal, for example, very strong ionic interactions exist between the positively charged Na^+ ions and the negatively charged Cl^- ions. This is reflected in the large positive enthalpy change when solid NaCl is separated into Na^+ and Cl^- ions:

$$NaCl(s) \longrightarrow Na^+(g) + Cl^-(g) \qquad (\Delta H^0_{298} = 785 \text{ kJ mol}^{-1})$$

Despite such strong ionic forces, solid NaCl nevertheless dissolves in water easily, with a small ΔH^0_{298} of only 4 kJ mol^{-1}:

$$NaCl(s) + \infty H_2O(l) \longrightarrow Na^+(aq) + Cl^-(aq) \qquad (\Delta H^0_{298} = 4 \text{ kJ mol}^{-1})$$

The reason is that there are strong interactions between the charged ions and the water molecules, so the net ΔH change is small.

The electronic distribution in an uncharged water molecule

is such that the bonding electrons are localized more on the oxygen atom than on the hydrogen atoms. Therefore, the oxygen atom is more negatively charged, and the hydrogen atoms are more positively charged. Because H_2O is bent, the geometric centers of the positive and negative charges do not coincide. We say that water has an electric dipole. The extent of charge separation is expressed in terms of the dipole moment. If two point charges $+e$ and $-e$ are separated by a distance r, the dipole moment is er. Experimentally, the dipole moment of H_2O is found to be equivalent to an electron and a unit positive charge separated by 0.38 Å (1 Å = 10^{-10} m). If water is modeled as three charges on the three atoms, the charge on the oxygen atom is -0.834 electronic charges, and each hydrogen atom is $+0.417$. Interactions between a charged ion and a neutral molecule with a dipole moment are called *ion–dipole interactions*. Ions can also interact with neutral molecules with zero dipole moment. Consider a molecule of CCl_4, for example. Though the bonding electrons are localized more on the chlorine atoms, the *permanent dipole moment* of the molecule is zero because the four Cl atoms are symmetrically located at the four corners of a tetrahedron, with the C atom occupying the center of the tetrahedron. However, if a charge is placed near a CCl_4 molecule, the charge will distort the electronic distribution in CCl_4. We say that the CCl_4 molecule becomes *polarized*. The centers of the positive and negative charges in the polarized CCl_4 molecule no longer coincide, and the molecule has now an *induced dipole moment*. Interactions between a charged ion and polarized molecules are called *charge-induced dipole interactions*.

Induced dipole–induced dipole interactions exist even between neutral molecules with no permanent dipole moments. This is because of fluctuations in the electronic distributions in the molecules. A molecule with no permanent dipole moment may acquire an instantaneous dipole moment because of a fluctuation. This instantaneous dipole can induce a dipole in a neighboring molecule. Interactions between such fluctuation dipole-induced dipoles are called *London interactions*. London interaction is always present and is always an attractive force between molecules. It is the only attractive force acting between identical rare gas atoms. It is responsible for the 8 kJ mol^{-1} enthalpy change necessary to vaporize solid argon to gaseous argon, for example. The force depends on the *polarizability* of the interacting molecules, which is a measure of how easy it is to distort the electron clouds of the molecule. In addition to the London interaction, uncharged molecules have *van der Waals interactions*. Van der Waals interactions include permanent dipole–permanent dipole interactions, permanent dipole–induced dipole attractions, and steric repulsions. The London–van der Waals interactions are usually nonspecific forces that contribute to the energies of all reactions. For example, the heat of vaporization of liquid *n*-butane, 21.5 kJ mol^{-1}, reflects mainly nonspecific London and van der Waals interactions in liquid *n*-butane. The London–van der Waals forces can become specific when careful fitting of molecules is involved. Binding of a particular substrate to an enzyme, antigen–antibody binding, and the function of specific membrane lipids may be dominated by London–van der Waals interactions.

The *hydrogen bond* is one of the important bonds that determines the three-dimensional structures of proteins and nucleic acids. A hydrogen atom covalently attached to one oxygen or nitrogen can form a weak bond to another oxygen or nitrogen. For O — H $\cdots$ O, N — H $\cdots$ N, and N — H $\cdots$ O hydrogen bonds, the bond enthalpy of 15–20 kJ mol^{-1} can be compared with values of about 400 kJ mol^{-1} for covalent bonds. This weak bond becomes even weaker in aqueous solution *because of competition between solute–solute hydrogen bonds and solute–water and water–water hydrogen bonds.* Between urea molecules in water the effective hydrogen-bond strength is just 5 kJ mol^{-1}. Urea can be considered as a model for a peptide bond, so this is the magnitude of the expected energy for breaking a hydrogen bond in a peptide in aqueous solution.

Hydrophobic Interactions

One type of interaction that is important in aqueous solutions is the *hydrophobic* (fear-of-water) interaction. Water molecules have a strong attraction for each other, primarily as a consequence of hydrogen-bond formation. The oxygen atom of most molecules of liquid water is hydrogen-bonded to two hydrogen atoms of two other H_2O molecules, and the hydrogen atoms of most molecules are hydrogen-bonded to the oxygen atoms of two other water molecules. Therefore, the molecules of liquid water form a mobile network: Most water molecules are primarily interacting, through hydrogen bonds, with four tetrahedrally oriented neighbors. The network is not a rigid one, and change of neighbors occurs rapidly because of thermal motions.

Let's consider what happens if a molecule such as propane (C_3H_8) is introduced into this network. A hole is created; some hydrogen bonds in the original network are broken. The C_3H_8 does not interact with water strongly; it does not form hydrogen bonds. The water molecules around the C_3H_8 molecule must orient themselves in a way that re-forms the hydrogen bonds that were disrupted by the hydrocarbon molecule. The net result is that water molecules around the C_3H_8 actually become more ordered. Since there is little change in the number of hydrogen bonds, the enthalpy change is small. The ordering of water molecules around the hydrocarbon molecule, however, is associated with a negative entropy change. These interpretations are consistent with experimental data:

$$C_3H_8(l) \longrightarrow C_3H_8(aq)$$

$$\Delta H_{298}^0 \approx -8 \text{ kJ mol}^{-1}$$

$$\Delta S_{298}^0 \approx -80 \text{ J K}^{-1} \text{ mol}^{-1}$$

$$\Delta G_{298}^0 \approx +16 \text{ kJ mol}^{-1}$$

Now let's consider two such hydrocarbon groups, R. Each separate group when exposed to liquid water will, from the data above, cause an unfavorable free-energy change. If the two groups cluster together, the disruptive effect on the solvent network will be less than the combined effects of two separate groups. Therefore, the association of the groups will be thermodynamically favored:

$$\underset{\text{separate}}{R(aq) + R(aq)} \longrightarrow \underset{\text{clustered}}{[R-R](aq)}$$

The clustering of the groups is not because they "like" each other, but because they are both "disliked" by water. The clustered arrangement of the hydrocarbons results in a decrease in the overall free energy of the system in comparison with the isolated hydrocarbons in water. Such hydrophobic interactions are important in many biological systems. For example, hydrocarbon groups in a water-soluble protein are usually found to cluster in the interior of the protein. Similarly, polar lipid molecules form bilayer sheets or membranes in water, in which the hydrocarbon portions are buried inside and the polar or charged portions are on the surface, exposed to the water. Such molecules are called *amphiphilic*. Hydrophobic interactions are characterized by low enthalpy changes and are entropy driven. We should note that hydrophobic interaction is a term that we use to describe the combined effects of London, van der Waals, and hydrogen-bonding interactions in certain processes in aqueous solutions; it is *not* a "force" different from the others we have discussed. In particular, there is no such thing as a hydrophobic bond.

Proteins and Nucleic Acids

Proteins are polypeptides of 20 naturally occurring amino acids, the structures of which are given in appendix table A.9. The amino acids are linked by amide bonds from the carboxyl group of one amino acid to the amino group of the next. The sequence of the amino acid residues in a polypeptide is called the *primary structure*. The polypeptides fold into different secondary structures held together by hydrogen bonds and other noncovalent interactions. The amide group

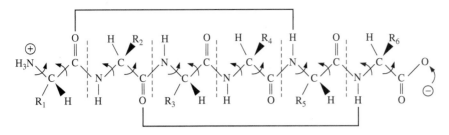

▲ FIGURE 3.4

A polypeptide chain. Residues are commonly numbered by starting from the amino terminal. The residues in the polypeptide chain are separated by dotted lines, and the arrows show the two bonds (on either side of each α-carbon) around which rotation can occur. The two long lines joining a carbonyl oxygen to an amide hydrogen 4 residue units away on its carboxyl terminal side represent hydrogen bond formation in an α-helix structure (see figure 3.5).

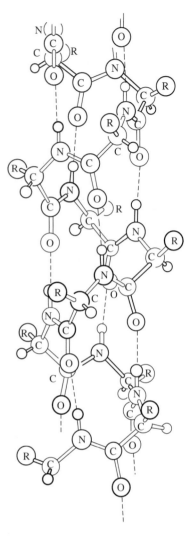

▶ FIGURE 3.5

The α-helix. The structure shown here is the right-handed conformation commonly found in proteins. [From Linus Pauling, *The Nature of the Chemical Bond* (Ithaca, NY: Cornell University Press, 1960, p. 500). Copyright © 1939 and 1940, 3d ed. © 1960 by Cornell University. Used by permission of Cornell University Press.]

is planar (because of partial double bond character of the central $C-N$ bond) and has a *trans* conformation as shown. The different folded conformations of a polypeptide depend on rotation about the two single bonds attached to the amide group, as shown in figure 3.4. A common secondary structural element in proteins is the α-helix shown in figure 3.5. It is a right-handed helix with hydrogen bonds between the $C=O$ of each amide to the $N-H$ of the amide 4 residue units away (figures 3.4 and 3.5).

Nucleic acids are polynucleotides of four naturally occurring nucleotides, with structures as given in table A.9 in the appendix. The nucleotides are composed of a sugar group connected to one of four bases and to a phosphate group. The nucleotides are linked by phosphodiester bonds [see figure 3.6(b)]. The sequence of the bases in the polynucleotide is the primary structure. In DNA two strands form a double helix held together by Watson–Crick base pairs as shown in Figure 3.6(a). The stacking of the base pairs, one above the other, plus the hydrogen bonds between bases provide the stabilizing enthalpy and free energy of the helix. Both cross-strand and same-

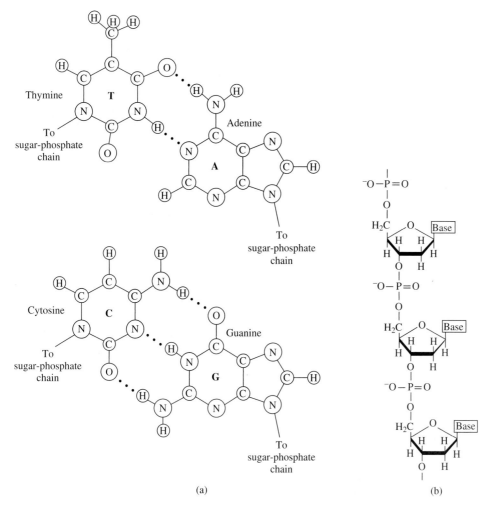

▲ FIGURE 3.6

(a) The Watson–Crick base pairs found in DNA. Adenine is complementary to thymine (A·T), and guanine is complementary to cytosine (G·C). In RNA, uracil substitutes for thymine (A·U). (b) The sugar–phosphate chain in RNA differs only in that the sugar is ribose instead of deoxyribose. In the structures shown in (a), a covalent bond, whether it is a single or a double bond, is represented by a solid line; a hydrogen bond is represented by a dotted line.

strand London–van der Waals interactions among the bases are important. The charged phosphate groups repel each other by Coulomb's law repulsion between like charges. The balance of these forces as well as interactions with the solvent, and other solutes such as positively charged ions, determine the conformation of the double helix. Two different double helices—secondary structures—are shown in figure 3.7. The B-form helix is the predominant structure found for DNA (deoxyribonucleic acid) in biological systems. The A-form helix is the structure found in double-stranded regions of RNA (ribonucleic acid).

Table 3.2 lists some measured enthalpies for biochemical reactions involving changes in shape (conformation) of molecules. By changes in conformation, we mean changes in the secondary structure of the molecule. (The primary struc-

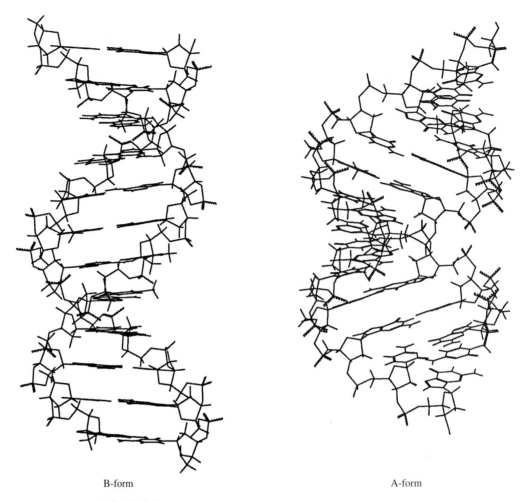

B-form   A-form

▲ FIGURE 3.7

Double-stranded, right-handed nucleic acid helices. The two polynucleotide strands in each double helix are antiparallel. One strand is as shown in figure 3.6(b); the other runs in the opposite direction. The base pairs are tilted relative to the helix axis in the A-form; they are nearly perpendicular to the helix axis in the B-form. DNA is mostly in the B-form in biological cells; double-stranded regions of RNA are in the A-form.

ture involves the covalent bonds.) Examples include the change in a polypeptide from a rigid helix to a flexible coil. Denaturation of proteins involves this type of change. The corresponding change in nucleic acids and polynucleotides is the change from a two-strand helix to two single strands. Hydrogen bonds between the amides, the nucleic acid bases, and the solvent are important, but so are London–van der Waals types of interactions. The magnitudes of ΔH^0 and ΔS^0 can help us understand the various interactions involved.

As mentioned earlier, the strength of a hydrogen bond between urea molecules in an aqueous environment is about 5 kJ mol^{-1}. From table 3.2, we see that this is consistent with the magnitude of ΔH^0 for the helix-coil transition of poly-L-glutamate in aqueous solution. In the α-helix, the peptide groups are hydrogen-bonded as shown in figure 3.5; in the coil, they are hydrogen-bonded to water. The enthalpy increase for denaturation of a pro-

TABLE 3.2 Thermodynamics of Biochemical Conformational Transitions and Noncovalent Reactions Studied Near Room Temperature and Neutral pH

Transition or noncovalent reaction	ΔH^0 (kJ mol^{-1})	ΔS^0 (J K^{-1} mol^{-1})	Reference*
Poly-l-glutamate helix-coil transition in 1.0 M KCl	4.5/amide	—	a
Lysozyme denaturation	≈450	—	a
RNA double-strand to single-strand transition in 1 M NaCl	≈40/base pair	≈104/base pair	b
DNA double-strand to single-strand transition in 1 M NaCl	≈35/base pair	≈88/base pair	b
Unstacking of bases in polyadenylic acid in 0.1 M KCl	36/nucleotide	113/nucleotide	a
Binding of Mg-ATP by tetrahydrofolate synthetase	31	182	c
Binding of Netropsin by poly dA · poly dT	−9.2	141	b

*(a) G. D. Fasman, ed., *Handbook of Biochemistry and Molecular Biology*, Vol. 1, 3d ed. (Cleveland: CRC Press, 1976); (b) Landolt-Börnstein Numerical Data and Functional Relationships in Science and Technology, Group VII, *Biophysics*, Vol. 1d, *Nucleic Acids* (1990); (c) N. P. Curthoys and J. C. Rabinowitz, *J. Biol. Chem.* 246(6942) (1971).

tein such as lysozyme depends greatly on the method used for denaturation (pH, urea, or temperature), but the large value of ΔH^0 clearly indicates many peptide–peptide hydrogen bonds are being broken.

For DNA and RNA double-strand to single-strand transitions, we need to break two hydrogen bonds per A·T base pair (in DNA) or A·U base pair (in RNA) and three hydrogen bonds per G·C base pair (see figure 3.6). The ΔH^0 for dissociating a G·C base pair is larger than for A·T or A·U as expected, but the magnitude of ΔH^0 depends on the sequence of bases. The main interactions that stabilize the double strands are the stacking of the base pairs on each other; therefore, the nearest neighbor sequences affect the magnitude of changes in thermodynamic quantities. In DNA, the range of ΔH^0 values for "melting" a base pair ranges from 30 kJ mol^{-1} for an A·T in a sequence of alternating –A–T–A–T–A–T–A– to 44 kJ mol^{-1} for a G·C in a sequence of alternating –G–C–G–C–G–C–. A measure of the thermodynamics of unstacking is illustrated by the data for polyadenylic acid in table 3.2. Polyadenylic acid unstacking refers to the transition between an ordered helical molecule with the adenine bases stacked on top of each other and a much more disordered molecule with the adenine bases not oriented relative to each other. This change in stacking does not involve hydrogen bonds directly, but it does have a ΔH^0 of 36 kJ per mole of nucleotide. This ΔH^0 is mainly London–van der Waals interactions among the bases and between the bases and the solvent.

The entropy changes for the double-strand to single-strand transition in nucleic acids are also consistent with the conclusions above. About 100 J K^{-1} mol^{-1} of entropy is gained for each base pair disrupted in DNA or RNA; this mainly corresponds to the increased possibility of rotation around single bonds in the nucleotides.

Two examples of binding of small molecules by a macromolecule are given in table 3.2. The ΔH^0 values mean that heat must be added to favor the binding of the Mg–ATP complex to the enzyme tetrahydrofolate synthetase, but heat is released when the antibiotic Netropsin binds to double-stranded polydeoxyadenylic acid·polythymidylic acid (poly dA·poly dT). The entropy increases for both, however, might be surprising because of the loss of translational entropy when two molecules form a complex. The explanation is that water must be released when the molecules are bound; there is thus a large net gain in translational entropy.

Use of Partial Derivatives in Thermodynamics

So far, eight thermodynamic variables of state have been considered: E, H, S, G, A, P, V, and T. We can select, based on convenience, a few parameters as independent variables and express the others as functions of these independent variables. For example, if our system is a fixed amount of a pure substance and if we choose P and T as independent variables, we can express V as a function of P and T. This is an example of an equation of state:

$$V = V(P, T)$$

The specific form of the function $V(P, T)$ depends on the nature of the substance. If it is an ideal gas,

$$V = V(P, T) = \frac{nRT}{P}$$

When we are interested in how V changes with T at constant P, we can take the derivative of V with respect to T, keeping P constant:

$$\frac{dV}{dT} = \frac{nR}{P} \qquad (P = \text{constant})$$

Instead of stating explicitly that $P = $ constant, we can use the notation of partial derivatives:

$$\left(\frac{\partial V}{\partial T}\right)_P = \frac{nR}{P} \tag{3.52}$$

$(\partial V / \partial T)_P$ means simply the derivative of V with respect to T at constant P. Similarly, instead of writing

$$\frac{dV}{dP} = nRT \frac{d}{dP}\left(\frac{1}{P}\right) \qquad (\text{at } T = \text{constant})$$

$$= -\frac{nRT}{P^2}$$

we can write

$$\left(\frac{\partial V}{\partial P}\right)_T = -\frac{nRT}{P^2} \tag{3.53}$$

It troubles some students in relations like Eqs. (3.52) and (3.53) that the partial derivative is actually a function of the variable that is held constant. What this means is that the slope of the curve of V as a function of T in Eq. (3.52), for example, has a different value at different pressures. However, at each pressure, the slope dV/dT is determined with the pressure held constant.

A partial derivative of particular importance in discussing chemical and physical equilibria (chapters 4 and 5) is partial molar Gibbs free energy, which is also termed the *chemical potential*. The chemical potential of a compound A in a solution containing n_A mol of A, n_B mol of B, and so on is defined by

$$\mu_A \equiv \left(\frac{\partial G}{\partial n_A}\right)_{T, P, n_B, n_C, \ldots} \tag{3.54}$$

Equation (3.54) describes the change in the Gibbs free energy that occurs upon an infinitesimal change dn_A in the number of moles of A, while keeping the temperature, pressure, and the number of moles of all other constituents constant. Component A may be either a solute or the solvent in the solution. For pure A, the chemical potential or partial molar Gibbs free energy of A is simply the Gibbs free energy per mole of A, $\overline{G}_A$. For a more rigorous expression of Eq. (3.38), the parameters $\overline{G}_C$, $\overline{G}_D$, and so on are to be replaced by the corresponding partial molal quantities μ_C, μ_D, and so on. (We will discuss this point in detail in the next chapter.)

Relations Among Partial Derivatives

For a state function used in thermodynamics, such as the volume V, it is generally true that if

$$V = V(P, T)$$

the total differential of V is

$$dV = \left(\frac{\partial V}{\partial P}\right)_T dP + \left(\frac{\partial V}{\partial T}\right)_P dT \qquad \textbf{(3.55)}$$

This equation states that for small changes, dV can be treated as the sum of how V changes with P (at constant T) and how V changes with T (at constant P). Also, the order of differentiation is not important:

$$\left[\frac{\partial}{\partial T}\left(\frac{\partial V}{\partial P}\right)_T\right]_P = \left[\frac{\partial}{\partial P}\left(\frac{\partial V}{\partial T}\right)_P\right]_T \qquad \textbf{(3.56)}$$

An example of the use of these equations follows.

EXAMPLE 3.10

Show that

$$\left(\frac{\partial V}{\partial T}\right)_P = -\left(\frac{\partial S}{\partial P}\right)_T$$

Solution

In the notation of partial derivatives, Eqs. (3.36) and (3.43) become

$$\left(\frac{\partial G}{\partial T}\right)_P = -S \qquad \textbf{(3.36)}$$

and

$$\left(\frac{\partial G}{\partial P}\right)_T = V \qquad \textbf{(3.43)}$$

Because the order of differentiation is unimportant for a thermodynamic state function,

$$\left[\frac{\partial}{\partial T}\left(\frac{\partial G}{\partial P}\right)_T\right]_P = \left[\frac{\partial}{\partial P}\left(\frac{\partial G}{\partial T}\right)_P\right]_T \qquad \textbf{(3.57)}$$

$$\left(\frac{\partial V}{\partial T}\right)_P = -\left(\frac{\partial S}{\partial P}\right)_T \qquad \textbf{(3.58)}$$

Equation (3.58) gives a general expression on how entropy depends on pressure. In the special case of an ideal gas, since

$$\left(\frac{\partial V}{\partial T}\right)_P = \frac{nR}{P}$$

we have

$$\left(\frac{\partial S}{\partial P}\right)_T = -\frac{nR}{P}$$

Integrating at constant temperature gives

$$\int_{S_1}^{S_2} dS = -\int_{P_1}^{P_2} \frac{nR}{P}\, dP = -nR \ln \frac{P_2}{P_1}$$

This is Eq. (3.21a), which we derived earlier by a different route.

Partial derivatives find many applications in thermodynamics, and we will illustrate here some of the results. The object is to obtain equations relating the state variables T, P, V, E, H, G, and S. We consider here a closed system with no external fields (such as gravitational or electrical fields) and no chemical reactions or phase changes. Therefore, the only reason that one of the above variables changes is because one or more of the other variables changes. We usually think of T, P, and V as the independent variables and E, H, G, and S as the dependent variables. Furthermore, because the equation of state tells us how P, V, and T are related, we need consider only how E, H, G, and S depend on two of the three variables P, V, and T.

For example, we have previously derived Eq. (3.35) for G as a function of P and T:

$$dG = V\, dP - S\, dT \qquad (3.35)$$

If we need to calculate the change of G resulting from a change of V, we can first find the change of P and T corresponding to this change of V and then use Eq. (3.35). To calculate $G(P_2, T_2) - G(P_1, T_1)$ for a change from P_1, T_1 to P_2, T_2, we must integrate Eq. (3.35):

$$G(P_2, T_2) - G(P_1, T_1) = \int_{P_1}^{P_2} V\, dP - \int_{T_1}^{T_2} S\, dT$$

If we need to know how ΔG for a chemical reaction or phase change depends on P and T, we use

$$\Delta G(P_2, T_2) - \Delta G(P_1, T_1) = \int_{P_1}^{P_2} \Delta V\, dP - \int_{T_1}^{T_2} \Delta S\, dT$$

where ΔV and ΔS are, respectively, the volume change and entropy change for the chemical reaction or phase change. To do the integration, we need to know ΔV as a function of P and ΔS as a function of T. For some systems, they are approximately constant.

We can obtain the corresponding equations for H as a function of P, T exactly analogous to the derivation for $G(P, T)$. We use

$$dH = \left(\frac{\partial H}{\partial T}\right)_P dT + \left(\frac{\partial H}{\partial P}\right)_T dP$$

The change of enthalpy with temperature at constant pressure was found to be the heat capacity at constant pressure [Eq. (2.38)]; thus,

$$\left(\frac{\partial H}{\partial T}\right)_P = C_P$$

We can derive $(\partial H/\partial P)_T$ from the identity $G = H - TS$ or $H = G + TS$:

$$dH = dG + T\,dS + S\,dT$$

$$= V\,dP - S\,dT + T\,dS + S\,dT$$

$$= V\,dP + T\,dS \tag{3.59}$$

We obtain the correct expression for $(\partial H/\partial P)_T$ by writing the derivative form of Eq. (3.59) and specifying that T is constant. Thus,

$$\left(\frac{\partial H}{\partial P}\right)_T = V + T\left(\frac{\partial S}{\partial P}\right)_T$$

We already have an expression for $(\partial S/\partial P)_T$ in terms of P, V, and T [Eq. (3.58)]; therefore,

$$\left(\frac{\partial H}{\partial P}\right)_T = V - T\left(\frac{\partial V}{\partial T}\right)_P$$

and

$$dH = C_P\,dT + \left[V - T\left(\frac{\partial V}{\partial T}\right)_P\right]dP \tag{3.60}$$

Equation (3.60) can be integrated to give $H(P_2, T_2) - H(P_1, T_1)$ for a finite change from P_1, T_1 to P_2, T_2.

For S, we have

$$dS = \left(\frac{\partial S}{\partial T}\right)_P dT + \left(\frac{\partial S}{\partial P}\right)_T dP$$

In Eq. (3.14), we used the integrated form of

$$\left(\frac{\partial S}{\partial T}\right)_P = \frac{C_P}{T}$$

Therefore, using Eq. (3.58),

$$dS = \frac{C_P}{T}\,dT - \left(\frac{\partial V}{\partial T}\right)_P dP \tag{3.61}$$

For E, the most useful variables to use are T and V instead of T and P. We write

$$dE = \left(\frac{\partial E}{\partial T}\right)_V dT + \left(\frac{\partial E}{\partial V}\right)_T dV$$

From Eq. (2.37),

$$\left(\frac{\partial E}{\partial T}\right)_V = C_V$$

To obtain $(\partial E / \partial V)_T$, we first consider a reversible path with PV work only:

$$dE = dq_{\text{rev}} + dw_{\text{rev}} \qquad \text{(first law, reversible path)}$$

$$= T \, dS - P \, dV \tag{3.62}*$$

Taking the derivative of Eq. (3.62) with respective to V at constant temperature gives

$$\left(\frac{\partial E}{\partial V} \right)_T = T \left(\frac{\partial S}{\partial V} \right)_T - P$$

We need to write $(\partial S / \partial V)_T$ in terms of P, V, and T. We do this from an analog of Eq. (3.35):

$$dA = -P \, dV - S \, dT \tag{3.63}$$

Applying the relation

$$\left[\frac{\partial}{\partial T} \left(\frac{\partial A}{\partial V} \right)_T \right]_V = \left[\frac{\partial}{\partial V} \left(\frac{\partial A}{\partial T} \right)_V \right]_T$$

we obtain

$$\left(\frac{\partial S}{\partial V} \right)_T = \left(\frac{\partial P}{\partial T} \right)_V$$

Therefore,

$$dE = C_V \, dT + \left[T \left(\frac{\partial P}{\partial T} \right)_V - P \right] dV \tag{3.64}$$

Note that for an ideal gas, PV = nRT means that the term in brackets in Eq. (3.64) is zero and $dE = C_V dT$. This conclusion was previously stated without proof.

To summarize our results, we can now calculate the change in E, H, S, and G for a system in which P_1, V_1, and T_1 change to P_2, V_2, and T_2, using the integrated forms of Eqs. (3.64), (3.60), (3.61), and (3.35). We need only be given two of these three variables P, V, and T because we can calculate the third from the equation of state.

Notice that we were able to derive many useful results from the following four equations:

$$dE = -P \, dV + T \, dS \tag{3.62}$$

$$dH = V \, dP + T \, dS \tag{3.59}$$

$$dG = V \, dP - S \, dT \tag{3.35}$$

$$dA = -P \, dV - S \, dT \tag{3.63}$$

Many more relations among partial derivatives can be obtained. Standard thermodynamics texts discuss the most useful ones.

---

*We note again, as we did in the derivation of Eq. (3.35), that all variables in Eq. (3.62) are path independent; thus, Eq. (3.62) is also valid for paths other than the special one chosen.

Summary
State Variables

Name	Symbol	Units	Definition
Entropy	S	$J\,K^{-1}$, cal deg^{-1}, eu	$dS \equiv dq_{rev}/T$
Gibbs free energy	G	J, erg, cal	$G \equiv H - TS$
Helmholtz free energy	A	J, erg, cal	$A \equiv E - TS$

Unit Conversions
Entropy

$$1 \text{ cal deg}^{-1} = 1 \text{ eu} = 4.184 \text{ J K}^{-1}$$

Free energy

$$1 \text{ cal} = 4.184 \text{ J} = 4.184 \times 10^7 \text{ erg}$$

General Equations
Efficiency for a Carnot-cycle heat engine

$$\text{efficiency} = 1 - \frac{T_{cold}}{T_{hot}} \tag{3.5}$$

Second law of thermodynamics

$$\Delta S \text{ (system)} + \Delta S \text{ (surroundings)} \geq 0 \tag{3.7}$$

$$\Delta S \text{ (isolated system)} \geq 0 \tag{3.8}$$

In the above equations, the equal sign applies for a reversible process and the other sign for an irreversible one.

Third law of thermodynamics

$$S_A(0 \text{ K}) \equiv 0 \tag{3.13}$$

A = any pure, perfect crystal

ΔG and a System's Capacity to Do Nonexpansion Work

$$\Delta G = w_{rev}^* \tag{3.28}$$

or

$$-\Delta G = -w_{rev}^*$$

The decrease in the Gibbs free energy $(-\Delta G)$ is the maximal amount of non-expansion work $(-w_{rev}^*)$ a system can do on the surroundings by a process at constant pressure and temperature.

Spontaneous Reactions at Constant T and P

$\Delta G < 0$	(spontaneous process)	(3.29)
$\Delta G > 0$	(not a spontaneous process)	(3.30)
$\Delta G = 0$	(system at equilibrium)	(3.31)

Changes in Entropy and Gibbs Free Energy

Pressure or temperature changes

Solids and Liquids. We assume in these equations that the volume of a solid or liquid is independent of T and P and that C_P and C_V are equal and do not depend on T and P.

$$S(P_2, T_2) - S(P_1, T_1) = n\overline{C}_P \ln \frac{T_2}{T_1} = n\overline{C}_V \ln \frac{T_2}{T_1}$$

n = number of moles
$\overline{C}_P$ = heat capacity per mole at constant P
$\overline{C}_V$ = heat capacity per mole at constant V

Gases. We assume that gases can be approximated by the ideal gas equation and that C_P and C_V are independent of T. Also, $PV = nRT$ and $\overline{C}_P = \overline{C}_V + R$.

$$S(P_2) - S(P_1) = -nR \ln \frac{P_2}{P_1} = nR \ln \frac{V_2}{V_1} \qquad \text{(constant } T\text{)} \qquad \textbf{(3.21a)}$$

$$G(P_2) - G(P_1) = nRT \ln \frac{P_2}{P_1} = -nRT \ln \frac{V_2}{V_1} \qquad \text{(constant } T\text{)} \qquad \textbf{(3.45)}$$

$$S(T_2) - S(T_1) = n\overline{C}_P \ln \frac{T_2}{T_1} \qquad \begin{array}{l}\text{(constant } P; C_P = n\overline{C}_P \\ \text{independent of } T\text{)} \quad \textbf{(3.15)}\end{array}$$

Phase changes at constant *T* and *P*

$$\Delta G \text{ (phase transition)} = 0 \qquad \text{(at equilibrium)}$$

Therefore,

$$\Delta S \text{ (phase transition)} = \frac{\Delta H \text{ (phase transition)}}{T \text{ (phase transition)}} \qquad \text{(at equilibrium)}$$

Chemical reactions

For a chemical reaction,

$$\Delta S(T_2) = \Delta S(T_1) + \int_{T_1}^{T_2} \Delta C_P \frac{dT}{T} \qquad \textbf{(3.18)}$$

$$\Delta C_P = n_C \overline{C}_{P,C} + n_D \overline{C}_{P,D} - n_A \overline{C}_{P,A} - n_B \overline{C}_{P,B}$$

$$\Delta G(T_2) = \Delta G(T_1) - (T_2 - T_1)\Delta S \qquad \text{(if } \Delta S \text{ is independent of } T\text{)}$$

$$\frac{\Delta G(T_2)}{T_2} = \frac{\Delta G(T_1)}{T_1} + \left(\frac{1}{T_2} - \frac{1}{T_1}\right)\Delta H \qquad \text{(if } \Delta H \text{ is independent of } T\text{)}$$

$$\frac{\Delta G(T_2)}{T_2} - \frac{\Delta G(T_1)}{T_1} = -\int_{T_1}^{T_2} \frac{\Delta H(T)}{T^2} dT \qquad \text{(if } \Delta H \text{ depends on } T\text{) } \textbf{(3.40)}$$

$$\Delta G(P_2) - \Delta G(P_1) = \Delta V(P_2 - P_1) \qquad \begin{array}{l}\text{(for solids and} \\ \text{liquids only, if } \Delta V \text{ is} \\ \text{independent of } P\text{)} \quad \textbf{(3.46)}\end{array}$$

$$\Delta V = n_C \overline{V}_C + n_D \overline{V}_D - n_A \overline{V}_A - n_B \overline{V}_B$$

$$\Delta G(P_2) - \Delta G(P_1) = \Delta nRT \ln \frac{P_2}{P_1} \qquad \begin{array}{l}\text{(at least one gaseous product} \\ \text{or reactant)}\end{array} \qquad \textbf{(3.47)}$$

$$\Delta n = \left(\begin{array}{c}\text{number of moles} \\ \text{of gaseous products}\end{array}\right) - \left(\begin{array}{c}\text{number of moles} \\ \text{of gaseous reactants.}\end{array}\right)$$

References

In addition to the textbooks listed in chapter 2, see

Atkins, P. W. 1994. *The Second Law,* Scientific American Library. New York: W. H. Freeman.

Craig, N. C. 1992. Entropy Analysis: An Introduction to Chemical Thermodynamics. New York: VCH Publishers.

Morowitz, H. J. 1996. *Entropy and the Magic Flute.* Oxford: Oxford University Press.

Saenger, W. 1984. *Principles of Nucleic Acid Structure.* New York: Springer-Verlag.

Von Baeyer, H. C. 1998. *Maxwell's Demon: Why Warmth Disperses and Time Passes.* New York: Random House.

Suggested Reading

Feynman, R. P., R. B. Leighton, and M. Sands. 1963. *The Feynman Lectures on Physics.* Reading, MA: Addison-Wesley.

Frautschi, S. 1982. Entropy in an Expanding Universe. *Science* 217:593–599.

McIver, R. T., Jr. 1980. Chemical Reactions Without Solvation. *Sci. Am.* 243 (November):186–196.

Wilson, S. S. 1981. Sadi Carnot. *Sci. Am.* 245 (August):134–145.

Problems

1. One mole of an ideal monatomic gas is expanded from an initial state at 2 atm and 400 K to a final state at 1 atm and 300 K.

   (a) Choose two different paths for this expansion, specify them carefully, and calculate w and q for each path.

   (b) Now calculate ΔE and ΔS for each path.

2. The temperature of the heat reservoirs for a Carnot-cycle (reversible) engine are $T_{hot} = 1200$ K and $T_{cold} = 300$ K. The efficiency of the engine is calculated to be 0.75, from Eq. (3.5).

   (a) If $w = -100$ kJ, calculate q_1 and q_3. Explain the meaning of the signs of w, q_1, and q_3.

   (b) The same engine can be operated in the reverse order. If $w = +100$ kJ, calculate q_1 and q_3. Explain the meaning of the signs.

   (c) Suppose it were possible to have an engine with a higher efficiency, say, 0.80. If $w' = -100$ kJ, calculate q_1' and q_3'. The superscript denotes quantities for this engine.

   (d) If we use all the work done by the engine in part (c) to drive the heat pump in part (b), calculate $(q_1 + q_1')$ and $(q_3 + q_3')$, the amount of heat transferred. Explain the signs. Note that the net effect of the combination of the two engines is to allow a

"spontaneous" transfer of heat from a cooler reservoir to a hotter reservoir, which should not happen.

   (e) Show that if it were possible to have a reversible engine with a lower efficiency operating between the same two temperatures, heat could also be transferred spontaneously from the cooler reservoir to the hotter reservoir. (*Hint:* Operate the engine with lower efficiency as a heat pump.)

3. The second law of thermodynamics states that entropy increases for spontaneous processes and that an increase in entropy is associated with transitions from ordered to disordered states. Living organisms, even the simplest bacteria growing in cultures, appear to violate the second law because they grow and proliferate spontaneously. They convert simple chemical substances into the highly organized structure of their descendants. Write a critical evaluation of the proposition that living organisms contradict the second law. Be sure to state your conclusion clearly and to present detailed, reasoned arguments to support it.

4. Consider the process where n_A mol of gas A initially at 1 atm pressure mix with n_B mol of gas B also at 1 atm to form 1 mol of a uniform mixture of A and B at a final total pressure of 1 atm, and all at constant temperature T. Assume that all gases behave ideally.

   (a) Show that the entropy change, $\Delta \bar{S}_{mix}$, for this process is given by $\Delta \bar{S}_{mix} = -X_A R \ln X_A - X_B R \ln X_B$, where X_A and X_B are the mole fractions of A and B, respectively. (*Hint:* Entropy changes are additive for components A and B; that is, $\Delta \bar{S}_{mix} = \Delta \bar{S}_A + \Delta \bar{S}_B$.)

(b) What can you say about the value (especially the sign) of $\Delta \bar{S}_{mix}$, and how does this correlate with the second law?

(c) Derive an expression for $\Delta \bar{G}_{mix}$ for the conditions of this problem and comment on values (and signs) of its terms.

5. A system consisting of 0.20 mol of supercooled gaseous water (95°C, 1 atm) partially condenses to liquid water; the process is done adiabatically at constant pressure.

(a) At equilibrium what is the temperature of the system?

(b) How many moles of gaseous water have condensed?

(c) Calculate the entropy change for the process.

6. The technology to build microscopic-size motors is now available. Can microscopic motors take advantage of fluctuations to beat the second law and do work at constant temperature? Consider the following fluctuation motor:

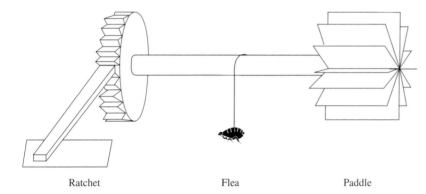

Ratchet   Flea   Paddle

Assume that the motor is so small that one gas molecule hitting a paddle can make it turn. The motor is placed in a box (at constant temperature) with a few gas molecules. A possible explanation of how the motor works is the following. Gas molecules randomly hitting the paddles will make the shaft turn clockwise sometimes and sometimes counterclockwise. The flea will randomly be raised and lowered by the fluctuations, *if we don't consider the ratchet.* However, the ratchet is held down by a spring, so it allows only clockwise rotations. Therefore, the flea will be raised because the ratchet has converted the random fluctuations into a directed rotation. Work is done (the mass of the flea times the acceleration of gravity times the distance it is raised) at constant temperature. The energy comes from the thermal energy of the gas, so the first law is not violated.

(a) Will the motor work? Discuss the action of the ratchet in converting random fluctuations into a directed clockwise motion.

(b) If the motor will not work as described, what simple change will allow it to work? Clearly, if the motor will work as described, you need not answer part (b).

This problem is discussed in *The Feynman Lectures on Physics*, Vol. 1.

7. For the following processes, determine whether each of the thermodynamic quantities listed is greater than, equal to, or less than zero for the system described (the system is shown in italic type in each case). Consider all gases to behave ideally. Indicate your reasoning and state explicitly any reasonable assumptions or approximations that you need to make.

(a) A sample of *gas* is carried through a complete Carnot cycle (isothermal expansion, adiabatic expansion, isothermal compression, and adiabatic compression—all reversible): ΔT, q, w, ΔE, ΔH, ΔS, and ΔG.

(b) A sample of hot *water* is mixed with a sample of cold *water* in a thermally insulated, closed container of fixed volume: w, q, ΔE, ΔH, and ΔS.

(c) An *ideal gas* expands adiabatically and reversibly: ΔV, ΔT, w, q, ΔE, ΔH, and ΔS.

(d) A flask of *liquid nutrient solution inoculated with a small sample of bacteria* is maintained for several days in a thermostat until the bacteria have multiplied 1000-fold: ΔT, w, q, ΔE, ΔH, and ΔG.

8. One suggestion for solving the fuel-shortage problem is to use electric power (obtained from solar energy) to electrolyze water to form $H_2(g)$ and $O_2(g)$. The hydrogen could be used as a pollution-free fuel.

(a) Calculate the enthalpy, entropy, and free-energy change on burning 1 kg of $H_2(g)$ to $H_2O(l)$ at 25°C and 1 atm.

(b) Calculate the enthalpy, entropy, and free-energy change on burning 1 kg of *n*-octane(*g*) to $H_2O(l)$ and $CO_2(g)$ at 25°C and 1 atm.

9. Use data from the tables A.5–7 in the appendix to answer the following questions.

   (a) A friend wants to sell you a catalyst that allows benzene(g) to be formed by passing $H_2(g)$ over carbon (graphite) at 25°C and 1 atm. Should you buy? Why?

   (b) Is the reaction $2NO(g) + O_2(g) \rightarrow 2NO_2(g)$ spontaneous at 25°C and 1 atm?

   (c) Is the reaction to form solid alanine, CH_3CHNH_2COOH, and liquid H_2O from $CH_4(g)$, $NH_3(g)$, and $O_2(g)$ spontaneous at 25°C and 1 atm?

10. For the following statements, choose the word or words inside the parentheses that serve to make a correct statement. Each statement has at least one, and may have more than one, correct answer.

   (a) According to the second law of thermodynamics, a spontaneous process, such as a balloon filled with hot gas cooling to the surroundings at constant pressure, will always occur (adiabatically, reversibly, irreversibly, without work done).

   (b) Associated with such a process, there is always an increase in entropy of (the system, the surroundings, the system plus the surroundings, none of these).

   (c) For the example given, the heat gained by the surroundings is just equal to the negative of the (internal energy change, enthalpy change, entropy change, Gibbs free-energy change) of the system.

   (d) To return a system to its initial state requires from the surroundings an expenditure of entropy whose magnitude is (greater than, equal to, less than) that which is gained during the spontaneous process.

11. Consider the reaction

$$C_2H_5OH(l) \longrightarrow C_2H_6(g) + \frac{1}{2}O_2(g)$$

   Calculate ΔH^0_{298}, ΔS^0_{298}, and ΔG^0_{298}. Estimate ΔE^0_{298}. State what further data would be needed to obtain ΔH^0 at 500°C and 1 atm.

12. The shells of marine organisms contain $CaCO_3$ largely in the crystalline form known as calcite. There is a second crystalline form of $CaCO_3$ known as aragonite.

   (a) Based on the thermodynamic and physical properties given for these two crystalline forms, would you expect calcite in nature to convert spontaneously to aragonite given sufficient time? Justify your answer.

   (b) Will the conversion proposed in part (a) be favored or opposed by increasing the pressure? Explain.

   (c) What pressure should be just sufficient to make this conversion spontaneous at 25°C?

   (d) Will increasing the temperature favor the conversion? Explain.

Properties at 298 K	CaCO₃ (calcite)	CaCO₃ (aragonite)
$\Delta \overline{H}^0_f$ (kJ mol^{-1})	−1206.87	−1207.04
$\Delta \overline{G}^0_f$ (kJ mol^{-1})	−1128.76	−1127.71
$\overline{S}^0$ (J K^{-1} mol^{-1})	92.88	88.70
$\overline{C}_P$ (J K^{-1} mol^{-1})	81.88	81.25
Density (g cm^{-3})	2.710	2.930

13. Consider the reaction that converts pyruvic acid ($CH_3COCOOH$) into gaseous acetaldehyde (CH_3CHO) and gaseous CO_2, which is catalyzed in aqueous solution by the enzyme pyruvate decarboxylase. Assume ideal gas behavior for the CO_2.

   (a) Calculate ΔG^0_{298} for this reaction.

   (b) Calculate ΔG for this reaction at 298 K and 100 atm. State any important assumptions needed in addition to ideal gas behavior.

14. Consider the reversible, isothermal, constant-pressure freezing of 1 mol of water at 0°C and 1 atm.

   (a) Calculate ΔE in kJ.

   (b) Calculate ΔH in kJ.

   (c) Calculate ΔS in J K^{-1}.

   (d) Calculate ΔG in kJ.

   (e) Calculate q in kJ. Is heat absorbed or evolved?

   (f) Calculate w in kJ. Is work done by the system or on the system?

15. For the following statements, choose the word or words inside the parentheses that serve(s) to make a correct statement. More than one answer may be correct. At 100°C, the equilibrium vapor pressure of water is 1 atm. Consider the process where 1 mol of water vapor at 1 atm pressure is reversibly condensed to liquid water at 100°C by slowly removing heat into the surroundings.

   (a) In this process, the entropy of the system will (increase, remain unchanged, decrease).

   (b) The entropy of the universe will (increase, remain unchanged, decrease).

   (c) Since the condensation process occurs at constant temperature and pressure, the accompanying free-energy change for the system will be (positive, zero, negative).

   (d) In practice, this process cannot be carried out reversibly. For the real process, compared with the ideal reversible one, different values will be observed for the (entropy change of the system, entropy change of the surroundings, entropy change of the universe, free-energy change of the system).

At a lower temperature of 90°C, the equilibrium vapor pressure of water is only 0.692 atm.

(e) If 1 mol of water is condensed reversibly at this temperature and pressure, the entropy of the system will (increase, remain unchanged, decrease).

(f) The entropy of the universe will (increase, remain unchanged, decrease).

(g) The free energy of the system will (increase, remain unchanged, decrease).

(h) Since the molar heat capacity of water vapor is only about one-half that of the liquid, the entropy change upon condensation at 90°C will be (more negative than, the same as, more positive than) that at 100°C.

16. Consider a system that undergoes a phase change from phase α to phase β at equilibrium temperature T_m and equilibrium pressure P_m. At these equilibrium conditions, the heat absorbed per mole of material undergoing this transition is q_m, and there is a molar volume change of ΔV_m. The molar heat capacities at constant pressure for the α and β phases are $C_{P,\alpha}$ and $C_{P,\beta}$; they are independent of temperature.

(a) Evaluate w, ΔE, ΔH, ΔS, and ΔG for converting 1 mol of the system from phase α to phase β at the equilibrium T_m and P_m. Your answer should be in terms of q_m, ΔV_m, etc.

(b) Evaluate ΔH and ΔS at temperature T^* different from T_m but at the same pressure P_m.

(c) The denaturation transition of a globular protein can be approximated as a phase change α to β. Calculate ΔG, ΔH, and ΔS for a heat of transition of $q_m = 638$ kJ mol^{-1} at $T_m = 70°C$, $P_m = 1$ atm; $C_{P,\alpha} - C_{P,\beta} = 8.37$ kJ mol^{-1} K^{-1}.

(d) Calculate ΔG, ΔH, and ΔS for the same transition at 37°C and 1 atm. Assume that $C_{P,\alpha} - C_{P,\beta}$ is independent of temperature.

(e) If ΔV_m for the transition is $+3$ mL mol^{-1}, will the equilibrium transition temperature T_m be increased or decreased at a pressure of 1000 atm? Explain.

(f) Write a thermodynamic cycle that would allow you to calculate the equilibrium transition temperature corresponding to a pressure of 1000 atm. Use the cycle to write an equation that in principle could be solved to calculate T_m at 1000 atm.

17. Table 3.2 gives thermodynamic data for transitions from ordered, helical conformations of polypeptides and polynucleotides to disordered states. The enthalpy changes and entropy changes are positive as expected; heat is absorbed and entropy is gained. However, when the synthetic polypeptide polybenzyl-l-glutamate undergoes a transition from an ordered helix to a disordered coil in an ethylene dichloride–dichloroacetic acid solvent at 39°C and 1 atm, $\Delta H^0 = -4.0$ kJ (per mole of amide) and $\Delta S^0 = -12$ J K^{-1} (per mole of amide).

(a) Give a plausible explanation for the experimental result that heat is released and entropy is decreased for this transition. Does increasing the temperature favor the helix-coil transition?

(b) Is the reaction spontaneous at 39°C? What thermodynamic criterion did you use to reach this conclusion?

(c) At what temperature (°C) will the helix-coil reaction be reversible? This temperature is often called the "melting" temperature of the helix. Assume that ΔH^0 and ΔS^0 are independent of temperature.

(d) Can a reaction occur in an *isolated system* that leads to a decrease in the entropy of the system? If it can, give an example; if it cannot, state why not.

18. The following thermodynamic data at 25°C have been tabulated for the gas-phase reaction shown below (dotted lines represent hydrogen bonds):

	ΔH_f^0 (kJ mol^{-1})	$\overline{S}^0$ (J K^{-1} mol^{-1})
HCOOH(g)	-362.63	251.0
(HCOOH)$_2$(g)	-785.34	347.7

(a) Write the equation for the standard heat of formation at 25°C of HCOOH(g) (specify the pressure and phase for each component).

(b) Calculate ΔH^0, ΔS^0, and ΔG^0 for the gas-phase dimerization at 298 K. Is the formation of dimer from monomers spontaneous under these conditions?

(c) Calculate the enthalpy change per hydrogen bond formed in the gas phase. Why is not a similar calculation useful to estimate the entropy or free energy of hydrogen bond formation?

19. An electrochemical battery is used to provide 1 milliwatt (mW) of power for a (small) light. The chemical reaction in the battery is

$$\frac{1}{2}N_2(g) + \frac{3}{2}H_2(g) \longrightarrow NH_3(g)$$

(a) What is the free-energy change for the reaction at 25°C, 1 atm?

(b) Calculate the free-energy change for the reaction at 50°C, 1 atm. State any assumptions made in the calculation.

(c) The limiting reactant in the battery is 100 g of H_2. Calculate the maximum length of time (seconds) the light can operate at 25°C.

20. Consider a fertilized hen egg in an incubator—a constant temperature and pressure environment. In a few weeks, the egg will hatch into a chick.

    (a) The egg is chosen as the system. Is the egg an open, an isolated, or a closed system? Define an isolated system.

    (b) In the fertilized egg, a highly ordered chick is formed. Does the entropy of the system increase or decrease? Does this violate the second law of thermodynamics? Explain in two or three sentences why the development of the chick is or is not consistent with the second law.

    (c) What happens to the energy of the system as the chick develops? What forms of energy contribute to the change in energy (if any) of the system?

    (d) Does the free energy of the system increase, decrease, or remain the same? How do you know?

21. Calculate the entropy change when

    (a) Two moles of $H_2O(g)$ are cooled irreversibly at constant P from 120°C to 100°C.

    (b) One mole of $H_2O(g)$ is expanded at constant pressure of 2 atm from an original volume of 20 L to a final volume of 25 L. You can consider the gas to be ideal.

    (c) One hundred grams of $H_2O(s)$ at −10°C and 1 atm are heated to $H_2O(l)$ at +10°C and 1 atm.

22. You are asked to evaluate critically the following situations. Some of the proposals or interpretations are reasonable, while others violate very basic principles of thermodynamics.

    (a) It is commonly known that one can supercool water and maintain it as a liquid at temperatures as low as −10°C. If a sample of supercooled liquid water is isolated in a closed, thermally insulated container, after a time it spontaneously changes to a mixture of ice and water at 0°C. Thus, it increases its temperature spontaneously with no addition of heat from outside; furthermore, some low-entropy ice is produced.

    (b) An inventor proposed a new scheme for heating buildings in the winter in arctic climates. Since freezing temperatures reach only a few feet down into the soil, he proposes digging a well and immersing a coil of copper tubing in the water. He will then connect the ends of the coil to the radiators in the building and use a heat pump to transfer heat into the building. What would be your advice to an attorney who is assigned to evaluate this patent application? Base your evaluation on the relevant thermodynamics.

    (c) On a hot summer's day, your laboratory partner proposes opening the door of the lab refrigerator to cool off the room.

    (d) A sample of air is separated from a large evacuated chamber, and the entire system is isolated from the surroundings. A small pinhole between the two chambers is opened, and roughly half the gas is allowed to effuse into the second chamber before the pinhole is closed. Because nitrogen effuses faster than oxygen, the gas in the second chamber is richer in nitrogen, and the gas in the first chamber is richer in oxygen than the original air. The gases have thus spontaneously unmixed (at least partially), and this is held to be a violation of the second law of thermodynamics.

    (e) Supercooled water at −10°C has a higher entropy than does an equal amount of ice at −10°C. Therefore, supercooled water cannot go spontaneously to ice at the same temperature in an isolated system.

    (f) A volume of an aqueous solution of hydrogen peroxide is placed in a cylinder and covered with a tight-fitting piston. A small amount of the enzyme catalase is placed on a probe and inserted through an opening in the base of the cylinder. The catalase catalyzes the decomposition of the hydrogen peroxide, and the oxygen gas formed serves to raise the piston. The catalase is then withdrawn and the hydrogen peroxide is re-formed, causing the piston to return to its initial position. The piston is connected to an engine, and net work is obtained indefinitely by cycles of simply inserting and withdrawing the catalase.

    (g) High temperatures can be achieved in practice by focusing the rays of the Sun on a small object, using a large parabolic reflector, such as is used in astronomical telescopes. Since the energy gathered increases as the square of the diameter of the reflector, it should be possible to produce temperatures higher than those of the Sun by using a large-enough reflector.

    (h) The maximum efficiency of a steam engine can be calculated using the second law of thermodynamics. If it operates between the boiling point of water and room temperature, the maximum efficiency is about

    $$\frac{373 - 293}{373} = \frac{80}{373} = 0.215 \quad \text{or} \quad 21.5\%$$

    Photosynthesis by green plants occurs almost entirely at ambient temperatures, yet publications report theoretical limits as high as 85% of the fraction of the light energy absorbed can be converted into chemical energy (free energy or net work). Clearly, such estimates must be wrong, or else the second law cannot apply.

23. (a) Consider 1 mol of liquid water to be frozen reversibly to ice at 0°C by slowly removing heat to the surroundings. In this process, the entropy of the system will (increase, remain unchanged, decrease).

(b) If the freezing process occurs at 0°C and a pressure of 1 atm, the accompanying change in the Gibbs free energy for the system will be (positive, zero, negative).

(c) For this process the enthalpy of the system will (increase, remain unchanged, decrease).

24. For the statements below, choose the word or words inside the parentheses that serve to make a correct statement. Each statement has at least one and may have more than one correct answer.

(a) For a sample of an ideal gas, the product PV remains constant as long as the (temperature, pressure, volume, internal energy) is held constant.

(b) The internal energy of an ideal gas is a function of only the (volume, pressure, temperature).

(c) The second law of thermodynamics states that the entropy of an isolated system always (increases, remains constant, decreases) during a spontaneous process.

(d) When a sample of liquid is converted reversibly to its vapor at its normal boiling point, (q, w, ΔP, ΔV, ΔT, ΔE, ΔH, ΔS, ΔG, none of these) is equal to zero for the system.

(e) If the liquid is permitted to vaporize isothermally and completely into a previously evacuated chamber that is just large enough to hold the vapor at 1 atm pressure, then (q, w, ΔE, ΔH, ΔS, ΔG) will

be smaller in magnitude than for the reversible vaporization.

25. Starting with the definition of enthalpy, $H \equiv E + PV$, and additional relations based on the first and second laws of thermodynamics, derive the following:

(a) dH as a function of T, S, V, and P.

(b) Equations for $(\partial H/\partial P)_S$ and $(\partial H/\partial S)_P$.

(c) The equation $(\partial T/\partial P)_S = (\partial V/\partial S)_P$.

(d) The result of part (a) starting from Eq. (3.35) and the definition $G \equiv H - TS$.

26. Earth's atmosphere behaves as if it is approximately isentropic—the molar entropy of air is a constant independent of altitude up to about 10 km. It is well known that pressure and temperature vary with altitude. Using the isentropic model, calculate the temperature of the atmosphere 10 km above Earth, where the pressure is found to be 210 torr. The temperature and pressure at the surface of Earth (sea level) are 25°C and 760 torr, respectively. Assume that air behaves like an ideal gas with $C_P = \frac{7}{2}R$. You may ignore gravitation influences. (The temperature measured during rocket flights above New Mexico is −50°C at 10 km above Earth.)

27. The temperature of a typical laboratory freezer unit is −20°C. If liquid water in a completely filled, closed container is placed in the freezer, estimate the maximum pressure developed in the container at equilibrium. The enthalpy of fusion of water may be taken as 333.4 kJ kg^{-1}, independent of temperature and pressure, and the densities of ice and liquid water at −20°C are 0.9172 and 1.00 g cm^{-3}, respectively.

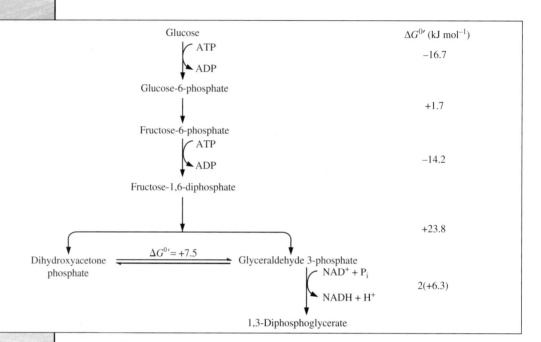

CHAPTER 4

Free Energy and Chemical Equilibria

Free energy is a thermodynamic variable of state that can predict whether a reaction can occur at constant temperature and pressure. It characterizes the maximum work that can be obtained from a reaction. It also specifies how the work depends on concentrations, temperature, and other variables. All the reactions in cells that make vital biochemicals, move muscles, and transport molecules are clarified by free energy.

Concepts

When a chemical reaction occurs in a closed system, the amounts of the reactants decrease, and the amounts of the products increase, until an equilibrium position is reached. At equilibrium, the reactants and products can interconvert, but the composition of the system remains unchanged. A triumph of thermodynamics is its ability to predict the properties of a system at equilibrium. We learned in the preceding chapter that at constant temperature and pressure, the change in the Gibbs free energy G of a system undergoing a *spontaneous* reaction must be negative. If ΔG is positive, only the reverse reaction can spontaneously occur; if it is zero, the system is at equilibrium. The Gibbs free energy G of an open system is dependent on the temperature, pressure, and the amounts of materials that make up the system. When a very small amount of a substance is added to the system at constant temperature and pressure without changing any of the other composition variables, the change in G per mole of the substance added is termed the *partial molar Gibbs free energy* of the added substance. We will see in this chapter that ΔG of a reaction at constant temperature and pressure is determined by the partial molar Gibbs free energies of the reactants and products. Thus, the partial molar Gibbs free energy of a substance determines its *chemical potential* in a reaction at constant temperature and pressure, just like the height of a boulder determines its potential energy in Earth's gravitational field. Because bioscientists are particularly interested in processes at constant temperature and pressure, the terms *partial molar Gibbs free energy* and *chemical potential* are used interchangeably in this book.

Many factors—such as temperature, pressure, and composition variables (partial pressures for gases in a gaseous mixture and concentrations for solutes in a solution)—influence the chemical potential of a substance. The chemical potential μ of a substance is invariably measured relative to the *standard* chemical potential μ^0 of the same substance under a chosen set of conditions, its *standard state*. By choosing appropriate standard states, $\mu - \mu^0$ of any substance can be expressed as a function of the more familiar variables: partial pressure if it is a gas, concentration in molarity or molality if it is a solute, and so on. For a component of a mixture of ideal gases, $\mu - \mu^0$ can be readily calculated from its partial pressure; for a solute of an ideal solution, from its concentration. For pure liquids or solids, $\mu - \mu^0$ is close to zero at moderate pressures. For real gases and solutions, the chemical potential of a substance is influenced by complex molecular interactions, but the difference $\mu - \mu^0$ can be experimentally determined.

Thermodynamics tells us that, in an equilibrium mixture at constant temperature and pressure, a simple relation exists between the chemical potentials of the reactants and products because ΔG is zero at equilibrium. This in turn shows that the composition of an equilibrium reaction is specified by the *standard* chemical potentials of the participants of the reaction. Thermodynamics clarifies the meaning of the equilibrium constants familiar to us and shows how these constants can be evaluated from tabulated thermodynamic data or adjusted when the temperature is changed. By establishing the relations for how the Gibbs free energy depends on various variables, thermodynamics also provides quantitative descriptions of *Le Châtelier's principle*, which states that perturbing a system at equilibrium results in changes that minimize the effect of the perturbation.

Applications

Thermodynamic analysis of reactions at equilibrium finds broad applications in chemical and biological systems. Although it is often difficult to describe a complex biological system in terms of equilibrium thermodynamics, we can usually separate a complex process into its components. Important components that can be analyzed thermodynamically include (1) metabolic reactions in which chemical bonds are broken and new ones formed, (2) dissociation of H^+ from acidic compounds and binding of H^+ to bases, (3) oxidation–reduction reactions in which electron transfer occurs, (4) interactions involving the aqueous medium in which metabolites and ionic species occur in the cytoplasm or other biological fluids, and (5) the assembly and disassembly of membranes and other multicomponent cellular structures. Applications of thermodynamic analysis to several cases of biological interest are illustrated in this chapter.

Chemical Potential (Partial Molar Gibbs Free Energy)
Gibbs Free Energy and the Chemical Potential

We introduced partial derivatives in chapter 3 to discuss changes in thermodynamic quantities when a system parameter is altered. Among these, the partial molar Gibbs free energy is particularly useful in discussing systems at constant temperature and pressure. As we will see shortly, the partial molar Gibbs free energy of a substance is called the *chemical potential* because it gives a measure of the *potential* of the substance in a chemical reaction at constant temperature and pressure. In the next chapter, we will see that the chemical potential is equally useful in discussing transformations that do not involve chemical changes, such as the evaporation of a liquid and the movement of ions across synthetic or natural membranes. For chemical reactions at constant temperature and volume, the chemical potential of a substance is given by the *partial molar Helmholtz free energy*. Because reactions in biological systems usually occur at constant temperature and pressure, in this and the next chapter the term *chemical potential* specifically refers to partial molar Gibbs free energy.

The chemical potential, or partial molar Gibbs free energy, of a substance A in a system containing other chemicals is

$$\mu_A \equiv \left(\frac{\partial G}{\partial n_A} \right)_{T,P,n_j \neq n_A} \tag{3.54}$$

The symbols T, P, and $n_j \neq n_A$ outside the parentheses specify that T, P, and the number of moles of all *other* chemicals present in the system are constant. Thus, the chemical potential of A, μ_A, describes the way in which the free energy of a system changes when the number of moles of A changes while all other variables are held constant. We see that this becomes equal to the free energy per mole of A, $\overline{G}_A$, if the system consists of pure A. In general, however, the free-energy change will depend on what else is in the system, and we must specify the composition of the system by specifying the values of n_j.

Consider first an *open system* made of substances A, B, C, and D. The free energy G of the system is a function of the temperature T, pressure P, the number of moles n_A, n_B, n_C, and n_D of A, B, C, and D, respectively:

$$G = G(T, P, n_A, n_B, n_C, n_D) \tag{4.1}$$

Applying the standard rules of differentiation gives

$$dG = \left(\frac{\partial G}{\partial T}\right)_{P,n_A,\ldots,n_D} dT + \left(\frac{\partial G}{\partial P}\right)_{T,n_A,\ldots,n_D} dP + \left(\frac{\partial G}{\partial n_A}\right)_{T,P,n_{j\neq A}} dn_A$$

$$+ \left(\frac{\partial G}{\partial n_B}\right)_{T,P,n_{j\neq B}} dn_B + \left(\frac{\partial G}{\partial n_C}\right)_{T,P,n_{j\neq C}} dn_C + \left(\frac{\partial G}{\partial n_D}\right)_{T,P,n_{j\neq D}} dn_D$$

$$= -S\,dT + V\,dP + \mu_A\,dn_A + \mu_B\,dn_B + \mu_C\,dn_C + \mu_D\,dn_D \qquad \textbf{(4.2)}$$

In the last step, we have substituted the various partial derivatives by their equivalents according to Eqs. (3.36), (3.43), and (3.54).

At constant temperature and pressure, Eq. (4.2) becomes

$$dG = \mu_A\,dn_A + \mu_B\,dn_B + \mu_C\,dn_C + \mu_D\,dn_D \qquad \textbf{(4.3)}$$

The Sum Rule for Partial Molar Quantities

Imagine that we start with a very small system containing substances A, B, C, and D and gradually add materials to it, at constant temperature and pressure, in such a way that infinitesimal amounts of the components (dn_A, dn_B, dn_C, and dn_D) are always added in proportion to their amounts in the mixture. In this way, the composition of the system, and hence the chemical potentials of the components, remain the same as n_A, n_B, n_C, and n_D are changed. Because the chemical potentials are kept constant during the additions, Eq. (4.3) can be readily integrated:

$$\int_0^G dG = \mu_A \int_0^{n_A} dn_A + \mu_B \int_0^{n_B} dn_B + \mu_C \int_0^{n_C} dn_C + \mu_D \int_0^{n_D} dn_D \qquad \textbf{(4.4)}$$

or

$$G = n_A\mu_A + n_B\mu_B + n_C\mu_C + n_D\mu_D \qquad \text{(constant } T, P) \qquad \textbf{(4.5)}$$

Equation (4.5) is termed the *sum rule for the partial molar Gibbs free energies*. From the way we arrived at this equation, we see that similar sum rules can be obtained for other partial molar quantities. The total volume V of a system containing components A, B, C, and D, for example, is equal to $n_A v_A + n_B v_B + n_C v_C + n_D v_D$, where v_A, v_B, v_C, and v_D are the partial molar volumes of the components.

Chemical Potential and Directionality of Chemical Reaction

We now consider a *closed system* consisting of four components A, B, C, and D that are undergoing a reversible chemical reaction:

$$aA + bB \rightleftharpoons cC + dD$$

As before, the Gibbs free energy G of the system is

$$G = G(T, P, n_A, n_B, n_C, n_D)$$

and

$$dG = -S\,dT + V\,dP + \mu_A\,dn_A + \mu_B\,dn_B + \mu_C\,dn_C + \mu_D\,dn_D \qquad \textbf{(4.2)}$$

In a closed system, however, the changes dn_A, dn_B, dn_C, and dn_D are interdependent and are related by the stoichiometry of the reaction:

$$\frac{dn_A}{a} = \frac{dn_B}{b} = -\frac{dn_C}{c} = -\frac{dn_D}{d}$$

Defining these ratios as $-d\alpha$ and substituting $dn_A = -a\,d\alpha$, $dn_B = -b\,d\alpha$, and so on into Eq. (4.2), we get

$$dG = -S\,dT + V\,dP + (c\mu_C + d\mu_D - a\mu_A - b\mu_B)\,d\alpha \qquad (4.6)$$

Here, $d\alpha$ is a measure of the extent of the reaction. A positive $d\alpha$ means dn_A and dn_B are negative and dn_C and dn_D are positive: The amounts of A and B are decreasing, and those of C and D are increasing. In other words, A and B are being converted to C and D, and the reaction is proceeding from left to right. Conversely, a negative $d\alpha$ means dn_A and dn_B are positive and dn_C and dn_D are negative, and the reaction is proceeding from right to left.

If the temperature and pressure are constant, $dT = 0$, $dP = 0$, and

$$dG = (c\mu_C + d\mu_D - a\mu_A - b\mu_B)\,d\alpha$$
$$= [(c\mu_C + d\mu_D) - (a\mu_A + b\mu_B)]\,d\alpha$$
$$= -[(a\mu_A + b\mu_B) - (c\mu_C + d\mu_D)]\,d\alpha \qquad (4.7)$$

If the reaction spontaneously proceeds from left to right at constant temperature and pressure, we know that dG must be negative. By definition, $d\alpha$ is positive if the reaction goes from left to right. It therefore follows from Eq. (4.7) that $[(a\mu_A + b\mu_B) - (c\mu_C + d\mu_D)]$ must be positive. In other words, if there is a net conversion of the reactants A and B into the products C and D at constant temperature and pressure, the sum of the chemical potentials of the reactants—each properly weighed according to the stoichiometry of the reaction—must be greater than the sum of the chemical potentials of the products similarly weighed. Thus, the chemical potentials are so named because they determine the directionality of chemical reactions at constant temperature and pressure. It also follows directly from Eq. (4.7) that if the system is at equilibrium,

$$[(a\mu_A + b\mu_B) - (c\mu_C + d\mu_D)] = 0$$

Note also that, from Eq. (4.7), $[(c\mu_C + d\mu_D) - (a\mu_A + b\mu_B)]$ is equal to $dG/d\alpha$, the change in the Gibbs free energy *per mole of the reaction* at constant temperature and pressure:

$$\Delta G(\text{per mole of reaction}) = \frac{dG}{d\alpha}$$
$$= [(c\mu_C + d\mu_D) - (a\mu_A + b\mu_B)] \qquad (4.8)$$

Two points should be clarified. First, "per mole of the reaction" is a quantity that depends on how we express the reaction, as we discussed in chapter 2 (see footnote on p. 49). For the oxidation of glucose to carbon dioxide and water, for example, we can express the reaction as

$$C_6H_{12}O_6 + 6O_2 \longrightarrow 6CO_2 + 6H_2O$$

Then, per mole of reaction refers to the conversion of 1 mol of glucose or 6 mol of oxygen or to the formation of 6 mol of CO_2 or 6 mol of water. If we express the same reaction as

$$\frac{1}{6}C_6H_{12}O_6 + O_2 \longrightarrow CO_2 + H_2O$$

then per mole of reaction refers to the conversion of $\frac{1}{6}$ mol of glucose or 1 mol of oxygen or to the formation of 1 mol of carbon dioxide or water. Recall that the Gibbs free energy is an extensive quantity, the value for ΔG (per mole of reaction) is dependent on how the stoichiometry of the reaction is specified. By convention, the change in free energy for a chemical reaction always refers to ΔG per mole of the particular reaction as specified. Therefore, in subsequent sections on chemical reactions, the quantity ΔG (per mole of reaction) will simply be termed ΔG.

Second, we should keep in mind that in the derivation of Eq. (4.8), ΔG for any reaction such as

$$C_6H_{12}O_6 + 6O_2 \longrightarrow 6CO_2 + 6H_2O$$

refers to $dG/d\alpha$—that is, *the change in the Gibbs free energy per mole of the reaction for an infinitesimal change $d\alpha$.* For an infinitesimal change, the chemical composition of the system is not significantly altered by the conversion. This simplifies things greatly; otherwise, as the reaction proceeds, the chemical composition of the system and therefore the chemical potentials of the reactants and products will keep on changing. An alternative way of thinking about ΔG per mole of the reaction is to imagine that the reaction occurs in a very large vessel containing large amounts of the reacting substances, so the net conversion of 1 mol of glucose does not significantly change the total number of moles of glucose, oxygen, carbon dioxide, or water in the reaction vessel; ΔG for the reaction as written can then be thought as the actual free-energy change when 1 mol of glucose combines with 6 mol of oxygen to form 6 mol of carbon dioxide and 6 mol of water under the particular reaction conditions.

Reactions of Gases: The Ideal Gas Approximation
Dependence of Chemical Potential on Partial Pressures

We start with reactions of gases because thermodynamically the gases are the simplest to treat, especially if the pressure is not too high so that, to a good approximation, the gases behave as ideal gases. For an ideal gas at a constant temperature, a simple relation exists between its chemical potential μ and pressure P. From Eq. (3.45), at a constant T the change in the Gibbs free energy G when the pressure of the gas changes from a pressure of P_1 to a pressure of P_2 is

$$G(P_2) - G(P_1) = nRT \ln\left(\frac{P_2}{P_1}\right) \qquad (T = \text{constant}) \qquad \textbf{(3.45)}$$

As we did before, we choose $P_1 = 1$ atm as the standard state and denote $G(P_1 = 1 \text{ atm})$ as G^0. The Gibbs free energy G at a pressure P is therefore related to P by

$$G - G^0 = nRT \ln\left(\frac{P}{1\text{ atm}}\right)$$

Dividing both sides by n and remembering that the chemical potential of a pure substance is just its molar free energy, we obtain

$$\mu = \mu^0 + RT \ln\left(\frac{P}{1\text{ atm}}\right) \tag{4.9}$$

where μ is the chemical potential of the ideal gas and μ^0 its chemical potential at the standard state of 1 atm. This relation also holds for an ideal gas A in a mixture of ideal gases. In a mixture, the pressure of A is the partial pressure P_A:

$$\mu_A = \mu_A^0 + RT \ln\left(\frac{P_A}{1\text{ atm}}\right) \tag{4.10}$$

EXAMPLE 4.1

Two gases, n_A mol of A and n_B mol of B, are initially in two separate flasks at the same temperature T and pressure P. A valve in the tube connecting the two flasks is then opened to allow complete mixing of the gases. Show that, from Eq. (4.10), the change in Gibbs free energy for mixing the two gases is

$$\Delta G_{mix} = n_A RT \ln X_A + n_B RT \ln X_B$$

where X_A and X_B are the mole fraction of A and B, respectively.

Solution

Before mixing, the total Gibbs free energy of the system is, from Eq. (4.5),

$$G(1) = n_A \mu_A(1) + n_B \mu_B(1)$$

$$= n_A\left[\mu_A^0 + RT \ln\left(\frac{P}{1\text{ atm}}\right)\right] + n_B\left[\mu_B^0 + RT \ln\left(\frac{P}{1\text{ atm}}\right)\right]$$

After mixing, the total Gibbs free energy of the system becomes

$$G(2) = n_A \mu_A(2) + n_B \mu_B(2)$$

$$= n_A\left[\mu_A^0 + RT \ln\left(\frac{P_A}{1\text{ atm}}\right)\right] + n_B\left[\mu_B^0 + RT \ln\left(\frac{P_B}{1\text{ atm}}\right)\right]$$

In the above equations, (1) and (2) denote the state before and after mixing, respectively, and P_A and P_B are the partial pressures of A and B, respectively, after mixing. The change in Gibbs free energy is therefore

$$\Delta G_{mix} = G(2) - G(1) = n_A RT \ln\left(\frac{P_A}{P}\right) + n_B RT \ln\left(\frac{P_B}{P}\right) \tag{4.11}$$

The partial pressure P_A is related to P by

$$P_A = \frac{n_A}{n_A + n_B}P = X_A P$$

or

$$\frac{P_A}{P} = X_A$$

Similarly,

$$\frac{P_B}{P} = X_B$$

Substituting these into Eq. (4.11) gives

$$\Delta G_{mix} = n_A RT \ln X_A + n_B RT \ln X_B$$

We have previously calculated the entropy change for this process and obtained

$$\Delta S_{mix} = -R[n_A \ln X_A + n_B \ln X_B] \tag{3.22}$$

The calculated ΔG_{mix} is equal to $-T\Delta S_{mix}$, as expected: ΔE as well as ΔH is zero when two ideal gases at the same temperature and pressure are mixed.

Equilibrium Constant

We now turn to one of the most important topics in thermodynamics: the relation between the equilibrium constant of a chemical reaction and the thermodynamic properties of the reactants and products in this reaction. For simplicity, we start by considering a reaction involving only gases, the *Haber process* of nitrogen fixation in which nitrogen and hydrogen react to form ammonia. Nitrogen is an essential element for the growth of all living organisms. Most organisms cannot use N_2 directly from the atmosphere but require its reduction to ammonia or oxidation to nitrite or nitrate first. The nitrogen-fixing bacteria, some of which live symbiotically with plants in root nodules, use a powerful reductant to convert N_2 metabolically to reduced nitrogen compounds, which can then be transferred to the plant. In the synthetic method developed by Fritz Haber, the reaction is

$$N_2 + 3H_2 \longrightarrow 2NH_3$$

From Eq. (4.5), the Gibbs free-energy change for the reaction is

$$\Delta G = 2\mu_{NH_3} - \mu_{N_2} - 3\mu_{H_2}$$

$$= 2\left[\mu_{NH_3}^0 + RT \ln\left(\frac{P_{NH_3}}{1 \text{ atm}}\right)\right] - \left[\mu_{N_2}^0 + RT \ln\left(\frac{P_{N_2}}{1 \text{ atm}}\right)\right]$$

$$- 3\left[\mu_{H_2}^0 + RT \ln\left(\frac{P_{H_2}}{1 \text{ atm}}\right)\right] \tag{4.12}$$

For simplicity, we will omit all "1 atm" terms because multiplying or dividing by 1 does not change the numerical value of a term. In doing so, however, we must keep in mind that all of the partial pressure terms become unitless numerical values of the actual partial pressures in units of atm. Equation (4.12) then becomes

$$\Delta G = (2\mu_{NH_3}^0 - \mu_{H_2}^0 - 3\mu_{H_2}^0) + RT[2 \ln P_{NH_3} - \ln P_{N_2} - 3 \ln P_{H_2}]$$

$$= \Delta G^0 + RT\left[\ln\left(\frac{P_{NH_3}^2}{P_{N_2} P_{H_2}^3}\right)\right] \tag{4.13}$$

In Eq. (4.13), ΔG^0 is equal to $(2\mu_{NH_3}^0 - \mu_{H_2}^0 - 3\mu_{H_2}^0)$ and is the standard Gibbs free energy per mole of the reaction at T when all participants of the reaction are in their standard states (1 atm partial pressure). At 298 K $\Delta G^0 = -32.90$ kJ from table A.5. In arriving at Eq. (4.13), we have also used the algebraic identity $a \ln x = \ln x^a$. We emphasize again that by dropping the "1 atm" terms in arriving at Eq. (4.13), we have implicitly specified that each partial pressure P_A in Eq. (4.13) is actually the ratio of P_A to 1 atm, which is the standard state. The numerical value of P_A can be used directly in Eq. (4.13) only if P_A is in units of atmospheres.

Equation (4.13) relates the free-energy change for any partial pressures, ΔG, to the standard free-energy change at 1 atm partial pressures, ΔG^0. It is important to note the presence or absence of the superscript. Its presence means standard conditions of 1 atm partial pressures of all participants of the reaction; its absence means different conditions, which must be specified. The superscript is less important for ΔH because enthalpy does not change very much with pressure, even for gases. However, for ΔG of a gas, it is vital.

For every chemical reaction at any temperature and pressure, there are partial pressures of products and reactants for which the system is at equilibrium. We may start with a mixture of nitrogen and hydrogen, or we may start with ammonia; if we wait long enough under the proper reaction conditions or if we use a catalyst for this reaction so that the wait is not too long, we will obtain an equilibrium mixture of nitrogen, hydrogen, and ammonia. We now ask the question, Are the partial pressures P_{N_2}, P_{H_2}, and P_{NH_3}, related when equilibrium is reached at constant temperature and total pressure?

Recall that when a system reaches equilibrium at constant temperature and pressure, $\Delta G = 0$. Therefore, from Eq. (4.13),

$$0 = \Delta G^0 + RT \ln \frac{(P^{eq}_{NH_3})^2}{(P^{eq}_{N_2})(P^{eq}_{H_2})^3}$$

or

$$\Delta G^0 = -RT \ln \frac{(P^{eq}_{NH_3})^2}{(P^{eq}_{N_2})(P^{eq}_{H_2})^3}$$

where the superscript "eq" specifies the quantity at equilibrium. The standard free-energy change ΔG^0 for a particular reaction is, however, a constant at any chosen temperature (because the states of the reactants and products are all specified to be their standard states and the Gibbs free energy is a state function). Therefore, at constant temperature and total pressure, the quotient $(P^{eq}_{NH_3})^2 / (P^{eq}_{N_2})(P^{eq}_{H_2})^3$ must be a constant, which we denote K:

$$\frac{(P^{eq}_{NH_3})^2}{(P^{eq}_{N_2})(P^{eq}_{H_2})^3} = K$$

When equilibrium is reached at a constant temperature and total pressure, thermodynamics tells us that the partial pressures $P^{eq}_{NH_3}$, $P^{eq}_{N_2}$, and $P^{eq}_{H_2}$ are not independent: they are related by the equation above. The constant K is called the equilibrium constant of the reaction and is directly related to ΔG^0 by

$$\Delta G^0 = -RT \ln K \qquad \text{(4.14)}$$

Note that at constant temperature and pressure the equilibrium constant of a reaction is related to the standard Gibbs free-energy change ΔG^0, and *not* ΔG itself, which is zero at equilibrium.

We can generalize Eqs. (4.13) and (4.14) for any reaction involving only gases that do not deviate too much from ideal gas behavior in terms of their free-energy dependence on partial pressure. For a reaction of ideal gases at temperature T,

$$aA + bB \longrightarrow cC + dD$$

$$\Delta G_T = \Delta G_T^0 + RT \ln Q \qquad \textbf{(4.15)}$$

$$Q = \frac{(P_C)^c (P_D)^d}{(P_A)^a (P_B)^b} \qquad \textbf{(4.16)}$$

where P_A, P_B, P_C, and P_D are the partial pressures of A, B, C, and D, respectively, each divided by 1 atm. The quotient Q is therefore the ratio of the numerical values of partial pressures (in atmospheres) of reactants and products, each raised to the power of its coefficient in the chemical equation. If we double the number of moles involved in the reaction, each coefficient is multiplied by 2, and Q is squared in Eq. (4.16). We see that Q will be large if product pressures are large or reactant pressures are small. A large Q means a positive (unfavorable) contribution to the free energy. Q is small if product pressures are small or reactant pressures are large. A small Q means a negative (favorable) contribution to the free energy. The following table gives quantitative values for the change of free energy with Q at 25°C. Values at any other temperature T can be calculated by multiplying those listed in the table by $(T/298)$.

Q	$\Delta G_{298} - \Delta G_{298}^0$ (kJ)
100	11.410
10	5.705
1	0
0.1	−5.705
0.01	−11.410

These tabulated values serve the purpose of emphasizing the distinction between standard Gibbs free-energy changes ΔG^0, for defined partial pressures, and actual free-energy changes ΔG, for any other partial pressures specified by Q.

For the reaction

$$aA + bB \longrightarrow cC + dD$$

at equilibrium

$$\Delta G^0 = -RT \ln K \qquad \textbf{(4.14)}$$

with

$$K = \frac{(P_C^{eq})^c (P_D^{eq})^d}{(P_A^{eq})^a (P_B^{eq})^b} \qquad \textbf{(4.17)}$$

where P_A^{eq}, P_B^{eq}, and so on are partial pressures, each divided by 1 atm, at equilibrium.

Equations (4.14) and (4.15) can also be combined to give

$$\Delta G_T = -RT \ln K + RT \ln Q$$

$$= RT \ln \frac{Q}{K} \qquad \textbf{(4.18)}$$

Equations (4.14) through (4.18) and their generalized forms, which are applicable to all substances, may be the most useful thermodynamic equations a biochemist learns. They relate the free-energy change for a reaction to experimentally measurable quantities. Determination of the standard Gibbs free energies of various substances permits calculation of ΔG^0 values for reactions involving these substances, which in turn permit calculation of the equilibrium constants; conversely, experimental determination of the equilibrium constant of a reaction allows us to calculate ΔG^0 of the reaction. Furthermore, Eq. (4.17), which defines the equilibrium constant, presents a ratio of equilibrium partial pressures that is constant for a chemical reaction at a given temperature.

EXAMPLE 4.2

What is the Gibbs free-energy change, relative to that under standard conditions, of forming 1 mol of NH_3 at 298 K if (a) 10.0 atm of N_2 and 10.0 atm of H_2 are reacted to give 0.0100 atm of NH_3 and (b) 0.0100 atm of N_2 and 0.0100 atm of H_2 are reacted to give 10.0 atm of NH_3? For part (b), NH_3 could be introduced with the reactants to give a high enough product partial pressure. The reaction is

$$N_2(g) + 3H_2(g) \longrightarrow 2NH_3(g) \qquad \Delta G^0_{298} = -32.90 \text{ kJ}$$

Solution

a. From Eq. (4.13),

$$\Delta G - \Delta G^0 = RT\left[\ln\left(\frac{P^2_{NH_3}}{P_{N_2}P^3_{H_2}}\right)\right]$$

$$= (8.314 \text{ J } K^{-1} \text{ mol}^{-1})(298 \text{ K})\ln\left[\frac{0.0100^2}{(10.0)(10.0^3)}\right]$$

$$= -4.56 \times 10^4 \text{ J mol}^{-1}$$

$$= -45.6 \text{ kJ mol}^{-1}$$

The free energy is decreased (relative to the standard free energy) by having high pressures of reactants and low pressures of products; the reaction is favored relative to standard conditions.

b. Substituting the partial pressures into Eq. (4.13) gives

$$\Delta G - \Delta G^0 = RT\left[\ln\left(\frac{P^2_{NH_3}}{P_{N_2}P^3_{H_2}}\right)\right]$$

$$= 5.70 \times 10^4 \text{ J mol}^{-1}$$

$$= 57.0 \text{ kJ mol}^{-1}$$

The free energy is increased by having high pressures of products and low pressures of reactants; the reaction is less favored relative to the standard conditions. The calculated free energies (with units of J mol^{-1}) are for the numbers of moles in the chemical reaction shown in the equation.

Nonideal Systems

Activity

It was clear to thermodynamicists that Eqs. (4.14) through (4.18) were simple and easy to use; therefore, it would be convenient if they applied to not only ideal gases but also real gases, liquids, solids, and solutions. Thermodynamicists decided that the way to ensure this is to *define* a new quantity, the activity, such that for any substance A its activity a_A is related to its chemical potential by

$$\mu_A = \mu_A^0 + RT \ln a_A \qquad (4.19)$$

It then follows, from Eq. (4.8), that for our usual reaction of $aA + bB \rightarrow cC + dD$,

$$\Delta G = [(c\mu_C + d\mu_D) - (a\mu_A + b\mu_B)] = \Delta G^0 + RT \ln\left(\frac{a_C^c a_D^d}{a_A^a a_B^b}\right)$$

$$= \Delta G^0 + RT \ln Q \qquad (4.20)$$

$$Q = \frac{a_C^c a_D^d}{a_A^a a_B^b} \qquad (4.21)$$

and the equilibrium constant is

$$K \equiv \frac{(a_C^{eq})^c (a_D^{eq})^d}{(a_A^{eq})^a (a_B^{eq})^b} \qquad (4.22)$$

These equations are of the same form as Eqs. (4.15), (4.16), and (4.17), but their application is no longer restricted to ideal gases. (When writing the expression for K, we often omit the superscript "eq." However, *equilibrium* activities are understood.)

Several comments may help clarify the use of the above equations. First, from the definition of the activity a [Eq. (4.19)], activities are unitless numbers. Second, the activity a_A of A is defined with respect to a standard state. The difference in chemical potential between two states, $\mu_A - \mu_A^0$, is a measurable quantity. Once the standard state is specified and the difference $\mu_A - \mu_A^0$ measured, the activity can be calculated from this difference: $a_A = e^{(\mu_A - \mu_A^0)/RT}$. It is also clear from its definition that a_A must be 1 when A is at its standard state. Third, to make full use of generalized eqs. (4.20) to (4.22), we need to relate the activities to parameters that have been widely used by chemists and biochemists, such as the partial pressures in the case of gases and the concentrations in the case of solutions. This is done by adopting various standard states, as we will see in the next section.

Standard States

In the following pages, we give the standard states that are commonly used in specifying activities. Some methods of measuring chemical potential differences and activities are discussed later. Students are often bewildered by the many different standard states. It may help to keep in mind that these standard states are chosen for convenience. The particular choices of standard states allow us to relate the activities to well-known experimental parameters. For ideal gases, the activity of a component is directly given by its partial pressure; for ideal solutions, the activity of a component is given by its concentration. For real gases and dilute solutions, the activity of a component remains a simple function of its partial pressure or concentration when convenient standard states are selected.

Ideal gases

The standard state for an ideal gas is the gas with a partial pressure equal to 1 atm. The activity of an ideal gas is defined as its actual partial pressure divided by 1 atm, its partial pressure in its standard state:

$$a_A = \frac{P_A}{1 \text{ atm}} \tag{4.23}$$

where P_A = partial pressure of ideal gas. The activity of an ideal gas is numerically equal to its partial presure in atmospheres. The standard conditions for a reaction involving only ideal gases are that each product and reactant has a partial pressure equal to 1 atm, as discussed in chapter 3.

Real gases

The activity of a real gas is a function of pressure, which we write

$$a_A = \frac{\gamma_A P_A}{1 \text{ atm}} \tag{4.24}$$

where γ_A is termed the activity coefficient and P_A is the partial pressure. We know that all gases become ideal at low-enough pressures, so γ_A must become 1 as the total pressure of the system approaches zero. Near atmospheric pressure, the activity coefficients of gases are close to 1, and real gases can be treated as ideal gases to a good approximation. From the definition of activity, for a real gas,

$$\mu_A - \mu_A^0 = RT \ln\left(\frac{\gamma_A P_A}{1 \text{ atm}}\right) = RT \ln \gamma_A + RT \ln\left(\frac{P_A}{1 \text{ atm}}\right)$$

Consider a system containing pure A. If $\mu_A - \mu_A^0$ is plotted as a function of $\ln(P_A/1 \text{ atm})$, at low pressure a straight line is obtained with a slope equal to RT (because $\gamma_A = 1$ at low pressures and therefore $\ln \gamma_A = 0$). If this straight line is extrapolated to $P_A = 1$ atm, $\mu_A - \mu_A^0 = 0$ and $\mu_A = \mu_A^0$, the chemical potential of the gas at its standard state. Thus, the standard state for a real gas is a hypothetical one: It is the *extrapolated state where the partial pressure is 1 atm but the properties are those extrapolated from low pressure.*

Pure solids or liquids

The standard state for a pure solid or liquid is the pure substance (solid or liquid) at 1 atm pressure. Therefore, the activity is equal to 1 for a pure solid or liquid at 1 atm. The Gibbs free energy of a solid or liquid changes only slightly with pressure [see Eq. (3.44)], so we can usually neglect the change and use 1 for the activity of a pure solid or liquid at any pressure:

$$a_A = 1 \text{ (pure solid or liquid)} \tag{4.25}$$

At high pressure, the activity of a pure solid or liquid can be calculated from the relation between Gibbs free energy and pressure [Eq. (3.43a)].

EXAMPLE 4.3

Calculate the activity a_{H_2O} of liquid water at 293 K and a pressure of (a) 10 atm and (b) 1000 atm.

Solution

At constant temperature, the Gibbs free energy of a substance is related to the pressure by

$$G(P_2) - G(P_1) = \int_{P_1}^{P_2} V \, dP \tag{3.43a}$$

This equation can also be written for 1 mol of the substance:

$$\overline{G}(P_2) - \overline{G}(P_1) = \int_{P_1}^{P_2} \overline{V}_A \, dP \tag{4.26}$$

where $\overline{G}$ and $\overline{V}_A$ are the Gibbs free energy and volume of 1 mol of water, respectively. Because the chemical potential μ_A of a pure substance A is equal to the Gibbs free energy per mole of A, $\overline{G}_A$, Eq. (4.26) becomes

$$\mu_A(P_2) - \mu_A(P_1) = \int_{P_1}^{P_2} \overline{V}_A \, dP \tag{4.27}$$

We set $P_1 = 1$ atm and let P_2 be any pressure P in atmospheres. At $P_1 = 1$ atm, $\mu_A(P_1) = \mu_A^0$. Thus,

$$\mu_A(P) - \mu_A^0 = \int_1^P \overline{V}_A \, dP \tag{4.28}$$

From page 32 of chapter 2, the molar volume of water is related to pressure by the empirical equation

$$\overline{V}_A = 18.07(1 - 45.9 \times 10^{-6} P) \, \text{mL mol}^{-1} \qquad (P \text{ in atm})$$

Substituting into Eq. (4.28) and integrating, we obtain

$$\mu_A(P) - \mu_A^0 = \int_1^P [18.07(1 - 45.9 \times 10^{-6} P)] \, dP$$

$$= \left[18.07(P - 1) - \frac{18.07 \times 45.9 \times 10^{-6}}{2}(P^2 - 1^2) \right] \quad (\text{mL atm mol}^{-1})$$

Thus,

$$\ln a_{H_2O} = \frac{\mu_A - \mu_A^0}{RT}$$

$$= \frac{18.07(P - 1) - \dfrac{18.07 \times 45.9 \times 10^{-6}}{2}(P^2 - 1^2)}{RT}$$

a. $P = 10$ atm. Substituting $P = 10$ atm, $R = 82.05$ mL atm mol^{-1} deg^{-1} and $T = 293$ deg into the equation gives $\ln a_{H_2O} = 0.00676$ or $a_{H_2O} = 1.007$, which is very close to the value 1.000, the activity of liquid water at 1 atm.

b. Similarly, at $P = 1000$ atm, $\ln a_{H_2O} = 0.732$ or $a_{H_2O} = 2.080$. At this high pressure, the activity deviates significantly from 1.

EXAMPLE 4.4

Calculate the equilibrium constant at 25°C for the decarboxylation of liquid pyruvic acid to form gaseous acetaldehyde and CO_2.

Solution

The reaction is

$$\underset{\substack{\parallel\\O}}{CH_3CCOOH}(l) \longrightarrow \underset{\substack{\parallel\\O}}{CH_3CH}(g) \; + \; CO_2(g)$$

From tables A.5–7 in the appendix,

$$\Delta G^0 = \Delta G^0_f(\text{acetaldehyde}) + \Delta G^0_f(CO_2) - \Delta G^0_f(\text{pyruvic acid})$$

$$= -133.30 + (-394.36) - (-463.38)$$

$$= -64.28 \text{ kJ}$$

From Eq. (4.14),

$$K = 10^{-\Delta G^0/2.303RT}$$

The factor $2.303RT = 5.708$ kJ mol^{-1} for 25°C; it will occur often, and it is worthwhile to make a note of this value for 25°C.

$$K = 10^{-\Delta G^0/5.708} \qquad (\Delta G^0 \text{ in kJ})$$

$$= 10^{64.28/5.708}$$

$$= 10^{11.26}$$

$$= 1.85 \times 10^{11}$$

Pyruvic acid, a liquid, is very unstable with respect to CO_2 and acetaldehyde at room temperature. The equilibrium constant is $K = (a_{\text{acetaldehyde}})(a_{CO_2})/(a_{\text{pyruvic acid}})$, where the activities are the values at equilibrium. The activity of pyruvic acid is of the order of 1; the activities of the gaseous products are related to their partial pressures according to Eq. (4.24). The very large equilibrium constant tells us that thermodynamically the reaction will go to completion; that is, all pyruvic acid will be converted to gaseous acetaldehyde and CO_2. If the reaction occurs rapidly, the sudden conversion of a condensed material to gaseous products can cause an explosion. A suitable catalyst is needed at room temperature, however, to accelerate this reaction.

Solutions

A *solution* is a homogeneous mixture of two or more substances. We can have solid solutions, liquid solutions, or very complicated mixtures of components as might be found in a living cell. The activity of each substance will depend on its concentration and the concentrations of everything else in the mixture. We write this dependence in a deceptively simple equation:

$$\text{activity} = (\text{activity coefficient})(\text{concentration}) \qquad \textbf{(4.29)}$$

The activity coefficient is not a constant; it incorporates all the complicated dependence of the activity of A on the concentrations of A, B, C, and so on. Another complication is that different concentration units are routinely used in Eq. (4.29). The commonly used concentration units and their standard states are discussed in the following section.

Mole fraction and solvent standard state

The standard state that uses mole fraction as a concentration unit is often called the solvent standard state. *The solvent standard state for a component of a solution defines the pure component as the standard state.* Let's choose the solvent standard state for component A in a solution. A useful concentration unit is the *mole fraction X*, the number of moles of species A divided by the total number of moles of all components present in the solution:

$$X_A = \frac{n_A}{n_T} \tag{4.30}$$

where n_A = number of moles of A
n_T = total number of moles of all components

The activity of A then, from Eq. (4.29), is

$$a_A = \gamma_A X_A \tag{4.31}$$

where a_A = activity of species A
γ_A = activity coefficient of A on the mole fraction scale
X_A = mole fraction of A

We choose the standard state so that the activity a_A becomes equal to the mole fraction X_A, as the mole fraction of A approaches unity. Mathematically, we write this as

$$\lim_{X_A \to 1} a_A = X_A$$

which is read "in the limit as X_A approaches 1, $a_A = X_A$." Because $\gamma = a_A/X_A$, we can express the same idea as

$$\lim_{X_A \to 1} \gamma = 1$$

In the limit as X_A approaches 1, $\gamma = 1$. Choosing $a_A \to X_A \to 1$ defines the standard state, the state where the activity is 1. For a liquid solution, the standard state for the solvent is defined as the pure liquid. The best example is dilute aqueous solutions, for which $a_{H_2O} = 1$ is used in introductory chemistry courses. We ignore the water in writing equilibrium constants in aqueous solutions. The logic is that we are using an activity on the mole fraction scale for the H_2O and that the solution is dilute, so $X_{H_2O} \approx 1$ and $a_{H_2O} \approx 1$. The mole fraction scale for activities is traditionally used for the *solvent* in a solution.

We now summarize the useful equations for a solvent in a solution:

Very dilute solution: $a_{solv} = 1$ (4.32)

Dilute solution or ideal solution: $a_{solv} = X_{solv}$ (4.33)

Real solution: $a_{solv} = \gamma X_{solv}$ (4.34)

Equation (4.33) defines an ideal solution: For any concentration, the solvent has the properties of the solvent in a dilute solution. To determine the activity coefficient and therefore the activity in Eq. (4.34), we have to measure the Gibbs free energy of the solvent in the solution. This can be done by methods involving measurements of the vapor pressure of the solvent in a solution, the freezing-point depression, the boiling-point elevation, or the osmotic pressure. These will be discussed in chapter 5.

Solute standard states

For solutions in which certain components never become very concentrated, such as dilute aqueous salt solutions, we define a solute standard state whose free energy can be obtained from measurements in dilute solutions. *The solute standard state for a component is defined as the extrapolated state where the concentration is equal to 1 molar (M) or 1 molal (m), but the properties are those extrapolated from very dilute solution.* The solute standard state is a hypothetical state in the sense that it corresponds to a 1 M or 1 m ideal solution. For real solutions, the ideal behavior is not approached except for concentrations that are much less than 1 M or 1 m. The definition of the solute standard state should become clearer after reading the following sections.

Molarity

Concentrations are commonly measured in *molarity,* with units of mol L^{-1}. (We use the symbol c for molarities.) The activity of solute B on the molarity scale and with a solute standard state is, from Eq. (4.29),

$$a_B = \gamma_B c_B \qquad\qquad (4.35)$$

where a_B = activity of species B

γ_B = activity coefficient of B on the molarity scale

c_B = concentration of B, mol L^{-1}, M

Note that the activity of B, a_B, for a molecule B in a solution will be different if Eq. (4.35) is used, rather than Eq. (4.31). The free energy per mole of B in a specified solution has a definite value, but different choices of standard states produce different values of a_B. The standard state for the molarity scale is chosen so that the activity becomes equal to the concentration as the concentration approaches zero:

$$\lim_{c_B \to 0} a_B = c_B$$

and

$$\lim_{c_B \to 0} \gamma = 1$$

The molarity scale for activities is used mainly for *solutes.*

To determine the activity coefficient and therefore the activity in Eq. (4.35), we measure the free energy of component B in the solution. This can be done indirectly through its effect on the vapor pressure of the solvent in solution. Depending on the properties of B, various methods also exist for measuring the free energy directly. If component B is volatile, its vapor pressure can be measured. Electrolytic cells provide easy methods for measuring certain activities (see "Galvanic Cells," pages 153–158).

The free energy in the standard state must also be measured to obtain a_B, so we must understand the definition of the standard state for the molarity scale. This is most easily seen in figure 4.1, in which a_B is plotted against c_B. This shows that the standard state for the molarity scale (solute standard state) is an extrapolated point. The dual requirement that $a_B = c_B$ in the limit

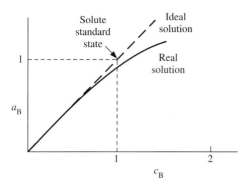

▲ FIGURE 4.1

Activity of a solute as a function of molarity for a real solution (solid curve) compared with that of an ideal solution (dashed line) extrapolated from very dilute conditions. The solute standard state lies on this extrapolated line, as shown.

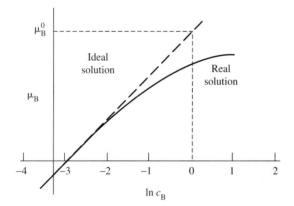

▲ FIGURE 4.2

Chemical potential of a solute plotted against the logarithm of molarity for a real solution (solid curve), compared with that of an ideal solution (dashed line) extrapolated from very dilute conditions.

as c_B approaches zero and that $a_B = 1$ in the standard state defines the extrapolated standard state. To obtain the free energy in the standard state, we measure μ_B as a function of $\ln c_B$ in a dilute solution and extrapolate linearly to $\ln c_B = 0$ ($c_B = 1$); we thus obtain μ_B^0, as shown in figure 4.2. The solute standard state is thus the state that has the properties of a very dilute solution extrapolated to a 1-M concentration. It is a hypothetical state rather than an actual solution that can be prepared.

If a solute is a strong electrolyte, in aqueous solution it dissociates completely into its component ions. Therefore, when we speak of the partial molar Gibbs free energy or chemical potential of NaCl in aqueous solution, we use the sum rule for the chemical potentials:

$$\mu_{NaCl} \equiv \mu_{Na^+} + \mu_{Cl^-}$$
$$= \mu_{Na^+}^0 + \mu_{Cl^-}^0 + RT \ln a_{Na^+} + RT \ln a_{Cl^-}$$
$$= \mu_{Na^+}^0 + \mu_{Cl^-}^0 + RT \ln(a_{Na^+} \cdot a_{Cl^-})$$

We can also express

$$\mu_{NaCl} = \mu^0_{NaCl} + RT \ln a_{NaCl} \tag{4.36}$$

in the usual way. A comparison of this expression with Eq. (4.30) shows that

$$a_{NaCl} = a_{Na^+} \cdot a_{Cl^-} \tag{4.37}$$

Similarly, for Na_2SO_4, the activity in aqueous solution is

$$a_{Na_2SO_4} = a^2_{Na^+} \cdot a_{SO_4^{2-}} \tag{4.38}$$

Molality

Another concentration unit that is frequently used is *molality, m,* with units of moles of solute per kilogram of solvent. The activity of B on the molality scale is

$$a_B = \gamma_B m_B \tag{4.39}$$

The discussion about molarity applies identically to molality. We have

$$\text{Dilute solution or ideal solution:} \quad a_B = m_B \tag{4.40}$$

$$\text{Real solution:} \qquad\qquad\qquad a_B = \gamma_B m_B \tag{4.41}$$

The standard state is an extrapolated state; μ^0_B is obtained by linearly extrapolating μ_B measured in dilute solution to $\ln m_B = 0$ ($m_B = 1$). Molality is used instead of molarity for the most accurate thermodynamic measurements. Because molalities are defined by weight, not volume, they can be measured quite accurately, and they do not depend on temperature. For dilute aqueous solutions, 1 L of solution contains about 1 kg of water and therefore the numerical values of molarity and molality are very close.

Biochemist's standard state

We have assumed in our discussion of activities and concentrations that we knew the concentration of each species involved in a system. For a molecule that dissociates in solution, this may be very difficult to determine. For example, a reaction may involve $H_2PO_4^-$; however, the species actually present in solution may include H_3PO_4, $H_2PO_4^-$, HPO_4^{2-}, and PO_4^{3-}. The distribution of these species will depend markedly on pH, so the concentration of $H_2PO_4^-$ may be difficult to specify. To simplify this situation, biochemists have chosen pH 7.0, which is near physiological pH, as their standard condition for a_{H^+}. This means that $a_{H^+} = 1$ for a concentration of $H^+ = 10^{-7}$ M. The activity of each other molecule is set equal to the *total concentration* of all species of that molecule at *pH 7.0.*

$$\text{Dilute solution:} \quad a_A = \sum_i^{\text{species}} c_{i,A} \qquad \text{(at pH 7.0)} \tag{4.42}$$

In our example, this sum over all species is the total concentration of phosphate added—that is, the concentration determined analytically. Knowledge of ionization constants is thus not needed, nor is it necessary to specify the concentration of each of the actual species involved in the reaction. When this biochemical standard state is used, the standard free energy is designated $\Delta G^{0\prime}$. $\Delta G^{0\prime}$ is the free energy change for a reaction at pH 7 when each product and reactant (except the H^+ ion) has a total 1-M concentration, but the solution is ideal. The equilibrium is actually measured in dilute solution, and the free energy is obtained by extrapolation to 1 M concentrations. The important difference between the biochemist's standard state and all others is that the equilibrium constant applies only at pH 7. For a reaction involving, for example, the hydrolysis of adenosine triphosphate (ATP), pH can be considered a variable analogous to temperature. The biochemist's standard state is then a practical and useful choice. The standard free energy and the equilibrium constant can be used directly from the tables at or near pH 7. For other pH's, one needs either to repeat the experiments to determine the new equilibrium concentrations or to use known ionization constants to calculate how the concentrations of reactive species depend on pH.

To illustrate the use of the biochemist's standard state, consider the hydrolysis of ethyl acetate to produce acetic acid and ethanol:

$$CH_3\overset{\overset{\displaystyle O}{\|}}{C}OCH_2CH_3 + H_2O \rightleftharpoons CH_3\overset{\overset{\displaystyle O}{\|}}{C}OH + CH_3CH_2OH$$

$$EtOAc + H_2O \rightleftharpoons HOAc + EtOH$$

The equilibrium constant is

$$K = \frac{(a_{HOAc})(a_{EtOH})}{(a_{EtOAc})(a_{H_2O})}$$

We could use the dilute solution standard states for ethanol, acetic acid, and ethyl acetate. This means their activities would be replaced by molarities in dilute solution. For water we use the pure liquid standard state and replace a_{H_2O} by 1 in dilute solution. These standard states allow us to calculate a standard free energy from the equilibrium constant, $\Delta G^0 = -RT \ln K$.

However, we may be interested in the reaction at other pH values because acetic acid is a weak acid and can dissociate into hydrogen ion and acetate; we want to consider the pH as an independent variable. We then write the reaction as

$$EtOAc + H_2O \rightleftharpoons OAc^- + H^+ + EtOH$$

$$K' = \frac{(a_{OAc^-})(a_{H^+})(a_{EtOH})}{(a_{EtOAc})(a_{H_2O})}$$

To use the biochemist's standard state, we measure the equilibrium at pH 7 and set $a_{H^+} = 1$, a_{OAc^-} = sum of the concentrations of OAc^- and HOAc, and the other activities are the same as before ($a_{H_2O} = 1$, a_{EtOH}, a_{EtOAc} = concentrations). The main difference thus is that we use $a_{H^+} = 1$ (not 10^{-7} M) and a_{OAc^-} represents the sum of the acetate species (OAc^- + HOAc). The calculated $\Delta G^{0'} = -RT \ln K'$ applies at pH 7.

Activity Coefficients of Ions

To simplify thermodynamic calculations, we often set activity coefficients equal to 1. For small uncharged molecules and for low concentrations of univalent ions in dilute aqueous solutions, this approximation may not be too bad. For multivalent ions such as Mg^{2+} or PO_4^{3-}, however, activity coefficients may be very different from 1, even at millimolar concentrations. In ionic solutions, the total number of positive charges is always equal to the total number of negative charges (electrical neutrality). Therefore, we cannot separately measure the activity coefficient of a positively or negatively charged ion; we can measure only the mean activity coefficient of an ion and its counterion. For HCl, $ZnSO_4$, or any 1–1 or 2–2 electrolyte, the mean ionic activity coefficient is

$$\gamma_\pm = (\gamma_+\gamma_-)^{1/2} \qquad (1\text{--}1, 2\text{--}2, \text{ or } n\text{--}n \text{ electrolyte}) \qquad \textbf{(4.43)}$$

For H_2SO_4 or any 1–2 electrolyte,

$$\gamma_\pm = (\gamma_+^2\gamma_-)^{1/3} \qquad (1\text{--}2 \text{ electrolyte}) \qquad \textbf{(4.44)}$$

For $LaCl_3$ or any 1–3 electrolyte,

$$\gamma_\pm = (\gamma_+\gamma_-^3)^{1/4} \qquad (1\text{--}3 \text{ electrolyte}) \qquad \textbf{(4.45)}$$

These equations are readily generalized to any type of salt. Figure 4.3 illustrates some measured activity coefficients for a 1–1 electrolyte (HCl), a 1–2 electrolyte (H_2SO_4), and a 2–2 electrolyte ($ZnSO_4$). The activity coefficients are much less than 1 for 0.1-M solutions; at concentrations above 1 M, the activity coefficients for some electrolytes increase and become even greater than 1.

To understand why activity coefficients are so different from 1 for electrolytes and why the ions with the larger charges have the smaller activity coefficients, we need to consider the interactions between the solutes. Activities are effective concentrations. In an HCl solution, the effective concentration of the H^+ ions is reduced by the surrounding Cl^- ions; the effective concentration of the Cl^- ions is reduced by the surrounding H^+ ions. Figure 4.3 shows that this decrease in effective concentrations of the ions is greater in H_2SO_4 solution, which has a doubly charged ion, SO_4^{2-}. Debye and Hückel (1923) calculated activity coefficients for individual ions using Coulomb's law for the interaction energy between charged particles. Their result applies only in

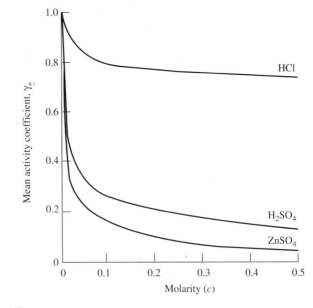

▲ FIGURE 4.3

Mean activity coefficient of various electrolytes at 25°C.

very dilute solution (less than 0.01 M), but it can provide a qualitative understanding of activity coefficients. For ions in water at 25°C,

$$\log \gamma_i = -0.509 Z_i^2 I^{1/2} \qquad (4.46)$$

where Z_i = charge on ion (± 1, ± 2, ± 3, etc.) and I = ionic strength = $\frac{1}{2}\Sigma_i\, c_i Z_i^2$. The numerical value, 0.509, depends on the temperature and the dielectric constant of the solvent. The dielectric constant is a measure of the strength of an electric field in a solvent compared to that in a vacuum; it is discussed in chapter 9.

The sum in the ionic-strength equation is over all ions in the solution. Ionic strength has been found to be a good measure of how different salts affect equilibria and reaction rates. The ionic strength rather than the concentration of the salt characterizes the effect; thus, the presence of a salt not directly involved in a reaction may nevertheless exert a significant effect. Note that 0.01 M NaCl, 0.0033 M MgCl$_2$, and 0.0025 M MgSO$_4$ all have ionic strength equal to 0.01.

The Equilibrium Constant and the Standard Gibbs Free Energies of the Reactants and Products

We now return to the relation between the equilibrium constant and the standard Gibbs free energies of the reactants and products. As mentioned earlier, an often-used method of determining free energies of reactions, particularly complicated biochemical reactions, is to measure the equilibrium constant for the reaction. The reverse is also true: Once standard free energies of reactants and products are obtained, it is easy to calculate the equilibrium constant and equilibrium concentrations of reactants and products. We repeat the appropriate equations (because they are so important) and give examples of how the various standard states are actually used.

$$\Delta G^0 = -RT \ln K \qquad (4.14)$$

Experimental determination of the equilibrium constant allows us to calculate the *standard* Gibbs free-energy change for the reaction, with all products and reactants in their standard states. The Gibbs free-energy change ΔG for the reaction at equilibrium is of course zero at constant temperature and pressure. The Gibbs free-energy change at arbitrary concentrations specified by Q is

$$\Delta G = \Delta G^0 + RT \ln Q \qquad (4.20)$$

We consider several reactions and see how the equilibrium constant and the standard Gibbs free energy depend on the choice of standard states. We begin with the dissociation of acetic acid, HOAc, into hydrogen ions, H$^+$, and acetate ions, OAc$^-$, in aqueous solutions:

$$\mathrm{HOAc}(aq) \rightleftharpoons \mathrm{H}^+(aq) + \mathrm{OAc}^-(aq)$$

The equilibrium constant is

$$K = \frac{(a_{\mathrm{H}^+})(a_{\mathrm{OAc}^-})}{(a_{\mathrm{HOAc}})} \qquad (4.47)$$

If we choose the molarity scale for all species, $a = \gamma c$ and

$$K = \frac{(c_{\mathrm{H}^+})(c_{\mathrm{OAc}^-})}{(c_{\mathrm{HOAc}})}\frac{\gamma_+\gamma_-}{\gamma_{\mathrm{HOAc}}} = K_c\frac{\gamma_+\gamma_-}{\gamma_{\mathrm{HOAc}}} = K_c\frac{\gamma_\pm^2}{\gamma_{\mathrm{HOAc}}}$$

$$K_c = \frac{(c_{\mathrm{H}^+})(c_{\mathrm{OAc}^-})}{(c_{\mathrm{HOAc}})} \qquad (4.48)$$

The activity coefficients γ_{HOAc}, γ_+, and γ_- are dependent on the concentrations of the solutes and approach 1 as the solution becomes very dilute. Therefore, for very dilute solutions, K and K_c are equal, and the experimental determination of one gives the other.

Before we elaborate further on the activity coefficients for more concentrated solutions, a comment here about the units of equilibrium constants may be helpful. For Eq. (4.47), it is clear that K is unitless because all activities are unitless. Thermodynamically speaking, K_c is also unitless because each concentration term is actually the concentration in M divided by 1 M, the chosen standard state, just like the case we discussed earlier for the formation of ammonia from nitrogen and hydrogen [see the section leading to Eq. (4.14)]. However, chemists and biochemists commonly express K_c in terms of the concentrations rather than the unitless ratios. In this practice, K_c would be associated with an apparent unit; for the example given above, K_c would have the unit of M. *Whenever this is done, we should view the apparent unit as a reminder of the chosen standard states of the reactants and products and keep in mind that K_c is actually a dimensionless quantity.*

Returning now to more concentrated solutions, we need to know the activity coefficients in order to calculate K_c from K, or vice versa. Although we cannot experimentally determine γ_+ and γ_- individually, a mean ionic activity coefficient $\gamma_\pm$, which for a 1–1 electrolyte like HOAc is equal to $(\gamma_+\gamma_-)^{1/2}$ [Eq. (4.43)], can be measured. The concentrations that require activity coefficients depend on the accuracy of the desired result and on the nature of the species in solution. As we have seen in the preceding section, ions are very nonideal, and in 0.1 M NaCl the mean ionic activity coefficient $\gamma_\pm$ is 0.78; in 0.01 M NaCl, it is 0.90. Therefore, 0.01-M solutions might be the upper limit for "dilute" solutions of 1–1 electrolytes.

From the thermodynamic equilibrium constant, we can calculate the standard free energy, ΔG^0 [Eq. (4.14)]:

$$HOAc(aq) \rightleftharpoons H^+(aq) + OAc^-(aq)$$

$$\Delta G^0 = \mu_{H^+}^0 + \mu_{OAc^-}^0 - \mu_{HOAc}^0$$

and the standard states of all species are the extrapolated 1-M hypothetical state—that is, 1-M concentration in water but with the properties of a very dilute solution. For the Gibbs free-energy change at any arbitrary concentration, we use Eq. (4.20). For example, for the reaction

$$HOAc(10^{-4}\ M, aq) \xrightarrow{\Delta G} H^+(10^{-4}\ M, aq) + OAc^-(10^{-4}\ M, aq)$$

the Gibbs free-energy change is

$$\Delta G = \Delta G^0 + RT \ln Q = \Delta G^0 + RT \ln \frac{(10^{-4})(10^{-4})}{10^{-4}}$$

$$= \Delta G^0 + RT \ln(10^{-4})$$

At 25°C this is

$$\Delta G = \Delta G^0 + (8.314)(298)\ln(10^{-4})$$

$$= \Delta G^0 - 22{,}820\ \text{J mol}^{-1}$$

This calculation tells us that dilution by a factor of 10^4 decreases the Gibbs free energy by 22,820 J mol^{-1} at 25°C, making the dissociation reaction that much more favorable. From Le Châtelier's principle, we expect that the lower the concentration, the more acetic acid dissociates; equation (4.20) enables us to quantitate the extent of this increase in dissociation.

A hydrolysis reaction illustrates the use of other standard states. Consider the hydrolysis of ethyl acetate:

$$H_2O + CH_3COOCH_2CH_3 \rightleftharpoons CH_3COOH + CH_3CH_2OH$$

$$H_2O + EtOAc \rightleftharpoons HOAc + EtOH$$

$$K = \frac{[a_{HOAc}][a_{EtOH}]}{[a_{H_2O}][a_{EtOAc}]}$$

We could choose the molarity standard state for all molecules, but the trouble with this choice is that in dilute aqueous solution the concentration of water would be near $55.5\,M$ $(1000/18)$ and the activity coefficient of the water would not be 1. A better choice would be the mole fraction for the water:

$$K = \frac{(c_{HOAc})(c_{EtOH})}{(X_{H_2O})(c_{EtOAc})} \cdot \frac{\gamma_{HOAc}\gamma_{EtOH}}{\gamma_{H_2O}\gamma_{EtOAc}}$$

In dilute aqueous solution, the mole fraction of H_2O approaches 1, and all the activity coefficients approach 1; hence, only concentrations appear in K. It should be clear that the numerical value of K will be very different, depending on our choice of standard state. Its value will depend on whether we use 55.5 or 1.0 for the activity of water. The corresponding standard free energies will also be different; they refer to different standard reaction conditions. The Gibbs free energy change ΔG for any particular concentration of reactants and products is a quantity that does not depend on choice of standard state.

Because one of the products, acetic acid, is ionizable, we might wish to choose the biochemist's standard state for HOAc. The concentration of the un-ionized species HOAc depends greatly on the total concentration of HOAc because of its dissociation into OAc^- and H^+. The equilibrium ratio of concentrations,

$$\frac{(c_{HOAc})(c_{EtOH})}{(c_{EtOAc})}$$

where c_{HOAc}, the concentration of the un-ionized species, will be difficult to measure, and be very dependent on the total concentration of HOAc. However, if we are mainly interested in the reaction near physiological conditions, we can simplify our problem by choosing the biochemist's standard state. We study the hydrolysis in a pH 7 buffer. We now find that the equilibrium ratio of concentrations, with c_{HOAc} equal the *total* concentration of HOAc, is nearly constant in dilute solution. The pH 7 buffer ensures that the acetic acid is essentially all in the form of acetate, and the HOAc concentration will be directly proportional to the acetate over a wide range of total concentration.

The decision among the possible standard states and equilibrium constants is simply to choose the most convenient for the system being studied.

Most of the time we will assume that the solutions are dilute enough so that we can set all activity coefficients equal to unity. However, we should realize that in concentrated solutions our calculations may be in error by factors of 2 or more because of activity coefficients. It is important to remember that the activity coefficient of a species depends on the concentrations of *all* species in solution. If one is studying the ionization constant of dilute acetic acid in water, activity coefficients of the ions are nearly 1; however, the addition of NaCl to 1-M concentration will have a strong effect on the activity coefficients of the ions in the dilute acid, even though NaCl does not play a direct role in the dissociation equilibrium. Later we will discuss methods of measuring activity coefficients.

Calculation of Equilibrium Concentrations: Ideal Solutions

One reason for studying equilibrium constants is to enable us to calculate concentrations at equilibrium. We may want to know how to make buffer solutions of any pH, how much product is obtained from a reaction, how much metal ion is bound in a complex, and so on. Some of these problems are very simple, but many of them require a computer for complete analysis. Here we discuss the general method, which in principle allows any equilibrium problem to be solved. We assume that we know the equilibrium constants for all of the equilibria involved, and we assume for the time being that all solutions are ideal; that is, we set all activity coefficients to 1.

The way most scientists solve equilibrium problems is to make intuitive approximations to simplify the arithmetic operations. The "intuition" comes either from having solved many similar problems or from a good memory of what we learned before. The method we present here is a general one; it applies when intuition fails on simple problems or when the problems become more difficult.

The usual requirement is to learn the concentrations of all species present in a mixture. The problem is essentially an algebraic one of finding simultaneous solutions for a number of equations. So the first requirement is to have as many equations as unknowns. The equations are equilibrium expressions, plus two types of conservation equations. The conservation equations are the ones we tend to use intuitively, but they are always included explicitly in our thinking. They are *conservation of mass*—the total mass of each element is not altered by any chemical reaction—and *conservation of charge*—the number of positive charges must always equal the number of negative charges in the mixture. Once the number of equations equals the number of unknowns, we look for approximations that allow a simple solution. A solution can always be obtained, however, even though the process is not necessarily simple.

Let's consider a few examples and see how the method works. Suppose a solution is prepared by adding c_A mol of acetic acid and c_S mol of sodium acetate to water to form 1 L of aqueous solution at 298 K. The solution will contain HOAc, OAc^-, Na^+, H^+, and OH^-. Even for this simple equilibrium, we have five species in addition to the solvent water. We therefore need five equations. We use brackets around a species to denote its concentration in molarity; the five equations are

$$\text{Mass balance:} \quad [Na^+] = c_S = \text{constant}$$

$$[HOAc] + [OAc^-] = c_A + c_S = \text{constant}$$

$$\text{Charge balance:} \quad [Na^+] + [H^+] = [OAc^-] + [OH^-]$$

$$\text{Equilibria:} \quad K_{HOAc} = \frac{[H^+][OAc^-]}{[HOAc]} = 1.8 \times 10^{-5}$$

$$K_{H_2O} = [H^+][OH^-] = 1.0 \times 10^{-14}$$

We are using the solute (1 M) standard state for H^+, OAc^-, HOAc, OH^-, and Na^+ and the mole fraction standard state for H_2O. These five equations allow us to solve any problem involving only these five species.

EXAMPLE 4.5

What are the concentrations of all species in pure water?

Solution

Mass balance: $\quad [Na^+] = 0, \quad [HOAc] = 0, \quad [OAc^-] = 0$

Charge balance: $\quad [H^+] = [OH^-]$

Equilibrium: $\quad [H^+][OH^-] = 1.0 \times 10^{-14}$

Therefore,

$$[H^+]^2 = 1.0 \times 10^{-14}$$

$$[H^+] = 1.0 \times 10^{-7}$$

$$[OH^-] = 1.0 \times 10^{-7}$$

EXAMPLE 4.6

What are the concentrations of all species in a 0.100 M HOAc solution?

Solution

Mass balance: $\quad [Na^+] = 0$

$$[HOAc] + [OAc^-] = 0.100$$

Charge balance: $\quad [H^+] = [OAc^-] + [OH^-]$

Equilibria: $\quad K_{HOAc} = \dfrac{[H^+][OAc^-]}{[HOAc]} = 1.80 \times 10^{-5}$

$$K_{H_2O} = [H^+][OH^-] = 1.00 \times 10^{-14}$$

Except for pure water or solutions close to pH 7, either $[H^+]$ or $[OH^-]$ will be negligible in the charge-balance equation. The solutions will be acidic or basic, so either $[OH^-]$ or $[H^+]$ can be ignored. Even for a solution at pH 6.5, the $[H^+]$ concentration is ten times the $[OH^-]$. We know that an acetic acid solution is acidic, so we ignore $[OH^-]$ in the charge-balance equation. Assuming that $[OAc^-] \gg [OH^-]$,

Charge balance: $\quad [H^+] \cong [OAc^-]$

Remember that we can sometimes ignore species in sums, but we can never ignore them in products. We can never ignore either $[H^+]$ or $[OH^-]$ in K_{H_2O}.

If we set $[H^+] = x$, we also have $[OAc^-] = x$ and $[HOAc] = 0.100 - x$:

$$K_A = \frac{(x)(x)}{0.100 - x} = 1.80 \times 10^{-5}$$

We can solve for x in the quadratic equation

$$x^2 + 1.8 \times 10^{-5}x - 1.8 \times 10^{-6} = 0$$

$$x = \frac{-1.8 \times 10^{-5} + \sqrt{3.24 \times 10^{-10} + 7.2 \times 10^{-6}}}{2}$$

$$= \frac{-1.8 \times 10^{-5} + 2.68 \times 10^{-3}}{2}$$

$$= 1.33 \times 10^{-3}$$

$$[H^+] = [OAc^-] = 1.3 \times 10^{-3}$$

$$[HOAc] = 0.100 - 1.33 \times 10^{-3} = 9.9 \times 10^{-2}$$

$$[OH^-] = \frac{1.0 \times 10^{-14}}{1.33 \times 10^{-3}} = 7.5 \times 10^{-12}$$

Note that our assumption that $[OAc^-] \gg [OH^-]$ is well verified.

Alternatively, we can solve the problem using successive approximations. The secret in solving by approximations comes in setting x equal to concentrations of species that are *not* the largest involved in the equilibrium. In this case, we have set $x = [H^+]$ or $[OAc^-]$. We do not set $x = [HOAc]$ because it is the species that we expect would have the largest concentration in 0.1-M acetic acid. Even before knowing the exact answer, we can guess that $[H^+]$ and $[OAc^-]$ will have to be much smaller than $0.100\ M$ to satisfy the equilibrium constant $K_{HOAc} = 1.8 \times 10^{-5}$.

As a first approximation when $x \ll 0.1$,

$$K_A = \frac{x^2}{0.1 - x} \cong \frac{x^2}{0.1} = 1.80 \times 10^{-5}$$

$$x \cong 1.34 \times 10^{-3}$$

Indeed, we see that our initial approximation that $x \ll 0.1$ is justified. To obtain a more precise answer, we can use this first solution to generate a second approximation:

$$\frac{x^2}{0.1 - 0.0013} = 1.80 \times 10^{-5}$$

$$x = 1.33 \times 10^{-3}$$

Clearly, no further approximations are needed.

EXAMPLE 4.7

What are the concentrations of all species in a $0.200\ M$ NaOAc solution?

Solution

NaOAc is the salt of a weak acid, and a hydrolysis reaction occurs when NaOAc is dissolved in water.

$$OAc^- + H_2O \rightleftharpoons HOAc + OH^-$$

Our general method gives

Mass balance: $[Na^+] = 0.200$

$$[HOAc] + [OAc^-] = 0.200$$

Charge balance: $[H^+] + [Na^+] = [OH^-] + [OAc^-]$

Equilibria: $K_{HOAc} = \dfrac{[H^+][OAc^-]}{[HOAc]} = 1.80 \times 10^{-5}$

$$K_{H_2O} = [H^+][OH^-] = 1.00 \times 10^{-14}$$

We recognize from the hydrolysis reaction that OH^- is produced. The solution will therefore be basic, and the term $[H^+]$ in the charge balance equation can be ignored. Furthermore, we know that $[Na^+] = 0.2000$.

Charge balance: $0.2000 = [OH^-] + [OAc^-]$

Comparing this with the mass-balance equation for $[HOAc]+[OAc^-]$, we see that

$$[HOAc] = [OH^-]$$

Now we set $x = [HOAc] = [OH^-]$ to facilitate making approximations. (The choice for x is between $[OAc^-]$ and $[HOAc] = [OH^-]$. The equilibrium expression for K_{HOAc} tells us that $[OAc^-]/[HOAc] \gg 1$ because $[H^+] \ll 10^{-7}$ for a basic solution.) Thus,

$$[OAc^-] = 0.200 - [HOAc] = 0.200 - x$$

$$[H^+] = \frac{10^{-14}}{x}$$

and

$$K_{HOAc} = \frac{(10^{-14}/x)(0.200 - x)}{x} = \frac{10^{-14}(0.200 - x)}{x^2} = 1.80 \times 10^{-5}$$

$$\frac{(0.200 - x)}{x^2} = 1.80 \times 10^{9}$$

We can solve the quadratic equation but it is easier to use successive approximation. As our first approximation, when $x \ll 0.200$,

$$\frac{0.200}{x^2} = 1.80 \times 10^{9}$$

$$x = 1.05 \times 10^{-5} \qquad \text{(consistent with assumption)}$$

No further approximations are needed:

$$[OH^-] = [HOAc] = 1.05 \times 10^{-5}$$

$$[H^+] = \frac{1.00 \times 10^{-14}}{1.05 \times 10^{-5}} = 9.5 \times 10^{-10}$$

$$[OAc^-] = 0.20$$

$$[Na^+] = 0.200$$

Note that $[Na^+] \gg [H^+]$, which justifies our early approximation in the charge-balance equation.

This general method works for any equilibrium. For more complicated examples, we must be sure that the equilibrium expressions are independent. In the preceding example, instead of the $K_A = K_{HOAc}$ and K_{H_2O}, we could have used K_B:

$$K_B = \frac{[HOAc][OH^-]}{[OAc^-]}$$

and K_{H_2O}, or K_B and K_A. However, we cannot count K_A, K_B, and K_{H_2O} as three independent equations; only two are independent because $K_A K_B$ always equals K_{H_2O}.

For buffer problems, the computations are usually simpler because both the acid and the salt (or the base and its corresponding salt) are present at an appreciable concentration. Furthermore, these concentrations are large compared with either $[H^+]$ or $[OH^-]$. In the case of a buffer involving acetic acid and sodium acetate, for example, we have

Mass balance: $[Na^+] = c_S = $ constant

$[HOAc] + [OAc^-] = c_A + c_S = $ constant

Charge balance: $[Na^+] + [H^+] = [OAc^-] + [OH^-]$

which becomes

$$[Na^+] = [OAc^-] = c_S$$

because $[H^+]$ and $[OH^-] \ll [Na^+]$ or $[OAc^-]$. Thus,

$$[HOAc] = c_A$$

and the equilibrium constant is

$$K_{HOAc} = \frac{[H^+][OAc^-]}{[HOAc]} = \frac{[H^+]c_S}{c_A}$$

This equation is often written in the form

$$pH = pK_A + \log \frac{c_S}{c_A} \tag{4.49}$$

where $pH = -\log[H^+]$
$pK_A = -\log K_A$

and it is sometimes called the *Henderson–Hasselbalch equation*. (Note that for real solutions, the proper definition is $pH = -\log a_{H^+}$.)

EXAMPLE 4.8

What amount of solid sodium acetate is needed to prepare a buffer at pH 5.00 from 1 L of 0.10 M acetic acid?

Solution

$$pH = pK_{HOAc} + \log \frac{c_S}{c_A}$$

$$5.00 = 4.75 + \log \frac{c_S}{0.10} = 5.75 + \log c_S$$

$$\log c_S = -0.75$$

$$c_S = 0.18 \text{ mol L}^{-1}$$

$$\text{wt of NaOAc} = (0.18 \text{ mol})(82.0 \text{ g mol}^{-1}) = 15 \text{ g}$$

Of course, this answer assumes that the solutions are ideal and that the addition of sodium acetate has a negligible effect on the volume. In practice, final adjustment of pH can be done by using a calibrated pH meter or other electrode system that measures a_{H^+} directly.

In the preceding examples, we have assumed that all activity coefficients of the solutes are close to 1. Similar calculations can be carried out, however, when the acitivity coefficients deviate substantially from 1. Example 4.9 illustrates such a case.

EXAMPLE 4.9

What are the concentrations of all species in a 0.200 M NaOAc solution if we take into account the mean ionic activity coefficients for the ions in the solution? The mean ionic activity coefficient for 1–1 electrolytes in this solution is $\gamma_\pm = 0.592$; we assume the activity coefficient of the uncharged HOAc is $\gamma_{HOAc} = 1$.

Solution

We have previously considered this problem in example 4.7 with the assumption that all activity coefficients are unity. The mass balance

$$[Na^+] = 0.200 \tag{1}$$

$$[HOAc] + [OAc^-] = 0.200 \tag{2}$$

and the charge balance

$$[H^+] + [Na^+] = [OH^-] + [OAc^-] \tag{3}$$

remain the same. The equilibrium constants K_{HOAc} and K_{H_2O} are

$$
\begin{aligned}
K_{HOAc} &= \frac{(a_{H^+})(a_{OAc^-})}{(a_{HOAc})} \\[6pt]
&= \frac{(c_{H^+})(c_{OAc^-})}{(c_{HOAc})} \frac{\gamma_+ \gamma_-}{\gamma_{HOAc}} \\[6pt]
&= \frac{[H^+][OAc^-]}{[HOAc]} \frac{\gamma_\pm^2}{\gamma_{HOAc}} \\[6pt]
&= 1.80 \times 10^{-5}
\end{aligned}
$$

$$
\begin{aligned}
K_{H_2O} &= \frac{(a_{H^+})(a_{OH^-})}{(a_{H_2O})} \\[6pt]
&= [H^+][OH^-]\gamma_\pm^2 \\[6pt]
&= 1.00 \times 10^{-14}
\end{aligned}
$$

Substituting $\gamma_\pm = 0.592$ we obtain

$$\frac{[H^+][OAc^-]}{[HOAc]} = \frac{1.80 \times 10^{-5}}{(0.592)^2} = 5.14 \times 10^{-5} \tag{4}$$

$$[H^+][OH^-] = \frac{1.00 \times 10^{-14}}{(0.592)^2} = 2.85 \times 10^{-14} \tag{5}$$

Solving Eqs. (1) through (5) the same way as described in example 4.7 gives

$$[Na^+] = 0.200 \ M$$

$$[OAc^-] = 0.200 \ M$$

$$[OH^-] = [HOAc] = 1.05 \times 10^{-5} \ M$$

$$[H^+] = 2.71 \times 10^{-9} \ M$$

The only change introduced by using activity coefficients is in $[H^+]$; previously (example 4.7) we obtained $[H^+] = 9.50 \times 10^{-10}$.

Temperature Dependence of the Equilibrium Constant

Once we know the equilibrium constant of a chemical reaction at one temperature, what can we say about it at another temperature? We show below that the temperature dependence of an equilibrium constant can be readily derived from the temperature dependence of the change in Gibbs free energy. From Eq. (3.39),

$$\frac{d\Delta G}{dT} = -\Delta S \qquad (P = \text{constant})$$

When all reactants and products are at their standard states, the above equation becomes

$$\frac{d\Delta G^0}{dT} = -\Delta S^0 \qquad (P = \text{constant}) \qquad \textbf{(4.50)}$$

Substituting $\Delta G^0 = -RT \ln K$ into the equation gives

$$-R \ln K - RT \frac{d \ln K}{dT} = -\Delta S^0$$

or

$$-RT \frac{d \ln K}{dT} = -\Delta S^0 + R \ln K$$

Multiplying both sides by T, we obtain

$$-RT^2 \frac{d \ln K}{dT} = -T\Delta S^0 + RT \ln K$$

$RT \ln K$ is equal to $-\Delta G^0$; therefore,

$$-RT^2 \frac{d \ln K}{dT} = -T\Delta S^0 - \Delta G^0$$

$$= -T\Delta S^0 - (\Delta H^0 - T\Delta S^0)$$

$$= -\Delta H^0 \qquad (P = \text{constant}) \qquad \textbf{(4.51)}$$

We can write Eq. (4.51) in an alternate form by noting that $d(1/T) = -dT/T^2$:

$$\frac{d \ln K}{d\left(\dfrac{1}{T}\right)} = -\frac{\Delta H^0}{R} \qquad (P = \text{constant})$$

or, in partial derivative form,

$$\left[\frac{\partial (\ln K)}{\partial \left(\dfrac{1}{T}\right)} \right]_P = -\frac{\Delta H^0}{R} \qquad\qquad \textbf{(4.52)}$$

Equation (4.52) is called the *van't Hoff equation*. Qualitatively, Eq. (4.52) tells us that if a reaction is exothermic under standard conditions (heat is given off, ΔH^0 is negative, and $-\Delta H^0$ is positive), then ln K and hence K itself increases with $1/T$. That is, as the temperature decreases and $1/T$ increases, ln K and therefore K becomes larger: An exothermic reaction is favored when the temperature is lowered. Conversely, an endothermic reaction is favored when the temperature is increased.

For example, we know from experience that a large amount of heat is generated upon mixing a 1-M acid solution (H^+) with 1-M base (OH^-) in the neutralization reaction to form water. The reverse reaction (the self-ionization of water) therefore absorbs heat:

$$H_2O(l) \longrightarrow H^+(aq) + OH^-(aq) \qquad \Delta H^0_{298} = 55.84 \text{ kJ}$$

Thus, we expect that increasing the temperature of water will result in an increase in K_w, the equilibrium constant for the self-ionization of water.

These predictions can of course also be made from Le Châtelier's principle. The usefulness of Eq. (4.52), however, goes beyond such qualitative conclusions. Quantitatively, if ΔH^0 is known as a function of temperature, then Eq. (4.52) can be used to calculate K at any temperature if K is known at one temperature. The equation can also be used to determine ΔH^0 from experimentally measured values of K at different temperatures. A plot of log K versus $1/T$ for the experimental values of K_w in the temperature range from 0° to 50°C is shown in figure 4.4. The plot gives a straight line, indicating that over the temperature range ΔH^0 is a constant. From Eq. (4.52), the slope of a plot of log K versus $1/T$ is $-\Delta H^0/2.303R$. The best-fit line through the data shown in Fig. 4.4 gives a value of $\Delta H^0 = 55.76$ kJ, which agrees well with the value 55.84 kJ calculated from the calorimetrically measured standard enthalpies of formation at 298 K. Measurements of equilibrium constants at different temperatures are widely used to determine ΔH^0 values from plots of ln K (or log K) versus $1/T$; the plot is known as a *van't Hoff plot*. If ΔH^0 changes

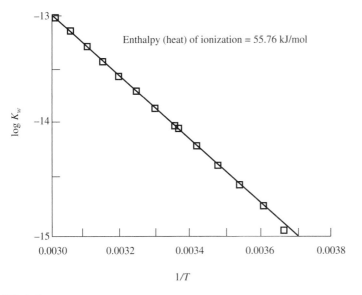

▲ FIGURE 4.4

A plot of the logarithm of the ionization constant of water K_w versus $1/T$. This is a van't Hoff plot. The slope is equal to $-\Delta H^0/2.303R$ [Eq. (4.52)]; the standard heat of ionization is thus found to be 55.76 kJ mol^{-1}. This is in excellent agreement with the value obtained from calorimetry.

with temperature, a van't Hoff plot will show some curvature. The slope will change with temperature, but the slope of a ln K versus $1/T$ plot at any temperature will equal $-\Delta H^0/R$ at that temperature (or $-\Delta H^0/2.303R$ if log K is plotted versus $1/T$).

When ΔH^0 is approximately a constant over the temperature range of interest, integration of Eq. (4.52) is straightforward:

$$\int_{K_1}^{K_2} d \ln K = -\frac{\Delta H^0}{R} \int_{T_1}^{T_2} d\frac{1}{T}$$

or

$$\ln \frac{K_2}{K_1} = -\frac{\Delta H^0}{R}\left(\frac{1}{T_2} - \frac{1}{T_1}\right) \qquad (\Delta H^0 = \text{constant}) \qquad \textbf{(4.53)}$$

This equation is often used to calculate an equilibrium constant K_2, at T_2 when K_1 and the standard enthalpy of the reaction, ΔH^0, are known. Alternatively, from the equilibrium constants measured at two different temperatures, one can calculate the enthalpy change.

EXAMPLE 4.10

The equilibrium constant for ionization of 4-aminopyridine is 1.35×10^{-10} at 0°C and 3.33×10^{-9} at 50°C. Calculate ΔG^0 at 0°C and 50°C as well as ΔH^0 and ΔS^0.

Solution

Values of ΔG^0 at the temperatures are readily calculated from Eq. (4.14):

$$\Delta G^0 = -RT \ln K$$

$$\Delta G^0(0°C) = -(8.314 \text{ J K}^{-1} \text{ mol}^{-1})(273 \text{ K}) \ln (1.35 \times 10^{-10})$$

$$= +51.58 \text{ kJ}$$

$$\Delta G^0(50°C) = -(8.314 \text{ J K}^{-1} \text{ mol}^{-1})(323 \text{ K}) \ln (3.33 \times 10^{-9})$$

$$= +52.42 \text{ kJ}$$

To obtain ΔH^0 and ΔS^0 from these data, we assume that ΔH^0 and ΔS^0 are independent of temperature. ΔH^0 or ΔS^0 can be first calculated from the data given, and the equation $\Delta G^0 = \Delta H^0 - T\Delta S^0$ can then be used to calculate the other quantity. If ΔH^0 is to be calculated first, we use equation (4.53):

$$\ln \frac{K_2}{K_1} = \ln \frac{1.35}{33.3} = -3.205$$

$$\Delta H^0 = -\frac{(8.314 \text{ J K}^{-1} \text{ mol}^{-1})(-3.205)}{5.67 \times 10^{-4} \text{ K}^{-1}}$$

$$= 47,000 \text{ J mol}^{-1} = 47.0 \text{ kJ mol}^{-1}$$

Because ΔH^0 is assumed to be a constant in obtaining Eq. (4.53), the calculated value is an average over the temperature range 0° and 50°C. The average value of ΔS^0 over the same temperature range is then obtained from

$$\Delta S^0 = \frac{\Delta H^0 - \Delta G^0}{T}$$

$$= \frac{47{,}000 - 51{,}580}{273}$$

$$= -16.78 \text{ J K}^{-1} \text{ mol}^{-1}$$

The same answer is obtained if values of ΔH^0 and ΔG^0 at $T = 323$ K are used:

$$\Delta S^0 = \frac{47{,}000 - 52{,}420}{323}$$

$$= -16.78 \text{ J K}^{-1} \text{ mol}^{-1}$$

We can also calculate first the average value of ΔS^0 from the ΔG^0 values at the two temperatures by integrating Eq. (4.50), assuming that ΔS^0 is a constant over the temperature range:

$$\int d\Delta G^0 = -\Delta S^0 \int dT \qquad\qquad \textbf{(4.50)}$$

$$\Delta G^0(T_2) - \Delta G^0(T_1) = -\Delta S^0(T_2 - T_1) \qquad\qquad \textbf{(4.54)}$$

From the calculated ΔG^0 values at the two temperatures, Eq. (4.54) gives

$$\Delta S^0 = -\frac{52.42 - 51.58}{323 - 273} \text{ kJ K}^{-1} \text{ mol}^{-1}$$

$$= -0.0168 \text{ kJ K}^{-1} \text{ mol}^{-1} = -16.8 \text{ J K}^{-1} \text{ mol}^{-1}$$

from which ΔH^0 can be obtained from the calculated ΔG^0 at either 323 K or 273 K:

$$\Delta H^0 = \Delta G^0 - T\Delta S^0$$

$$= 52{,}420 - 323 \times 16.8$$

$$= 46{,}994 \text{ J mol}^{-1}$$

$$= 46.99 \text{ kJ mol}^{-1}$$

$$= 51{,}580 - 273 \times 16.8$$

$$= 46{,}994 \text{ J mol}^{-1}$$

$$= 46.99 \text{ kJ mol}^{-1}$$

In calculating the temperature dependence of the equilibrium constant, the assumption that ΔH^0 is constant [Eq. (4.53)] is usually adequate for a small temperature range (less than 25°). For a larger temperature range, we should consider the temperature dependence of ΔH^0, which can be obtained from Eq. (2.43b) and the temperature-dependent heat capacity values, $C_P(T)$, for the reactants and products. Heat capacity data for many compounds are tabulated in a number of compilations of thermodynamic data.

Galvanic Cells

So far we have used free energies and presented tables of standard free energies without discussing the easiest and most accurate method of measuring free energies. We will show that if a chemical reaction can be made to occur reversibly in a galvanic cell, the maximum voltage of the cell is a direct measure of the Gibbs free energy of the reaction. Conversely, from the known Gibbs free-energy change of a reaction, we can calculate the maximum voltage that can be obtained when the reaction takes place in an electrochemical cell. As we will see, the use of galvanic cells in the measurements of Gibbs free energies is widely applicable to many types of chemical and biochemical reactions, as well as processes such as the solubility of a salt or concentration differences across osmotically active membranes.

For a reversible process at constant T and P, we have shown that the Gibbs free-energy change is equal to the useful work w_{rev}^* [Eq. (3.28)], that is, work the surroundings does on the system excluding the pressure–volume work. In a galvanic cell, w_{rev}^* is the reversible electric (rev elec) work:

$$-\Delta G = -w_{\text{rev elec}} \tag{4.55}$$

We use minus signs in the equation to emphasize that the *decrease* in Gibbs free energy is equal to the reversible electrical work done *by* the system on the surroundings. Electrical work is equal to the voltage times the current multiplied by the time [Eq. (2.8)]. Because current multiplied by time is charge, the electrical work done by a chemical reaction in a galvanic cell is the voltage of the cell multiplied by the amount of charge transferred in the reaction. For n electrons transferred, the Gibbs free energy is

$$\Delta G \text{ (electron volts)} = -n\mathscr{E} \tag{4.56}$$

where n = number of electrons involved in the reaction and $\mathscr{E}$ = reversible voltage of cell (the sign convention for a cell will be discussed shortly). An electron volt (eV) is a unit of energy. One mole of electrons moving 1 cm (centimeter) in a field of 1 V cm^{-1} acquires an energy of 96.485 kJ mol^{-1}.

$$\Delta G \text{ (kJ)} = -96.485n\mathscr{E} \tag{4.57}$$

The Gibbs free energy is also often expressed by

$$\Delta G \text{ (J)} = -n\mathscr{E}F \tag{4.58}$$

where F = faraday = 96,485 coulombs (C) mol^{-1} of electrons. When a reaction occurs spontaneously in a galvanic cell, the Gibbs free-energy change must be negative, and the voltage, by convention, is positive.

Because the reversible voltage is proportional to Gibbs free energy, it follows from $(\partial G/\partial T)_P = -S$ that the entropy change for the reaction in the cell depends on the temperature dependence of the voltage:

$$\Delta S \text{ (J K}^{-1}) = 96,485n\left(\frac{\partial\mathscr{E}}{\partial T}\right)_P \tag{4.59}$$

Also, from the relation that $\Delta H = \Delta G + T\Delta S$ at constant temperature,

$$\Delta H \text{ (kJ)} = 96,485n\left[-\mathscr{E} + T\left(\frac{\partial\mathscr{E}}{\partial T}\right)_P\right] \tag{4.60}$$

Therefore, the thermodynamic characteristics of the cell reaction can be obtained from measurements of the reversible voltage of a galvanic cell at several temperatures.

As an example, consider the reaction between $H_2(g)$ and $O_2(g)$ to produce $H_2O(l)$. The equation is

$$H_2(g) + \frac{1}{2}O_2(g) \longrightarrow H_2O(l)$$

For a galvanic cell, we need to make electrical contact with the chemical species involved. Platinum metal electrodes are usually used for this purpose because they are chemically inert and can be prepared to give a catalytic surface ("platinum black") that aids in attaining nearly reversible conditions. We also need ions present in water to conduct the electrical current. We can use "inert" ions that do not participate in this reaction, such as Na^+ and Cl^-, or we can use ions such as H^+ or OH^- that do. One such arrangement, which uses HCl as the electrolyte, is shown in figure 4.5. $H_2(g)$ is provided through a gas bubbler at one Pt electrode, and $O_2(g)$ is provided at a second Pt electrode. The electrodes, in contact with the bubbling gases, are placed in an aqueous solution of HCl. We use a porous barrier or a salt bridge between the two electrode compartments, which prevents mixing of the bulk solutions on the two sides yet allows diffusion of aqueous ions between the two sides. The reaction to form water is both highly spontaneous ($\Delta G < 0$) and exothermic ($\Delta H < 0$) under normal conditions.

In the galvanic cell shown in figure 4.5, the oxidation takes place at the left electrode:

Oxidation: $H_2(g, P_{H_2}) \longrightarrow 2H^+(c_{H^+}) + 2e^-$

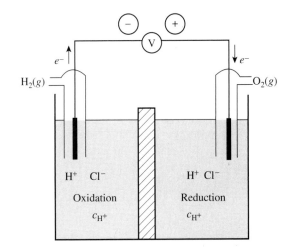

▲ FIGURE 4.5

A galvanic cell in which $H_2(g)$ and $O_2(g)$ form water by oxidizing the H_2 at one electrode and reducing the O_2 at the other electrode. The electrodes are inert conductors such as platinum. They allow transfer of the electrons through the external wires and the potentiometer; the direction of the flow of the electrons is shown. The electrical connection in the solution is provided by the ions and the semipermeable porous barrier (the crosshatched partition in the figure). The voltage measures the difference in chemical potentials of products and reactants—the ΔG of the reaction.

Thus, the left terminal is negative; electrons are being produced. The reduction takes place at the right electrode:

$$\text{Reduction:} \quad 2e^- + \frac{1}{2}O_2(g, P_{O_2}) + 2H^+(c_{H^+}) \longrightarrow H_2O(l)$$

Thus, the right terminal is positive; electrons are being removed. In addition to preventing diffusive mixing, the porous barrier maintains a path for electrical conduction (by ion diffusion). Such cells are often described using an abbreviated notation:

$$\text{Pt} \mid H_2(g, P_{H_2}) \mid H^+(c_{H^+}) \parallel H^+(c_{H^+}) \mid O_2(g, P_{O_2}) \mid \text{Pt}$$

Each vertical line represents a boundary between two different phases, and the double line represents the porous barrier or salt bridge. By convention, the electrode where the oxidation reaction (production of electrons) occurs is placed on the left and the electrode where reduction (removal of electrons) occurs is placed on the right. An easy way to remember this convention is to associate the r's: reduction, right.

The cell reaction for this cell is the sum of the two half reactions written above, balanced so that the electrons cancel in the cell reaction:

$$H_2(g) + \frac{1}{2}O_2(g) \longrightarrow H_2O(l)$$

Note that H^+ does not appear in the cell reaction for this cell.

To determine the Gibbs free-energy change for the reaction at the gas pressures provided, P_{H_2} and P_{O_2}, we use the measured reversible voltage of the cell in Eqs. (4.56) through (4.58) with $n = 2$. It is easy to measure the reversible voltage of the cell accurately with a potentiometer (represented by V in the diagram, which measures the voltage when the current flow approaches zero). The dependence of the reversible voltage on the partial pressures or concentrations of the reactants and products can be measured, and by extrapolation the reversible voltage for standard conditions can be determined. This provides the standard Gibbs free-energy change for the reaction.

Standard Electrode Potentials

Reversible voltages have been determined for many reactions at standard conditions. Each reaction of course includes both an oxidation and a reduction half-cell reaction. If we *define* the voltage of one half-cell reaction as zero, all others can be obtained relative to this arbitrary choice. Chemists have chosen the reduction of H^+ to H_2 gas under standard conditions to have a standard electrode potential of zero:

$$H^+(a = 1) + e^- \longrightarrow \frac{1}{2}H_2(g, P = 1 \text{ atm}) \qquad \mathscr{E}^0 = 0.000 \text{ V}$$

With this choice, the standard reduction potentials given in table 4.1 were obtained.

To calculate the standard potential (and therefore the standard free energy) for a chemical reaction from table 4.1, it is necessary to combine the appropriate potentials for the half-cell reactions. The value of $\mathscr{E}^0_{cell}$ is $\mathscr{E}^0$ for the reduction half-cell reaction minus $\mathscr{E}^0$ for the oxidation half-cell reaction. By convention for galvanic cells, this is equal to $\mathscr{E}^0$ for the right-hand side of

TABLE 4.1 Standard reduction electrode potentials at 25°C

Oxidant/Reductant	Electrode reaction	$\mathscr{E}^0$ (V)*	$\mathscr{E}^{0\prime}$ (V)† (pH 7)
Li^+/Li	$Li^+ + e^- \rightarrow Li$	−3.045	
Na^+/Na	$Na^+ + e^- \rightarrow Na$	−2.714	
Mg^{2+}/Mg	$Mg^{2+} + 2e^- \rightarrow Mg$	−2.363	
$OH^-/H_2/Pt$	$2H_2O + 2e^- \rightarrow H_2 + 2OH^-$	−0.8281	
Zn^{2+}/Zn	$Zn^{2+} + 2e^- \rightarrow Zn$	−0.7628	
Acetate/acetaldehyde	$OAc^- + 3H^+ + 2e^- \rightarrow CH_3CHO + H_2O$		−0.581
Fe^{2+}/Fe	$Fe^{2+} + 2e^- \rightarrow Fe$	−0.4402	
Gluconate/glucose	$C_6H_{11}O_7^- + 3H^+ + 2e^- \rightarrow C_6H_{12}O_6 + H_2O$		−0.44
Spinach ferredoxin	$Fd[Fe(III)] + e^- \rightarrow Fd[Fe(II)]$		−0.432
CO_2/formate	$CO_2 + 2H^+ + 2e^- \rightarrow HCO_2^- + H^+$	−0.20	−0.42
$NAD^+/NADH^\ddagger$	$NAD^+ + H^+ + 2e^- \rightarrow NADH$	−0.105	−0.320
Fe^{3+}/Fe	$Fe^{3+} + 3e^- \rightarrow Fe$	−0.036	
$H^+/H_2/Pt$	$2H^+ + 2e^- \rightarrow H_2$	0	−0.421
Acetoacetate/ β-hydroxybutyrate	$CH_3COCH_2CO_2^- + 2H^+ + 2e^- \rightarrow$ $CH_3CHOHCH_2CO_2^-$		−0.346
Mn hematoporphyrin IX	$He[Mn(III)] + e^- \rightarrow He[Mn(II)]$		−0.342
$NADP^+/NADPH$	$NADP^+ + H^+ + 2e^- \rightarrow NADPH$		−0.324
Horseradish peroxidase	$HRP[Fe(III)] + e^- \rightarrow HRP[Fe(II)]$		−0.271
$FAD/FADH_2^\S$	$FAD + 2H^+ + 2e^- \rightarrow FADH_2$		−0.219
Acetaldehyde/ethanol	$CH_3CHO + 2H^+ + 2e^- \rightarrow CH_3CH_2OH$		−0.197
Pyruvate/lactate	$CH_3COCO_2^- + 2H^+ + 2e^- \rightarrow$ $CH_3CHOHCO_2^-$		−0.18
Oxaloacetate/malate	$^-O_2CCOCH_2CO_2^- + 2H^+ + 2e^- \rightarrow$ $^-O_2CCHOHCH_2CO_2^-$		−0.166
Fumarate/succinate	$^-O_2CCH{=}CHCO_2^- + 2H^+ + 2e^- \rightarrow$ $^-O_2CCH_2CH_2CO_2^-$		+0.031
Myoglobin	$Mb[Fe(III)] + e^- \rightarrow Mb[Fe(II)]$		+0.046
Dehydroascorbate/ ascorbate	$C_6H_7O_7^- + 2H^+ + 2e^- \rightarrow C_6H_9O_7^-$		+0.058
Ubiquinone	$UQ + 2H^+ + 2e^- \rightarrow UQH_2$		+0.10
$AgCl/Ag/Cl^-$	$AgCl + e^- \rightarrow Ag + Cl^-$	+0.2223	
Calomel	$\frac{1}{2}Hg_2Cl_2 + e^- \rightarrow Hg + Cl^-$	+0.268	
Cytochrome c	$Cyt[Fe(III)] + e^- \rightarrow Cyt[Fe(II)]$		+0.254
Cu^{2+}/Cu	$Cu^{2+} + 2e^- \rightarrow Cu$	+0.337	
$I_2/I^-/Pt$	$I_2 + 2e^- \rightarrow 2I^-$	+0.5355	
$O_2/H_2O_2/Pt$	$O_2 + 2H^+ + 2e^- \rightarrow H_2O_2$	+0.69	+0.295
$Fe^{3+}/Fe^{2+}/Pt$	$Fe^{3+} + e^- \rightarrow Fe^{2+}$	+0.771	
Ag^+/Ag	$Ag^+ + e^- \rightarrow Ag$	+0.799	
$NO_3^-/NO_2^-/Pt$	$NO_3^- + 2H^+ + 2e^- \rightarrow NO_2^- + H_2O$	+0.94	+0.421
$Br_2/Br^-/Pt$	$Br_2 + 2e^- \rightarrow 2Br^-$	+1.087	
$O_2/H_2O/Pt$	$O_2 + 4H^+ + 4e^- \rightarrow 2H_2O$	+1.229	+0.816
$Cl_2/Cl^-/Pt$	$Cl_2 + 2e^- \rightarrow 2Cl^-$	+1.359	
$Mn^{3+}/Mn^{2+}/Pt$	$Mn^{3+} + e^- \rightarrow Mn^{2+}$	+1.4	
$Ce^{4+}/Ce^{3+}/Pt$	$Ce^{4+} + e^- \rightarrow Ce^{3+}$	+1.61	
$F_2/F^-/Pt$	$F_2 + 2e^- \rightarrow 2F^-$	+2.87	

*$\mathscr{E}^0$ refers to the solute standard state with unit activity for all species.
†$\mathscr{E}^{0\prime}$ refers to the biochemist's standard state with pH 7.
‡NAD^+ is nicotinamide adenine dinucleotide.
§FAD is flavin adenine dinucleotide.

the cell minus $\mathcal{E}^0$ for the left-hand side of the cell. For example, let's consider the reaction

$$Zn(s) + Cu^{2+}(a = 1) \longrightarrow Zn^{2+}(a = 1) + Cu(s)$$

The half-cell reactions and their potentials are

$$Cu^{2+}(a = 1) + 2e^- \longrightarrow Cu(s) \qquad \mathcal{E}^0 = +0.337 \text{ V}$$

$$Zn(s) \longrightarrow Zn^{2+}(a = 1) + 2e^- \qquad \mathcal{E}^0 = -(-0.763 \text{ V})$$

$$\overline{\mathcal{E}^0_{cell} = +1.100 \text{ V}}$$

The standard reversible voltage is +1.100 V, and the standard Gibbs free energy is obtained from Eqs. (4.56) through (4.58):

$$\Delta G^0 = -2.20 \text{ eV} = -212 \text{ kJ}$$

Because table 4.1 gives reduction potentials, the sign of the potential must be changed for the oxidation half-cell reaction. However, the voltages must *not* be multiplied by 2 even though there are two electrons involved. The voltage of the cell is independent of the number of electrons involved. The Gibbs free energy does depend on the number of electrons, and therefore n appears explicitly in Eqs. (4.56) through (4.58). Free energy is extensive, but voltage is intensive.

If the reaction as written does not occur spontaneously, the calculated voltage is found to be negative, and the standard Gibbs free energy is positive.

Concentration Dependence of $\mathcal{E}$

Table 4.1 is analogous to a table of standard free energies. The standard free energies of a great many chemical reactions can be obtained from the table. What if we are interested in the free energies at arbitrary concentrations? We know the relation between standard Gibbs free energy and Gibbs free energy:

$$\Delta G = \Delta G^0 + RT \ln Q \qquad (4.20)$$

Substituting Eq. (4.58) for ΔG and ΔG^0, we obtain the *Nernst equation*:

$$\mathcal{E} = \mathcal{E}^0 - \frac{RT}{nF} \ln Q \qquad (4.61)$$

For the reaction

$$aA + bB \longrightarrow cC + dD$$

$$Q = \frac{(a_C)^c (a_D)^d}{(a_A)^a (a_B)^b}$$

At 25°C the Nernst equation can be written explicitly as

$$\mathcal{E} = \mathcal{E}^0 - \frac{0.0591}{n} \log Q \qquad (4.62)$$

The Nernst equation can be used to calculate the voltage and Gibbs free energy for any concentrations of reactants and products in a cell. At equilibrium $\Delta G = 0$; therefore, $\mathcal{E} = 0$, also. Thus, the standard potential gives the equilibrium constant for the reaction in the cell:

$$\mathcal{E}^0 = \frac{RT}{nF} \ln K \qquad (4.63)$$

We can solve for K and, at 25°C,

$$K = 10^{n\mathcal{E}^0/0.0591} \qquad (4.64)$$

Biochemical Applications of Thermodynamics

Knowledge of equilibrium constants, free energies, and their dependence on concentration and temperature is very important in biochemistry and biology. Here we will discuss a few simple examples. Table 4.2 and figure 4.6

TABLE 4.2 Ionization constants and enthalpies (heats) of ionization at 25°C			
Compound	Ionizing species*	pK†	ΔH^0 (kJ mol^{-1})
Acetic acid	$HOAc \rightarrow H^+ + OAc^-$	4.76	−0.25
Adenosine	$AH^+ \rightarrow A + H^+$	3.55	15.9
5′-Adenylic acid			
(adenosine phosphate)	$pAH^+ \rightarrow pA + H^+$	3.7	17.6
	$pA \rightarrow pA^- + H^+$	6.4	−7.5
	$pA^- \rightarrow pA^{2-} + H^+$	13.1	45.6
Adenosine triphosphate (ATP)	$pppAH^+ \rightarrow pppA + H^+$	4.0	15.5
	$pppA \rightarrow pppA^- + H^+$	7.0	−5.0
Alanine	$^+H_3NRCOOH \rightarrow {^+}H_3NRCOO^- + H^+$	2.35	2.9
	$^+H_3NRCOO^- \rightarrow H_2NRCOO^- + H^+$	9.83	45.2
Ammonia	$NH_4^+ \rightarrow NH_3 + H^+$	9.24	52.2
Aspartic acid	$^+H_3NRHCOOH \rightarrow {^+}H_3NRHCOO^- + H^+$	2.05	7.5
	$^+H_3NRHCOO^- \rightarrow {^+}H_3NR^-COO^- + H^+$	3.87	4.2
	$^+H_3NR^-COO^- \rightarrow H_2NR^-COO^- + H^+$	10.3	38.5
Carbonic acid	$H_2CO_3 \rightarrow HCO_3^- + H^+$	6.36	7.66
	$HCO_3^- \rightarrow CO_3^{2-} + H^+$	10.24	14.85
Fumaric acid	$R(COOH)_2 \rightarrow {^-}OOCRCOOH + H^+$	3.10	0.4
	$^-OOCRCOOH \rightarrow {^-}OOCRCOO^- + H^+$	4.6	−2.9
Glycine	$^+H_3NRCOOH \rightarrow {^+}H_3NRCOO^- + H^+$	2.35	3.92
	$^+H_3NRCOO^- \rightarrow H_2NRCOO^- + H^+$	9.78	44.2
Histidine	$^+H_3NRH^+COOH \rightarrow {^+}H_3NRH^+COO^- + H^+$	1.82	—
	$^+H_3NRH^+COO^- \rightarrow {^+}H_3NRCOO^- + H^+$	6.00	29.9
	$^+H_3NRCOO^- \rightarrow H_2NRCOO^- + H^+$	9.16	43.6
Hydrocyanic acid	$HCN \rightarrow CN^- + H^+$	9.21	43.5
Lysine	$^+H_3NRH^+COOH \rightarrow {^+}H_3NRH^+COO^- + H^+$	2.18	0.30
	$^+H_3NRH^+COO^- \rightarrow {^+}H_3NRCOO^- + H^+$	8.95	12.8
	$^+H_3NRCOO^- \rightarrow H_2NRCOO^- + H^+$	10.53	11.6
Phenol	$ROH \rightarrow RO^- + H^+$	9.98	23.6
Phosphoric acid	$H_3PO_4 \rightarrow H_2PO_4^- + H^+$	2.15	−7.95
	$H_2PO_4^- \rightarrow HPO_4^{2-} + H^+$	7.21	4.15
	$HPO_4^{2-} \rightarrow PO_4^{3-} + H^+$	12.35	14.7
Pyruvic acid	$RCOOH \rightarrow RCOO^- + H^+$	2.49	12.1
Tyrosine	$^+H_3NRHCOOH \rightarrow {^+}H_3NRHCOO^- + H^+$	2.20	—
	$^+H_3NRHCOO^- \rightarrow {^+}H_3NR^-COO^- + H^+$	9.11	—
	$^+H_3NR^-COO^- \rightarrow H_2NR^-COO^- + H^+$	10.05	25.1
Water	$H_2O \rightarrow OH^- + H^+$	14.00	55.84

*The ionizable groups corresponding to the pK's in the table are shown in figure 4.6. Glycine is just like alanine with the methyl group replaced by a hydrogen. 5′-Adenylic acid is shown in the figure; adenosine is the same except that the phosphate is lacking. In ATP the single phosphate (OPO_3H_2) of adenylic acid is replaced by a triphosphate $(OPO_2)^-(OPO_2)^-(OPO_3H_2)$.
†pK = −log K. The K values are thermodynamic equilibrium constants on the molarity scale. $\Delta G^0 = -RT \ln K$.
Source: Reprinted from *Handbook of Biochemistry and Molecular Biology,* 3d ed., *Physical and Chemical Data,* Vol. 1 (Boca Raton, Fla.: CRC Press, 1976); D. D. Wagman et al., eds., 1982, "The NBS Tables of Chemical Thermodynamic Properties," *J. Phys. Chem. Ref. Data* 11, Suppl. 2.

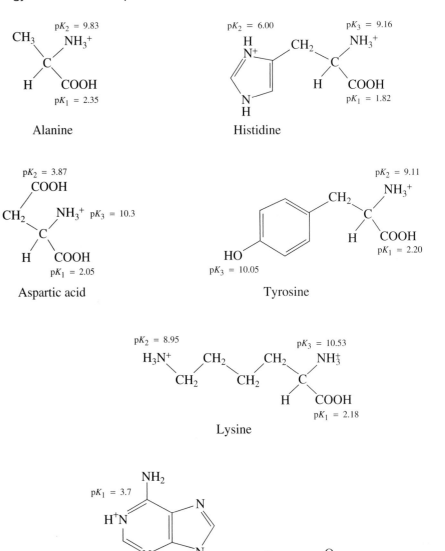

▲ FIGURE 4.6

The pK values from table 4.2 of several amino acids and of a nucleotide; p$K = -\log K$.

show equilibrium constants at 25°C and heats of ionization for various acids. A great deal of information about the acids is summarized in a small space. For example, some biochemical systems (such as human beings) work best at 37°C (98.5°F). Table 4.2 provides $\Delta H°$ values that allow calculation of pK's at 37°C from the 25°C data. We see that acetic acid is one of the few acids listed whose pK is the same at 37°C as at 25°C, because its $\Delta H \sim 0$. We usually consider that a neutral aqueous solution has a pH of 7, but this is true only at 25°C.

EXAMPLE 4.11

What is the pH of pure water at 37°C?

Solution

The necessary equation is

$$\ln \frac{K_2}{K_1} = -\frac{\Delta H^0}{R}\left(\frac{1}{T_2} - \frac{1}{T_1}\right) \tag{4.53}$$

The ionization constant of water is 1.00×10^{-14} at 25°C, its $\Delta H^0 = 55.84$ kJ (see table 4.2), $K_1 = 10^{-14}$, $T_1 = 298$, and $T_2 = 310$. Then,

$$\ln \frac{K_2}{10^{-14}} = \frac{-55,840}{8.314}\left(\frac{1}{310} - \frac{1}{298}\right)$$

$$= +0.87$$

$$\frac{K_2}{10^{-14}} = e^{+0.87} = 2.4$$

$$K_2 = 2.4 \times 10^{-14}$$

This is the ionization constant of water at 37°C. The H^+ and OH^- in pure water is surely low enough so that the activities of the ions equal concentrations: $[H^+] = [OH^-]$. At 37°C,

$$[H^+][OH^-] = 2.4 \times 10^{-14}$$

$$[H^+] = 1.55 \times 10^{-7} \, M$$

$$pH = -\log[H^+]$$

$$= 6.81$$

EXAMPLE 4.12

A buffer is prepared by adding 26.8 mL of 0.200 M HCl to 50.0 mL of 0.200 M tris-(hydroxymethyl)aminomethane, a weak base known as Tris and widely used as a buffer by biochemists. Tris has a pK = 8.3 at 20°C: Its pK changes with temperature with a temperature coefficient of $\Delta pK/\Delta T = -0.029$ K^{-1}. The mixture is then diluted with pure water to a total volume of 200 mL. (a) What is the pH of the buffer solution at 20°C? (b) What is the "buffer capacity" of this solution, expressed as the change in pH that would occur upon the addition of 1 mmol of strong acid or base to 1 L of the buffer? (c) What would be the pH of the buffer solution of part (a) at 37°C?

Solution

The equilibrium involved for Tris buffer is

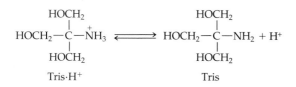

Tris·H⁺   Tris

a. Addition of the HCl solution (26.8 mL of 0.200 M HCl contains 5.36 mmol of HCl) to the Tris solution (50.0 mL of 0.200 M Tris contains 10.00 mmol of Tris) causes the conversion of 5.36 mmol of Tris to Tris$\cdot$H$^+$. Thus, the concentrations in the final 200 mL of buffer solution are

$$[\text{Tris}\cdot\text{H}^+] = \frac{5.36 \text{ mmol}}{200 \text{ mL}} = 26.8 \times 10^{-3} \, M$$

$$[\text{Tris}] = \frac{(10.00 - 5.36) \text{ mmol}}{200 \text{ mL}} = 23.2 \times 10^{-3} \, M$$

Using Eq. (4.49)

$$\text{pH} = 8.3 + \log \frac{23.2 \times 10^{-3}}{26.8 \times 10^{-3}} = 8.3 - 0.063 = 8.2$$

If a more precise pH is desired, final adjustment of the buffer can be accomplished by the addition of a small volume of one or the other stock solution, monitoring the buffer solution with an accurate (calibrated) pH meter.

b. The *buffer capacity* is a measure of the resistance to change in pH upon addition of H$^+$ or OH$^-$. Pure water has no buffer capacity. Addition of 1 mmol of HCl (or NaOH) to 1 L of water changes the pH from 7 to 3 (or 11). However, for 1 L of the buffer solution of part (a), addition of 1.0 mmol of HCl, for example, would result in the following changes

$$[\text{Tris}\cdot\text{H}^+] = (26.8 + 1.0) \times 10^{-3} = 27.8 \times 10^{-3} \, M$$

$$[\text{Tris}] = (23.2 - 1.0) \times 10^{-3} = 22.2 \times 10^{-3} \, M$$

Thus,

$$\text{pH} = 8.3 + \log \frac{22.2 \times 10^{-3}}{27.8 \times 10^{-3}} = 8.3 - 0.097 = 8.2$$

The change is only $\Delta\text{pH} = 0.097 - 0.063 = 0.034$ pH units. The buffer is still nominally at pH 8.2. This illustrates the ability of buffer solutions to resist changes in pH when, for example, H$^+$ is produced or consumed in metabolic reactions. Note that the buffer capacity is a direct function of the buffer concentration. This 50 mM Tris buffer solution has about ten times the buffer capacity of a 5 mM buffer solution at pH 8.2.

c. The pK for Tris decreases by 0.029 per degree (K) rise in temperature. Thus, at 37°C

$$\text{p}K_{37} = \text{p}K_{20} - (0.029)(37 - 20) = 8.3 - 0.49 = 7.8$$

and the pH of the solution prepared as described above would be

$$\text{pH}_{37} = \text{p}K_{37} - 0.063 = 7.7$$

This is a very significant change with temperature and should be kept in mind in all biochemical investigations. The temperature coefficient can be related to an enthalpy change upon dissociation, as tabulated for many weak acids in table 4.2. Beginning with Eq. (4.53), we can readily derive the equation

$$\Delta H_T^0 = \frac{2.303 R T_1 T_2}{T_1 - T_2} \log \frac{K_2}{K_1} = \frac{2.303 R T_1 T_2}{T_2 - T_1} (\text{p}K_1 - \text{p}K_2)$$

For Tris,

$$\Delta H_{298}^0 = \frac{(2.303)(8.314 \text{ J K}^{-1} \text{ mol}^{-1})(298 \text{ K})(299 \text{ K})}{1 \text{ K}} (+0.029)$$

$$= 49 \text{ kJ mol}^{-1}$$

From the pK values given in table 4.2 or those that we calculate at other temperatures, we can calculate the activities and concentrations of various ionized species present in solution. For example, consider histidine in solution at 25°C. The equilibria are

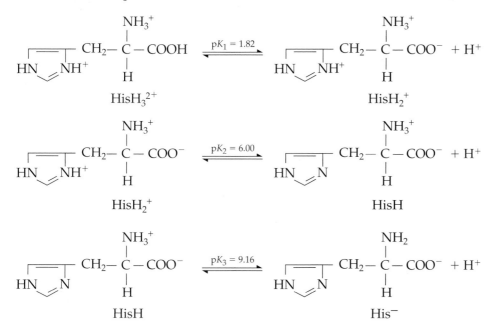

We often need to know the concentration of each species at a given pH. Each ionic species has different chemical reactivity and physical properties; therefore, the properties of the solution will depend on the concentration of each species.

EXAMPLE 4.13

Calculate the concentration of each species in a 0.10-M solution of histidine at pH 7.

Solution

The three equilibrium expressions are

$$K_1 = 1.51 \times 10^{-2} = \frac{[H^+][HisH_2^+]}{[HisH_3^{2+}]}$$

$$K_2 = 1.00 \times 10^{-6} = \frac{[H^+][HisH]}{[HisH_2^+]}$$

$$K_3 = 6.92 \times 10^{-10} = \frac{[H^+][His^-]}{[HisH]}$$

We assume that activities are equal to concentrations. If necessary, we could measure or estimate activity coefficients to obtain more accurate concentrations. The mass-balance equation is

$$[HisH_3^{2+}] + [HisH_2^+] + [HisH] + [His^-] = 0.100 \ M$$

We now have four equations and four unknowns, so we can solve for the unknowns. As often happens, it will be convenient to solve by successive approximations. First, we find the ratios of species from the three equilibrium constants and the value of $[H^+] = 10^{-7} \ M$.

$$\frac{[\text{HisH}_2^+]}{[\text{HisH}_3^{2+}]} = 1.51 \times 10^5$$

$$\frac{[\text{HisH}]}{[\text{HisH}_2^+]} = 10.0$$

$$\frac{[\text{His}^-]}{[\text{HisH}]} = 6.92 \times 10^{-3}$$

From the ratios of species, we see that a good first approximation will be to ignore $[\text{HisH}_3^{2+}]$ and $[\text{His}^-]$ in the mass-balance equation. That is, we need consider only the two species involved in the equilibrium whose pK's are closest to the pH of the solution.

The first-approximation mass balance is, assuming that $[\text{HisH}_3^{2+}] + [\text{His}^-] \ll [\text{HisH}] + [\text{HisH}_2^+]$,

$$[\text{HisH}] + [\text{HisH}_2^+] = 0.100 \ M$$

but

$$[\text{HisH}] = [\text{HisH}_2^+] \cdot (10.0)$$

$$[\text{HisH}_2^+](10.0) + [\text{HisH}_2^+] = 0.100$$

$$[\text{HisH}_2^+] = \frac{0.100}{11.0} = 9.1 \times 10^{-3} \ M$$

$$[\text{HisH}] = 10[\text{HisH}_2^+] = 9.1 \times 10^{-2} \ M$$

$$[\text{His}^-] = (6.92 \times 10^{-3})(9.1 \times 10^{-2}) = 6.3 \times 10^{-4} \ M$$

$$[\text{HisH}_3^{2+}] = \frac{9.1 \times 10^{-3}}{1.51 \times 10^5} = 6.0 \times 10^{-8} \ M$$

We have ignored activity coefficients, so the concentrations are at best accurate to $\pm 10\%$. If the second approximation does not change values by more than 10%, we can stop at the first approximation.

The second-approximation mass balance is

$$[\text{HisH}] + \text{HisH}_2^+] = 0.100 - [\text{His}^-] - [\text{HisH}_3^{2+}]$$

$$= 0.0994 \ M$$

The first approximation is good enough; the concentrations are

$$[\text{HisH}_2^+] = 9.1 \times 10^{-3} \ M$$

$$[\text{HisH}] = 9.1 \times 10^{-2} \ M$$

$$[\text{His}^-] = 6.3 \times 10^{-4} \ M$$

$$[\text{HisH}_3^{2+}] = 6.0 \times 10^{-8} \ M$$

Note that the sum of the last two species is much less than the first two.

Using the methods illustrated in example 4.12 we calculated the concentrations of all histidine species as a function of pH as shown in figure 4.7. The pK's correspond to the pH values where two species cross—where their concentrations are equal. Note that the concentration of each species reaches a maximum near 0.1 M (the total concentration of all histidine species).

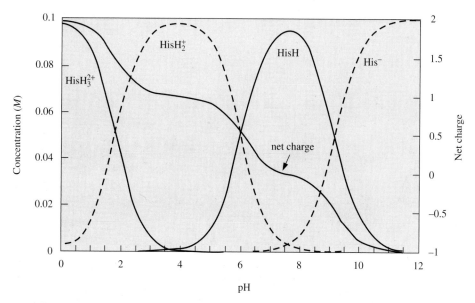

▲ FIGURE 4.7

The concentrations of histidine species as a function of pH; the values were calculated as shown in example 4.12. Note the sharp change in concentration of species near their pK's (1.82, 6.00, 9.16), and the result that no more than two histidine species are present in significant concentrations at each pH. The net charge on an average histidine molecule is shown at each pH. The net charge decreases in steps from +2 to +1 (pH 4) to 0 (pH 8) to −1. This is analogous to what happens to a protein molecule as a function of pH. It will have a net positive charge at low pH, a zero charge at its isoelectric point, and a negative charge at high pH.

These methods can of course be used to calculate the concentrations of species needed to produce a buffer of a given pH and salt concentration. For common buffers, the algebra has already been done, and buffer tables can be found in such handbooks as the *Handbook of Biochemistry and Molecular Biology* (CRC Press, Boca Raton, Florida).

Thermodynamics of Metabolism

Metabolism refers to the process by which cells use energy from their environment to synthesize the building blocks of their macromolecules. Hundreds of reactions are involved, and the path to a single product can involve many intermediates. One of the advantages of thermodynamic analysis is that we can focus on the beginning and the end of any process and not worry about the steps in-between. However, we can also learn about the driving force for each of the steps.

Figure 4.8 presents a scheme for glycolysis, which is the metabolic process that converts glucose into pyruvate, and produces adenosine triphosphate, ATP. In aerobic organisms, pyruvate then enters the mitochondria, where it is completely oxidized to CO_2 and H_2O in the citric acid cycle. These reactions produce additional ATP. Under less aerobic conditions, in muscle during active exercise, pyruvate is converted to lactate. Under anaerobic conditions in yeast, pyruvate is transformed into ethanol.

The overall oxidative reaction (combustion) of glucose with $O_2(g)$ to form $CO_2(g)$ and $H_2O(l)$ is associated with a large negative Gibbs free-energy

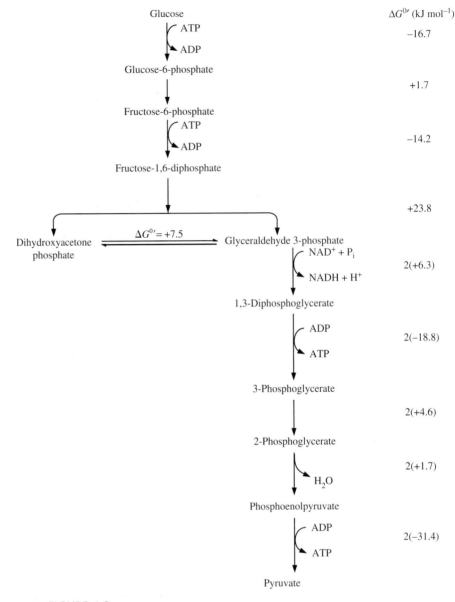

▲ FIGURE 4.8

The glycolytic path for the conversion of glucose to pyruvate in cells. The $\Delta G^{0\prime}$ for reactions from glyceraldehyde-3-phosphate to pyruvate are multiplied by 2 because 2 mol of these species are formed per mole of glucose.

change. Using ΔG_f^0 values at 298 K from tables A.5–7 in the appendix, we obtain for the reaction

$$C_6H_{12}O_6(s) + 6O_2(g) \longrightarrow 6CO_2(g) + 6H_2O(l) \qquad (\Delta G_{298}^0 = -2878.41 \text{ kJ})$$

α-D-Glucose

Furthermore, most of this free-energy change arises from the enthalpy contribution, $\Delta H_{298}^0 = -2801.6$ kJ, and relatively little from the entropy, $-T\Delta S_{298}^0 = -77.2$ kJ. Thus, when glucose is burned directly in oxygen, nearly all the

TABLE 4.3   Standard free energies of reaction at 25°C, pH 7 for steps in the metabolism of glucose*

Reaction	$\Delta G^{0\prime}$ (kJ mol^{-1})
D-Glucose + ATP → D-glucose-6-phosphate + ADP	−16.7
D-Glucose-6-phosphate → D-fructose-6-phosphate	+1.7
D-Fructose-6-phosphate + ATP → D-fructose-1,6-diphosphate + ADP	−14.2
Fructose-1,6-diphosphate → dihydroxyacetone phosphate + glyceraldehyde-3-phosphate	+23.8
Dihydroxyacetone phosphate → glyceraldehyde-3-phosphate	+7.5
Glyceraldehyde-3-phosphate + phosphate + NAD$^+$ → 1,3-diphosphoglycerate + NADH +H$^+$	+6.3
1,3-Diphosphoglycerate + ADP → 3-phosphoglycerate + ATP	−18.8
3-Phosphoglycerate → 2-phosphoglycerate	+4.6
2-Phosphoglycerate → phosphoenolypyruvate + H$_2$O	+1.7
2-Phosphoenolpyruvate + ADP → pyruvate + ATP	−31.4
Pyruvate + NADH + H$^+$ → lactate + NAD$^+$	−25.1
Pyruvate → acetaldehyde + CO$_2$	−19.8
Acetaldehyde + NADH +H$^+$ → ethanol +NAD$^+$	−23.7

*An important reaction in many of these steps is the hydrolysis of ATP: ATP + H$_2$O → ADP + phosphate. $\Delta G^{0\prime} = -31.0$ kJ mol^{-1}. $\Delta H^{0\prime} = 24.3$ kJ mol^{-1}.
Source: C. K. Mathews and K. E. van Holde, 1990, *Biochemistry* (Redwood City, Calif.: Benjamin/Cummings).

chemical potential is released as heat. By contrast, in living cells a significant fraction of the chemical potential is retained as chemical bond energy in the form of ATP and other synthesized molecules.

The free energies at 25°C and pH 7 for the steps in the metabolic degradation (catabolism) of glucose are given in table 4.3 and illustrated in figure 4.8. The biochemist's standard state is used in this table. This means that the pH is 7, the activity of water is equal to 1, metabolites are solutes in aqueous solutions, and the activities of all solutes are replaced by their total concentrations in mol L^{-1}. For example, ATP in the reaction refers to the sum of all ionized species of ATP present at pH 7. Standard states of all solutes have a total concentration equal to 1 M, but the solution is ideal.

The first step in the metabolism of glucose (table 4.3) is the formation of glucose-6-phosphate. The direct reaction of glucose with phosphate has a relatively large positive standard free-energy change and will not occur significantly under physiological conditions.

$$\text{glucose + phosphate} \longrightarrow \text{glucose-6-phosphate} + H_2O \qquad \textbf{(1)}$$

$$(\Delta G^{0\prime} = +14.3 \text{ kJ})$$

In the first step of glycolysis, however, this reaction is coupled to the hydrolysis of ATP to form adenosine diphosphate, ADP, which is thermodynamically favorable.

$$\text{ATP} + H_2O \longrightarrow \text{ADP + phosphate} \qquad (\Delta G^{0\prime} = -31.0 \text{ kJ}) \qquad \textbf{(2)}$$

The sum of these two reactions has a negative standard free energy, and that process occurs spontaneously. Adding (1) and (2), we obtain

$$\text{glucose} + \text{ATP} \longrightarrow \text{glucose-6-phosphate} + \text{ADP} \quad \textbf{(3)}$$

$$(\Delta G^{0\prime} = -16.7 \text{ kJ})$$

as shown in table 4.3. In this manner the strongly spontaneous hydrolysis of ATP is coupled to the otherwise nonspontaneous glucose phosphorylation. This reaction is typical of the role played by ATP in metabolism.

All reactions in the glycolytic path are enzyme catalyzed; the first step is catalyzed by the enzyme hexokinase. The enzyme speeds the reaction and serves to control the glycolytic path in response to the cell's demand for metabolic energy, but the enzyme does not affect the thermodynamics of the process. The free-energy changes are associated with differences in chemical potential between reactants and products; they do not indicate how fast the processes occur.

The second step in the metabolism of glucose is the isomerization of glucose-6-phosphate to fructose-6-phosphate catalyzed by the enzyme phosphoglucose isomerase. This reaction is also nonspontaneous in the standard state, but the positive free-energy change is relatively small, $\Delta G^{0\prime} = +1.7$ kJ mol^{-1}. The equilibrium constant associated with this free-energy change at 25°C is 0.50, which means that significant amounts of product are formed under glycolytic conditions. Furthermore, the glycolytic process is drawn on by the third step in the process, which is the attachment of a second phosphate group to fructose-6-phosphate to form fructose-1,6-diphosphate. This step, like the first one, is driven by phosphate transfer from ATP and has a $\Delta G^{0\prime}$ that is strongly negative (-14.2 kJ mol^{-1}).

In the fourth step in the glycolytic path, the six-carbon sugar fructose-1,6-diphosphate is split into two three-carbon compounds, dihydroxyacetone phosphate and glyceraldehyde-3-phosphate, catalyzed by the enzyme aldolase. This reaction has a large positive free-energy change,

fructose-1,6-diphosphate $\longrightarrow$

$$\text{dihydroxyacetone phosphate} + \text{glyceraldehyde-3-phosphate} \quad \textbf{(4)}$$

$$(\Delta G^{0\prime}_{298} = +23.8 \text{ kJ})$$

which would make it very nonspontaneous under standard conditions. But the concentrations of these metabolites in the cell are actually much less than 1 M, and this results in a very small value for the reaction quotient, $Q \approx 10^{-4}$, for this reaction. The effect of such a small reaction quotient is to cause the actual free energy, ΔG, to be significantly less positive. At 37°C each factor of 0.1 in the reaction quotient decreases the free energy by about 6 kJ mol^{-1} ($RT \ln 0.1 = -5.9$ kJ mol^{-1}), so the effective value for the free-energy change under actual metabolic conditions is $\Delta G \approx 0$. This is a direct consequence of the fact that in reaction (4) there are more product species than reactant species—all at low concentrations. Le Châtelier's principle tells us that a decrease in overall concentration for such a reaction will favor products. Even though the system is distinctly not at equilibrium under physiological conditions, that is the direction in which the spontaneous process is driven under low-concentration conditions. An example of this behavior was presented in example 4.2.

Dihydroxyacetone phosphate is in equilibrium with glyceraldehyde-3-phosphate. The free-energy change for the reaction

dihydroxyacetone phosphate $\rightleftharpoons$ glyceraldehyde-3-phosphate

$(\Delta G_{298}^{0\prime} = +7.5 \text{ kJ})$

gives an equilibrium constant $K = 0.05$. Nevertheless, the next reactions in the glycolytic path drain the pool of glyceraldehyde-3-phosphate sufficiently that essentially all dihydroxyacetone phosphate becomes converted to glyceraldehyde-3-phosphate in the continuation of glycolysis.

The next step results in the simultaneous oxidation and phosphorylation of glyceraldehyde-3-phosphate to 1,3-diphosphoglycerate. The oxidation is carried out by the biological oxidant nicotinamide adenine dinucleotide (NAD), and the phosphate is taken up directly from the pool of "inorganic" phosphate, P_i:

glyceraldehyde-3-phosphate $+ NAD^+ + P_i \longrightarrow$

1,3-diphosphoglycerate $+ NADH + H^+$

$(\Delta G_{298}^{0\prime} = +6.3 \text{ kJ})$

Although this reaction has a positive standard free-energy change, the process continues because of the next step in glycolysis, which is the transfer of one of the phosphates of 1,3-diphosphoglycerate to ADP to generate ATP and 3-phosphoglycerate. This reaction has a strongly negative standard free-energy change:

1,3-diphosphoglycerate $+ ADP \longrightarrow$ 3-phosphoglycerate $+ ATP$

$(\Delta G_{298}^{0\prime} = -18.8 \text{ kJ})$

The remaining three steps on the path to pyruvate involve two isomerization reactions and a final step in which the remaining phosphate is transferred to ADP to form ATP. No new principles are involved in these steps. The overall process by which one molecule of glucose is metabolized all the way to pyruvate can be summarized by the reaction

glucose $+ 2NAD^+ + 2ADP + 2P_i \longrightarrow$

2pyruvate $+ 2NADH + 2ATP + 2H^+ + 2H_2O$

$(\Delta G_{298}^{0\prime} = -80.6 \text{ kJ})$

In the overall glycolytic process to form pyruvate from glucose, there is a net formation of 2 mol of ATP and 2 mol of the physiological reductant NADH per mole of glucose consumed. In this way a significant fraction of the available chemical potential is retained as chemical energy for use in other biosynthetic or energy-requiring reactions. For example, under strenuous exercise in muscle cells, NADH serves to reduce pyruvate to lactate according to the reaction

pyruvate $+ NADH + H^+ \longrightarrow$ lactate $+ NAD^+$ $(\Delta G_{298}^{0\prime} = -25.1 \text{ kJ})$

EXAMPLE 4.14

The enzyme aldolase catalyzes the conversion of fructose-1,6-diphosphate (FDP) to dihydroxyacetone phosphate (DHAP) and glyceraldehyde-3-phosphate (G3P).

Under physiological conditions in red blood cells (erythrocytes), the concentrations of these species are [FDP] = 35 μM, [DHAP] = 130 μM, and [G3P] = 15 μM. Will the conversion occur spontaneously under these conditions?

Solution

The standard free-energy change for this reaction is

$$FDP \longrightarrow DHAP + G3P \qquad (\Delta G^{0\prime}_{298} = +23.8 \text{ kJ})$$

and

$$Q = \frac{[DHAP][G3P]}{[FDP]} = \frac{(130 \times 10^{-6})(15 \times 10^{-6})}{(35 \times 10^{-6})}$$

$$= 55.7 \times 10^{-6}$$

$$\Delta G = +23{,}800 + (8.314)(298) \ln (55.7 \times 10^{-6})$$

$$= +23{,}800 - 24{,}300$$

$$= -500 \text{ J} = -0.5 \text{ kJ}$$

Under the actual conditions in the cell, the free energy of the reaction is negative, and the reaction will occur spontaneously.

Biological Redox Reactions

Oxidation–reduction reactions are essential for energy storage and conversion in biological organisms. For example, the pyruvate produced as a product of glycolysis (figure 4.8) undergoes oxidative decarboxylation to form acetyl coenzyme A, which then enters the *citric acid cycle.* In the citric acid cycle, the acetate is transferred to oxaloacetate to form the six-carbon molecule, citrate. Citrate then undergoes a series of at least eight reactions that involve progressive oxidation of two of the carbon atoms to CO_2 and return of the remaining four-carbon portion as oxaloacetate for reentry into the cycle. Coupled to the oxidation of the acetate to CO_2 is the reduction of 3 mol of NAD^+ to NADH and the production of 1 mol of reduced flavin adenine dinucleotide ($FADH_2$). This reducing power generated in the citric acid cycle is subsequently used for biosynthetic reactions and for the formation of ATP by oxidative phosphorylation (respiration) in mitochondria.

The step that completes the citric acid cycle is the oxidation of malate to oxaloacetate coupled with the reduction of NAD^+ to NADH by the enzyme malate dehydrogenase:

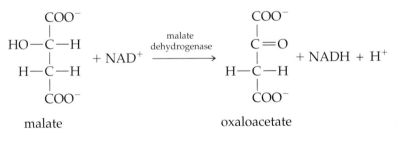

To calculate the $\Delta G^{0\prime}$ associated with this reaction, we can use $\mathscr{E}^{0\prime}$ values for the corresponding redox half-reactions listed in table 4.1:

$$\mathscr{E}^{0\prime} \text{ (volts)}$$

$$NAD^+ + H^+ + 2e^- \longrightarrow NADH \qquad -0.320$$

$$\text{Oxaloacetate} + 2H^+ + 2e^- \longrightarrow \text{malate} \qquad -0.166$$

$$\mathscr{E}^{0\prime} = -0.320 - (-0.166) = -0.154 \text{ V}$$

$$\Delta G^{0\prime} = -(2)(96.485)(-0.154) = 29.7 \text{ kJ}$$

This step in the cycle is not spontaneous under standard conditions. However, it will become spontaneous under conditions where the demand for NADH is high; that is, when $[NADH]/[NAD^+]$ is low. Under these conditions, the malate will undergo conversion to oxaloacetate, which serves to turn on the citric acid cycle. This is an example of metabolic regulation whereby supply and demand are kept in reasonable balance.

A principal source of cellular ATP is the process of *oxidative phosphorylation* (respiratory electron transport) carried out in mitochondria.*

This is a series of electron transfer reactions catalyzed by three membrane-associated enzyme complexes whereby NADH is oxidized, ultimately by O_2. Each complex contributes to the formation of a transmembrane proton gradient that is coupled to the formation of ATP. The overall redox reaction is

$$2NADH + 2H^+ + O_2(g) \longrightarrow 2NAD^+ + 2H_2O$$

It is useful to analyze the thermodynamics of this process in terms of each of the three enzyme complexes.

NADH-Q Reductase

NADH, which is a soluble reductant produced in the citric acid cycle and elsewhere, is imported into the mitochondria. It undergoes oxidation at the enzyme complex NADH-Q reductase, which contains several reducible groups. Initially, NADH reduces the flavin, (FMN), which in turn reduces an iron–sulfur (FeS) center. Finally, the electrons are passed on to reduce ubiquinone (UQ). We can sketch the coupled oxidation–reduction reactions catalyzed by this complex as follows:

NADH ⤸⤷ FMN ⤸⤷ FeS$_{red}$ ⤸⤷ UQ

NAD$^+$ ⤸⤷ FMNH$_2$ ⤸⤷ FeS$_{ox}$ ⤸⤷ UQH$_2$

NADH-Q reductase

The UQH_2 is then released to go on to the next complex. In the course of this process, H^+ ions are discharged outside the mitochondrial membranes into an intermembrane space, and H^+ is taken up from the internal space. This is the origin of one contribution to the pH gradient that results in ATP formation.

The net reaction for this portion of the electron transfer (per two electrons) is

$$NADH + H^+ + UQ \longrightarrow NAD^+ + UQH_2$$

$$\mathscr{E}^{0\prime} = 0.10 - (-0.32) = 0.42 \text{ V}$$

$$\Delta G^{0\prime} = -(2)(96.485)(0.42) = -81 \text{ kJ}$$

---

*Mitochondria are found in eukaryotic cells. They are membrane-surrounded structures in which oxidative metabolism occurs.

Cytochrome Reductase

UQH_2 diffuses to a second mitochondrial membrane-bound complex, cytochrome reductase. UQH_2 releases its electrons to an FeS center in the enzyme complex, which passes them one at a time to water-soluble cytochrome c. For the overall reaction,

$$QH_2 + 2Cyt\, c_{ox} \longrightarrow Q + 2H^+ + 2Cyt\, c_{red}$$

$$\mathscr{E}^{0\prime} = 0.254 - 0.10 = 0.154\ V$$

$$\Delta G^{0\prime} = -(2)(96.485)(0.154) = -29.7\ kJ$$

Cytochrome c Oxidase

The third mitochondrial membrane complex, cytochrome c oxidase, receives electrons one at a time from Cyt c_{red}, and transmits them to molecular oxygen, which requires four electrons overall to become reduced to two water molecules. The cytochrome c oxidase complex accomplishes this one-electron to four-electron transfer through the mediation of two heme centers. Each heme is associated with a copper ion near the heme iron. The net reaction (per two electrons transferred) is

$$2Cyt\, c_{red} + \frac{1}{2}O_2 + 2H^+ \longrightarrow 2Cyt\, c_{ox} + H_2O$$

$$\mathscr{E}^{0\prime} = 0.816 - 0.254 = 0.562\ V$$

$$\Delta G^{0\prime} = -(2)(96.485)(0.562) = -108.5\ kJ$$

For each of these three stages of mitochondrial electron transport, there is a strongly negative value of $\Delta G^{0\prime}$. This free energy is not dissipated, however. A portion of it is retained as chemical potential associated with a transmembrane proton gradient, which in turn provides the proton motive force that drives the formation of ATP. We will examine the mechanism and the thermodynamics of this coupling in chapter 5.

Double Strand Formation in Nucleic Acids

The formation of a double-stranded helix from two single strands of DNA is crucial in identifying and locating specific sequences of nucleotides in a DNA. A probe oligonucleotide (a single strand of DNA typically containing 20 to 40 nucleotides) is added to the single-stranded target DNA to form a double helix with its complementary sequence. The double helix has Watson–Crick complementary base pairs in which adenine (A) pairs with thymine (T) and guanine (G) pairs with cytosine (C). This technique is used to locate genes in chromosomes, to diagnose infectious or genetic diseases, and to amplify DNA by the polymerase chain reaction (PCR). In PCR amplification, the reaction mixture contains a minute amount of single- or double-stranded DNA to be amplified, a pair of oligonucleotides called the *primers*, a DNA polymerase stable at a temperature high enough to denature the DNA in its double-stranded form, and deoxynucleoside triphosphates that the polymerase uses in DNA synthesis. Samples for PCR are placed in an instrument programmed to cycle the incubation temperature through various settings. In each cycle, the temperature is first raised to denature double-stranded DNA, and then lowered to allow the binding of one primer to one DNA strand and the other primer to the complementary strand. The DNA polymerase then extends the 3′ ends of the bound primers to copy the DNA strands. Following this extension, a second cycle is started. As each cycle doubles the number of DNA molecules, rapid amplification can be obtained.

Knowledge of the equilibrium constant for formation of a DNA double strand of any sequence at any temperature is of great interest. Systematic studies were done by mixing complementary single strands in 1 M NaCl at pH 7 (Allawai and Santa Lucia 1997). The amount of double helix—and thus the equilibrium constant—was determined from the decrease in ultraviolet light absorbance when a DNA double helix forms (see chapter 10). The standard free-energy changes were calculated from the equilibrium constants; calorimetric measurements provided the enthalpy changes.

When a series of similar reactions is studied, one looks for simple patterns in the data. The obvious question to ask for formation of DNA double strands is, Does the thermodynamics just depend on the number of $A \cdot T$ base pairs and the number of $G \cdot C$ base pairs formed? The answer is no; this simple assumption does not fit the data. The next approximation is to assume that the sequence is important only through the base-pair nearest neighbors. This approximation is valid. We can rationalize it by noting that the main interaction involved in forming a double strand is hydrogen bonding between base pairs, plus London–van der Waals interaction between neighboring (stacked) base pairs. Thus, ΔG^0, ΔH^0, ΔS^0—and also K as a function of temperature—can be calculated from ten Watson–Crick nearest-neighbor parameters, plus parameters for the formation of the first base pair of the double strand as shown in table 4.4. The general expression is

$$\Delta G^0 = \Delta G^0(\text{initiation}) + \Sigma\, \Delta G^0(\text{nearest neighbors}) \qquad (4.65)$$

and similar expressions for ΔH^0 and ΔS^0. Consider the reaction

5′–ATAGCA–3′

$+$ $\rightleftharpoons$ 5′–ATAGCA–3′
                                 $\cdot\ \cdot\ \cdot\ \cdot\ \cdot\ \cdot$
                                 3′–TATCGT–5′

5′–TGCTAT–3′

The double strand contains antiparallel complementary strands; the structures of the molecules are shown in figures 3.6 and 3.7 and table A.9 in the appendix. To calculate the thermodynamic values, we use Eq. 4.65. The ΔG^0 (initiation) is positive—unfavorable. It mainly represents the loss of entropy for bringing the two strands together; the enthalpy of forming the first base pair is small. The ΔG^0 (nearest neighbors) are all negative—favorable; they characterize the hydrogen bonding and stacking interactions of adding successive base pairs. It does not matter which base pair is assumed to form first. However, it is convenient to start at the left-hand end for counting nearest neighbors. For the example given, the result is

$$\Delta G^0 = \Delta G^0(\text{initiation}) + \Delta G^0\!\begin{pmatrix} AT \\ \cdot\ \cdot \\ TA \end{pmatrix} + \Delta G^0\!\begin{pmatrix} TA \\ \cdot\ \cdot \\ AT \end{pmatrix}$$

$$+\ \Delta G^0\!\begin{pmatrix} AG \\ \cdot\ \cdot \\ TC \end{pmatrix} + \Delta G^0\!\begin{pmatrix} GC \\ \cdot\ \cdot \\ CG \end{pmatrix} + \Delta G^0\!\begin{pmatrix} CA \\ \cdot\ \cdot \\ GT \end{pmatrix}$$

TABLE 4.4 Thermodynamic parameters* for calculating double strand stability in DNA (pH 7, 1 M NaCl)

	ΔG^0 (kJ mol^{-1}) at 37°C	ΔH^0 (kJ mol^{-1})	ΔS^0 (J K^{-1} mol^{-1})
$5'-A \quad \longrightarrow \quad -A-A$ $3'-T \qquad\qquad\quad -T-T$	−4.2	−33.1	−93.0
$5'-A \quad \longrightarrow \quad -A-T$ $3'-T \qquad\qquad\quad -T-A$	−3.7	−30.2	−85.4
$5'-T \quad \longrightarrow \quad -T-A$ $3'-A \qquad\qquad\quad -A-T$	−2.4	−30.2	−89.2
$5'-A \quad \longrightarrow \quad -A-C$ $3'-T \qquad\qquad\quad -T-G$	−6.0	−35.2	−93.8
$5'-C \quad \longrightarrow \quad -C-A$ $3'-G \qquad\qquad\quad -G-T$	−6.0	−35.6	−95.0
$5'-A \quad \longrightarrow \quad -A-G$ $3'-T \qquad\qquad\quad -T-C$	−5.4	−32.7	−87.9
$5'-G \quad \longrightarrow \quad -G-A$ $3'-C \qquad\qquad\quad -C-T$	−5.4	−34.3	−93.0
$5'-C \quad \longrightarrow \quad -C-G$ $3'-G \qquad\qquad\quad -G-C$	−9.1	−44.4	−113.9
$5'-G \quad \longrightarrow \quad -G-C$ $3'-C \qquad\qquad\quad -C-G$	−9.3	−41.0	−102.2
$5'-C \quad \longrightarrow \quad -C-C$ $3'-G \qquad\qquad\quad -G-G$	−7.7	−33.5	−83.3

Initiation: The bringing together of two strands to form a duplex involves a loss of entropy and an unfavorable free energy: ΔG^0(initiation) = +8.1 kJ mol^{-1}, ΔS^0(initiation) = −23.4 J K^{-1} mol^{-1}, ΔH^0(initiation) = +0.8 kJ mol^{-1}.
*Data are from H. T. Allawai and J. SantaLucia Jr., 1997, *Biochemistry* 36, 10581–10594.

Using Table 4.4,

$$\Delta G^0 \text{ (kJ)} = 8.1 - 3.7 - 2.4 - 5.4 - 9.3 - 6.0 = -18.7 \text{ kJ}$$

$$(K_{310} = e^{-\Delta G^0/RT} = e^{-18,700/(8.314)(310)} = 1416)$$

For equilibrium constants at other temperatures, calculate ΔH^0 and ΔS^0 from table 4.4; then ΔG^0 can be calculated at any temperature by assuming ΔH^0 and ΔS^0 are independent of temperature: $\Delta G_T^0 = \Delta H^0 - T\Delta S^0$.

Ionic Effect on Protein–Nucleic Acid Interactions

Macromolecules often carry many charged groups. A double-stranded DNA, for example, has two phosphate groups every 3.4 Å along the molecule, with each phosphate carrying a negative charge. In aqueous solution, the negatively charged phosphate groups of the polyelectrolyte interact with positive-charged small ions that are present. We call these small ions *counterions*. For simplicity, we consider a DNA solution containing only Na^+ as the counterions.

The Na^+ counterions interact with DNA in two ways. Because of the very large negative-charge density of the DNA, there is a direct condensation of a fraction of the positive counterions on the polyelectrolyte. The remaining unneutralized negative charges are screened by a mobile ion atmosphere (Manning 1979). We let ψ be the effective fraction of each phosphate charge neutralized by Na^+ ions condensed on the DNA. When a protein L binds to a particular site of the DNA, a number of Na^+ originally bound to the DNA are displaced. If L interacts with n phosphate groups of the DNA, $n\psi$ of Na^+ originally condensed on the DNA are released. We express the reaction as

$$P + L \underset{}{\overset{K}{\rightleftharpoons}} PL + n\psi Na^+$$

The equilibrium constant K, which is independent of the Na^+ concentration in the solution, is

$$K = \frac{(a_{PL})(a_{Na^+})^{n\psi}}{(a_P)(a_L)}$$

$$\cong \frac{[PL][Na^+]^{n\psi}}{[P][L]}$$

where [L] = concentration of protein, M
    [P] = concentration of DNA, mol phosphate L^{-1}
    [PL] = concentration of complex, M

The concentration quotient

$$\frac{[PL]}{[P][L]} \equiv K_{obs}$$

is usually measured in media containing different amounts of Na^+. We note that

$$K_{obs} = K[Na^+]^{-n\psi}$$

Clearly, the observed apparent equilibrium constant K_{obs} will be strongly dependent on $[Na^+]$ if $n\psi$ is large. For double-stranded DNA, the value of ψ is near 0.88. Therefore, from a plot of $\log K_{obs}$ versus $\log [Na^+]$, n can be calculated from the slope; this tells us how many phosphate groups on the DNA the protein L interacts with (see problem 29 and Record et al. 1981).

Summary
Chemical Potential (Partial Molar Gibbs Free Energy)

Gibbs free energy as a function of temperature, pressure, and the composition variables

$$G = G(T, P, n_A, n_B, n_C, \ldots) \tag{4.1}$$

where $n_A, n_B, n_C, \ldots$ = the number of moles of A, B, C, and so on.

The chemical potential (partial molar Gibbs free energy) of a substance A

$$\mu_A \equiv \left(\frac{\partial G}{\partial n_A} \right)_{T,P,n_j \neq n_A} \tag{3.54}$$

The sum rule for chemical potentials

At constant T and P, the Gibbs free energy of a system is equal to the sum of the products of the number of moles of each component times its chemical potential:

$$G = n_A\mu_A + n_B\mu_B + \cdots \tag{4.5}$$

Chemical potentials of reactants and products at equilibrium

For a chemical reaction,

$$aA + bB \longrightarrow cC + dD$$

at equilibrium

$$[(a\mu_A + b\mu_B) - (c\mu_C + d\mu_D)] = 0$$

or

$$a\mu_A + b\mu_B = c\mu_C + d\mu_D$$

At equilibrium, the sum of the chemical potentials of the reactants, each weighted according to the stoichiometry of the reaction, is equal to the sum of the chemical potentials of the products similarly weighted.

Chemical potential and activity

$$\mu_A = \mu_A^0 + RT \ln a_A \tag{4.19}$$

or

$$a_A = e^{(\mu_A - \mu_A^0)/RT}$$

where a_A is the activity of a substance A at temperature T and μ_A and μ_A^0 are its chemical potential and standard chemical potential, respectively, at temperature T.

Standard States and Activities

Ideal gases
Standard state is gas at 1 atm pressure. The activity of an ideal gas A is

$$a_A = \frac{P_A}{1 \text{ atm}} \tag{4.23}$$

where P_A is the partial pressure of the gas.

Real gases
Standard state is a hypothetical state where the pressure is 1 atm, but the properties are those extrapolated from low pressures. The activity of a real gas is

$$a_A = \frac{\gamma_A P_A}{1 \text{ atm}} \tag{4.24}$$

where γ_A, the activity coefficient, is dependent on pressure.

Pure solids and liquids
Standard state is the pure substance at 1 atm. The activity is equal to 1 for a pure solid or liquid at 1 atm, and it changes only slightly with pressure.

$$a_A = 1 \tag{4.25}$$

Solvent of a solution
Standard state is the pure solvent at 1 atm. The activity is

$$a_A = \gamma_A X_A \tag{4.31}$$

where X_A = mole fraction of A.

Solutes in solution
Standard state of a solute is a hypothetical one where the concentration of the solute is equal to 1 molar (M) or 1 molal (m), but the properties are those extrapolated from very dilute solutions.

$$a_B = \gamma_B c_B \qquad (\gamma_B \longrightarrow 1 \quad \text{as} \quad c_B \longrightarrow 0) \tag{4.35}$$

$$a_B = \gamma_B m_B \qquad (\gamma_B \longrightarrow 1 \quad \text{as} \quad m_B \longrightarrow 0) \tag{4.41}$$

Activity coefficients of ions
The Debye–Hückel limiting law for the activity coefficient of an ion in water at 25°C:

$$\log \gamma_{\text{ion}} = -0.509\, Z_i^2 I^{1/2} \tag{4.46}$$

Z_i = charge on ion ($\pm 1, \pm 2, \pm 3$, etc.)

I = ionic strength = $\dfrac{1}{2} \sum_i c_i Z_i^2$

Biochemists' standard state
At pH 7, $a_{H^+} = 1$; the activity of each of the other species is set equal to the *total concentration* of all species of that molecule at pH 7.0:

$$\text{Dilute solution:} \quad a_A = \sum_i^{\text{species}} c_{i,A} \qquad \text{(at pH 7.0)} \tag{4.42}$$

Gibbs Free-Energy Change and Equilibrium Constant for a Chemical Reaction

Chemical reactions

$$aA + bB \longrightarrow cC + dD$$

$$\Delta G = \Delta G^0 + RT \ln Q \tag{4.20}$$

ΔG = Gibbs free-energy change for the reaction at T and concentrations specified in Q

ΔG^0 = standard Gibbs free-energy change for the reaction at T with all reactants and products at their standard states

$$Q = \frac{(a_C)^c(a_D)^d}{(a_B)^b(a_A)^a}$$

a_A, a_B, a_C, a_D = activities of reactants and products at the concentrations specified

$$\Delta G^0 = -RT \ln K \tag{4.14}$$

$K \equiv$ equilibrium constant

$$= \frac{(a_C^{eq})^c(a_D^{eq})^d}{(a_B^{eq})^b(a_A^{eq})^a} \tag{4.22}$$

a_A^{eq}, etc. = activities of reactants and products at equilibrium

$$R = 8.3144 \text{ J K}^{-1} \text{ mol}^{-1}$$

$$2.3026RT = 5708 \text{ J at 298 K} \qquad (K = e^{-\Delta G^0/RT} = 10^{-\Delta G^0/2.303RT})$$

Temperature dependence of equilibrium constant

The van't Hoff equation:

$$\left[\frac{\partial(\ln K)}{\partial\left(\dfrac{1}{T}\right)} \right]_P = -\frac{\Delta H^0}{R} \tag{4.52}$$

$$\ln \frac{K_2}{K_1} = -\frac{\Delta H^0}{R}\left(\frac{1}{T_2} - \frac{1}{T_1} \right) \qquad (\Delta H^0 = \text{constant}) \tag{4.53}$$

Galvanic Cells

$$\Delta G = -n\mathscr{E} \text{ (in eV)} = -96.485n\mathscr{E} \text{ (in kJ)} = -n\mathscr{E}F \text{ (in J)} \quad \textbf{(4.56) to (4.58)}$$

$$(F = 96,485 \text{ C mol}^{-1})$$

$$\Delta H(\text{kJ}) = 96.485n\left[-\mathscr{E} + T\left(\frac{\partial\mathscr{E}}{\partial T}\right)_P \right] \tag{4.60}$$

$$\Delta S(\text{J K}^{-1}) = 96,485n\left(\frac{\partial\mathscr{E}}{\partial T}\right)_P \tag{4.59}$$

The Nernst equation:

$$\mathscr{E} = \mathscr{E}^0 - \frac{RT}{nF} \ln Q \qquad\qquad \textbf{(4.61)}$$

$$\mathscr{E}^0 = \frac{RT}{nF} \ln K \qquad\qquad \textbf{(4.63)}$$

Mathematics Needed for Chapter 4

Review partial derivatives in chapter 3.

References

The biochemistry texts listed in chapter 1 describe metabolism in depth.

Suggested Reading

Allawai, H. T., and J. SantaLucia Jr. 1997. Thermodynamics and NMR of Internal G·T Mismatches in DNA. *Biochemistry* 36:10581–10594.

Manning, G. S. 1979. Counterion Binding in Polyelectrolyte Theory. *Acc. Chem. Res.* 12:443–447.

Moser, C. C., J. M. Keske, K. Warnke, R. S. Farid, and P. L. Dutton. 1992. Nature of Biological Electron Transfer. *Nature* 355:796–802.

Record, M. T. Jr., S. J. Mazur, P. Melançon, J.-H. Roe, S. L. Shaner, and L. Unger. 1981. Double Helical DNA: Conformation, Physical Properties and Interactions with Ligands. *Annu. Rev. Biochem.* 50:997–1024.

SantaLucia, J. Jr., Allawi, H. T., and Seneviratne, P. A. 1996. Improved Nearest-Neighbor Parameters for Predicting DNA Duplex Stability. *Biochemistry* 35:3555–3562.

Turner, D. T., N. Sugimoto, and S. M. Freier. 1988. RNA Structure Prediction. *Annu. Rev. Biophys. Biophys. Chem.* 17:167–193.

Problems

1. A key step in the biosynthesis of triglycerides (fats) is the conversion of glycerol to glycerol-1-phosphate by ATP:

$$\text{glycerol} + \text{ATP} \xrightarrow[\alpha\text{-glycerol kinase}]{\text{Mg}^{2+}}$$
$$\text{glycerol-1-phosphate} + \text{ADP}$$

At a steady state in the living cell, $(\text{ATP}) = 10^{-3}\,M$ and $(\text{ADP}) = 10^{-4}\,M$. The maximum (equilibrium) value of the ratio (glycerol-1-P)/(glycerol) is observed to be 770 at 25°C and pH 7.

   (a) Calculate K and $\Delta G^{0\prime}_{298}$ for the reaction.

   (b) Using the value of $\Delta G^{0\prime}_{298}$ for the reaction ADP + $P_i \rightarrow$ ATP + H_2O, together with the answer to part (a), calculate $\Delta G^{0\prime}_{298}$ and K for the reaction

$$\text{glycerol} + \text{phosphate} \longrightarrow$$
$$\text{glycerol-1-phosphate} + H_2O$$

2. In the frog muscle rectus abdominis, the concentrations of ATP, ADP, and phosphate are $1.25 \times 10^{-3}\,M$, $0.50 \times 10^{-3}\,M$, and $2.5 \times 10^{-3}\,M$, respectively.

   (a) Calculate the Gibbs free-energy change $\Delta G^{0\prime}$ for the hydrolysis of ATP in muscle. Take the temperature and pH of the muscle as 25°C and 7, respectively.

   (b) For the muscle described, what is the maximum amount of mechanical work it can do per mole of ATP hydrolyzed?

   (c) In muscle, the enzyme creatine phosphokinase catalyzes the following reaction:

$$\text{phosphocreatine} + \text{ADP} \longrightarrow \text{creatine} + \text{ATP}$$

The standard Gibbs free energy of hydrolysis of phosphocreatine at 25°C and pH 7 is $-43.1\ \text{kJ mol}^{-1}$:

$$\text{phosphocreatine} + H_2O \longrightarrow \text{creatine} + \text{phosphate}$$

Calculate the equilibrium constant of the creatine phosphokinase reaction.

3. An important step in the glycolytic path is the phosphorylation of glucose by ATP, catalyzed by the enzyme hexokinase and Mg^{2+}:

$$\text{glucose} + \text{ATP} \xrightarrow[\text{hexokinase}]{\text{Mg}^{2+}} \text{glucose-6-P} + \text{ADP}$$

In the absence of ATP, glucose-6-P is unstable at pH 7, and in presence of the enzyme glucose-6-phosphatase, it hydrolyzes to give glucose:

$$\text{glucose-6-P} + H_2O \xrightarrow[\text{G-6-phosphatase}]{} \text{glucose} + \text{phosphate}$$

(a) Using data from table 4.3, calculate $\Delta G^{0\prime}$ at pH 7 for the hydrolysis of glucose-6-P at 298 K.

(b) If the phosphorylation of glucose is allowed to proceed to equilibrium in the presence of equal concentrations of ADP and ATP, what is the ratio (glucose-6-P)/(glucose) at equilibrium? Assume a large excess of ATP and ADP; that is, (ATP) = (ADP) $\gg$ [(glucose) + (glucose-6-P)].

(c) In the absence of ATP (and ADP), calculate the ratio (glucose-6-P)/(glucose) at pH 7, if phosphate = $10^{-2}\ M$.

4. In the red blood cell, glucose is transported into the cell against its concentration gradient. The energy for this transport is supplied by the hydrolysis of ATP:

$$ATP + H_2O \longrightarrow ADP + P_i \qquad (\Delta \overline{G}^0 = -31.0\ kJ)$$

Assume that the overall transport reaction is 100% efficient and given by

$$ATP + H_2O + 2glucose(out) \longrightarrow$$
$$2glucose(in) + ADP + P_i$$

(a) At 25°C, under conditions where [ATP], [ADP], and [P_i] are held constant at $1 \times 10^{-2}\ M$ by cell metabolism, find the maximum value of

$$\frac{[glucose(in)]}{[glucose(out)]}$$

Assume all activity coefficients are equal to 1.

(b) If the stoichiometry of transport were 1 mol of glucose transported per mole of ATP hydrolyzed, what would be the maximum concentration gradient of glucose under the same conditions as part (a)?

(c) In an actual cell, the glucose inside the cell may have an activity coefficient much less than 1 due to nonideal behavior. Would this increase or decrease the maximum *concentration* gradient obtainable? (Assume that all other activity coefficients are equal to 1.)

5. The following reactions can be coupled to give alanine and oxaloacetate:

glutamate + pyruvate $\rightleftharpoons$ ketoglutarate + alanine

$$(\Delta G^{0\prime}_{303} = -1004\ J\ mol^{-1})$$

glutamate + oxaloacetate $\rightleftharpoons$
ketoglutarate + aspartate

$$(\Delta G^{0\prime}_{303} = -4812\ J\ mol^{-1})$$

(a) Write the form of the equilibrium constant for the reaction

pyruvate + aspartate $\rightleftharpoons$ alanine + oxaloacetate

and calculate the numerical value of the equilibrium constant at 30°C.

(b) In the cytoplasm of a certain cell, the components are at the following concentrations: pyruvate = $10^{-2}\ M$, aspartate = $10^{-2}\ M$, alanine = $10^{-4}\ M$, and oxaloacetate = $10^{-5}\ M$. Calculate the Gibbs free-energy change for the reaction of part (a) under these conditions. What conclusion can you reach about the direction of this reaction under cytoplasmic conditions?

6. (a) Use the following equilibrium constants to calculate ΔH^0 from the slope of $\ln K$ vs. (1/T) for the reaction 3-phosphoglycerate to 2-phosphoglycerate at pH 7.

T(°C)	0	20	30	45
K	0.1535	0.1558	0.1569	0.1584

(b) Calculate $\Delta \overline{G}^0$ at 25°C.

(c) What is the concentration of each isomer of phosphoglycerate when 0.150 M phosphoglycerate reaches equilibrium at 25°C?

7. Oxides of sulfur are important in atmospheric pollution, arising particularly from burning coal. Use the thermodynamic data at 25°C given in the table below to answer the following questions.

(a) In air, the oxidation of SO_2 can occur: $\frac{1}{2}O_2(g)$ + $SO_2(g) \rightarrow SO_3(g)$. Calculate the standard Gibbs free energy for this reaction at 25°C.

(b) Find the equilibrium ratio of partial pressures of $SO_3(g)$ to $SO_2(g)$ in air at 25°C; the partial pressure of $O_2(g)$ is 0.21 atm.

(c) $SO_3(g)$ can react with $H_2O(g)$ to form sulfuric acid, $H_2SO_4(g)$. Air that is in equilibrium with liquid water at 25°C has a partial pressure of $H_2O(g)$ of 0.031 atm. Find the equilibrium ratio of partial pressures of $H_2SO_4(g)$ to $SO_3(g)$ in air at 25°C.

Compound	ΔH^0_f (kJ mol^{-1})	S^0 (J K^{-1} mol^{-1})
$O_2(g)$	0	205.1
$H_2O(g)$	−241.8	188.7
$SO_2(g)$	−296.8	248.2
$SO_3(g)$	−395.7	256.8
$H_2SO_4(g)$	−814.0	156.9

8. An important metabolic step is the conversion of fumarate to malate. In aqueous solution, an enzyme (fumarase) allows equilibrium to be attained.

fumarate + H_2O $\rightleftharpoons$ malate

at 25°C the equilibrium constant $K = (a_M/a_F) = 4.0$. The activity of malate is a_M, and the activity of fumarate is a_F, defined on the molarity concentration scale ($a = c$ in dilute solution).

(a) What is the standard Gibbs free-energy change for the reaction at 25°C?

(b) What is the Gibbs free-energy change for the reaction at equilibrium?

(c) What is the Gibbs free-energy change when 1 mol of 0.100 M fumarate is converted to 1 mol of 0.100 M malate?

(d) What is the Gibbs free-energy change when 2 mol of 0.100 M fumarate are converted to 2 mol of 0.100 M malate?

(e) If $K = 8.0$ at 35°C, calculate the standard enthalpy change for the reaction; assume that the enthalpy is independent of temperature.

(f) Calculate the standard entropy change for the reaction; assume that ΔS^0 is independent of temperature.

9. Use the Gibbs free energy of hydrolysis of ATP under standard conditions at 25°C, 1 atm, to answer the following questions.

(a) Calculate the ΔG for the reaction when $[ATP] = 10^{-2} M$, $[ADP] = 10^{-4} M$, and $[phosphate] = 10^{-1} M$.

(b) Calculate the maximum available work under the conditions of part (a) when 1 mol of ATP is hydrolyzed. This work can be used, for example, to contract a muscle and raise a weight.

(c) Calculate ΔG and the maximum available work if $[ATP] = 10^{-7} M$, $[ADP] = 10^{-1} M$, and $[phosphate] = 2.5 \times 10^{-1} M$.

10. What is the pressure of $CO(g)$ in equilibrium with the $CO_2(g)$ and $O_2(g)$ in the atmosphere at 25°C? The partial pressure of $O_2(g)$ is 0.2 atm, and the partial pressure of $CO_2(g)$ is 3×10^{-4} atm. CO is extremely poisonous because it forms a very strong complex with hemoglobin. Should you worry?

11. For a reaction, $aA + bB \rightarrow cC + dD$,

$$\Delta G_T^0 = c\mu_C^0 + d\mu_D^0 - a\mu_A^0 - b\mu_B^0$$

and

$$\Delta G_T^0 = -RT \ln K$$

Combining these two expressions and dividing each term by RT gives

$$\ln K = -\frac{1}{R}\left(\frac{c\mu_C^0}{T} + \frac{d\mu_D^0}{T} - \frac{a\mu_A^0}{T} - \frac{b\mu_B^0}{T}\right)$$

Show that

$$\frac{d}{dT}\left(\frac{\mu_i^0}{T}\right) = -\frac{\overline{H}_i^0}{T^2}$$

where $\overline{H}_i^0$ is the standard partial molal enthalpy of species i. (Note that at constant T, $\Delta G^0 = \Delta H^0 - T\Delta S^0$, and $\mu_i^0 = H_i^0 - TS_i^0$.)

12. (a) From the ionization constant, calculate the standard Gibbs free-energy change for the ionization of acetic acid in water at 25°C.

(b) What is the Gibbs free-energy change at equilibrium for the ionization of acetic acid in water at 25°C?

(c) What is the Gibbs free-energy change for the following reaction in water at 25°C? (Assume that activity coefficients are all equal to 1.)

$$H^+(10^{-4} M) + OAc^-(10^{-2} M) \longrightarrow HOAc(1M)$$

(d) What is the Gibbs free-energy change for the following reaction in water at 25°C? (Assume that activity coefficients are all equal to 1.)

$$H^+(10^{-4} M) + OAc^-(10^{-2} M) \longrightarrow HOAc(10^{-5} M)$$

(e) What is the Gibbs free-energy change for transferring 1 mol of acetic acid from an aqueous solution of $1 M$ concentration to an aqueous solution of $10^{-5} M$? (Assume that activity coefficients are all equal to 1.)

13. You want to make a pH 7.0 buffer using NaOH and phosphoric acid. The sum of the concentrations of all phosphoric acid species is 0.100 M. The equilibrium constants for concentrations given in mol L^{-1} for the following equilibria are

$$H_3PO_4 \rightleftharpoons H^+ + H_2PO_4^- \qquad (K_1 = 7.1 \times 10^{-3})$$

$$H_2PO_4^- \rightleftharpoons H^+ + HPO_4^{2-} \qquad (K_2 = 6.2 \times 10^{-8})$$

$$HPO_4^{2-} \rightleftharpoons H^+ + PO_4^{3-} \qquad (K_3 = 4.5 \times 10^{-13})$$

(a) Write the equation that specifies that the solution is electrically neutral.

(b) Calculate the concentrations of all species in the buffer.

(c) Use data in table 4.2 to calculate K_2 at 37°C.

14. You made a pH 9.0 buffer solution at 25°C by mixing NaOH and histidine (HisH) to give a solution that is 0.200 M in total concentration of histidine.

(a) Calculate the concentrations of all species at 25°C.

(b) What is the pH of the buffer at 40°C? The two most important ΔH^0 values for histidine are given in table 4.2; assume that $\Delta H^0 = 0$ for the first ionization. You can ignore the volume change of the solution.

(c) Calculate the concentrations of all species at 40°C.

15. An important reaction in visual excitation is the activation of an enzyme that catalyzes the hydrolysis of guanosine triphosphate:

$$GTP + H_2O \longrightarrow GDP + P_i \qquad (K'_{298} = 1.9 \times 10^5)$$

(a) Typical concentrations of GTP, GDP, and P_i in the retinal rod cell are 50 mM, 5 mM, and 15 mM, respectively. What is $\Delta G^{0\prime}_{298}$ for the above reaction in the rod cell?

(b) If the GTP concentration were suddenly doubled, by how much would $\Delta G^{0\prime}_{298}$ change?

(c) The solution described in part (a) is allowed to come to equilibrium. What are the final concentrations of GTP, GDP, and P_i?

16. In general, native proteins are in equilibrium with denatured forms:

$$\text{protein (native)} \rightleftharpoons \text{protein (denatured)}$$

For ribonuclease (a protein), the following concentration data for the two forms were experimentally determined for a total protein concentration of 1×10^{-3} mol L^{-1}:

Temperature (°C)	Native	Denatured
50	9.97×10^{-4} mol L^{-1}	2.57×10^{-6} mol L^{-1}
100	8.6×10^{-4} mol L^{-1}	1.4×10^{-4} mol L^{-1}

Determine ΔH^0 for the denaturation reaction, assuming it to be independent of temperature.

17. A single-stranded oligonucleotide that has complementary ends can form a base-paired hairpin loop.

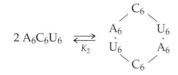

$$\text{AAAAAACCCCCCUUUUUU} \underset{}{\overset{K_1}{\rightleftharpoons}}$$

(a) At 25°C, the equilibrium constant $K_1 = 0.86$. What are the concentrations of loop and single strand at equilibrium if the initial concentration of single strands is 1.00 mM? Will increasing the initial concentration of oligonucleotide increase or decrease the fraction of hairpin loop? Explain.

(b) At 37°C, the equilibrium constant $K_1 = 0.51$. Calculate ΔH^0, ΔS^0, and ΔG^0 at 37°C.

(c) As the concentration of oligonucleotide increases, another reaction becomes possible. A double-stranded molecule with an internal loop can form:

$$2\,A_6C_6U_6 \underset{K_2}{\rightleftharpoons}$$

At 25°C, $K_2 = 1.00 \times 10^{-2}$ M^{-1} (and $K_1 = 0.86$). Calculate the concentrations of all three species (single strand, SS; double strand, DS; and hairpin loop, H) at equilibrium in a solution of initial concentration of single strands = 0.100 M.

18. Cytochromes are iron–heme proteins in which a porphyrin ring is coordinated through its central nitrogens to an iron atom that can undergo a one-electron oxidation–reduction reaction. Cytochrome f is an example of this class of molecules, and it operates as a redox agent in chloroplast photosynthesis. The standard reduction potential $\mathscr{E}^{0\prime}$ of cytochrome f at pH 7 can be determined by coupling it to an agent of known $\mathscr{E}^{0\prime}$, such as ferricyanide/ferrocyanide:

$$\text{Fe(CN)}_6^{3-} + e^- \longrightarrow \text{Fe(CN)}_6^{4-} \qquad (\mathscr{E}^{0\prime} = 0.440 \text{ V})$$

In a typical experiment, carried out spectrophotometrically, a solution at 25°C and pH 7 containing a ratio

$$\frac{[\text{Fe(CN)}_6^{4-}]}{[\text{Fe(CN)}_6^{3-}]} = 2.0$$

is found to have a ratio $[\text{Cyt } f_{red}]/[\text{Cyt } f_{ox}] = 0.10$ at equilibrium.

(a) Calculate $\mathscr{E}^{0\prime}$ (reduction) for cytochrome f.

(b) On the basis of the standard reduction potential $\mathscr{E}^{0\prime}$ for the reduction of O_2 to H_2O at pH 7 and 25°C, is oxidized cytochrome f a good enough oxidant to cause the formation of O_2 from H_2O at pH 7?

19. Consider the following reaction, in which two electrons are transferred:

$$2 \text{ cytochrome } c \text{ (ferrous)} + \text{pyruvate} + 2H^+ \longrightarrow$$
$$2 \text{ cytochrome } c \text{ (ferric)} + \text{lactate}$$

(a) What is $\mathscr{E}^{0\prime}$ for this reaction at pH 7 and 25°C?

(b) Calculate the equilibrium constant for the reaction at pH 7 and 25°C.

(c) Calculate the standard Gibbs free-energy change for the reaction at pH 7 and 25°C.

(d) Calculate the Gibbs free-energy change (at pH 7 and 25°C) if the lactate concentration is five times the pyruvate concentration and the cytochrome c (ferric) is ten times the cytochrome c (ferrous).

20. The cell Ag(s), AgI(s)|KI(10^{-2} M)| |KCl(10^{-3} M)|Cl$_2$(g, 1 atm), Pt(s) has the voltage 1.5702 V at 298 K.

(a) Write the cell reaction.

(b) What is ΔG_{298}?

(c) What is ΔG^0_{298}?

(d) Calculate the standard reduction potential for the half-cell on the left.

(e) Calculate the solubility product of AgI; $K_{AgI} = (a_{Ag^+})(a_{I^-})$

(f) The cell has a potential of 1.5797 V at 288 K. Estimate ΔS^0_{298} for the reaction.

21. Ferredoxins (Fd) are iron- and sulfur-containing proteins that undergo redox reactions in a variety of microorganisms. A particular ferredoxin is oxidized in a one-electron reaction, independent of pH, according to the equation

$$Fd_{red} \rightleftharpoons Fd_{ox} + e^-$$

To determine the standard potential of Fd_{red}/Fd_{ox}, a known amount was placed in a buffer at pH 7.0 and bubbled with H_2 at 1 atm pressure. (Finely divided platinum catalyst was present to ensure reversibility.) At equilibrium, the ferredoxin was found spectrophotometrically to be exactly one-third in the reduced form and two-thirds in the oxidized form.

(a) Calculate K', the equilibrium constant, for the system

$$\frac{1}{2}H_2 + Fd_{ox} \rightleftharpoons H^+ + Fd_{red}$$

(b) Calculate $\mathscr{E}^{0\prime}$ for the Fd_{red}/Fd_{ox} half-reaction at 25°C.

22. The conversion of β-hydroxybutyrate, β-HB⁻, to acetoacetate, AAc⁻, is an important biochemical redox reaction that uses molecular oxygen as the ultimate oxidizing agent:

$$\beta\text{-HB}^- + \frac{1}{2}O_2(g) \longrightarrow AAc^- + H_2O$$

(a) Using the standard reduction potentials given in table 4.1, calculate $\Delta G^{0\prime}$ and the equilibrium constant for this system at pH 7 and 25°C.

(b) In a solution at pH 7 and 25°C saturated at 1 atm with respect to dissolved air (which is 20% oxygen), what is the ratio of [ACc⁻] to [β-HB⁻] at equilibrium?

23. Consider the reaction

$$CH_3CH_2OH(aq) + \frac{1}{2}O_2(g) \longrightarrow CH_3\overset{\overset{\displaystyle O}{\|}}{C}H(aq) + H_2O$$

ethanol   acetaldehyde

(a) Calculate $\mathscr{E}^{0\prime}$ for this reaction at 25°C.

(b) Calculate the standard Gibbs free energy (in kJ) for the reaction at 25°C.

(c) Calculate the equilibrium constant at 25°C for the reaction.

(d) Calculate $\mathscr{E}$ for the reaction at 25°C when: a(ethanol) = 0.1, P_{O_2} = 4 atm, a(acetaldehyde) = 1, and $a(H_2O)$ = 1.

(e) Calculate ΔG for the reaction in part (d).

24. Magnesium ion and other divalent ions form complexes with adenosine triphosphate, ATP:

$$ATP + Mg^{2+} \longrightarrow complex$$

(a) Describe an electrochemical cell that would allow you to measure the activity of Mg^{2+} at any concentration in a 0.100 M ATP solution.

(b) Describe how you could measure the thermodynamic equilibrium constant for binding Mg^{2+} by ATP with an electrochemical cell.

25. Photosystem 1, in higher plants, converts light into chemical energy. Energy, in the form of photons, is absorbed by a chlorophyll complex, P700, which donates an electron to A. The electron is passed down an electron-transport chain, at the end of which $NADP^+$ is reduced. The reduction potentials of P700⁺, A, and $NADP^+$, at pH 7.0, 25°C, are 0.490 V, −0.900 V, and −0.350 V, respectively.

(a) Calculate, at pH 7.0, 25°C, $\mathscr{E}^{0\prime}$ of the reaction

$$P700 + A \longrightarrow P700^+ + A^-$$

(b) What is $\Delta G^{0\prime}$, in kJ, for the same reaction?

(c) At pH 7.0, 25°C, find $\Delta G^{0\prime}$, in kJ, for the reaction

$$NADP^+ + H_2(g) \longrightarrow NADPH + H^+$$

26. Certain dyes can exist in oxidized or reduced form in solution. The half-reaction for one such dye, methylene blue (MB), is

$$\underset{\text{(blue)}}{MB(ox)} + 2H^+ + 2e^- \rightleftharpoons \underset{\text{(colorless)}}{MB(red)} \quad (\mathscr{E}^0 = +0.4\ V)$$

As indicated, the oxidized form is blue, and the reduced form is colorless. From the color of the solution, the relative amounts of the two forms can be estimated.

(a) Write the equation for the half-cell reduction potential of methylene blue in terms of [MB(ox)], [MB(red)], [H⁺], and $\mathscr{E}^{0\prime}$.

(b) A very small amount of MB(ox) is added to a solution containing an unknown substance. The pH = 7.0 ([H⁺] = 1×10^{-7} M). From the color of the solution, it was estimated that the ratio of concentrations [MB(red)]/[MB(ox)] = 1×10^{-3} at equilibrium. Assuming that all activity coefficients are equal to 1, determine the half-cell potential of the unknown substance in solution.

27. In the presence of oxygen, most living cells make ATP by oxidative phosphorylation, which takes place in mitochondria. One of the major substrates that is oxidized is NADH. The overall reaction for this energy-yielding process is

$$NADH + H^+ + \frac{1}{2}O_2 \longrightarrow NAD^+ + H_2O$$

(a) Use table 4.1 to calculate the Gibbs free-energy change of the above reaction as it takes place in mitochondria at pH 7 and 25°C. The concentrations of reactants and products are [NADH] = 1 mM, [NAD$^+$] = 2 mM, and P_{O_2} = 0.1 atm.

(b)   Given that 3 mol of ATP is made from P_i and ADP for every mole of NADH oxidized, what fraction of the Gibbs free-energy change of part (a) is used in making ATP in mitochondria during oxidative phosphorylation? [ADP] = 1 mM, [ATP] = 3 mM, and P_i = 10 mM.

28. Consider the following half-cell reactions and their standard reduction potentials at 298 K and pH 7.0 in aqueous solution:

$$2H^+ + \frac{1}{2}O_2 + 2e^- \longrightarrow H_2O \quad (\mathcal{E}^0 = +0.816 \text{ V})$$

$$2H^+ + \text{cystine} + 2e^- \longrightarrow 2\text{cysteine} \quad (\mathcal{E}^0 = -0.34 \text{ V})$$

Although not required for the problem, the structures of cystine and cysteine are as follows:

Cystine:
$$\begin{array}{c} CH_2-S-S-CH_2 \\ | \qquad\qquad | \\ CH-NH_3^+ \quad CH-NH_3^+ \\ | \qquad\qquad | \\ CO_2^- \qquad\quad CO_2^- \end{array}$$

Cysteine:
$$\begin{array}{c} CH_2-SH \\ | \\ CH-NH_3^+ \\ | \\ CO_2^- \end{array}$$

(a)   If you prepare a 0.010-M solution of cysteine at pH 7.0 and let it stand in contact with air at 298 K, what will be the ratio of [cystine]/[cysteine] at equilibrium? The partial pressure of oxygen in the air is 0.20 atm. The activity coefficients may be taken as unity.

(b)   What is ΔG for the reaction when the activities of the reactants and products are the equilibrium values?

29. The binding of oligomers (lys)$_n$ of the amino acid lysine, where n indicates the number of lysines in each oligomer, to a synthetic double-stranded RNA poly-(rA·rU) has been studied by S. A. Latt and H. A. Sober [*Biochemistry* 6:3293–3306 (1967)]. The dependence of the apparent binding constant K_{obs} on [NaCl] in the medium is tabulated as follows:

[NaCl], M	0.06	0.10	0.25	0.39
log K_{obs}, $n = 4$	4.62	—	2.39	1.60
log K_{obs}, $n = 5$	—	4.71	2.79	1.88

Plot log K_{obs} against log [Na$^+$] for each of the oligolysines and interpret the data. The value of ψ for poly-(rA·rU) is close to that of duplex DNA.

30. Self-complementary oligonucleotides can form double-stranded helices in aqueous solution stabilized by Watson–Crick base pairs. For example,

$$2(5'-CGCGATATCGCG-3') \rightleftharpoons \begin{array}{c} 5'-CGCGATATCGCG-3' \\ 3'-GCGCTATAGCGC-5' \end{array}$$

Derive an expression for the equilibrium constant K as a function of c, the initial concentration of single strands and f, the fraction of single strands that are in the double-stranded helix at equilibrium. The equilibrium is

$$2S \rightleftharpoons D$$

$$K = \frac{[D]}{[S]^2}$$

Note that

$$f = \frac{2[D]}{2[D] + [S]} = \frac{2[D]}{c}$$

31. (a)   Use nearest-neighbor values (table 4.4) for the thermodynamics of double strand formation to calculate ΔG^0(25°C), ΔH^0, and ΔS^0 for

$$\begin{array}{c} 5'-GGGCCC-3' \\ 3'-CCCGGG-5' \end{array} \text{ and } \begin{array}{c} 5'-GGTTCC-3' \\ 3'-CCAAGG-5' \end{array}$$

(b)   The melting temperature, T_{melt}, is defined as the temperature where f, the fraction of single strands in the helix at equilibrium, is $\frac{1}{2}$. Calculate the melting temperature for each double helix above at 1.0 × 10^{-4} M initial concentrations of *each* single strand (the initial concentration of strands, c, is 1.0 × 10^{-4} M for the helix on the left and 2.0 × 10^{-4} M for the helix on the right). Note that the oligonucleotides in the double helix on the right are not self-complementary; its equilibrium constant in terms of f and c is slightly different from that in problem 30.

32. Genomic DNA sequences are detected by hybridization with a radioactively labeled probe that is complementary to the target sequence and forms a double strand.

(a)   For the equilibrium

$$\text{probe} + \text{target} \longrightarrow \text{double strand}$$

express K, the equilibrium constant for the association of the probe with the target, in terms of c_{probe}, the initial probe concentration, c_{target}, the initial concentration of the target sequence, and f, the fraction of target hybridized to the probe to form a double strand. Assume that the probe is present in a large excess.

(b)   The target sequence is

$$5'-GGGGAATCA-3'$$

Calculate ΔG^0 (use table 4.4) and K at 25°C for the formation of a double strand with the above sequence at pH 7.0, 1 M NaCl.

(c)   Find the melting temperature of the probe from the target, defined as the temperature at which half of the target is hybridized to the probe. Assume that c_{probe} = 1.00 × 10^{-4} M and c_{target} = 1.00 × 10^{-8} M.

(d) Does the melting temperature as defined in part (c) go up, down, or remain the same when the following changes are made? Give reasons for your answers.

(1) The probe concentration is doubled.

(2) The target concentration is doubled.

(3) The probe contains a single base mismatch.

(4) The salt concentration is decreased.

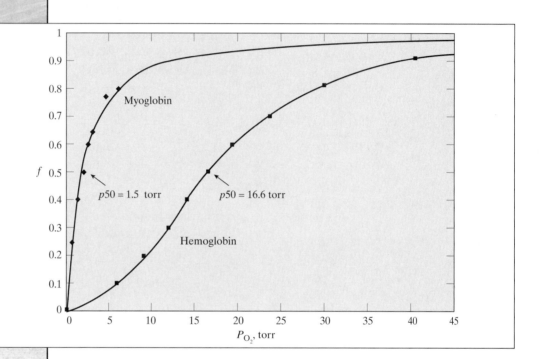

CHAPTER 5

Free Energy and Physical Equilibria

Free energy not only characterizes chemical reactions, it applies to all sorts of processes (at constant temperature and pressure), and it sometimes leads to counterintuitive results. For example, the vapor pressure of benzene (vapor pressure is the pressure of the gas in the vapor in contact with the liquid) is the same for an aqueous solution saturated with benzene as it is for pure benzene—even though the aqueous solution is very dilute in benzene.

Concepts

Biological organisms are highly inhomogeneous; different regions of each cell have different concentrations of molecules and different biological functions. The storage and expression of genes occur in the nucleus; protein synthesis occurs in the cytoplasm. The ATP to drive many of these processes comes from oxidative phosphorylation in mitochondria. Transport of O_2 and essential nutrients into cells and the removal of CO_2 and waste products are required for all these processes. In this chapter, we focus on the physical chemistry underlying the equilibrium and transport of molecules among different regions of a system. Examples of these regions include cell compartments separated by membranes, the inside and outside of a membrane, a liquid or solid in contact with a gas, and two immiscible liquids. In such systems the chemical potential (partial molar free energy) is the thermodynamic property that tells us whether two or more compartments are in equilibrium with one another (chemical potential is the same in all compartments at equilibrium) or whether they are not in equilibrium (unequal chemical potentials).

If a species has a different chemical potential in two phases (or compartments) in contact with one another, it will move to the phase with the lower chemical potential. This will continue until the species reaches equilibrium, and its chemical potential becomes the same in all phases. This simple idea allows us to distinguish and characterize active and passive transport in cells, the equilibrium concentrations of molecules separated by semipermeable membranes, and the equilibria of molecules between solids, liquids, and gases that determine freezing points, solubilities, boiling points, and osmotic pressure.

Applications
Membranes and Transport

Living cells are separated from their surroundings by membranes and contain a variety of subcellular particles—*organelles*—also enclosed by membranes. Most biological membranes consist of a lipid bilayer that contains proteins and other molecules that serve as recognition sites, signal transmitters, or ports of entrance and exit. Membranes are so thin, having thicknesses of only one or two molecules, that they are often considered to be two-dimensional phases. The thermodynamic properties of membranes are then described in terms of surface properties, such as the surface chemical potential and the surface tension or pressure. Membranes not only separate the contents of a compartment from its surroundings but also permit the controlled transport of molecules and signals between the inside and outside. Differences between the inside and outside of a cell influence the exchange of metabolites and electrical signals, the flow of heat, and changes in shape. Temperature differences cause heat flow, pressure differences cause changes in shape, and electrochemical potential differences cause molecular transport and electrical signals.

Ligand Binding

Noncovalent interactions that bind ligands like O_2 to hemoglobin, substrates to enzymes, and complementary strands of DNA or RNA to one another underlie essential dynamic processes in living cells. Equilibrium dialysis provides a method of exploring the binding between macromolecules and small ligand molecules. In equilibrium dialysis a semipermeable membrane allows a ligand to reach equilibrium between two phases, one of which contains a macromolecule. The difference in concentrations of the ligand on opposite sides of the membrane depends on the interaction of ligand and macromolecule. This provides a very useful and easy method for studying equilibrium binding constants.

Colligative Properties

The chemical potential must be the same for each component present in two or more phases at equilibrium with one another. The component can be the solvent or each of the solutes in a solution. The phases are solids, liquids, gases, or solution compartments separated by a semipermeable membrane. Any change in a property such as temperature, pressure, or activity in one phase that results in a change in chemical potential must be accompanied by an equal change in chemical potential in the other phases, for the system to remain in equilibrium. This fundamental fact allows us to explain the *colligative properties*, such as the freezing-point lowering, the boiling-point elevation, the vapor-pressure lowering, and the increase of osmotic pressure when a solute is dissolved in a solvent. Colligative properties play essential roles in biological cells. Osmotic pressure that results from the activities of components present in the cytoplasm needs to be balanced by a suitable external pressure lest the cell rupture and burst. Colligative properties are used to determine the concentrations and molecular weights of solutes in solution; they can be used to measure association and dissociation equilibrium constants of biopolymers.

Phase Equilibria

The transfer of a chemical from one phase to another is illustrated by the evaporation of liquid water into the vapor phase; the heat removed from our bodies by evaporation of sweat is essential to survival in hot climates. Another example is the transport of ions from inside a cell to outside, which is vital to nerve conduction. We can consider the inside and the outside of the cell as two different phases.

Equilibrium is a state where nothing seems to happen—nothing apparently changes with time. For living systems to exist, they must be out of equilibrium; dynamic processes need to occur to maintain the living state. When an organism dies, it approaches closer to equilibrium. The usefulness of considering equilibrium in connection with living organisms is that it helps define the direction of dynamic processes. One of the consequences of the second law of thermodynamics (chapter 3) is that spontaneous processes result in the system moving toward a state of equilibrium. For an open system, spontaneous processes are accompanied by a decrease in chemical potential, $\Delta\mu_{T,P} \leq 0$. At equilibrium, $\Delta\mu_{T,P} = 0$.

One-Component Systems

A pure substance can occur as a solid, liquid, or gas, depending on the temperature and pressure. For many substances, several solid phases with different structures may occur under different conditions. Some compounds form ordered liquid–crystalline phases that are distinct from both solids and isotropic (unordered) liquids.

If a system consists of two or more phases of the same substance in equilibrium at the same temperature and pressure, then the difference in chemical potential between the substance in the different phases is zero. Therefore,

$$\mu(\text{phase } 1) = \mu(\text{phase } 2) = \mu(\text{phase } 3) \qquad (T, P \text{ constant}) \qquad \textbf{(5.1)}$$

The chemical potential depends on the activity of the substance in each phase:

$$\mu = \mu^0 + RT \ln a = \mu^0 + RT \ln \gamma c \qquad \textbf{(4.19)}$$

Therefore, if the same standard state is used for a substance in two different phases, its activity in the two phases will be the same at equilibrium:

$$a(\text{phase } 1) = a(\text{phase } 2) \qquad \textbf{(5.2)}$$

The equilibrium between liquid (l) water and its gaseous vapor (g) is illustrated in figure 5.1. Both phases are at the same temperature and pressure; hence, the chemical potential of water is the same in both phases: $\mu_{H_2O}(l) = \mu_{H_2O}(g)$. If the temperature is raised, the water will establish a new equilibrium having a new (higher) vapor pressure and a new chemical potential that still will be identical for the two phases. The chemical potentials do not depend on the amount of liquid or gaseous water, only on both phases being present and at equilibrium.

Boiling Point and Freezing Point

The history of mountaineering and the thermodynamics of gas–liquid equilibria have always been closely linked. The reason is that the boiling point of water, or any liquid, depends on the pressure of its vapor in equilibrium with it. In the 1850s, intrepid mountaineers on first ascents carried 760-mm-long mercury manometers to record the pressure and thus the altitude at the peak. The scientist-mountaineer carried a thermometer to measure the boiling point of water at the peak and calculated the pressure from the measured boiling point. Present-day mountaineers who climb above 15,000 ft routinely carry a pressure cooker for meals and an aneroid barometer to measure altitudes.

It is straightforward to derive the equation relating the boiling point of a pure liquid to the pressure of its gas (g) phase in equilibrium—the vapor pressure. Consider the reaction

$$A(\text{pure liquid}) \longrightarrow A(g)$$

The equilibrium constant is

$$K = \frac{a_A(g)}{a_A(l)}$$

Gases are always much more nearly ideal than liquids because the interactions of the molecules in the gas phase are always weaker than the interactions of the same molecules in the liquid phase. In liquids, molecules are in

H$_2$O (g)
Temperature, T
Vapor pressure, P_{H_2O}

H$_2$O (l)
Temperature, T

▲ FIGURE 5.1

Water in a closed flask at equilibrium, involving liquid and vapor at a uniform temperature and pressure

direct contact; in gases (except for very high pressures) they are often far apart. Therefore, it is appropriate to approximate the gas as an ideal gas and to replace the activity of the gas by the pressure of the gas in atmospheres:

$$a_A(g) = P(\text{atm})$$

We can ignore the effect of pressure on the activity of the liquid and arbitrarily take the activity as unity:

$$a_A(l) = 1$$

Therefore, the equilibrium constant is just the equilibrium vapor pressure,

$$K = P$$

The temperature dependence of the equilibrium constant [Eq. (4.53)] gives

$$\ln \frac{P_2}{P_1} = \frac{-\Delta \overline{H}_{vap}}{R}\left(\frac{1}{T_2} - \frac{1}{T_1}\right) \tag{5.3}$$

where $\Delta \overline{H}_{vap} = \overline{H}_g - \overline{H}_l$ = molar enthalpy (heat) of vaporization. This equation is known as the *Clausius–Clapeyron equation.* These properties are illustrated for water in figure 5.2.

The pressures P_2 and P_1 are the equilibrium vapor pressures of a liquid at temperatures T_2 and T_1, respectively. The normal boiling point, T_{boil}, is defined as the temperature where the vapor pressure is 1 atm. Therefore, for a pure liquid whose molar enthalpy of vaporization and normal boiling point are known, one can use Eq. (5.3) to obtain the equilibrium vapor pressure at other temperatures.

EXAMPLE 5.1

At 20,320-ft altitude at the summit of Mt. Denali, pure water boils at only 75°C. We note in passing that "hot" tea will be tepid and weak at this altitude. What is the atmospheric pressure?

Solution

The enthalpy of vaporization of water depends on temperature; for Eq. (5.3), we use an approximate value of $\Delta \overline{H}_{vap} = 42{,}000$ J mol^{-1}, $P_1 = 1$ atm, and $T_1 = 373$ K (the normal boiling point of water):

$$\ln P = \frac{-42{,}000}{8.314}\left(\frac{1}{348} - \frac{1}{373}\right)$$

$$P = 0.38 \text{ atm}$$

This is the pressure of the water in equilibrium with the boiling liquid. Liquid boils when the vapor pressure of the liquid equals that of the surrounding atmosphere; therefore, the atmospheric pressure is 0.38 atm. The effect of this reduced pressure on the equilibrium binding of oxygen to hemoglobin in the lungs is another problem in phase equilibria that concerns mountaineers.

The Clausius–Clapeyron equation can be used to calculate a normal boiling point temperature from two measured vapor pressures at different temperatures. First, the enthalpy of vaporization is calculated using Eq. (5.3); then the temperature corresponding to a vapor pressure of 1 atm is calculated. An easy way to apply the Clausius–Clapeyron equation is to plot ln P

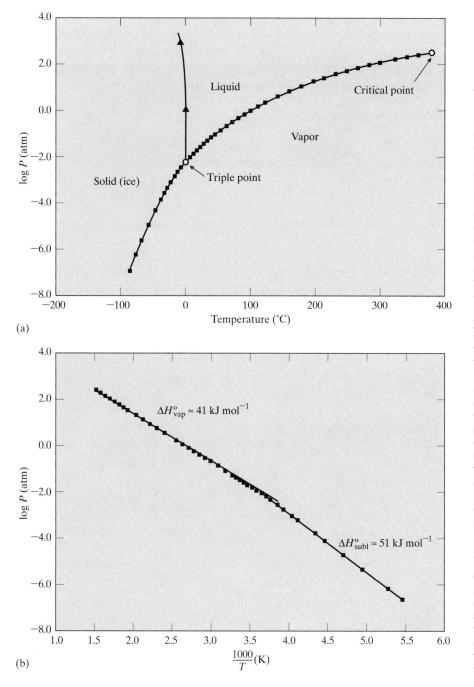

(a)

(b)

◀ FIGURE 5.2

Phase relations for pure water over a range of pressure and temperature. (a) The vapor pressure of water, on a logarithmic scale, is plotted against temperature for ice (lower left) and liquid water (upper right), as well as the pressure dependence of the freezing or melting temperature (upper center). The three curves meet at the *triple point:* the only point where three phases of water can be in equilibrium with one another. The triple point for water occurs at 0.0075°C and 4.579 torr. The curve at the upper right ends at the *critical point,* beyond which there is no longer a distinction between liquid and vapor phases. The critical point for water occurs at 647.3 K and 217.6 atm. The critical volume is 56.0 cm³ mol⁻¹. The nearly vertical curve showing the transition between ice and liquid extends to greater than 2000 atm. At still higher pressures, different solid phases of water occur. (b) Phase relations for solid–vapor and liquid–vapor equilibria of water plotted according to the Clausius–Clapeyron equation, Eq. (5.3). From the slopes of the curves, $\Delta \overline{H}^0_{subl}$ and $\Delta \overline{H}^0_{vap}$, respectively, can be calculated. The slight curvature of the liquid–vapor data results from the difference in heat capacity between liquid and gaseous water. The $\Delta \overline{H}^0_{subl}$ and $\Delta \overline{H}^0_{vap}$ values shown on the plot are averages over a wide temperature range. Values given in table 2.2 are more precisely determined from the tangents to the curves at the temperature indicated.

versus $1/T$. The slope is equal to $-\Delta \overline{H}_{vap}/R$. The vapor pressure at any temperature, or the boiling point at any pressure, can be read from the graph.

In using Eq. (5.3), we should remember its limitations. The main one is the assumption that $\Delta \overline{H}_{vap}$ is independent of temperature. Over a wide-enough temperature range, $\ln P$ versus $1/T$ will be a curve; the slope at any point will give $\Delta \overline{H}_{vap}$ at that temperature. The assumption of gas ideality and the neglect of pressure dependence of the liquid activity are usually valid. We can increase the activity and therefore increase the vapor pressure of a liquid by applying an external pressure. However, this effect is small compared with the large effect of temperature on the vapor pressure.

For the effect of pressure on the freezing- or melting-point temperature, we need to use a somewhat different and even more general approach. Consider a solid and a liquid in equilibrium at temperature T and pressure P. Then,

$$\overline{G}_s = \overline{G}_l$$

at T and P. Now increase the temperature by an amount dT and the pressure by an amount dP so as to maintain the solid and liquid in equilibrium. Under the new conditions,

$$\overline{G}_s + d\overline{G}_s = \overline{G}_l + d\overline{G}_l$$

at $T + dT$, $P + dP$. Subtracting the first equation from the second, we obtain

$$d\overline{G}_s = d\overline{G}_l$$

This states that, if we change the pressure of a solid substance in equilibrium with its liquid, we must also change the temperature so as to keep the increment in free energy the same for the two phases. Otherwise, the one with higher free energy will be converted into the one with lower free energy.

Using

$$d\overline{G} = -\overline{S}\,dT + \overline{V}\,dP \qquad \textbf{(3.35)}$$

we obtain

$$-\overline{S}_s\,dT + \overline{V}_s\,dP = -\overline{S}_l\,dT + \overline{V}_l\,dP$$

which can be rearranged to give

$$\frac{dT}{dP} = \frac{\Delta\overline{V}_{\text{fus}}}{\Delta\overline{S}_{\text{fus}}}$$

$$\Delta\overline{V}_{\text{fus}} = \overline{V}_1 - \overline{V}_s \qquad \Delta\overline{S}_{\text{fus}} = \overline{S}_1 - \overline{S}_s$$

Because the process is reversible,

$$\Delta\overline{S}_{\text{fus}} = \frac{\Delta\overline{H}_{\text{fus}}}{T}$$

and

$$\frac{dT}{dP} = \frac{T\Delta\overline{V}_{\text{fus}}}{\Delta\overline{H}_{\text{fus}}} \qquad \textbf{(5.4)}$$

This is an example of a general expression that can be applied to equilibrium between any two phases. For example, Eq. (5.3) can be derived as a special case for liquid–vapor equilibrium.

The effect of external pressure on melting points is not large, so we can write Eq. (5.4) as

$$\frac{T_2 - T_1}{P_2 - P_1} = T\frac{\Delta\overline{V}_{\text{fus}}}{\Delta\overline{H}_{\text{fus}}} \qquad \textbf{(5.5)}$$

where T = average of T_2 and T_1
$\Delta\overline{V}_{\text{fus}} = \overline{V}(l) - \overline{V}(s)$ = molar volume change of fusion
$\Delta\overline{H}_{\text{fus}} = \overline{H}(l) - \overline{H}(s)$ = molar enthalpy (heat) of fusion

We must be careful with units in this equation. If P is in atm, $\Delta \overline{V}_{fus}$ in m^3 mol^{-1} and $\Delta \overline{H}_{fus}$ in J mol^{-1}, the conversion factor of 9.869×10^{-6} m^3 atm J^{-1} must be used. The effect of pressure on a melting point depends on the signs and magnitudes of two thermodynamic variables: $\Delta \overline{H}_{fus}$ and $\Delta \overline{V}_{fus}$. The heat of fusion is always positive; heat is always absorbed on melting. The volume of fusion, however, can be positive or negative, although it is usually positive. A notable exception is water, where the volume decreases on melting.

Solutions of Two or More Components

For solutions containing two or more components present in two or more phases at equilibrium, we can generalize the equations in the previous section.

$$\mu_A(\text{phase } 1) = \mu_A(\text{phase } 2) = \cdots \tag{5.6}$$

$$\mu_B(\text{phase } 1) = \mu_B(\text{phase } 2) = \cdots \tag{5.7}$$

Vapor pressure

For equilibrium between liquid and vapor, there is equality of the chemical potentials of each component in the two phases. For example, if A is the solvent in a liquid solution,

$$\mu_A(\text{solution}) = \mu_A(g, P_A)$$

$$\mu_A^{\bullet}(\text{pure solvent}) = \mu_A^{\bullet}(g, P_A)$$

where P_A = vapor pressure of A for solution
$P_A^{\bullet}$ = vapor pressure of pure solvent A

Note that $\mu_A^{\bullet}$ and $P_A^{\bullet}$ refer to the pure solvent A at the temperature of the system. These are, in general, different from μ_A^0 and P_A^0 in the standard state, where $P_A^0 = 1$ atm. For liquid solvents below their boiling temperature, $P_A^{\bullet}$ will be less than 1 atm. For an ideal gas, we can use Eq. (3.45) to obtain

$$\mu_A(g, P_A) - \mu_A^{\bullet}(g, P_A^{\bullet}) = RT \ln \frac{P_A}{P_A^{\bullet}}$$

Any nonideality in the gas at pressures near 1 atm can be ignored, because it will be small compared to nonidealities in solution. Therefore,

$$\mu_A(\text{solution}) - \mu_A^{\bullet}(\text{pure solvent}) = RT \ln \frac{P_A}{P_A^{\bullet}}$$

We can now relate vapor pressures to the activity of the solvent. From Eqs. (4.19) and (4.23),

$$\mu_A(\text{solution}) - \mu_A^{\bullet}(\text{pure solvent}) = RT \ln a_A$$

Therefore, at equilibrium with gas,

$$RT \ln \left(\frac{P_A}{P_A^{\bullet}} \right) = RT \ln a_A$$

$$a_A = \frac{P_A}{P_A^{\bullet}} \tag{5.8}$$

All of these must be measured at the same temperature. Vapor-pressure measurements thus provide a simple method for measuring the activity of the solvent.

The activity of the solvent can be written

$$a_A = \gamma_A X_A \tag{5.9}$$

which leads to Raoult's law for ideal solutions, where $\gamma_A \equiv 1$:

$$P_A = X_A P_A^\bullet \tag{5.10}$$

It can be shown that when Raoult's law applies to the solvent in a solution, Henry's law must apply to the solute.

Henry's law

Consider as an example the oxygen O_2 that is necessary to support life in multicellular organisms (fish, animals, humans) and many microorganisms. Oxygen is supplied to the atmosphere by photosynthesis and is consumed by a variety of biological and nonbiological processes. At present, the concentration of O_2 in the atmosphere is about 20%, and lakes and oceans also store a significant amount as dissolved O_2. At equilibrium at 25°C, the mole fraction of dissolved O_2 in water, X_{O_2}, is 4.7×10^{-6}. Although this seems like a very small number, the amount of O_2 dissolved in 1 L of water is about 3% of that contained in 1 L of air at 1 atm and 25°C.

The equilibrium described by the expression

$$O_2(g) \rightleftharpoons O_2(aq)$$

is governed by *Henry's law*. Henry found experimentally that the solute (O_2) vapor pressure is proportional to the solute mole fraction in solution. Henry's law is stated

$$P_B = k_B X_B \tag{5.11}$$

where P_B = vapor pressure of solute B
$\qquad X_B$ = mole fraction of B in the liquid solution
$\qquad k_B$ = Henry's law constant for solute B in solvent A

Thus, Henry's law shows that gas solubility is directly proportional to the gas pressure. Henry's law constants for some common gases in water are given in table 5.1. Among these gases, He is the least soluble, and C_2H_2 and CO_2 are

TABLE 5.1 Henry's Law Constants, k_B (atm) = P_B/X_B, for Aqueous Solutions at Three Temperatures

Gas	0°C	25°C	37°C
He	133×10^3	141×10^3	140×10^3
N_2	51×10^3	85×10^3	99×10^3
CO	35×10^3	58×10^3	68×10^3
O_2	26×10^3	43×10^3	51×10^3
CH_4	23×10^3	39×10^3	47×10^3
Ar	24×10^3	39×10^3	46×10^3
CO_2	0.72×10^3	1.61×10^3	2.16×10^3
C_2H_2	0.72×10^3	1.34×10^3	1.71×10^3

Source: Data from A. H. Harvey, *AIChE Journal* 42 (1996): 1491–1494.

the most soluble. Gas solubility is also a function of temperature and depends on the solvent. Oxygen is about 20% less soluble at 37°C (physiological temperature) than at 25°C. By contrast, O_2 is 30 times more soluble in the liquid perfluorooctylbromide (perflubron, PFB). PFB is undergoing testing as a liquid breathing medium for certain infants and adults subject to lung failure and as a temporary blood substitute during coronary operations.

Raoult's law

In a complete description of the two-phase (gas, liquid) system containing two components (O_2, water), we note that the volatile solvent (water) will be present also in the vapor phase. At equilibrium, the vapor pressure of the solvent above the solution is given by *Raoult's law:*

$$P_A = X_A P_A^{\bullet} \tag{5.10}$$

Note that this expresses the same proportionality between vapor pressure and mole fraction as does Henry's law, but the proportionality constant is different. For dilute solutions, where these laws hold best, X_A is typically near 1.0. It is reasonable that in this limit the solvent vapor pressure will approach that of pure solvent, $P_A^{\bullet}$, which is the proportionality constant for Raoult's law. By contrast, solute B in the same solution is surrounded almost entirely by A molecules, which is typically a very different environment from that in pure liquid B. As a consequence, Henry's law constants, k_B, are typically different in different solvents and different from $P_B^{\bullet}$. For very concentrated solutions, the distinction between solvent and solute is arbitrary. Furthermore, nonideal-solution behavior results in the failure of ideal solution relations like Eqs (5.10) and (5.11).

EXAMPLE 5.2

Use Raoult's law to calculate the vapor pressure of water saturated with air at 1 atm pressure and 25°C. For this purpose, we can consider dry air to be a mixture of 78.08% N_2 (by volume), 20.95% O_2, and 0.93% Ar. Using Henry's law and the data in table 5.1, we calculate the total mole fraction of dissolved air:

$$X_{air} = X_{N_2} + X_{O_2} + X_{Ar} = \frac{P_{N_2}}{k_{N_2}} + \frac{P_{O_2}}{k_{O_2}} + \frac{P_{Ar}}{k_{Ar}}$$

$$X_{air} = \frac{0.7808}{86 \times 10^3} + \frac{0.2095}{43 \times 10^3} + \frac{0.0093}{40 \times 10^3} = 14.18 \times 10^{-6}$$

The vapor pressure of pure water at 25°C obtained from standard tables is 23.756 torr = 0.03126 atm. The *vapor-pressure lowering* for air-saturated water at 25°C and 1 atm is

$$\Delta P_{H_2O} = P_{H_2O}^{\bullet} - P_{H_2O} = P_{H_2O}^{\bullet}(1 - X_{H_2O}) = P_{H_2O}^{\bullet} X_{air}$$

$$\Delta P_{H_2O} = (23.756 \text{ torr})(14.18 \times 10^{-6}) = 3.3 \times 10^{-4} \text{ torr}$$

Thus, the effect of the dissolved air is to *decrease* the water vapor pressure by only 1.4×10^{-3}%. This is a completely negligible change from the vapor pressure of pure water. The presence of saturated water vapor in the air (the definition of 100% humidity) does, however, change the percent composition of the humid air, which contains $(23.756/760) \times 100 = 3.126$ mol % H_2O. The percentages of the other gases relative to dry air need to be decreased accordingly.

Vapor-pressure lowering is a practical method for determining the amount of a solute in solutions. Normal saline solution (0.9% by wt NaCl), for example, exhibits a vapor-pressure lowering relative to that of pure water of 0.27 torr, which is readily measured using a sensitive manometer. Another example of the use of this technique is in determining the molecular mass of a protein in solution. If a weighed sample of a pure protein is dissolved in a known volume of water, then measurement of the vapor-pressure lowering of the solution allows the mole fraction of the protein to be calculated by the method of example 5.2. Knowing both the weight concentration and the mole fraction allows us to calculate the molecular mass of the protein in the solution.

For a system like air-saturated water to be in equilibrium, each component must have the same chemical potential in each phase. If the concentration of O_2 in the liquid-water phase is so low that the chemical potential μ of the dissolved O_2 is less than that of the O_2 in the gas phase, then O_2 will be transferred from the gas phase until equilibrium is reached. At that point, $\Delta\mu$ for the transfer is zero, and no further change occurs. Therefore, at equilibrium

$$\mu_{O_2}(aq) = \mu_{O_2}(g)$$

This is also true for water in the system and for nitrogen and argon. If we have a more complex system at equilibrium, then the same relation holds for each component in each phase. For example, perflubron (PFB) is a nonpolar liquid that is almost completely insoluble in water. If you mix air, water, and PFB together in a flask, three phases appear: a gas phase on top, an aqueous phase in the middle, and a liquid PFB phase at the bottom (figure 5.3). This arrangement occurs because the gas has the lowest density and the PFB has the greatest density. At equilibrium,

$$\mu_{O_2}(\text{PFB}) = \mu_{O_2}(aq) = \mu_{O_2}(g)$$
$$\mu_{H_2O}(\text{PFB}) = \mu_{H_2O}(aq) = \mu_{H_2O}(g)$$
$$\mu_{\text{PFB}}(\text{PFB}) = \mu_{\text{PFB}}(aq) = \mu_{\text{PFB}}(g)$$
$$\mu_{N_2}(\text{PFB}) = \mu_{N_2}(aq) = \mu_{N_2}(g)$$
$$\mu_{Ar}(\text{PFB}) = \mu_{Ar}(aq) = \mu_{Ar}(g)$$

The first equation states that the chemical potential (free energy per mole) of O_2 in the PFB solution is equal to the chemical potential of O_2 in the aqueous solution, which is equal to the chemical potential of O_2 in the gas phase. The second equation makes the same statement for water.

The chemical potential depends on the activity of each species in each phase:

$$\mu_A = \mu_A^0 + RT \ln a_A = \mu_A^0 + RT \ln \gamma_A c_A \tag{4.19}$$

Therefore, if the same standard state is used for a component in two different phases, its activity in the two phases will be the same at equilibrium. For phase equilibrium (same standard state in both phases),

$$a_A(\text{phase 1}) = a_A(\text{phase 2})$$

For the air, water, and PFB mixture, for example, we can choose pure O_2 gas at 1 atm as the standard state for all three phases. The activities of O_2 are

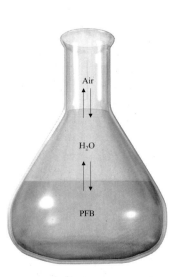

▲ FIGURE 5.3

A flask containing the immiscible liquids water and perflubron (PFB) is in equilibrium with air. (The labels indicate only the major component in each phase.) For the many components present in this system, the chemical potential of each component is the same in all three phases at equilibrium.

equal in the three phases at equilibrium, but the concentrations are very different. In the gas phase, as we saw in example 5.2, the mole fraction of O_2 is $(0.2095)(1 - 0.03126) = 0.2030$; in the aqueous phase, the mole fraction of O_2 is 4.7×10^{-6}, and in the PFB phase the mole fraction of O_2 is 1.4×10^{-4}. These values are for total pressure 1 atm and temperature 298 K. It is important to emphasize that the activities of O_2 are the same in all three phases because we have chosen the same standard state for O_2 for all three phases.

Equilibrium Dialysis

If we had to rely only on normal saline solution flowing in our veins and arteries, the amounts of O_2 transferred from our lungs to muscle cells would be far too little to support life as we know it. The dissolved oxygen concentration is too low, even if we breathe pure O_2. Red blood cells (erythrocytes) solve this problem for us by packaging hemoglobin (Hb) molecules, which are proteins containing a heme (Fe protoporphyrin IX, see appendix table A.9) prosthetic group that binds O_2 strongly. Muscle cells contain a related protein, myoglobin (Mb), that binds O_2 and stores it until needed. Red blood cells are packed full of Hb and contained by a plasma membrane boundary. The plasma membrane is freely permeable to small, neutral molecules like O_2 but is impermeable to large protein molecules like Hb and to many charged ions. Of the dozen or so proteins associated with lipids in the membrane, one is an anion channel that enables bicarbonate to be exchanged for chloride. A second function of the red blood cells is to carry the product of oxidative metabolism, bicarbonate, from the muscle cells back to the lungs, where it is converted to CO_2 gas and exhaled. Despite all of this complexity, red blood cells are small enough to squeeze through the tiniest capillaries in the circulatory system.

The operation of the red bood cell is illustrative of a technique called *equilibrium dialysis* that is a laboratory method of studying the binding of a ligand (small molecule like O_2) by a macromolecule (protein, nucleic acid, etc.) quantitatively (See figure 5.4). Materials made from reformulated cellulose are typically used instead of a biological membrane. These sheet polymers contain small pores that allow the free passage of small molecules, including ions, but prevent the large macromolecules from passing. Tubes of cellulose can be filled with 100 mL or so of a macromolecular solution and tied off. The closed specimen solution in the dialysis bag can then be immersed in a beaker of solution containing the ligand. Immediately, the ligand starts to diffuse across the dialysis membrane, and this continues until equilibrium is reached. Stirring the contents speeds the approach to equilibrium.

Once equilibrium is reached, all concentrations remain constant. Portions of the contents of the dialysis bag and of the outside medium may then be removed for analysis. In a "control" where no macromolecule is present in the dialysis bag, the concentrations of all species will be the same inside and outside the bag. In a different "control" where a protein like albumin that does not bind O_2 is in the bag, then the O_2 concentration at equilibrium will be the same inside and outside, although the protein is inside and not outside. If a protein like myoglobin is in the bag, then the total concentration of O_2 (in mol L^{-1}) inside will be significantly larger than the concentration of O_2 outside, where Mb is absent. The presence of the Mb inside serves to concentrate the O_2 by binding it. By a series of quantitative measurements, it is possible to determine both the binding equilibrium constant and the number

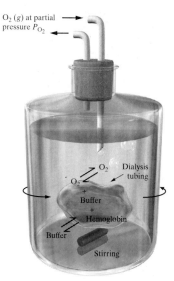

▲ FIGURE 5.4

Equilibrium dialysis of a solution of hemoglobin against a buffer saturated with O_2 gas at a partial pressure P_{O_2}. The hemoglobin solution is contained inside a closed dialysis bag and is in the same buffer as is present outside the bag. The system is well stirred to provide a more rapid approach to equilibrium. The dialysis bag material is permeable to water, O_2, and buffer components, but not to hemoglobin.

of ligand binding sites per macromolecule. A convenient way of treating the data to extract these values is through the Scatchard equation (discussed in the next section).

Consider the equilibrium involving a macromolecule, M, with a single binding site for a ligand, A.

$$M + A \overset{K}{\rightleftharpoons} M \cdot A$$

$$K = \frac{[M \cdot A]}{[M][A]}$$

To relate these concentrations to the measurable quantities,

$$c_M = [M \cdot A] + [M] = \text{total concentration of M(inside)}$$

$$c_A(\text{outside}) = [A] \tag{5.12}$$

$$c_A(\text{inside}) = [A] + [M \cdot A]$$

$$c_A(\text{bound}) = c_A(\text{inside}) - c_A(\text{outside}) = [M \cdot A] \tag{5.13}$$

An assumption inherent in these equations is that the concentration of free (unbound) A inside is the same as the total concentration of A outside. This is a consequence of the requirement that the chemical potential and therefore the activity of A must be the same inside and outside. Combining the relations above, we obtain an expression for the equilibrium constant in terms of measurable quantities:

$$K = \frac{c_A(\text{bound})}{\{c_M - c_A(\text{bound})\}c_A(\text{outside})}$$

The Scatchard Equation

The equation for K can be simplified and generalized by introducing ν, the average number of ligand molecules bound per macromolecule at equilibrium.

$$\nu = \frac{c_A(\text{bound})}{c_M} \tag{5.14}$$

where ν = average number of A's bound per macromolecule
$c_A(\text{bound})$ = concentration of A bound
c_M = concentration of macromolecule

Writing the expression for K in terms of ν (divide the numerator and denominator by c_M) we obtain

$$K = \frac{\nu}{(1 - \nu)c_A(\text{outside})} = \frac{\nu}{(1 - \nu)[A]}$$

or

$$\frac{\nu}{[A]} = K(1 - \nu)$$

The equilibrium constant was written with the assumption that only one molecule of A was bound per macromolecule; this means that ν can vary only from 0 (no A bound) to 1 (each macromolecule has bound an A). However, many macromolecules have more than one binding site for a ligand. (Hb for example has four binding sites for O_2.) Now ν can vary from 0 to N,

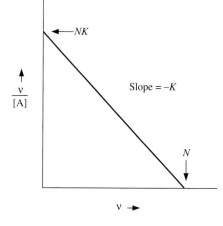

◀ FIGURE 5.5

Scatchard plot of the binding of substrate A to a macromolecule containing N binding sites, where ν is the average number of A molecules bound per macromolecule and K is the equilibrium binding constant.

the number of binding sites on each macromolecule. If the sites are identical and independent, it is very easy to generalize the equation. If there are N *identical and independent binding sites* per macromolecule, this means that the N sites have the same binding equilibrium constant K and that binding at one site does not change the binding at another site. Then we can replace ν in the equation for the equilibrium constant above by ν/N.

$$\frac{\nu/N}{[A]} = K\left(1 - \frac{\nu}{N}\right) \tag{5.15}$$

By rearranging, we obtain the *Scatchard equation:*

$$\frac{\nu}{[A]} = K(N - \nu) \tag{5.16}$$

The Scatchard equation (5.16) is often used to study binding to a macromolecule. The binding can be measured by any suitable method, but equilibrium dialysis is often convenient. The value of [A], which we designated as $c_A(\text{free})$, is the concentration on the solvent side of the dialysis membrane at equilibrium. The value of ν comes from the concentration of A on the macromolecule side at equilibrium [see Eqs. (5.13) and (5.14)]. A plot of $\nu/[A]$ versus ν should give a straight line with slope of $-K$, y-intercept of NK, and x-intercept of N, as shown in figure 5.5. If a Scatchard plot does not give a straight line, this indicates that the binding sites are not identical or not independent. (The four binding sites for O_2 in Hb are neither identical nor independent.) The consequences of interactions between adjacent binding sites are discussed more fully in the next section.

Data describing the binding of O_2 to Mb are shown in figure 5.6. Taking advantage of the different visible absorption spectra of myoglobin with and without O_2 bound, the fraction of Mb having O_2 bound can be determined experimentally for a range of O_2 pressures. These values are plotted directly in figure 5.6(a). Two features of this plot are noteworthy: (1) The fraction of binding sites occupied increases toward a maximum with increasing pressure of O_2. (2) The pressure of O_2 at which the Mb sites are half-saturated, designated $p50$, is a measure of the strength of the binding. The $p50$ value for Mb at 30°C is 1.5 torr. Data obtained for O_2 binding to Mb at several temperatures are plotted as Scatchard plots in figure 5.6(b). Despite the scatter in the

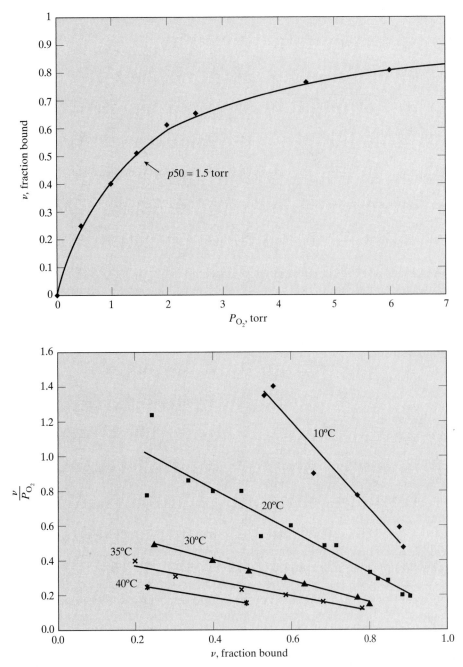

▶ FIGURE 5.6

Binding of O_2 to human myoglobin. (a) Data obtained at 30 °C are plotted as the fraction of binding sites occupied, ν, against the pressure of O_2 (torr). (b) Scatchard plots of the binding of O_2 to Mb at several temperatures. [Data from A. Rossi-Fanelli and E. Antonini, *Arch. Biochem. Biophys. 77* (1958): 478–492.]

experimental data for individual temperatures, it is evident that they can be fitted by a set of straight lines. The fact that the straight lines extrapolate reasonably close to $\nu = 1$ at $\nu/P_{O_2} = 0$ is consistent with a picture that each Mb molecule has a single and necessarily independent binding site. From the point where each line crosses the vertical corresponding to $\nu = 0.5$, we see that the $p50$ values increase with increasing temperature. The studies are consistent with the mechanism

$$Mb + O_2 = Mb \cdot O_2$$

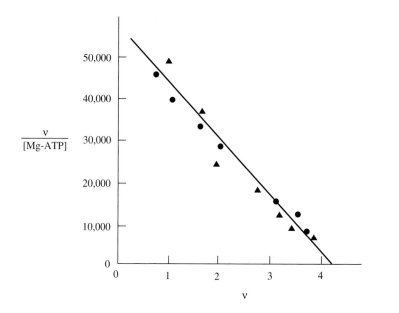

◀ FIGURE 5.7

Binding of Mg–ATP by tetrahydrofolate synthetase from *C. cylindrosporum*. The circles and triangles
represent two sets of data obtained
by two different techniques. Measurements were carried out at 23°C
and a pH of 8.0. [Data from N. P.
Curthoys and J. C. Rabinowitz,
J. Biol. Chem. 246 (1971): 6942.]

EXAMPLE 5.3

The formyltetrahydrofolate synthetases can utilize the energy of hydrolysis of ATP
(to ADP and phosphate) for the formation of a carbon–nitrogen bond between
(*l*)-tetrahydrofolate and formate. Binding of the Mg complex of ATP to such an enzyme from *C. cylindrosporum* has been measured; the data are plotted in Scatchard
form in figure 5.7.

The data fit the identical-and-independent-sites model well. The slope of the
plot gives an equilibrium constant $K = 1.37 \times 10^4 \, M^{-1}$. The intercept gives $N = 4.2$.
Since the enzyme is known to have four identical subunits (each has a molecular
weight of 60,000), the binding results are consistent with one Mg–ATP binding
site per subunit.

In the discussion leading to Eq. (5.16) we defined the quantities ν and N
as follows:

$$\nu = \text{average number of A's bound to macromulecule}$$

$$N = \text{number of binding sites on macromolecule}$$

A macromolecule is made of monomer units, such as the peptide units in a
protein. A DNA molecule is a polynucleotide consisting of nucleotides as its
monomer building blocks. Sometimes it is more convenient to define the
binding parameters on a *per monomer unit basis* as follows:

$$r = \text{number of bound A } \textit{per monomer unit of the macromolecule}$$

$$n = \text{number of binding sites } \textit{per monomer unit of the macromolecule}$$

The Scatchard equation is unchanged by these new definitions:

$$\frac{r}{[A]} = K(n - r) \qquad\qquad \textbf{(5.17)}$$

Thus, the Scatchard plot can also be presented with r and n as the parameters.

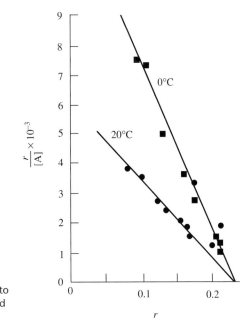

▶ FIGURE 5.8

Scatchard plot of the binding of ethidium to DNA in a medium containing 3 M CsCl and 0.01 M Na$_3$EDTA.

In figure 5.8, data are plotted for the binding of the trypanocide drug ethidium (ethidium is sometimes used to treat animals infected by parasitic trypanosomes) to a double-stranded DNA. Note that the data are represented reasonably well by the identical-and-independent-sites model. The r intercept gives $n \approx 0.23$, or approximately one binding site per four nucleotides (two base pairs). This is consistent with the fact that ethidium binds to DNA by *intercalation*. Planar, aromatic molecules like ethidium slide between two adjacent base pairs in DNA so that the hydrophobic parts are stacked on the base pairs; this is intercalation. Note also that at the lower temperature the line is steeper, indicating a greater binding constant. This temperature dependence of the binding constant yields a negative $\Delta \overline{H}^0$ (calculated to be ~ -25 kJ mol^{-1}) for the binding of the drug. The fairly strong interaction ($K = 2.66 \times 10^4\ M^{-1}$ and $5.60 \times 10^4\ M^{-1}$ at 20°C and 0°C, respectively) between the drug and DNA is evidence that the target of the drug *in vivo* is the DNA of the trypanosome.

Cooperative Binding and Anticooperative Binding

In the previous section, a macromolecule with many binding sites was assumed to have identical and noninteracting sites. Frequently, however, although the sites are identical, there is significant interaction among them. The binding of a ligand to one site affects the binding of ligands to other sites. The first ligand bound affects the binding of the next one, which affects the binding of the next one, and so on. In *cooperative binding*, the first ligand bound makes it easier for the next one to be bound. This may be caused by a conformational change in a multisubunit protein that makes it easier for successive ligands to be bound. The limiting case of cooperative binding is *all-or-none binding*. In all-or-none binding the first ligand increases the binding of the other ligands so much that all N ligands are bound at once. As the concentration of ligand is increased, the number of ligands bound increases sharply to the maximum possible. The macromolecule has either no ligands bound or N ligands bound.

In *anticooperative binding,* each succeeding ligand is bound less strongly than the previous one. An example of anticooperative binding occurs, for example, in the titration of a protein which has ten carboxyl groups from ten glutamic acid side chains. The carboxylate groups with pK's near 4 have the same intrinsic binding constant for hydrogen ions; however, they can interact coulombically. As the pH is lowered toward 4, the first proton is attracted by ten negative charges; it is bound strongly. The next proton is not attracted as strongly, and so forth. Each proton bound makes it harder to bind the next proton. For some macromolecules and ligands, the anticooperativity may be so strong that the binding of a ligand prevents the binding of another ligand at a neighboring site. This is *excluded-site binding*—the binding at one site excludes binding at another site. An example of excluded-site binding is the binding of the antibiotic drug actinomycin to DNA. An actinomycin molecule consists of a heterocyclic ring phenoxazone, with two cyclic pentapeptides attached to the ring. It bonds to double-stranded DNA by intercalating the phenoxazone ring between two adjacent base pairs and tucking the peptides into the narrow groove of the DNA helix. One guanine–cytosine $(G \cdot C)$ base pair must be present at each binding site.

A useful way to plot binding data and to learn about the cooperativity (or anticooperativity) of binding is to plot fraction of sites bound, f, versus concentration of free ligand. A vital example of cooperative binding is the binding of oxygen by hemoglobin. Hemoglobin has four heme binding groups that bind four oxygen molecules cooperatively. The binding is cooperative so that hemoglobin releases most of its bound oxygen at the low oxygen pressure in the tissues but binds the maximum amount of oxygen in the lungs. Figure 5.9 compares the binding of oxygen to myoglobin with only one heme-binding site and to hemoglobin with four interacting binding sites. The cooperative-binding curve for hemoglobin shows a characteristic *sigmoidal* shape; sigmoidal means "shaped like the letter S." Myoglobin, which has only one binding site so it cannot bind cooperatively, has a binding curve characteristic of a molecule with any number of identical and noninteracting sites. That is, the data for myoglobin are represented by the Scatchard equation; a linear Scatchard plot is obtained.

A *Hill plot* is a plot to quantitatively assess the cooperativity of binding. The Scatchard equation was derived for no interaction among sites. In the form of Eq. (5.15) it can be rewritten as

$$\frac{\nu/N}{(1 - \nu/N)} = \frac{f}{(1 - f)} = K[A] \qquad (5.18)$$

where $f = \nu/N$ is the fraction of sites occupied. This equation represents the binding curve for myoglobin or for any number of noninteracting and identical sites. For cooperative binding, the form of the equation becomes

$$\frac{f}{(1 - f)} = K[A]^n \qquad (5.19)$$

n = Hill coefficient
K = a constant, not the binding constant for one ligand

The Hill coefficient, n, varies from 1 for no cooperativity to N for all-or-none binding of N ligands. The way to determine the Hill coefficient or coopera-

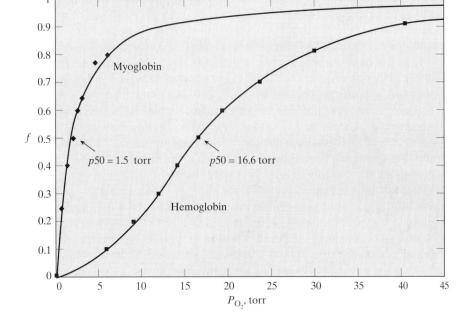

▶ **FIGURE 5.9**

Fraction of hemes of Mb or Hb occupied by oxygen, f, as a function of the pressure of oxygen. Myoglobin has only one heme group and therefore cannot show cooperativity. Hemoglobin has four hemes; the binding is cooperative. [Data for Mb from figure 5.6; data for Hb from P. Astrup, K. Engel, J. W. Severinghaus, and E. Munson, *Scandinavian J. Clin. & Laboratory Investigation* 17 (1965): 515–523.]

tivity coefficient is to take the logarithm of Eq. (5.19) and plot $\log[f/(1 - f)]$ against $\log[A]$:

$$\log \frac{f}{(1 - f)} = n \log[A] + \log K \qquad (5.20)$$

Clearly, the slope of the plot gives n, the cooperativity coefficient. This is illustrated in figure 5.10 for Mb and Hb. For Mb, the cooperativity coefficient is 1, as it must be for a single site. The slope found is typically 2.8 for the cooperative binding of oxygen by Hb, which is greater than 1 but less than the all-or-none value of 4.

EXAMPLE 5.4

A formulation of cooperativity in the binding of oxygen to Hb was presented by Monod, Wyman, and Changeux in 1965. This model proposes that Hb exists in two different conformational states that are in equilibrium with one another but differ in their affinity for O_2. The weakly liganding form, designated T for *tense*, requires a structural constraint. The predominant form for liganded Hb is designated R for *relaxed*, to reflect the release of the structural constraint upon binding O_2. For the cooperative binding of O_2 by Hb at pH 7.0 in phosphate buffer, the equilibrium between the T and R states is defined by the expression $L = [T_0]/[R_0]$, where $[T_0]$ and $[R_0]$ are the concentrations with no ligand bound. In addition, the intrinsic affinity of the T state for O_2 is nearly 100-fold smaller than that of the R state; thus, $c = K_R/K_T = 0.014$, where K_R and K_T are the dissociation constants for the R and T states, respectively. Using this model, we can formulate expressions for the successive O_2-binding steps. Because there are four independent sites for O_2 to bind to the R state and to the T state, we write

$$[Hb(O_2)] = 4\left(\frac{[R_0][O_2]}{K_R} + \frac{[T_0][O_2]}{K_T}\right) = \frac{4[R_0][O_2]}{K_R}(1 + Lc)$$

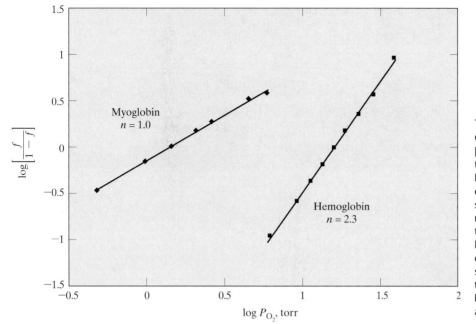

Comparison of the Hill plots [Eq. (5.20)] for the binding of O_2 to Mb and Hb. The slope of each line is the Hill cooperativity coefficient. For no cooperativity, the slope is 1; for maximum (all-or-none) cooperativity the slope is N, the total number of binding sites. Hemoglobin typically has a Hill coefficient of 2.8 for four binding sites, although the value is closer to 2.3 for the data shown. The binding is cooperative, but not all-or-none.

Similarly, there are six ways to bind two O_2 ligands, four ways to bind three, and only one way to bind four. (A description of this treatment of multiple independent binding equilibrium is given in chapter 11.) Therefore

$$[Hb(O_2)_2] = 6\left(\frac{[R_0][O_2]^2}{K_R^2} + \frac{[T_0][O_2]^2}{K_T^2}\right) = \frac{6[R_0][O_2]^2}{K_R^2}(1 + Lc^2)$$

$$[Hb(O_2)_3] = \frac{4[R_0][O_2]^3}{K_R^3}(1 + Lc^3) \qquad \text{and}$$

$$[Hb(O_2)_4] = \frac{R_0[O_2]^4}{K_R^4}(1 + Lc^4)$$

Noting that $[Hb] = [R_0] + [T_0] = [R_0](1 + Lc)$, we can now write expressions for the macroscopic equilibrium constants $K_1 \ldots K_4$:

$$\left.\begin{aligned}
K_1 &= \frac{[Hb(O_2)]}{[Hb][O_2]} = \frac{4(1 + Lc)}{K_R(1 + L)} \\[2mm]
K_2 &= \frac{[Hb(O_2)_2]}{[Hb(O_2)][O_2]} = \frac{3(1 + Lc^2)}{2K_R(1 + Lc)} \\[2mm]
K_3 &= \frac{[Hb(O_2)_3]}{[Hb(O_2)_2][O_2]} = \frac{2(1 + Lc^3)}{3K_R(1 + Lc^2)} \\[2mm]
K_4 &= \frac{[Hb(O_2)_4]}{[Hb(O_2)_3][O_2]} = \frac{1(1 + Lc^4)}{4K_R(1 + Lc^3)}
\end{aligned}\right\} \qquad \textbf{(5.21)}$$

The fraction of sites occupied is

$$f = \frac{\alpha(1 + \alpha)^3 + L\alpha(1 + c\alpha)^3}{(1 + \alpha)^4 + L(1 + c\alpha)^4} \qquad \textbf{(5.22)}$$

where $\quad \alpha = [O_2]/K_R \qquad \textbf{(5.23)}$

We use the values of $p50$ from the experimental data of Astrup et al., part of which are shown in figure 5.9, to determine a value $L = 12{,}550$ and the corresponding values of α. This is done by numerical approximation using Eq (5.22) and results

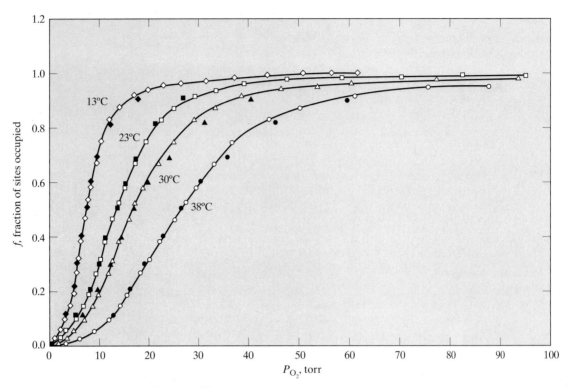

▲ FIGURE 5.11

Plots showing the binding of O_2 to Hb and comparison with the Monod–Wyman–Changeaux (MWC) model. Experimental data (filled symbols) from Astrup et al. (1965) for human Hb at four temperatures. Calculated values (open symbols) determined using the MWC model were obtained using Eq. (5.22) with parameters $c = 0.014$ and $L = 12{,}550$, together with $p50$ values taken from the experiments at each temperature. Then, Eq. (5.23) allows us to calculate K_R (torr) = 13.6 (13°C), 25.2 (23°C), 33.2 (30°C), and 53.4 (38°C).

in values $K_1 K_R = 0.056$, $K_2 K_R = 0.029$, $K_3 K_R = 0.199$, and $K_4 K_R = 0.242$. Finally, we determine K_R at each temperature using the experimental values of $p50$ and the corresponding values of α in Eq. (5.23). The experimental data are shown in figure 5.11 at four temperatures using solid symbols and the calculated values from the MWC model using the corresponding open symbols. Qualitatively, this model accounts for the sigmoidal behavior that results from relatively weak binding, compared with Mb for example, at low oxygen pressure and then the rapid increase in affinity as the first sites of each Hb become occupied, reflecting the cooperative interactions between the Hb subunits. The fact that the calculated and observed data do not agree perfectly results from several factors: The experimental data are for human Hb at several temperatures, whereas the value for c was chosen to be optimal for horse Hb at 19°C. Better fits can be obtained by optimizing L and c for each data set to which they are being fit. An alternative model proposed by Koshland, Nemethy, and Filmer (1966) can be made to fit these data about equally well, so the decision between the models must be made on the basis of other studies.

Free Energy of Transfer Between Phases
Transfer between two different solvents

The equality of chemical potential for a solute in equilibrium in different phases leads to some very useful but surprising conclusions. Consider a solid solute A added in excess to liquid 1, to form a saturated solution. In a separate container, solute A is added to a different liquid 2, to form a saturated solution in the second liquid. A saturated solution is one in which the solution has come to equilibrium with a solid phase. We know that for the solute A,

$$\mu_A(s) = \mu_A(\text{saturated soln. } 1) = \mu_A(\text{saturated soln. } 2)$$

Thus, for two saturated solutions of a solute at the same T and P but in different solvents, the chemical potential of the solute is the same in the solutions *even if they are not in contact with each other.* The presence of the solid solute in the saturated solutions provides the common feature that allows us to make a thermodynamic connection for the chemical potential of solute A in different solvents at the same temperature and pressure, even if the solubilities are very different and the saturated solutions are not in contact with each other. This is extremely useful. We use the activity of the solute in each saturated solution to calculate the standard chemical potential for the solute in that solvent, $\mu_A^0(1)$ or $\mu_A^0(2)$, relative to that of solid A, $\mu_A^0(s)$. The solute standard states correspond to solutions with activities $a_A(1) = 1.0$ or $a_A(2) = 1.0$, even if those activities (concentrations, for ideal solutions) are greater than the actual concentrations in the saturated solutions. With these standard chemical potentials, we can readily calculate $\Delta\mu_A^0 = \mu_A^0(2) - \mu_A^0(1)$ for transferring the solute from solvent 1 to solvent 2.

Relatively hydrophobic solutes, like the fatty acid compounds $C_nH_{2n+1}COOH$, with $n \geq 5$, are more soluble in a hydrocarbon like liquid n-heptane than in a dilute aqueous buffer. For palmitic acid, A $= C_{15}H_{31}COOH$, solubility measurements lead to the result that $\mu_A^0 = \mu_A^0(\text{heptane}) - \mu_A^0(aq) = -38$ kJ mol^{-1}. For the series of fatty acids compounds having a number of carbon atoms n_C, where $n_C = 8$ (octanoic), $n_C = 10$ (decanoic), $n_C = 12$ (dodecanoic), and so on, to $n_C = 22$ (behenic), the solubility data can be used to generate the relation (in kJ mol^{-1}):

$$\mu_A^0(\text{heptane}) - \mu_A^0(aq) = 17.82 - 3.45\, n_C \qquad \textbf{(5.24)}$$

This equation gives the standard free-energy change in transferring 1 mol of fatty acid from liquid water to liquid heptane. The positive value for the first term on the right side of Eq. (5.24) reflects the preference of the polar COOH group for water as solvent. The addition of CH_2 groups to increase the length of the hydrocarbon part of the molecule then results in a decrease of the strength of the interaction with water relative to that with heptane by 3.45 kJ mol^{-1} per CH_2 group. For $n_C \geq 6$, the fatty acids are more soluble in hydrocarbons, whereas the shorter chain carboxylic acids are more soluble in water. Figure 5.12 shows that the experimental data follow closely a linear relation between μ_A^0 and n_C.

Another useful application of this approach is to determine the relative hydrophobicities of the side chains of the amino acids that are the constituents of proteins. Many structures of water-soluble proteins are now known in detail, based on X-ray crystallography (chapter 12) or high-resolution nuclear magnetic resonance (chapter 10). An almost uniform characteristic of these structures is that charged or polar amino acid side chains are exposed to the aqueous medium at the surface of the protein molecules, whereas nonpolar hydrocarbon side chains are buried in the interior. The aqueous pro-

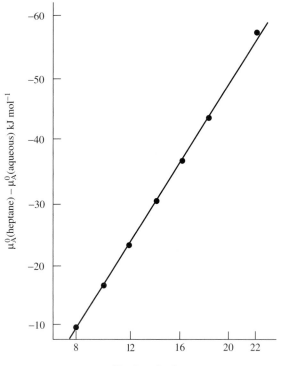

▲ FIGURE 5.12

The standard free energies of transfer from aqueous solution to heptane solution of aliphatic carboxylic acids as a function of number of carbons in the fatty acid. The negative values of the free-energy differences mean that the fatty acids prefer to be in the nonpolar hydrocarbon solution rather than in the aqueous solution. The data were measured at 23°C to 25°C; a least-squares fit to the data provides the parameters for [Eq. (5.24)]. [Data are from figure 3-1 of C. Tanford, *The Hydrophobic Effect*, 2d ed. (New York: Wiley, 1980).]

tein solution has a lower free energy this way. For proteins that are embedded in lipid membranes, the hydrocarbon side chains are on the outside of the protein, and the polar groups are hidden inside.

Amino acid side chains can be characterized as hydrophilic (polar) or as hydrophobic (hydrocarbon-like). A quantitative evaluation is provided by a *hydropathy index* based on measurements of the solubility of the different amino acid side chains in polar versus nonpolar solvents. One such index is presented in table 5.2.

Transfer between the same solvents

There is a change in free energy when a molecule is transferred from a solution at one concentration to one at a different concentration in the same solvent. Examples include transfer of molecules from inside a cell to outside or transfer across a dialysis membrane. We can even treat the diffusion of a molecule from high to low concentration in a poorly mixed solution by the same method. For these transfers, the *standard* free energies are equal because the solvent is the same, and we keep T and P constant. Because the concentrations in the different phases do not correspond to equilibrium and

TABLE 5.2 A Hydropathy Index for the Side Chains of the Amino Acids in Proteins*

Side chain	Hydropathy index	Side chain	Hydropathy index
Isoleucine	4.5	Tryptophan	−0.9
Valine	4.2	Tyrosine	−1.3
Leucine	3.8	Proline	−1.6
Phenylalanine	2.8	Histidine	−3.2
Cysteine/cystine	2.5	Glutamic acid	−3.5
Methionine	1.9	Glutamine	−3.5
Alanine	1.8	Aspartic acid	−3.5
Glycine	−0.4	Asparagine	−3.5
Threonine	−0.7	Lysine	−3.9
Serine	−0.8	Arginine	−4.5

*The hydropathy index is the relative hydrophobicity-hydrophilicity of the side-chain groups in proteins at neutral pH. Positive values characterize hydrophobic groups which tend to be buried in water-soluble proteins; negative values characterize hydrophilic groups which appear on the surface of the proteins. The relative values were deduced from free energies of transfer of side-chain molecules from water to organic solvents or from water to the gas phase. See J. Kyte and R. F. Doolittle, *J. Mol. Biol.* 157 (1982): 105–132.

the standard free-energy difference $\Delta\mu_A^0$ is zero, the free energy per mole for transferring A from one phase to another is

$$\Delta\mu = \mu_A(\text{phase 2}) - \mu_A(\text{phase 1})$$

$$= RT \ln \frac{a_A(\text{phase 2})}{a_A(\text{phase 1})} \qquad (5.25)$$

If the activity coefficients for A are the same in the two phases, the free energy of transfer per mole is

$$\Delta\mu = RT \ln \frac{c_A(\text{phase 2})}{c_A(\text{phase 1})} \qquad (5.26)$$

EXAMPLE 5.5

The concentration of creatine in blood is 20 mg L^{-1} and in urine is 750 mg L^{-1}. Calculate the free energy per mole required for its transfer from blood to urine at 37°C.

Solution

For the transfer from blood to urine

$$\Delta\mu = RT \ln\left(\frac{750}{20}\right) = (8.314 \text{ J K}^{-1} \text{ mol}^{-1})(310 \text{ K})\ln 37.5 = 9.3 \text{ kJ mol}^{-1}$$

Equations (5.25) and (5.26) apply to uncharged molecules and to molecules in the absence of an electric field. If there is a potential difference, $\phi_2 - \phi_1 = V$ volts, between the two phases, an additional term is required for charged molecules. The associated free-energy change is equal to the electrical work [Eqs. (4.57) and (4.58)] of transferring an ion of charge Z against a voltage difference V. The corresponding free energy for 1 mol of ions is $N_0 Z V$ (where N_0 is Avogadro's number and the units are electron volts, eV) or ZFV [where $F = 96{,}485$ J $(\text{eV})^{-1}$, the Faraday conversion factor, and the units are joules]. The total free energy of transfer for a charged species in the presence of a field is therefore

$$\Delta\mu = \mu_A(\text{phase 2}) - \mu_A(\text{phase 1}) + ZFV$$

$$= RT \ln \frac{a_A(\text{phase 2})}{a_A(\text{phase 1})} + ZFV \tag{5.27}$$

where Z = charge on ion, $\pm 1, \pm 2, \pm 3, \ldots$
    F = Faraday = $96{,}485$ J $(\text{eV})^{-1}$
    $V \equiv \phi_2 - \phi_1$ is the potential difference in volts between the two phases; ϕ_1 and ϕ_2 are the electric potential of phase 1 and phase 2, respectively. The sign of V is positive when ϕ_2 is electrically positive relative to ϕ_1.

These equations will be useful in understanding active transport in biological cells. The transport of ions and metabolites across membranes is characterized by Eqs. (5.25) through (5.27). Equation (5.27) can be rearranged in the form

$$\Delta\mu = (\mu_A + ZF\phi)(\text{phase 2}) - (\mu_A + ZF\phi)(\text{phase 1})$$

It is convenient to define

$$\mu_{A,\text{tot}} \equiv \mu_A + ZF\phi \tag{5.28}$$

as the total chemical potential of A in the presence of the electric potential ϕ. The condition for equilibrium between the two phases in the presence of a field is then, at constant T and P,

$$\mu_{A,\text{tot}}(\text{phase 2}) = \mu_{A,\text{tot}}(\text{phase 1})$$

This is the first example we have given of an external field affecting the free energy of a species. We will discuss the effects of other external fields, such as centrifugal fields and gravitational fields, in the next chapter. The important idea to remember is that the free energy depends on the interactions of molecules with other molecules and with their surroundings. If the surroundings include electric fields, or other fields, we must consider the *total* chemical potential in calculating the free energy.

Donnan Effect and Donnan Potential

We described equilibrium dialysis as an easy way to measure binding of a small molecule ligand by a macromolecule. The method becomes more complicated if the macromolecule and ligand are charged. The requirement that the solutions on each side of the dialysis membrane must be electrically neutral means that there can be an apparent increase in binding of a ligand with the opposite charge to that of the macromolecule and a decrease in binding of a ligand with the same charge. These effects depend on the net charge on the macromolecule and are not caused by binding at specific sites. The effect of the net charge of the macromolecule on the apparent binding of a ligand can be minimized by using high concentrations of a salt not involved in binding.

Thus, the charged macromolecule and ligands are not the main contributors to the total concentration of ions in the solutions.

When equilibrium (except for the macromolecule) is reached for charged species, a voltage is developed across the membrane. The asymmetric distribution of ions caused by the charged macromolecule is called the *Donnan effect*, and the transmembrane potential is called the *Donnan potential.* We can use thermodynamics to quantitatively calculate these effects. For simplicity, we consider the dialysis of a macromolecule with a net charge of Z_M and concentration c_M, against a NaCl solution. The requirement of electrical neutrality outside the membrane means that

$$c_{Na^+}(\text{outside}) = c_{Cl^-}(\text{outside}) = c \qquad (5.29)$$

where c is the concentration of NaCl outside the membrane at equilibrium. At equilibrium, because the membrane is permeable to NaCl,

$$\mu_{NaCl}(\text{inside}) = \mu_{NaCl}(\text{outside})$$

But

$$\mu_{NaCl} = \mu^0_{NaCl} + RT \ln a_{NaCl} \qquad (4.36)$$

We make the simplifying assumption that the activity coefficients are approximately equal inside and outside the membrane; thus,

$$c_{Na^+}(\text{inside})c_{Cl^-}(\text{inside}) = c_{Na^+}(\text{outside})c_{Cl^-}(\text{outside}) \qquad (5.30)$$

From Eq. (5.29), we see that

$$c_{Cl^-}(\text{inside}) = \frac{c^2}{c_{Na^+}(\text{inside})} \qquad (5.31)$$

But for electrical neutrality inside, we have

$$c_{Cl^-}(\text{inside}) = Z_M c_M + c_{Na^+}(\text{inside}) \qquad (5.32)$$

We equate Eqs. (5.31) and (5.32) and solve the following quadratic equation for $c_{Na^+}(\text{inside})$:

$$c^2 = Z_M c_M c_{Na^+}(\text{inside}) + c^2_{Na^+}(\text{inside})$$

The positive solution of the quadratic equation is

$$c_{Na^+}(\text{inside}) = \frac{-Z_M c_M + \sqrt{(Z_M c_M)^2 + 4c^2}}{2}$$

The ratio r of $c_{Na^+}(\text{inside})$ to $c_{Na^+}(\text{outside})$ is

$$r = \frac{c_{Na^+}(\text{inside})}{c}$$

$$= \frac{-Z_M c_M}{2c} + \sqrt{\left(\frac{Z_M c_M}{2c}\right)^2 + 1} \qquad (5.33)$$

From Eq. (5.31), we can also calculate the ratio of Cl^- inside to outside:

$$\frac{c_{Cl^-}(\text{inside})}{c_{Cl^-}(\text{outside})} = \frac{c}{c_{Na^+}(\text{inside})} = \frac{1}{r}$$

The equations above show that for a positively charged macromolecule the concentration of positive ions inside will be less than that outside ($r < 1$), and the concentration of negative ions inside will be greater than outside.

▶ FIGURE 5.13

The origin of the Donnan potential, $V = \phi_{out} - \phi_{in}$, across a membrane (dialysis tubing) separating a solution that contains a positively charged macromolecule M^+ in a solution containing NaCl as electrolyte from a similar electrolyte solution outside, but without the macromolecule. The analysis of the concentration differences and the origin of the Donnan potential are described in the text.

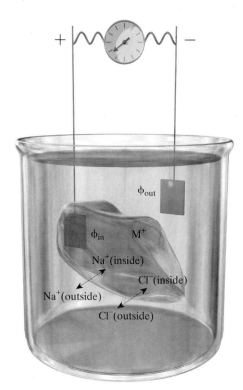

For a negatively charged macromolecule, $r > 1$, and the concentration of positive ions is greater inside than outside. For example, for a macromolecule with a net positive charge of 10 ($Z_M = 10.0$) at a concentration of 1 mM ($c_M = 1.00 \times 10^{-3}$) dialyzed against 0.10 M NaCl ($c = 0.100$), the value of $r = 0.95$ from Eq. (5.33). This means that the ratio of Na^+ inside to outside is 0.95, and the ratio of Cl^- inside to outside is $1/r = 1.053$. Increasing the positive charge or concentration of a macromolecule will make r decrease; increasing the concentration of NaCl will make r approach 1.

The asymmetry in Na^+ or Cl^- concentration on the two sides of the membrane is surprising because the membrane is permeable to Na^+ or Cl^- and we might think that the relations $\mu_{Na^+}(\text{inside}) = \mu_{Na^+}(\text{outside})$ and $\mu_{Cl^-}(\text{inside}) = \mu_{Cl^-}(\text{outside})$ would predict equal concentrations of Na^+ and Cl^- on the two sides of the membrane. There is, however, a potential difference V, the Donnan potential, across the membrane, as illustrated in figure 5.13. From Eq. (5.27) and the requirement $\Delta G = 0$ when the inside and outside are in equilibrium with each other, we obtain

$$RT \ln \frac{a_{Na^+}(\text{inside})}{a_{Na^+}(\text{outside})} + ZFV = 0$$

or

$$V = -\frac{RT}{ZF} \ln \frac{a_{Na^+}(\text{inside})}{a_{Na^+}(\text{outside})}$$

$$\approx -\frac{RT}{ZF} \ln \frac{c_{Na^+}(\text{inside})}{c_{Na^+}(\text{outside})}$$

$$\approx -\frac{RT}{ZF} \ln r \tag{5.34}$$

From the example above with $r = 0.95$, the charge $Z = 1$ on Na^+, $T = 298$ K, $R = 8.314$ J K^{-1}, and $F = 96,485$ J $(eV)^{-1}$, we can calculate a transmembrane voltage of $+1.3$ mV. The plus sign means that the electrical potential is higher inside than outside the membrane because the positively charged macromolecule is inside. We of course calculate the same value for the voltage if we use the ratio of Cl^- concentration inside and outside.

Membranes

When we think of phases, we think of the usual solid, liquid, or gaseous states. Whenever there are two phases in contact, there is also a surface, or interface, between them. This surface has properties different from those in the two bulk phases and therefore will have different behavior. The surface has thermodynamic properties specified by its free energy, enthalpy, and so on, just as bulk phases have. However, there are differences. A surface is two-dimensional instead of three-dimensional. This means that concentration units for a surface are mol cm^{-2} instead of mol cm^{-3}.

Biological cells are small volumes surrounded by *membranes*. (Membranes are present inside the cells, also.) The membranes typically contain lipids, proteins, glycolipids, and other amphiphilic molecules. Membranes constitute a different phase from the rest of the cell, and for some purposes they can be thought of as a surface phase. The membranes of course have a finite thickness and definite volumes, but it may be more useful to think about their concentrations in molecules cm^{-2} instead of molecules cm^{-3}.

Lipid Molecules

Oil droplets or particles of greasy dirt are hydrophobic. When they are suspended in water, the surface or interface is attractive to amphiphilic molecules (see chapter 3). *Amphiphilic molecules* consist of two parts or regions, one of which interacts primarily with the hydrophobic phase and the other with the aqueous phase. Amphiphilic molecules locate preferentially at the interface. This happens to molecules that have a hydrocarbon tail and a polar head; they are called surface-active molecules, or *surfactants*. Sodium dodecylsulfate (see figure 5.14), a principal component of many commercial detergents, orients at the surface of dirt particles and makes them soluble in water. The hydrocarbon tail is attracted to the oily dirt "phase," and the polar sulfate head faces the water (see figure 5.15). The hydrocarbon part is hydrophobic; the polar group is hydrophilic, which is typical of amphiphilic molecules.

The structures of several surfactants are shown in figure 5.14. There is a large variety of hydrophilic head groups: some are polar owing to exposed —OH or —NH₂; others are negatively charged (phosphatidylglycerol, phosphatidylserine, or phosphatidylethanolamine) or zwitterionic (phosphatidylcholine), bearing both positive and negative charges at pH 7. There is also a variety among the hydrocarbon tails: Chain lengths range between 12 and 18 or more carbon atoms, double bonds or large multiple-ring structures as in cholesterol may be present. In addition there is variability in the number of hydrocarbon tails (but usually one or two) attached to a head group. Such variations in composition are important for defining the properties of the surfaces, especially in membranes and vesicles. They determine the thermal stability, fluidity, and curvature of the membranes, as well as the interactions

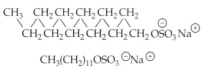

$CH_3(CH_2)_{11}OSO_3 \ominus Na \oplus$

Sodium dodecylsulfate

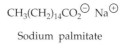

$CH_3(CH_2)_{14}CO_2^\ominus \ Na^\oplus$

Sodium palmitate

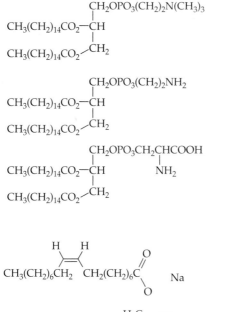

Dipalmitoyl phosphatidylcholine, DPPC
(Dipalmitoyl lecithin)

Dipalmitoyl phosphatidylethanolamine

Dipalmitoyl phosphatidylserine

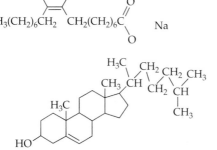

Sodium oleate

Cholesterol

▲ FIGURE 5.14

Some common surface active molecules.

with proteins and other molecules associated with or embedded in the membrane. An illustration of the variety of forms and interactions exhibited by lipids and other surfactants is shown in figure 5.15.

Lipid Bilayers

A typical cell membrane consists of a lipid bilayer into which are incorporated other amphiphilic molecules, such as cholesterol, sphingolipids (found in brain tissue), proteins, and glycolipids (molecules containing sugar or saccharide residues). One can think of the lipid bilayer as a kind of two-dimensional solvent; however, its thickness cannot be ignored. As illustrated in figure 5.15, the bilayer is arranged as a sandwich with the hydrophobic portions of the lipids turned inward and the hydrophilic portions on the outer surfaces in

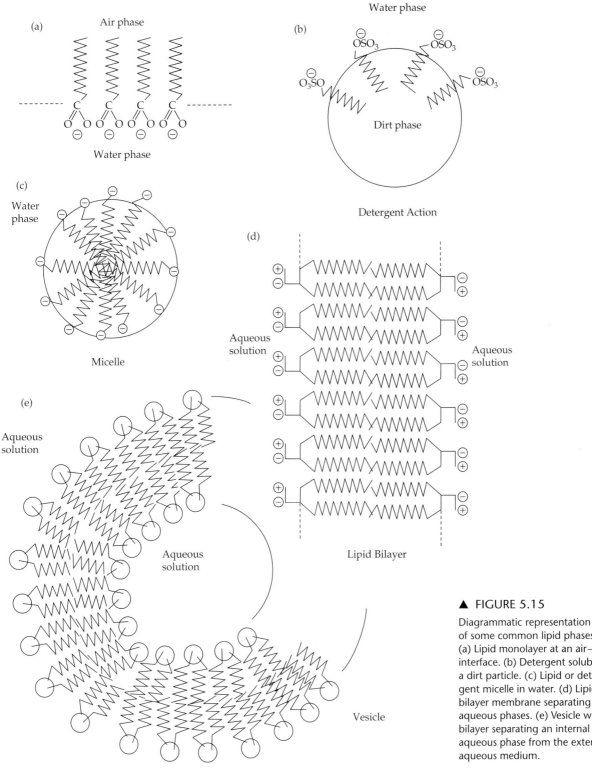

(a) Air phase

Water phase

(b) Water phase

$\overset{\ominus}{O}SO_3$ $\overset{\ominus}{O}SO_3$

$O_3\overset{\ominus}{S}O$ $\overset{\ominus}{O}SO_3$

Dirt phase

Detergent Action

(c) Water phase

Micelle

(d) Aqueous solution

Aqueous solution

Lipid Bilayer

(e) Aqueous solution

Aqueous solution

Vesicle

▲ FIGURE 5.15

Diagrammatic representation
of some common lipid phases.
(a) Lipid monolayer at an air–water
interface. (b) Detergent solubilizing
a dirt particle. (c) Lipid or detergent micelle in water. (d) Lipid
bilayer membrane separating two
aqueous phases. (e) Vesicle with
bilayer separating an internal
aqueous phase from the external
aqueous medium.

contact with the aqueous phases. Thus, the bilayer serves as a boundary between two aqueous phases that, in biological cells, typically differ from one another in solute concentrations, ionic strength, pH, and so on. (A soap bubble is a bilayer that separates two regions of air.)

Membrane proteins, glycolipids, and other components associated with the lipid bilayer serve important functions, such as gating the transport of molecules across the bilayer from one aqueous phase to the other, receiving and transmitting signals, defining the electrical properies of the cell, decorating the surface to identify the cell type, and many others. Most membrane proteins span the bilayer; that is, they have a dumbbell shape with a hydrophobic waist and polar ends that extend into the aqueous environment on either side. The polar ends are often not identical in structure, which helps define distinctions between inside and outside the cell or organelle. Although the lipid bilayer serves effectively to block the passive transport of ions or of large molecules and even small-molecule metabolites, nevertheless water moves readily across the hydrophobic interior of the bilayer. The mechanism of this water transport is not well understood, but it illustrates the dynamic character of the lipid molecules in the bilayer.

Phase Transitions in Lipids, Bilayers, and Membranes

Based on measurements of the two-dimensional diffusion of small molecules in membranes of living cells, the lipid bilayers are thought to have a fluidity something like that of olive oil. The behavior is liquidlike in two dimensions. As the temperature is decreased, transitions to a solidlike membrane may occur. Under these conditions, the two-dimensional mobility is greatly decreased. The actual situation is in fact more complex than this.

Lipid-phase transitions can be studied using a technique called differential thermal analysis (DTA), using a differential scanning calorimeter. In this method, heat is added to the sample electrically at a constant rate, and the temperature is continuously monitored using a small probe. As long as the lipid remains a single solidlike phase, the temperature rises smoothly and with a rate determined by the heat capacity of the lipid. When the temperature reaches the "melting point" of the lipid, extra heat is required to induce the transition from solidlike to liquidlike behavior. Once the transition is completed, the temperature once again rises smoothly in proportion to the heat capacity. Figure 5.16 shows a plot of the excess specific heat associated with the transition in the lipid dipalmitoylphosphatidylcholine (DPPC). The peak of the curve at 41.55°C indicates the transition temperature, T_{trans}, for this lipid. The width of the peak, which becomes broader as the heating rate is increased, is a measure of the sharpness of the transition. Very sharp peaks are associated with first-order phase transitions, where the transition from one phase to the other is highly cooperative.

Commonly, the behavior is more complex than for DPPC. There are typically three or four peaks for a given lipid preparation, each peak occurring at a different temperature. For anhydrous lipid crystals with charged headgroups like phosphatidylethanolamine or phosphatidylcholine, the true melting transition observed visually in a melting point capillary occurs at 200°C or higher. The enthalpy of this high-temperature melting is very small in comparison with earlier transitions that occur between 0° and 70°C, depending on the specific headgroup and hydrocarbon tails. It is known from

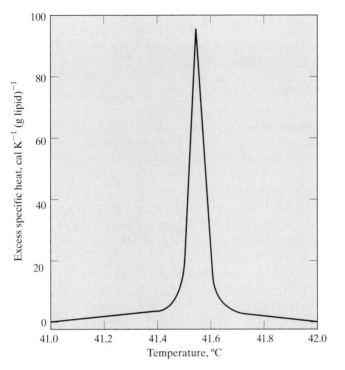

▲ FIGURE 5.16

Differential scanning calorimetry of a suspension of dipalmitoylphosphatidylcholine (0.508 mg g^{-1}) in water. [Data from N. Albon and J. M. Sturtevant, *Proc. Nat. Acad. Sci., USA* 75 (1978): 2258–2260.]

measurements using X-ray diffraction (see chapter 12) or nuclear magnetic resonance (see chapter 10) that the melting occurs in several stages. The lowest-temperature transition involves melting of only the hydrocarbon tails; the headgroups still remain in a reasonably well-ordered two-dimensional lattice. The hydrocarbon tails in the low-temperature solidlike phase are arranged in an ordered all-trans configuration. This is a structure in which there is maximum van der Waals interaction between the hydrocarbon tails because they are fully aligned along their length. Starting at low temperature with a solidlike lipid and adding heat, the thermal energy of motion is increased until disorder is introduced to the hydrocarbon layer. The relatively large enthalpy change associated with this is responsible for the first transition peak in the DTA curve at temperature T_{trans}. Investigations of lipid bilayer structures show that during this first transition the thickness of the bilayer structure decreases because the tails from the opposite faces of the sandwich begin to interdigitate. The headgroups, especially when water of hydration and ionic countercharges are present, are held together in a stronger hydrogen-bonded and ionic lattice that typically requires much higher temperatures before it is disrupted. This is a very important property of membrane lipids in the region of physiological temperature: The lipids are liquidlike in the hydrocarbon regions, permitting rapid lateral diffusion of incorporated small molecules, whereas the ordered headgroups stabilize the bilayer structure and prevent the lipid from simply melting into spherical droplets. That happens only above 200°C. Other more subtle changes in lipid conformation produce additional transitions between the lowest and the highest transition temperatures. For example, the headgroups are typically found to lie parallel to the membrane surface with the hydrocarbon tails pointing toward the center of the sandwich. The average angle of the hydrocarbons with respect to the surface of the bilayer changes during some of the minor transitions.

The mobility of molecules such as DPPC in bilayers has also been investigated using techniques such as deuterium nuclear magnetic resonance and electron paramagnetic spin labeling (see chapter 10). It is evident from these studies that mobility increases significantly from outside to inside the bilayer. The headgroups are held relatively rigidly in their network, whereas the hydrocarbon tails exhibit more mobility. Even this mobility is graded from the regions close to the headgroup (mostly rotation about an axis normal to the bilayer surface in this region) to much greater liquidlike flexibility of the terminal lipid region in the interior of the bilayer sandwich.

Factors that influence the transition temperatures for lipids include (1) the length of the hydrocarbon tail, (2) the nature of the headgroup, (3) the presence of unsaturation (double bonds) in the hydrocarbons, and (4) additional molecules like cholesterol that are incorporated into the bilayer. In general, transition temperatures reflecting the hydrocarbon fluidity that is important for biological function (in the $-30°$ to $100°C$ temperature range) are increased for longer hydrocarbon tails (C12 through C18 and higher), are decreased by the presence of double bonds, and are increased by admixture of cholesterol, which is a rigid multiring hydrophobic molecule. In these respects, the properties of the bulk lipid materials mirror those of the bilayers. Some biological organisms, including single-celled bacteria and flowering plants like oleander, are able biochemically to alter the composition of their membrane lipids in response to large changes in growth temperature. In this fashion, membranes of cells maintain their fluidity when grown at temperatures so low that the membranes of the high-temperature grown organism would solidify. This is an example of thermal adaptation.

Surface Tension

Because of intermolecular attractions, the molecules at the surface (exposed to air) of a bulk liquid are attracted inward. This creates a force in the surface that tends to minimize the surface area. If the surface is stretched, the free energy of the system is increased. The *free energy per unit surface area,* or the force per unit length on the surface, is called the *surface tension.* Note that the units for energy per unit area and force per unit length are identical. The SI units are millinewtons per meter ($mN\ m^{-1}$), equal to $mJ\ m^{-2}$. Data in table 5.3 illustrate the range of surface tension that can occur.

What happens when a substance is dissolved in the liquid or added to the surface? The surface tension typically either decreases or does not change very much. It never increases greatly. There is a thermodynamic reason for this, but before stating it let's consider what would happen if we could greatly increase the surface tension of water. Water drops from your faucet could become the size of basketballs because the size of the drop is directly proportional to the surface tension. You might have to cut water with a knife and

TABLE 5.3   Surface Tensions of Pure Liquids in Air

Substance	Surface tension (mN m^{-1})	Temperature (°C)
Platinum	1819	2000
Mercury	487	15
Water	71.97	25
	58.85	100
Benzene	28.9	20
Acetone	23.7	20
n-Hexane	18.4	20
Neon	5.2	−247

chew it well before swallowing. Water would not wet anything, so processes depending on capillary action would not work. We are saved from these possible catastrophes. Any substance that tends to raise the surface tension of a liquid raises the free energy of the surface. The substance therefore concentrates less at the surface.

The surface tension, or the surface free energy, is just $(\partial G/\partial A)_{T,P}$, where A is the surface area. Substances that lower the surface tension also lower the free energy of the surface; they preferentially migrate to the surface. Thus, substances that lower the surface tension concentrate at the surface and give large decreases in surface tension, but substances that raise the surface tension avoid surfaces and give only small increase in surface tension. The quantitative expression for this is called the *Gibbs adsorption isotherm*:

$$\Gamma = -\frac{1}{RT}\frac{d\gamma}{d\ln a} \cong -\frac{1}{RT}\frac{d\gamma}{d\ln c} \qquad (5.35)$$

where Γ = adsorption (excess concentration) of solute at surface, mol m^{-2}
γ = surface tension, N m^{-1}
R = gas constant = 8.314 J K^{-1} mol^{-1}
a = activity of solute in bulk solution
c = concentration of solute in bulk solution (any units can be used)

The sign of the excess surface concentration, Γ, is opposite to the sign of the change of the surface tension with concentration (or activity) of solute in the solution. Figure 5.17 shows the change in surface tension of water when LiCl or ethanol is added. Ionizing salts are almost the only solutes that raise the surface tension of water.

There are many naturally occurring surfactants in plants and animals. Dipalmitoyl lecithin (see figure 5.14) forms a layer that lowers the surface tension at the surface of the alveoli* in the lung and allows one to breathe. A large surface area is necessary for the efficient exchange of gases in the lung. Dipalmitoyl lecithin lowers the surface tension of water to nearly zero and allows large aqueous surface areas to exist in the lung. Premature babies lack

---

*The alveoli are the smallest air compartments found in the lung.

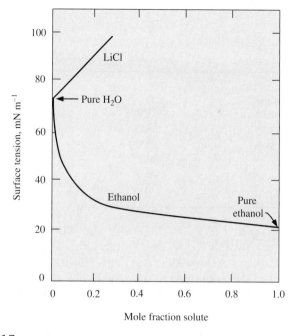

▲ FIGURE 5.17

Effect of solute concentration on surface tension in aqueous solution at 20°C.

this vital surfactant and have great difficulty breathing effectively. The use of perfluorocarbon liquid breathing medium is being tested to alleviate this condition, as mentioned earlier in this chapter.

The surface-tension decrease caused by surface-active molecules can be measured with a Langmuir film balance, illustrated in figure 5.18. One adds some surfactant, usually in a volatile or aqueous solvent, to the surface of water between a movable barrier and a float sensor. As the barrier is slowly moved to compress the film, an increasing force must be applied to the float from the opposite direction to prevent the float from being displaced from its null position. The surface pressure necessary to maintain the film in the compressed state is just the difference between the surface tension γ_0 of the water beyond the float and the surface tension γ of the coated water.

$$P_{\text{surface}}(\text{mN m}^{-1}) = \gamma_0 - \gamma$$

The surface concentration in mol m^{-2} depends on the surface pressure. The experiment is the two-dimensional analog of a pressure versus volume experiment in three dimensions. It is traditional to plot surface pressure versus area per molecule in Å^2 molecule^{-1}. The area occupied by the film at various stages of compression is determined using a calibrated distance scale associated with the movable barrier. By knowing the amount of surfactant added, one can calculate the area per molecule, which is just the reciprocal of the surface concentration. Some representative results are shown in figure 5.19. The experiment typically begins with an expanded film, corresponding to large values of the area per molecule. Under these conditions, the surfactant molecules are far from one another, on average, giving a film with low surface pressure, corresponding to a surface tension equal to that of pure water.

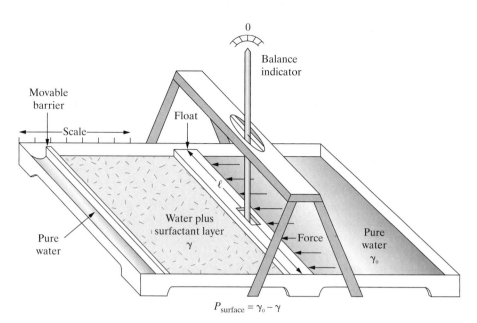

▲ FIGURE 5.18

A Langmuir film balance measures surface tension. The force per unit length F/ℓ required to compress a surface containing a surfactant layer is known as the surface pressure $P_{surface}$. The float is connected to a torsion wire (not shown) to measure the force.

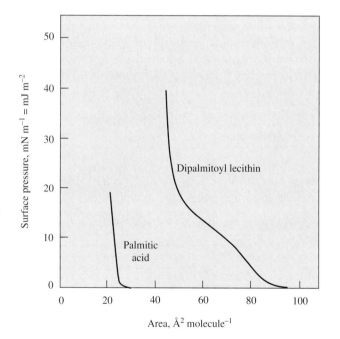

▲ FIGURE 5.19

Surface pressure versus area curves for two amphiphilic molecules (surfactants) at an air–water interface. The molecular areas correspond to average cross sections of the molecules projected on the surface. The steeply rising portions of the curves occur when the molecules in the surface layer have been compressed into a compact arrangement.

As the film is compressed, the surface pressure rises. In some cases, disconti-
nuities appear in the plots owing to phase transitions that occur in some surfactants. The high surface pressures at low area per molecule (high surface
concentration) result from the low surface tensions produced by the surface-
active molecules. The maximum possible surface pressure, equal to the surface tension of pure water, occurs when the surface tension of the film falls to
zero. This is implicit in the preceding equation. The experiment typically
ends with the collapse of the film, whereupon the surface pressure drops
abruptly to zero.

Surface Free Energy

Amphiphilic molecules containing hydrocarbon chains prefer the surface of
an aqueous solution, which lowers the surface tension. The longer the chain,
the higher is the surface concentration of the amphiphile. To understand the
thermodynamic consequences of the interactions of lipids in an aqueous environment, we need to look more closely at the free energy of transfer of molecules between bulk phases.

Hydrophobic effect

Hydrocarbons are almost completely insoluble in water, and long hydrocar-
bon-chain amphiphiles typically have low solubility. The thermodynamic
consequence is that for most nonpolar molecules the free energy is positive
for transfer from an organic solvent to water. Contrary to what one might expect, the enthalpy of transfer for small hydrocarbons is negative. A large drop
in entropy occurs when these small hydrocarbons enter water, accounting
for the positive free-energy change.

To explore this situation thermodynamically, consider a liquid hydrocarbon phase in equilibrium with a water phase. The mole fraction of hydrocarbon in the hydrocarbon phase, $X_H(H)$, is close to 1. Therefore,

$$\mu_H(H) = \mu_H^0(H) + RT \ln a_H(H) = \mu_H^0(H) + RT \ln \gamma_H(H)X_H(H)$$

where $\gamma_H(H) \to 1$ as $X_H(H) \to 1$ defines the standard state for the hydrocarbon in the hydrocarbon phase.

In this limit,

$$\mu_H(H) = \mu_H^0(H) + RT \ln X_H(H) \tag{5.36}$$

For hydrocarbon in the water phase, the mole fraction $X_H(W)$ is close to zero.

$$\mu_H(W) = \mu_H^0(W) + RT \ln a_H(W) = \mu_H^0(W) + RT \ln \gamma_H(W)X_H(W)$$

where $\gamma_H(W) \to 1$ as $X_H(W) \to 0$ defines the standard state for the hydrocarbon in the water phase.

In this limit,

$$\mu_H(W) = \mu_H^0(W) + RT \ln X_H(W) \tag{5.37}$$

The two terms in the chemical potential reflect two different origins, termed
unitary and *cratic*. The cratic contribution, $RT \ln X_H$, is statistical in nature
and arises from the entropy of mixing solute and solvent. The unitary contributions to the chemical potential are $\mu_H^0(H)$ and $\mu_H^0(W)$. They result from
expressing the solute in mole fraction units. Thus, $\Delta\mu^0 = \mu_H^0(W) - \mu_H^0(H)$
contains only the changes of the internal energy and the entropy of the

TABLE 5.4 Thermodynamic Data for the Transfer of Hydrocarbons to Water at 25°C

Hydrocarbon	$\mu_H^0(W) - \mu_H^0(H)$ (kJ mol^{-1})	$\overline{H}_H^0(W) - \overline{H}_H^0(H)$ (kJ mol^{-1})	$\overline{S}_H^0(W) - \overline{S}_H^0(H)$ (J K^{-1} mol^{-1})
C_2H_6	16.3	−10.5	−88
C_3H_8	20.5	−7.1	−92
C_4H_{10}	24.7	−3.3	−96
C_5H_{12}	28.7	−2.1	−105
C_6H_{14}	32.4	0	−109
C_6H_6	19.2	+2.1	−59
$C_6H_5-CH_3$	22.6	+1.7	−71
$C_6H_5-C_2H_5$	25.9	+2.0	−79
$C_6H_5-C_3H_7$	28.9	+2.1	−88

Source: Data compiled in C. Tanford, *The Hydrophobic Effect: Formation of Micelles and Biological Membranes,* 2d ed. (New York: Wiley, 1980).

hydrocarbon and its interactions with solvent as the hydrocarbon environment changes from hydrocarbon to water. But $\Delta\mu^0 = \mu_H^0(W) - \mu_H^0(H) = RT \ln X_H(W) - RT \ln X_H(H) = RT \ln X_H(W)$ provides a useful measurement of the unitary contribution; $X_H(W)$ is the solubility of the hydrocarbon in water, which is typically very small. The separation into unitary enthalpic and entropic terms is obtained from the relation $\Delta\mu^0 = \Delta\overline{H}^0 - T\Delta\overline{S}^0$ where the unitary enthalpy ΔH^0 can be obtained experimentally from calorimetric measurements or from the temperature dependence of $\Delta\mu^0$. Data for several hydrocarbons have been collected by Tanford (1980), and a selection is shown in table 5.4. The values for $\Delta\overline{H}^0$ are significantly negative for small hydrocarbons, but they rise as the size of the hydrocarbon molecule increases and are positive for aromatic molecules. The corresponding entropies are negative for both aliphatic and aromatic hydrocarbons and do not show a marked dependence on the size of the molecule. In every case, the values for $\Delta\mu^0$ are positive. In parallel studies, it has been shown that aliphatic alcohols behave similarly to the aliphatic hydrocarbons. Furthermore, investigations of the partial molar volume show that $\Delta\overline{V}^0 < 0$ for these hydrophobic or amphiphilic molecules. This means that the molecules occupy less volume in the aqueous environment than in the pure organic phase. Clearly, this behavior is somehow tied to the counterintuitive decrease in entropy (increase in ordering) that occurs when amphiphiles are added to water.

Explanations for the behavior just described are usually made in terms of the effect on the structure of liquid or "solvent" water. As we have seen, there is evidence from thermodynamic and physical measurements on pure water that the hydrogen bonding in water gives rise to an organized but fluctuating network of water molecules in the liquid state. The decrease in entropy associated with introducing amphiphiles must then result in a further ordering of the water of solvation, at least in the immediate vicinity of the introduced solute molecule. These have been described as "microscopic icebergs"; however, that may be an exaggeration of the degree of increased structure. In any event, the solvation water has a decreased entropy and apparently a decreased volume in comparison with bulk water.

Silver (1985) summarizes the consequences of this hydrophobic effect for phospholipids: At extremely low concentrations, translational entropy (cratic) ensures dispersion of the lipid as a homogeneously distributed solute in water. Above a small threshold of concentration, phospholipids form aggregates, including bilayers, to lower the total free energy. In this fashion, the hydrophobic effect serves to keep membranes intact in an aqueous environment. This may be demonstrated in a dramatic way by taking a red blood cell suspension and adding an organic solvent such as acetone to the medium. The acetone destabilizes the cell membranes and spills out the hemoglobin, presumably by disrupting the hydrogen-bonding networks that are normally present in the aqueous phase. The hydrophobic effect's contribution to the stability of a lipid bilayer in water is equivalent to a lateral pressure squeezing the phospholipids together and thus preventing the hydrocarbon chains from coming in contact with water.

Vapor Pressure and Surface Tension

If small droplets and a much larger drop of water are placed in a closed container saturated with water vapor, the small droplets eventually disappear, and the large one becomes larger. This phenomenon is a result of surface tension; the small droplets have a higher vapor pressure than the large one. When water or any other substance is finely divided, the surface effect becomes significant, and we need to include the term $\gamma \, dA$ in the expression for the change in free energy associated with change in droplet size. We replace

$$dG = -S \, dT + V \, dP + \mu_{H_2O} \, dn_{H_2O} \tag{5.38}$$

for bulk water by

$$dG = -S \, dT + V \, dP + \gamma \, dA + \mu_{H_2O} \, dn_{H_2O} \tag{5.39}$$

for the small droplets, where $\gamma \, dA$ is the change in free energy when the surface area is changed. The free energy per mole of bulk water (water in the interior of the liquid) is μ_{H_2O}. Equation (5.39) shows that the free energy of a liquid is increased by increasing its surface area; thus, a liquid will minimize its free energy by decreasing its surface area. This means a liquid will form a sphere if there are no other forces acting on it, such as gravity or other surfaces. Combining two spheres into one lowers the total surface area and therefore the total free energy. Thus, small spheres will tend to combine to form larger spheres. Mercury with its higher surface tension illustrates this better than water. We usually ignore the effect of surface area and the contribution of surface free energy to the total free energy; the effects are too small. However, significant effects can occur as shown by the higher vapor pressure of small—micron-sized—droplets. We can combine the terms $\gamma \, dA + \mu_{H_2O} \, dn_{H_2O}$ of Eq. (5.39) to obtain the *total chemical potential* of the liquid water; this takes into account the effect of the surface on the water molecules. Water in small droplets (with a high surface to volume ratio) has a higher total chemical potential and higher vapor pressure than bulk water and can therefore transfer spontaneously into a larger drop.

Biological Membranes

The composition and orientation of the molecules in biological membranes determine their function. Along with the diversity in composition comes a richness of properties that correlates with the variety of membrane functions, such as the rate of transport of water, ions, or other solutes across the surface. Usually, membranes contain an assortment of lipid molecules with diverse chemical structures, together with proteins and sometimes polysaccharides. The lipids are typically fatty acid esters that differ in the length of the fatty acid chain, the degree of unsaturation, the charge or polarity of the esterifying group, and the number of fatty acids esterified per molecule. The proteins may be intrinsic or integral, in the sense that they are incorporated directly into the membrane structure, or extrinsic, if they are attached to the membrane surface or interact strongly with it. Other components, such as cholesterol, may also be present. Electron micrographs giving two different views of chloroplast membranes are shown in figure 5.20. A model known as the Fluid Mosaic Model of membrane structure was developed by Singer and Nicholson (1972). Not only is the biological membrane characterized by a high degree of fluidity,

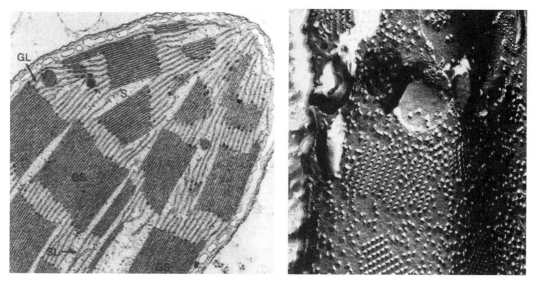

▲ FIGURE 5.20

Electron micrographs of plant chloroplasts. (a) A section through a chloroplast from corn. The section is fixed using glutaraldehyde and osmium tetroxide. The heavy osmium atoms show the location of double bonds in lipid molecules, with which the osmium reacts. The outer chloroplast membrane is seen to be a single stained bilayer; the internal thylakoid membranes, which occur in some regions in stacks called grana, GS, and in other regions separately as stroma lamellae, SL, show doubled staining because these are enclosed vesicles that separate an inner lumenal space from an outer stromal space, S. (Photograph courtesy of L. K. Shumway, Washington State University.) (b) A freeze-etch picture of a freeze-fractured surface that exposes surface features of the thylakoid membranes from spinach chloroplasts. The surfaces are shadowed using a heavy metal that is opaque to electrons. Several types of membrane-associated particles are seen on the different surfaces exposed. Some of these particles are photosynthetic pigment proteins; others are ATPase complexes involved in energy transduction. (Photograph courtesy of Roderic Park, University of California, Berkeley.)

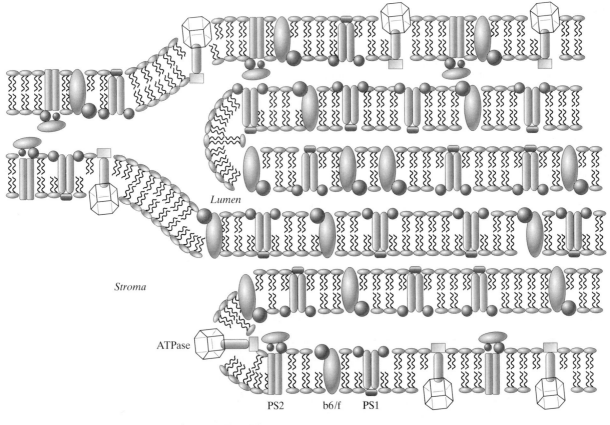

▲ FIGURE 5.21

The Fluid Mosaic Model of membrane structure applied to chloroplast thylakoids. Immersed in the lipid bilayer of the internal membranes in the grana stacks (right side) are photosystem 1 (PS1) and cytochrome b6/f/Rieske (b6/f) protein complexes. The external membranes in the grana stacks and the stroma lamellae (left side) also contain photosystem 2 (PS2) and the phosphorylation coupling factor (ATPase). Each complex has a particular orientation between the lumenal and the stromal faces of the thylakoid membrane.

but it is also constituted of a heterogenous set of membrane-associated proteins. Application of this model to the chloroplast membrane is shown in figure 5.21.

The variety of membrane composition results in a range of physical properties. For example, the mobility of the molecules in two dimensions in a membrane may be that of a typical liquid or that of a solid. As we have discussed, it is relatively common to encounter phase transitions in membranes or artificial surface films, and these transitions are closely analogous to their three-dimensional counterparts. The transition temperature or melting point, for example, is sensitive to the lipid composition; the thermodynamic analysis of two-dimensional solutions can be carried out just as for three-dimensional systems.

The relation between the intrinsic proteins and the lipids is more complex and is not yet well understood. Many proteins are amphiphilic and interact with both the hydrophobic lipids and the polar aqueous interface. These hydrophobic and hydrophilic sites are located in different regions of the protein molecule, resulting in a strong orientation with respect to the membrane surface. Apart from this orientation, the protein may behave much like a solute in the lipid solvent; essentially a two-dimensional solution is formed. There is even a two-dimensional analog of precipitation, in which the solution becomes saturated with respect to a protein constituent and the protein separates as a distinct phase. Electron microscopy (see chapter 12) of membrane surfaces provides a useful method for detecting and characterizing the phase changes.

EXAMPLE 5.6

A spherical cell 1 μm in diameter divides into two spherical cells. Calculate the work necessary to increase the surface area if its surface tension is 12.3×10^{-3} N m^{-1}. Assume that the volume of the daughter cells taken together is the same as that of the parent.

Solution

The surface area of the parent cell is $A_1 = \pi d^2 = 3.14 \times 10^{-12}$ m^2. The volume of the parent cell is $V_1 = \pi d^3/6 = 0.524 \times 10^{-18}$ m^3 and that of each daughter cell is $V_2 = 0.262 \times 10^{-18}$ m^3. Rearranging the equations for the area and surface of a sphere, we obtain $A_2 = (36\pi V_2^2)^{1/3} = 1.980 \times 10^{-12}$ m^2 and $2A_2 = 3.96 \times 10^{-12}$ m^2 for the two daughter cells. Thus, the increase in surface area associated with the cell division is $\Delta A = (3.96 - 3.14) \times 10^{-12}$ $m^2 = 0.82 \times 10^{-12}$ m^2. The work necessary to carry out this process is given by $w = \Delta G = \gamma \Delta A = (12.3 \times 10^{-3}$ N $m^{-1}) (0.82 \times 10^{-12}$ $m^2) = 1.0 \times 10^{-14}$ N m $= 1.0 \times 10^{-14}$ J. To put this in perspective, we may compare it with the free energy involved in the hydrolysis of ATP, for which $\Delta G \sim 25$ kJ $mol^{-1} = 4 \times 10^{-19}$ J molecule^{-1}. Thus, the energy from about 25,000 ATP molecules is required to provide the work involved in forming the new membrane associated with a dividing cell.

Active and Passive Transport

Frogs, seaweed, and other organisms that live in contact with water have semi-permeable skins. Water and some ions and small molecules pass through the skins; macromolecules generally do not. The frog or the seaweed can selectively concentrate certain molecules inside itself and selectively exclude or excrete other molecules. How do they do it? The thermodynamic question is, If a molecule can easily pass through the skin, how can the inside concentration be maintained at a value that is different from the outside concentration? The answer is to ensure that the free-energy change for transporting the molecule inside is negative. For example, the presence of a protein inside seaweed that strongly binds iodide ion ensures that the iodide concentration in the seaweed is always higher than in the seawater. If the concentration of *free* iodide is the same inside and outside, the bound iodide would account for the concentrating effect of the seaweed. This effect, known as *passive transport*, does not depend on whether the seaweed is alive or dead; that is, metabolism is not involved in the concentrating effect. The presence of the specific binding protein lowers the chemical potential of the iodide, which therefore concentrates in the seaweed. In equilibrium dialysis, we used the transport of ligands like O_2 through a dialysis membrane to measure binding to a macromolecule like myoglobin or hemoglobin on one side of the membrane. This is clearly passive transport.

There is another kind of transport, known as *active transport*, that is closely dependent on active cellular metabolism. Active transport is defined as the transport of a substance from a lower to a higher chemical potential. Because the total free-energy change of the process must be negative, active transport is tied to a chemical reaction that has a negative free-energy change. In a biological system, this means that metabolism is occurring and driving the pump. Therefore, an experimental test of whether active transport is involved is to poison the metabolic activity to see whether the transport also stops.

A well-known example of active transport is the concentration of K^+ ions and the depletion of Na^+ ions that occur inside most animal cells. A protein complex called the *sodium–potassium pump* uses the free energy of hydrolysis of ATP to pump Na^+ ions out of the cell and K^+ ions into the cell. The net reaction for the active transport is thought to be

$$\left.\begin{array}{c} 3Na^+(\text{inside}) \\ + \\ 2K^+(\text{outside}) \end{array}\right\} + ATP \longrightarrow ADP + \text{phosphate} + \left\{\begin{array}{c} 3Na^+(\text{outside}) \\ + \\ 2K^+(\text{inside}) \end{array}\right.$$

This Na–K ATPase is inhibited by cardiotonic steroids, such as digitoxigenin, obtained from the foxglove plant, or ouabain, a dart poison obtained from African trees. Cardiotonic steroids have a strong effect on the heart. An illustration involving representative concentration differences across a typical animal cell membrane is shown in figure 5.22. There is also a voltage difference of about -70 mV across the cell membrane; the inside is negative relative to the outside as shown. This membrane potential arises from the action of the Na–K ATPase pump and the fact that the typical cell membrane is more permeable to K^+ than to Na^+. Thus, the magnitude of the potential is determined largely by the ratio $a_{K^+}(\text{outside})/a_{K^+}(\text{inside})$.

A mechanism for the reaction is given in figure 5.23. The phosphorylation of the sodium–potassium pump by ATP causes a conformational change in the protein that exchanges bound Na^+ ions for K^+ ions outside the cell. Subsequent hydrolysis of the phosphate group from the sodium–potassium pump causes a conformational change that exchanges bound K^+ ions for Na^+ ions inside the cell. The molecular machine is now ready for another cycle. The details of the mechanism are not known, but it is important to make sure that the process is consistent with the laws of thermodynamics. If it is not, we know that we have left something out.

The free energy for the net reaction must be negative. This means that the positive free energy of actively transporting ions against concentration and voltage gradients must be more than balanced by the negative free energy of ATP hydrolysis.

Let's divide the process into three parts: the transport of Na^+ ions out, the transport of K^+ ions in, and the hydrolysis of ATP. The free energy per mole for the transport of Na^+ ions is

$$\Delta\mu(\text{transport of } Na^+ \text{ out}) = RT \ln \frac{a_{Na^+}(\text{outside})}{a_{Na^+}(\text{inside})} + ZFV \qquad \textbf{(5.27)}$$

We replace ratios of activities by ratios of concentrations and use a temperature of 37°C. For an answer in kilojoules,

$$\Delta\mu(\text{transport of } Na^+ \text{ out}) = (0.008314)(310) \ln \frac{140}{10} + (1)(96.485)(+0.07)$$

$$\Delta\mu(\text{transport of } Na^+ \text{ out}) = +13.6 \text{ kJ mol}^{-1}$$

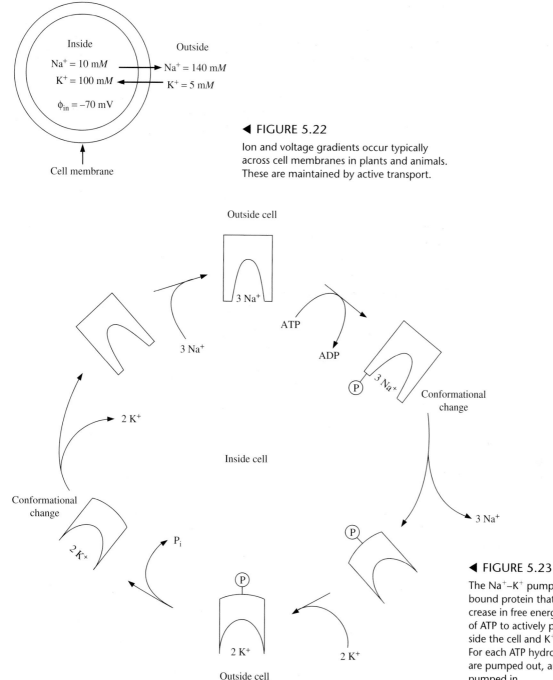

◀ FIGURE 5.22

Ion and voltage gradients occur typically across cell membranes in plants and animals. These are maintained by active transport.

◀ FIGURE 5.23

The Na$^+$–K$^+$ pump is a membrane-bound protein that uses the decrease in free energy of hydrolysis of ATP to actively pump Na$^+$ outside the cell and K$^+$ inside the cell. For each ATP hydrolyzed, 3Na$^+$ are pumped out, and 2K$^+$ are pumped in.

The last term has the factor $+0.07$ because the voltage is $\phi_{out} - \phi_{in} = 0.0 - (-0.07)$. The free energy per mole for the transport of K$^+$ ions is

$$\Delta\mu(\text{transport of K}^+ \text{ in}) = (0.008314)(310) \ln \frac{100}{5} + (1)(96.485)(-0.07)$$

$$\Delta\mu(\text{transport of K}^+ \text{ in}) = +1.0 \text{ kJ mol}^{-1}$$

The reason for difference in free energy for moving Na^+ ions out and K^+ ions in is that the electrical potential difference of -70 mV favors transport of positive ions in; thus, the favorable effect of the electric field nearly cancels the effect of the concentration gradient. The total free-energy cost of transporting 3 mol of Na^+ ions out and 2 mol of K^+ ions in is

$$\Delta G(\text{total ion transport}) = 3\Delta\mu(\text{transport of } Na^+ \text{ out})$$
$$+ 2\Delta\mu(\text{transport of } K^+ \text{ in})$$
$$\Delta G(\text{total ion transport}) = +42.8 \text{ kJ} \tag{5.40}$$

The free energy for the hydrolysis of ATP ($ATP \rightarrow ADP + P$) is

$$\Delta\mu = \Delta\mu^0 + RT \ln Q$$
$$= \Delta\mu^0 + RT \ln \frac{[ADP][P]}{[ATP]}$$

$\Delta\mu^0$ is -31.0 kJ for ATP hydrolysis at 298 K and -31.3 kJ at 310 K. The ratio of ATP to ADP in cells is commonly 100 or larger. The concentration of phosphate varies, but estimates of Q range from 10^{-2} to 10^{-3}. This gives

$$\Delta\mu(\text{ATP hydrolysis}) = -31.3 + (0.008314)(310)\ln Q$$
$$= -31.3 - 11.8 \quad \text{or} \quad = -31.3 - 17.8$$
$$= -43.1 \text{ to } -49.1 \text{ kJ} \tag{5.41}$$

Clearly, the free energy of ATP hydrolysis is sufficient to account for the active transport of the ions; the negative free energy values of Eq. (5.41) are larger than the positive value of Eq. (5.40). The free-energy calculation does not prove the mechanism. It only shows that the mechanism is a possible one that does not violate thermodynamic principles. This is a very important test, however, because if the calculation did not give a net negative free energy, the mechanism would have been immediately disproved.

The mechanism illustrated in figure 5.23 is only one of several that are involved in ion translocation. In excitable membranes, such as those in the sensory nervous system of animals, signal transmission in a neuron occurs as a triggered consequence of the flow of ions across the plasma membrane. These membranes contain separate Na^+ and K^+ channels, each of which in turn exhibits a response to the concentration of activators like Ca^{2+}, to the transmembrane potential, to the time dependence of these parameters, and to inhibitors like tetrodotoxin (the puffer fish poison). Using the techniques of modern electrophysiology, the currents associated with single channels can be monitored, as illustrated in figure 5.24. By recording the current flowing across a small region of membrane-containing neuronal channels, the pattern of opening and closing of individual channels is seen as pulses of current that occur intermittently and with a range of duration. When one channel is open, a fixed magnitude of ionic current (upward deflection) flows; when two channels are open simultaneously, twice as much current flows. When the channel gates close, the current falls to zero. The effect of applying a depolarizing transmembrane potential on the magnitude of current flowing in an open channel is also seen in figure 5.24.

An important aspect of photosynthetic energy conversion involves H^+ gradients (ΔpH) across the thylakoid membrane (see figure 5.21). Light acti-

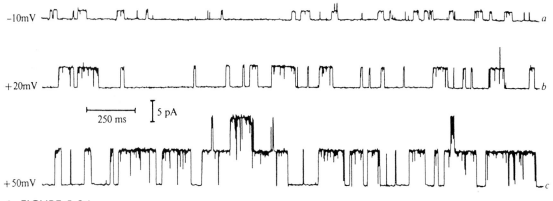

-10mV   *a*

+20mV   *b*

$\overline{}$ $\boxed{}$ 5 pA
250 ms

+50mV   *c*

▲ FIGURE 5.24

Single-channel currents observed from K^+ channels in cultured rat myotubes. After the patch of membrane has been held at the potential given at the left for several seconds, the stationary responses shown are recorded. Depolarization increases the current magnitude along with the probability and duration of channel opening. [Data from B. S. Pallotta, K. L. Magleby, and J. N. Barrett, *Nature (London)* 293 (1981): 471–474.]

vation of PS1 and PS2 pumps H^+ ions from the stromal to the lumenal phase. A transmembrane electric field $\Delta\phi$ (positive inside) also results from the electron transport induced by PS1 and PS2. Both the ΔpH and the $\Delta\phi$ provide the source of chemical potential to power the ATPase, which couples ADP with phosphate to produce high chemical potential ATP. The ATP in turn provides a source of chemical potential for a variety of biochemical reactions, including those involved in converting CO_2 to sucrose, cellulose, and other plant materials.

Colligative Properties

The *normal boiling point* of a liquid is the temperature at which the pure liquid is in equilibrium with its vapor at 1 atm. The *normal freezing point* is the temperature at which the liquid and solid are in equilibrium at 1 atm. The boiling point and freezing point of a solution depend on the concentration of any solutes dissolved in the liquid. The solubility of gases and solids in a solvent depends on the temperature and pressure. These properties involve phase equilibria between gas and liquid or between solid and liquid. They can be understood by straightforward thermodynamic analysis.

The osmotic pressure of a solution—the pressure that must be applied to a solution to keep it in equilibrium with pure solvent—depends on the concentrations of the solutes dissolved. If a solution is separated from pure solvent by a semipermeable membrane (permeable to solvent but not to solutes), solvent will flow through the membrane into the solution as long as equilibrium is not attained. This effect is vital in transporting water from the roots up to the leaves of trees. It also causes rupture of red blood cells in contact with pure water or with a solution that is too dilute (hypotonic).

The properties described above are called *colligative* because they are tied together by a common origin. In dilute solution, they depend only on the number of solute molecules, not on the kinds of solute molecules. This means that colligative properties can be used to count the number of molecules present in a solution. We can easily prepare a solution containing a known weight of solute molecules; thus, we can use a colligative property to determine their molecular weight.

For sufficiently dilute solutions, the vapor pressure, boiling point, freezing point, and osmotic pressure of a solution are linearly dependent on the concentration of solutes present. Furthermore, they are all closely related to one another. If one is measured, the others can be calculated. For example, if the osmotic pressure of a solution is measured, the vapor pressure of the solvent is immediately known. For dilute or ideal solutions, from the measured boiling-point rise, the freezing-point lowering can be calculated. We find that in ideal solutions the colligative properties depend only on the number of solute molecules. So if we measure the molar concentration of the solute in solution, we can calculate any colligative property. The effective link between the measured colligative properties is the activity of the solvent. For a real solution, we can obtain the activity of the solvent in the solution from one measured colligative property. Then any other colligative property at the same temperature can be calculated.

Raoult's law

Adding a solute to a solvent decreases the vapor pressure of the solvent. Raoult found experimentally that the vapor pressure of the solvent in a solution is directly proportional to its mole fraction. *Raoult's law* is stated:

$$P_A = X_A P_A^{\bullet} \qquad (5.10)$$

where P_A = vapor pressure of component A in solution
$\quad X_A$ = mole fraction of A
$\quad P_A^{\bullet}$ = vapor pressure of pure A

Raoult's law is like the ideal gas law, in that it applies approximately to many solutions and it applies exactly to very dilute solutions. Therefore, for a dilute solution, you can always calculate the vapor pressure of the solvent P_A simply by knowing its mole fraction X_A and the vapor pressure of pure solvent $P_A^{\bullet}$, at the same temperature. If Raoult's law applies, the solution is defined to be ideal.

Henry's law

The solutes in the solution may also be volatile. What about their vapor pressures? Henry found experimentally that the solute vapor pressure is proportional to the solute mole fraction, but the proportionality constant is not, in general, the vapor pressure of the pure solute. *Henry's law* is stated:

$$P_B = k_B X_B \qquad (5.11)$$

where P_B = vapor pressure of component B
$\quad X_B$ = mole fraction of B
$\quad k_B$ = Henry's law constant for component B in solvent A

You can think of Henry's law as specifying the pressure dependence of the solubility of a gas in a solution. It says that the gas solubility should be directly proportional to pressure:

$$X_B = \frac{P_B}{k_B} \qquad (5.42)$$

The solubility also depends on temperature and the nature of the solvent. These must be specified for each value of the Henry's law constant k_B. Table 5.1 gives Henry's law constants for various gases in water at several temperatures in order of increasing solubility of the gas. Using Eq. (5.42), we obtain the solubility as mole fraction, but because these are dilute aqueous solutions, it is easy to see that the molarity is just 55.6 ($= 1000/18$) times the mole fraction; for dilute aqueous solutions, $c = 55.6X$.

EXAMPLE 5.7

Divers get "the bends" when bubbles of N_2 gas form in their bloodstream if they rise too rapidly from a deep dive. Calculate the solubility (mol L^{-1}) of N_2 in water (this is roughly equal to the solubility in blood serum) at sea level and at a depth of 300 ft below the surface of the ocean.

Solution

First, we need to find the pressure at a depth of 300 ft. We know that a column of mercury 760 mm ($= 2.49$ ft) high exerts a pressure of 1 atm. An equal column of seawater will exert a pressure proportional to the ratio of the density of seawater (1.01) to mercury (13.6); therefore, the pressure 300 ft deep in the ocean is

$$P = \frac{300 \text{ ft}}{2.49 \text{ ft atm}^{-1}} \left(\frac{1.01}{13.6} \right) + 1 \text{ atm}$$

$$= (8.9 + 1)\text{atm} = 9.9 \text{ atm}$$

The solubilities are obtained from Eq. (5.42) and $c = 55.6X$.

$$c_{N_2} = \frac{55.6 P_{N_2}}{k_{N_2}} = \frac{55.6 P_{N_2}}{86 \times 10^3}$$

$$= 0.65 \times 10^{-3} \text{ mol L}^{-1} \text{ at 1 atm}$$

$$= 6.4 \times 10^{-3} \text{ mol L}^{-1} \text{ at 9.9 atm}$$

The diver should rise slowly to allow plenty of time for the extra dissolved nitrogen in her blood to equilibrate with the decreasing external pressure.

Boiling-point rise and freezing-point lowering

When a small amount of solute is added to a solvent, the boiling point of the solution is raised, and the freezing point of the solution is lowered, relative to those of the pure solvent. The changes in the boiling point and the freezing point are found to be directly proportional to the concentration of the solute in dilute solution:

$$T_0 - T_{\text{freeze}} = K_{\text{freeze}} m \qquad \textbf{(5.43)}$$

$$T_{\text{boil}} - T_0 = K_{\text{boil}} m \qquad \textbf{(5.44)}$$

where T_0 = freezing point or boiling point temperature of pure solvent (it depends on pressure)

$T_{\text{freeze}}, T_{\text{boil}}$ = freezing point and boiling point temperature of solution

$K_{\text{freeze}}, K_{\text{boil}}$ = constants that depend only on the solvent ($K_{\text{freeze}} = 1.86$ and $K_{\text{boil}} = 0.51$ for water)

m = molality of solution = number of moles of solute in 1000 g of solvent

Because K_{freeze} and K_{boil} depend only on the solvent, Eqs. (5.43) and (5.44) can be used to measure total concentration of solutes in complicated mixtures of solutes in solution. For example, the freezing point of seawater immediately gives a measure of the total number of moles of ions and other solutes of all kinds present in the seawater. In using Eqs. (5.43) and (5.44), we must remember that the solutions should be dilute and that only the pure solvent must freeze or evaporate. The solutes should not be volatile at the boiling point or soluble in the solid solvent at the freezing point for the simple equations to apply.

We have used three different concentration scales, so it is well to know the relations among them. From the definition of mole fraction X, molality m, and molarity c, we can convert from one to the other. Here we give the relations that apply to dilute solutions (m or $c < 0.1$):

$$c = \rho_0 m \tag{5.45}$$

$$X_B = \frac{M_A m}{1000} \tag{5.46}$$

where X_B = mole fraction of solute
$\quad M_A$ = molecular weight of solvent
$\quad m$ = molality of solute
$\quad c$ = molarity of solute
$\quad \rho_0$ = density of solvent

We note that for dilute aqueous solution,

$$X_B = \frac{m}{55.6}$$

$$c = m \tag{5.47}$$

The change of the boiling point or freezing point of a dilute solution is inversely proportional to the molecular weight of the solute. Equations (5.43) and (5.44) can be written:

$$M_B = \frac{wt_B}{kg\ solvent} \cdot \frac{K_{boil}}{\Delta T_{boil}} \tag{5.48}$$

$$M_B = \frac{wt_B}{kg\ solvent} \cdot \frac{K_{freeze}}{\Delta T_{freeze}} \tag{5.49}$$

where M_B = molecular weight of solute
$\quad \Delta T_{boil}$ = increase in boiling-point temperature
$\quad \Delta T_{freeze}$ = decrease in freezing-point temperature
$wt_B/kg\ solvent$ = weight of solute per 1000 g of solvent

These methods for measuring molecular weights are limited to temperatures near the boiling point or freezing point of the solvent. As a consequence, they have limited applicability to purely biological systems, but in special cases they may be quite useful. A study of the molal concentration of total solute particles in different cells of the kidney was made by putting a kidney slice on the cold stage of a microscope. The freezing point determined for each separate cell immediately gave the total molality of its contents.

The activity of the solvent is the link that connects the seemingly unrelated colligative properties. We use the solvent standard state for the activity of the solvent, so the chemical potential of the standard state μ^0 is chosen to be that of the pure solvent. When a nonvolatile solute is added to a solvent, the vapor pressure of the solution decreases, and therefore the normal boiling point is raised. How much higher must the temperature of the solution be so that its vapor pressure is 1 atm, the same as that of the pure solvent at the normal boiling point T_0? Because the vapor pressure is decreased in the solution, we need to raise the temperature from T_0 to T_{boil}, which we choose so that $P_A = 1$ atm. The vapor pressure of the pure solvent at temperature T_{boil} would be $P_A/a_A = (1 \text{ atm})/a_A$. Now we can use Eq. (5.3), which gives the vapor pressure as a function of temperature. Let $P_1 = 1$ atm at $T_1 = T_0$ and $P_2 = (1 \text{ atm})/a_A$ at $T_2 = T_{boil}$. After rearranging terms, we obtain

$$\ln a_A = \left(\frac{\Delta \overline{H}_{vap}}{R} \right) \left(\frac{1}{T_{boil}} - \frac{1}{T_0} \right) \tag{5.50}$$

where a_A = activity of solvent in solution—if the solution is ideal, $a_A = X_A$
$\quad \Delta \overline{H}_{vap}$ = enthalpy of vaporization of solvent
$\quad T_{boil}$ = boiling point of solution
$\quad T_0$ = boiling point of pure solvent

Equation (5.50) looks different from the familiar Eq. (5.44) relating boiling point and concentration in dilute solution, but Eq. (5.50) reduces to the familiar result as follows. For dilute (two component) solutions, we have the relations

$$\frac{1}{T_{boil}} - \frac{1}{T_0} = \frac{T_0 - T_{boil}}{T_{boil}T_0} \cong \frac{T_0 - T_{boil}}{T_0^2}$$

$$\ln a_A \cong -\frac{n_B}{n_A}$$

$$\ln a_A \cong -\frac{mM_A}{1000} \tag{5.51}$$

where a_A = activity of solvent
$\quad m$ = molality of solute (moles of solute per 1000 g of solvent)
$\quad M_A$ = molecular weight of solvent

Therefore, from Eqs. (5.50) and (5.51),

$$T_{boil} - T_0 = \frac{RT_0^2 M_A}{1000 \, \Delta \overline{H}_{vap}} m = K_{boil}m \tag{5.52}$$

The boiling-point rise is directly proportional to the molality, just as in Eq. (5.44). The proportionality constant depends on the properties of only the solvent. In fact, Eq. (5.52) allows us to calculate values of K_{boil} from known thermodynamic properties of the solvent. For water, we can use $T_0 = 373$ K, $M_A = 18.0$, and $\Delta \overline{H}_{vap} = 40.66$ kJ mol^{-1} at 373 K to calculate a value of $K_{boil} = 0.51$ K m^{-1} (m = molal). This agrees well with the experimental value of 0.52 K m^{-1}, obtained by extrapolating measured values to infinite dilution.

Adding a solute to a solution decreases the activity of the solvent and thus lowers the melting point. The temperature of the solution must be lowered to keep it in equilibrium with the pure solid solvent. The equation is

$$\ln a_A = \frac{\Delta \overline{H}_{fus}}{R}\left(\frac{1}{T_0} - \frac{1}{T_f}\right) \tag{5.53}$$

where a_A = activity of pure solvent—if the solution is ideal, $a_A = X_A$
$\Delta \overline{H}_{fus}$ = enthalpy of fusion per mole
T_{freeze} = freezing point of solution
T_0 = freezing point of pure solvent

For a dilute solution,

$$T_0 - T_{freeze} = \frac{RT_0^2 M_A}{1000\ \Delta \overline{H}_{fus}}m \tag{5.54}$$

$$\Delta T_{freeze} = K_{freeze}m \tag{5.44}$$

Again, for water at $T_0 = 273$ K, $\Delta \overline{H}_{fus} = 6.007$ kJ mol^{-1}, and we calculate $K_{freeze} = 1.875$ K m^{-1}, compared with the experimental value 1.852 K m^{-1}.

Solubility

When the solid in equilibrium with a solution is pure solute instead of pure solvent, we think of the equilibrium as a *solubility*. We can then derive an equation for the temperature dependence of solubility. For an ideal solution,

$$\ln \frac{X_B(T_2)}{X_B(T_1)} = \frac{-\Delta \overline{H}_{sat}}{R}\left(\frac{1}{T_2} - \frac{1}{T_1}\right) \tag{5.55}$$

where $X_B(T_2), X_B(T_1)$ = solubility of solute at temperatures T_2, T_1
$\Delta \overline{H}_{sat} = \overline{H}_B(\text{solution}) - \overline{H}_B(\text{solid})$ = molar enthalpy (heat) of solution of solute in saturated solution

This equation is most useful for slightly soluble solutes because then the solutes are dilute enough for the assumption of ideality to be valid. For dilute solutions, the mole fraction is proportional to molarity and molality, so the ratio of solubilities in any convenient units can be used in Eq. (5.55). Usually, we expect solubility to increase with increasing temperature. However, we now see that the change of solubility with temperature actually depends on the sign of the heat of solution. Some ionic salts decrease in solubility with increasing temperature.

Osmotic Pressure

It has happened that a hospital patient died because pure water was accidentally injected into the veins. The osmotic pressure of the solution inside the blood cells is high enough to cause them to break under these circumstances. Osmotic pressure is defined as the pressure that must be applied to a solution to keep its solvent in equilibrium with pure solvent at the same temperature. If a solution is separated from a solvent by a semipermeable membrane (such as the membrane of a red blood cell), solvent will flow into the solution side until equilibrium is reached. At constant temperature, equilibrium can be attained by increasing the pressure on the solution or by diluting the solution

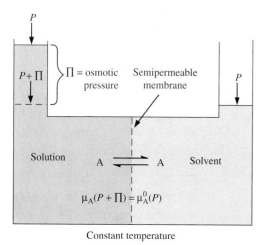

$P + \Pi$ $\Pi =$ osmotic   Semipermeable P
pressure   membrane

Solution   A $\rightleftharpoons$ A   Solvent

$$\mu_A(P + \Pi) = \mu_A^0(P)$$

Constant temperature

◀ FIGURE 5.25

Osmometer showing conditions at
equilibrium. A is a solvent molecule.

until it is essentially pure solvent. A red blood cell placed in pure water or in
a dilute salt solution will burst before equilibrium is reached.

Osmotic pressure is a colligative property closely related to the vapor
pressure, freezing point, and boiling point. The activity of the solvent is the
parameter that is most easily related to the osmotic pressure.

Adding a solute to a solution lowers the activity of the solvent; by raising the pressure on the solution, the activity of the solvent is increased. Van't
Hoff found empirically that for dilute solutions the osmotic pressure Π is directly proportional to the concentration of the solute. Because of the importance of osmotic pressure in biology, we will derive the relevant equations.

Figure 5.25 shows the condition at equilibrium for a solution in contact
with pure solvent through a semipermeable membrane that allows only solvent to pass. The equilibrium occurs in an osmometer in which the extra
pressure Π necessary for equilibrium is produced by the hydrostatic pressure of a water column on the solution side. At equilibrium, there will be no
net flow of solvent across the membrane, so the chemical potential of the
pure solvent at pressure P must be equal to the chemical potential of the solvent in solution at pressure $P + \Pi$:

$$\mu_A(\text{solution}, P + \Pi) = \mu_A(\text{solvent}, P)$$

The temperature is constant throughout. If we let $P = 1$ atm, then at
equilibrium,

$$\mu_A(\text{solution}, 1 + \Pi) = \mu_A^0(\text{solvent})$$

Because of the presence of the solute in solution, the chemical potential of
A at 1 atm pressure is decreased by an amount that can be calculated using
Eq. (4.19):

$$\Delta\mu_{osm} = \mu_A(\text{solution}, 1 \text{ atm}) - \mu_A^0(\text{solvent}) = RT \ln a_A$$

To compensate for this change and restore equilibrium with the solvent, we
need to increase the chemical potential of A by an equal amount by applying
external pressure Π to the solution. Remembering that

$$\overline{G}_2 - \overline{G}_1 = \int_{P_1}^{P_2} \overline{V} \, dP = \overline{V}(P_2 - P_1) = \Delta\mu_{osm} \qquad \textbf{(3.44)}$$

we obtain

$$\Delta\mu_{osm} = RT \ln a_A = \int_{1+\Pi}^{1} \overline{V}_A \, dP = -\overline{V}_A \Pi$$

and

$$\ln a_A = \frac{-\Pi\overline{V}_A}{RT} \tag{5.56}$$

For dilute solutions, we can use the following approximation for $\ln a_A$:

$$\ln a_A = \ln X_A = \ln(1 - X_B) \cong -X_B = -\frac{n_B}{n_A + n_B} \cong -\frac{n_B}{n_A}$$

Then,

$$\Pi = \frac{RT}{n_A\overline{V}_A}n_B \tag{5.57}$$

Furthermore, in dilute solutions, $n_A\overline{V}_A$ is essentially equal to the volume of solution (the volume of solute is negligible). Therefore, for dilute solutions,

$$\Pi V = n_B RT \tag{5.58}$$

$$\Pi = cRT \tag{5.58a}$$

where Π = osmotic pressure, atm
V = volume of solution, L
n_B = number of moles of solute in solution
R = gas constant = 0.08205 L atm K^{-1} mol^{-1}
c = concentration of solute, mol L^{-1}

Equation (5.58a) is the relation that was first found empirically by van't Hoff. These equations state that the osmotic pressure of a solution is directly proportional to the concentration of solute in dilute solution. Equation (5.58) is written to look like the ideal gas equation, to make it easy to remember. However, Π is not the pressure of the solute, but rather the pressure that must be applied to the solution to keep solvent from flowing in through a semipermeable membrane.

To gain an appreciation for the magnitude of osmotic pressure, note that according to Eq. (5.58) a solution with $c = 0.5$ mol L^{-1} at 310 K (37°C) will have an osmotic pressure of 12.7 atm. This is typical of the osmotic concentration of the cytoplasm inside animal cells. When such cells come into contact with pure water, the osmotic pressure may rupture the cell membranes. This happens in the hemolysis of red blood cells when the blood serum is diluted with pure water.

Osmotic-pressure measurement is the most useful colligative method for determining molecular weights. It is accurate for molecular weights approaching 1 million, and small molecule impurities, buffers, or salts do not interfere if suitable membranes are used. The molecular weight of only the solutes that do not pass the semipermeable membrane is obtained. Equation (5.58) can be rewritten as

$$\Pi = \frac{w}{M}RT \tag{5.59}$$

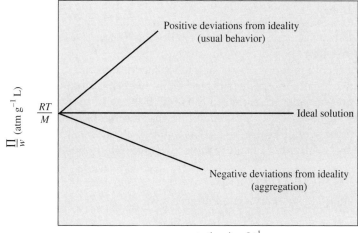

▲ FIGURE 5.26

Dependence of osmotic pressure on weight concentration.

where Π = osmotic pressure, atm
$\quad w$ = concentration of solution, g L^{-1}
$\quad R$ = gas constant = 0.08205 L atm K^{-1} mol^{-1}
$\quad M$ = molecular weight
$\quad T$ = absolute temperature

Usually, a series of concentrations is measured, and extrapolation is made to zero concentration. A plot of Π/w versus w gives RT/M as the intercept; the slope characterizes deviations from ideality (see figure 5.26). A positive slope is usually found for these plots; it can be understood in terms of the activity coefficient of the solute. A negative slope indicates an increase in molecular weight with increasing concentration. This occurs if the solute undergoes aggregation. For an ideal solution, $\Pi/w = RT/M$ at any concentration; for real ionic solutions, this relation begins to fail for concentrations greater than 0.01 M. Osmotic pressure, as well as the other colligative methods, is used to study aggregation in solution at higher concentration.

As derived earlier, the osmotic pressure of a solution is a measure of the activity of the solvent:

$$\ln a_{\mathrm{A}} = \frac{-\Pi \overline{V}_{\mathrm{A}}}{RT} \qquad (5.56)$$

Both vapor pressure and osmotic pressure can be measured at any temperature. They are therefore more useful methods than boiling-point or freezing-point measurements when working with biological molecules.

Molecular-Weight Determination

The molecular weight is an important property of any molecule, but is particularly important for proteins, nucleic acids, and other macromolecules. Just learning the rough size of the molecule may be sufficient for some purposes (M = 1000? 10,000? 1,000,000?), but usually we need more accurate molecular weights. The most accurate molecular weights come from precise knowledge of the amino acid sequence of a protein or the base sequence of a nucleic acid. Nevertheless, sequence information is not informative about posttranslational processing or aggregation of proteins; these are detectable by measurements based on colligative properties. Osmotic pressure is the most useful colligative property to use for macromolecules.

Vapor-pressure lowering, freezing-point lowering, boiling-point rise, and osmotic pressure are all related to the mole fraction of solvent in dilute solution. This means that they all depend on the number of solute molecules (or ions, in the case of electrolytes) added to the solvent. If we divide the weight of solute by the number of solute molecules, we obtain the molecular weight.

Vapor-Pressure Lowering

In dilute solution, we can write Raoult's law,

$$P_A = X_A P_A^\bullet \tag{5.10}$$

because

$$X_B = 1 - X_A = \frac{P_A^\bullet - P_A}{P_A^\bullet} = \frac{n_B}{n_A + n_B} \cong \frac{n_B}{n_A}$$

where X_B = mole fraction of solute
$P_A^\bullet$ = vapor pressure of pure solvent
P_A = vapor pressure of solvent in solution
$P_A^\bullet - P_A$ = vapor pressure lowering of solvent
n_A, n_B = number of moles of solvent and solute, respectively

The number of moles of solute n_B is equal to wt_B/M_B; therefore,

$$M_B = \frac{wt_B}{n_A} \cdot \frac{P_A^\bullet}{P_A^\bullet - P_A} \tag{5.60}$$

For a solution of known weight concentration, vapor-pressure measurements provide the molecular weight of the solute. The solution should be dilute and ideal, so measurements are made as a function of concentration and extrapolated to zero concentration. Inaccuracies in measuring vapor pressures put an upper limit on the molecular weight that can be measured. For very high molecular weights, the vapor-pressure lowering is so small that slight temperature differences between solvent and solution can ruin the results.

A more significant problem for high-molecular-weight solutes is the effect of impurities. Because the vapor pressure depends on the mole fraction of solvent, each solute molecule contributes equally to decreasing the vapor pressure. In dilute solution, a large molecule has the same effect on vapor pressure as a small molecule. On a weight basis, the small molecule produces a much larger effect. One milligram of solute with a molecular weight of 100 lowers the vapor pressure of solution as much as 1 g of solute with a 100,000 molecular weight. It is clear that vapor-pressure lowering cannot be used to measure the molecular weight of a protein dissolved in a buffer solution; the buffer ions would dominate the vapor-pressure lowering.

EXAMPLE 5.8

To determine the molecular weight of creatine (see example 5.5), a sample was purified and used to prepare a solution containing 0.1 g L^{-1}. This solution exhibited an osmotic pressure against pure water of 13.0 torr at 25°C. What is the molecular weight of creatine? Using this result, we can calculate the vapor pressure lowering of water above the solution.

Solution

An osmotic pressure of 13.0 torr results from a solution with a concentration $c = \Pi/RT = (13/760)/(0.08205)(298) = 7.0 \times 10^{-4}$ mol L^{-1}. Thus, the molecular weight of creatine is $(0.1 \text{ g } L^{-1})/(7.0 \times 10^{-4} \text{ mol } L^{-1}) = 143$ g mol^{-1}.

The vapor pressure lowering of this solution is $\Delta P/P = X_{creatine} = (7.0 \times 10^{-4})/(55.6) = 1.26 \times 10^{-5}$. The vapor pressure of pure water at 25°C is 23.8 torr. Therefore, the vapor pressure lowering of this solution is $\Delta P = 1.26 \times 10^{-5}(23.8) = 3 \times 10^{-4}$ torr. Clearly, the osmotic pressure provides a much more sensitive measure of concentration or molecular weight for such solutions than does vapor pressure lowering.

Summary
Phase Equilibrium
Equilibrium conditions

$$T(\text{phase 1}) = T(\text{phase 2})$$

$$P(\text{phase 1}) = P(\text{phase 2})$$

$$\mu_A(\text{phase 1}) = \mu_A(\text{phase 2}) \tag{5.1}$$

Free energy of transport of n mol of an uncharged molecule from phase 1 to phase 2

$$\Delta G = nRT \ln \frac{a_A(\text{phase 2})}{a_A(\text{phase 1})} \cong nRT \ln \frac{c_A(\text{phase 2})}{c_A(\text{phase 1})} \tag{5.25), (5.26}$$

Free energy of transport of n mol of an ion with charge Z from phase 1 to phase 2

$$\Delta G = n \left[RT \ln \frac{a_A(\text{phase 2})}{a_A(\text{phase 1})} + FZV \right] \tag{5.27}$$

F = Faraday = 96,485 J $(eV)^{-1}$
V = potential difference between phases

Scatchard equation for binding to a macromolecule

$$\frac{\nu}{[A]} = K(N - \nu) \tag{5.16}$$

ν = average number of A's bound to macromolecule
N = number of binding sites on macromolecule

Scatchard equation for binding per monomer unit of a macromolecule

$$\frac{r}{[A]} = K(n - r) \tag{5.17}$$

r = number of bound A per monomer unit of the macromolecule
n = number of binding sites per monomer unit of the macromolecule

Hill equation to assess the cooperativity of binding

$$\frac{f}{(1-f)} = K[A]^n \tag{5.19}$$

f = fraction of sites bound
n = Hill coefficient; it varies from 1 for no cooperativity
    to N for all-or-none binding of N ligands
K = a constant, not the binding constant for one ligand

Solutions

Gibbs adsorption isotherm for adsorption at a surface of a solution

$$\Gamma = -\frac{1}{RT}\frac{\partial \gamma}{\partial \ln c} \tag{5.35}$$

Γ = adsorption (excess concentration) of solute at surface, mol m^{-2}
γ = surface tension, N m^{-1} = J m^{-2}
R = gas constant = 8.314 J K^{-1} mol^{-1}
c = concentration of solute in bulk solution

Raoult's law for solvent

$$P_A = X_A P_A^{\bullet} \tag{5.10}$$

P_A = vapor pressure of solvent in solution with mole fraction X_A
$X_A = \dfrac{n_A}{n_A + n_B}$
$P_A^{\bullet}$ = vapor pressure of pure solvent

Henry's law for solute

$$P_B = k_B X_B \tag{5.11}$$

k_B = Henry's law constant (from tables)

Clausius–Clapeyron equation for the change of the boiling point with pressure and the change of the vapor pressure with temperature

$$\ln\frac{P_2}{P_1} = \frac{-\Delta\overline{H}_{vap}}{R}\left(\frac{1}{T_2} - \frac{1}{T_1}\right) \tag{5.3}$$

$\Delta\overline{H}_{vap}$ = molar enthalpy of vaporization

Pressure dependence of the freezing point

$$\frac{T_2 - T_1}{P_2 - P_1} = T\frac{\Delta\overline{V}_{fus}}{(9.869 \text{ cm}^3 \text{ atm J}^{-1})\,\Delta\overline{H}_{fus}} \tag{5.5}$$

$\Delta\overline{V}_{fus}$ = molar volume (cm^3 mol^{-1}) of fusion
$\Delta\overline{H}_{fus}$ = molar heat (J mol^{-1}) of fusion
    T = average of T_2 and T_1

P must be in atmospheres to be consistent with the unit-conversion factor.

Concentration units

$$\text{mole fraction:} \quad X_B = \frac{n_B}{n_A + n_B}$$

molality: m = moles of solute per 1000 g of solvent

molarity: c = moles of solute per liter of solution

Dilute solutions

$$X_B = \frac{M_A m}{1000} \tag{5.46}$$

$$c = \rho_0 m \tag{5.45}$$

M_A = molecular weight of solvent of density ρ_0

Boiling-point rise

$$T_{\text{boil}} - T_0 = K_{\text{boil}} m = \frac{M_A R T_0^2}{1000 \, \Delta \overline{H}_{\text{vap}}} m \tag{5.52}$$

For H_2O, $K_{\text{boil}} = 0.51$.

Freezing-point lowering

$$T_0 - T_{\text{freeze}} = K_{\text{freeze}} m = \frac{M_A R T_0^2}{1000 \, \Delta \overline{H}_{\text{fus}}} m \tag{5.54}$$

For H_2O, $K_{\text{freeze}} = 1.86$.

Osmotic pressure (in atm)

$$\Pi = \frac{n_B R T}{V} = cRT \tag{5.58, 5.58a}$$

$R = 0.08205 \text{ L atm K}^{-1} \text{ mol}^{-1}$
V = volume of solution containing n_B moles solute, L
c = concentration, mol L^{-1}

Osmotic-pressure molecular weight

$$M = \frac{wRT}{\Pi} \tag{5.59}$$

where w = concentration of solution, g L^{-1}

Vapor-pressure-lowering molecular weight

$$M = \frac{\text{wt of solute}}{\text{moles of solvent}} \cdot \frac{P_A^{\bullet}}{P_A^{\bullet} - P_A} \tag{5.60}$$

P_A = vapor pressure of solution
$P_A^{\bullet}$ = vapor pressure of pure solvent

Boiling-point and freezing-point molecular weights

$$M = \frac{\text{wt}_B}{\text{kg solvent}} \cdot \frac{K_{\text{boil}}}{\Delta T_{\text{boil}}} \tag{5.48}$$

$$M = \frac{\text{wt}_B}{\text{kg solvent}} \cdot \frac{K_{\text{freeze}}}{\Delta T_{\text{freeze}}} \tag{5.49}$$

$\Delta T_{\text{boil}}, \Delta T_{\text{freeze}}$ = change in boiling point or freezing point
$K_{\text{boil}}, K_{\text{freeze}}$ = constants from tables

The activity of the solvent is obtained from colligative properties.

Osmotic pressure, Π

$$\ln a_A = \frac{-\Pi \overline{V}_A}{RT} \tag{5.56}$$

$\overline{V}_A$ = molar volume of solvent

Vapor pressure

$$a_A = \frac{P_A}{P_A^\bullet} \tag{5.8}$$

Boiling point and freezing point

$$\ln a_A = \left(\frac{\Delta \overline{H}_{\text{vap}}}{R} \right) \left(\frac{1}{T_{\text{boil}}} - \frac{1}{T_0} \right) \tag{5.50}$$

$$\ln a_A = \left(\frac{\Delta \overline{H}_{\text{fus}}}{R} \right) \left(\frac{1}{T_0} - \frac{1}{T_{\text{freeze}}} \right) \tag{5.53}$$

$\Delta \overline{H}_{\text{vap}}, \Delta \overline{H}_{\text{fus}}$ = molar heats of vaporization, fusion

References

Branton, D., and D. Deamer. 1972. *Membrane Structure*. New York: Springer-Verlag.

Gennis, R. D. 1989. *Biomembranes: Molecular Structure and Function*. New York: Springer Verlag.

Hille, B. 1992. *Ionic Channels of Excitable Membranes*, 2d ed. Sunderland, MA: Sinauer.

Koshland, D. L. Jr., G. Nemethy, and D. Filmer, 1966. Comparison of Experimental Binding Data and Theoretical Models in Proteins Containing Subunits. *Biochemistry* 5:365–385.

Monod, J., J. Wyman, and J. P. Changeux. 1965. On the Nature of Allosteric Transitions: A Plausible Model. *J. Mol. Biol.* 12:88–118.

Nicholls, D. G. 1982. *Bioenergetics, An Introduction to Chemiosmotic Theory*. New York: Academic Press.

Silver, B. L. 1985. *The Physical Chemistry of Membranes, an Introduction to the Structure and Dynamics of Membranes*. Boston: Allen & Unwin.

Tanford, C. 1980. *The Hydrophobic Effect: Formation of Micelles and Biological Membranes*, 2d ed. New York: Wiley.

Wyman, J., and S. J. Gill, 1990. *Binding and Linkage, Functional Chemistry of Biological Macromolecules*. Mill Valley, CA: University Science Books.

Suggested Reading

Dickerson, R. E., and I. Geis. 1983. *Hemoglobin: Structure, Function, Evolution, and Pathology.* New York: Benjamin/Cummings.

Linder, M. E., and A. G. Gilman. 1992. G Proteins. *Sci. Am.* 267:56–65. (G proteins are membrane-bound proteins that cause cellular responses to chemical messengers.)

Phillips, J. L. 1998. *The Bends: Compressed Air in the History of Science, Diving and Engineering.* New Haven, CT: Yale University Press.

Singer, S. J., and G. L. Nicholson. 1972. The Fluid Mosaic Model of the Structure of Cell Membranes. *Science* 175:720–731.

Ward, P., L. Greenwald, and O. E. Greenwald. 1980. The Buoyancy of the Chambered Nautilus. *Sci. Am.* 243:190–203.

Wolfenden, R. 1983. Waterlogged Molecules. *Science* 222:1087–1093.

Internet

White, S. 1999. *Membrane Proteins and Bilayers.* [Biophysical Society, Online Biophysics Textbook.]   Available: http://biosci.cbs.umn.edu/biophys/OLTB/membrane .html (requires Adobe Acrobat Reader).

Problems

1. (a) How many kilograms of water can be evaporated from a 1-m^2 surface of water in a lake on a hot, clear summer day, if it is assumed that the limiting factor is the Sun's heat of 4 J min^{-1} cm^{-2} for an 8-hour day? Assume that the air and water temperature remains constant at 40°C.

    (b) What volume of air is needed to hold this much water vapor at 40°C?

    (c) When the Sun sets, the air temperature drops to 20°C. Assuming that the excess water vapor is precipitated as rain, calculate the weight of the rain and the amount of heat released when the water condenses. When it starts raining, do you expect the temperature of the air to rise or fall? Explain.

    (d) Calculate the vapor pressure of liquid water at 200°C. Assume that the heat of vaporization does not change from its value at 100°C.

    (Table 2.2 is useful for this problem.)

2. The vapor pressure of pure, solid pyrene at 25°C is P_1 atm. The solubility at 25°C of pyrene in water is 1.0×10^{-4} M; the solubility of pyrene in ethanol at 25°C is greater than 10^{-4} M. Neither water nor ethanol dissolves significantly in pyrene.

    (a) What is the vapor pressure of pyrene above a saturated solution of pyrene in water at 25°C?

    (b) If ethanol is added to the saturated aqueous solution, more pyrene dissolves. When equilibrium with the solid pyrene is reached again, will the vapor pressure of the pyrene above the new solution be greater than, less than, or the same as in part (a)?

    (c) The solubility of pyrene at 25°C is 1.1×10^{-3} M in an aqueous solution of 0.10 M cytosine. This increase in solubility is due to the formation of a complex between pyrene and cytosine. Calculate the values of the equilibrium constants for the following reactions at 25°C, where P = pyrene and C = cytosine. Use concentration in M instead of activities.

    $$P(\text{solution}) + C(\text{solution}) = P \cdot C(\text{solution}) \quad (1)$$

    $$P(\text{solid}) + C(\text{solution}) = P \cdot C(\text{solution}) \quad (2)$$

3. (a) Which of the following has the lowest standard chemical potential at 25°C: $CH_4(l)$, $CH_4(s)$, $CH_4(g)$, or $CH_4(aq)$? Explain your answer.

    (b) Calculate the solubility in mol L^{-1} for CH_4 in water at 25°C when the CH_4 gas pressure is 1 atm.

    (c) Will the solubility be greater or less at 0°C? Explain your answer.

    (d) Henry's law constants for nearly all of the gases listed in table 5.1 increase with increasing temperature. What can you conclude from this about an important thermodynamic property (for example, enthalpy, heat capacity, entropy, or free energy) associated with the dissolution of gases in water?

4. The standard free energy for the reaction of oxygen binding to myoglobin

    $$\text{myoglobin} + O_2 \longrightarrow \text{oxymyoglobin}$$

    is $\Delta G^0 = -30.0$ kJ mol^{-1} at 298 K and pH 7. The standard state of O_2 is the dilute solution, molarity scale; therefore the concentration of O_2 must be in M.

    What is the ratio (oxymyoglobin)/(myoglobin) in an aqueous solution at equilibrium with a partial pressure of oxygen $P_{O_2} = 30$ torr? Assume ideal behavior of O_2 gas, and for the protein in solution.

5. According to the chemiosmotic theory, an electrochemical proton gradient is used to synthesize ATP in

mitochondria. The enzyme that does this is located on the inside of the mitochondrial membrane. The oxidation of carbohydrates and fats is used to pump protons outside the mitochondrial membrane until the steady-state membrane potential is -140 mV and the pH gradient is ΔpH = 1.5. Inside the mitochondrion, pH = 7.0, [ATP] = 1 mM, $[P_i]$ = 2.5 mM, [ADP] = 1 mM, and T = 298 K.

(a) How much chemical potential is required to synthesize ATP inside the mitochondria?

(b) How much free energy is made available by moving 1 mol of protons from the outside to the inside? Is this enough to drive ATP synthesis?

(c) How many protons must be translocated per ATP synthesized?

6. In living biological cells, the concentration of sodium ions inside the cell is kept at a lower concentration than the concentration outside the cell, because sodium ions are actively transported from the cell.

(a) Consider the following process at 37°C and 1 atm.

1 mol NaCl(0.05 M inside) $\longrightarrow$

1 mol NaCl(0.20 M outside)

Write an expression for the free-energy change for this process in terms of activities. Define all symbols used.

(b) Calculate $\Delta\mu$ for the process. You may approximate the activities by concentrations in M. Will the process proceed spontaneously?

(c) Calculate ΔG for moving 3 mol of NaCl from inside to outside under these conditions.

(d) Calculate $\Delta\mu$ for the process if the activity of NaCl inside is equal to that outside.

(e) Calculate $\Delta\mu$ for the process at equilibrium.

(f) The standard free energy for hydrolysis of ATP to ADP (ATP + H_2O → ADP + phosphate) in solution is ΔG^0 = -31.3 kJ mol^{-1} at 37°C, 1 atm. The free energy of this reaction can be used to power the sodium-ion pump. For a ratio of ATP to ADP of 10, what must be the concentration of phosphate to obtain -40 kJ mol^{-1} for the hydrolysis? Assume that activity coefficients are 1 for the calculation.

(g) If the ratio of ATP to ADP is 10, what is the concentration of phosphate at equilibrium? Assume ideal solution behavior. What do you conclude from your answer?

7. The binding of a ligand A to a macromolecule was studied by equilibrium dialysis. In each measurement a 1.00×10^{-6} M solution of the macromolecule was dialyzed against an excess amount of a solution containing A. After equilibrium was reached, the total concentra-

tion of A on each side of the dialysis membrane was measured. The following data were obtained:

Total concentration of A (M)

The side without macromolecules	The side with $1.00 \times 10^{-6}\,M$ macromolecules
0.51×10^{-5}	0.67×10^{-5}
1.02×10^{-5}	1.28×10^{-5}
2.01×10^{-5}	2.34×10^{-5}
5.22×10^{-5}	5.62×10^{-5}
10.50×10^{-5}	10.91×10^{-5}
20.00×10^{-5}	20.47×10^{-5}

Using these data, calculate ν for each value of [A]. Make a Scatchard plot and evaluate the intrinsic equilibrium constant K and the total number of sites per macromolecule, if the independent-and-identical-sites model appears to be applicable.

8. Plot $\nu/[A]$ versus ν for the following cases:

(a) There is a total of ten identical and independent sites per polymer and the intrinsic binding constant is $K = 5 \times 10^5\,M^{-1}$.

(b) There is a total of ten independent sites per polymer. Nine of the ten sites are identical, with $K = 5 \times 10^5\,M^{-1}$. The binding constant for the tenth site is $5 \times 10^6\,M^{-1}$. For more than one type of site, the Scatchard equation can be generalized to

$$\frac{\nu}{[A]} = \sum_i \frac{N_i K_i}{1 + K_i[A]}$$

[Note that the deviation from linearity is small for part (b). Because experimental error in such measurements is usually appreciable, be cautious in concluding from a linear Scatchard plot that the sites are identical.]

(c) Repeat the calculation and plot of part (b) for five identical sites of each type.

9. The enzyme tetrahydrofolate synthetase has four identical and independent binding sites for its substrate ATP. For an equilibrium dialysis experiment, a solution of the enzyme was prepared. First the osmotic pressure of this solution was measured to be 2.4×10^{-3} atm at 20°C, using an osmometer. Then the enzyme solution was placed in a dialysis bag, and the binding of ATP to the enzyme was measured by equilibrium dialysis at the same temperature. In one run, after the binding equilibrium was established, the concentration of free ATP outside the bag was found to be 1.0×10^{-4} M, and the total ATP concentration inside the bag was found to be 3.0×10^{-4} M.

(a) On the basis of the information above, calculate K, the equilibrium constant for the binding of ATP to the enzyme at 20°C.

(b) The equilibrium dialysis experiment was repeated for a series of outside ATP concentrations at two different temperatures (20°C and 37°C). The following Scatchard plots were obtained:

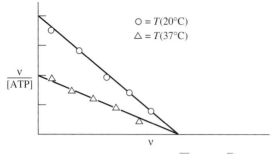

Using this information, calculate $\Delta \overline{H}^0$ and $\Delta \overline{S}^0$ for the binding of ATP to its site on the enzyme. (Assume that $\Delta \overline{H}^0$ and $\Delta \overline{S}^0$ are independent of temperature.)

10. (a) Using Henry's law constants in table 5.1, calculate the solubility (in M) of each gas in water at 25°C if $P_{O_2} = 0.2$ atm, $P_{N_2} = 0.75$ atm, and $P_{CO_2} = 0.05$ atm.

(b) What will be the vapor pressure at 25°C of the water in this solution if Raoult's law holds? The vapor pressure of pure water at 25°C is 23.756 torr.

11. (a) Assume that Raoult's law holds and calculate the boiling point of a 2-M solution of urea in water.

(b) Urea actually forms complexes in solution in which two or more molecules hydrogen bond to form dimers and polymers. Will this effect tend to raise or lower the boiling point?

12. When cells of the skeletal vacuole of a frog were placed in a series of aqueous NaCl solutions of different concentrations at 25°C, it was observed microscopically that the cells remained unchanged in 0.7 wt % NaCl solution, shrank in more concentrated solutions, and swelled in more dilute solutions. Water freezes from the 0.7% salt solution at −0.406°C.

(a) What is the osmotic pressure of the cell cytoplasm relative to that of pure water at 25°C?

(b) Suppose that sucrose (mol wt 342) was used instead of NaCl (mol wt 58.5) to make the isoosmotic solution. Estimate the concentration (wt %) of sucrose that would be sufficient to balance the osmotic pressure of the cytoplasm of the cell. Assume that sucrose solutions behave ideally.

13. On the planet Taurus II, ammonia plays a role similar to that of water on Earth. Ammonia has the following properties: Normal (1 atm) boiling point is −33.4°C, where its heat of vaporization is 1368 kJ kg^{-1}. Normal freezing point is −77.7°C.

(a) Estimate the temperature at which the vapor pressure of NH$_3$(l) is 60 torr.

(b) Calculate the entropy of vaporization of NH$_3$ at its normal boiling point. Compare this value with that predicted by Trouton's rule ($\Delta S^0_{vap} \cong 88$ J K^{-1} mol^{-1}) for typical liquids. Give a likely reason for any discrepancy.

(c) Ammonia undergoes self-dissociation according to the reaction

$$2NH_3(l) \rightleftharpoons NH_4^+(am) + NH_2^-(am)$$

where (am) refers to the "ammoniated" ions in solution. At −50°C, the equilibrium constant for this reaction is

$$K = (NH_4^+)(NH_2^-) = 10^{-30}$$

Calculate $\Delta \mu^0_{223}$ for the self-dissociation. What is the standard state for which $a_{NH_3} = 1$?

(d) When 1 mol of ammonium chloride, NH$_4$Cl, is dissolved in 1 kg of liquid ammonia, the boiling point at 760 torr is observed to occur at −32.7°C. What conclusions can you reach about the nature of this solution? State the evidence for your answer.

14. One of the important factors responsible for geologic evolution (erosion in particular) is the pressure developed by the freezing of water trapped in enclosed spaces in rock formations. Estimate the maximum pressure that can be developed by water freezing to ice on a cold night ($T = -10$°C). The densities of ice and water at −10°C may be taken as 0.9 g cm^{-3} and 1.0 g cm^{-3}, respectively.

15. Proteins can be hydrolyzed to amino acids by heating in very dilute aqueous NaOH solution. The protein itself is also very dilute. The reaction is carried out at 120°C by heating the solution in a sealed tube that was evacuated prior to sealing. What internal pressure must the tube withstand at this temperature? The normal boiling point and $\Delta \overline{H}^0_{vap}$ of water may be taken as 100°C and 40.66 kJ mol^{-1}, respectively.

16. Indicate whether each of the following statements is true or false as it is written. Alter each false statement to make it correct.

(a) If two aqueous solutions containing different nonvolatile solutes exhibit exactly the same vapor pressure at the same temperature, the activities of water are identical in the two solutions.

(b) The most important reason that foods cook more rapidly in a pressure cooker is that equilibrium is shifted in the direction of products at the higher pressure.

(c) If solutions of a single solute are prepared at equal molalities but in different solvents, the freezing-point lowering will be the same for the different solutions, assuming that each behaves ideally.

(d) If solutions of a single solute are prepared at equal mole fractions but in different solvents, the fractional vapor-pressure lowering will be the same for the different solutions, assuming that each behaves ideally.

(e) If two liquids, such as benzene and water, are not completely miscible in one another, the mixture can never be at equilibrium.

17. A solution of the sugar mannitol (mol wt 182.2) is prepared by adding 54.66 g of mannitol to 1000 g of water. The vapor pressure of pure liquid water is 17.54 torr at 20°C. Mannitol is nonvolatile and does not ionize in aqueous solutions.

(a) Assuming that aqueous mannitol solutions behave ideally, calculate the vapor-pressure lowering (the difference between the vapor pressure of pure water and that of the solution) for the above solution at 20°C.

(b) The *observed* vapor-pressure lowering of the mannitol solution above is 0.0930 torr. Calculate the activity coefficient (based on mole fraction) of water in this solution.

(c) Calculate the osmotic pressure of the mannitol solution of part (b) when it is measured against pure water and compare it with the osmotic pressure of the ideal solution.

18. (a) Assuming that glucose and water form an ideal solution, what is the equilibrium vapor pressure at 20°C of a solution of 1.00 g of glucose (mol wt 180) in 100 g of water? The vapor pressure of pure water is 17.54 torr at 20°C.

(b) What is the osmotic pressure, in torr, of the solution in part (a) versus pure water?

(c) What is the activity of water as a solvent in such a solution?

(d) What would be the osmotic pressure versus pure water of a solution containing both 1.00 g of glucose and 1.00 g of sucrose (mol wt 342) in 100 g of water at 20°C?

19. Workers in underwater caissons or diving suits necessarily breathe air at greater than normal pressure. If they are returned to the surface too rapidly, N_2, dissolved in their blood at the previously higher pressure, comes out of solution and may cause emboli (gas bubbles in the bloodstream), bends, and decompression sickness. Blood at 37°C and 1 atm pressure dissolves 1.3 mL of N_2 gas (measured for pure N_2 at 37°C and 1 atm) in 100 mL of blood. Calculate the volume of N_2 likely to be liberated from the blood of a caisson worker returned to 1-atm pressure after prolonged exposure to air pressure at 300 m of water (below the surface). The

total blood volume of the average adult is 3.2 L; density of mercury is 13.6 g mL^{-1}; air contains 78 vol % N_2.

20. Hexachlorobenzene, C_6Cl_6, is a solid compound that is somewhat volatile but not very soluble in water; water is not soluble in C_6Cl_6. Solid C_6Cl_6 and liquid water are allowed to reach equilibrium at 20°C to form a solid phase, an aqueous phase, and a gaseous phase. The vapor pressure of pure solid C_6Cl_6 is 1.0×10^{-2} torr at 20°C and 1.0×10^{-1} torr at 40°C.

(a) A sample of the equilibrium aqueous phase was removed and the equilibrium concentration of C_6Cl_6 was found to be mole fraction $X = 1.00 \times 10^{-5}$ and molarity $c = 55.6 \times 10^{-5}$. What are the vapor pressures of H_2O and C_6Cl_6 for this phase at 20°C?

(b) C_6Cl_6 binds to proteins. A protein is added to the aqueous solution, and equilibrium with solid C_6Cl_6 is established. State whether each of the following increases, decreases, or remains unchanged compared to part (a):

(1) The vapor pressure of C_6Cl_6 for the aqueous solution

(2) The vapor pressure of water for the aqueous solution

(3) The concentration of dissolved (free) C_6Cl_6 in the aqueous solution

(c) What is the mole fraction of C_6Cl_6 in the gas phase at equilibrium at 20°C?

(d) Calculate the heat of vaporization of solid C_6Cl_6.

21. The solubilities of two amino acids in two solvents at 25°C are given below; they are the concentrations present in saturated solutions.

	In water	In ethanol
Glycine	3.09 M	4.04×10^{-4} M
Valine	0.60 M	1.32×10^{-3} M

(a) Calculate the standard free energy of transfer $\Delta\mu^0$ of 1 mol of glycine from the solid to the aqueous solution at 25°C. The standard state for the solid is the pure solid; the standard state in the solution corresponds to $a = 1$ extrapolated from a dilute solution; the molarity concentration scale is used. Consider the solutions to be ideal.

(b) Calculate the standard free energy of transfer $\Delta\mu^0$ of 1 mol of glycine from ethanol to the aqueous solution at 25°C.

(c) Assume that the effects of the backbone and side chain are simply additive (glycine essentially has no side chain) and calculate the standard free energy of transfer of 1 mol of the $(CH_3)_2CH-$ side chain of valine from water to ethanol at 25°C.

(d) Ethanol can be considered to mimic the interior of a protein. Will the mutation of a glycine to a valine in the interior of a protein favor the folding of the protein if interaction with the solvent is the dominant effect? Explain.

(e) Assume that the protein-folding equilibrium is a two-state transition: folded or unfolded. Calculate the change in equilibrium constant due to the mutation from glycine to valine for folding at 25°C, if interaction with the solvent is the dominant effect.

22. The human red-blood-cell membrane is freely permeable to water but not at all to sucrose. The membrane forms a completely closed bag, and it is found by trial and error that adding the cells to a solution of a particular concentration of sucrose results in neither swelling nor shrinking of the cells. In a separate experiment, the freezing point of that particular sucrose solution is found to be −0.56°C.

(a) If the exterior of the cell contains only KCl solution, estimate the KCl concentration. Assume that the membrane is also impermeable to KCl.

(b) If these cells are suspended in distilled water at 0°C, what would be the internal hydrostatic pressure at equilibrium, assuming that the cell did not change volume?

23. (a) What is the osmotic pressure in atm of a 0.1-M solution of urea in ethanol compared with that of pure ethanol at 27°C? State what assumptions you make in the calculation.

(b) What is the osmotic pressure in atm of a 0.1-M solution of NaCl in water compared with that of pure water at 27°C? State what assumptions you make in the calculation.

(c) An aqueous solution of osmotic pressure 5 atm is separated from an aqueous solution of osmotic pressure 2 atm. Which way will the solvent flow? How large a pressure must be applied to prevent this solvent flow?

(d) What is the free-energy change for transferring 1 mol of water from a solution of osmotic pressure 5 atm to a solution of osmotic pressure 2 atm?

24. A protein solution containing 0.6 g of protein per 100 mL of solution has an osmotic pressure of 22 mm of water at 25°C and at a pH where the protein has no net charge (the isoelectric point). What is the molecular weight of the protein?

25. A climber has carefully measured the boiling point of water at the top of a mountain and found it to be 95°C. Calculate the atmospheric pressure at the mountaintop. The boiling point of water at 1 atm pressure is 100°C, and the heat absorbed in vaporizing 1 mol of water at constant pressure is 40.66 kJ mol^{-1}.

26. A cell membrane at 37°C is found to be permeable to Ca^{2+} but not to anions, and analysis shows the inside concentration to be 0.100 M and the outside concentration to be 0.001 M in Ca^{2+}.

(a) What potential difference in volts would have to exist across the membrane for Ca^{2+} to be in equilibrium at the stated concentrations? Assume that activity coefficients are equal to 1. Give the sign of the potential inside with respect to that outside.

(b) If the measured inside potential is +100 mV with respect to the outside, what is the minimum (reversible) work required to transfer 1 mol of Ca^{2+} from outside to inside under these conditions?

27. The water in freshwater lakes is supplied by transfer from the oceans. Consider the oceans to be a 0.5 M NaCl solution and freshwater lakes to be a 0.005 M $MgCl_2$ solution. For simplicity, consider the salts to be completely dissociated and the solutions sufficiently dilute so that all the equations for dilute solutions apply:

(a) Calculate the osmotic pressure of the ocean water and of the lake water at 25°C in equilibrium with pure water.

(b) Give a simple relation between the osmotic pressure and the freezing-point depression of a solution by which you could calculate ΔT_{freeze} without knowing the freezing-point depression constant for the solvent.

(c) How much free energy is required to transfer 1 mol of pure water from the ocean to the lake at 25°C? (In the actual process, the Sun supplies this energy.)

(d) Which solution, the ocean or the lake, has the highest vapor pressure of water?

(e) The observed water vapor pressure at 100°C for 0.5 M NaCl is 747.7 torr. What is the activity of water at this temperature? The vapor pressure of pure water at 100°C is 760 torr.

28. For each of the following processes state whether ΔG, ΔH, and ΔS of the system increase, decrease, or do not change. If there is not enough information given, state what experimental measurements need to be made to allow you to deduce the direction of the change.

(a) A system contains a 1 M sucrose solution and pure water separated by a membrane permeable to water only. A small amount of water is transferred through the membrane from the pure water to the sucrose solution at constant temperature and pressure.

(b) Liquid water at 100°C, 1 atm pressure is evaporated to water vapor at 100°C, 1 atm pressure. The system is the water.

(c) A system contains pure solid water at 0°C suspended in a 0.01 M KCl solution at the same temperature. The pressure on the system is 1 atm. Some of the liquid water is frozen to ice at constant temperature and pressure. The system is the solid water plus the KCl solution.

(d) Water vapor at 1 atm pressure and 25°C is converted to liquid water at the same temperature and pressure along a reversible path. The system is the water.

29. An aqueous solution containing 100 mg L^{-1} of pyruvic acid is separated from pure H_2O by a membrane semipermeable to water, all at 25°C.

(a) Calculate the osmotic pressure that you expect for this solution. State any important assumptions that you make.

(b) The osmotic pressure observed is 36.9 torr. Use this value to estimate the pK for the acid dissociation of pyruvic acid at 25°C.

(c) What fractional vapor-pressure lowering relative to pure water do you expect for the solution of pyruvic acid at 80°C?

30. Which of the following has the largest (most positive) value for the:

(a) Molar entropy increase?

$H_2O(l, 0°C) \rightarrow H_2O(l, 100°C)$

$H_2O(s, 0°C) \rightarrow H_2O(l, 0°C)$

$H_2O(l, 100°C) \rightarrow H_2O(g, 1 \text{ atm}, 100°C)$

$H_2O(g, 1 \text{ atm}, 100°C) \rightarrow H_2O(g, 0.1 \text{ atm}, 100°C)$

(b) Free energy of transfer of 1 mol of the substance from inside to outside a cell, all at 37°C?

$NaCl(a = 0.010 \text{ inside}) \rightarrow NaCl(a = 0.140 \text{ outside})$

$KCl(a = 0.100 \text{ inside}) \rightarrow KCl(a = 0.005 \text{ outside})$

$Glucose(a = 0.10 \text{ inside}) \rightarrow Glucose(a = 0.20 \text{ outside})$

$NaCl(a = 0.10 \text{ inside}) \rightarrow NaCl(a = 0.20 \text{ outside})$

31. A 1.0 M sucrose solution and a 0.5 M sucrose solution (both in water) in separate open vessels are placed next to each other in a sealed container at constant temperature. What, if anything, will occur? Explain your answer briefly.

32. The measured osmotic pressure of seawater is 25 atm at 273 K.

(a) What is the activity of water in seawater at 273 K? Take the partial molar volume of water in seawater to be 0.018 M^{-1}.

(b) The vapor pressure of pure water at 273 K is 4.6 torr. What is the vapor pressure of water in seawater at this temperature?

(c) What temperature would you expect to find for seawater in equilibrium with the polar ice caps? The melting point of pure water at 1 atm pressure is 0°C, and $\Delta \overline{H}^0_{\text{fus}}$ may be taken as 5.86 kJ mol^{-1}, independent of temperature.

33. The following device is a candidate for a perpetual motion machine:

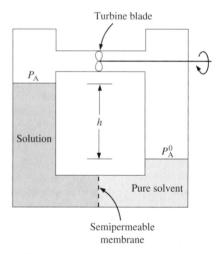

(1) All parts of the machine are at the *same* temperature T, equal to that of the heat bath surrounding the machine.

(2) The compartment at the left contains a solution, separated from the compartment at the right (containing pure solvent) by a semipermeable membrane, permitting solvent transfer only. A pressure difference $\Delta P = \rho g h$ balances the osmotic pressure difference Π across the membrane.

(3) The vapor pressure $P^{\bullet}_A$ above pure solvent is higher than the pressure P_A above the solution (Raoult's law). There is therefore a pressure difference $\Delta P = P^{\bullet}_A - P_A$ in the tube, and the flow of vapor in the upper connecting tube, from right to left, drives a turbine blade.

(4) Vapor moving through the connecting tube condenses at the surface of the solution. It tends to dilute

the solution, therefore reducing its osmotic pressure. This means that solvent will flow back through the semipermeable membrane, restoring the original concentration of the solution.

(5)   The solvent is therefore set into perpetual motion, flowing as vapor from right to left in the upper part and from left to right in the lower part, yielding energy.

Analyze the operation of this machine carefully. First, decide whether this machine works or fails. Then defend your answer as quantitatively and clearly as possible.

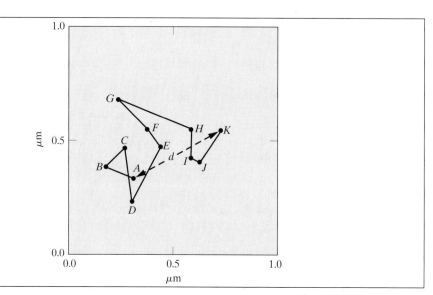

CHAPTER 6

Molecular Motion and Transport Properties

Although we have emphasized molecules in the first five chapters, thermodynamics really deals with properties, such as energy, that don't depend on whether molecules exist. Now we start describing what molecules do and how fast they move in outer space, in Earth's gravity, in a centrifuge, or in an electric field. We can sequence DNA by cutting it and measuring the molecular motions of the fragments, we can learn the sizes of proteins and viruses, and we can study their interactions.

Concepts

We have learned how thermodynamic properties—energy and entropy—are related to the energy of each molecule, to molecular interactions, and to the randomness of the collection of molecules (chapters 2 and 3). Now we will learn about properties that depend directly on how fast molecules move.

The random motion of molecules is dependent on the temperature. At 0 K, molecular motion ceases, except for quantum mechanical zero-point vibration. As the temperature is raised, molecules increasingly vibrate, rotate, and translate—they move. It is easiest to analyze their motion in the gas phase where molecular interactions are least important. For ideal gases, the velocity of a molecule depends only on the temperature and the mass of the molecule. All molecules with the same mass in a gas of constant temperature do not have the same velocity. Each molecule is continually colliding with other molecules and with the walls of the container; all molecules are constantly changing direction, speeding up, and slowing down. However, given the molecular weight and the temperature, we can calculate the probability that a molecule will have a particular velocity, and we can calculate the average velocity for all the molecules. The *Maxwell–Boltzmann theory* allows us to understand molecular velocities and kinetic energies.

In a liquid, the random motion of molecules depends on the temperature, but because collisions are so important in liquids, molecular motion also depends on the sizes and shapes of molecules. The larger the volume of a molecule, the slower is its motion; spherical molecules move faster than those with asymmetric shapes of the same volume. The random motion of molecules is *diffusion;* by observing the diffusion, we learn about size and shape.

If a force is applied to a molecule, its motion is no longer random; it is directed. On Earth's surface, all molecules experience the force of gravity pulling them toward the center of Earth. They tend to sediment to the bottom of a container unless their random motion—their diffusion—keeps the solution mixed. Light molecules in solution remain uniformly mixed; heavy molecules sediment to the bottom. Earth's gravitational field is not very strong, so only large particles, such as viruses or chromosomes, will sediment significantly. By using a centrifugal field, which can be many orders of magnitude larger than Earth's field, any molecule can be sedimented. *Sedimentation velocity* is the motion of a particle in a gravitational or centrifugal field; its rate depends on the mass plus the size and shape of the particle. *Sedimentation equilibrium* refers to the distribution of molecules in a gravitational or centrifugal field after equilibrium has been reached. The balance between the sedimenting force in one direction and the randomizing effect of diffusion means that a *concentration gradient* is formed; there is a higher concentration near the bottom of the container. The concentration gradient at sedimentation equilibrium depends on the molecular weight of the molecules.

The *viscosity* of a fluid is its resistance to flow. When a fluid flows, one part of the fluid moves faster than another part. For example, in a fluid moving through a tube, the part of the fluid at the wall is stationary, while the part in the middle of the tube is flowing most rapidly. The molecules in different parts of the fluid thus have different average velocities in the direction of flow. The resistance to flow, the viscosity, depends on how strongly the molecules interact. How much do the stationary molecules at the surface of the tube slow down the faster-moving molecules toward the middle? For example, the intermolecular interaction provided by the three hydrogen-bond donors and acceptors in glycerol, [$HOCH_2CH(OH)CH_2OH$], makes it more than 1000 times more viscous than water. When macromolecules are added to a solvent, the viscosity increases. The amount of increase depends on the concentration of the molecules and their size and shape.

Electrophoresis is the motion of charged molecules in an electric field. It depends on the charge on the molecule, plus its size and shape. For proteins and nucleic acids with many ionizable groups, the charge on the molecule can be adjusted by changing the pH of the solution. In aqueous solutions, it is very difficult to quantitate the effective electric field and effective charge on a macromolecule because of all the other ions present. Thus, electrophoresis is used to measure relative sizes and shapes of molecules rather than absolute values.

All the properties we have described depend on the sizes and shapes of macromolecules; the properties are called *transport properties*. Diffusion and sedimentation measure mass transport; viscosity measures momentum transport; and electrophoresis measures charge transport. We care about transport properties because they tell us what the molecules look like in solution. Are they big or small, compact or extended, rigid or flexible? Furthermore, diffusion and flow are important methods of transport of molecules in biological cells, animals, and plants. We need to know how fast molecules can move from one place to another in a living cell, for example, and how much their speed changes when the temperature or molecular size changes.

Applications

Transport properties are used to analyze, separate, and purify cellular particles, proteins, and nucleic acids. Sedimentation provides fractionation based on differences in sedimentation coefficients, which depend on the mass of the particle, its shape, and its buoyancy—its density relative to the solvent. Gel electrophoresis is used to separate native proteins that differ by as little as one charge or denatured proteins that differ by one peptide unit. Two-dimensional gel electrophoresis has the ability to resolve nearly every protein present in a human cell. The changes in the pattern of proteins produced during differentiation of cells can be followed. Gel electrophoresis can separate nucleic acids that differ by one nucleotide, and thus determine their sequence, or can separate DNA fragments that are millions of base pairs in length. Both of these abilities are crucial for the Human Genome Project.

The absolute molecular weights of macromolecules can be determined by a combination of sedimentation and diffusion or by sedimentation equilibrium. Relative molecular weights can be determined easily by gel electrophoresis. The molecular weights of macromolecules are a key to their identification. The interactions of macromolecules to form higher-order structures can be characterized by the molecular weight of the complex.

The shapes of macromolecules and how they change in different environments can be followed by any of the transport properties. Thus, all sorts of reactions of macromolecules are studied by measurements of diffusion, sedimentation, viscosity, or electrophoresis.

Kinetic Theory
Brownian Motion and Random Molecular Motion

In 1827 the English botanist Robert Brown observed through a microscope the motion of tiny bits of pollen suspended in water. The pollen grains danced randomly around like water bugs on the surface of a pond. This *Brownian motion* is similar to the motion of the water molecules themselves or to motions of atoms in a gas. The random translational motion seen by Brown is the result of the thermal energy that every particle has, whether it is a pollen grain, a macromolecule, a solvent molecule, or a gas atom. We will learn that the average translational kinetic energy—energy of motion—of every particle depends only on the temperature. Collisions between the water molecules and the pollen ensure that the average translational kinetic energies of both are identical.

We cannot use a microscope to follow individual water molecules moving at random, but individual fluorescent molecules have been observed moving—undergoing Brownian motion—in a lipid bilayer (Schmidt et al. 1996). A highly fluorescent dye (tetramethyl rhodamine) was covalently attached to a phospholipid molecule in a lipid bilayer on a glass slide. The slide was illuminated by an intense laser beam, and the fluorescence was imaged by a microscope and photodetector. The lipid bilayer ensured that the dye moved essentially in two dimensions, not three, so that it stayed in focus during the measurement of its path. The position of the fluorescently labeled molecule was detected every 35 ms (millisecond) to determine its path. The trajectories of 531 individual molecules were analyzed; a representative path is shown in figure 6.1.

The fluorescently labeled lipid molecule moves randomly under the influence of many collisions. At equal-time intervals, a laser pulse was used to determine the position of the molecule, indicated by $A, B, C, \ldots, K$. This random path is very similar to the path of a gas molecule that travels in a straight line until it collides with another molecule and changes direction. In figure 6.1 the lipid molecule collides many times between the points where it is observed; it does not travel in a straight line between points. However, the

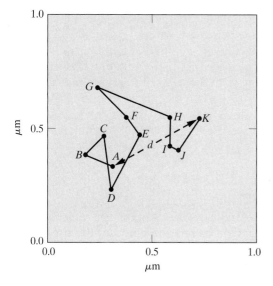

◀ FIGURE 6.1

The path of a fluorescently labeled phospholipid molecule undergoing Brownian motion in a lipid bilayer. The molecule is first observed at position A, and then at successive time intervals of 35 ms it is found at $B, C, \ldots, K$. The straight-line distance traveled is directly proportional to the square root of time. Einstein showed that the mean-square displacement $\langle d^2 \rangle$ is directly proportional to the time and to the diffusion coefficient D of the molecule.

random paths of molecules in a dilute gas, and molecules in a liquid, have many characteristics in common. The straight-line distance, d, traveled between two points is proportional to the square root of time. The actual path taken between points A and K in figure 6.1 is long and complicated, but the average of the square of the distance $\langle d^2 \rangle$ is directly proportional to time. The molecule is diffusing randomly away from where it started, with its progress measured by how far away it is moving per second. Usually, diffusion is measured by following how the concentration of many molecules changes with time. The macroscopic measurement of diffusion, characterized by the diffusion coefficient D, can be related to the molecular motion. Einstein in fact showed that the mean-square distance moved by a molecule along any axis, labeled x or y, is

$$\langle x^2 \rangle = \langle y^2 \rangle = 2Dt \tag{6.1}$$

For random motion in two dimensions in a membrane, or three dimensions in solution, Eq. (6.1) is generalized to

$$\langle d^2 \rangle = \langle x^2 \rangle + \langle y^2 \rangle = 4Dt \qquad \left(\begin{array}{c} \text{two-dimensional} \\ \text{diffusion} \end{array} \right) \tag{6.1a}$$

and

$$\langle d^2 \rangle = \langle x^2 \rangle + \langle y^2 \rangle + \langle z^2 \rangle = 6Dt \qquad \left(\begin{array}{c} \text{three-dimensional} \\ \text{diffusion} \end{array} \right) \tag{6.1b}$$

Comparison of many paths as illustrated in figure 6.1 with a macroscopic measurement of the diffusion coefficient D gave reasonable agreement. This is one example of how a molecular property can be related to a macroscopic one. As we might expect, small molecules diffuse faster than large molecules; spherical molecules diffuse faster than asymmetric molecules; and both diffuse faster as the temperature increases. To understand these types of methods better, we consider gases first because they are the simplest. However, the concepts and methods we learn are applicable to more complicated environments like liquids and solutions.

In a gas at ordinary pressures, molecules move freely in a volume much greater than their size. Molecules in a dilute gas are thus usually far apart; the motion of one molecule is little affected by the others unless it collides with another molecule. For the time being, we will not consider molecular collisions. We assume that the molecules are so small or that the gas is so dilute that collisions between molecules are highly improbable. Molecules in such a gas are nevertheless colliding with the walls of the container. The average translational kinetic energy of each molecule is the same and depends on the temperature of the walls. The pressure exerted on the walls by collisions of the molecules depends on the temperature, the volume, and the number of molecules.

Velocities of Molecules, Translational Kinetic Energy, and Temperature

We first relate the pressure of a gas to the collisions of the gas molecules on the walls of the container. This provides a relation between the pressure–volume product of the gas and the velocities of the gas molecules. Comparing this expression with the ideal gas equation, we learn how the average translational kinetic energy and the average velocity of the molecules depend on temperature. This is a very important result that also applies to liquids and solids. In condensed phases, the molecules are continually colliding and interacting. There is constant interchange of kinetic and potential energy. However, every molecule, independent of its size, has the same average translational kinetic energy. We can even consider the motion of each atom in each molecule in the sample and find that the average translational kinetic energy of all atoms depends only on the temperature of the sample.

To relate molecular velocities to kinetic energy and to pressure, we need to use two definitions from elementary physics. First, translational kinetic energy (U) equals one-half mass (m) times velocity (u) squared:

$$U = \frac{1}{2}mu^2 \qquad (6.2)$$

Second, Newton's second law [force (f) equals mass times acceleration (a)] can also be written as force equals the derivative of the momentum (mu) with respect to time (t), because acceleration is the time derivative of the velocity (u):

$$f = ma = \frac{d(mu)}{dt} \qquad (6.3)$$

Force, acceleration, and velocity are all vectors; they have both magnitude and direction. We therefore use a **boldface** notation for them. Usually, we will refer to the magnitude of the vectors; then they will be written as plain text. The word *speed* will be used for the magnitude of the velocity; speed is always positive.

We can now calculate the pressure (force per unit area) exerted by one molecule of mass m and velocity u on the walls of a cubic container of length l; the area of each wall is $A = l^2$ (figure 6.2). The velocity u can be resolved into its three Cartesian components (u_x, u_y, and u_z) along the three axes. Let's first focus on the collisions between this molecule and face A. The molecule will pass many times between the two walls perpendicular to the x-axis. If the x-component of the velocity is $\pm u_x$, because each round-trip spans a distance of $2l$, the number of times the molecule hits face A per second is $u_x/2l$. Now consider the momentum change per collision. Each time the molecule approaches wall A, it has a perpendicular velocity component u_x and a cor-

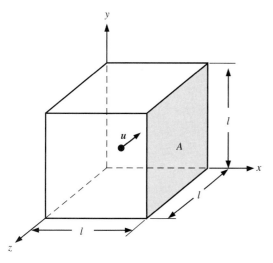

◀ FIGURE 6.2

Molecule of mass m and velocity u in a cubic container of side l. To calculate the pressure, we need to now the number of collisions between the molecule and the shaded face labeled A. The velocity is shown as a vector; it has components along three perpendicular directions. To calculate the pressure, we need to use the x-component of the velocity u_x.

responding momentum mu_x. As it departs in the opposite direction after a collision, the velocity component perpendicular to the wall changes to $-u_x$, with a corresponding momentum $-mu_x$. Thus, the momentum change along the x-direction is $2mu_x$ per collision.

The total momentum change along the x-axis per second, which according to Newton's second law [Eq. (6.3)] is the force exerted by the molecule along the x-axis, is then

$$f_x = \left(\begin{array}{c} \text{momentum change} \\ \text{per collision} \end{array} \right) \times \left(\begin{array}{c} \text{number of collisions} \\ \text{per second} \end{array} \right)$$

$$= 2mu_x \frac{u_x}{2l}$$

$$= \frac{mu_x^2}{l} \qquad (6.4)$$

The pressure caused by one molecule is equal to the force exerted by that molecule divided by the area, f_x/l^2:

$$P = \frac{mu_x^2}{l^3}$$

If there are N identical molecules in the container, each will have the same mass m, but a different velocity, u_i. The total force in the direction perpendicular to the wall A exerted by all the molecules is

$$F_x = \frac{mu_{x1}^2}{l} + \frac{mu_{x2}^2}{l} + \cdots + \frac{mu_{xN}^2}{l}$$

$$= \frac{m}{l} \sum_{i=1}^{N} u_{xi}^2$$

The total force per unit area, in the direction normal to wall A, is by definition the pressure P. Thus,

$$P = \frac{F_x}{l^2}$$

$$= \frac{m}{l^3} \sum_{i=1}^{N} u_{xi}^2$$

$$= \frac{m}{V} \sum_{i=1}^{N} u_{xi}^2 \qquad (6.5)$$

where $V = l^3$ is the volume of the container.

We define the mean-square velocity (the mean of the squares of the velocities) of the molecules in the x-direction as

$$\langle u_x^2 \rangle = \frac{1}{N} \sum_{i=1}^{N} u_{xi}^2 \qquad (6.6)$$

Equation (6.5) then becomes

$$P = \frac{Nm}{V} \langle u_x^2 \rangle \qquad (6.7)$$

We now have an expression for pressure in terms of the mean-square velocity in the x-direction, but there is nothing special about the x-direction. We need an expression in terms of molecular velocities, not just one component of the velocity. The square of the magnitude of any vector is the sum of the squares of its components, so that for each velocity u_i,

$$u_i^2 = u_{xi}^2 + u_{yi}^2 + u_{zi}^2$$

By adding all these equations and dividing each side by N, we obtain

$$\frac{1}{N}\sum_{i=1}^{N} u_i^2 = \frac{1}{N}\sum_{i=1}^{N} u_{xi}^2 + \frac{1}{N}\sum_{i=1}^{N} u_{yi}^2 + \frac{1}{N}\sum_{i=1}^{N} u_{zi}^2 \qquad \textbf{(6.8)}$$

We define the mean-square speed $\langle u^2 \rangle$ as

$$\langle u^2 \rangle = \frac{1}{N}\sum_{i=1}^{N} u_i^2$$

Therefore, Eq. (6.8) becomes

$$\langle u^2 \rangle = \langle u_x^2 \rangle + \langle u_y^2 \rangle + \langle u_z^2 \rangle$$

Since the molecules are moving randomly, there is no bias for the x-, y-, or z-directions. Therefore,

$$\left. \begin{array}{c} \langle u_x^2 \rangle = \langle u_y^2 \rangle = \langle u_z^2 \rangle \\[2mm] \langle u_x^2 \rangle = \dfrac{1}{3}\langle u^2 \rangle \end{array} \right\} \qquad \textbf{(6.9)}$$

Substituting Eq. (6.9) into (6.7), we obtain

$$P = \frac{1}{3}\frac{Nm}{V}\langle u^2 \rangle$$

or

$$PV = \frac{1}{3}Nm\langle u^2 \rangle \qquad \textbf{(6.10)}$$

We thus have derived an equation relating the pressure of a dilute (ideal) gas to the properties of individual molecules—namely, to their masses and the average speed of the N molecules.

According to the ideal gas law,

$$PV = nRT = \frac{N}{N_0}RT \qquad \textbf{(6.11)}$$

where $n = N/N_0$, N_0 is Avogadro's number, and n is the number of moles of gas. Combining Eqs. (6.10) and (6.11) gives

$$RT = \frac{1}{3}N_0 m\langle u^2 \rangle$$

$$= \frac{2}{3}\left(\frac{1}{2}N_0 m\langle u^2 \rangle\right) \qquad \textbf{(6.12)}$$

Note that $\frac{1}{2}m\langle u^2 \rangle$ is the average translational kinetic energy per molecule [Eq. (6.2)] and that $\frac{1}{2}N_0 m\langle u^2 \rangle$ is the translational kinetic energy U_{tr} of 1 mol of gas. Therefore,

$$RT = \frac{2}{3}\langle U_{tr} \rangle$$

and

$$\langle U_{tr} \rangle = \frac{3}{2}RT \tag{6.13}$$

Note that $\langle U_{tr} \rangle$ is a function of T only. Also, since $N_0 m$ is the molecular weight M, Eq. (6.12) gives

$$\langle u^2 \rangle = 3\frac{RT}{M} \tag{6.14}$$

Equation (6.14) is an important result. It states that for a gas consisting of molecules with molecular weight M, the mean-square speed is a function of T only. In deriving Eq. (6.14), we have neglected molecular collisions; the same result is obtained if collisions are considered.

The root-mean-square speed (the square root of the mean-square velocities)

$$\langle u^2 \rangle^{1/2} = \left(\frac{3RT}{M}\right)^{1/2} \tag{6.15}$$

is one measure of the average speed of the molecules.

EXAMPLE 6.1

Calculate the root-mean-square speed of oxygen and of geraniol vapor at 20°C. Geraniol, an alcohol responsible for the fragrance of roses, has a molecular weight of 154.2.

Solution

The molecular weight of O_2 is 32.0. To obtain useful units (m s^{-1}) for the velocity, we must express the gas constant R as 8.314 J K^{-1} mol^{-1}. Since 1 J = 1 kg m^2 s^{-2}, R = 8314 g m^2 s^{-2} K^{-1} mol^{-1}.

From Eq. (6.15),

$$\langle u^2 \rangle^{1/2} = \left(\frac{3RT}{M}\right)^{1/2}$$

$$= \left[3(8314 \text{ g m}^2\text{ s}^{-2}\text{ K}^{-1}\text{ mol}^{-1})\frac{(293 \text{ K})}{32 \text{ g mol}^{-1}}\right]^{1/2}$$

$$= (2.284 \times 10^5 \text{ m}^2\text{ s}^{-2})^{1/2}$$

$$= 478 \text{ m s}^{-1}$$

Similarly, for geraniol,

$$\langle u^2 \rangle^{1/2} = 218 \text{ m s}^{-1}$$

This example shows that the speeds are rather high—several tenths of kilometers per second (1 km s^{-1} ≅ 2200 mph). The values that we have calculated for oxygen and geraniol are of the same order of magnitude as the speed of sound in air (~330 m s^{-1}).

The average translational kinetic energy of each atom in a molecule in a solid, liquid, or gas depends only on the absolute temperature. It is surprising but true that the average translational kinetic energy of each atom does *not* depend on whether it is in a gas, liquid, or solid. Therefore, the equation that we derived for an ideal gas [Eq. (6.14)] still applies. For an average atom in a molecule in any sample (averaged over all the atoms in the sample), the average translational kinetic energy is

$$\langle U_{\text{tr}} \rangle = \frac{3}{2} kT \qquad \qquad \textbf{(6.13)}$$

k = Boltzmann constant = R/N_0
  = $1.380 \times 10^{-23} \, \text{J K}^{-1} \, \text{molecule}^{-1}$

The motion of an atom includes its vibrational motion, plus the translation and rotation of the molecule of which it is a part. The fact that collisions, interactions, and even bonding the atoms into molecules do not change the atom's average kinetic energy is very surprising and very useful. As long as the molecular and atomic motions can be treated classically, the simple equations for average energies and average speeds are correct. They are applicable to a lipid molecule in a membrane, as well as the motions and interactions of proteins and nucleic acids in solution.

To emphasize one of the main conclusions of this section, we repeat the fact that every molecule, and every atom in the molecule, in any environment has an average kinetic energy (its energy of motion) equal to $\frac{3}{2} kT$ per atom, or $\frac{3}{2} RT$ per mole of atoms.

Maxwell–Boltzmann Distribution of Velocities

We have found that the mean-square speed of all molecules in a gas is simply related to the absolute temperature. The squares of the speeds—equal to the squares of the velocities—are pertinent because kinetic energy depends on velocity squared. Other measurable properties depend on the mean speed $\langle u \rangle$. For example, the rate of molecular collisions, which is crucial for the rate of of a chemical reaction, depends on the mean speed. It is easy to see that unless all speeds are equal, the root-mean-square speed $\langle u^2 \rangle^{1/2}$ is different from the mean speed $\langle u \rangle$. For any distribution of speeds, the root-mean-square speed $\langle u^2 \rangle^{1/2}$ must be greater than the mean speed $\langle u \rangle$. Similarly, the root-mean-cube speed $\langle u^3 \rangle^{1/3}$ is greater than $\langle u^2 \rangle^{1/2}$ and $\langle u \rangle$. If we know the molecular speeds u_i, for all molecules at any one time, we can calculate any average speed such as the mean, the median, the most probable, the root-mean-square, and so on. Ludwig Boltzmann of entropy fame and James Clerk Maxwell (best known for his electromagnetic equations) independently derived equations for the probability of finding a molecule with any chosen speed.

To help understand their derivation, let's consider a large number, such as Avogadro's number, of (dilute) gas molecules in a container. In a thought experiment, we can instantaneously measure the speed and thus the translational kinetic energy of each molecule at any time. For example, we can count the number of molecules that have nearly zero speed. We must specify some range of speeds because there is always some uncertainty in any measurement. So we count the number of molecules with speed between $0 \, \text{m s}^{-1}$ and $0.1 \, \text{m s}^{-1}$, between $100 \, \text{m s}^{-1}$ and $100.1 \, \text{m s}^{-1}$, and so on. These numbers will be independent of time as long as the conditions, such as temperature, number of molecules, and the like, do not change. Instead of the numbers of molecules within each range of speeds, it is more useful to know the *fraction* of molecules within each range of speeds. It does not matter whether the container has 1 mol of O_2 molecules or 10 mol; the distribution of speeds will be the same for the dilute gas molecules. Because the molecules are identical, the fraction of molecules within a certain range of speeds is equal to the probability that any one molecule will be in this range. For example, we can say that 10% of the molecules have speed less than 50 m s^{-1} or, equally correctly, say that the probability of finding a molecule with speed less than 50 m s^{-1} is 0.10.

Let's see how Maxwell and Boltzmann derived the speed distribution for molecules in dilute gases, using statistics and probability. Boltzmann derived the general result that the probability of finding a molecule with energy E_i is proportional to $e^{-E_i/kT}$. Here k is Boltzmann's constant, and T is the absolute temperature.

$$P_i \propto g_i e^{-E_i/kT}$$

$P_i =$ the probability of finding a molecule with energy E_i
$g_i =$ degeneracy, the number of different states of a
     molecule having the same energy E_i

We see that the probability of of a molecule having energy E_i depends on the exponential of E_i/kT, as well as on a new property called the *degeneracy*. We will learn more about degeneracies in chapter 9 on quantum mechanics. The magnitude of the degeneracy depends on the type of energy. For electronic energy levels, the degeneracies are small integers, such as 1, 2, or 3 as described for the hydrogen atom in chapter 9. For translational energies, the degeneracies are very large and increase rapidly with energy.

The ratio of the number of molecules in two different energy levels according to Boltzmann is thus

$$\frac{P_j}{P_i} = \frac{N_j}{N_i} = \frac{g_j e^{-E_j/kT}}{g_i e^{-E_i/kT}} = \frac{g_j}{g_i} e^{-(E_j-E_i)/kT} \tag{6.16}$$

$N_j/N_i =$ the ratio of the number of molecules with energies E_j and E_i
     at equilibrium

A probability, like the fraction of the number of molecules, must be less than 1. Thus, the Boltzmann most-probable energy distribution (the distribution at equilibrium) is

$$P_j = \frac{N_j}{N} = \frac{g_j e^{-E_j/kT}}{\sum_i g_i e^{-E_i/kT}} \tag{6.17}$$

$P_j =$ the probability of finding a molecule with energy E_j
$N_j/N =$ the fraction of molecules with energy E_j

By dividing by the sum over all the exponential terms (which are each proportional to the probability of finding a molecule with energy E_i), we make sure that the sum over all probabilities is equal to 1:

$$\sum_1 P_i = 1$$

The Boltzmann equation is very important in all fields of science. It applies to any type of energy, not just translational kinetic energy. It applies to vibrational, rotational, and electronic energies, for example. It applies to quantized energy states (as required to treat electronic energies), as well as to energies that can be treated classically, such as translational energies. For interacting molecules, as found in liquids, solids, and high-pressure gases, the same equation applies except now the energies E_i refer to the *entire* collection of molecules. We can in principle calculate the probability of finding a protein in a particular folded state in solution by calculating the energies of possible folded and unfolded states of the protein in the solvent. (We will discuss this further in chapter 9.)

Maxwell derived a special case of the Boltzmann equation [Eq. (6.16)] specifically for translational kinetic energies with E_i replaced by $\frac{1}{2}mu_i^2$. Furthermore, he considered a continuous distribution of energies and speeds. Although all energies are quantized, as we will describe in chapter 9, the translational energy levels are so closely spaced that they can be treated as if they were continuous. For a continuous distribution of energies, the sum over energies in Eq. (6.17) can be replaced by an integral from zero to infinity. The Maxwell–Boltzmann distribution of speeds is obtained by replacing E by $\frac{1}{2}mu^2$ in the exponent and using the degeneracies pertinent to translational energies in three dimensions:

$$dP(u) = F(u)\,du = \frac{u^2 e^{-mu^2/2kT}\,du}{\displaystyle\int_0^\infty u^2 e^{-mu^2/2kT}\,du} \qquad (6.18)$$

$dP(u)$ = the probability of finding a molecule with speed
between u and $u + du$

$F(u)\,du$ = the fraction of molecules with speed between u and $u + du$

The sum over all P_j in Eq. (6.17) and the integral over $dP(u)$ in Eq. (6.18) give values of 1, as appropriate for probabilities:

$$\sum_i P_i = \int_0^\infty dP(u) = 1$$

To obtain the Maxwell–Boltzmann equation, we need to evaluate the integral in Eq. (6.18). From a table of integrals we find that

$$\int_0^\infty x^2 e^{-ax^2}\,dx = \frac{1}{4\pi}\left(\frac{\pi}{a}\right)^{3/2}$$

Replacing a by $(m/2kT)$ we obtain the integral in Eq. (6.18):

$$\int_0^\infty u^2 e^{-mu^2/2kT}\,du = \frac{1}{4\pi}\left(\frac{2\pi kT}{m}\right)^{3/2} \qquad (6.19)$$

The Maxwell–Boltzmann equation for the probability of finding a molecule in a dilute gas with speed between u and $u + du$ is then

$$dP(u) = F(u)\,du = 4\pi\left(\frac{m}{2\pi kT}\right)^{3/2} u^2 e^{-mu^2/2kT}\,du \qquad (6.20)$$

We notice that the probability depends on the mass m of the molecule and the absolute temperature. A plot of $F(u)$ from Eq. (6.20) versus u is shown in figure 6.3. You can see that as the temperature increases, the distribution of

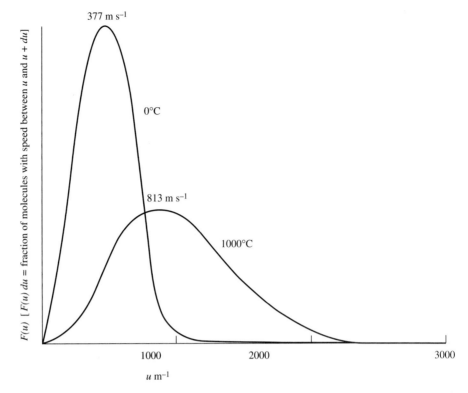

▶ FIGURE 6.3

The Maxwell–Boltzmann distribution of speeds at two temperatures for O_2 gas. Notice that as the temperature increases the most probable speed increases (the maximum in the curve), but the spread of the speeds increases more markedly. This means that the fraction of molecules with the highest speeds increases much faster with temperature than the average speeds. This is one reason why temperature has such a large effect on reaction rates.

speeds broadens and shifts to higher speeds. The molecules move faster and have a wider spread of speeds. Decreasing the mass of each molecule has a similar effect.

The Maxwell–Boltzmann distribution [Eq. (6.20)] and the curves shown in figure 6.3 give us a tremendous amount of information about the speeds of molecules in dilute (ideal) gases. The most probable speed, $u = (2kT/m)^{1/2}$, is calculated from the derivative of Eq. (6.20) with respect to u. A maximum or minimum occurs when the derivative is equal to zero. Obviously, we can calculate any average speed we need. The mean speed $\langle u \rangle$ and the mean-square speed $\langle u^2 \rangle$ are obtained from

$$\langle u \rangle = \int_0^\infty u\, F(u)\, du \tag{6.21}$$

$$\langle u^2 \rangle = \int_0^\infty u^2\, F(u)\, du \tag{6.22}$$

Substitution of $F(u)$ from Eq. (6.20) into these integrals and the use of a table of integrals gives the results

$$\langle u \rangle = \left(\frac{8kT}{\pi m} \right)^{1/2} \tag{6.23}$$

$$\langle u^2 \rangle = \frac{3kT}{m} \tag{6.24}$$

We had obtained Eq. (6.24) earlier with k/m replaced by the equivalent R/M in Eq. (6.14).

We have described molecular velocities in the gas phase in great detail because they provide a simple example of the ability to quantitatively relate molecular properties to macroscopic properties. In the rest of this chapter and in succeeding chapters, we will treat more complex systems and systems more relevant to biochemical applications. However, the ideas introduced here—such as the Boltzmann distribution of energies—can be applied directly.

Molecular Collisions

The speeds of molecules in the gas phase are very large. Figure 6.3 shows ranges of hundreds to thousands of m s^{-1}; example 6.1 gives a root-mean-square speed of 218 m s^{-1} at room temperature for the fragrance of roses. These high speeds might make you think that the fragrance from a bottle of perfume opened in the center of a room, in still air, would reach the corners in a fraction of a second. This does not happen because of molecular collisions. As we saw in figure 6.1, the molecules do not travel far before colliding and changing their direction. They move fast, but they move randomly and do not migrate very quickly from where they start.

Let's think about the number of collisions per second z that one molecule undergoes in a gas containing N molecules in a volume V. What does the number of collisions depend on? The larger the concentration of molecules, the more collisions will occur, so z will be proportional to N/V. Surely, the faster the molecules are moving, the more collisions will occur. This means that z depends on the speeds of the molecules. The number of collisions must depend on the sizes of the molecules; larger molecules have a higher probability of colliding. When two molecules collide, the size of each molecule is important. We consider the simplest case of two identical, spherical molecules each with diameter σ (sigma) The number of collisions per second for one molecule z is thus proportional to σ^2. Putting all this together and using the mean speed $\langle u \rangle$ as a measure of the molecular speeds, we obtain the result that

$$z \text{ is proportional to } \frac{N}{V}\sigma^2\langle u \rangle$$

To make sure that our reasoning makes sense, we check that the units of the expression are correct. The units of z are s^{-1}, so the product of $(N/V)\sigma^2\langle u \rangle$ must also equal s^{-1}, which it does. A detailed derivation shows that the mean speed is the correct one to use and provides the proportionality constant. The number of collisions per second for one molecule is

$$z = \sqrt{2}\pi\frac{N}{V}\sigma^2\langle u \rangle \qquad \textbf{(6.25)}$$

Using Eq. (6.23) for $\langle u \rangle$, we obtain

$$z = 4\sqrt{\pi}\frac{N}{V}\sigma^2\left(\frac{RT}{M}\right)^{1/2} \qquad \textbf{(6.26)}$$

Since there are N/V molecules per unit volume, there are Nz/V molecules per unit volume undergoing collisions per unit time. The total number of collisions per unit volume per unit time Z by all the molecules present is then

$$Z = \left(\frac{N}{V}\right)\frac{z}{2} \qquad \textbf{(6.27)}$$

The factor of 2 in the denominator results from the fact that each collision involves two molecules. Combining Eqs. (6.27) and (6.26), we obtain for the total number of collisions per volume per second

$$Z = 2\sqrt{\pi}\left(\frac{N}{V}\right)^2 \sigma^2 \left(\frac{RT}{M}\right)^{1/2} \tag{6.28}$$

The total number of molecular collisions in a container is seen to be proportional to the square of the concentration, the square of the molecular diameter, the square root of the absolute temperature, and to be inversely proportional to the square root of the molecular weight.

Mean Free Path

The mean free path l is defined as the average distance a molecule travels between two successive collisions with other molecules. Because $\langle u \rangle$ is the average distance a molecule travels per unit time and z is the number of intermolecular collisions it encounters during that time, the mean free path is given by

$$l = \frac{\langle u \rangle}{z} = \frac{1}{\sqrt{2}\pi(N/V)\sigma^2} \tag{6.29}$$

using Eq. (6.25).

EXAMPLE 6.2

For N_2 gas at 1 atm and 25°C, calculate (a) the number of collisions each N_2 molecule encounters in 1 s, (b) the total number of collisions in a volume of 1 m^3 in 1 s, and (c) the mean free path of an N_2 molecule. The diameter of an N_2 molecule can be taken as 3.74 Å (3.74×10^{-10} m).

Solution

a. For N_2 gas at 1 atm and 25°C, the number of moles per unit volume is P/RT, and the number of molecules per unit volume is

$$\frac{N}{V} = \frac{N_0 P}{RT}$$

$$= \frac{(6.023 \times 10^{23} \text{ molecules mol}^{-1})(1 \text{ atm})}{(0.08205 \text{ L atm K}^{-1} \text{ mol}^{-1})(298 \text{ K})}$$

$$= 2.46 \times 10^{22} \text{ molecules L}^{-1}$$

$$= 2.46 \times 10^{25} \text{ molecules m}^{-3}$$

From Eq. (6.26), the number of collisions each molecule encounters per unit time is

$$z = 4\sqrt{\pi}(2.46 \times 10^{25} \text{ m}^{-3})(3.74 \times 10^{-10} \text{ m})^2$$

$$\times \left[(8.31) \text{ J K}^{-1} \text{ mol}^{-1} \left(\frac{298 \text{ K}}{0.028 \text{ kg mol}^{-1}} \right) \right]^{1/2}$$

$$= 7.27 \times 10^9 \text{ (J m}^{-2} \text{ kg}^{-1})^{1/2}$$

but $1\,J = 1\,kg\,m^2\,s^{-2}$; therefore,

$$z = 7.27 \times 10^9\,s^{-1}$$

b. From Eq. (6.27), the total number of collisions is

$$Z = \left(\frac{N}{V}\right)\frac{z}{2}$$

$$= \frac{(2.46 \times 10^{25}\,m^{-3})(7.27 \times 10^9\,s^{-1})}{2}$$

$$= 8.94 \times 10^{34}\,\text{collisions}\,m^{-3}\,s^{-1}$$

c. From Eq. (6.29), the mean free path is

$$l = \frac{1}{\sqrt{2}\pi(2.46 \times 10^{25}\,m^{-3})(3.74 \times 10^{-10}\,m)^2}$$

$$= 6.53 \times 10^{-8}\,m$$

Thus, at 1 atm and 25°C, a typical N_2 molecule goes a distance almost 200 times its diameter between collisions.

Diffusion
The Random Walk and Diffusion in a Gas

A molecule in a gas may have a high speed (hundreds of meters per second), but—because its movement is random—after 1 s it does not end up very far from where it started. A gas molecule moves in a straight line for a short distance and then collides with another molecule and changes direction. This kind of path is illustrated by the random walk in two dimensions of a lipid molecule in a membrane in figure 6.1. The lipid molecule has many collisions between the points labeled in the figure, but the important characteristics of the three-dimensional path of a gas molecule, the two-dimensional path of the lipid in a membrane, and a one-dimensional random path are all the same. This type of motion is called *Brownian motion,* and the path is a *random walk,* or *random flight.* A characteristic of a random path is that the total path is large compared with the distance from the beginning to the end of the path. This straight-line distance is called the displacement d.

The problem of the random walk occurs in many different areas of science. The shape of a flexible polymer such as polyethylene or a random polypeptide (a denatured protein), for example, can be calculated from a random walk model. The question is, What is the average distance between the beginning and end of a path of N random steps of length l? We can deduce the answer easily for a random walk in one dimension; a detailed derivation is given in chapter 11. Think of a person who takes a step randomly either to the right or left along a line; he flips a coin to decide whether to step right or left. How far does he get after N steps of unit length? Table 6.1 gives the answers for a small number of steps. In the table, r means a step right and l means a step left.

TABLE 6.1 A Random Walk in One Dimension

No. of steps N	Possible paths	Possible locations, d_i	Mean displacement, $\langle d \rangle$	Mean-square displacement, $\langle d^2 \rangle$
1	r	+1	0	1
	l	−1		
2	r r	+2	0	2
	r l, l r	0, 0		
	l l	−2		
3	r r r	+3	0	3
	r r l, r l r, l r r	+1, +1, +1		
	r l l, l r l, l l r	−1, −1, −1		
	l l l	−3		

In table 6.1, we calculated the means by averaging over all possible paths for a given number of steps, such as the eight possibilities for three steps. The mean displacement

$$\langle d \rangle = \frac{\sum_i^{\text{paths}} d_i}{\text{no. of paths}} \tag{6.30}$$

and the mean-square displacement

$$\langle d^2 \rangle = \frac{\sum_i^{\text{paths}} d_i^2}{\text{no. of paths}} \tag{6.31}$$

are calculated for the possible distances from the origin after 1, 2, and 3 steps. The mean displacement is always zero because there is equal probability of moving left or right. We find that the mean-square displacement is always equal to the number of steps. If we do the calculation in three dimensions—to produce what is sometimes called a *random flight*—we get exactly the same result. So for N random steps in any direction of length l, the mean-square displacement is

$$\langle d^2 \rangle = N l^2 \tag{6.32}$$

and the root-mean-square displacement is

$$\langle d^2 \rangle^{1/2} = \sqrt{N} l \tag{6.33}$$

We can use this equation to calculate the average end-to-end distance in a random polymer of N monomer units. The monomers in a real polymer (for example, the CH_2 units in polyethylene) are not arranged randomly because of the tetrahedral bond angles between carbons, so the measured average end-to-end distance is slightly larger, but the square root dependence on the number of monomer units is correct.

For the random diffusion of a molecule in a gas, the random walk equation equivalent to Eq. (6.32) is

$$\langle d^2 \rangle = z l^2 \tag{6.34}$$

Here $\langle d^2 \rangle$ is the mean-square displacement of the molecule per second, z is the number of collisions per second, and l is the mean free path. Using Eq. (6.29) for the mean free path in a dilute gas, we obtain

$$\langle d^2 \rangle = \frac{\langle u \rangle^2}{z} \qquad\qquad (6.35)$$

The square root of $\langle d^2 \rangle$ is the root-mean-square displacement,

$$\langle d^2 \rangle^{1/2} = \sqrt{z}\, l \qquad\qquad (6.36)$$

$$= \frac{\langle u \rangle}{\sqrt{z}} \qquad\qquad (6.36a)$$

which is a measure of the average displacement per unit time.

We mentioned earlier that if a bottle of perfume is opened in the middle of a windless room, the scent takes much longer to reach the other parts of the room than one might expect from the high value of $\langle u \rangle$. The explanation lies in Eq. (6.35), which tells us that the mean-square displacement is inversely proportional to the number of collisions per second z. Although the mean speed $\langle u \rangle$ is large, every molecule has many collisions per second, and each collision changes its direction. At atmospheric pressure, z is a large number ($\sim 10^{10}\,s^{-1}$), as we found in example 6.2.

EXAMPLE 6.3

Consider a molecule of N_2 gas at 1 atm and 25°C.
a. Estimate how far it will move in 1 s away from where it started.
b. What is the total distance traveled along the zigzag path in 1 s?

Solution

a. The root-mean-square displacement $\langle d^2 \rangle^{1/2}$ is a measure of its displacement. From Eq. (6.36) and using the results of example 6.2,

$$\langle d^2 \rangle^{1/2} = \sqrt{z}\, l$$

$$= (7.27 \times 10^9)^{1/2}(6.53 \times 10^{-8}\,m)$$

$$= 5.57 \times 10^{-3}\,m = 0.557\,cm$$

b. The total distance traveled per second is

$$zl = (7.27 \times 10^9)(6.53 \times 10^{-8}\,m)$$

$$= 475\,m$$

Note that these distances differ by a factor of almost 10^5 for this example.

Diffusion Coefficient and Fick's First Law

The diffusion of the molecules that produce the scent of a perfume is an example of diffusion in the gas phase. We can calculate the rate of diffusion in a dilute gas from the mean speed and the number of collisions per second of the molecules. In a liquid, there are many more collisions per second than in a gas, and it is more difficult to calculate their number. A macroscopic treatment of diffusion is more convenient for liquids, although the random motion of individual molecules can be measured, as seen in figure 6.1. The rate of diffusion depends on the size and shape of the molecule and on the properties—such as viscosity—of the solvent.

We first define the measurable macroscopic properties that characterize diffusion. *Diffusion* occurs whenever there is a concentration difference in a container; diffusion will eventually make the concentration uniform. Diffusion is a slow process compared to mixing caused by stirring, for example, but over small distances—as occur in biological cells—diffusion can be an effective method for transporting molecules.

One way to measure diffusion is to measure the movement of molecules caused by a concentration difference. Think of molecules in solution crossing a 1-cm^2 cross-sectional area in the yz-plane (figure 6.4). The flux J_x is defined as the net amount of solute that diffuses through this unit area, per unit time, in the x-direction. Units of flux, for example, are mol cm^{-2} s^{-1}. If there is no concentration gradient—no difference in concentration—in the x-direction, you expect equal numbers of molecules crossing this area from the left and from the right. Therefore, the net flux J_x is zero. Suppose the concentration decreases as x increases; this means that the concentration gradient dc/dx is negative. The concentration decreases with x; therefore, there are more molecules per unit volume to the left of the area than to the right. You expect, then, that more molecules per unit time diffuse across the area from the left than from the right. In other words, there is a net transport of material in the direction opposite to the concentration gradient, as shown in figure 6.4. The steeper the concentration gradient, the larger is the net flux. These considerations led to the equation for *Fick's first law:*

$$J_x = -D\left(\frac{dc}{dx}\right) \qquad (6.37)$$

where D, the proportionality constant, is called the *diffusion coefficient.* The negative sign indicates that the net transport by diffusion is in a direction opposite to the concentration gradient. The units of D are cm^2 s^{-1}; the concentration gradient is in mol cm^{-4}.

As diffusion occurs, the solution becomes more uniform; the concentration gradient dc/dx decreases. As time goes on, the system gradually approaches homogeneity, and dc/dx approaches zero. Thus, Eq. (6.37) is given in terms of the instantaneous flux at any time t. Equation (6.37), Fick's first law of diffusion, has been shown experimentally to be correct.

▶ FIGURE 6.4

Flux J_x is the net diffusional transport of material per unit time (mol cm^{-2} s^{-1}), in the x-direction, across a unit cross-sectional area perpendicular to x. The concentration c decreases with increasing x; the flux J_x is in the direction opposite to the concentration gradient.

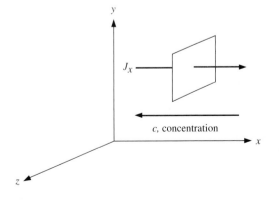

Fick's Second Law

It is also possible to describe how the concentration in a gradient changes with time. In a uniform concentration gradient, where dc/dx is constant for all values of x, Eq. (6.37) tells us that the flux J_x is the same at all positions and, therefore, c will not change with time. (The flux into every volume element from one side is exactly equal to that going out the other side.) This describes a steady-state flow of material by diffusion.

However, if the concentration gradient is not the same everywhere, the concentration will change with time. The change of concentration with time $(\partial c/\partial t)$ at position x is proportional to the gradient in the flux $(\partial J/\partial x)$ at that point. Intuitively, if more material is diffusing into the volume element from the left than is diffusing out to the right, then $(\partial J/\partial x) < 0$, and the concentration inside should increase with time, $(\partial c/\partial t) > 0$. This logic leads to *Fick's second law*:

$$\left(\frac{\partial c}{\partial t}\right)_x = D\left(\frac{\partial^2 c}{\partial x^2}\right)_t \tag{6.38}$$

We use the notation of partial derivatives because the concentration depends on both time t and distance x. Fick's second law states that the change in concentration with time at any position x, $(\partial c/\partial t)_x$, is proportional to the second derivative of the concentration with respect to x at time t, $(\partial^2 c/\partial x^2)_t$. The proportionality constant is the diffusion coefficient D, assumed to be independent of concentration. An illustration of the application of Fick's second law is shown in figure 6.5.

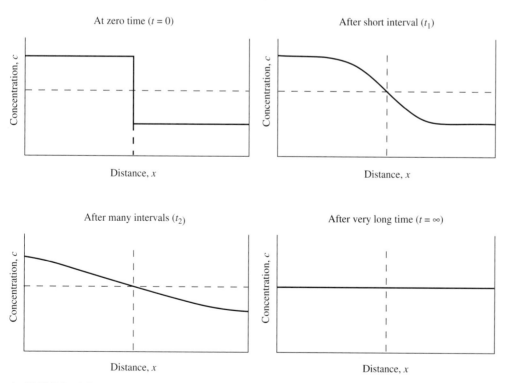

▲ FIGURE 6.5

Diffusion of a material with a concentration gradient that is initially a step. Profiles are shown at time zero (t_0) and at successive times, t_1, t_2, and t_∞. The exact behavior is described by Fick's second law [Eq. (6.38)].

Determination of the Diffusion Coefficient

In principle, D can be determined by the use of either Fick's first law [Eq. (6.37)] or Fick's second law [Eq. (6.38)]. A simple way to measure a diffusion coefficient is to use Fick's first law and to measure the amount of material that is transferred through unit area per unit time—the flux J. We use a porous glass disk of thickness Δx to separate two solutions of two different concentrations, as shown in figure 6.6. The rate of transfer of material (mol s^{-1} or g s^{-1}) across the disk can be measured using a radioactive label, for example. The effective area of the porous disk is determined by calibrating the disk with a substance of known diffusion coefficient. Dividing the rate of transfer of material by the effective area of the porous disk gives the flux J. The diffusion coefficient is obtained from Fick's first law [Eq. (6.37)]:

$$D = -J\left(\frac{\Delta x}{c_2 - c_1}\right) \tag{6.39}$$

J = flux with units of mol cm^{-2} s^{-1} or g cm^{-2} s^{-1}
c_2, c_1 = concentrations with units of mol cm^{-3} or g cm^{-3}
Δx = thickness of porous glass disk, in cm
D = diffusion coefficient with units of cm^2 s^{-1}

The concentrations are kept constant during the experiment by using large volumes on each side and by stirring near the porous disk. Although Fick's second law [Eq. (6.38)] was written for a concentration-independent diffusion coefficient, D in general varies with concentration. In the experiment shown in figure 6.6, the measured value of D corresponds to the average of the two concentrations. By changing these concentrations, we can obtain D at any desired concentration, and we can extrapolate D to zero concentration.

To use Fick's second law [Eq. (6.38)] to measure a diffusion coefficient, we measure the concentration c or the concentration gradient dc/dx as a function of time. For example, the experiment shown in figure 6.5 can provide a diffusion coefficient. A step concentration gradient is prepared at the beginning of the experiment; then the diffusion of the gradient is followed as a function of time. From measured values of the concentration c as a function of position x and time t, the derivatives needed to evaluate D from Eq. (6.38) can be obtained. An integrated form of Eq. (6.38) is most useful. The concentration is usually measured optically, using the absorbance or refractive index of the solution. Thus, pictures are taken of the concentration gradient at different times and analyzed in terms of Fick's second law.

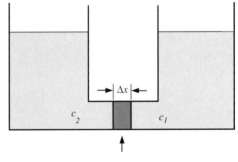

▶ FIGURE 6.6

Measurement of a diffusion coefficient D by measuring the diffusion through a porous glass disk. The disk's effective porous area can be calibrated using a molecule with a known diffusion coefficient. The two solutions are stirred well, and the concentrations are kept essentially constant during the experiment. The flux (amount transferred per unit area per unit time) is measured. D is calculated from Fick's first law, Eq. (6.37).

Porous glass disk with effective area A and thickness Δx

Relation Between the Diffusion Coefficient and the Mean-Square Displacement

In our discussion of the molecular basis of diffusion, we used the quantity $\langle d^2 \rangle$, which is the average of the square of the displacement of a molecule in 1 s. Because diffusion is due to the random motion of molecules, we expect that the larger the value of $\langle d^2 \rangle$, the larger will be the diffusion coefficient D. That is, the faster the molecules can move from one point to another, the greater the flux J for a given concentration gradient. Einstein in 1905 was the first to derive the equation (Eq. 6.1) linking random molecular motion to a macroscopically measured diffusion coefficient. We can derive it from Fick's second law [Eq. (6.38)], which describes how the concentration at position x changes with time t. We start with a very thin layer of a solution containing w grams of solute per unit area sandwiched between layers of solvent. Thus, all the molecules start at $x = 0$ at time $t = 0$, as illustrated in figure 6.7(a).

Equation (6.38) describes how the concentration $c(x, t)$ changes as diffusion progresses. It can be solved to give c in g cm^{-3} at any position x and time t:

$$c = \frac{w}{(4\pi Dt)^{1/2}} e^{-x^2/4Dt} \tag{6.40}$$

You can check that Eq. (6.40) is correct by taking its second derivative with respect to x and comparing it with the first derivative with respect to t. The concentration profiles, plotted in figure 6.7(b), are distributions in x at two different times. That is, during time t some solute molecules have moved far away from their original positions at $x = 0$, but others have hardly moved away at all. Each concentration profile is that of a Gaussian distribution in x at any time; we can use it to calculate any average property of the distribution. The mean-square displacement $\langle x^2 \rangle$ can be evaluated directly from this distribution using Eq. (6.22) to give

$$\langle x^2 \rangle = \int_{-\infty}^{\infty} \frac{1}{(4\pi Dt)^{1/2}} x^2 e^{-x^2/4Dt} = 2Dt \tag{6.41}$$

$$D = \frac{\langle x^2 \rangle}{2t}$$

The diffusion coefficient is one-half of the mean-square displacement per unit time.

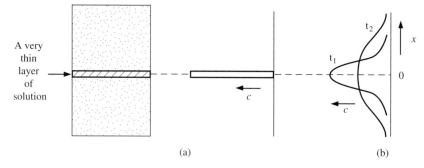

(a)   (b)

▲ FIGURE 6.7

Diffusion from a very thin layer of solution: (a) initial state; (b) concentration profiles after time t_1 and t_2. The concentration is a Gaussian distribution in x at any time t [Eq. (6.40)].

EXAMPLE 6.4

The diffusion coefficient of the oxygen-storage protein myoglobin (Mb) is $D_{20,w} = 11.3 \times 10^{-7} \, \text{cm}^2 \, \text{s}^{-1}$, at 20°C in water. Estimate a mean value for the time required for an Mb molecule to diffuse a distance of 10 μm along a particular direction, which is the order of the size of a cell.

Solution

The mean-square displacement $\langle x^2 \rangle$ is a measure of the displacement of an average molecule along direction x. From Eq. (6.41),

$$t = \frac{\langle x^2 \rangle}{2D} = \frac{(10 \times 10^{-4} \, \text{cm})^2}{2(11.3 \times 10^{-7} \, \text{cm}^2 \, \text{s}^{-1})} = 0.44 \, \text{s}$$

Although the diffusion coefficient of Mb in a cell is smaller than in water because of the higher viscosity of cytoplasm, nevertheless the diffusion of even relatively large molecules such as Mb (mol wt 17,000) occurs rapidly across dimensions comparable to the size of a cell.

For the Gaussian distribution [Eq. (6.40)] shown in figure 6.7, the diffusion coefficient D can be measured from the width of the distribution at half maximum. The concentration c is maximum at $x = 0$. For c equal to half its maximum value, Eq. (6.40) shows that $x = \pm 2(Dt \ln 2)^{1/2}$; the full width at half maximum is $4(Dt \ln 2)^{1/2}$. Thus, the spread with time t of the initially sharp distribution is a direct measure of the diffusion coefficient D. Note the appearance of the square root of time $\sqrt{t}$, which is related to the square root of the number of steps, or of collisions, expected for random walks.

A diffusion coefficient can be measured from the disappearance of any concentration gradient caused by diffusion. One way to create a concentration gradient and to measure the concentration of the diffusing species as a function of time is by photobleaching (Axelrod et al. 1976). Proteins embedded in a cell membrane can often diffuse in the plane of the membrane. A laser is focused on a small spot on the membrane of a living cell, and an intense laser pulse is used to photochemically destroy absorbing groups in the proteins in the spot. The laser (with much lower intensity) can now be used to measure the return by diffusion of neighboring protein molecules with unchanged absorbing groups. Thus, a hole is produced in the apparent color of a membrane, and the time that it takes for the hole to disappear provides a diffusion coefficient for the absorbing species. By making the hole, or spot, a few microns in size, the diffusion experiment need take only a few seconds. (See example 6.4 for the time it takes for myoglobin to diffuse 10 μm.)

Fluorescence is a more sensitive method for detecting membrane proteins than is absorption, so fluorescence is usually used to follow the motion of the molecules. The terms *fluorescence photobleaching recovery* (FPR) and *fluorescence recovery after photobleaching* (FRAP) have been coined for this method.

Determination of the Diffusion Coefficient by Laser Light Scattering

The preceding methods start with a nonequilibrium system, and the diffusion coefficient is evaluated from the way the system moves toward equilibrium. But since the diffusion coefficient D is related to the random motion of the molecules, which occurs in equilibrium as well as nonequilibrium systems, one should be able to measure D for a system at equilibrium by monitoring the random motion of molecules directly. One such method is *laser light scattering*, sometimes called *quasi-elastic light scattering*.

There is random motion of molecules in a homogeneous solution. If a beam of monochromatic light of frequency ν_0 passes through the solution, the scattered light is no longer monochromatic. Some molecules will be moving toward the light and some away; this causes a Doppler broadening of the scattered light. The intensity of the scattered light, when plotted as a function of frequency, has a maximum at ν_0, and its width at half-height is directly proportional to D (for molecules whose characteristic dimensions are small compared with the wavelength of the light). For a typical protein molecule with a D of 10^{-6} cm^2 s^{-1}, the width of the spectrum of the scattered light is of the order of 10^4 hertz (Hz). This is still an extremely sharp line, considering the fact that ν_0 is of the order of 10^{15} Hz. The availability of highly monochromatic light from laser sources has made it possible to measure diffusion coefficients of macromolecules by this method. It provides a rapid method of measuring diffusion coefficients without the need to wait for the return of a concentration gradient to a uniform concentration.

Some representative results are given in table 6.2. For small rigid proteins, the precision of the measurements is about ±3%, and the agreement with other literature values is quite good. The value for tobacco mosaic virus is less reliable because the virus is not small compared with the wavelength of light. Nevertheless, for suitable systems, the light-scattering method is rapid and convenient.

Diffusion Coefficient and Molecular Parameters

An important application of transport measurements is to provide information on the size and shape of macromolecules. According to Newton, a force acting on any object causes an acceleration. However, in a viscous medium, there is a counterforce, called a *frictional force*, which is proportional to the velocity of the object, but in the opposite direction:

$$\text{frictional force} = -f\boldsymbol{u}$$

where f is called the *frictional coefficient*. Because the frictional force increases with $\boldsymbol{u}$, a velocity is soon reached at which the frictional force balances the

TABLE 6.2 Diffusion Coefficients for Macromolecules Measured Using Laser Light Scattering (LS) or Conventional (Lit) Methods

	pH	Salt	$D \times 10^7$ (cm^2 s^{-1}) LS	Lit
Bovine serum albumin	6.8	0.5 M KCl	6.7 ± 0.1	6.7
Ovalbumin	6.8	0.5 M KCl	7.1 ± 0.2	8.3
Lysozyme	5.6	—	11.5 ± 0.3	11.6
Tobacco mosaic virus	7.2	0.01 M Na phosphate	0.40 ± 0.02	0.3

Source: S. B. Dubin, J. H. Lunacek, and G. B. Benedek. 1967, *Proc. Natl. Acad. Sci. USA* 57:1164–71.

driving force, and the acceleration becomes zero. This velocity is called the *terminal velocity*. In a viscous medium, the terminal velocity is quickly reached, so it is the frictional coefficient that characterizes how fast a molecule can move under the forces acting in diffusion, sedimentation, electrophoresis, and other transport measurements. Although Newton's law—force equals mass times acceleration—is of course acting on each of the particles in the solution, the observed effect is that the driving force equals frictional coefficient f times terminal velocity u. For a particle in a viscous medium,

$$\text{force} = fu \qquad (6.42)$$

The units of f are (force/velocity) equal to (mass $\times$ acceleration/velocity); we will use g s^{-1}. The frictional coefficient of a particle, such as a protein or nucleic acid, is the molecular property that characterizes how fast the particle moves under a driving force. A larger frictional coefficient means a smaller velocity. By measuring the frictional coefficient, we can learn about sizes and shapes of molecules and about their interactions with the surrounding medium.

In a diffusion experiment, the driving force is the average of the random forces that comes from the thermal kinetic energy of the surrounding molecules. A measure of the average velocity of motion is the diffusion coefficient. The basis for relating the diffusion coefficient to molecular parameters is the equation

$$D = \frac{kT}{f} \qquad (6.43)$$

where k, the Boltzmann constant, is equal to the gas constant R divided by Avogadro's number, N_0. This equation is another one of the equations, derived by Einstein as part of the work for his PhD thesis, that relates a macroscopic measurement, the diffusion coefficient, to a molecular property—the frictional coefficient. The frictional coefficient can be related to the size and shape of a molecule. For a sphere or spherical macromolecule of radius r, Stokes found that

$$f = 6\pi\eta r \qquad (6.44)$$

where η is the viscosity coefficient, a measure of how viscous the surrounding medium is. (We discuss viscosity in more detail in a later section.)

If a macromolecule of known volume is spherical and unsolvated, we can calculate its radius and then calculate its frictional coefficient from the Stokes equation [Eq. (6.44)]. This calculated frictional coefficient can be compared with the measured frictional coefficient determined from the diffusion coefficient [Eq. (6.43)]. The calculated f is usually smaller than the measured f. The explanation for this is that the macromolecule is either solvated (which increases its radius), or nonspherical, or both. In the following sections, we learn that both increase the frictional coefficient.

Solvation

Some solvent is usually associated with a macromolecule in solution. This *solvation* increases the effective or hydrodynamic volume of the macromolecule and therefore increases its frictional coefficient. Properties such as the diffusion coefficient and the sedimentation coefficient are directly related to the frictional coefficient.

We can measure the partial specific volume of a macromolecule in solution and use it to calculate the hydrodynamic volume of the macromolecule. The partial specific volume $\bar{v}_2$ is the change in volume of a solution when w_2 grams of solute are added:

$$\bar{v}_2 \equiv \left(\frac{\partial V}{\partial w_2} \right)_{T,P,w_1} \tag{6.45}$$

The partial specific volume can be measured just as implied by Eq. (6.45). The volume of a solvent is measured and then a known weight of solute is added. The partial specific volume $\bar{v}_2$ is the change in volume of the solution ΔV divided by the added weight Δw. The partial specific volume can depend on the concentration of the solute, the composition of the solvent, and temperature and pressure. We specify that these parameters are held constant by the subscripts T, P, w_1, in the definition of $\bar{v}_2$. For an impermeable object, such as a glass bead, the increase in the volume of the solution is just equal to the volume of the bead. Therefore, the partial specific volume $\bar{v}_2$ is equal to the specific volume, v_2 (volume per gram) of the bead. It will be independent of solvent composition and of solute concentration. For a macromolecule, the partial specific volume may be very dependent on such variables as pH, salt concentration, and macromolecule concentration. In neutral aqueous solutions, $\bar{v}_2$ is 0.7–0.75 cm^3 g^{-1} for most proteins, and $\bar{v}_2$ is 0.4–0.5 cm^3 g^{-1} for DNA.

To calculate the hydrodynamic volume of an unhydrated molecule (think of a glass bead), we just multiply the partial specific volume $\bar{v}_2$ by the weight per molecule to obtain the volume of the molecule. However, for a hydrated molecule, we must add the volume of the solvent bound to the molecule; these bound water molecules increase the hydrodynamic volume of the molecule. Values of hydration δ_1 range from about 0.2 to 0.6 (gram of water per gram of protein) for proteins and from about 0.5 to 0.7 for DNA. The volume of water bound per gram of macromolecule is δ_1/ρ with ρ the density of the solvent. The hydrodynamic volume of the macromolecule is obtained by adding the volume of bound water per gram to the partial specific volume of the molecule and then multiplying the sum by the weight of the macromolecule.

Shape Factor

For a nonspherical macromolecule, the frictional coefficient is greater than that for a sphere of the same volume. It seems reasonable that any deviation from a spherical shape for the same size particle will increase its frictional coefficient because of the increased surface area in contact with the solvent. Computer calculations can be done to calculate the frictional coefficient for any shape particle, and tables exist for frictional coefficients of simple shapes. Figure 6.8 illustrates the increase in frictional coefficient as a sphere is changed to an ellipsoid of revolution with the same volume. An oblate ellipsoid of revolution is generated by rotating an ellipse about its minor axis; a prolate ellipsoid of revolution is generated by rotating an ellipse about its major axis. Figure 6.8 shows that as the axial ratio increases the frictional coefficient increases faster for a prolate ellipsoid than for an oblate ellipsoid. Proteins and nucleic acids are not spheres or ellipsoids, but figure 6.8 shows that for small deviations from spheres the frictional coefficient calculated from the Stokes equation [Eq. (6.44)] is a reasonable approximation.

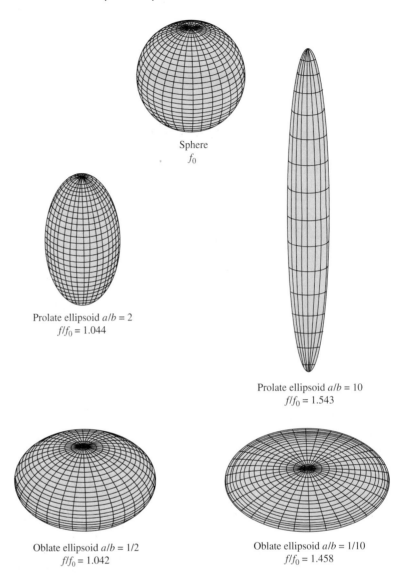

Sphere
f_0

Prolate ellipsoid $a/b = 2$
$f/f_0 = 1.044$

Prolate ellipsoid $a/b = 10$
$f/f_0 = 1.543$

Oblate ellipsoid $a/b = 1/2$
$f/f_0 = 1.042$

Oblate ellipsoid $a/b = 1/10$
$f/f_0 = 1.458$

▶ FIGURE 6.8

The frictional coefficients of ellipsoids of revolution relative to the frictional coefficients of spheres of the same volume. Prolate ellipsoids are shaped like footballs; oblate ellipsoids are shaped like pancakes. The axis of revolution is designated by *a* for each molecule. Note that for slight deviations from a sphere—axial ratios of less than 2 to 1—the increase of *f* is less than 5%. The frictional coefficient increases faster for a prolate ellipsoid than for an oblate ellipsoid as the axial ratio increases.

EXAMPLE 6.5

The diffusion coefficient D of ribonuclease from bovine pancreas, an enzyme that digests RNA, has been measured in a dilute buffer at 20°C. The value is $D = 13.1 \times 10^{-7}$ cm^2 s^{-1}.

a. Calculate the frictional coefficient f.

b. Assuming that the protein molecule is a sphere, calculate its hydrodynamic radius from the Stokes equation [Eq. (6.44)].

Solution

a. From Eq. (6.43),

$$f = \frac{kT}{D}$$

$$k = 1.380 \times 10^{-16} \text{ erg K}^{-1} \text{ molecule}^{-1}$$

$$= 1.380 \times 10^{-16} \text{ g cm}^2 \text{ s}^{-2} \text{ K}^{-1} \text{ molecule}^{-1}$$

Therefore,

$$f = \frac{(1.380 \times 10^{-16} \text{ g cm}^2 \text{ s}^{-2} \text{ K}^{-1})(298 \text{ K})}{13.1 \times 10^{-7} \text{ cm}^2 \text{ s}^{-1}}$$

$$= 3.09 \times 10^{-8} \text{ g s}^{-1}$$

b. From Eq. (6.44),

$$r = \frac{f}{6\pi\eta}$$

The viscosity coefficient for a dilute aqueous buffer is approximately equal to that of water, which is, at 20°C, 1.00 centipoise (cP) or 1.00×10^{-2} g s^{-1}cm^{-1}; this is equal to 1.00 mPa s (see table 2.2). Thus,

$$r = \frac{3.09 \times 10^{-8} \text{ g s}^{-1}}{6\pi \; 1.00 \times 10^{-2} \text{ g s}^{-1} \text{ cm}^{-1}}$$

$$= 1.64 \times 10^{-7} \text{ cm}$$

$$= 16.4 \text{ Å}$$

Thus, from a macroscopic measurement of concentration of a solution versus time, we can measure D and thus learn the radius of the diffusing molecules. Hydration and a nonspherical shape can affect the calculated radius but probably not by more than 20 to 30%. Of course, if we knew that the molecule was a long rod, we would not use the Stokes equation for a sphere to calculate its radius.

Diffusion Coefficients of Random Coils

Our discussions in the preceding section apply to macromolecules that are reasonably rigid. However, very flexible macromolecules, such as DNA molecules with many thousands of base pairs, are different. The DNA in solution, with its length much greater than its diameter, resembles somewhat a loose ball of thread. As the DNA moves through the solution, it carries a large amount of solvent with it. Therefore, the DNA molecule acts hydrodynamically as a highly hydrated sphere. The radius of this sphere is expected to be directly proportional to the average three-dimensional size of the molecule.

The root-mean-square end-to-end distance in a random polymer of N units is given by Eq. (6.33):

$$\langle d^2 \rangle^{1/2} = \sqrt{N} l$$

The root-mean-square radius of a random polymer is also proportional to the square root of the number of monomer units. Thus, the root-mean-square radius is proportional to the square root of the molecular weight of a macromolecule if the macromolecule is a very flexible coil. DNA molecules with molecular weights M in the millions act as flexible coils; therefore, their frictional coefficients are expected to be proportional to $M^{1/2}$, and their diffusion coefficients are expected to be proportional to $M^{-1/2}$. This is found to be approximately true.

Sedimentation

Sedimentation is used to separate, purify, and analyze all sorts of cellular species ranging from individual proteins and nucleic acids to viruses, chromosomes, mitochondria, and so forth. Let's first consider sedimentation in a gravitational field. The free fall of a particle of mass m and partial specific

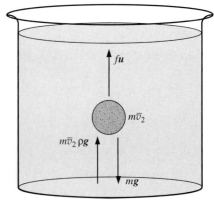

▶ FIGURE 6.9

Free fall of a particle, of mass m and specific volume $\bar{v}_2$, through a viscous medium of density ρ. The forces acting on the particle are, in one direction, the gravitational force mg, and in the other direction, the buoyant force $m\bar{v}_2\rho g$ and the frictional force fu, where u is the velocity of the particle and f is the frictional coefficient.

volume $\bar{v}_2$ through a viscous medium of density ρ is illustrated in figure 6.9. The driving force that causes the particle to sink is the difference between the gravitational force and the buoyant force:

$$\text{driving force} = mg - m\bar{v}_2\rho g$$

$$= m(1 - \bar{v}_2\rho)g$$

where g is the gravitational acceleration.

At time zero, when the particle is released, its velocity is zero. As it falls, it picks up speed. Let u be its velocity at time t. As the particle moves through the medium with a velocity u, it experiences a frictional force proportional to u and to the frictional coefficient f. From Newton's law, we obtain

$$m(1 - \bar{v}_2\rho)g - fu = m\frac{du}{dt}$$

where du/dt is the acceleration.

Since the frictional force increases with u, a velocity is soon reached at which the frictional force balances the driving force, and the acceleration becomes zero. This velocity is called the terminal velocity, u_t:

$$m(1 - \bar{v}_2\rho)g - fu = 0$$

or

$$u_t = \frac{m(1 - \bar{v}_2\rho)}{f}g \tag{6.46}$$

In a centrifugal field, the centrifugal acceleration is $\omega^2 x$, where ω is the angular velocity of the centrifuge in units of rad s^{-1} and x is the distance from the center of rotation. By analogy to the free fall of a sphere in a gravitational field, a molecule (of mass m and partial specific volume $\bar{v}_2$) sedimenting through a viscous medium in a centrifugal field also reaches a terminal velocity u_t in the x-direction:

$$u_t = \frac{m(1 - \bar{v}_2\rho)}{f}\omega^2 x \tag{6.47}$$

The centrifugal acceleration $\omega^2 x$ now replaces the gravitational acceleration g in Eq. (6.46).

The quantity $u_t/\omega^2 x$, which is the velocity per unit acceleration, is called the *sedimentation coefficient*. The symbol s is usually used for the sedimentation coefficient. The dimensions of s are (cm sec^{-1})/(cm sec^{-2}) or sec. (Here we will abbreviate seconds as sec to avoid confusion with the sedimentation coefficient.) A convenient unit for s is the *svedberg*, named in honor of T. Svedberg, who pioneered much research on sedimentation in an ultracentrifuge. One svedberg, or 1 S, is defined as 10^{-13} sec. From the definition of s and Eq. (6.47), we have

$$s = \frac{m(1 - \bar{v}_2 \rho)}{f} \qquad \textbf{(6.48)}$$

Determination of the Sedimentation Coefficient

Two methods can be used to measure the sedimentation coefficients of macromolecules. In one, a homogeneous solution is spun in an ultracentrifuge. As the macromolecules move down the centrifugal field, a solution–solvent boundary is generated. By following the movement of the boundary with time, the sedimentation coefficient can be calculated. This method is called *boundary sedimentation* and is illustrated in figure 6.10.

The generation of a boundary means that a concentration gradient is also generated, which, according to Eq. (6.37), results in diffusional transport of solute molecules. Thus, transport by sedimentation is opposed by transport by diffusion. If the macromolecules are very large or the centrifugal acceleration is very high, transport by sedimentation is much greater than transport by diffusion. Under such conditions, the boundary is very sharp. If transport by diffusion is significant, the boundary broadens as it moves downfield. As

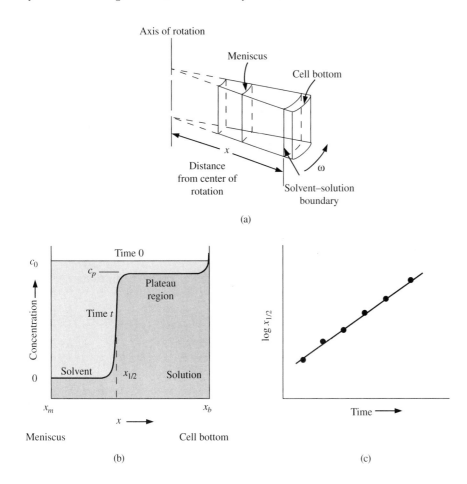

(a)

(b)

(c)

◀ FIGURE 6.10

Boundary sedimentation. (a) The compartment of the centrifuge cell containing the solution is sector shaped, with the center of the sector at the axis of rotation. The dimensions of the cell are exaggerated for clarity. (b) Concentration as a function of the distance from the center of rotation. At zero time, the concentration (c_0) is uniform across the cell. At time t, owing to sedimentation of the macromolecular solute molecules, a sharp boundary has been generated, with solvent to its left and solution to its right. The concentration in the plateau region c_p is independent of position; c_p is lower than c_0 because of the sector shape of the cell compartment and because the centrifugal field increases with increasing x. It can be shown that $c_p/c_0 = (x_m/x_{1/2})^2$, where x_m is the position of the meniscus and $x_{1/2}$ is the position of the boundary. (c) Plot of $\log x_{1/2}$ versus t gives a straight line. The sedimentation coefficient can be obtained from the slope of this line [Eq. (6.49)].

an approximation, we can assume that the diffusion process does not affect the rate of movement of the boundary. It is usually sufficiently accurate to calculate s from the position of the midpoint of the boundary, $x_{1/2}$:

$$s = \frac{u_t}{\omega^2 x} = \frac{dx_{1/2}/dt}{\omega^2 x_{1/2}} = \frac{1}{\omega^2}\frac{d \ln x_{1/2}}{dt} = \frac{2.303}{\omega^2}\frac{d \log x_{1/2}}{dt} \qquad \textbf{(6.49)}$$

For a more rigorous discussion of the calculation of s, a more advanced text should be consulted. Several suggestions are listed at the end of this chapter.

EXAMPLE 6.6

Escherichia coli DNA ligase, an enzyme that catalyzes the formation of a phosphodiester bond from a pair of 3'-hydroxyl and 5'-phosphoryl groups in a double-stranded DNA, has a molecular weight of 74,000 and a $\bar{v}_2$ of 0.737 cm^3 g^{-1} at 20°C. Boundary sedimentation in a dilute aqueous buffer (0.02 M potassium phosphate, 0.01 M NH$_4$Cl, and 0.2 M KCl, pH 6.5) at 20.6°C and a speed of rotation of 56,050 rpm gave the following results (data courtesy of P. Modrich):

Time (min)	$x_{1/2}$ (cm)	log $x_{1/2}$
0	5.9110	0.7717
20	6.0217	0.7797
40	6.1141	0.7863
60	6.2068	0.7929
80	6.3040	0.7996
100	6.4047	0.8065
120	6.5133	0.8138
140	6.6141	0.8205

a. Calculate s.

b. Calculate the frictional factor f in the dilute aqueous buffer; the density of the buffer at 20.6°C is 1.010 g cm^{-3}.

Solution

a. A plot of log $x_{1/2}$ versus t gives a straight line with a slope of 3.42×10^{-4} min^{-1}. The angular velocity ω is

$$\omega = (56{,}050 \text{ revolutions min}^{-1})(2\pi \text{ rad revolution}^{-1})(1 \text{ min}/60 \text{ sec})$$

$$= 5.870 \times 10^3 \text{ rad sec}^{-1}$$

(note that the unit radian is defined as an arc length divided by a radius and is therefore dimensionless)

From Eq. (6.49),

$$s = \frac{2.303}{\omega^2}(3.42 \times 10^{-4} \text{ min}^{-1})$$

$$= \frac{2.303(3.42 \times 10^{-4} \text{ min}^{-1})}{(5.870 \times 10^3 \text{ sec}^{-1})^2(60 \text{ sec min}^{-1})}$$

$$= 3.81 \times 10^{-13} \text{ sec}$$

$$= 3.81 \text{ S}$$

b. From Eq. (6.48),

$$f = \frac{m(1 - \bar{v}_2\rho)}{s}$$

$$m = \frac{74{,}000}{6.023 \times 10^{23}} = 1.23 \times 10^{-19} \text{ g}$$

Therefore,

$$f = \frac{(1.23 \times 10^{-19} \text{ g})(1 - 0.737 \times 1.010)}{3.81 \times 10^{-13} \text{ sec}}$$

$$= 8.24 \times 10^{-8} \text{ g sec}^{-1}$$

A second method of sedimentation is called *zone sedimentation*. A thin layer of a solution of macromolecules is placed on top of a solvent at the beginning of centrifugation. As centrifugation continues, the macromolecules sediment through the solvent as a zone or band. However, because the density of the macromolecular solution will be greater than that of the solvent, an instability will occur. The sedimenting zone will be denser than the solvent below it and will tend to mix. To avoid such instability, a solvent gradient is usually imposed so that the net density gradient in the direction of the centrifugal field is always positive. For example, a preformed sucrose gradient can be used, as illustrated in figure 6.11. Alternatively, if a concentrated salt solution is used as the medium for sedimentation measurements, the density gradient generated by the sedimentation of the salt is sufficient to provide the stabilization.

Standard Sedimentation Coefficient

To facilitate comparison of sedimentation coefficients measured in solvents with different values of solvent density ρ and solvent viscosity η it is common practice to standardize the measured s. From Eq. (6.48),

$$s = \frac{m(1 - \bar{v}_2\rho)}{f}$$

we know that s is directly proportional to the buoyancy factor and inversely proportional to the frictional coefficient f. The frictional coefficient is directly proportional to the solvent viscosity, and the buoyancy factor depends on the solvent density. Therefore, we can calculate a standard value of a sedimentation coefficient, chosen to be the value at 20°C in water, $s_{20,w}$, by multiplying the measured s by ratios of viscosities and buoyancy factors:

$$s_{20,w} = s\left(\frac{\eta}{\eta_{20,w}}\right)\frac{(1 - \bar{v}_2\rho)_{20,w}}{(1 - \bar{v}_2\rho)} \tag{6.50}$$

To use Eq. (6.50) to convert the measured s to $s_{20,w}$, we must measure the density and viscosity of the solvent we use for the sedimentation experiments. For the calculation to be valid for different solvents, it is necessary that the solvent *not* change the shape or solvation of the macromolecule significantly. In Eq. (6.50) we assume that the only effect on the frictional coefficient is through the solvent viscosity. If two different solvents give very different values of $s_{20,w}$, we know that the macromolecule has different conformations in the two solvents.

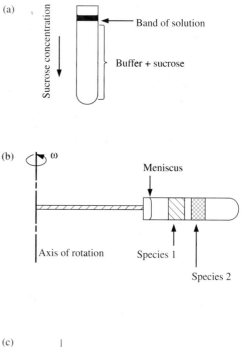

(a)

Band of solution

Buffer + sucrose

Sucrose concentration

(b)

ω

Meniscus

Axis of rotation

Species 1

Species 2

▶ FIGURE 6.11

Zone sedimentation through a preformed density gradient. (a) At time = 0, a layer of solution is placed on a preformed gradient. In this illustration, a sucrose gradient (for example, a linear gradient from 5 to 20% sucrose) is employed. The buffer concentration is constant along the tube. (b) After the tube has spun in a centrifuge at an angular speed ω for a certain time t, the macromolecular species have sedimented. In this example, the original solution contained two macromolecular species with different sedimentation coefficients. (c) The concentration profile after time t. A typical measurement technique involves the use of radioactively labeled macromolecules. The centrifuge is stopped after time t, a hole is punched through the bottom of the tube, and fractions are collected. The radioactivity of each fraction is then determined.

(c)

Species 1

Species 2

Concentration

Distance from center of rotation

EXAMPLE 6.7

Convert the value of s obtained in example 6.6(a) to $s_{20,w}$. At 20°C the ratio of the viscosity of the buffer η_b to that of water η_w has been measured to be $\eta_b/\eta_w = 1.003$.

Solution

For a dilute buffer, it is sufficiently accurate to assume that its temperature dependence of viscosity is the same as that of water. Thus,

$$\frac{\eta_b}{\eta_w} = 1.003$$

From a handbook, we obtain $\eta_{20.6,w} = 0.9906$ cP and $\eta_{20,w} = 1.0050$ cP. Therefore,

$$\frac{\eta_{20.6,w}}{\eta_{20,w}} = \frac{1.003 \times 0.9906}{1.0050} = 0.989$$

We assume for the protein that v_2 at 20.6°C is the same as v_2 at 20°C. The density of water at 20°C is 0.9982 g cm^{-3} from table 2.2. Thus,

$$\frac{(1 - \bar{v}_2\rho)_{20,w}}{(1 - \bar{v}_2\rho)} = \frac{1 - 0.737 \times 0.9982}{1 - 0.737 \times 1.010} = 1.034$$

Substituting these values into Eq. (6.50), we obtain

$$s_{20,w} = 3.81 \text{ S} \times 0.989 \times 1.034 = 3.90 \text{ S}$$

Determination of Molecular Weights from Sedimentation and Diffusion

Equation (6.48) states that the sedimentation coefficient s depends on the mass m, the frictional coefficient f, and the buoyancy factor $(1 - \bar{v}_2\rho)$ of a molecule. But f can be obtained from a measured diffusion coefficient:

$$f = \frac{kT}{D}$$

Combining Eqs. (6.48) and (6.43) and replacing the molecular mass by the mass per mol of molecules ($M = mN_0$), we obtain

$$M = \frac{RTs}{D(1 - \bar{v}_2\rho)} \tag{6.51}$$

Thus, the molecular weight of a molecule can be obtained by measuring its sedimentation coefficient s, its diffusion constant D, and its partial specific volume $\bar{v}_2$. This provides an absolute method for determining molecular weights.

EXAMPLE 6.8

The following data have been obtained for ribosomes from a paramecium: $s_{20,w} = 82.6$, $D_{20,w} = 1.52 \times 10^{-7} \text{cm}^2 \text{sec}^{-1}$, and $\bar{v}_{20} = 0.61 \text{ cm}^3 \text{g}^{-1}$ (Reisner, Rowe, and Macindoe, *J. Mol. Biol.* 32, 587 [1968]). Calculate the molecular weight of the ribosomes.

Solution

From Eq. (6.51),

$$M = \frac{(8.314 \text{ J K}^{-1} \text{ mol}^{-1})(293 \text{ K})(82.6 \times 10^{-13} \text{ s})}{(1.52 \times 10^{-7} \text{ cm}^2 \text{ s}^{-1})(10^{-2} \text{ m cm}^{-1})^2(1 - 0.61 \times 1.00)}$$

$$= 3.4 \times 10^3 \text{ J s}^2 \text{ m}^{-2}$$

$$= 3.4 \times 10^3 \text{ J s}^2 \text{ m}^{-2} \left(\frac{\text{kg m}^2 \text{ s}^{-2}}{\text{J}}\right)\left(\frac{10^3 \text{ g}}{\text{kg}}\right)$$

$$= 3.4 \times 10^6 \text{ g mol}^{-1}$$

In a given medium for a family of macromolecules of the same $\bar{v}_2$, we expect s to be proportional to M/f. We discussed previously that, for random coils like large DNA molecules, f is proportional to $M^{1/2}$, so s is proportional to $M/M^{1/2}$ or $M^{1/2}$. Experimentally, for large DNA molecules, s is found to be proportional to $M^{0.44}$, close to our expectation. For spherical, unsolvated molecules of the same partial specific volume, f is proportional to r and therefore to $M^{1/3}$, and s is expected to be proportional to $M^{2/3}$.

Determination of Molecular Weights from Sedimentation Equilibrium

Suppose a homogeneous two-component solution (a solute plus a solvent) is spun in an ultracentrifuge. However, we do not spin the centrifuge fast enough to sediment all the macromolecules to the bottom of the tube. Nevertheless, because of sedimentation, a concentration gradient is generated. Diffusion then sets in. Because transport by sedimentation and by diffusion go in opposite directions, an equilibrium concentration gradient can be generated by centrifugation, in which transport by sedimentation exactly balances transport by diffusion. Similar concentration gradients develop in Earth's atmosphere and in the oceans, although convective currents disrupt the balance between gravitational and diffusive transport.

The concentration at equilibrium in a gravitational field or a centrifugal field does not depend on frictional coefficients, sedimentation coefficients, or diffusion coefficients of the molecules. At equilibrium, thermodynamics, not kinetics, is controlling. We describe the equilibrium concentration gradient for an ideal solution or for real solutions at low concentrations. A rigorous thermodynamic derivation is given, for example, in Cantor and Schimmel (1980).

The concentration of molecules at equilibrium in Earth's gravitational field will be higher at the bottom of a container than at the top because of the molecules' lower potential energy in the field at the bottom. For a molecule in an external field, the total chemical potential (the sum of the chemical potential plus the potential energy in the external field) determines the concentration at equilibrium. The total chemical potential for a molecule in Earth's gravitational field is $\mu + M_{eff}gx$. Here M_{eff} is the effective molecular weight $M(1 - \bar{v}_2\rho)$, g is the acceleration of gravity, and x is the distance from the bottom of the container.

To derive the equation for the equilibrium concentrations in a gravitational or centrifugal field, we will use the Boltzmann equation [Eq. (6.16)] for the probability of finding a molecule with energy E_i. The ratio of the number of molecules with energy E_j to the number with energy E_i (when the degeneracy factors are equal) is

$$\frac{P_j}{P_i} = \frac{e^{-E_j/kT}}{e^{-E_i/kT}} \tag{6.52}$$

The ratio of probabilities is equal to the ratio of concentrations. We can now write the equation as

$$\frac{c_j}{c_i} = e^{-(E_j - E_i)/kT} \tag{6.53}$$

For energies per mole instead of energies per molecule,

$$\frac{c_j}{c_i} = e^{-(\bar{E}_j - \bar{E}_i)/RT} \tag{6.53a}$$

For a molecule in a solvent of density ρ, the energy for a mole of molecules in a gravitational field with acceleration g is

$$E = M(1 - \bar{v}_2\rho)gx$$

Substituting this expression into Eq. (6.53b) we obtain

$$\frac{c_2}{c_1} = e^{-M(1-\bar{v}_2\rho)g(x_2-x_1)/RT} \tag{6.54}$$

Equation (6.54) states that at equilibrium the concentration of any molecule will vary exponentially with position above the bottom of the container. In the laboratory for distances of a few centimeters, the effect is large only for high molecular weights (M greater than 10^8). To obtain equilibrium, it is important not to stir the solution, nor to allow temperature gradients to cause convective mixing.

For large distances, the exponential decrease of concentration with height above Earth's surface is important for all molecules. The change of pressure with altitude is approximately characterized by Eq. (6.54). With the buoyancy factor set equal to 1 and using the molecular weight of oxygen gas, we calculate reasonable values for the oxygen pressure as a function of altitude. We of course should also take into account the change of temperature with altitude, and even the change of g with altitude.

To determine molecular weights of most molecules, we need stronger fields than gravity to establish measurable concentration gradients, so we use centrifuges. The centrifugal acceleration for a centrifuge spinning at ω radians s^{-1} ($\omega = 2\pi$ revolutions per second) is $\omega^2 x$, where x is measured from the center of rotation. The force acting toward the bottom of the cell is $m(1 - \bar{v}_2\rho)\omega^2 x$. The potential energy decreases with increasing x; the energy is the negative integral of the force with respect to x. Thus, the energy for 1 mol of molecules in a centrifugal field with acceleration $\omega^2 x$ is

$$E = -M\frac{(1 - \bar{v}_2\rho)\omega^2 x^2}{2}$$

Substituting this expression into Eq. (6.53a), we obtain

$$\frac{c_2}{c_1} = e^{+M(1-\bar{v}_2\rho)\omega^2(x_2^2-x_1^2)/2RT} \tag{6.55}$$

The difference in sign between Eqs. (6.54) and (6.55) is due to the different convention for measuring x. In both cases, the concentration increases toward the bottom of the cell, where the bottom in a centrifuge cell is farthest from the center of rotation. By choosing the speed of rotation ω, we can sediment molecules of any molecular weight. Note that the buoyancy factor can be positive or negative; this means that some molecules sink, but others, such as lipoproteins, float.

To determine a molecular weight from a sedimentation equilibrium experiment we take the logarithm of Eq. (6.55):

$$\ln\left(\frac{c_2}{c_1}\right) = \frac{M(1 - \bar{v}_2\rho)\omega^2(x_2^2 - x_1^2)}{2RT} \tag{6.56}$$

We see that a plot of $\ln c$ versus x^2 is a straight line with slope equal to

$$\frac{M(1 - \bar{v}_2\rho)\omega^2}{2RT}$$

Thus, to obtain an absolute molecular weight we use

$$M = \frac{2RT}{(1 - \bar{v}_2\rho)\omega^2}(\text{slope of } \ln c \text{ vs. } x^2)$$

Absolute molecular weights are emphasized because with gel electrophoresis methods (see following sections) we can measure only relative molecular weights. The most accurate molecular weights are obtained from the known amino acid sequence of a protein or the nucleotide sequence of a nucleic acid. These provide good standards. However, for proteins containing more than one polypeptide, or if association or dissociation of macromolecules occurs, then sedimentation-diffusion or sedimentation-equilibrium can provide the molecular weights of the complexes.

Density-Gradient Centrifugation

A second application of equilibrium centrifugation involves the spinning of a concentrated salt solution, such as CsCl, at high speed. A concentration gradient is generated, which results in a *density gradient*, because the density of a CsCl solution increases with increasing concentration. If a macromolecular species is also present in the solution, it will form a band at a point in the salt gradient where the macromolecules are "buoyant." The buoyancy term $(1 - \bar{v}_2\rho)$ is zero at the value of solvent density $\rho = 1/\bar{v}_2$ in the density gradient. The macromolecules above this point sink; those below it float. This means that macromolecules that were uniformly distributed throughout the cell before sedimentation concentrate at this position when the density gradient forms. This is illustrated in figure 6.12. The higher the molecular weight of the macromolecular species, the sharper will be the band that forms in the

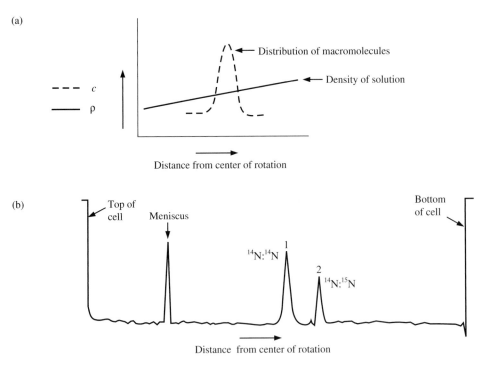

▲ FIGURE 6.12

Density-gradient centrifugation. (a) A macromolecular species in a concentrated salt solution of an appropriate density is spun in an ultracentrifuge. The solution was initially homogeneous. After a certain time, equilibrium is reached. The concentration of the salt—and, consequently, the density of the solution—increases with increasing distance from the center of rotation. The macromolecular species forms a band at a position at which the solvated molecules are buoyant. (b) An actual tracing of two DNA species in a CsCl solution. The initial homogenous solution has a density of 1.739 g cm^{-3}. After 17 hr at 44,770 rpm and 25°C, the DNA species form two sharp bands. Species 1 is a bacterial virus DNA with a molecular weight of 20×10^6. Species 2 is the same DNA except that it contains a heavier isotope of nitrogen (^{15}N rather than the usual ^{14}N). The substitution of ^{14}N by ^{15}N increases the buoyant density of this DNA by 0.012 g cm^{-3}.

density gradient. Many biological macromolecules have "buoyant densities" sufficiently different that they can be resolved by the density-gradient centrifugation method.

The original experiment that showed that DNA replicated by making a new complementary strand for each original strand in the parent DNA was done by using density-gradient centrifugation. Bacteria with DNA containing ^{14}N were grown in a medium containing ^{15}N. The DNA was isolated as a function of time and analyzed in a CsCl density gradient as shown in figure 6.12. The original DNA had both strands labeled with ^{14}N; this DNA is designated as ^{14}N:^{14}N. After one generation, DNA appears with one ^{14}N strand and one ^{15}N strand (^{14}N:^{15}N). It is not until the second generation that ^{15}N:^{15}N DNA appears.

Viscosity

When an external force F is applied to a solid particle to make it move through a liquid with a velocity u_t, the molecules of the liquid at the interface with the particle move at the same velocity u_t because of the attractive forces between the two. Far away from the particle, the liquid remains stationary. Thus, the movement of the particle through the liquid generates a velocity gradient in the medium.

Whenever a velocity gradient is maintained in a liquid, momentum is constantly transferred in a direction opposite to the direction of the velocity gradient. This is illustrated in figure 6.13. Here the liquid is moving in the direction x. The x-components of the average velocities of molecules in different layers are represented by arrows of different lengths. The velocity gradient is in the y direction; that is, the x-component of the velocity u_x increases with increasing y. The molecules are also moving in the y- and z-directions. However, because there is no net flow in the y- and z-directions, the movement of the molecules in these directions is random. Now consider a unit cross-sectional area in the xz-plane. In 1 s, a certain number of molecules will move across this area from below, and an equal number of molecules will

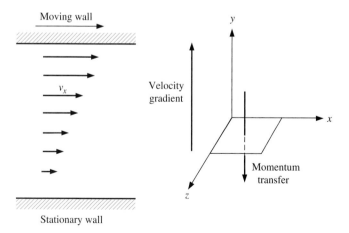

▲ FIGURE 6.13

Uniform velocity gradient produced in a fluid placed between a moving wall and a parallel stationary wall. The velocity gradient is a vector in the direction of increasing velocity y and is perpendicular to the direction of flow x. Momentum transfer occurs in the direction $-y$, opposite to the velocity gradient.

move across this area from above. But since molecules from below have lower u_x and consequently lower momentum in the x-direction, there is a net transfer of momentum in the x-direction from above to below—that is, in a direction opposite to the velocity gradient. The steeper the velocity gradient, the larger is the net momentum transfer. Mathematically, we write

$$J_{mu} \propto -\frac{du_x}{dy}$$

or

$$J_{mu} = -\eta\frac{du_x}{dy} \qquad (6.57)$$

where J_{mu} is the rate of momentum transfer per unit time per unit cross-sectional area.

The quantity η (eta) in Eq. (6.57) is called the *viscosity coefficient*, or viscosity. Its SI units are $kg\ m^{-1}\ s^{-1} \equiv Pa\ s$, but viscosities are often given in poise [1 poise (P) = $1\ g\ cm^{-1}\ s^{-1}$; $P = 10^{-1}\ Pa\ s$; $1\ cP = 1\ mPa\ s$]. If η is a constant independent of du_x/dy, the fluid is called a *Newtonian fluid*. If η is itself dependent on du_x/dy, the fluid is *non-Newtonian*.

Thus, when a particle moves through a stationary fluid under an external force F, a velocity gradient is generated, and the velocity gradient in turn imposes a viscous drag F' on the particle. The final velocity gradient is such that F and F' balance each other, and the particle moves at the terminal velocity u_t.

Measurement of Viscosity

In our discussion of the free fall of a particle through a viscous medium, we obtained

$$u_t = \frac{m(1 - \bar{v}\rho)g}{f} \qquad (6.46)$$

where we have dropped the subscript 2 on $\bar{v}$ for simplicity. Substituting the Stokes equation for a sphere, $f = 6\pi\eta r$, we obtain, upon rearranging,

$$\eta = \frac{m(1 - \bar{v}\rho)g}{6\pi r u_t} \qquad (6.58)$$

Thus, η for a fluid can be determined by measuring u_t for a sphere falling through it. Note also that, from measurements using the same particle, the relative viscosities of two liquids η_2 and η_1 can be calculated from

$$\frac{\eta_2}{\eta_1} = \left(\frac{1 - \bar{v}\rho_2}{1 - \bar{v}\rho_1}\right)\frac{u_{t1}}{u_{t2}} \qquad (6.59)$$

where the subscripts 1 and 2 refer to the quantities for liquids 1 and 2, respectively.

A more convenient method is to measure the volume rate of flow through a capillary. A simple viscometer, called an Ostwald viscometer, is shown in figure 6.14. Here the time t required for the liquid level to drop from mark 1 to mark 2 is measured. The relative viscosities of two liquids are measured from the ratio of their flow times, t_1 and t_2, and the ratio of their densities, ρ_1 and ρ_2.

$$\frac{\eta_2}{\eta_1} = \frac{\rho_2 t_2}{\rho_1 t_1} \qquad (6.60)$$

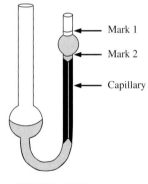

Mark 1

Mark 2

Capillary

▲ FIGURE 6.14

Simple Ostwald viscometer. Liquid is drawn up initially into the arm on the right to above mark 1. Upon release, it begins to flow back under the influence of gravity, but it is restricted by viscosity primarily in the capillary region. The elapsed time t is measured from when the meniscus passes mark 1 until it passes mark 2.

Viscosities of Solutions

Adding macromolecules to a solvent increases the viscosity of the solution. The viscosity of a solution of macromolecules depends on concentration and the size and shape of the macromolecule. The specific viscosity of a solution is defined as

$$\eta_{sp} \equiv \frac{\eta' - \eta}{\eta} \tag{6.61}$$

where η is the viscosity of the solvent and η' is the viscosity of the solution. The specific viscosity is unitless. To separate the effect of concentration on viscosity from that of molecular size and shape, the intrinsic viscosity $[\eta]$ is defined as the limit of the specific viscosity divided by concentration, as the concentration approaches zero.

$$[\eta] \equiv \lim_{c \to 0} \frac{\eta_{sp}}{c} \tag{6.62}$$

The units of $[\eta]$ are the reciprocal of concentration c. We will use c in $g\,cm^{-3}$; thus, the units of $[\eta]$ are $g^{-1}\,cm^3$. The intrinsic viscosity can be related to molecular properties by

$$[\eta] = \nu(\bar{v}_2 + \delta_1 v_1^0) \tag{6.63}$$

where ν is a unitless shape factor that equals 2.5 for spheres and increases rapidly for nonspherical shapes. Prolate ellipsoids have higher values than oblate ellipsoids; for an axial ratio of 10, the value of ν is about 15 for prolate ellipsoids and about 8 for oblate ellipsoids. The intrinsic viscosity depends on the volume of the particle, so it depends on its partial specific volume $\bar{v}_2$ as well as its hydration δ_1 and the specific volume of the solvent v_1^0.

Measurement of the viscosity of a solution of macromolecules is a convenient method to study denaturation of a protein or the assembly of several molecules into a large particle. The change of viscosity as a function of time, pH, or temperature is a measure of the change in shape or volume of the molecules.

Electrophoresis

Biological macromolecules are usually charged; therefore, they will move in the presence of an electric field. Obviously, molecules with a net positive charge will move toward the negative electrode, and molecules with a net negative charge will move away from the negative electrode. The velocity of motion depends on the magnitude of the electric field E, the net charge on the molecule, and the size and shape of the molecule as characterized by its frictional coefficient f. The net charge on the molecule is designated by Ze where Z is the number of electronic charges e.

For a particle in a nonconducting solvent, the velocity u of migration $(m\,s^{-1})$ in an electric field is

$$u = \frac{ZeE}{f} \tag{6.64}$$

where Z = the (unitless) number of charges
$\quad e = 1.6022 \times 10^{-19}\,C$ (coulomb)
$\quad E$ = the electric field in volt m^{-1}
$\quad f$ = the frictional coefficient in $kg\,s^{-1}$

The velocity per unit electric field, u/E, is called the *electrophoretic mobility μ*. The electrophoretic mobility and Eq. (6.64) can be used to measure the charge of a particle in a nonconducting medium. The charge of the electron was originally measured this way with a charged oil drop in air. However, biological macromolecules are found in aqueous solutions with other ions, buffers, and so on. Therefore, the charged macromolecule is surrounded by an atmosphere of small ions. This ion atmosphere greatly complicates the interpretation of electrophoretic mobility. The shielding effect of the ion atmosphere reduces the electric field experienced by the macromolecule. Furthermore, when the macromolecule moves, it drags its ion atmosphere with it. Therefore, the frictional coefficient is changed. The complications make it very difficult to interpret electrophoretic mobility quantitatively as we have done for the other hydrodynamic measurements. Electrophoresis is a powerful analytical tool, however, and macromolecules with very small differences in their properties can be resolved. One example is the separation of normal hemoglobin and hemoglobin from patients who suffer from sickle-cell anemia. Sickle-cell hemoglobin S differs from normal hemoglobin A by a single change of valine for glutamic acid in each of the two β-chain peptides. As a consequence, the proteins differ by 2 charges per molecule, which is sufficient to effect a clear separation by electrophoresis.

In the following sections, we describe some applications of electrophoresis to the characterization of macromolecules.

Gel Electrophoresis

Nearly all electrophoresis of macromolecules is done in gels. We want to separate molecules according to charge or size or shape; the gel greatly reduces mixing of the molecules by convection and by diffusion.

A gel is a three-dimensional polymer network dispersed in a solvent. A variety of gels have been used. For example, an agarose gel consists of an aqueous medium and a polysaccharide obtained from agar. An acrylamide gel consists of an aqueous medium and a copolymer of acrylamide ($CH_2{=}CH{-}CO{-}NH_2$) and N,N'-methylenebisacrylamide ($CH_2{=}CH{-}CO{-}NH{-}CH_2{-}NH{-}CO{-}CH{=}CH_2$). In this gel, the water-soluble acrylamide, with its reactive double bond, can polymerize into a linear chain:

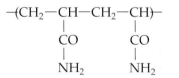

Adding the bisacrylamide, which has two double bonds, results in the formation of cross-links between different chains. The degree of cross-linking can be controlled by the ratio of the bis compound to acrylamide. In a typical gel, over 90% of the space is occupied by the aqueous medium, but the presence of the three-dimensional polymer network prevents convectional flow.

The gel contributes additional factors that affect the electrophoretic mobility: (1) The path traveled by a macromolecule through the porous gel is considerably longer than the length of the gel; (2) the gel imposes additional frictional resistance to the macromolecules as they move through pores of comparable size; (3) the macromolecules will interact with charged groups on the gel network; and (4) some pores in the gel will be too small for a macromolecule to enter. The combination of these factors further improves the resolution of electrophoresis.

DNA Sequencing

Gel electrophoresis in a denaturing solvent provides a rapid, effective method to determine the sequence of a single strand of DNA. The mobility of a nucleic acid is determined by the number of nucleotide phosphate groups in the strand when electrophoresis is done in a gel containing a high concentration of a neutral reagent (such as urea) that disrupts base pairing. This simple method has been combined with brilliant logic to invent a quick method for determining the base sequence of a nucleic acid. The problem was to determine the sequence of four bases in a single strand of nucleic acid. The solution was to make a break in the strand after only one type of base (such as guanine) but not at every location of this base. If one end of the strand is labeled, for example, with ^{32}P, we end up with strands of different length, each with ^{32}P on one end and guanine on the other. Measuring the chain length by gel electrophoresis gives us the positions of all the guanines. Repeating the procedure for each of the other bases gives the base sequence. This Nobel Prize–winning method was invented independently by Maxam and Gilbert [*Proc. Natl. Acad. Sci. USA* 74 (1977): 560–564] and by Sanger and colleagues [*Proc. Natl. Acad. Sci. USA* 74 (1977): 5463–5467]. Figure 6.15 shows an example of the procedure.

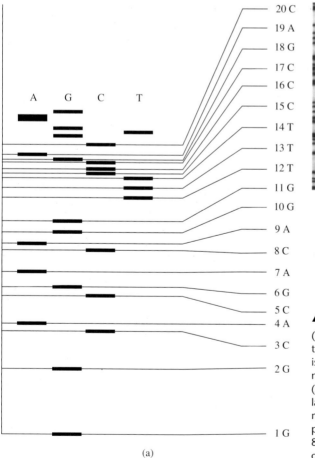

▲ FIGURE 6.15

(a) Schematic illustration of a sequencing gel electrophoresis pattern. The sequence of the first 20 bases of a single strand of DNA is shown. The strand is labeled at the 5′-hydroxyl group of the first nucleotide with radioactive ^{32}P. The strand is broken at adenines (A), guanines (G), cytosines (C), or thymines (T) and placed in four lanes at the top of a denaturing gel of 20% acrylamide, 0.67% methylene bisacrylamide, and 7 *M* urea. After electrophoresis (the positive electrode is at the bottom of the gel), autoradiography for 8 h produced the pattern shown. The sequence is now simply read off. (b) An actual gel pattern.

Sanger's method uses an enzyme, DNA polymerase, to make a complement of the DNA being sequenced. A primer (an oligonucleotide complementary to the DNA sequence) is used to specify where the DNA polymerase starts synthesis. The enzyme adds deoxynucleoside triphosphates (dNTPs) to the primer to form the newly synthesized strand; it is the one actually sequenced. The breaks in the strand, which give the different length fragments, are made by adding a few percent of a dideoxynucleoside triphosphate (ddNTP) to the dNTP mixture. Incorporation of a ddNTP into the new strand stops the polymerization at that point. For example, to obtain the positions of the guanines in the newly synthesized strand, 10% ddGTP is added. Most of the time, dGTP, which allows the chain to continue, is incorporated, but 10% of the time, ddGTP, which stops the chain, is incorporated. Thus, a ladder of chains with increasing lengths is obtained. All strands end in G; all start at the primer. Electrophoresis as shown in figure 6.15 gives the chain lengths. Repeating the procedure with each of the other bases completes the sequence. By using fluorescent derivatives of the ddNTPs, we can do the gel electrophoresis in a single lane instead of four separate lanes. The lane is scanned by a photodetector. Thus, the position of the band gives the chain length; its fluorescence characteristics identify the base. This method allows many different experiments to be done on a single gel and provides a more efficient recording of the sequence data.

The Human Genome Project has obtained a nearly complete sequence of a human genome: sequences for all 24 human chromosomes—about 3×10^9 base pairs in all. Hunkapiller, Kaiser, Koop, and Hood (1991) discuss methods and improvements in automation and data acquisition that helped make this possible. Electrophoresis in capillaries and in micron-sized channels etched in silicon wafers allow rapid sequence analysis of tiny amounts of DNA. For the latest information on the progress made in sequencing the human genome and in interpreting the results go to the National Center for Biotechnology Information, National Library of Medicine site at *www.ncbi .nlm.nih.gov.*

Double-Stranded DNA

Double-stranded DNA molecules can be separated according to molecular weight by electrophoresis in either polyacrylamide gels or agarose gels. No denaturant, such as urea, is added, so the gels are called *native gels*; the DNA runs in its native form. DNA has a uniform-charge density because each base pair adds two phosphate charges; therefore, in free solution the electrophoretic mobility is essentially independent of molecular weight. In gels the DNA molecules have to wander through the pores of the gel, and good separation can be obtained in the range of 10–100,000 base pairs (bp). The concentration of the agarose, or the cross-linking of the polyacrylamide, is optimized to give the best separation for each size range. Figure 6.16 illustrates the excellent separation of six DNA fragments in the range of 100–1000 bp.

DNA Fingerprinting

Although all humans have similar sequences of their DNA—the similarity is what will be found through the Human Genome Project—we are all unique in our total DNA sequence. Identical twins start out with identical sequences

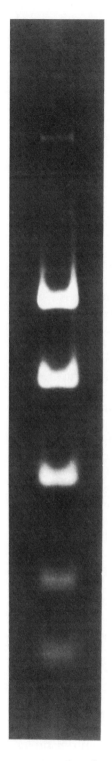

1100 bp

420 bp

290 bp

160 bp

90 bp

67 bp

◀ FIGURE 6.16

Electrophoresis of double-stranded DNA fragments on 3.5% acrylamide gel (acrylamide/bisacrylamide ≈ 30 wt/wt). The mixture originally started as a thin band at the top of the picture and was electrophoresed toward the positive electrode. After electrophoresis, the gel was immersed in a staining solution containing a fluorescent dye, ethidium, and then photographed under UV light. The sizes of the fragments are given in terms of base pairs (bp). One base pair corresponds to a molecular weight of ~660. (Unpublished data of T-S. Hsieh and J. C. Wang.)

at conception, but mutations during growth and development cause slight divergences of their sequences. A DNA sample obtained from skin, blood, semen, hair, or the like, left at a crime scene, for example, can be compared with the DNA from a suspect to see if they match.

Only about 2% of human DNA codes for proteins; much of the rest is simple repeating sequences. The number of repetitions of the sequence and the sequences differ from person to person. A region of the DNA where this occurs has been called the *hypervariable fingerprint region* (Jeffreys, Wilson, and Thein 1986). Restriction enzymes cut double-stranded DNA at specific sequences (such as CAATTG). They produce a distinct set of different length fragments of the fingerprint DNA; the fragments are separated by native gel electrophoresis (see figure 6.16). This method is referred to as RFLP analysis: *restriction fragment length polymorphism analysis;* it can distinguish members of a single family, but not identical twins. In criminal cases, prosecutors, defense lawyers, and jurors are having to learn about the reproducibility of gel electrophoresis patterns and to judge when similar patterns mean guilt or innocence.

Conformations of Nucleic Acids

We have been describing the separation of nucleic acids based solely on their molecular weights—the number of nucleotides in single strands on denaturing gels or the number of base pairs in double strands on native gels. But nucleic acids of the same molecular weight will have different electrophoretic mobilities when they have different shapes. An obvious example is linear double-stranded DNA versus circular double-stranded DNA. The compact circular DNA travels faster than the linear molecule. Many natural DNAs occur as circles or twisted circles (figure 8's and so forth). These superhelical DNAs differ only in their topology—the number of twists in the DNA before the ends are joined; they are called *topological isomers*, or *topoisomers.* The topoisomers can be separated and quantitated by gel electrophoresis. The concentrations of DNA topoisomers are used to assay for topoisomerases, enzymes that wind and unwind the DNA during its replication. These enzymes are targets for anticancer drugs as discussed by Liu (1989), because the most rapidly dividing cells require the largest amounts of these enzymes.

The DNA double helix is a flexible linear rod that is approximately linear for a few hundred base pairs; because of its flexibility, however, it acts more like a random coil when it is thousands of base pairs in length. However, certain sequences of DNA, such as repeating A·T pairs, cause the DNA to bend. Many chemicals that react with DNA, including carcinogens (such as *N*-acetylaminofluorene) and anticancer drugs [such as *cis*-diamminedichloroplatinum (II)] also cause the DNA to bend. All these bent DNAs have anomalous electrophoretic mobilities and can be distinguished from linear molecules of the same size; this is reviewed by Leng (1990).

RNA molecules are synthesized as single strands, which then fold into intramolecular double-strand regions with single-stranded loops. The simplest folded structure is called a *hairpin*—a single stem and loop. More complex structures are formed by further folding and interactions of different kinds of loops and stems. These structures are vital for the correct interactions of RNA molecules with proteins and with other RNAs. One example is transfer RNA, which adds each amino acid to the growing polypeptide chain during protein synthesis. The folding of RNA molecules can be easily monitored by native gel electrophoresis; mutations can be made in the RNA to determine how the sequence determines the folding (Jacques and Susskind 1991).

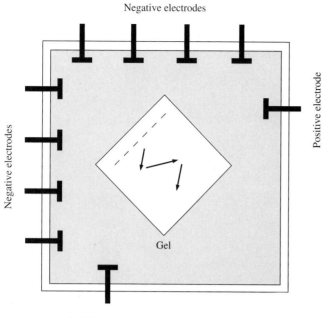

Negative electrodes

Negative electrodes

Positive electrode

Gel

Positive electrode

◀ FIGURE 6.17

A pulsed-field gel electrophoresis (PFGE) apparatus in which the DNA molecules move alternately down and to the right in the figure. The gel is actually horizontal and is immersed in a buffer solution. The net motion of the DNA is along the diagonal of the apparatus and is approximately straight in the gel. The angle between alternate electric fields ranges from 100° to 150°; this improves resolution. Each electrical pulse continues from 1 s to several minutes, depending on the size of the molecules.

Pulsed-Field Gel Electrophoresis

In the usual gel electrophoresis, the gel pores act like a sieve; different-sized DNA molecules pass through with different mobilities. For DNA molecules with more than 50,000 bp, the usual gel electrophoresis method does not work. The DNA can no longer pass through the pores in its randomly coiled form. To travel through the gel, the DNA is distorted by the field and moves along its long axis through the pores. It travels like a snake. Once the DNA is stretched out, its mobility becomes nearly independent of length and no separation according to molecular weight is obtained. The solution to this problem is to make the DNA molecules change direction during electrophoresis. If an electric field is applied perpendicular to the direction of motion of the DNA, the molecule must change shape to allow it to travel in the new direction. The time necessary for this to occur depends on the size of the DNA. By using a pulsed electric field that alternates in direction, Schwartz and Cantor (1984) were able to separate DNA molecules in the million base-pair range. Figure 6.17 shows one configuration of a *pulsed-field gel electrophoresis* (PFGE) apparatus; many other arrangements of electrodes have been used. The key variable for obtaining optimum separation is the length of each pulse; the longer the molecules, the longer are the pulses required. Individual DNA molecules can be seen moving through a gel by using acridine-labeled DNA and a fluorescence microscope. Computer modeling of this motion allows calculation of the optimum pulse length to separate a given size range (see Smith, Heller, and Bustamante 1991).

An example of the effect of pulse length in *field-inversion gel electrophoresis* (FIGE) is shown in figure 6.18. The sign of the electric field is changed periodically to obtain improved separation of the DNA molecules. A field that is positive for 0.50 s and negative for 0.25 s gives good separation in the size range of 10,000–50,000 bp. A field that is positive for 3 s and negative for 1 s gives good separation in the size range of 50,000–200,000 bp.

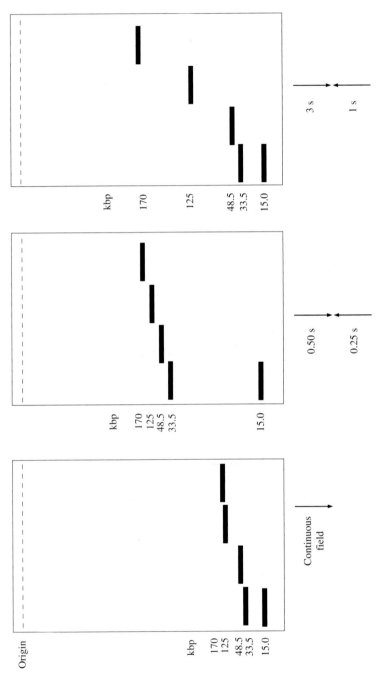

▲ FIGURE 6.18

Separation of DNA molecules by field-inversion gel electrophoresis (FIGE). The efficiency of separation is increased by alternately reversing the direction of the field. Three separate experiments are shown on DNA molecules ranging in size from 15.0 kilobase pairs (kbp) to 170 kbp. The DNA molecules are T4 bacteriophage, 170 kbp; T5 bacteriophage, 125 kbp; lambda bacteriophage, 48.5 kbp; two restriction enzyme fragments of lambda bacteriophage, 15.0 and 33.5 kbp. The molecules start at the top of the figure (labeled "Origin") and move down in a 1% agarose gel. In a continuous field applied for 4 h all sizes move with similar mobilities. In a field of 0.50 s forward and 0.25 s back for 12 h, the two restriction fragments are well separated. In a field of 3 s forward and 1 s back for 12 h, the lambda, T4, and T5 DNAs are well separated (Carle, Frank, and Olson 1986).

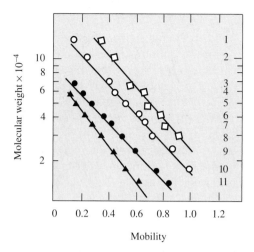

◀ FIGURE 6.19

The electrophoretic mobilities of proteins in SDS–polyacrylamide gel electrophoresis (SDS–PAGE). The logarithms of the molecular weights are linear in the mobilities of the proteins, as stated in Eq. (6.65). The slope and intercept of the linear function depends on the amount of cross-linking and the concentration of the gel. The acrylamide concentrations illustrated are 15% (▲), 10% (●), 7.5% (○), and 5% (□). The weight ratio of acrylamide to methylenebisacrylamide is 37:1. The numbers 1–11 refer to the following proteins: β-galactosidase, phosphorylase a, serum albumin, catalase, fumarase, aldolase, glyceraldehyde–phosphate dehydrogenase, carbonic anhydrase, trypsin, myoglobin, and lysozyme. [From K. Weber, J. R. Pringle, and M. Osborn, *Methods Enzymol.* 26 (1972): 3.]

Protein Molecular Weights

Nucleic acids can be separated according to molecular weight by electrophoresis because there is one charged phosphate group for each nucleotide monomer unit. For proteins, the number of charges depends on their amino acid compositions and the pH of the buffer. Furthermore, a polypeptide chain of a specific length can fold into different shapes with different frictional coefficients. Therefore, to use electrophoresis to determine protein molecular weight, it is necessary to denature the protein and to introduce a charge on each peptide. This is done by adding an anionic detergent, sodium dodecylsulfate (SDS), and 2-mercaptoethanol; the latter disrupts sulfur–sulfur linkages in proteins. The combined action of these reagents causes the proteins to unfold. Furthermore, for most proteins at an SDS concentration greater than 10^{-3} M, a nearly constant amount of SDS is bound per unit weight of protein (approximately 0.5 detergent molecule is bound per amino acid residue). Thus, the charge of the protein–dodecylsulfate complex is due primarily to the charges of the bound dodecylsulfate groups, making the charge per unit weight approximately the same for most proteins.

Under these conditions, the mobilities of the SDS-treated proteins are determined by their molecular weights. If M is the molecular weight of a protein and x is the distance migrated in the gel (proportional to the mobility), the relation

$$\log M = a - bx \tag{6.65}$$

is usually observed, where a and b are constants for a given gel at a given electric field. A set of proteins of known molecular weight must be used for calibration in the determination of a protein of unknown molecular weight. An example is shown in figure 6.19.

We emphasize that Eq. (6.65) is dependent on two factors: (1) a constant amount of bound detergent per unit weight of protein and (2) charges due to bound detergent dominate those carried by the protein itself. Deviations from this type of relation occur when a protein binds an abnormal amount of dodecylsulfate (such as glycoproteins) or carries a large number of charges (such as histones).*

---

*A *glycoprotein* is a protein that has oligosaccharide groups attached. *Histones* are highly positively charged proteins rich in lysine and arginine residues.

Protein Charge

For native proteins, the mobility depends on the net charge and the frictional coefficient. The net charge depends on the amino acid composition and the charges of any ligands bound covalently or reversibly. The net charge is always a function of pH; therefore, by judicious choice of pH, different proteins can be separated. All proteins will be positively charged at low enough pH because the carboxyl groups of aspartic and glutamic acid will be neutral (COOH), but the amino groups of lysine and arginine will be positive (NH_3^+). At high enough pH, all proteins will be negatively charged because now the carboxyls are negative (COO^-) and the aminos are neutral (NH_2). This means that at some pH each protein must be neutral. This pH is the *isoelectric point*—the pH at which the protein has zero mobility.

Isoelectric focusing (IEF) uses the different isoelectric points of proteins to provide an effective separation method. Buffers are used to establish a pH gradient in the gel, with high pH at the negative electrode and low pH at the positive electrode. We can start the sample on the high-pH side near the negative electrode. The negatively charged proteins will move toward the positive electrode until each one reaches its isoelectric pH and stops moving. Similarly, if we start the proteins at the low-pH side near the positive electrode, each will move to its isoelectric point. Combining IEF in one direction followed with SDS electrophoresis to separate by molecular weight at right angles to the first separation makes a very powerful analytical method. Figure 6.20 shows an example of the hundreds of different proteins that can be detected from a single cell.

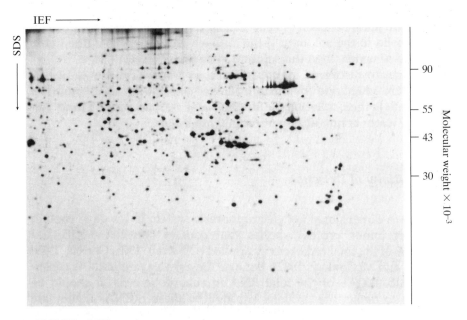

▲ FIGURE 6.20

A two-dimensional gel electrophoresis pattern of acidic proteins (isoelectric points are between pH 7.5 and 4.5), separated by isoelectric focusing (IEF) in the first dimension and SDS–PAGE in the second dimension. The proteins are radioactively labeled by growing cells on ^3H or ^{35}S media; the radioactivity is detected by placing an X-ray sensitive film on the gel. [From R. Bravo, *Two-Dimensional Gel Electrophoresis of Proteins* ed. J. D. Celis and R. Bravo (New York: Academic Press, 1984), Fig. 6, p. 22.]

Macromolecular Interactions

Gel electrophoresis can be used to study interactions between macromolecules. To determine the interaction between a protein and a nucleic acid, we do gel electrophoresis of the mixture. For example, to learn which restriction fragment of DNA binds to a protein, we compare the gel pattern of the ^{32}P-labeled DNA with and without protein. The fragment that is retarded in the presence of the protein is the one that is bound. Similar methods can be used to study the interaction of two proteins or two nucleic acids.

If the kinetics of the macromolecular reactions is slow enough so that re-equilibration does not occur during electrophoresis, equilibrium constants for reaction can be measured by gel electrophoresis. Mixtures are equilibrated before the mixture is applied to the gel; then electrophoresis is done to separate the individual species. If the electrophoresis does not change the equilibrium concentrations, the concentration of each species determined after separation determines the equilibrium constant. For example, the binding of an RNA enzyme—a ribozyme—to its substrate was measured by this method (Pyle, McSwiggen, and Cech 1990). It is well to emphasize that the gel electrophoresis method for measuring equilibrium binding is valid only if the kinetics of dissociation is slow. Otherwise, as the complex is moved away from its substrate, it will dissociate; the apparent equilibrium constant will be a complicated function of the rate of separation and the rate of reaction.

A careful study by gel electrophoresis of equilibria and kinetics of the *lac* repressor* protein binding to its operator DNA was done by Fried and Crothers (1981). A 203-bp operator sequence binds from one to eight repressor proteins with increasing concentration of protein. The DNA fragment and the complexes containing 1, 2, 3, . . . , up to 8 bound repressor molecules each migrate as a separate band in the polyacrylamide gel. At low protein concentrations, only free DNA and a complex with one repressor bound are seen. As the protein concentration increases, complexes with higher ratios of protein to DNA appear. The molecular weights of the complexes, estimated from the electrophoretic mobilities of the bands, are consistent with 1 DNA : n repressor ($n = 1, . . . , 8$). Kinetics of dissociation were measured by mixing excess unlabeled DNA with radioactively labeled DNA in a complex. At various times, the aliquots were analyzed by gel electrophoresis. The amount of radioactivity remaining in the complex was followed versus time to obtain the rate of dissociation.

Size and Shape of Macromolecules

Transport properties—diffusion, sedimentation, viscosity, and electrophoresis—tell us about sizes and shapes of macromolecules. Molecular weights can be obtained from sedimentation and diffusion, from sedimentation at equilibrium, and from gel electrophoresis. Properties for some representative macromolecules and particles are given in table 6.3. From the diffusion coefficient or the sedimentation coefficient, the frictional coefficient f can be calculated. We can use the Stokes equation ($f = 6\pi\eta r$) to calculate for com-

---

*A repressor is a protein that binds to a DNA sequence called an *operator* and prevents RNA polymerase from binding and synthesizing messenger RNA (mRNA). The *lac* repressor controls the synthesis in *E. coli* of the RNA that codes for enzymes involved in the metabolism of β-galactosides.

TABLE 6.3 Transport and Related Properties of Some Proteins

	$s_{20,w} \times 10^{13}$ (s)	$D_{20,w} \times 10^7$ (cm^2 s^{-1})	$\bar{v}_{20,w}$ (cm^3 g^{-1})	M	f/f_0
Ferricytochrome c (bovine heart)	1.91	13.20	0.707	12,744	1.077
Ribonuclease (bovine pancreas)	2.00	13.10	0.707	13,690	1.066
Myoglobin (horse heart)	2.04	11.30	0.741	16,951	1.105
Lysozyme (chicken egg white)	1.91	11.20	0.703	14,314	1.210
Chymotrypsinogen (bovine pancreas)	2.54	9.50	0.721	25,666	1.193
Immunoglobulin G (human)	6.6–7.2	4.00	0.739	156,000	1.513
Myosin	6.4	1.1	0.728	570,000	3.6
Tobacco mosaic virus	192	0.44	0.73	40,000,000	2.6

Sources: Selected from Cantor and Schimmel (1980) and from other sources.

parison the frictional coefficient f_0 for an unsolvated spherical molecule with molecular weight M and partial specific volume $\bar{v}_2$ that are the same as those of the macromolecule of interest. If f/f_0 is close to 1, we conclude that the macromolecule is approximately spherical and only lightly solvated. If f/f_0 is much greater than 1, the molecule is either highly solvated, asymmetric, or both. Additional information is provided by the intrinsic viscosity. The viscosity shape factor ν contains information that complements that obtained from the ratio f/f_0.

Since the transport properties are sensitive to the sizes and shapes of molecules, they can be used to study interactions between molecules and conformational changes of molecules. Some examples are interactions between subunits of a macromolecular structure (such as a multisubunit protein or a ribosome), conformational change of a protein due to the binding of a substrate, circularization of a linear DNA, unfolding of a protein chain, and antibody–antigen interactions.

More detailed knowledge on the structural features and functional aspects of macromolecules can be obtained only by other physical–chemical methods, some of which we will discuss in later chapters. The understanding of how molecules function in biological processes is usually achieved by a combination of many methods.

Summary
Kinetic Theory

Average translational kinetic energy:

$$\langle U_{\text{tr}} \rangle = \frac{3}{2} kT \tag{6.13}$$

$\langle U_{tr}\rangle =$ the average translational kinetic energy of an
atom in a molecule in any medium
$k =$ Boltzmann's constant
$T =$ absolute temperature

Mean-square velocity:

$$\langle u^2\rangle = \frac{3RT}{M} \qquad (6.14)$$

$\langle u^2\rangle \equiv$ average of the square of the velocities (m s^{-1}) of
molecules in a gas of molecular weight M (kg) at T K
$R =$ gas constant 8.314 J K^{-1} mol^{-1}
1 J $=$ 1 kg m^2 s^{-2}

Boltzmann most-probable distribution of energies:

$$P_i = \frac{N_i}{N} = \frac{g_i e^{-E_i/kT}}{\sum\limits_i g_i e^{-E_i/kT}} \qquad (6.17)$$

$P_i =$ the probability of finding a molecule with energy E_i and
degeneracy g_i
$N_i/N =$ the fraction of molecules with energy E_i and degeneracy g_i

The sum is over all possible energies.

Collision frequency:

$$z = 4\sqrt{\pi}\frac{N}{V}\sigma^2\left(\frac{RT}{M}\right)^{1/2} \qquad (6.26)$$

$z =$ number of collisions a molecule encounters per second
$\dfrac{N}{V} =$ number of molecules per unit volume
$\sigma =$ diameter of a molecule (each molecule is considered
to be a hard sphere)

Total number of collisions Z per unit volume per second:

$$Z = \left(\frac{N}{V}\right)\frac{z}{2} \qquad (6.27)$$

Mean free path:

$$l = \frac{1}{\sqrt{2}\pi(N/V)\sigma^2} \qquad (6.29)$$

Root-mean-square displacement:

$$\langle d^2\rangle^{1/2} = \sqrt{N}l \qquad (6.33)$$

where $\langle d^2\rangle =$ the root-mean-square displacement
after N random steps of length l

Diffusion

Fick's first law:

$$J_x = -D\left(\frac{\partial c}{\partial x}\right)_t \tag{6.37}$$

J_x = net amount of material that diffuses in the x-direction
   per second, across an area 1 cm^2 perpendicular to x

D = diffusion coefficient, which is directly related to the
   mean-square displacement per unit time; the units are cm^2 s^{-1}

$\left(\dfrac{\partial c}{\partial x}\right)_t$ = concentration gradient; it is the change of concentration with
   respect to x at a specified time t

Fick's second law:

$$\left(\frac{\partial c}{\partial t}\right)_x = D\left(\frac{\partial^2 c}{\partial x^2}\right)_t \tag{6.38}$$

$\left(\dfrac{\partial c}{\partial t}\right)_x$ = change of concentration versus time at a given position x

$\left(\dfrac{\partial^2 c}{\partial x^2}\right)_t$ = second derivative of the concentration c with respect to the
   position x, at a given time t

Diffusion coefficient D and frictional coefficient f:

$$D = \frac{kT}{f} \tag{6.43}$$

k = Boltzmann constant $\equiv R/N_0$
   = 1.380×10^{-16} g cm^2 s^{-2} K^{-1} molecule^{-1}
   = 1.380×10^{-23} J K^{-1} molecule^{-1}

Sedimentation

$$s = \frac{m(1 - \bar{v}_2\rho)}{f} \tag{6.48}$$

$s \equiv \dfrac{u_t}{\omega^2 x}$ = velocity of sedimentation per unit centrifugal acceleration

m = mass of the macromolecule
$\bar{v}_2$ = partial specific volume of the macromolecule
ρ = density of the solution
f = frictional coefficient

The dimension of s is seconds. A convenient unit for s, the svedberg (S), is
equal to 10^{-13} sec (1 S = 10^{-13} sec).

$$s_{20,w} = s\left(\frac{\eta}{\eta_{20,w}}\right)\frac{(1 - \bar{v}_2\rho)_{20,w}}{1 - \bar{v}_2\rho} \tag{6.50}$$

$s_{20,w} \equiv$ sedimentation coefficient corrected to give the expected value
   at 20°C in water

The subscript in $s_{20,w}$ refers to quantities in water at 20°C.

Frictional Coefficient and Molecular Parameters

The frictional coefficient f can be obtained from either s or D. For a sphere of radius r,

$$f = 6\pi\eta r \qquad (6.44)$$

η = viscosity of medium

For an unsolvated, spherical molecule of mass m and partial specific volume $\bar{v}_2$, its radius r is

$$r = \left(\text{volume}\cdot\frac{3}{4\pi}\right)^{1/3}$$

$$= \left(\frac{3m\bar{v}_2}{4\pi}\right)^{1/3}$$

The frictional coefficient f_0 for this unsolvated sphere is

$$f_0 = 6\pi\eta\left(\frac{3m\bar{v}_2}{4\pi}\right)^{1/3}$$

The deviation of f/f_0 from unity for a macromolecule can be due to either solvation or asymmetric shape.

Combination of Diffusion and Sedimentation

$$M = \frac{RTs}{D(1 - \bar{v}_2\rho)} \qquad (6.51)$$

Molecular weight M can be calculated from s, D, and the partial specific volume $\bar{v}_2$. ρ is the density of the solution.

Equilibrium centrifugation:

$$M = \frac{2RT}{\omega^2(1 - \bar{v}_2\rho)}\frac{d\ln c}{d(x^2)} \qquad (6.56)$$

$\omega = 2\pi\nu$ = angular speed, rad s^{-1} (ν = revolutions s^{-1})
c = concentration
x = distance from the center of rotation

When centrifuged to equilibrium, the molecular weight of a macromolecule can be calculated from the slope of a plot of $\ln c$ versus x^2.

Viscosity

$$J_{mu} = -\eta\frac{du_x}{dy} \qquad (6.57)$$

J_{mu} = rate of transfer of momentum in the x-direction per second across a cross-sectional area 1 cm^2 in the xz-plane

$\dfrac{du_x}{dy}$ = velocity gradient, or the rate of change of the x-component of the velocity with respect to y

η = viscosity coefficient

Newtonian fluid: η is independent of du_x/dy; non-Newtonian fluid: η is dependent on du_x/dy.

Solutions of macromolecules are often non-Newtonian, and viscosity measurements are usually done at several values of du_x/dy and extrapolated to $du_x/dy = 0$.

Specific viscosity:

$$\eta_{sp} = \frac{\eta' - \eta}{\eta} \tag{6.61}$$

$\eta' \equiv$ viscosity coefficient of a solution of a macromolecular species
$\eta \equiv$ viscosity coefficient of solvent

Intrinsic viscosity is the limiting value of η_{sp}/c as c approaches 0, where c is the concentration of macromolecules in g cm^{-3}:

$$[\eta] \equiv \lim_{c \to 0} \frac{\eta_{sp}}{c} \tag{6.62}$$

Relation between $[\eta]$ and molecular parameters:

$$[\eta] = \nu(\bar{v}_2 + \delta_1 v_1^0) \qquad \text{(rigid molecules)} \tag{6.63}$$

$\nu \equiv$ shape factor; $\nu = 2.5$ for a sphere
$\bar{v}_2 \equiv$ partial specific volume of the macromolecule
$\delta_1 \equiv$ grams of solvent molecules hydrodynamically associated with each gram of macromolecule
$v_1^0 \equiv$ specific volume of the solvent; for a dilute aqueous solution, $v_1^0 = 1$ cm^3 g^{-1}

Very flexible coils (random coils):

$$[\eta] \text{ approximately } \propto M^{1/2}$$

Electrophoresis
Electrophoretic mobility:

$$\mu = \frac{u}{E}$$

u = velocity of charged particle, cm s^{-1}
E = electric field strength, V cm^{-1}

$$\frac{ZeE}{f} \tag{6.64}$$

Z = number of charges on particle
f = frictional coefficient
e = electronic charge

Electrophoretic mobility $\equiv \mu/E$, or velocity per unit electric field.

Gel Electrophoresis
Molecular weight:

$$\log M = a - bx \tag{6.65}$$

M = molecular weight of protein
x = distance traveled in SDS gel (proportional to electrophoretic mobility)
a, b = parameters established by measuring reference proteins of known molecular weights

References

The following books are introductory texts similar in level to this chapter.

Andrews, A. T. 1986. *Electrophoresis: Theory, Techniques, and Biochemical and Clinical Applications.* Oxford, England: Oxford University Press.

Chang, R., 2000. *Physical Chemistry for the Chemical and Biological Sciences.* Sausalito, CA: University Science Books.

Harding, S. E., A. J. Rowe, and J. C. Horton, eds. 1992. *Analytical Ultracentrifugation in Biochemistry and Polymer Science.* Cambridge, England: Royal Society of Chemistry.

Schuster, T. M. A., and T. M. Laue, eds. 1994. *Modern Analytical Ultracentrifugation: Acquisition and Interpretation of Data for Biological and Synthetic Polymer Systems.* New York: Springer-Verlag.

Van Holde, K. E., W. C. Johnson, and P. S. Ho. 1998. *Principles of Physical Biochemistry.* Upper Saddle River, NJ: Prentice Hall.

The material in this chapter is covered at a more advanced level in

Cantor, C. R., and P. R. Schimmel. 1980. *Biophysical Chemistry.* Part II: *Techniques for the Study of Biological Structure and Function.* San Francisco: Freeman.

Celis, J. D., and R. Bravo, eds. 1984. *Two-Dimensional Gel Electrophoresis of Proteins.* Orlando, FL: Academic Press.

Suggested Reading

Axelrod, D., D. E. Koppel, J. Schlessinger, E. Elson, and W. W. Webb. 1976. Mobility Measurement by Analysis of Fluorescence Photobleaching Recovery Kinetics. *Biophys. J.* 16:1055–69.

Bauer, W. R., F. H. C. Crick, and J. H. White. 1980. Supercoiled DNA. *Sci. Am.* 243:118.

Carle, G. F., M. Frank, and M. V. Olson. 1986. Electrophoretic Separation of Large DNA Molecules by Periodic Inversion of the Electric Field. *Science* 232:65–68.

Hunkapiller, T., R. J. Kaiser, B. F. Koop, and L. Hood. 1991. Large-Scale and Automated DNA Sequence Determination. *Science* 254:59–67.

Fried, M., and D. M. Crothers. 1981. Equilibria and Kinetics of the *lac* Repressor-Operator Interactions by Polyacrylamide Gel Electrophoresis. *Nucleic Acids Research* 9:6505–25.

Jacques, J.-P., and M. M. Susskind. 1991. Use of Electrophoretic Mobility to Determine the Secondary Structure of a Small Antisense RNA. *Nucleic Acids Res.* 19:2971–77.

Jeffreys, A. J., V. Wilson, and S. L. Thein. 1986. Individual-Specific "Fingerprints" of Human DNA. *Nature* 316:76–79.

Kheterpal, I., and R. A. Mathies. 1999. Capillary Array Electrophoresis DNA Sequencing. *Analytical Chemistry* 71:31A–37A.

Leng, M. 1990. DNA Bending Induced by Covalently Bound Drugs: Gel Electrophoresis and Chemical Probe Studies. *Biophys. Chem.* 35:155–63.

Liu, L. 1989. DNA Topoisomerase Poisons as Anticancer Drugs. *Annu. Rev. Biochem.* 58:351–75.

Pyle, A. M., J. A. McSwiggen, and T. R. Cech. 1990. Direct Measurement of Oligonucleotide Substrate Binding to Wild-Type and Mutant Ribozymes from *Tetrahymena. Proc. Natl. Acad. Sci. USA* 87:8187–91.

Righetti, P. G. 1990. Recent Developments in Electrophoretic Methods. *J. Chromatography* 516:3–22.

Schmidt, Th., G. J. Schütz, W. Baumgartner, H. J. Gruber, and H. Schindler. 1995. Characterization of Photophysics and Mobility of Single Molecules in a Fluid Lipid Membrane. *J. Phys. Chem.* 99:17662–8.

Schmidt, Th., G. J. Schütz, W. Baumgartner, H. J. Gruber, and H. Schindler. 1996. Imaging of single molecule diffusion. *Proc. Natl. Acd. Sci. USA* 93, 292–9.

Schwartz, D. C., and C. R. Cantor. 1984. Separation of Yeast Chromosome-Sized DNAs by Pulsed-Field Gradient Gel Electrophoresis. *Cell* 37:67–75.

Smith, S. B., C. Heller, and C. Bustamante. 1991. Model and Computer Simulations of the Motion of DNA Molecules During Pulse-Field Gel Electrophoresis. *Biochemistry* 30:5264–74.

Problems

1. The collisional diameter σ of a H_2 molecule is about 2.5 Å, or 2.5×10^{-8} cm. For H_2 gas at 0°C and 1 atm, calculate the following:

   (a) The root-mean-square velocity

   (b) The translational kinetic energy of 1 mol of H_2 molecules

   (c) The number of H_2 molecules in 1 cm^3 of the gas

   (d) The mean free path

   (e) The number of collisions each H_2 molecule encounters in 1 s

   (f) The total number of intermolecular collisions in 1 s in 1 cm^3 of the gas

2. A proton in a magnetic field can be in only one of two energy levels; each level has a degeneracy of 1. The energy spacing ΔE corresponds to a transition frequency ν of 100 MHz, using the equation $\Delta E = h\nu$,

with h = Planck's constant. Use the Boltzmann distribution to calculate the absolute temperature when the following occur:

(a) The proton is in the lower state only.

(b) The probability of finding the proton in either level is equal.

(c) The probability of finding the proton in the lower level is 1.000015 times the probability of finding it in the upper level. (Remember that $e^x = 1 + x + \cdots$ when x is small.)

3. The following data were reported for human immunoglobulin G (IgG) at 20°C in a dilute aqueous buffer:

$$M = 156{,}000$$
$$D_{20,w} = 4.0 \times 10^{-7} \text{ cm}^2 \text{ s}^{-1}$$
$$\bar{v}_2 = 0.739 \text{ cm}^3 \text{ g}^{-1}$$

(a) Use the diffusion coefficient to calculate f, the frictional coefficient.

(b) Use the Stokes equation to calculate f_0, the frictional coefficient of an unhydrated sphere of the same M and $\bar{v}_2$ as IgG. The volume of a sphere is $\frac{4}{3}\pi r^3$.

(c) If IgG is not significantly hydrated, use figure 6.8 to estimate the dimensions of a prolate ellipsoid that best fit the data.

4. Diffusion, sedimentation, and electrophoresis can be used to characterize, identify, or separate macromolecules.

(a) Describe one experimental method to measure a diffusion coefficient, one to measure a sedimentation coefficient, and one to measure an electrophoretic mobility. For each method, state what is measured and how it is used to obtain the desired parameter.

(b) Describe what you could learn about a protein by applying the three methods.

(c) Describe what you could learn about a nucleic acid by applying the three methods.

5. The sedimentation coefficient of a certain DNA in 1 M NaCl at 20°C was measured by boundary sedimentation at 24,630 rpm. The following data were recorded:

Time, t (min)	Distance of boundary from center of rotation, x (cm)
16	6.2687
32	6.3507
48	6.4380
64	6.5174
80	6.6047
96	6.6814

(a) Plot log x versus t. Calculate the sedimentation coefficient s.

(b) The partial specific volume of the sodium salt of DNA is 0.556 cm³ g⁻¹. The viscosity and density of 1 M NaCl and the viscosity of water can be found in the International Critical Tables; the values are 1.104 cP, 1.04 g cm⁻³, and 1.005 cP, respectively. Calculate $s_{20,w}$ for the DNA.

6. For a bacteriophage T7, the following data at zero concentration have been obtained [(Dubin et al. 1970, J. Mol. Biol. 54:547)]:

$$s^0_{20,w} = 453 \text{ S}$$
$$D^0_{20,w} = 6.03 \times 10^{-8} \text{ cm}^2 \text{ s}^{-1}$$
$$\bar{v}_2 = 0.639 \text{ cm}^3 \text{ g}^{-1}$$

(a) Calculate the molecular weight of the bacteriophage.

(b) Phosphorus and nitrogen analyses of the bacteriophage show that 51.2% by weight of the bacteriophage is DNA. Calculate the molecular weight of T7 DNA. Each bacteriophage contains one DNA molecule.

7. Two *spherical* viruses of molecular weights M_1 and M_2, respectively, happen to have the same partial specific volume $\bar{v}_2$. Neglecting hydration, what are the expected values for the ratios s_2/s_1, D_2/D_1, and $[\eta]_2/[\eta]_1$, where s_1, D_1, and $[\eta]_1$ are the sedimentation coefficient, diffusion coefficient, and intrinsic viscosity, respectively, of the virus of molecular weight M_1, and s_2, D_2, and $[\eta]_2$ are the corresponding parameters of the virus of molecular weight M_2?

8. For a rodlike particle with length L and diameter d, its hydrodynamic properties are similar to those of a prolate ellipsoid of the same length and volume. Show that

$$\frac{L}{d} = \left(\frac{3}{2}\right)^{1/2}\frac{a}{b}$$

where a and b are the long and short semiaxes of the ellipsoid, respectively.

9. The following questions can be answered by the application of the equations discussed in this chapter.

(a) Some lipoproteins sediment in a centrifugal field and others float. What is the partial specific volume of a lipoprotein that neither sinks nor floats in a solution of density 1.125 g cm⁻³?

(b) Calculate the sedimentation coefficient in seconds of a parachutist who is falling toward Earth at a velocity of 2 m s⁻¹ (about 4.5 miles per hour).

(c) The intrinsic viscosity of a solution of spherical viruses is 1.5 cm³ g⁻¹. The molecular weight is 5×10^6 g mol⁻¹. What is the volume of each solvated virus particle?

(d) A mutant protein with molecular weight 100,000 differs from the normal protein by the replacement of an alanine amino acid (side chain —CH₃) by an aspartic acid (side chain —CH₂COO⁻). This small change has a negligible effect on the molecular weight and frictional coefficient. How would you separate the two proteins from each other?

(e) The rate of a gas-phase reaction is directly proportional to the total number of collisions per second that occur in the container. State by what factor the rate of the reaction increases when the concentration of molecules doubles; when the collision cross section (σ^2) doubles; when the absolute temperature doubles; when the molecular weight doubles.

10. Nucleosome particles are composed of DNA wrapped around a "core" of protein. Nucleosome particles were analyzed using ultracentrifugation and dynamic light scattering. In a solution at 20°C, the diffusion constant was measured to be 4.37×10^{-7} cm^2 s^{-1}. In the same solution, boundary centrifugation was done with angular speed of rotation of $\omega = 2\pi(18,100)$ radians per minute to obtain the following data for the position of the boundary measured from the center of rotation.

Time (minutes)	Boundary position (cm)
0	4.460
80	4.593
160	4.713
240	4.844

The density of the solvent was 1.02 g cm^{-3}, and the partial specific volume of the nucleosome was determined to be 0.66 cm^3 g^{-1}.

(a) What is the molecular weight of a nucleosome particle?

The following gel pattern was found when the DNA in nucleosomes was cut with an enzyme, all protein was removed, and then gel electrophoresis of the DNA was done.

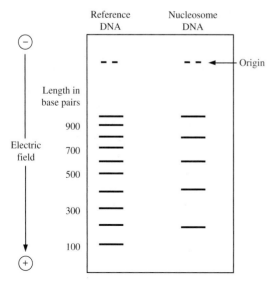

We can assume that the shortest fragment found corresponds to a piece of DNA that had been wrapped around a single nucleosome core, and longer fragments came from DNA associated with 2, 3, 4, ..., n nucleosome particles. One DNA base pair has molecular weight 660 g mol^{-1}.

(b) How many DNA base pairs are wrapped around each protein core?

(c) What is the molecular weight of a single protein in the nucleosome if (as other evidence suggests) this "nucleosome core" is made up of eight protein molecules that bind together, each of about the same molecular weight?

11. Use the data for lysozyme given in table 6.3 to answer the following questions:

(a) Calculate the frictional coefficient in g s^{-1}, f, and the frictional coefficient, f_0, if lysozyme were an unhydrated sphere.

(b) Under certain conditions, lysozyme dimerizes. If the ratio $f_{dimer}/f_{monomer} = 1.6$, calculate $s_{20,w}$ for the lysozyme dimer.

12. Consider a mixture of two proteins with molecular weights of 20,000 and 200,000. For simplicity of calculation, both may be approximated as unhydrated spheres with $\bar{v}_2 = 0.740$ cm^3 g^{-1}.

(a) Calculate the ratio of the sedimentation coefficients of the proteins in the same medium (density = 1.05 g cm^{-3}) at the same temperature.

(b) Consider the following experiments. A mixture of the two proteins is placed on top of a centrifuge tube filled with a dilute aqueous buffer containing a linear gradient of sucrose ranging from 5% at the top to 20% at the bottom, and the tube is spun in a centrifuge, as illustrated in the following diagram. The top and bottom of the liquid column are 4.0 and 8.0 cm from the axis of rotation, respectively. When the larger protein has sedimented a distance of 3.0 cm, how far has the smaller protein traveled? [Use your answer to part (a) in the calculation and neglect the viscosity and density gradients that are associated with the sucrose concentration gradient.] Based on your calculation, is this an effective method of separation for the two proteins?

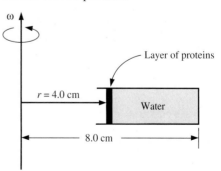

(c) As the proteins are sedimenting, the initial concentrated thin layer of protein molecules will spread out due to diffusion. If the distance over which the protein molecules spread due to diffusion is much greater than the distance separating the two proteins, this method will not be useful as a separation approach since the protein layers will overlap. To see if this might be a problem, calculate the average distances the two proteins move due to diffusion at 298 K if the sedimentation experiment lasts 12 h. Take the average viscosity of the medium as 1.5 cp. Is the "spreading distance" due to diffusion significant compared to the separation of the proteins due to sedimentation?

(d) At sedimentation equilibrium, do you expect to find a gradual variation in the concentration of protein along r near the bottom of the tube?

13. Consider a small spherical molecule of radius 2 Å (2×10^{-8} cm).

(a) At 298 K, what is the average time required for such a molecule to diffuse across a phospholipid bilayer 40 Å thick? The viscosity of the bilayer interior is about 0.10 P.

(b) At a constant temperature, the average time to diffuse across a bilayer hydrophobic interior depends only on the radius of the diffusing molecule. Hence, all molecules with the same size will cross in the same time. However, it is well known that for the same concentration gradient the flux of nonpolar molecules across bilayers is much greater than that for polar molecules of the same size. Are these statements contradictory? Explain your answer.

14. The pK_a values of various groups in proteins are tabulated as follows:

		pK_a
	α-COOH	~2
	α-NH$_3^+$	~9
Side chain	—COOH	~4
Side chain	—NH$_3^+$	~11
	Histidine	~6

A biochemist studying the properties of the enzyme hexokinase isolates and purifies two different mutants (I and II) for this enzyme from bacteria. She then performs isoelectric focusing and runs an SDS–gel electrophoresis for these mutant enzymes as well as for the normal enzyme. Her gels after protein staining are:

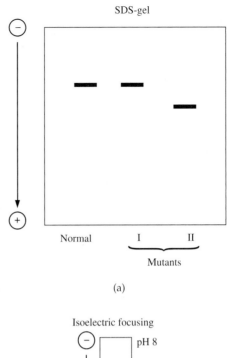

(a)

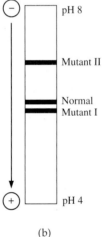

(b)

(a) On the basis of the gel results, what can you say about the possible kinds of changes in the amino acid composition of these two mutants? Explain briefly.

(b) You have a small unidentified peptide. You determine its molecular weight by SDS–gel electrophoresis to be 3000 and its isoelectric pH to be 10. With a specific protease, you cleave a *single* peptide bond in your protein. You find that your cleaved peptide still has approximately the same molecular weight on an SDS–gel electrophoresis, but its isoelectric point is reduced to about 8. How would you explain your results?

15. For prolate ellipsoids of large axial ratios—say, with $a/b > 20$—it can be shown that the viscosity shape factor $\nu \approx 0.207(a/b)^{1.723}$. A certain rodlike macromole-

cule is believed to form an end-to-end dimer. Show that the intrinsic viscosity of the dimer is expected to be 3.32 times that of the monomer.

16. The DNA from an animal virus, polyoma, has a sedimentation coefficient $s_{20,w}$ of 20 S. Digesting the DNA very briefly with pancreatic DNase I, an enzyme that introduces single-chain breaks into a double-stranded DNA, converts it to a species sedimenting at 16 S. This reduction in s could be due to either a reduction in molecular weight or a conformational change of the DNA (so that its frictional coefficient is increased). How would you design an experiment to decide between these possibilities?

17. The following data have been obtained for human serum albumin:

$$s_{20,w} = 4.6\,\text{S}$$

$$D_{20,w} = 6.1 \times 10^{-7}\,\text{cm}^2\,\text{s}^{-1}$$

$$[\eta] = 4.2\,\text{cm}^3\,\text{g}^{-1}$$

$$\bar{v}_2 = 0.733\,\text{cm}^3\,\text{g}^{-1}$$

Calculate the molecular weight of this protein.

18. (a) Proteins A and B with molecular weight of 16,500 and 35,400 move 4.60 cm and 1.30 cm, respectively, during electrophoresis and in SDS polyacrylamide gel. What is the molecular weight of protein C that moves 2.80 cm under the same conditions in the same gel?

(b) The sedimentation coefficient of a DNA molecule is 22.0×10^{-13} sec. A dilute solution of the DNA is spun in a centrifuge at 40,000 rpm starting from a distance of 6.0 cm from the axis of rotation. How far will the DNA move in 20 min?

19. In 6 M guanidine hydrochloride and in the presence of 2-mercaptoethanol, it is generally believed that complete unfolding of proteins occurs. Could you test whether this is true by viscosity measurements? Give a brief and concise discussion. Some experimental data are listed below in the following table (taken from C. Tanford, 1968, *Adv. Protein Chem.* 23:121):

20. The O_2-carrying protein, hemoglobin, contains a total of four polypeptide chains: two α chains and two β chains per molecule. The hemoglobin of a certain person from Boston, designated hemoglobin M Boston, differs from normal hemoglobin (hemoglobin A) in that a histidine residue in each of the α chains of hemoglobin A is substituted by a tyrosine residue. From the ionization constants given in table 4.2, do you expect the electrophoretic mobilities at pH 7 of the two proteins, hemoglobin M Boston and hemoglobin A, to differ? Give your reasons. What would be a reasonable pH to use to separate the two by electrophoresis? Which is more negatively charged at this pH?

21. The protein β-lactalbumin (molecular weight = 14,000) has been studied under different solution conditions by dynamic light scattering. At pH 7 and 40°C in a dilute solution, the diffusion constant was found to be $14.25 \times 10^{-7}\,\text{cm}^2\,\text{s}^{-1}$.

(a) If the viscosity of the solvent is 0.0101 P, estimate the diameter of the protein, assuming that it is spherical in shape.

(b) At pH <2, this protein becomes inactive and changes spectral characteristics. Under these conditions, the diffusion constant was found to be $12.80 \times 10^{-7}\,\text{cm}^2\,\text{s}^{-1}$. Calculate the change in volume of the protein (assume all else remained the same).

(c) At intermediate pH, there is an equilibrium between these two forms of the protein. Calculate the ratio of sedimentation coefficients ($s_{\text{high pH}}/s_{\text{low pH}}$) for these two forms. Assume the solvent has a density of $1.04\,\text{g cm}^{-3}$.

22. A protein involved in light harvesting for photosynthesis in bacteria was investigated using sedimentation equilibrium and SDS–gel electrophoresis. Two centrifuge experiments were done, the first in 1 M NaCl solution and the second in 1 M NaSCN. The samples were spun at 20°C at 21,380 rpm until an equilibrium concentration gradient was established. The partial specific volume of the protein $\bar{v}_2$ was found to be 0.709 cm^{-3} g; the density ρ for the salt solutions was 1.20 g cm^3. Ignore any possible effects of density gradients that are established due to the salt in the so-

◄ PROBLEM 19
Data

Protein	M^* molecular weight	In dilute aqueous buffer	In 2 M guanidine · HCl and in the presence of 2-mercaptoethanol
		$[\eta]$ (cm^3 g^{-1})	
Ribonuclease (cow)	13,690	3.3	16.6
Myoglobin (horse)	17,568	3.1	20.9
Chymotrypsinogen (cow)	25,666	2.5	26.8
Serum albumin (cow)	66,296	3.7	52.2

*These proteins have only one polypeptide chain per molecule; therefore, the molecular weight does not change upon unfolding of the protein.

lutions. The concentration profile was determined and used to generate the following graph; c = protein concentration at position x in the centrifuge cell. Below is a drawing of the denaturing SDS–gel electrophoresis.

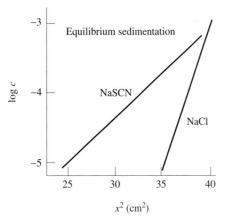

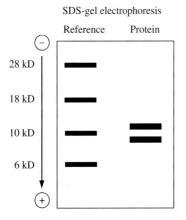

SDS-gel electrophoresis

(a)   From the centrifuge data, calculate the effective molecular weight of the light-harvesting protein in each of the two salt solutions.

(b)   From the SDS–gel electrophoresis data, estimate the molecular weights of the two polypeptides that make up the light-harvesting protein.

(c)   How can you explain the two sets of data? Give a concise explanation of what each observation indicates and how the observations fit together.

23.   The enzyme aspartate transcarbamylase (ATCase) has a molecular weight of 310 kD and undergoes a change in shape upon binding of substrate or inhibitors. This change has been characterized by ultracentrifugation. In the absence of any ligands, a sedimentation coefficient of $s_{20,w}$ = 11.70 S was measured. The solvent density was 1.00 g cm^3, the viscosity was 1.005 cP, and the specific volume of the protein was 0.732 $cm^3\ g^{-1}$.

(a)   Find the radius of the protein in Å, assuming that it is a sphere.

(b)   Upon adding a ligand that bound to the enzyme, the sedimentation coefficient increased by 3.5%. What is the radius of the ligated enzyme?

(c)   From X-ray diffraction, it was shown that the presence of ligands causes a contraction of the enzyme of about 12 Å along one axis. Why did the radius not decrease by 6 Å? (Explain briefly.)

24.   T4 is a large bacterial virus. The virus has a symmetric (approximately spherical) head group, which contains DNA. These head particles were studied and were found to have the following parameters: $s_{20,w}$ = 1025 S; $D_{20,w}$ = 3.60 × 10^{-8} cm^2 sec^{-1}; $\bar{v}_2$ = 0.605 cm^3 g^{-1} at 20°C. Calculate the following:

(a)   The molecular weight of the head group

(b)   The volume in cm^3 of the head group from $\bar{v}_2$ (this assumes no hydration)

(c)   The frictional coefficient of the head group from the diffusion coefficient

(d)   The volume of the head group from the frictional coefficient in part (c) and the Stokes equation. The difference between the calculated volume in part (b) and the measured volume in part (d) is caused by hydration of the head group.

25.   A small spherical virus has a molecular weight of 1.25 × 10^6 g mol^{-1} and a diameter of 100 Å; it is not significantly hydrated. Calculate the intrinsic viscosity in units of cm^3 g^{-1} of an aqueous solution of this virus.

26.   For each of the following changes, state whether the sedimentation coefficient of the particles will increase, decrease, remain the same, or whether it is impossible to tell. Also give an equation or a one- or two-sentence explanation that supports your answer.

(a)   The temperature of the solvent is increased from 20°C to 30°C.

(b)   The long axis of the particle (which is a prolate ellipsoid) is cut in half.

(c) ^{15}N is substituted for ^{14}N in the particle.

(d)   Provide answers for the diffusion coefficient of the particle in parts (a)–(c).

27.   An important step in the blood coagulation process is the enzymatic conversion of the protein prothrombin to the lower-molecular-weight clotting agent, thrombin, via cleavage of a portion of the prothrombin polypeptide.

prothrombin $\longrightarrow$ thrombin + cleaved peptide

The values of $s_{20,w}$ and $D_{20,w}$ for prothrombin have been found to be 4.85 × 10^{-13} sec and 6.24 × 10^{-7} cm^2 sec^{-1}, respectively. The diffusion coefficient for thrombin is 8.76 × 10^{-7} cm^2 sec^{-1}. Assume that both thrombin and prothrombin are unhydrated spheres with $\bar{v}_2 = 0.70$ cm^3 g^{-1} and calculate the molecular weights of prothrombin, thrombin, and the cleaved peptide.

28. (a) Gel electrophoresis is used to determine the sequences of DNA molecules. Describe how this is done. In particular, describe how the positions of bands on a gel are related to the sequence of the DNA.

    (b) Gel electrophoresis is used to determine molecular weights of proteins. Describe how this is done and mention the limitations to this method.

    (c) Describe one other method that could be used to determine the molecular weight of a protein. State what is measured and how this is related to molecular weight.

    (d) Describe how an equilibrium-binding constant can be measured by gel electrophoresis.

29. Three transport properties of proteins are easy to measure: s (sedimentation coefficient), D (diffusion coefficient), and μ (electrophoretic mobility). State what happens to each of these three quantities for the following changes. Be quantitative and give equations to justify your answers.

    (a) The frictional coefficient of the protein doubles, but the molecular weight is constant.

    (b) The charge on the protein doubles, but nothing else changes.

    (c) The protein changes from a prolate ellipsoid to an oblate ellipsoid of the same axial ratio; nothing else changes.

    (d) The amount of hydration of the protein doubles.

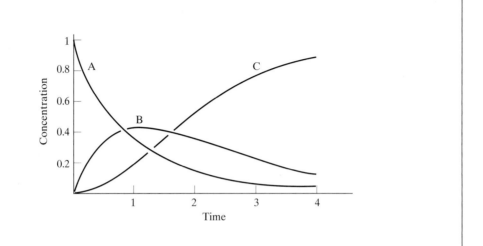

CHAPTER 7

Kinetics: Rates of Chemical Reactions

Knowing how fast molecules move and how often they collide, we can begin to learn how fast they react. Thermodynamics tells us whether a reaction can occur; kinetics tells us when it will occur—in a hundred years or in a millisecond. Kinetics helps us to influence rates—to speed them and to slow them.

Concepts

A chemical reaction may occur when two molecules collide. For the reaction to actually occur, the molecules must have enough energy to break the covalent bonds of the reactants and to form the new bonds of the products. Thus, increasing the concentrations of reactants and increasing the temperature nearly always increase the rate of a chemical reaction. The higher concentrations mean more frequent collisions; the higher temperature provides more energy per collision and also more collisions per second. However, not all reaction rates increase with increasing temperature and concentration. We need to understand all of this.

Chemical kinetics includes two distinct parts: experiments and theory. First, we must do experiments to measure the dependence of the rate on concentrations of reactants and, possibly, of products. (There is often product inhibition of the rate, and there may be catalysis by the product.) We change the solvent environment. Concentrations of molecules not directly involved in the reaction can have a large effect on the rate. Added salts, metal ions, [H$^+$], and so forth can act as catalysts or inhibitors of the reaction. We determine the effect of temperature and any other pertinent variables. The rate may depend on whether the reaction is done in a glass container or a plastic container. It may depend on whether it is done in a dark room or in sunlight. Joseph Priestley accidentally learned that plants produce oxygen only in sunlight, not in the dark. We have since learned much more about photosynthesis and photochemical reactions in general. The important lesson is that many things may have large effects on the rate of a reaction. A spark introduced into a container of H_2 and O_2 gas changes the rate of formation of H_2O by more than 10 orders of magnitude: A stable gas mixture is converted into a bomb.

Once we have determined the effects of many variables on the rate, the next step is to explain the effects in terms of what the molecules are doing. The theoretical explanation is called a *mechanism* of the reaction. A mechanism is one or more equations showing which molecules react and which bonds break or form. The equations represent elementary reactions. Elementary reactions involve one or two molecules, at most three molecules, that collide and react. The mechanism is distinct from the *stoichiometric reaction*. The stoichiometry tells us the number of moles of reactants that produce the number of moles of products. The stoichiometry tells us nothing about the dependence of rate on concentration; the mechanism is an explanation of the experimental rate data. For a chemical reaction, there is only one stoichiometric reaction; however, many mechanisms are consistent with the rate data. Each mechanism is a hypothesis to explain the kinetic data; the mechanism is a guide to design further experiments to better understand the results.

An important concept in understanding the mechanism of a reaction and the temperature dependence of a rate is the *transition state*. For a reaction in which one species is converted to another, the structure of the transition state is assumed to be intermediate between the two species. It is an unstable structure (*unstable* means that it exists only for the time of a molecular vibration) that quickly transforms to either stable species. The transition state corresponds to an energy maximum between two stable species that exist in energy minima. The *transition state energy* is the energy difference between the transition state and a reactant; its magnitude characterizes the temperature dependence of the rate of the reaction. A small transition state energy means that little energy is required to transform reactant to product and the rate does not depend much on temperature. A large transition state energy produces a large temperature dependence of the rate.

Applications

Kinetics—the study and understanding of the rate of change of anything—is useful in all areas of medicine, biology, and biochemistry. The effect of food supply, predators, and climate on the number of animals in a population follows the same differential equations used for treating molecular reactions in the atmosphere or in a single biological cell. Bacterial growth rates, radioactive decay, biosynthesis of deoxynucleotides, viral infectivity, and antibiotic cures all depend on the rates of reactions. We will learn how to measure rates of reaction and to study the variables that affect the rates. For some purposes, this may be enough. For example, if we use an enzyme as a catalyst, it may be sufficient to assay its activity every day. When the activity gets too low, we throw out the old solution and make a new batch. The rate of denaturation of the enzyme is important, but the mechanism may be too complicated to understand. We are satisfied with the empirical rate data.

For most reactions, we do want to understand what is happening. Why do some reaction rates increase by a factor of 2 for a 10-degree rise in temperature, but some do not change or even decrease? When should a salt not directly involved in the reaction affect the rate? The energy and structure of the transition state is important in answering these questions. It may be difficult to generalize the concept of a mechanism and a transition state to the complex rates involved in predator–prey dependence. However, the idea is to propose a hypothesis that explains the data. The test of the hypothesis is how well it explains other data for different animals under different conditions.

Kinetics

Chemical kinetics is the study of rates of reactions. Some reactions, such as that between hydrogen gas and oxygen gas in an undisturbed clean flask, occur so slowly as to be unmeasurable. Radioisotopes of some nuclei have very long lifetimes; the carbon isotope $^{14}_{6}C$ decays so slowly that half of the initial amount is still present after 5770 years. Other radionuclei have half-lives that are orders of magnitude longer than this. Processes such as the growth of bacterial cells are slow but easily measurable. The rate of reaction between H^+ and OH^- to form water is so fast that its study requires special techniques, such as temperature-jump kinetics. The frontier is expanding to include still faster processes, and reaction times of fractions of a picosecond (1 ps = 10^{-12} s) are currently being studied. Nuclear reactions involving species with lifetimes shorter than 10^{-20} s are known. Clearly, the methods of observation are very different to include processes over such an enormous range of time.

The first chemical reaction rate studied quantitatively involved a compound of biological origin. In 1850 L. Wilhelmy reported that the hydrolysis of a solution of sucrose to glucose and fructose occurred at a rate that decreased steadily with time but always remained proportional to the concentration of sucrose remaining in the solution. He followed the reaction indirectly by measuring the change with time of the optical rotation—the rotation of the plane of polarization of light passing through the solution. The phenomenon of optical rotation results from the molecular chirality of sugars such as sucrose, glucose, and fructose; *chirality* (handedness) refers to the fact that these molecules are structurally (and chemically) distinct from their mirror-image molecules (see chapter 10). Subsequent work demonstrated that systems at equilibrium are not static but are undergoing transformations between reactants and products in both directions and at equal rates.

The role of *catalysts,* substances that increase the rates of reactions without themselves being consumed, was recognized early from the influence of hydrogen ion on the rate of sucrose hydrolysis. The early "ferments" used to convert sugars from grain or grapes into beverages contained enzyme catalysts that greatly sped up the rates of these processes. Biological organisms contain thousands of different enzymes—protein molecules that selectively catalyze all the reactions in living cells. In the hydrogen–oxygen gas mixture, the introduction of a trace of finely divided platinum catalyst leads to a violent explosion. Because catalysts have no effect on the position of equilibrium, we can conclude that the hydrogen–oxygen mixture in the absence of a catalyst is not at equilibrium.

The ability of increasing temperature to speed most chemical reactions was put on a quantitative basis by Arrhenius (1889). The mixture of hydrogen and oxygen, which is stable indefinitely at room temperature, explodes on heating it to a temperatures in excess of 400°C. In other cases, such as the processes that occur in biological cells, an increase in temperature causes the enzyme-catalyzed reactions to cease altogether.

Light or electromagnetic radiation serves as a "reagent" in some biological processes that are essential to our survival. Photosynthesis and vision are just two of the most obvious examples. Other processes that are not chemical in nature, such as the nuclear reactions that provide the energy source of the Sun and have been the major source of heat within Earth, can nevertheless be described using the methods of chemical kinetics. Population dynamics, ecological changes and balance, atmospheric pollution, and biological waste disposal are just a few of the relatively new applications of this powerful approach.

The methods of chemical kinetics or reaction-rate analysis were developed for the resolution and understanding of the relatively simple systems encountered by chemists. These approaches are also valuable in analyzing the much more complex processes of biology. The reason that the methods work is often because one or a few steps control the rate of an extensive chain of reactions. All the steps involved in metabolism—cell division and replication, muscular contraction, and so on—are subject to the same basic principles, as are the elementary reactions of the chemist.

The rate or velocity, v, of a reaction or process describes how fast it occurs. Usually, the rate is expressed as a change in concentration per unit time,

$$v = \frac{dc}{dt} = \text{rate of reaction}$$

but it may alternatively express the change of a population of cells with time, the increase or decrease in the pressure of a gas with time, or a change in the absorption of light by a colored solution with time. In general, the rate of a process depends in some way on the concentrations or amounts involved;

the rate is a function of the concentrations. This relation is known as the *rate law:*

$$v = f(\text{concentrations})$$

The rate law may be simple (v = constant, for example) or complex, but it gives important information about the mechanism of the process. One of the main objectives of research in kinetics is the determination of the rate law.

Rate Law

Substances that influence the rate of a reaction can be grouped into two categories:

1. Those whose concentration changes with time during the course of the reaction:

   Reactants—decrease with time.

   Products—increase with time.

   Intermediates—increase and then decrease during the course of the reaction. An example is substance C in the following two-step reaction:

   $$A \longrightarrow C \longrightarrow B$$

2. Those whose concentrations do not change with time:

   Catalysts (both promoters and inhibitors), including enzymes and active surfaces.

   Intermediates in a steady-state process, including reactions under flowing conditions.

   Components that are buffered by means of equilibrium with large reservoirs.

   Solvents and the environment in general.

These influences do not change during a single run, but they can be changed from one experiment to the next. The concentrations of these components frequently do influence the rates of reactions.

Order of a Reaction

It is important in kinetics to learn immediately the vocabulary that kineticists use. We need to distinguish the stoichiometric reaction, the order of the reaction, and the mechanism of the reaction. It is essential to understand these terms. The *stoichiometry* of the reaction describes how many moles of each reactant are needed to form each mole of products. Only ratios of moles are significant. For example,

$$H_2 + \frac{1}{2}O_2 = H_2O$$

$$2H_2 + O_2 = 2H_2O$$

are both correct stoichiometric reactions. The *mechanism* of a reaction describes how the molecules react to form products. The mechanism is, in general, a set of *elementary reactions* consistent with the stoichiometric reaction.

For the reaction of H_2 and O_2 in the gas phase, the reaction is probably a chain involving H, O, and OH radicals:

$$H_2 \longrightarrow 2H$$

$$H + O_2 \longrightarrow OH + O$$

$$OH + H_2 \longrightarrow H_2O + H$$

$$O + H_2 \longrightarrow OH + H$$

Each step in the mechanism describes an elementary reaction; the four reactions constitute the proposed mechanism.

The kinetic *order* of a reaction describes the way in which the rate of the reaction depends on the concentration. Consider a reaction whose stoichiometry is

$$A + B \longrightarrow P$$

For many such reactions, the rate law is of the form

$$v = kc_A^m c_B^n c_P^q \qquad (7.1)$$

where the concentrations c_A, c_B, and c_P, are raised to powers m, n, and q, that are usually integers or zero ($c_A^0 = 1$) but may be nonintegral as well. The order of the reaction with respect to a particular component A, B, and P, is just the exponent of its concentration. Because the rate may depend on the concentrations of several species, we need to distinguish between the order with respect to a particular component and the overall order, which is the sum of the exponents of all components. Some representative examples are listed in table 7.1.

If the concentration of a component is unchanged during the course of the reaction, it is frequently omitted in the rate-law expression. A more complete rate law for the first reaction tabulated in table 7.1 is

$$v = k'[\text{sucrose}][H^+][H_2O]$$

TABLE 7.1 Rate Laws and Kinetic Order for Some Reactions

Stoichiometric reaction	Rate law	Kinetic order
Sucrose + H_2O $\xrightarrow{H^+}$ fructose + glucose	$v = k[\text{sucrose}]$	1
L-Isoleucine $\rightarrow$ D-isoleucine	$v = k[\text{L-isoleucine}]$	1
$^{14}_6C \rightarrow {}^{14}_7N + \beta^-$	$v = k[^{14}_6C]$	1
2Proflavin $\rightarrow$ proflavin dimer	$v = k[\text{proflavin}]^2$	2
p-Nitrophenyl acetate + $2OH^-$ $\xrightarrow{\text{pH 9}}$ p-nitrophenolate$^-$ + acetate$^-$ + H_2O	$v = k[p\text{-nitrophenyl acetate}][OH^-]$	2 (overall)
Hemoglobin $\cdot 3O_2 + O_2 \rightarrow Hb \cdot 4O_2$	$v = k[Hb \cdot 3O_2][O_2]$	2 (overall)
$H_2 + I_2 \rightarrow 2HI$	$v = k[H_2][I_2]$	2 (overall)
$H_2 + Br_2 \rightarrow 2HBr$	$v = \dfrac{k[H_2][Br_2]^{1/2}}{k' + [HBr]/[Br_2]}$	Complex
$CH_3CHO \rightarrow CH_4 + CO$	$v \cong k[CH_3CHO]^{3/2}$	$\frac{3}{2}$ (approx.)
$C_2H_5OH \xrightarrow{\text{Liver enzymes}} CH_3CHO$	$v = $ constant	0

However, H^+ is a catalyst, and its concentration is constant during a run; the concentration of H_2O, the solvent, is also little changed because it is present in vast excess. Therefore, the terms $[H^+]$ and $[H_2O]$ are omitted in the rate law (called pseudo-first order) given in the table. When the reaction is carried out in the presence of different concentrations of H^+ or with an added inert solvent, the first-order dependence of the reaction on $[H^+]$ and on $[H_2O]$ is seen.

Table 7.1 illustrates that there is no simple relation between the stoichiometry and the rate law. It is never possible to deduce the order of the reaction by inspection of the stoichiometric equation. Kinetic experiments must be done to measure the order of the reaction. The reaction of H_2 and I_2 is first order with respect to each reactant over a wide range of conditions, whereas the similar reaction of $H_2 + Br_2$ exhibits a complex rate law that cannot be described by a single "order" under all conditions. Note that in this case the rate depends on the concentration of a product as well as on reactants. This comes about because of a reverse step in the mechanism that becomes important as the product concentration builds up. Reaction orders may be non-integral, as in the case of the thermal decomposition of acetaldehyde, and they may be significantly different during the initial stages, when the reaction is getting under way, or at the end, when other complications set in. The significance of a zero-order reaction is that the rate is constant and independent of the concentration of the reactants. This is characteristic of reactions catalyzed by enzymes, such as liver alcohol dehydrogenase, under the special condition where the enzyme is saturated with reactants (called *substrates* in enzyme reactions). The role of enzymes and other catalysts, such as H^+ in the sucrose hydrolysis reaction, is detected by observing that a change in the catalyst concentration produces change in the experimental rate constant for the reaction, even though the catalyst concentration does not vary with time during the course of any single experiment. Even these preliminary comments indicate that the rate law contains important information about the mechanism of the reaction.

Experimental Rate Data

The rate law for a reaction must be determined from experimental data. We may simply want to know how the rate depends on concentrations for practical reasons, or we may want to understand the mechanism of the reaction. For example, if we are inactivating a virus by reaction with formaldehyde, it is vital for us to know how the rate depends on the concentrations of virus and formaldehyde. If the rate is first order in virus concentration, the rate will be one-tenth as fast for 10^5 viruses per milliliter as for 10^6 viruses per milliliter. If the rate is second order in virus concentration (unlikely), the rate will be only one-hundredth as fast. Knowledge of the rate law is necessary to determine the time of treatment with formaldehyde needed to ensure that all the viruses are inactivated before they are used to immunize a population. We also must know how temperature, pH, and solvent affect the rate.

There are many possible ways to obtain the rate data. A usual method is to obtain concentrations of reactants and products at different times during the reaction. For the virus inactivation, we would presumably measure live virus versus time by an infectivity assay. In general, any analytical method that determines concentration can be used. If the time required to perform the analysis is long relative to the rate of the reaction, *quenching*, or sudden stopping of the reaction, is necessary. An enzyme-catalyzed reaction, for example, can be quenched by cooling the reaction mixture quickly, by the addition of an agent that denatures the enzyme or by the addition of a chelating compound if a multivalent metal ion is necessary for catalytic activity. In a *quenched-flow* apparatus, reactants in two syringes are forced through a mixing chamber to initiate the reaction. The mixture flows through a tube into a second mixing chamber, where it is mixed with a stopping reagent. The reaction time in such an experiment is the time it takes for the solution to flow from the first to the second mixing chamber and can be as short as several milliseconds.

If there is a physical property that changes significantly as the reaction proceeds, it can be used to follow the reaction. If the reactants and products absorb light at characteristic wavelengths, the absorbance can be related to the extent of the reaction (see chapter 10). The concentration of H^+ can be easily monitored with a glass electrode, but different techniques are needed when the pH changes rapidly with time. Reactions between charged species can be followed by the electrical conductivity of the solution.

Very fast reactions require special experimental approaches. The *relaxation methods,* such as temperature-jump methods, will be described later in this chapter.

Zero-Order Reactions

A *zero-order reaction* corresponds to the rate law:

$$\frac{dc}{dt} = k_0 \tag{7.2}$$

where c is the concentration of a product and k_0 is a constant. The units of the rate constant k_0 for a zero-order reaction are obviously concentration per time, such as $M\,s^{-1}$. This expression can readily be integrated by writing it in the differential form with the variables c and t separated on each side of the equality:

$$dc = k_0\,dt$$

Integrating both sides, we obtain

$$\int dc = k_0 \int dt + \text{constant}$$

where the rate constant k_0 is placed outside the integral sign because it does not depend on time. Once the initial and final conditions are specified, the equation can be written as a definite integral. If the concentration is c_1 at time t_1 and c_2 at time t_2, then

$$\int_{c_1}^{c_2} dc = k_0 \int_{t_1}^{t_2} dt$$

$$c_2 - c_1 = k_0(t_2 - t_1)$$

If we integrate from c_0 at time zero to c at time t, we obtain another commonly used form of a zero-order equation, where c_0 refers to the initial concentration at zero time and the concentration is c at any later time t. Then,

$$c - c_0 = k_0 t \tag{7.3}$$

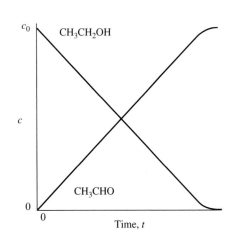

▶ FIGURE 7.1

Plot of concentration versus time for a zero-order reaction. For ethanol, the equation is $c = c_0 - k_0 t$. For acetaldehyde, the equation is $c = k_0 t$. The magnitude of the slope of each straight line is equal to the rate constant k_0. The order must eventually change from being zero order as the concentration of CH_3CH_2OH approaches zero.

This is the equation for a straight line giving the dependence of c on t; the slope is the rate constant k_0.

This behavior is illustrated by the conversion of ethanol to acetaldehyde by the enzyme liver alcohol dehydrogenase (LADH). The oxidizing agent is nicotinamide adenine dinucleotide (NAD^+) and the reaction can be written

$$CH_3CH_2OH + NAD^+ \xrightarrow{\text{LADH}} CH_3CHO + NADH + H^+$$

In the presence of an excess of alcohol over the enzyme and with the NAD^+ buffered via metabolic reactions that rapidly restore it, the rate of this reaction in the liver is zero order over most of its course:

$$v = -\frac{d[CH_3CH_2OH]}{dt} = \frac{d[CH_3CHO]}{dt} = k_0 \tag{7.4}$$

The negative sign is used with the reactant, ethanol, because its concentration decreases with time; the concentration of the product, acetaldehyde, increases with time. This behavior is illustrated in figure 7.1.

The reaction cannot be of zero order for all times; because obviously the reactant concentration cannot become less than zero, and the product concentration cannot become larger than the initial reactant concentration. For the oxidation of alcohol by LADH the reaction is zero order only while alcohol is in excess.

First-Order Reactions

A *first-order reaction* corresponds to the rate law:

$$\frac{dc}{dt} = k_1 c \tag{7.5}$$

The units of k_1 are time^{-1}, such as s^{-1}. There are no concentration units in k_1, so it is clear that we do not need to know absolute concentrations; only relative concentrations are needed. An elementary step in a reaction of the form

$$A \longrightarrow B$$

has a rate law of the form

$$v = -\frac{d[A]}{dt} = \frac{d[B]}{dt} = k_1[A] \tag{7.6}$$

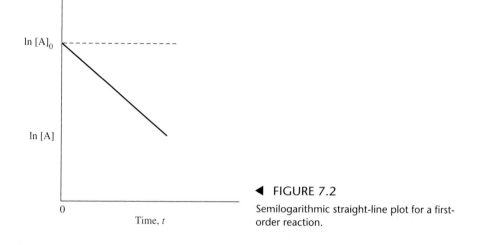

◀ FIGURE 7.2

Semilogarithmic straight-line plot for a first-order reaction.

where k_1 is the rate constant for the reaction and [A] and [B] are concentrations. The rate of the reaction can be expressed in terms of either the rate of disappearance of reactant, $-d[A]/dt$, or the rate of formation of product, $d[B]/dt$. The stoichiometric equation assures us that these two quantities will always be equal to one another. To solve the rate-law expression, we choose the form involving the smallest number of variables:

$$-\frac{d[A]}{dt} = k_1[A] \tag{7.7}$$

Here, time is one variable, and the concentration of A is the other. Dividing both sides by [A], we obtain

$$\frac{d[A]}{[A]} = -k_1\,dt$$

In this form, the variables are *separated* in the sense that the left side depends only on [A] and the right side only on t. Once the variables are separated, the equation can be integrated, separately, on each side:

$$\int \frac{d[A]}{[A]} = \int -k_1\,dt = -k_1 \int dt$$

$$\ln[A] = -k_1 t + C \tag{7.8}$$

where C is a constant of integration. This states that for a first-order reaction, the *logarithm* of the concentration will be a linear function of time, as shown in figure 7.2. To evaluate the constant C, we need to know one concentration at one time. For example, if $[A]_0$ is the value of the concentration initially when $t = 0$, it too must satisfy Eq. (7.8). Substitution gives

$$\ln[A]_0 = C$$

which provides an alternative form of Eq. (7.8):

$$\ln \frac{[A]}{[A]_0} = -k_1 t \tag{7.9}$$

A more general form involving any two points during the course of a first-order reaction is

$$\ln \frac{[A]_2}{[A]_1} = -k_1(t_2 - t_1) \tag{7.10}$$

By taking the exponential of each side of Eq. (7.9), we obtain (remember that $e^{\ln x} = x$)

$$[A] = [A]_0 e^{-k_1 t} = [A]_0 10^{-k_1 t/2.303} \tag{7.11}$$

This says that the concentration of A decreases exponentially with time for a first-order reaction. It starts at an initial value $[A]_0$, since $e^0 = 1$, and reaches zero only after infinite time! Strictly speaking, the reaction is never "finished." Before worrying about the philosophical implications of this, be assured that the inability to detect any remaining [A] will occur in finite time even using the most sensitive analytical methods.

Let's consider the stability of the antibiotic penicillin as a practical example. Assume that our job is to learn how long penicillin remains active when it is stored at room temperature. We want to determine its activity versus time. The general structure of penicillin is

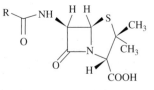

TABLE 7.2 The Number of Units of Penicillin Present After Storage at 25°C. Penicillin Units Are Proportional to the Concentration of Penicillin; They Are a Measure of the Antibiotic Effectiveness of the Penicillin Solution.

Time (weeks)	Penicillin units*	ln (penicillin units)
0	10,100	9.220
1.00	8180	9.009
2.00	6900	8.839
3.00	5380	8.590
4.00	4320	8.371
5.00	3870	8.261
7.00	2190	7.692
8.00	2000	7.601
9.00	1790	7.490
10.00	1330	7.193
11.00	1040	6.947
12.00	898	6.800
13.00	750	6.620
14.00	572	6.349
15.00	403	5.999
16.00	403	5.999
17.00	314	5.749
18.00	279	5.631
19.00	181	5.198
20.00	167	5.118

*The precision of the data produces only three significant figures for this column.

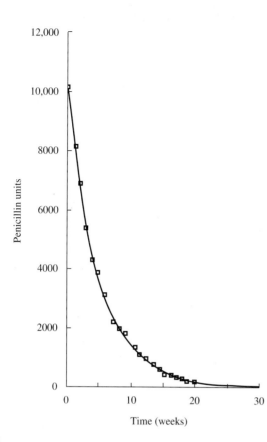

▲ FIGURE 7.3

The concentration (in units of antibiotic activity) of penicillin plotted versus time (in weeks). The concentration is not linear in time; thus, the data are not consistent with a zero-order reaction.

Ultraviolet absorbance, infrared absorbance, or nuclear magnetic resonance could be used to measure its concentration, but we want to know if it will kill bacteria; therefore, we must measure its antibiotic properties. We use a standard assay to count the number of bacterial colonies that survive after treatment with successive dilutions of the penicillin solution under standardized conditions. We thus obtain the number of standard units of penicillin in the sample. The penicillin is in a well-buffered solution at pH 7 kept at 25°C. The mean values for data measured in triplicate are given in table 7.2.

A plot of penicillin units versus time is shown in figure 7.3. The amount of penicillin decreases rapidly with time; an exponential decrease is a good guess.

$$[\text{penicillin}] = [\text{penicillin}]_0 e^{-kt}$$

We note that after three to four weeks, the penicillin is only half as active as it was originally. The data clearly are not consistent with zero-order kinetics; the concentration is not linear in time. To test for first-order kinetics, the natural logarithm (or the base 10 logarithm) of the penicillin units is plotted versus time, as shown in figure 7.4. A linear plot is found; this indicates that the

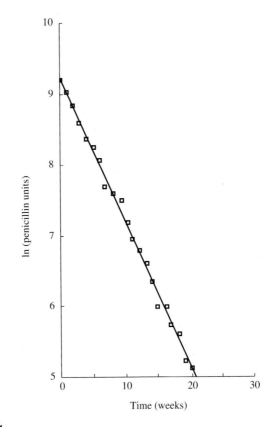

▲ FIGURE 7.4

The ln (concentration) of penicillin plotted versus time. The data are consistent with first-order kinetics; the slope of the line is equal to minus the rate constant k_1 for the reaction.

data are consistent with first-order kinetics. A best least-squares line through the data (a linear regression) has the equation

$$\ln (\text{penicillin units}) = 9.237 - 0.2059 \text{ time (in weeks)}$$

Comparison with Eq. (7.9) shows that the slope of the line is the first-order rate constant; $k_1 = 0.2059$ week^{-1}. The intercept is ln (penicillin units) extrapolated to zero time; therefore, ln [penicillin]$_0$ = 9.237 and [penicillin]$_0$ = $e^{9.237}$ = 10,270. This is consistent with the measured value of 10,100.

A common parameter used to describe the kinetics of first-order rate processes is the *half-life* $t_{1/2}$. This is simply the time required for half the initial concentration to react; we saw that for the data in figure 7.3 this was about 3.5 weeks. For a quantitative determination, we put [A] = $\frac{1}{2}$[A]$_0$ into Eq. (7.11):

$$\frac{1}{2}[A]_0 = [A]_0 e^{-k_1 t_{1/2}}$$

$$\frac{1}{2} = e^{-k_1 t_{1/2}}$$

Taking logarithms of both sides, we have

$$\ln \frac{1}{2} = -k_1 t_{1/2}$$

Since $\ln \frac{1}{2} = -0.6931$ and therefore $\ln 2 = +0.6931$, we see that

$$t_{1/2} = \frac{\ln 2}{k_1} = \frac{0.6931}{k_1} \tag{7.12}$$

For the penicillin data, $k_1 = 0.2059$ week^{-1}, and the half-life is $t_{1/2} = 3.37$ weeks.

EXERCISE 7.1

From the definition of half-life show that

$$\frac{[A]}{[A]_0} = 2^{-t/t_{1/2}} \tag{7.13}$$

and that $t_{1/2}$ is independent of the initial concentration of reactant. Just combine Eqs. (7.11) and (7.12) to obtain Eq. (7.13); then show that the time for the concentration to change from $[A]_0$ to $\frac{1}{2}[A]_0$ is the same as the time for the concentration to change from $\frac{1}{2}[A]_0$ to $\frac{1}{4}[A]_0$.

Another parameter used to characterize first-order kinetic data is the *relaxation time* τ; it is the reciprocal of the rate constant k. From Eq. (7.11),

$$\frac{[A]}{[A]_0} = e^{-t/\tau} \tag{7.14}$$

$$\tau = \frac{1}{k_1} \tag{7.15}$$

The relaxation time τ is the time required for the concentration to decrease to $e^{-1} = 0.3679$ times its initial value (at $t = \tau$, $e^{-\tau/\tau} = e^{-1}$).

Note that the half-life, the relaxation time, and the first-order rate constant depend only on ratios of concentrations. We do not need to know actual concentrations to characterize the kinetics of first-order reactions. We can measure any property proportional to concentration, such as the optical absorbance or the mutagenic ability. It is only for first-order reactions that the rate constant (with units of time^{-1}) does not contain units of concentration. A familiar example is radioactive decay, which is always first order. It does not matter whether there is a large or small amount of a radioactive isotope—the half-life is the same. For all other rate processes, the rate constant does contain units of concentration, and half-lives and relaxation times depend on initial concentrations.

We can now characterize the loss of activity of the penicillin at one temperature and one buffer solution by a single number—the half-life of the first-order reaction is 3.37 weeks. It is useful to be able to characterize the data taken over a 20-week period by one number instead of a table of many numbers, such as table 7.2. How can knowledge of kinetics help further? The fact that most reactions are faster at higher temperatures suggests that penicillin will remain active longer when stored in a refrigerator at 4°C or in liquid nitrogen at 77 K. Because it might take years to measure the stability at low temperatures, it will be quicker to measure the rate of decomposition at higher temperatures and use the Arrhenius equation, which we will discuss shortly, to extrapolate to 4°C. We would need to make measurements in frozen solutions at low temperatures to extrapolate to 77 K.

A proposed mechanism can help us understand the data and may help us formulate a better buffer solution for the penicillin. Penicillin has a lactam structure—an amide ring. The strained four-member ring is easily hydrolyzed to give an inactive compound. The stoichiometric reaction is

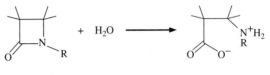

Lactam ring

By analogy with other ester and amide hydrolysis mechanisms, we propose

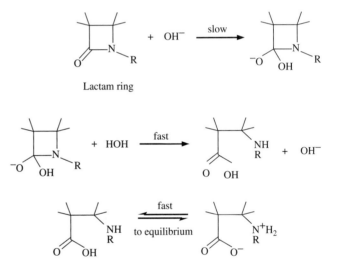

Lactam ring

The reaction starts with a hydroxide ion adding to the carbonyl carbon of the lactam. This is proposed to be the slowest step in the reaction; therefore, we expect that the rate of the reaction will be directly proportional to the concentration of OH^-. The next two steps are much faster than the first step, so they do not affect the kinetics. We say that the first step is rate determining. In a buffer with a pH of 6 instead of 7, the proposed mechanism predicts that the rate of decomposition should be one-tenth as fast. The half-life would be 33.7 weeks instead of 3.37 weeks. Of course, the rate of penicillin decomposition at pH 6 would have to be measured, and the effect of injection into patients of pH 6 solution would need to be determined, but clearly knowledge of kinetics has been helpful.

We will discuss reaction mechanisms in much more detail later in this chapter. Here we just want to give an idea of how kinetic measurements can be used.

EXAMPLE 7.1

Carbon dioxide in the atmosphere contains a small but readily detectable amount of the radioactive isotope $^{14}_{6}C$. This isotope is produced by high-energy neutrons (in cosmic rays) that transform nuclei of nitrogen atoms by the process

$$^{14}_{7}N + ^{1}_{0}n \longrightarrow ^{14}_{6}C + ^{1}_{1}H$$

The $^{14}_{6}C$ nucleus is unstable and decays by the first-order process

$$^{14}_{6}C \longrightarrow ^{14}_{7}N + ^{0}_{-1}\beta$$

with a half-life of 5770 years. The production and decay of ^{14}C leads to a nearly constant steady-state concentration. The amount of ^{14}C present in a carbon-containing sample can be determined by measuring its radioactivity (the rate of production of high-energy electrons, $^{0}_{-1}\beta$ particles).

In the atmosphere, CO_2 contains a nearly constant amount of ^{14}C. Once CO_2 is "fixed" by photosynthesis, however, it is taken out of the atmosphere, and new ^{14}C is no longer added to it. The level of radioactivity then decreases by a first-order process with a 5770-year half-life.

A sample of wood from the core of an ancient bristlecone pine in the White Mountains of California shows a ^{14}C content that is 54.9% as great as that of atmospheric CO_2. What is the approximate age of the tree?

Solution

Since the process is first order, we need know only the *ratio* of the present radioactivity to the original value. Assume that the level of ^{14}C in the atmosphere has not changed over the life of the tree. (This is not strictly true. Corrections can be made, but they are small.) Thus,

$$\frac{[^{14}C]}{[^{14}C]_0} = 0.549$$

Using Eqs. (7.9) and (7.12), we obtain

$$\ln \frac{[^{14}C]}{[^{14}C]_0} = -\frac{0.693}{t_{1/2}}t$$

where t is the age of the wood. Therefore,

$$\ln (0.549) = -\frac{0.693}{5770 \text{ yr}}t$$

$$t = \frac{(0.600)(5770 \text{ yr})}{0.693} = 4990 \text{ yr}$$

We conclude that the carbon in the tree was taken from the atmosphere approximately 4990 years ago.

Second-Order Reactions

A *second-order reaction* corresponds to the rate law

$$v = k_2 c^2 \quad \text{or} \quad v = k_2 c_A c_B$$

The units of k_2 are concentration^{-1} time^{-1}, such as $M^{-1}\,s^{-1}$.

It is useful to separate the treatment of second-order reactions into two major classifications, depending on whether the rate law depends on (I) the second power of a single reactant species or (II) the product of the concentrations of two different reagents.

Class I (A + A → P)

$$v = k_2[A]^2$$

Although the stoichiometric equation may involve either one or several components, the rate law for many reactions depends only on the second power of a single component. Some examples are:

$$2\ \text{proflavin} \longrightarrow [\text{proflavin}]_2\ ;\ v = k_2[\text{proflavin}]^2$$

$$\underset{\text{ammonium cyanate}}{NH_4OCN} \longrightarrow \underset{\text{urea}}{NH_2CONH_2}\ ;\ v = k_2[NH_4OCN]^2$$

$$2\underset{\text{hexanucleotide*}}{A-A-G-C-U-U} \xrightarrow{k_1} \underset{\text{hexanucleotide dimer}}{\begin{matrix} A-A-G-C-U-U \\ \cdot\ \ \cdot\ \ \cdot\ \ \cdot\ \ \cdot\ \ \cdot \\ U-U-C-G-A-A \end{matrix}}\ ;\ v = k_2[A_2GCU_2]^2$$

In each case, the rate law is of the form

$$v = -\frac{d[A]}{dt} = k_2[A]^2 \tag{7.16}$$

The variables can be separated,

$$\frac{-d[A]}{[A]^2} = k_2\,dt$$

and each side integrated, to give

$$\frac{1}{[A]} = k_2t + C$$

Since $[A] = [A]_0$ when $t = 0$, the constant of integration is $1/[A]_0$, and we obtain an integrated form of the second-order rate equation:

$$\frac{1}{[A]} - \frac{1}{[A]_0} = k_2t \tag{7.17}$$

An alternative form is

$$\frac{1}{[A]_0 - x} - \frac{1}{[A]_0} = k_2t \tag{7.18}$$

where x is the concentration of A that has reacted. Note that for second-order kinetics one expects a linear relation between the reciprocal of the reactant concentration and time. This is in contrast to first-order kinetics where the logarithm of concentration is linear in time.

The half-life of any reaction is the time required for half of $[A]_0$ to react. For a second-order (class I) reaction, we use Eq. (7.17):

---

*See the appendix for the structure of the nucleotides.

$$\frac{1}{\frac{1}{2}[A]_0} - \frac{1}{[A]_0} = k_2 t_{1/2}$$

$$\frac{1}{[A]_0} = k_2 t_{1/2}$$

$$t_{1/2} = \frac{1}{k_2[A]_0} \qquad \textbf{(7.19)}$$

For second-order reactions, the half-life depends on the initial concentration as well as on the rate constant. It is only first-order reactions where the half-life (or any other fractional life) is independent of initial concentration.

Class II (A + B → P)

$$v = k_2[A][B] \qquad \textbf{(7.20)}$$

A reaction that is second order overall may be first order with respect to *each* of the two reactants. Some examples are

$$CH_3COOC_2H_5 + OH^- \longrightarrow CH_3COO^- + C_2H_5OH ;$$

$$v = k_2[CH_3COOC_2H_5][OH^-]$$

$$NO(g) + O_3(g) \longrightarrow NO_2(g) + O_2(g) ; v = k_2[NO][O_3]$$

$$H_2O_2 + 2Fe^{2+} + 2H^+(excess) \longrightarrow 2H_2O + 2Fe^{3+}; v = k_2[H_2O_2][Fe^{2+}]$$

Again we see that the overall stoichiometric equation is not a valid indicator of the rate law. We may expect that there are significant underlying differences in the mechanisms of these reactions.

[a ≠ b]   In general, for reactions of the type

$$A + B \longrightarrow \text{products}$$

exhibiting class II kinetics, the initial concentrations of the reactants, a and b, need not be the stoichiometric ratio. For this reason, we rewrite the rate law

$$v = k_2[A][B] \qquad \textbf{(7.20)}$$

in terms of the reaction variable x, where x = the concentration of each species reacted. For a stoichiometric reaction of A + B → products

$$[A] = a - x; \qquad [A]_0 = a$$
$$[B] = b - x; \qquad [B]_0 = b$$

Following substitution into Eq. (7.20) and separation of variables, we obtain

$$\frac{dx}{(a - x)(b - x)} = k_2 \, dt \qquad \textbf{(7.21)}$$

A table of integrals gives the result

$$\frac{1}{a - b} \ln \frac{b(a - x)}{a(b - x)} = k_2 t \qquad (a \neq b) \qquad \textbf{(7.22)}$$

or, alternatively,

$$\frac{1}{[A]_0 - [B]_0} \ln \frac{[B]_0[A]}{[A]_0[B]} = k_2 t \qquad (7.23)$$

After separating the constant terms involving only initial concentrations, we see that class II second-order reactions exhibit a linear relation between $\ln([A]/[B])$ and time.

[a = b]   Note that when $a = b$ (when the initial concentrations are stoichiometric), then Eqs. (7.22) and (7.23) do not apply. In this case, however, the method of class I is appropriate because the values of [A] and [B] will be in a constant ratio throughout the entire course of the reaction. That is, if $a = b$, then [A] = [B] at all times and

$$v = k_2[A][B] = k_2[A]^2$$

which can be integrated exactly as in class I to give Eq. (7.17).

EXAMPLE 7.2

Hydrogen peroxide reacts with ferrous ion in acidic aqueous solution according to the reaction

$$H_2O_2 + 2\,Fe^{2+} + 2H^+ \longrightarrow 2H_2O + 2Fe^{3+}$$

From the following data, obtained at 25°C, determine the order of the reaction and the rate constant.

$$[H_2O_2]_0 = 1.00 \times 10^{-5}\,M$$
$$[Fe^{2+}]_0 = 1.00 \times 10^{-5}\,M$$
$$[H^+]_0 = 1.0\,M$$

Time (min)	0	5.3	8.7	11.3	16.2	18.5	24.6	34.1
$10^5 \times [Fe^{3+}]\,M$	0	0.309	0.417	0.507	0.588	0.632	0.741	0.814

Solution

Because the H^+ concentration is in huge excess, it will not change significantly during the course of the reaction. As a consequence, we can write the rate expression as

$$v = k[H_2O_2]^i[Fe^{2+}]^j$$

recognizing that this is the simplest general form. Note also that the initial concentration of H_2O_2 is twice the amount needed to react with all of the Fe^{2+} present.

1. Test for first order in $[Fe^{2+}]$, where $i = 0$ and $j = 1$. In this case, we expect from Eq. (7.9) that

$$\ln \frac{[Fe^{2+}]_0}{[Fe^{2+}]} = kt \quad \text{or} \quad \frac{1}{t} \ln \frac{[Fe^{2+}]_0}{[Fe^{2+}]} = k$$

2. Test for first order in $[H_2O_2]$, where $i = 1$ and $j = 0$. Then

$$\frac{1}{t} \ln \frac{[H_2O_2]_0}{[H_2O_2]} = k$$

The data for testing proposals 1 and 2 are as follows:

Time (min)	0	5.3	8.7	11.3	16.2	18.5	24.6	34.1
$10^5 \times [Fe^{2+}]\,(M)$	1.00	0.691	0.583	0.493	0.412	0.368	0.259	0.186
$\ln \dfrac{[Fe^{2+}]_0}{[Fe^{2+}]}$	0	0.370	0.540	0.707	0.887	1.000	1.351	1.682
(1) $\dfrac{10^3}{t} \ln \dfrac{[Fe^{2+}]_0}{[Fe^{2+}]}$	—	69.8	62.0	62.6	54.7	54.0	54.9	49.3
$10^5 \times [H_2O_2]\,(M)$	1.00	0.845	0.791	0.741	0.706	0.684	0.630	0.593
(2) $\dfrac{10^3}{t} \ln \dfrac{[H_2O_2]_0}{[H_2O_2]}$	—	31.8	26.9	26.5	21.5	20.5	18.8	15.3

The values in the rows designated (1) and (2) are not constant, but show a trend, indicating that neither of these predictions is valid for the reaction.

3. Test for second order overall, where $i = 1$ and $j = 1$. Because the stoichiometric coefficients are not unity, we need to modify the development of Eq. (7.22). For a reaction with stoichiometry A + 2B → product, we define

$$v = -\frac{d[A]}{dt} = k_2[A][B] = -\frac{1}{2}\frac{d[B]}{dt}$$

where $[A] = a - x$, $[A]_0 = a$ and $[B] = b - 2x$, $[B]_0 = b$. The equation analogous to Eq. (7.21) is now

$$\frac{dx}{(a-x)(b-2x)} = k_2\,dt$$

or multiplying the equation by 2,

$$\frac{dx}{(a-x)(b/2-x)} = 2k_2\,dt$$

This is readily solved by noting that with respect to Eq. (7.21) we have just replaced b by $b/2$ and k_2 by $2k_2$. We simply make the same replacements in the solution, Eq. (7.22), to obtain the result

$$\frac{1}{a-\dfrac{b}{2}}\ln\frac{\dfrac{b}{2}(a-x)}{a\left(\dfrac{b}{2}-x\right)} = 2k_2t$$

or

$$\frac{1}{2a-b}\ln\frac{b(a-x)}{a(b-2x)} = k_2t$$

or

$$\frac{1}{2[H_2O_2]_0 - [Fe^{2+}]_0}\ln\frac{[Fe^{2+}]_0[H_2O_2]}{[H_2O_2]_0[Fe^{2+}]} = k_2t$$

Since $[Fe^{2+}]_0/[H_2O_2]_0 = 1.0$ from the initial conditions, we expect the relation

$$\frac{1}{t}\ln\frac{[H_2O_2]}{[Fe^{2+}]} = \text{constant}$$

to hold if the rate law is first order with respect to each reactant.

Proposal 3 is tested by the following tabulation:

Time (min)	0	5.3	8.7	11.3	16.2	18.5	24.6	34.1
$\dfrac{[H_2O_2]}{[Fe^{2+}]}$	1.00	1.223	1.357	1.503	1.714	1.859	2.432	3.188
$(3)\ \dfrac{1}{t}\ln\dfrac{[H_2O_2]}{[Fe^{2+}]}(\text{min}^{-1})$	—	0.0380	0.0351	0.0361	0.0332	0.0335	0.0361	0.0340

The entries in the bottom row are approximately constant and show that the results do correspond to this rate law. Therefore,

$$v = k_2[H_2O_2][Fe^{2+}]$$

The value of k_2 can be determined from

$$k_2 = \frac{1}{2[H_2O_2]_0 - [Fe^{2+}]_0}\left[\frac{1}{t}\ln\frac{[H_2O_2]}{[Fe^{2+}]}\right]$$

The average value of $(1/t)\ln([H_2O_2]/[Fe^{2+}]_0)$ from the table is 0.035 min^{-1}. Therefore,

$$k_2 = \frac{0.035\ \text{min}^{-1}}{(2\times 10^{-5} - 1\times 10^{-5})\ M} = 3.5\times 10^3\ M^{-1}\ \text{min}^{-1}$$

We could of course have solved this problem by graphical rather than numerical methods.

Renaturation of DNA as an Example of a Second-Order Reaction

A double-stranded DNA is made of two antiparallel chains with nucleotide sequences that are complementary to each other. If a linear DNA fragment is heated or subjected to high pH, the two chains separate, and this disordered product is termed the *denatured*, or *coiled, form*. If the temperature or the pH is then lowered so that the double helix is again the stable form, pairing of the bases between chains of complementary sequences occurs, and the chains will *reassociate*, or *renature*. If we designate the strands with complementary sequences as A and A', the renaturation reaction can be written as

$$A + A' \xrightarrow{k} AA'$$

which has been found experimentally to be second order. The most striking feature of renaturation kinetics is that when DNA from different sources is first broken down to about the same size (by sonication, for example) and then the rates of renaturation are measured at the same DNA concentrations in moles of nucleotides per liter, the rates are found to span a range of several orders of magnitude (figure 7.5). The following analysis shows that the rate is expected to be inversely proportional to the *sequence complexity* of the DNA.

Figure 7.6 illustrates that the rate of renaturation depends on the repetitiveness of the DNA sequence. When a long DNA molecule with no repetitive sequences is fragmented into pieces longer than 20 base pairs (bp), each piece should be a unique sequence. For example, a DNA 10^6 bp long can be broken

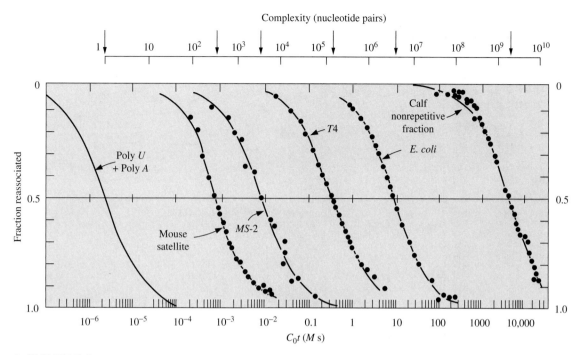

▲ FIGURE 7.5

Reassociation rates in solutions containing 0.18 M Na$^+$ for several double-stranded DNAs (mouse satellite—part of the mouse genome that has highly repetitive sequences, T4 bacteriophage, *E. coli* bacteria, and part of the calf genome). Data for two double-stranded RNAs (MS-2 viral RNA and a synthetic RNA, poly A · poly U) are also shown. The large nucleic acids were first broken into double-stranded fragments of about 400 base pairs by sonication. The fragments were then denatured into single strands of about 400 nucleotides by heating the solution to 90°C for a few minutes. The solution was cooled quickly and the fraction reassociated to double strands was measured versus time t. The rate of reassociation (renaturation) to double strands depends on the concentration of the single strands C_0 and on the number of different sequences in the nucleic acids—the sequence complexity. C_0 is expressed in units of moles of nucleotides per liter. The complexity of the samples, indicated by arrows over the upper scale, range from 1 for poly A · poly U to over 10^9 for the nonrepetitive fraction of calf DNA. (From R. Britten and D. Kohne, 1968, *Science* 161:529.)

into essentially 10^6 (actually $10^6 - 19$) pieces of length 20. The number of possible DNA sequences that are 20 bp long is 4^{20} (any of the 4 bases can be in each position). Because $4^{20} = 10^{12}$, which is much, much larger than 10^6, each piece is likely to be unique. When the pieces are denatured, the concentration of each unique strand is very small, and the probability of finding its Watson–Crick complementary strand is small. The rate of renaturation will be very slow [figure 7.6(a)]. However, if there are repetitive sequences, some of the fragmented pieces can be identical, and it will be easy for a denatured strand to find its complement [figure 7.6(b)]. The analysis can be done quantitatively.

Let C_0 be the total concentration (in moles of nucleotide per liter) of all the single strands before any renaturation occurs. This total concentration can be measured easily from the UV absorbance of the solution. The rate of renaturation will depend on $[A]_0$, the initial concentration (in moles of nucleotide per liter) of fragment A that is complementary to fragment A'. We see from figure 7.5 that $[A]_0$ can vary from $C_0/2$ for poly dA · poly dT [figure 7.6(b)] to $C_0/2N$ for a DNA of N bp of a nonrepeating sequence.

DNA with nonrepetitive sequence of 10^6 base pairs

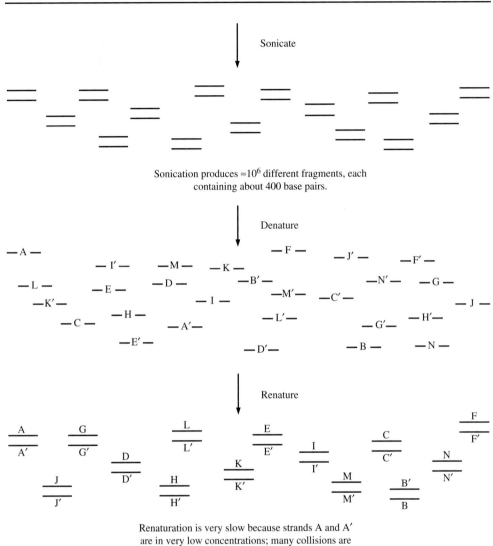

Renaturation is very slow because strands A and A′
are in very low concentrations; many collisions are
needed to find complementary partners.

(a)

▲ FIGURE 7.6

Renaturation (or reassociation) rates depend on the sequence complexity of the DNA. In (a) a complex DNA, such as *E. coli* DNA with a nonrepeating sequence, is fragmented to give essentially all different fragments. The concentration of each fragment is very small compared with the total concentration of all fragments. After the double-stranded fragments are denatured, the low concentrations of the complementary single strands lead to a slow rate of renaturation. In (b) all the fragments of poly dA · poly dT are the same. The concentrations of the single strands are high, and the rate of renaturation is much faster.

Therefore, we define N as the sequence complexity—the total number of base pairs in the smallest repeating sequence in the DNA. The concentration $[A]_0$ is proportional to C_0/N. If the DNA is that of the bacterium *E. coli*, for example, there is very little repetition of the sequence, and N is the same as the number of base pairs per genome, or about 3×10^6; if the DNA is polyde-

DNA with repeating sequence of 10^6 base pairs—poly dA · poly dT

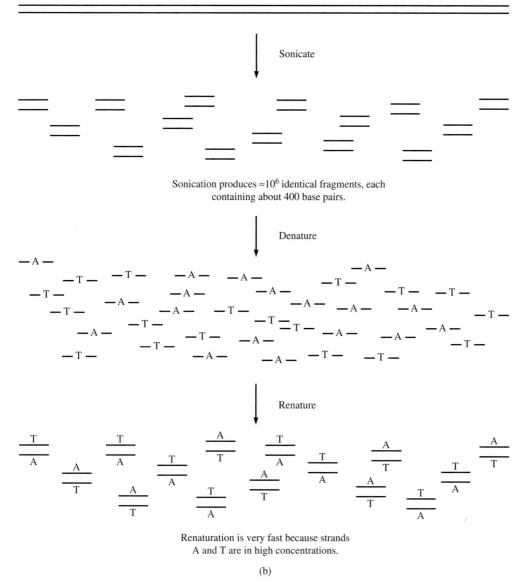

Sonicate

Sonication produces $\approx 10^6$ identical fragments, each containing about 400 base pairs.

Denature

Renature

Renaturation is very fast because strands A and T are in high concentrations.

(b)

▲ FIGURE 7.6   (*continued*)

oxyadenylate (poly dA) in one strand and polydeoxythymidylate (poly dT) in the other, then N is 1, since the smallest repeating sequence is just a single base pair.

The rate of renaturation is

$$\frac{-d[A]}{dt} = \frac{-d[A']}{dt} = k[A][A'] = k[A]^2$$

The DNA is originally double stranded, so the concentrations of A and A' must be equal. The equation is the same as Eq. (7.16), and the half-life of the reaction is, from Eq. (7.19),

$$t_{1/2} = \frac{1}{k[A]_0} \tag{7.19}$$

But $[A]_0$ is proportional to C_0/N, and so

$$[A]_0 \propto \frac{C_0}{N}$$

Substituting into the equation above gives

$$t_{1/2} \propto \frac{N}{kC_0}$$

or

$$C_0 t_{1/2} \propto N$$

Thus, the half-life of renaturation times the total initial denatured strand concentration is proportional to the sequence complexity of the DNA. A long half-life means that each single strand will make many collisions before it finds its complement; the sequence is complex. The quantity $C_0 t_{1/2}$ can thus be used to measure the complexity of the nucleotide sequence of the genome of an organism. The method is obviously applicable to double-stranded RNA as well. Figure 7.5 shows that the sequence complexity varies from 1 for the synthetic poly A·poly U to 10^9 for the nonrepetitive fraction of calf thymus DNA.

Reactions of Other Orders

It is easy to generalize the treatment of class I reactions of order n, where n is any positive or negative number except $+1$. The rate law is then

$$v = k[A]^n \tag{7.24}$$

which can be integrated to give

$$\frac{1}{n-1}\left[\frac{1}{[A]^{n-1}} - \frac{1}{[A]_0^{n-1}}\right] = kt \qquad (n \neq 1) \tag{7.25}$$

The half-life will be

$$t_{1/2} = \frac{2^{n-1} - 1}{(n-1)k[A]_0^{n-1}} \tag{7.26}$$

For example, the thermal decomposition of acetaldehyde is $\frac{3}{2}$ order over most of the course of the reaction. A linear plot should be obtained, according to Eq. (7.25), if $1/[A]^{1/2}$ is plotted versus time. The half-life will be given by

$$t_{1/2} = \frac{\sqrt{2} - 1}{\frac{1}{2}k[A]_0^{1/2}}$$

Determining the Order and Rate Constant of a Reaction

We have discussed various simple orders that can represent reaction kinetics, and we have given specific illustrations. Here we mention some general methods that can be used to determine the order and rate constant for a reaction. We need to know the stoichiometry of the reaction, and we need to be able to measure the concentration of a product or reactant. To determine the order of the reaction with respect to each reactant, we actually need only to measure a property that is directly proportional to concentration. Except

TABLE 7.3   The Characteristics of Simple-order Reactions

Reaction order	Linear plot	Rate proportional to	Units of k
Zero-order	c vs. t	$dc/dt = k$	M time^{-1}
First-order	$\ln c$ vs. t	$dc/dt = kc$	time^{-1}
Second-order (I)	$1/c$ vs. t	$dc/dt = kc^2$	M^{-1} time^{-1}
Second-order (II)	$\ln(c_A/c_B)$ vs. t	$dc_A/dt = kc_A c_B$	M^{-1} time^{-1}
nth-order ($n \neq 1$)	$1/c^{n-1}$ vs. t	$dc/dt = kc^n$	$M^{-(n-1)}$ time^{-1}

for first-order reactions, we do need to know concentrations to obtain rate constants. The characteristics of the different order reactions are given in table 7.3.

Direct data plots

A plot of concentration versus time gives a clue to the order. If the plot is linear, a zero-order reaction is indicated. If a curved plot results, other plots, such as log c versus time or $1/c$ versus time, can be tried. Look for a linear plot that indicates the order and whose slope is proportional to the rate constant.

Rate versus concentration plots (differentiation of the data)

From measurements of the slopes of plots of concentration versus time, you can determine the rate of the reaction at different times. These rates can then be plotted versus the corresponding reactant concentration to try to deduce the form of the rate law. This approach requires many experimental points or a continuous curve.

Method of initial rates

The rate is measured during the earliest stages of the reaction. The concentrations do not change much with time, and they can be replaced in the rate law by the initial concentrations. This method is particularly useful when the rate law is not of the simplest form. For example, in many enzyme-catalyzed reactions, the method of initial rates has been used to find that

$$v = \frac{k[S]}{K + [S]}$$

where k and K are constants and [S] is the substrate concentration. (We discuss this Michaelis–Menten type of kinetics in chapter 8.)

Changes in initial concentrations

The influence of different initial concentrations on the initial rate, the half-life, or some other parameter can be determined. The method measures the order of the reaction with respect to a particular component directly. To find the order with respect to component A in the expression

$$v = k[A]^a[B]^b[C]^c$$

we measure the initial rate v_1 with initial concentrations A_0, B_0, and C_0:

$$v_1 = k[A_0]^a[B_0]^b[C_0]^c$$

We then double the concentration of A_0, keeping the others constant, and measure the new initial rate v_2:

$$v_2 = k(2[A_0])^a[B_0]^b[C_0]^c$$

The ratio of these two rates leads to the value of a,

$$\frac{v_2}{v_1} = 2^a$$

because all other factors cancel.

Methods of reagents in excess

We can decrease the kinetic order of a reaction by choosing one or more of the reactant concentrations to be in large excess. Normally, the concentration of a reactant decreases steadily during the course of a reaction. This complicates the kinetic analysis, especially if the rate law is not a simple one. By using a large excess of one component, its concentration is maintained nearly constant, and the corresponding term in the rate law now hardly changes during the course of the reaction. In the example above, we could choose B and C to be 1 M and A to be 0.01 M. Concentrations of B and C would change only slightly (depending on the stoichiometry) during the complete reaction of A. We could essentially study the kinetics with respect to A independently of the B and C kinetics.

Once the order of the reaction is known with respect to each component, the value of the rate constant can be calculated. It is important to specify what rate and what species are being described. Uncertainty can occur when the stoichiometric coefficients for the reaction are not all the same. For example, the reaction

$$N_2O_5 \longrightarrow 2NO_2 + \tfrac{1}{2}O_2$$

exhibits first-order kinetics. If we express concentrations of these substances in mol L^{-1}, the rate of disappearance of reactant N_2O_5 is half as great as the rate of formation of the product NO_2 and twice as great as the rate of formation of O_2. Expressed mathematically, this is

$$-\frac{d[N_2O_5]}{dt} = \frac{1}{2}\frac{d[NO_2]}{dt} = 2\frac{d[O_2]}{dt}$$

The coefficients in this rate equation are simply the reciprocals (apart from the signs) of the stoichiometric coefficients. The standard convention is that for the following stoichiometric equation

$$mM + nN \longrightarrow pP + qQ$$

the rate of the reaction is written as

$$v = -\frac{1}{m}\frac{d[M]}{dt} = -\frac{1}{n}\frac{d[N]}{dt} = \frac{1}{p}\frac{d[P]}{dt} = \frac{1}{q}\frac{d[Q]}{dt}$$

The rate-law expression

$$v = k[M]^i[N]^j \cdots$$

will serve to define the relation of the rate constant to the rate of the reaction. (Remember that the exponents i and j in the rate law are not related to the stoichiometric coefficients.)

Reaction Mechanisms and Rate Laws

The form of the rate law determined from the kinetics of a reaction does not tell us the actual mechanism, although it gives important clues. A very complex process, such as the growth and multiplication of bacterial cells, exhibits simple first-order kinetics. Growth, as measured by the increase in the number of cells, occurs in proportion to the number of cells present at any given time and with a rate constant determined by the turnaround time for the complex events in the cell cycle. We cannot hope for this single parameter, the rate constant, to give us detailed information about the individual steps of DNA replication, transcription into RNA, synthesis of protein, mitotic division, and so on that are involved in the cell cycle. The rate law does, however, make a quantitative statement about how fast the cells grow.

Typically, a complex reaction sequence or mechanism is made up of a set of elementary reactions. Consider, for example, the process by which the ozone layer of the upper atmosphere is maintained. This ozone layer acts as an important shield, absorbing potentially harmful UV radiation from the Sun and protecting organisms on the surface of Earth from dangerous biological consequences. Ozone is formed from atmospheric oxygen by a process that can be described by the stoichiometric equation

$$3O_2 \longrightarrow 2O_3$$

Ozone formation starts with a photochemical step resulting from the absorption of short-wavelength UV radiation by O_2, followed by a second elementary reaction in which the oxygen atoms produced in the first step react with O_2 to form ozone:

$$O_2 \xrightarrow{\;h\nu\;\text{(below 242 nm)}\;} O + O$$

$$O + O_2 + M \longrightarrow O_3 + M$$

where M is a third body (any other gas molecule) whose function is to remove excess energy from the excited ozone initially formed. The protective role of ozone results from its absorption of somewhat longer wavelength UV light not absorbed by O_2, and this radiation leads to the decomposition of ozone by the elementary reaction

$$O_3 \xrightarrow{\;h\nu\;\text{(190 to 300 nm)}\;} O_2 + O$$

The balance in the ozone concentration is achieved by these three elementary steps operating together, supplemented by lesser contributions of additional reactions involving other elements.

Johnston (1975) pointed out that the flights of supersonic transports in the upper atmosphere posed a threat to the stability of the protective ozone layer. This comes about because of the role of oxides of nitrogen (NO_x) in the exhaust gases of the jets. The oxides catalyze the decomposition of ozone:

$$O + O_3 \xrightarrow{\;NO_x\;} 2O_2$$

This process can be understood in terms of the elementary reactions involved, which are

$$NO + O_3 \longrightarrow NO_2 + O_2$$

$$NO_2 + O \longrightarrow NO + O_2$$

Not only does this show in molecular detail how NO_x produces the ozone decomposition, but it also shows that NO and NO_2 are not used up in the process. Together they serve a true catalytic function.

Another class of catalyst for the destruction of ozone is chlorofluorocarbons (Molina and Rowland 1974; Anderson, Toohey, and Brune 1991). Chlorine atoms are produced by the photochemical decomposition of, for example, dichlorodifluoromethane (Freon 12):

$$CCl_2F_2 \longrightarrow CClF_2 + Cl$$

The chlorine atoms catalyze the decomposition of ozone molecules by oxygen atoms:

$$Cl + O_3 \longrightarrow ClO + O_2$$

$$O + ClO \longrightarrow Cl + O_2$$

The net reaction is that ozone is converted to oxygen and the chlorine atoms are not used up. Because of the environmental dangers caused by the loss of ozone, the use of chlorofluorocarbons in aerosol cans has been banned in the United States.

It often happens that more than one mechanism can be written consistent with a given rate law. The best one can be chosen only if we learn more about the intermediates, the energetics of the process, and the rates of elementary steps determined from other reactions. The concept of elementary reactions is important to chemical kinetics for the same reason that the study of chemical bonds or of functional groups is important in simplifying and organizing the chemistry of the hundreds of thousands of different organic compounds. The kinetic parameters of an elementary step are the same regardless of the overall reaction in which it is participating.

Elementary reactions are described in terms of their *molecularity:* the number of reactant particles involved in the elementary reaction. Reactions between two particles are called *bimolecular.* They may be thought of in terms of a collision between the two reactants to form a transition state. A *transition state* is an unstable arrangement of atoms intermediate between reactants and products. The transition state exists for no longer than the time it takes for a molecular vibration (of order 10 femtoseconds). For example, the reaction of ozone with nitric oxide in the gas phase occurs as a simple bimolecular process. The transition state is the transient species in which an O—O bond is being broken as an O—N bond forms:

$$O_3 + NO \xrightarrow{\text{collision}} \left[\begin{array}{c} \overset{+}{\underset{}{O}} \\ O \diagup \ddots O^- \cdots \overset{+}{N}\!-\!O^- \\ \text{transition state} \end{array} \right] \longrightarrow O_2 + NO_2$$

Since the process is bimolecular, it follows that it must exhibit second-order kinetics. The converse is not true, however; many second-order reactions do not have simple bimolecular mechanisms but are more complex. *The form of the rate law alone cannot predict the mechanism but the mechanism directly provides the rate law.* This is an important distinction to learn.

From the example given above, we can define *molecularity* as the number of particles that come together to form the transition state in an elementary reaction. Note that the term *molecularity* applies only to elementary reactions and not to the overall mechanism. *Unimolecular reactions* are those that involve only a single reactant particle—for example, radioactive decay. True unimolecular processes are rare because most reactions require some *activation energy* in order to proceed. This energy is commonly picked up by collisions with the surroundings (other molecules, walls of the vessel). *Termolecular reactions* involve collisions of three molecules and are more likely to occur at high pressures or in liquid solution.

The simplest examples of reaction mechanisms are those that involve only a single elementary reaction. In this case, the rate law can be written by inspection. The rate is proportional to the concentration of each of the species involved in forming the transition state, and the exponents are determined by the numbers of each kind of particle. For example, if the reaction

$$2A + B \longrightarrow \text{products}$$

is *elementary*, its rate will be given by

$$v = k[A]^2[B]$$

The real problem occurs when the overall reaction is complex and consists of several connected elementary reactions. The overwhelming majority of reactions and processes in nature are complex.

A few examples will illustrate some important methods of dealing with complex reactions. One general method is to see whether one elementary reaction controls the overall kinetics—that is, to look for the *rate-determining step.* For certain mechanisms, one elementary reaction is found to dominate the kinetics of the complex reaction. When this is so, the analysis of the mechanism is greatly simplified. We can write the rate for the rate-determining step and equate it to the overall rate. In the following examples, we discuss the general case first and then indicate the simplification that occurs if there is a rate-determining step.

Parallel Reactions

Parallel reactions are of the following type:

$$A \begin{array}{c} \overset{k_1}{\nearrow} B \\ \underset{k_2}{\searrow} C \end{array}$$

The substance A can decompose by either of two paths, giving rise to different products, B and C. The rate expressions are

$$-\frac{d[A]}{dt} = k_1[A] + k_2[A] = (k_1 + k_2)[A] \tag{7.27}$$

$$\frac{d[B]}{dt} = k_1[A] \tag{7.28}$$

$$\frac{d[C]}{dt} = k_2[A] \tag{7.29}$$

The solution to the first equation, which has the form of the first-order rate law, Eq. (7.7), is

$$\ln \frac{[A]}{[A]_0} = -(k_1 + k_2)t$$

$$[A] = [A]_0 \exp[-(k_1 + k_2)t] \tag{7.30}$$

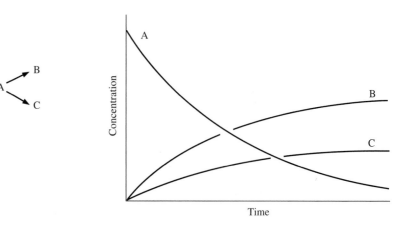

▶ **FIGURE 7.7**

Kinetic curves for reactant and products for two first-order reactions in parallel.

Thus, [A] decays exponentially, as illustrated in figure 7.7. To find out how [B] and [C] increase with time, we need to solve Eqs. (7.28) and (7.29). Because each of these equations contains three variables, it cannot be solved as it stands. However, we can substitute for A using Eq. (7.30). Thus,

$$\frac{d[B]}{dt} = k_1[A] = k_1[A]_0 \exp[-(k_1 + k_2)t]$$

and

$$\frac{d[C]}{dt} = k_2[A] = k_2[A]_0 \exp[-(k_1 + k_2)t]$$

We now separate variables and integrate. If $[B]_0 = [C]_0 = 0$, the solutions are

$$[B] = \frac{k_1[A]_0}{k_1 + k_2}\{1 - \exp[-(k_1 + k_2)t]\} \tag{7.31}$$

$$[C] = \frac{k_2[A]_0}{k_1 + k_2}\{1 - \exp[-(k_1 + k_2)t]\} \tag{7.32}$$

Note that B and C are formed in a constant ratio, $[B]/[C] = k_1/k_2$ throughout the course of the reaction. The extension of this method to three or more parallel reactions is straightforward.

EXAMPLE 7.3

Radioactive decay frequently occurs simultaneously by two separate first-order paths. For example $^{64}_{29}Cu$ is unstable against decay, emitting either an electron or a positron. The scheme can be written as follows:

$$^{64}_{29}Cu \begin{array}{c} \xrightarrow{k_-} {}^{64}_{30}Zn + \beta^- \quad (62\%) \\ \xrightarrow{k_+} {}^{64}_{28}Ni + \beta^+ \quad (38\%) \end{array}$$

The decay of $^{64}_{29}Cu$ occurs exponentially with a single half-life of 12.80 h. Calculate the rate constants for the two paths, k_+ and k_-.

Solution

From Eq. 7.30 we see that $^{64}_{29}$Cu decays as a first-order process with rate constant equal $k_+ + k_-$. Thus,

$$k_+ + k_- = \frac{\ln 2}{t_{1/2}} = \frac{\ln 2}{12.80 \text{ h}} = 0.05415 \text{ h}^{-1}$$

But

$$k_+/k_- = \frac{38}{62}$$

Combining the two equations for k_+ and k_-,

$$k_+ = \frac{0.05415}{1 + \frac{62}{38}} = 0.0206 \text{ h}^{-1}$$

$$k_- = \frac{0.05415}{1 + \frac{38}{62}} = 0.0336 \text{ h}^{-1}$$

(handwritten margin notes:)
$$\frac{k_+ + k_-}{1 + \frac{k_-}{k_+}} = \frac{k_+ + k_-}{\frac{k_+ + k_-}{k_+}} = k_+ + k_- \times \frac{k_+}{k_+ k_-}$$

In parallel reactions, if one step is much faster than the others, this fast step primarily determines the rate of reaction. This is very different from the case of series reactions, in which the slow step determines the rate, as we will see in the next section. In the example above, if 99% of the reaction goes through the β^- path, we can ignore the β^+ path in calculating the decay. We can see this quantitatively in Eq. (7.30). When $k_1 \gg k_2$, the decay of [A] with time depends only on k_1 and $[A] \cong [A]_0 \exp(-k_1 t)$.

Series Reactions (First Order)

Series reactions are of the type

$$A \xrightarrow{k_1} B \xrightarrow{k_2} C$$

Compound A reacts to form B, which then goes on to form C. The rate expressions are

$$v_1 = -\frac{d[A]}{dt} = k_1[A]$$

$$[A] = [A]_0 \exp(-k_1 t)$$

$$\frac{d[B]}{dt} = k_1[A] - k_2[B] = k_1[A]_0 \exp(-k_1 t) - k_2[B] \qquad \textbf{(7.33)}$$

This differential equation is slightly more difficult to solve, but we show the result. If $[B]_0 = 0$,

$$[B] = \frac{k_1[A]_0}{k_2 - k_1}[\exp(-k_1 t) - \exp(-k_2 t)] \qquad \textbf{(7.34)}$$

Similarly,

$$v_2 = \frac{d[C]}{dt} = k_2[B]$$

$$= \frac{k_1 k_2 [A]_0}{k_2 - k_1}[\exp(-k_1 t) - \exp(-k_2 t)]$$

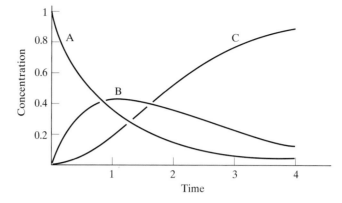

▶ FIGURE 7.8

Kinetic curves for reactant, A, intermediate, B, and product, C, for two first-order reactions in series (A → B → C).

Variables [C] and t can be separated here to give, for $[C]_0 = 0$,

$$[C] = [A]_0 \left\{ 1 - \frac{1}{k_2 - k_1} [k_2 \exp(-k_1 t) - k_1 \exp(-k_2 t)] \right\} \quad (7.35)$$

Where such complex formulas are involved, it is important to get some intuitive feeling for how these reactions behave. Figure 7.8 illustrates the pattern for series first-order reactions. As the concentration of A decreases, it does so in a simple exponential fashion. If there is no B present initially, the rate of the second reaction will be negligible at the beginning, $[v_2]_0 = 0$. As B is formed by the first step, its concentration rises, reaches a maximum, and then declines to zero as the second step removes it from the reaction mixture. In accordance with the rate law, the curve for [C] starts out with zero slope, where $[v_2]_0 = 0$. The rising portion of curve [C] then undergoes an inflection (the point where $d^2[C]/dt^2 = 0$) when [B] has its maximum value. Finally, [C] approaches a limiting value as the reaction nears completion. These properties are useful diagnostics for series reactions.

When two or more reactions occur in series, the slow reaction is the rate-determining step. It dominates in the kinetic control of the overall process. The analogy with a city water-distribution system is appropriate. It does not matter how large the reservoir, water mains, and distribution pipes are. If the smallest cross section of the whole system is the $\frac{1}{2}$-in. pipe coming into your house, that will limit the maximum possible flow, and the only practical way to increase the flow is to replace the small pipe with one of larger diameter.

For the series reactions

$$A \xrightarrow{k_1} B \xrightarrow{k_2} C$$

we can consider the two cases $k_1 \gg k_2$ and $k_1 \ll k_2$. The first situation means that the first step is much faster than the second, and the second is the rate-determining step. Once the reaction is started, A will be converted to B rapidly in comparison with the subsequent reaction of B to C. During most of the course of the reaction, B undergoes a first-order conversion to C, which is controlled by the rate constant k_2. If the formation of C is our measure of the rate of the reaction, its appearance is nearly identical to that of a single-step (elementary) first-order reaction. This is illustrated schematically in figure 7.9. We arrive at the same conclusion from examination

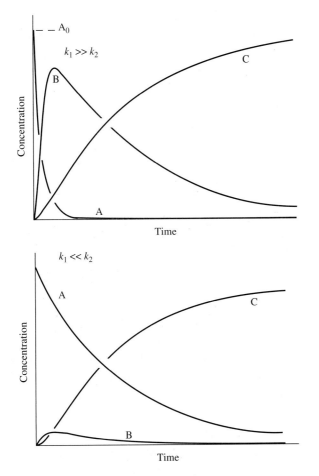

◀ FIGURE 7.9

Two first-order reactions in series
(A → B → C), with the *second* as
the rate-determining step.

◀ FIGURE 7.10

Two first-order reactions in series
(A → B → C), with the *first* as the
rate-determining step.

of Eq. (7.35) in the limit where $k_1 \gg k_2$. Under those conditions, Eq. (7.35) reduces to

$$[C] = [A]_0[1 - \exp(-k_2 t)]$$

for $k_1 \gg k_2$, which has the same form as the simple first-order formation kinetics.

The second situation, where $k_1 \ll k_2$, starts out with a slow conversion of A to B and is followed by a very rapid reaction for B going on to C. In this case, the concentration of B remains small throughout the course of the reaction, and C appears essentially as A disappears. This is illustrated in figure 7.10. The rate of the reaction can be measured either by the formation of C or by the disappearance of A. Thus, Eq. (7.33) applies to the overall reaction.

The chief reason for emphasizing rate-limiting steps in series reactions is because complex processes in biology frequently have rate-limiting steps. For our analysis, it does not matter how many steps are involved in the series mechanism; all that is required is that one step should be appreciably slower than any of the others. If the rate-limiting step is not first order, as in the example treated here, it must be treated using the kinetic equations for the appropriate order.

EXAMPLE 7.4

A complex but illustrative example of the kinetic behavior of series reactions occurs in the clotting of blood. The final step in the process is the conversion of fibrinogen to fibrin (the clotting material) catalyzed by the proteolytic (protein cutting) enzyme thrombin. Thrombin is not normally present in the blood but is converted from a precursor, prothrombin, by the action of another activated proteolytic enzyme and calcium ion. The overall process involves a cascade of at least eight such steps:

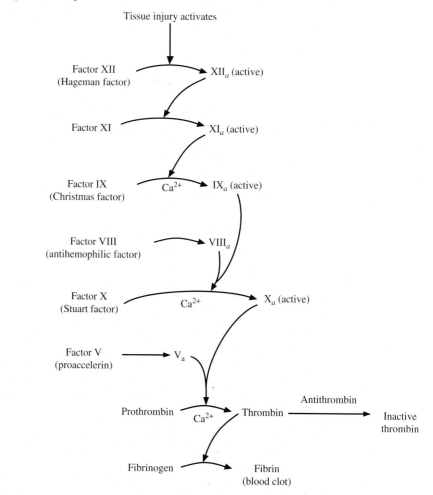

A consequence of this cascade of steps is the fact that the thrombin concentration, as measured by the decrease in clotting time for blood samples, increases dramatically following wounding and then decreases again once the clot is formed. The behavior, shown in figure 7.11, is very much like that of an intermediate in a series of reactions. In the proposed mechanism, the injury sets off a cascade of reactions involving a series of proteolytic enzymes, each of which activates a target protein. This cascade phenomenon serves to accelerate the mobilization of the clotting mechanism so that it can be switched from fully off to fully on in less than 1 min. The conversion of prothrombin to thrombin, catalyzed by active proaccelerin and Ca^{2+}, is the immediate cause of clotting. The level of thrombin in the blood must not remain high, however, or it will cause fibrin to form in the circulatory system and block the normal flow of blood. Antithrombin is an additional factor that inactivates thrombin, usually within a few minutes after it has reached its maximum level. In this way, the action of thrombin is restricted to the period when it is critically needed.

Although this overall process is much more elaborate than the simple example of series reactions considered later, a similarity is seen when we inspect the

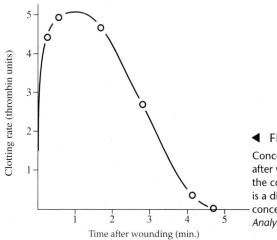

◀ FIGURE 7.11

Concentration of active thrombin after wounding. The clotting rate for the conversion of fibrinogen to fibrin is a direct measure of the thrombin concentration. (From R. Biggs, 1953, *Analyst* 78:84.)

formation and disappearance of thrombin as rate-limiting steps. For this purpose, prothrombin plays the role of A, thrombin is B, and inactive thrombin is C. The appearance of active proaccelerin turns on the first step (A → B), and the subsequent formation of antithrombin turns on the second step (B → C). The rate of clotting of fibrinogen serves (kinetically speaking) as a simple indicator of the level of thrombin present at any time in the course of this process.

Equilibrium and Kinetics

All reactions approach equilibrium. Therefore, in principle, for every forward reaction step, there is a backward step. In practice, we can sometimes ignore the backward step because either the equilibrium constant is very large or the concentrations of products are kept very small. Nevertheless, it is important to know the general relation between kinetic rate constants (k) and thermodynamic equilibrium constants (K).

Let's consider an elementary first-order reversible reaction:

$$A \underset{k_{-1}}{\overset{k_1}{\rightleftharpoons}} B$$

The rate of disappearance of A is

$$-\frac{d[A]}{dt} = k_1[A] - k_{-1}[B]$$

At equilibrium, $-d[A]/dt = 0$; therefore,

$$\frac{[B]^{eq}}{[A]^{eq}} = \frac{k_1}{k_{-1}} = K$$

EXERCISE 7.2

Show that for the elementary first-order reversible reaction

$$A \underset{k_{-1}}{\overset{k_1}{\rightleftharpoons}} B$$

the approach to equilibrium concentrations starting from any concentrations of A and B is a first-order reaction with a rate constant equal to the sum of the rate constants for the forward and backward reactions.

$$\ln \frac{[A] - [A]^{eq}}{[A]_0 - [A]^{eq}} = \ln \frac{[B] - [B]^{eq}}{[B]_0 - [B]^{eq}} = -(k_1 + k_{-1})t$$

where $[A]_0$ and $[B]_0$ are the concentrations of A and B, respectively, at $t = 0$.
Hint: Let $x = [A] - [A]^{eq} = [B]^{eq} - [B]$, then

$$-\frac{d[A]}{dt} = k_1[A] - k_{-1}[B]$$

$$= k_1([A]^{eq} + x) + k_{-1}(x - [B]^{eq})$$

But because

$$-\frac{d[A]^{eq}}{dt} = k_1[A]^{eq} - k_{-1}[B]^{eq} = 0$$

we therefore obtain

$$-\frac{d[A]}{dt} = -\frac{dx}{dt} = (k_1 + k_{-1})x$$

$$\ln \frac{x}{x_0} = -(k_1 + k_{-1})t$$

From the definition of x, we obtain the equation we set out to prove:

$$\ln \frac{[A] - [A]^{eq}}{[A]_0 - [A]^{eq}} = \ln \frac{[B] - [B]^{eq}}{[B]_0 - [B]^{eq}} = -(k_1 + k_{-1})t$$

Often there is more than one path for the reaction of A to form B. To be consistent with the principles of equilibrium thermodynamics, we must apply the principle of *microscopic reversibility*. The principle requires that at equilibrium the reactions between any pair of reactant, intermediate, or product species must occur with equal frequency in both directions, that is, the following mechanism is *not* possible:

Each elementary step in the reaction must be reversible to be consistent with the principle of microscopic reversibility. Thus, we must formulate the mechanism as follows:

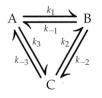

The relation to thermodynamics requires further that

$$K = \frac{[B]^{eq}}{[A]^{eq}} = \frac{[B]^{eq}}{[C]^{eq}} \frac{[C]^{eq}}{[A]^{eq}}$$

and thus

$$K = \frac{k_1}{k_{-1}} = \frac{k_2}{k_{-2}} \frac{k_3}{k_{-3}} \tag{7.36}$$

The six rate constants are therefore not independent.

Complex Reactions

For many reactions, particularly those involving enzymes as catalysts, a series of reversible steps is involved. An example is the following set of coupled elementary reactions:

$$A + B \underset{k_{-1}}{\overset{k_1}{\rightleftharpoons}} X$$

$$X \underset{k_{-2}}{\overset{k_2}{\rightleftharpoons}} P + Q$$

In this case, the exact solution to the rate equations is complex because two of the elementary reactions are bimolecular and a total of five molecular species are involved. It is useful to learn some approximations that can be applied to these complex reactions.

Initial-rate approximation

At the beginning of a reaction, the product concentrations are usually very small or zero. For the set of the above opposed reactions, the step designated -2 can be neglected in an analysis of *initial* velocities. This simplifies the kinetic expressions considerably. As the extent of reaction increases, the experimental results will begin to differ from the prediction of the approximate theory.

Prior-equilibrium approximation

This approach is valuable when steps 1 and -1 are rapid in comparison with step 2:

$$A + B \underset{k_{-1}}{\overset{k_1}{\rightleftharpoons}} X \qquad \text{(fast, equilibrium)}$$

$$X \overset{k_2}{\rightleftharpoons} P + Q \qquad \text{(slow)}$$

A, B, and X rapidly attain a state of quasi equilibrium, such that

$$v_1 = v_{-1}$$

and

$$k_1[A][B] = k_{-1}[X]$$

We can write this as an equilibrium expression:

$$K = \frac{k_1}{k_{-1}} = \frac{[X]}{[A][B]}$$

Step 2 is the rate-limiting step, and the rate of formation of product is given by

$$v = \frac{d[P]}{dt} = \frac{d[Q]}{dt} = k_2[X]$$

Substituting from above, we see that the rate

$$v = \frac{k_2 k_1}{k_{-1}}[A][B] \tag{7.37}$$

can be expressed in terms of reactant concentrations only. This is particularly valuable if the relative concentration of X remains very small ($k_{-1} \gg k_1$), but is just as valid when X is fairly large ($k_{-1} \sim k_1$). The important criterion for the prior-equilibrium approximation to apply can be written

$$v \cong v_2 \ll v_1 \cong v_{-1}$$

This should be read: "The overall rate of the reaction is limited by the slow step 2, and the rate is much slower than the forward and reverse reactions of step 1, which are essentially in equilibrium."

Steady-state approximation

It frequently happens that an intermediate is formed that is very reactive. As a consequence, it never builds up to any significant concentration during the course of the reaction:

$$A + B \xrightarrow{k_1} X \qquad (\text{slow})$$
$$X + D \xrightarrow{k_2} P \qquad (\text{fast})$$

To a first approximation, X reacts as rapidly as it is formed:

$$v_1 = v_2$$
$$k_1[A][B] = k_2[X][D] \tag{7.38}$$

Hence,

$$v = \frac{d[P]}{dt} = k_1[A][B]$$

in terms of reactant concentrations only. A more general way of formulating the steady-state approximation is to consider all steps involving formation and disappearance of the reactive intermediate and to set the sum of their rates equal to zero. In this example, X is formed in step 1, and it disappears in step 2. Thus,

$$\frac{d[X]}{dt} = k_1[A][B] - k_2[X][D] \cong 0 \tag{7.39}$$

This gives the same result [Eq. (7.38)]. However, this second approach is more general. The significance of this equation is not that [X] is constant during the course of the reaction. Such is never the case. It is true, however, that the *slope, d*[X]/*dt*, of the curve of [X] versus time is much smaller than those typical of reactants or products. For example, this can be seen for component B for the scheme shown in figure 7.10. It is this nearly zero slope throughout the course of the reaction that is the characteristic of the steady-state condition.

The steady-state approximation can be applied in many chemical and biochemical reactions. Care must be taken that the intermediate satisfies the criterion that $d[X]/dt \cong 0$. Note that this is not necessarily the case where prior equilibrium is involved. It is also not true for the intermediate B illustrated in figure 7.8.

Deducing a Mechanism from Kinetic Data

There is no straightforward way to obtain a mechanism from kinetic data. There is always more than one mechanism that will be consistent with the kinetic data. Therefore, why propose a mechanism at all? We like to think that we understand a reaction and that we can guess what the molecules are doing. A mechanism gives us some basis for predicting what should happen for other reactions.

The best way to illustrate how to hypothesize a mechanism is to discuss specific examples. The only general advice that can be given is to think of a simple and plausible mechanism and then calculate the kinetics and see if they are consistent with the data. A unique rate equation can always be obtained from a proposed mechanism. If the rate law is simple first or second order, assume a unimolecular or bimolecular rate-determining step. Then, if necessary, add other steps to agree with the stoichiometry. For example, suppose for the stoichiometric reaction

$$A + B \longrightarrow C$$

the rate law is found to be

$$-\frac{d[A]}{dt} = k[A][OH^-]$$

We can assume the mechanism

$$A + OH^- \xrightarrow{k_1} M^-$$

$$M^- + B \xrightarrow{k_2} C + OH^-$$

The first elementary reaction postulates that $A + OH^-$ can react to form an intermediate, M^-; this gives the correct rate law. The second elementary reaction must be added to account for the stoichiometry. From the proposed mechanism, we can now predict rate laws for [B] and [C] and experimentally check the predictions. Another mechanism consistent with the data is

$$A + OH^- \overset{K}{\rightleftharpoons} N^- \qquad \text{(fast to equilibrium)}$$

$$N^- \xrightarrow{k_1} P^- \qquad \text{(slow, rate determining)}$$

$$P^- + B \xrightarrow{k_2} C + OH^- \qquad \text{(fast)}$$

The rate-determining step is

$$-\frac{d[N^-]}{dt} = k_1[N^-]$$

but

$$\frac{[N^-]}{[A][OH^-]} = K$$

Therefore,

$$-\frac{d[N^-]}{dt} = k_1 K[A][OH^-]$$

Because all other elementary reactions are fast, each time N^- reacts to form P^-, A and B react and C is formed. The rate laws are then

$$-\frac{d[A]}{dt} = -\frac{d[B]}{dt} = \frac{d[C]}{dt} = k_1 K[A][OH^-]$$

We can do other kinetic experiments to decide between the two mechanisms and can make attempts to detect the postulated intermediates. Many other mechanisms could also be written. It should be clear that proposing reasonable mechanisms requires practice and experience.

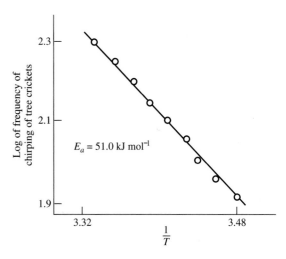

► FIGURE 7.12

Arrhenius plot of the chirping rate of tree crickets at different temperatures. The linearity of this plot suggests that the complex biological process is governed by a rate-limiting step having the activation energy shown. The chirping frequency changes by over a factor of 2 in the temperature range from 285 K to 300 K (54°F to 81°F). (From K. J. Laidler, 1972, *J. Chem. Educ.* 49:343.)

Temperature Dependence

It seems almost intuitive that if you want a reaction to proceed faster, you should heat it. It came as a surprise when some chemical reactions were found to have negative temperature coefficients; that is, the reaction was *slower* at higher temperature. In biology this is a common occurrence. Few organisms or biochemical processes can withstand temperatures above 50°C for very long because of irreversible structural changes that alter the mechanisms or destroy the organism.* Processes such as coagulation of the protein egg albumin (egg white) occur during cooking, and they occur more rapidly at high temperatures. This seems normal. However, if a protein is first carefully denatured, it can be renatured more rapidly at a lower temperature than at a higher one! There is a kinetic effect in addition to the shift in equilibrium constant for this process.

Observations by Arrhenius and others showed that for most chemical reactions there is a logarithmic relation between the rate constant and the reciprocal of the temperature in kelvins. We can therefore write the Arrhenius result as

$$\ln k = -\left(\frac{E_a}{RT}\right) + \ln A \tag{7.40}$$

or

$$k = A \exp\left(-\frac{E_a}{RT}\right) \tag{7.41}$$

The *activation energy* for the reaction is E_a and A is called the pre-exponential factor. This is an empirical equation that represents the behavior of many chemical systems and even some complex biological processes (figure 7.12). The usual way of obtaining the activation energy is to plot $\ln k$ versus $1/T$. The slope of the line is equal to $-E_a/R$, with the units

---

*There are, however, *thermophiles*—organisms that live at temperatures of 100 °C or higher in hot pools in Yellowstone Park and in hot deep-sea vents.

of E_a dependent on the value of R. Once E_a is known, Eq. (7.41) can be solved for A. An alternative method of obtaining E_a and A is illustrated in example 7.5.

EXAMPLE 7.5

The decomposition of urea in 0.1 M HCl occurs according to the reaction

$$NH_2CONH_2 + 2H_2O \longrightarrow 2NH_4^+ + CO_3^{2-}$$

The first-order rate constant for this reaction was measured as a function of temperature, with the following results:

Expt.	Temperature (°C)	k (min^{-1})
1	61.0	0.713×10^{-5}
2	71.2	2.77×10^{-5}

Calculate the activation energy E_a and the A factor in Eq. (7.41) for this reaction.

Solution

This problem can be solved either numerically or graphically. The numerical solution is given here. Since $\ln k = -E_a/RT + \ln A$, we will first calculate $\ln k$ and $1/T$ for each experiment.

Expt.	$\ln k$	$\dfrac{1}{T}$ (K)$^{-1}$
1	-11.85	2.922×10^{-3}
2	-10.49	2.904×10^{-3}

We can write

$$\ln k_2 - \ln k_1 = -\frac{E_a}{R}\left(\frac{1}{T_2} - \frac{1}{T_1}\right)$$

Combining experiments 1 and 2 in this way, we obtain

$$E_a = -\frac{R(\ln k_2 - \ln k_1)}{1/T_2 - 1/T_1}$$

$$= -\frac{0.008314 \; kJ \; k^{-1} \, mol^{-1} \, (-10.49 + 11.85)}{(2.904 - 2.992) \times 10^{-3} \, K^{-1}}$$

$$= 128.5 \; kJ \; mol^{-1}$$

Using the data from either experiment, we can calculate a value for A from

$$\ln A = \ln k + \frac{E_a}{RT}$$

Using the data of experiment 2,

$$\ln A = -10.49 + \frac{128.5}{(0.008314)(344.4)}$$

$$= -10.49 + 44.84 = 34.35$$

Therefore,

$$A = 8.28 \times 10^{14} \text{ min}^{-1}$$

$$= 1.38 \times 10^{13} \text{ s}^{-1}$$

Note that the value for A is in the range of infrared vibrational frequencies. This means that with enough energy—at very high temperatures—the reaction occurs about as rapidly as the atoms can move.

To understand the meaning of the activation energy, consider the elementary reaction $M + N \rightarrow P$ with an energy barrier between reactants and products, as illustrated in figure 7.13. In this figure, we plot the energy versus the *reaction coordinate*. The reaction coordinate is a convenient way to represent the change of the reactants into products as a reaction takes place. It is not a simple coordinate, such as the x-coordinate for a point in space; instead, it can represent the positions of all the significant atoms in the reacting molecules. For example, in the reaction

$$OH^- + CH_3Br \longrightarrow CH_3OH + Br^-$$

the reaction coordinate represents the concerted motions of the OH^- approaching the central carbon, while the Br^- is leaving and the three hydrogen atoms are also moving. We can plot any property versus the reaction coordinate, but energy or free energy is particularly useful.

The molecules M and N must have sufficient energy to react. Because the process must be reversible at the microscopic level, the top of the barrier will be the same for the reverse reaction. The energy of the products of a reaction is usually different from that of the reactants; the energy difference, ΔE, can be positive or negative.

$$\Delta E = E_P - (E_M + E_N)$$

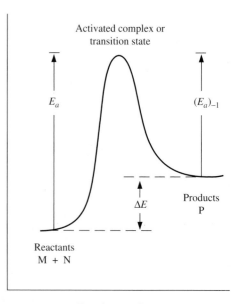

▶ **FIGURE 7.13**

Energy changes that occur throughout the course of a representative reaction. The reaction coordinate is a variable that measures the progress from reactants to transition state to product. E_a is the activation energy for the forward reaction, and $(E_a)_{-1}$ is the activation energy for the reverse reaction.

Therefore, the activation energy for the reverse reaction is usually not the same as for the forward reaction. The relation

$$(E_a)_{\text{forward}} - (E_a)_{\text{reverse}} = \Delta E \tag{7.42}$$

does hold, however, at least for elementary reactions.

As the temperature is increased, the fraction of molecules having energy greater than E_a increases, and the rate of reaction correspondingly increases. In some cases, correlations can be made between the activation energy and the energy of a bond that must be broken in order for a reaction to proceed.

Very detailed theories of chemical kinetics exist that can predict rate constants for simple, gas-phase reactions. We discuss more general theories that are useful mainly in helping us understand the factors that affect the kinetics.

In the *collision theory* of chemical kinetics, we assume that, to react, molecules must collide (including collisions at the walls of the vessel in many cases). The rate of a reaction whose mechanism is

$$M + N \longrightarrow P$$

is given by

$$v = k[\text{M}][\text{N}] = A[\text{M}][\text{N}]\exp\left(-\frac{E_a}{RT}\right) \tag{7.43}$$

At high temperatures, the exponential approaches unity, and $k \cong A$:

$$\lim_{T \to \infty} A \exp\left(-\frac{E_a}{RT}\right) = A$$

The maximum rate of this reaction at high temperature would be $A[\text{M}][\text{N}]$, which is therefore interpretable as a collision frequency for the molecules M and N. The exponential factor resembles a Boltzmann probability distribution (see chapter 6). It can be considered to represent the fraction of molecules with proper orientation that have sufficient energy to react.

Examined in detail, the collision theory has serious shortcomings. Collision frequencies of molecules in the gas phase can be calculated quite well using kinetic theory of gases (see chapter 6). From Eq. (7.43), theoretical values for the rate constant of a bimolecular gas-phase reaction can be calculated. These are on the order of 10^{11} mol^{-1} L s^{-1} for small molecules, somewhat dependent on the molecular radii. Experimentally, values of A range widely, from 10^2 to 10^{13} mol^{-1} L s^{-1} for reactions that are thought to be relatively simple. Although a few reactions proceed much faster than the rates calculated from simple collision theory, many are orders of magnitude slower. Explaining these results by collision theory is difficult.

Transition-State Theory

An alternative model was introduced in 1935 by Eyring and others. Its significant new feature was that it considered the reactive intermediate, or *transition state** (an unstable species at a free-energy maximum), to be just like a stable molecule, except for motion of atoms along the reaction coordinate between reactants and products. Thus, the transition state is a molecule lasting

---

*The transition state was originally called an *activated complex*.

only a few molecular vibrations. When the transition-state theory was proposed, it was not possible to study a transition state because of its short lifetime. However, transition states in the gas phase now have been characterized spectroscopically (Zewail and Bernstein 1988; Gruebele and Zewail 1990) using laser pulses of femtoseconds (1 fs = 10^{-15} s) to picoseconds (1 ps = 10^{-12} s) in length.

In the transition-state theory, every elementary reaction

$$\text{M} + \text{N} \xrightarrow{k} \text{P}$$

is rewritten as

$$\text{M} + \text{N} \underset{}{\overset{K^{\ddagger}}{\rightleftharpoons}} \text{MN}^{\ddagger} \xrightarrow{k^{\ddagger}} \text{P}$$

It is assumed that reactants are in rapid equilibrium with the transition state, $\text{MN}^{\ddagger}$. Here $K^{\ddagger}$ is the equilibrium constant between the transition state and reactants, and $k^{\ddagger}$ is a universal rate constant:

$$K^{\ddagger} = \frac{[\text{MN}^{\ddagger}]}{[\text{M}][\text{N}]} \tag{7.44}$$

$$\frac{d[\text{P}]}{dt} = k^{\ddagger}[\text{MN}^{\ddagger}] \tag{7.45}$$

Neither $K^{\ddagger}$ nor $k^{\ddagger}$ are experimentally measurable from Eqs. (7.44) and (7.45) because the concentration of the transition state is not measurable. However, $K^{\ddagger}$ and $k^{\ddagger}$ can be related to concentrations of reactants and to measurable rates. Combining Eqs. (7.44) and (7.45), we obtain

$$\frac{d[\text{P}]}{dt} = k^{\ddagger}K^{\ddagger}[\text{M}][\text{N}] \tag{7.46}$$

The experimental rate expression is

$$\frac{d[\text{P}]}{dt} = k[\text{M}][\text{N}]$$

So we have related an experimental rate constant k to properties of the transition state:

$$k = k^{\ddagger}K^{\ddagger} \tag{7.47}$$

But every equilibrium constant is related to the standard free energy of the reaction

$$\Delta G^{\ddagger} = -RT \ln K^{\ddagger}$$

and transition-state theory shows that the universal rate constant $k^{\ddagger}$ is equal to

$$k^{\ddagger} = \frac{k_{\text{B}}T}{h} \tag{7.48}$$

where k_{B} is the Boltzmann constant* and h is Planck's constant. Combining Eqs. (7.47) and (7.48), we obtain

$$k = \frac{k_{\text{B}}T}{h}K^{\ddagger} = \frac{k_{\text{B}}T}{h}\exp\left(-\frac{\Delta G^{\ddagger}}{RT}\right) \tag{7.49}$$

---

*In this section only, we use k_{B} for the Boltzmann constant to distinguish it from rate constants.

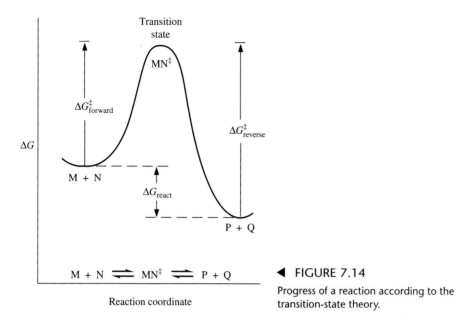

M + N $\rightleftharpoons$ MN‡ $\rightleftharpoons$ P + Q

Reaction coordinate

◀ FIGURE 7.14

Progress of a reaction according to the transition-state theory.

ΔH is

This equation permits a calculation of the free energy of activation, $\Delta G^{\ddagger}$, from a measured rate constant. The rate constant must have units (s^{-1}) consistent with Boltzmann's and Planck's constants; the concentration units (usually M) specify the standard state for the free energy. The same derivation could be given for the reverse reaction. So for any reaction, the picture shown in figure 7.14 holds. The free-energy difference between products and reactants is equal to the difference in activation free energies:

$$\Delta G_{react} = \Delta G^{\ddagger}_{forward} - \Delta G^{\ddagger}_{reverse} \qquad (7.50)$$

Additional insight comes if we consider the distinction between energy (enthalpy) and entropy contributions. At constant temperature,

$$\Delta G^{\ddagger} = \Delta H^{\ddagger} - T\Delta S^{\ddagger} \qquad (7.51)$$

where $\Delta H^{\ddagger}$ and $\Delta S^{\ddagger}$ are the activation enthalpy and entropy, respectively. Substituting into Eq. (7.49),

$$k = \frac{k_B T}{h} \exp\left(\frac{\Delta S^{\ddagger}}{R}\right)\exp\left(-\frac{\Delta H^{\ddagger}}{RT}\right) \qquad (7.52)$$

From this equation we can see that transition-state theory predicts a slightly different temperature dependence from that of Arrhenius, although the main temperature dependence for both are in the exponential terms. According to Arrhenius [Eq. (7.40)], a plot of ln k versus $1/T$ gives a straight line with slope equal to $-E_a/R$. In transition-state theory, a plot of ln k versus $1/T$ has a slope of $-(\Delta H^{\ddagger} + RT)/R$. This is seen by taking the derivative of ln k with respect to $1/T$. For most reactions, $\Delta H^{\ddagger}$ is much larger than RT, so the slope is nearly equal to $-\Delta H^{\ddagger}/R$, analogous to the Arrhenius theory. Thus,

$$\Delta H^{\ddagger} \approx E_a \qquad (7.53)$$

The value of $\Delta S^{\ddagger}$ can be obtained from Eq. (7.52) and $\Delta H^{\ddagger}$; it can also be calculated from the Arrhenius pre-exponential term A. Substituting Eq. (7.53) into Eq. (7.52) and comparing the result with Eq. (7.41), we obtain

$$\Delta S^{\ddagger} \approx R\left(\ln \frac{Ah}{k_B T}\right) \qquad (7.54)$$

In Eq. (7.54) we use an average T in the temperature range of interest.

The chief advantage of the transition-state theory is that it relates kinetic rates to thermodynamic properties of reactants and a transition intermediate. This can help us understand some of many factors that influence the rates. Entropies of activation are typically negative for simple gas-phase reactions. They reflect the loss in randomness in the transition state relative to the reactants.

Consider the bimolecular dimerization of butadiene in the gas phase:

$$2C_4H_6 \xrightarrow[-\Delta S^{\ddagger} = +68 \text{ J K}^{-1} \text{ mol}^{-1}]{} (C_4H_6 \cdots C_4H_6)^{\ddagger} \longrightarrow C_8H_{12}$$

The fairly large negative entropy of activation corresponds to the loss of three translational degrees of freedom in going from two independent reactant particles to the single one in the associated transition state.

EXAMPLE 7.6

Calculate the entropy of activation at 71.2°C for the reaction in example 7.5 involving the decomposition of urea.

Solution

Using the approximation that $\Delta H^{\ddagger} = E_a$, we use Eq. (7.54), remembering that A must have units of s^{-1}. Therefore, for $A = 1.38 \times 10^{13} \text{ s}^{-1}$,

$$\Delta S^{\ddagger} = R\left(\ln A - \ln \frac{k_B T}{h}\right)$$

$$= (8.314 \text{ J K}^{-1} \text{ mol}^{-1})\left[30.256 - \ln \frac{(1.318 \times 10^{-23})(344.4)}{6.626 \times 10^{-34}}\right]$$

$$= (8.314 \text{ J K}^{-1} \text{ mol}^{-1})(30.256 - 29.602)$$

$$= 5.44 \text{ J K}^{-1} \text{ mol}^{-1}$$

The entropy of activation for this reaction is small. This is consistent with our conclusion in example 7.5 that, in the absence of an energy barrier, this reaction occurs as rapidly as the atoms can move. For reactions in which the solvent water is able to interact less favorably with the transition state than with the reactant, a positive entropy of activation is often observed.

Electron Transfer Reactions: Marcus Theory

Theoretical calculations for the energies along possible reaction paths have been done for reactions of diatomic or triatomic molecules in the gas phase. This leads to an energy surface for the reaction and predictions of the reactivity of each vibrational and rotational state of the molecules. The predictions can be tested by molecular beam experiments, which provide kinetic data for molecules colliding with different amounts of kinetic energy (translational, vibrational, and rotational) and different orientations relative to each other (Lee 1987). Reaction cross sections as a function of energy and angle of collision are measured. Values of the rate constant, its temperature dependence, and thus $\Delta H^{\ddagger}$ and $\Delta S^{\ddagger}$ have been calculated for reactions in the gas phase. These parameters represent combinations of the detailed reaction cross sections that have been obtained by molecular beam experiments.

Theoretical calculations for solution reactions are in general much more difficult, but success has been obtained for charge-transfer reactions such as an electron transfer from Fe^{2+} to Ce^{4+} to produce Fe^{3+} and Ce^{3+} in aqueous solution. Electron-transfer reactions do not involve bond breakage or formation, but they are crucial in photosynthesis and in biochemical oxidation–reduction reactions. In 1956 Marcus derived a theory for the rates of electron-transfer reactions based on some simple ideas (see Nobel lecture, Marcus 1993). He thought about the simplest electron-transfer reaction—one in which the products and reactants are the same but one of the reactants is radioactively labeled (designated by *):

$$Fe^{2+} + {}^*Fe^{3+} \longrightarrow Fe^{3+} + {}^*Fe^{2+}$$

or

$$Ce^{4+} + {}^*Ce^{3+} \longrightarrow Ce^{3+} + {}^*Ce^{4+}$$

What controls the rates of these reactions? Clearly, the free-energy change for both reactions is zero because there is no net change. What about the free energies of activation that characterize the rates? The motions of electrons are very fast compared with motions of atoms, so Marcus reasoned that the reorientation of water molecules around the ions must be the rate-limiting process. The radius of Fe^{3+} is smaller than that of Fe^{2+}, and the larger charge on Fe^{3+} orients the surrounding water molecules more tightly. Therefore, the orientations of the water molecules must be different for reactant and product ions, and the transition state (where the electron transfer occurs) must correspond to an unstable configuration of water molecules that is in between the stable configurations for products and reactants. Starting from these simple reactions, Marcus derived an expression for the free energy of activation $\Delta G^{\ddagger}$ of electron-transfer reactions in general:

$$\Delta G^{\pm} = \frac{\lambda}{4}\left(1 + \frac{\Delta G^{\circ}}{\lambda}\right)^2 \qquad (7.55)$$

ΔG° is the standard free enrgy of the reaction, and the factor λ is a reorganization energy with contributions from the solvent and from the reactants (when the reactants are molecules, not spherical metal ions). Note that for the ionic reactions above where the standard free-energy change $\Delta G^{\circ} = 0$, the free energy of activation is simply $\lambda/4$. If the solvent is treated as a continuum with refractive index n and dielectric constant ε, the Marcus solvent reorientation energy has a simple form:

$$\lambda_{\text{solv}} = (\Delta e)^2 \left(\frac{1}{2a_1} + \frac{1}{2a_2} - \frac{1}{R}\right)\left(\frac{1}{n^2} - \frac{1}{\varepsilon}\right) \qquad (7.56)$$

Δe = charge transferred
a_1, a_2 = radii of the ions
R = center-to-center distance between the reactants
n = refractive index of solvent
ε = dielectric constant (relative permittivity) of solvent

Equation (7.56) has the form of the Coulomb energy of interaction of two charges q at distance r in a dielectric medium ($q^2/\varepsilon r$). To calculate λ, it is convenient to measure charge in electrons transferred ($\Delta e = 1, 2, \ldots,$) and distances in Å (10^{-10} m). To obtain λ in kJ mol^{-1}, the results must be multiplied by 1389. For water, the refractive index is about 1.33, and the dielectric constant is about 80; both are unitless. Marcus' Nobel Prize address (1993) describes what led him to the problem and its solution; he also discusses applications to biochemical reactions.

Ionic Reactions and Salt Effects

The effect of the nonideal behavior of reactants in solution can be treated by extending the transition-state theory to include activities and activity coefficients explicitly. Equation (7.45) is written as

$$v = k^{\ddagger}c^{\ddagger} = \frac{k_B T}{h}c^{\ddagger} \tag{7.57}$$

and Eq. (7.44) can be rewritten as

$$K^{\ddagger} = \frac{[MN^{\ddagger}]}{[M][N]} = \frac{c^{\ddagger}\gamma^{\ddagger}}{c_M\gamma_M c_N\gamma_N} \tag{7.58}$$

where the c's are concentrations and the γ's are activity coefficients. Combining Eqs. (7.57) and (7.58) and substituting into Eq. (7.46), we obtain the result

$$v = \frac{k_B T}{h}K^{\ddagger}\frac{\gamma_M\gamma_N}{\gamma^{\ddagger}}c_M c_N \tag{7.59}$$

In this case, the rate constant can be written as

$$k = \frac{k_B T}{h}K^{\ddagger}\frac{\gamma_M\gamma_N}{\gamma^{\ddagger}} = k_0\frac{\gamma_M\gamma_N}{\gamma^{\ddagger}} \tag{7.60}$$

where k_0 is the rate constant extrapolated to infinite dilution of all species in solution; at infinite dilution, all activity coefficients = 1.

For reactions involving ionic species, we can use the Debye–Hückel limiting law (chapter 4) to express the dependence of the rate constant k on ionic strength. For each ionic species, we use

$$\log\gamma_i = -0.51Z_i^2\sqrt{I}$$

$$I = \frac{1}{2}\sum_i^{\text{all species}} c_i Z_i^2 \tag{7.61}$$

where the constant 0.51 applies to water solutions at 25°C, Z_i is the charge on species i, and I is the ionic strength. The charge relation for the reaction $M + N \rightarrow MN^{\ddagger}$ is clearly $Z_M + Z_N = Z^{\ddagger}$, and we can write the logarithm of Eq. (7.60) as

$$\log k = \log k_0 - 0.51[Z_M^2 + Z_N^2 - (Z_M^2 + Z_N^2)^2]\sqrt{I}$$

$$= \log k_0 + 2(0.51)Z_M Z_N\sqrt{I} \tag{7.62}$$

Equation (7.62) predicts that a plot of $\log k$ versus $\sqrt{I}$ for dilute ionic solutions should give a straight line with a slope $2(0.51)Z_M Z_N \cong Z_M Z_N$. The slope then gives direct information about the charges of the reacting species. This has been verified for a variety of ionic reactions in dilute aqueous solutions. For example, when the rate of decomposition of H_2O_2 in the presence of HBr was studied as a function of ionic strength, it was found that the data fit Eq. (7.62) with a slope of -1. The stoichiometric reaction

$$H_2O_2 \longrightarrow H_2O + \frac{1}{2}O_2$$

does not involve ionic species, but the kinetics imply that H^+ and Br^- are involved in the formation of the transition-state species.

EXAMPLE 7.7

The acid denaturation of CO–hemoglobin has been studied at pH 3.42 in a formic acid–sodium formate buffer as a function of sodium formate ($Na^+ + {}^-OOCH$) concentration (Zaiser and Steinhardt 1951). The half-lives for the first-order denaturation are as follows:

NaOOCH (M)	0.007	0.010	0.015	0.020
$t_{1/2}$ (min)	20.2	13.6	8.1	5.9

Determine whether these results follow the Debye–Hückel limiting law in the dependence of the rate constant on ionic strength and, if so, what is the apparent charge on the CO–hemoglobin?

Solution

Since NaOOCH is an uni-univalent electrolyte, the ionic strength I is equal to the molar concentration c of sodium formate (we ignore the small H^+ concentration). For a first-order reaction,

$$k = \frac{\ln 2}{t_{1/2}} = \frac{0.693}{t_{1.2}}$$

Therefore,

$\sqrt{I}$	0.084	0.100	0.122	0.141
k (min^{-1})	0.0343	0.0510	0.0856	0.117
$\log k$	-1.465	-1.293	-1.068	-0.930

A plot of these data is shown in figure 7.15. Although the data show some curvature, there is a reasonably good fit to a straight line with a slope of 9. Since the denaturation occurs by reaction with monovalent H^+, this suggests a charge of approximately $+9$ for the CO–hemoglobin at pH 3.42.

Isotopes and Stereochemical Properties

The rate law, the activation energy, the ionic effects, and so on all provide important clues to the mechanism of a reaction. A number of other experimental and theoretical approaches have been devised; the use of isotopes and the

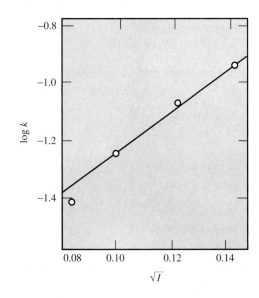

▶ FIGURE 7.15

Data for the rate of acid denaturation of CO–hemoglobin. (From E. M. Zaiser and J. S. Steinhardt, 1951, *J. Am. Chem. Soc.* 73:5568.)

examination of the stereochemical properties of reactants and products are two of the most powerful experimental methods to provide the structural details of a reaction. In the hydrolysis of an ester,

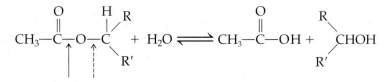

the rate law cannot tell us whether the bond cleaved is indicated by the solid or dashed arrow. When the reaction is carried out in ^{18}O- (a stable isotope of the more abundant ^{16}O) labeled H_2O, ^{18}O is found only in the acid and not in the alcohol produced. Thus, the cleavage must be occurring at the position indicated by the solid arrow. In the reaction between acetaldehyde bound to an enzyme,

and a nucleophilic (electron-donating) reagent X^-, the stereospecificity of the product is different depending on the direction of attack by X^-. The nucleophile can approach from above or below the plane containing the four atoms:

The two products are depicted as follows:

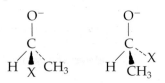

They are not superimposable but are mirror images (chapter 10). Therefore, from the stereospecificity of the product, we can deduce from which side of the plane the attacking group approaches. Kineticists have made extensive use of isotopic labeling and stereospecificity in deducing mechanisms; the two examples given provide a glimpse of how this might be achieved.

Very Fast Reactions

The rates of many reactions can be measured by mixing two solutions and measuring the change of concentrations with time. The concentrations are measured continuously by absorbance or other spectroscopic method, or the reaction is quenched and an appropriate analytical method is used. However, some reactions are too fast for this technique. Consider the very simple and very important reactions

$$H^+ + OH^- \longrightarrow H_2O$$

$$CH_3COO^- + H^+ \longrightarrow CH_3COOH$$

The kinetics of neutralization in aqueous solution occur in the reaction mechanisms for many reactions. Yet the kinetics of these reactions cannot be studied by mixing sodium hydroxide and hydrochloric acid or sodium acetate and hydrochloric acid. The reaction is over long before the solutions can be mixed thoroughly; therefore, the apparent rate of reaction is actually a measure of the rate of mixing. Methods that do not require mixing two solutions are needed for these very fast reactions. *Relaxation methods* allow you to measure a rate by perturbing an equilibrium and following the return of the system to equilibrium.

Relaxation Methods

Consider a reversible system:

$$A \underset{k_{-1}}{\overset{k_1}{\rightleftharpoons}} B$$

At equilibrium, the rates of the forward and reverse reactions are the same. The position of equilibrium is represented in terms of an equilibrium constant,

$$K = \frac{k_1}{k_{-1}} = \frac{[B]^{eq}}{[A]^{eq}}$$

which is the ratio of the rate constants k_1 and k_{-1}. The usual measurements of the "average" composition of the system at equilibrium can provide a value for only the ratio, not for the individual rate constants. Clearly, k_1 and k_{-1} can range from very large to very small and still give the same ratio (the equilibrium constant K). If we change the temperature or pressure of the system so as to attain a new equilibrium state, the new state will in general have a different value for the equilibrium constant. We are still completely ignorant of the individual rate constants, however.

Several approaches are possible to extract the desired rate constants and mechanistic information. One category involves the use of relaxation methods, in which the system at equilibrium is suddenly perturbed in such a way that it finds itself out of equilibrium under the new conditions. It then relaxes to a state of equilibrium under the new conditions, and the rate of relaxation is governed by the rate constants and mechanism of the process. Examples of relaxation processes include:

1. Temperature jump, which can be achieved by the discharge of an electric capacitor, the use of a rapid laser pulse, or by plunging the sample into hot water to produce a sudden increase in temperature of a system
2. Pressure jump, where a restraining diaphragm is suddenly ruptured to release pressurization of the system
3. Flash- or laser-pulse photolysis, where a pulse of light produces a sudden change in the population of electronically excited states of absorbing molecules. In cases where the excited state is more acidic than the ground state, this results in a jump in $[H^+]$

In each case, the system suddenly finds itself out of equilibrium at the instant of the perturbation, and the experimenter monitors the approach to a new equilibrium state under the altered conditions.

A second major category of methods for dealing with systems at equilibrium involves the study of fluctuations about the equilibrium state. Because of both the dynamic and statistical nature of systems at equilibrium, the system and many of its properties undergo constant and rapid fluctuations about some average or median state (see chapter 3). Any normal (long-term) measurements always give the same result if the system is truly at equilibrium, and these results serve to characterize the average equilibrium state. On a short-enough time scale, however, we observe that the properties of the system exhibit a kind of Brownian motion, in the sense that they are constantly moving one way or the other with respect to their average equilibrium value. This can be true for such properties as the following:

1. Refractive index, where fluctuations result in variations in scattering of light from laser beams
2. Concentration, which can be detected as fluctuations in the absorption or fluorescence of particular species
3. Position, involved in scattering of fine laser beams off single particles
4. Pressure, where fluctuations can produce "noise" in an acoustic detector

In each case, there is no need to perturb the system artificially; we simply take advantage of the fluctuations that occur naturally because of the particulate and statistical nature of matter.

Relaxation Kinetics

As an example of relaxation kinetics, consider an equilibrium of the type

$$A + B \underset{k_{-1}}{\overset{k_1}{\rightleftharpoons}} P$$

The rate expression is

$$\frac{d[P]}{dt} = k_1[A][B] - k_{-1}[P] \tag{7.63}$$

At equilibrium, $d[P]/dt$ is zero, and

$$K = \frac{k_1}{k_{-1}} = \frac{[\overline{P}]}{[\overline{A}][\overline{B}]} \tag{7.64}$$

where the bar over the concentration terms emphasizes the equilibrium values. Now generate a nearly discontinuous change (typically within 10^{-6} s, but it can be as short as 10^{-8} s) of temperature or pressure. Changing temperature or pressure will in general cause the equilibrium concentrations to change; we can use thermodynamics to tell us how large a change to expect. From chapter 3, we remember that

$$\left(\frac{\partial \Delta G}{\partial T}\right)_P = -\Delta S \quad \text{and} \quad \left(\frac{\partial \Delta G}{\partial P}\right)_T = \Delta V \qquad \textbf{(7.65)}$$

Combining these equations with $\Delta G^0 = -RT \ln K$, we obtain

$$\left(\frac{\partial \ln K}{\partial T}\right)_P = \frac{\Delta H^0}{RT^2} \quad \text{and} \quad \left(\frac{\partial \ln K}{\partial P}\right)_T = -\frac{\Delta V^0}{RT} \qquad \textbf{(7.66)}$$

Thermodynamic standard states for solutions include the requirement that the pressure is 1 atm. Therefore, we note that the pressure dependence of K refers to change in equilibrium concentrations, not activities. So Eq. (7.66) tells us that the equilibrium constant K will be different at the new temperature or pressure immediately after the pulse. How much different will depend on the enthalpy change of the reaction for a T jump or the volume change of the reaction for a P jump. The system finds itself suddenly out of equilibrium under the new conditions, and we can follow the concentration changes during the subsequent relaxation to the new equilibrium state.

Conditions are usually chosen so that the T jump or P jump produces relatively small displacements from equilibrium; that is, a jump of 5 to 10 degrees or 100 to 1000 atm is used. This simplifies not only the mathematics but also the thermodynamic analysis of the results. Suppose, following the sudden perturbation, the system relaxes so as to increase the concentration of A by a small amount $\Delta[A]$, B by $\Delta[B]$, and to decrease P by $\Delta[P]$, as shown in figure 7.16. The perturbation is small enough so that $(\Delta[A])^2$ is negligible compared with $\Delta[A]$ and so on. The values of $\Delta[P]$, $\Delta[A]$, and $\Delta[B]$ are all defined as positive, and because of the stoichiometry, they are all equal. It is then convenient to rewrite the instantaneous, time-dependent concentrations as

$$[P] = [\overline{P}] + \Delta[P]$$
$$[A] = [\overline{A}] - \Delta[A] = [\overline{A}] - \Delta[P]$$
$$[B] = [\overline{B}] - \Delta[B] = [\overline{B}] - \Delta[P]$$

The values of $[\overline{P}]$, $[\overline{A}]$, and $[\overline{B}]$ refer to the equilibrium concentrations at the final temperature or pressure. Therefore,

$$\frac{d[P]}{dt} = \frac{d([\overline{P}] + \Delta[P])}{dt} = \frac{d(\Delta[P])}{dt}$$

because $d[\overline{P}]/dt = 0$. Combining this result with Eq. (7.63), we obtain

$$\frac{d(\Delta[P])}{dt} = \frac{d[P]}{dt} = k_1[A][B] - k_{-1}[P]$$

$$= k_1([\overline{A}] - \Delta[P])([\overline{B}] - \Delta[P]) - k_1([\overline{P}] + \Delta[P])$$

$$= k_1[\overline{A}][\overline{B}] - k_1[\overline{P}] - k_1([\overline{A}]\Delta[P] + [\overline{B}]\Delta[P] - (\Delta[P])^2) - k_{-1}\Delta[P]$$

$$\qquad \textbf{(7.67)}$$

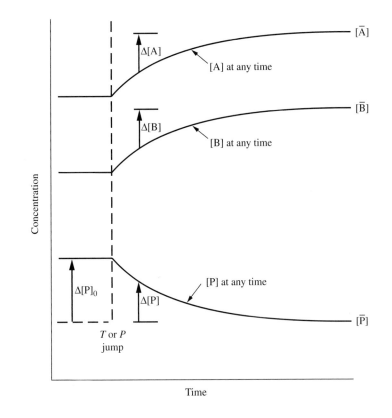

Effect of temperature or pressure jump on concentrations of reactants and products for a system initially at equilibrium or in a steady state. The reaction is A + B $\rightleftharpoons$ P, and the perturbation causes the concentration of P to decrease slightly while A and B increase slightly. After a short time, a new equilibrium is reached with concentrations $[\overline{A}]$, $[\overline{B}]$, $[\overline{P}]$ at the higher temperature or pressure.

$$-\frac{d(\Delta[P])}{dt} = \{k_1([\overline{A}] + [\overline{B}]) + k_{-1}\}\Delta[P] \tag{7.68}$$

To achieve Eq. (7.68), we dropped the term involving $(\Delta[P])^2$ because it is small for small displacements from equilibrium, and we have used Eq. (7.63) to show that the first two terms of Eq. (7.67) sum to zero. The resulting Eq. (7.68) is a simple first-order equation because $\{k_1([A] + [B]) + k_{-1}\}$ is independent of time. If we rewrite Eq. (7.68) as

$$-\frac{d(\Delta[P])}{dt} = \frac{\Delta[P]}{\tau}$$

it can be integrated to give

$$\Delta[P] = \Delta[P]_0 e^{-t/\tau} \tag{7.69}$$

where τ is the relaxation time for the process. For this example,

$$\tau = \frac{1}{k_{-1} + k_1([\overline{A}] + [\overline{B}])} \tag{7.70}$$

Results for other examples are presented in table 7.4.

A very important and at first surprising result of analyses of each of the examples listed in table 7.4 is that the relaxation kinetics is *always* simple first order (exponential in time), regardless of the number of molecules involved as reactants or products. This is true because the perturbations produce only small changes in the equilibrium concentrations, and we can ignore squared terms. The behavior is illustrated in figure 7.16 for the case where the perturbation shifts the equilibrium in favor of reactants. The reverse situation is also found, depending on the sign of ΔH^0 or ΔV^0.

TABLE 7.4 Relaxation Times for Reactions Involving Single Steps

Mechanism	Relaxation time*
$A \underset{k_{-1}}{\overset{k_1}{\rightleftharpoons}} B$	$\tau = \dfrac{1}{k_1 + k_{-1}}$
$A + B \underset{k_{-1}}{\overset{k_1}{\rightleftharpoons}} P$	$\tau = \dfrac{1}{k_{-1} + k_1([\overline{A}] + [\overline{B}])}$
$A + B + C \underset{k_{-1}}{\overset{k_1}{\rightleftharpoons}} P$	$\tau = \dfrac{1}{k_{-1} + k_1([\overline{A}][\overline{B}] + [\overline{B}][\overline{C}] + [\overline{A}][\overline{C}])}$
$A + B \underset{k_{-1}}{\overset{k_1}{\rightleftharpoons}} P + Q$	$\tau = \dfrac{1}{k_1([\overline{A}] + [\overline{B}]) + k_{-1}([\overline{P}] + [\overline{Q}])}$
$2A \underset{k_{-1}}{\overset{k_1}{\rightleftharpoons}} A_2$	$\tau = \dfrac{1}{4k_1[\overline{A}] + k_{-1}}$

*$[\overline{A}]$, $[\overline{B}]$, etc., represent the equilibrium concentrations after the perturbation.
Source: M. Eigen and L. De Maeyer, in *Investigation of Rates and Mechanisms of Reactions,* 3d ed., vol. 6, part 2, ed. G. G. Hammes (New York: Wiley-Interscience, 1974), chapter 3.

An experimental trace is shown in figure 7.17, for the dimerization of proflavin, according to the reaction

$$2P \underset{k_{-1}}{\overset{k_1}{\rightleftharpoons}} P_2$$

Because of the fast kinetics involved, it was necessary to use a pulsed laser to provide the temperature jump. For dimerizations, it is possible to simplify the relaxation expression by writing it in terms of the total concentration $[P]_t$ of monomers and dimers. Turner and colleagues (1972) used the expression

$$\frac{1}{\tau^2} = k_{-1}^2 + 8k_1 k_{-1}[P]_t \qquad \textbf{(7.71)}$$

The analysis according to Eq. (7.71) does not require a prior knowledge of the equilibrium constant to determine both rate constants. The data should

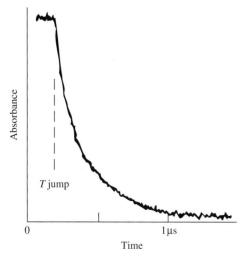

◀ FIGURE 7.17

Relaxation of proflavin dimerization. Total concentration was 4.64×10^{-3} *M,* pH 7.8, and final temperature was 6°C. Absorbance was monitored at 455 nm. [From D. H. Turner, G. W. Flynn, S. K. Lundberg, L. D. Faller, and N. Sutin, 1972, *Nature* 239:215.]

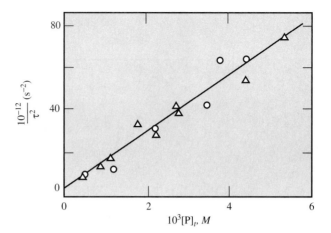

▶ FIGURE 7.18

Plot of relaxation data for proflavin dimerization as a function
of total proflavin concentration,
$[P]_t$; $\bigcirc$ = pH 7.8; $\triangle$ = pH 4.0; temperature = 25°C. [From D. H. Turn-
er, G. W. Flynn, S. K. Lundberg,
L. D. Faller, and N. Sutin, 1972,
Nature 239:215.]

fall on a straight line when plotted according to Eq. (7.71). The plot shown
in figure 7.18 demonstrates not only that this is observed but also that the
rate constants are independent of pH between 4 and 7.8. From the slope
and intercept of the line shown in figure 7.18, $k_1 = 8 \times 10^8\,M^{-1}\,s^{-1}$ and $k_{-1} =
2.0 \times 10^6\,s^{-1}$ at 25°C were obtained. The dimerization rate constant corre-
sponds to a process that is almost at the limit placed by the diffusion of the
monomeric species in aqueous solution, as discussed in the next section.

EXERCISE 7.3

Use the results of this kinetic analysis to determine the equilibrium constant for
proflavin dimerization at 25°C.

EXAMPLE 7.8

The dimerization of the decanucleotide A_4GCU_4 has been studied by Pörschke,
Uhlenbeck, and Martin (1973). The letters A, G, C, and U represent the bases adenine, guanine, cytosine, and uracil, and this oligonucleotide forms base-paired
double-stranded helices, which are models for nucleic acids. The two strands are
antiparallel; therefore, the oligonucleotide is self-complementary.

$$2A_4GCU_4 \underset{k_{-1}}{\overset{k_1}{\rightleftharpoons}} \begin{array}{c} A-A-A-A-G-C-U-U-U-U \\ \cdot\ \cdot\ \cdot\ \cdot\ \cdot\ \cdot\ \cdot\ \cdot\ \cdot\ \cdot \\ U-U-U-U-C-G-A-A-A-A \end{array}$$

a. A temperature jump was applied to a solution containing $7.45 \times 10^{-6}\,M$
oligonucleotide at pH 7.0 to give a final temperature of 32.4°C:

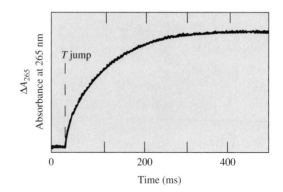

From the response in the UV absorption shown in the figure, the following data can be extracted:

Time (ms)	0	50	100	150	200	250	300	∞
$100 \times \Delta A_{265}$	0	2.0	3.2	3.9	4.28	4.47	4.57	4.70

The difference in absorbance at 265 nm (ΔA_{265}) is directly proportional to the concentration of product at time t minus the concentration at zero time. Use the results to calculate the relaxation time constant under these conditions.

b. Similar experiments were carried out at 23.3°C and pH 7.0 for a series of concentrations of single-stranded oligomer. The observed relaxation times were as follows:

τ (ms)	455	370	323	244
$[\overline{M}]$ (μM)	1.63	2.45	3.45	5.90

Determine k_1 and k_{-1} for double-strand (dimer) formation and dissociation, respectively, under these conditions.

Solution

a. First, test the data for first-order relaxation. The absorbance should approach its value at infinite time (its equilibrium value) exponentially:

$$\frac{\Delta A_\infty - \Delta A_t}{\Delta A_\infty} = e^{-t/\tau}$$

This is equivalent to Eq. (7.69).

Time (ms)	0	50	100	150	200	250	300
$100 \times (\Delta A_\infty - \Delta A_t)$	4.7	2.7	1.5	0.8	0.42	0.23	0.13

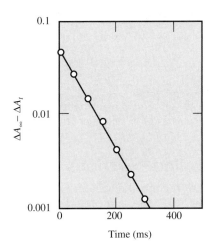

The data fit a linear first-order plot (use semilog graph paper) with a slope of 5.19 s^{-1}. The relaxation time is

$$\tau = \frac{1}{(2.30)(5.19\ \text{s}^{-1})}$$

$$= 0.0838\ \text{s}$$

$$= 84\ \text{ms}$$

b. We use the appropriate expression from table 7.4,

$$\tau = \frac{1}{4k_1[\overline{M}] + k_{-1}}$$

where $[\overline{M}]$ is the concentration of oligonucleotide. Thus, a plot of $1/\tau$ versus $[\overline{M}]$ should give a straight line.

$[\overline{M}]$ (μM)	1.63	2.45	3.45	5.90
$1/\tau (s^{-1})$	2.20	2.70	3.10	4.10

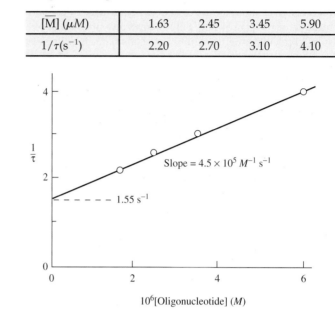

The straight line corresponds to the equation

$$\frac{1}{\tau} = 4k_1[\overline{M}] + k_{-1}$$

Therefore, from the plot shown,

$$4k_1 = 4.5 \times 10^5 \, M^{-1}\,s^{-1}$$
$$k_1 = 1.12 \times 10^5 \, M^{-1}\,s^{-1}$$
$$k_{-1} = 1.55 \, s^{-1}$$

In this case, the dimerization step is several powers of 10 slower than the diffusion limit, discussed in the following section.

Diffusion-Controlled Reactions

The concept of a diffusion-controlled reaction is useful for interpreting the kinetics of fast reactions. Imagine a simplified system where reactant species M and N initially move about in the system independently of one another and far enough apart so that they do not interact. In the gas phase, the intervening space is mostly empty, and the molecules would come together, react, and the product(s) depart without interference from other collisions.

The situation in the liquid (or solid) phase is quite different. The reactant species are always in contact with the solvent or other solutes, and they are constantly being bumped by their neighbors (figure 7.19). The motion of each reactant molecule is a random-walk process, as described earlier. Because the spaces between molecules in the liquid phase are small (typically 10% or so of the molecular

Reactants

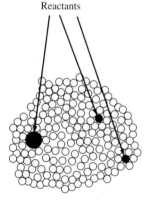

▲ FIGURE 7.19

Schematic diagram of the environment of two reactant species in tight "cages" in liquid solution.

of the molecular diameters), there results a *cage effect* in which the molecule makes many collisions with its neighbors before it moves away to a new site, where it encounters new neighbors. The motion is similar to that of people in a crowded room. The time required to move across the room is very great when the room is crowded, but a nearly empty room is easily traversed.

The consequence of random motions of molecules is described in terms of their diffusion coefficients (see chapter 6). Because the diffusion coefficient D is defined as the proportionality between the flux or velocity of material flow J_x, in units of (concentration) $\times$ (cm s^{-1}), across a unit area in response to a concentration gradient dc/dx, we can write the equation for Fick's first law:

$$J_x = -D\frac{dc}{dx} \qquad (6.37)$$

The diffusion coefficient is a measure of the mobility of molecules and thus characterizes how often they can collide. Diffusion coefficients for small molecules are typically on the order of 10^{-5} cm^2 s^{-1}. Large molecules have smaller diffusion coefficients compared with those of small species, but the numerical values depend on the nature of both the diffusing species and the medium (solvent) through which they are moving. Hydrogen ions in water diffuse with unusual ease, because lacking electrons they have the smallest radius of any ion and because they can "hop" from one water molecule to another.

The effect of diffusion on kinetics can be developed mathematically so as to calculate an encounter frequency for two solute species M and N. Putting the encounter frequency into the rate expression, we obtain a value for the A factor in the Arrhenius expression

$$A_{\text{diffusion}} = \frac{4\pi(r_{\text{MN}})(D_{\text{M}} + D_{\text{N}})N_0}{1000} \qquad (7.72)$$

for bimolecular reactions in liquid solution. Here r_{MN} is the encounter distance of the pair of reacting molecules M and N and is typically several angstroms. N_0 is Avogadro's number and 1000 converts cm^3 to L. For diffusion coefficients $D_{\text{M}} = D_{\text{N}} = 1.5 \times 10^{-5}$ cm^2 s^{-1} and $r_{\text{MN}} = 4$ Å $= 4 \times 10^{-8}$ cm, the value of $A_{\text{diffusion}}$ is about 10^{10} L mol^{-1} s^{-1}. For a reaction with zero activation energy, this would be the value of the rate constant if every encounter led to reaction. It constitutes an approximate upper bound for reactions that are limited only by the frequency of encounters. These are known as *diffusion-controlled reactions*.

Rate constants for a variety of second-order reactions are listed in table 7.5. Note that these are all rate constants at room temperature. Where activation energies are not zero, the Arrhenius A factors will be somewhat larger than the rate constants given in the table. Despite this limitation, it is clear that many reactions involving protonation or reaction with hydroxide operate near the diffusion limit. Because of electrostatic attractions, reactions involving oppositely charged ions or ions with polar molecules are faster than calculations from simple diffusion theory would predict. While a few enzyme–substrate complex formation reactions appear to be diffusion limited, many others are slower by several orders of magnitude. It should be obvious that this difference is at least partly associated with the appreciable negative entropy of activation [Eq. (7.54)] that is typically found for such reactions.

TABLE 7.5 Rate Constants for Some Second-order Reactions in Liquid Solution

Reaction		$k\ (M^{-1}\,s^{-1})$
$H^+ + OH^-$		1.3×10^{11}
$H^+ + NH_3$		4.3×10^{10}
$H^+ + $ imidazole		1.5×10^{10}
$OH^- + NH_4^+$		3.4×10^{10}
$OH^- + $ imidazole$^+$		2.3×10^{10}
Enzyme–substrate complex formation	ribonuclease + cytidine 3′-phosphate	6×10^7
	lactate dehydrogenase + NADH	1×10^9
	creatine kinase + ADP	2.2×10^7
	aspartate aminotransferase	
	$\quad + \alpha$-methylaspartate	1.2×10^4
	$\quad + \beta$-erythrohydroxyaspartate	3.1×10^6
	$\quad + $ aspartate	$>1 \times 10^8$
	catalase + H_2O_2	5×10^6
$O_2 + $ hemoglobin		4×10^7

EXAMPLE 7.9

Glucose binds to the enzyme hexokinase from yeast with a rate constant $k = 3.7 \times 10^6\ M^{-1}\,s^{-1}$. Estimate the rate constant for the diffusion-limited reaction and compare this calculation with the experimental value.

Solution

To use Eq. (7.72), we need to know the diffusion coefficients of glucose and hexokinase and to estimate the radius of a reactive encounter.

$$D(\text{glucose}) = 0.673 \times 10^{-5}\ \text{cm}^2\,\text{s}^{-1}$$

$$D(\text{hexokinase}) = 2.9 \times 10^{-7}\ \text{cm}^2\,\text{s}^{-1}$$

We assume that the encounter distance is of the order of 5×10^{-8} cm. Then, from Eq. (7.72),

$$A_{\text{diff}} = \frac{4\pi(5 \times 10^{-8}\ \text{cm})(0.673 + 0.029)(10^{-5}\ \text{cm}^2\,\text{s}^{-1})(6.0 \times 10^{23})}{1000}$$

$$= 2.6 \times 10^9\ M^{-1}\,s^{-1}$$

which is nearly 1000 times faster than the observed rate constant. Thus, the reaction is not diffusion limited under these conditions.

Photochemistry and Photobiology

Light in our environment has both highly beneficial and potentially hazardous effects. We survive on planet Earth because of the process of photosynthesis, which uses solar energy to form the chemical compounds we and other animals eat. Our eyes employ rhodopsin to convert light stimuli into neural signals to the brain, enabling us to see. Solar irradiation of the upper atmosphere induces photochemistry that results in the formation of a complex distribution of oxygen, nitrogen, and hydrogen compounds. One consequence is the presence of the ozone shield, which screens out harmful UV radiation. The reason for concern about UV light is that its photons are of sufficiently high energy to induce mutations in both microorganisms and higher organisms. Such mutations are more likely to be harmful than beneficial. Furthermore, prolonged exposure to sunlight rich in UV components is known to lead to an increase in the incidence of skin cancer. Ultraviolet germicidal lamps are potent lethal weapons for microorganisms and are used for sterilizing food, drugs, serum solutions, and operating rooms. Ordinary fluorescent light has been used therapeutically for babies suffering from hyperbilirubinemia, a malfunctioning of heme metabolism. Even this short list gives an indication of the wide variety of roles played by light in biology.

To understand the role of light in photoreactions, we must realize that only radiation *absorbed* by the system is effective in producing photochemical change. The quantitative measure of light is its intensity I, usually given in photons per square meter per second ($m^{-2} s^{-1}$), einsteins $m^{-2} s^{-1}$, or $J m^{-2} s^{-1}$ ($= W m^{-2}$). The relation among these quantities can be derived from the *Planck equation*,

$$\varepsilon = h\nu = \frac{hc}{\lambda} \tag{7.73}$$

which states that the energy of a single photon ε is proportional to its frequency ν (in s^{-1}) or inversely proportional to its wavelength λ (in m); h is Planck's constant (6.626×10^{-34} J s), and c is the velocity of light (3.0×10^{8} m s^{-1} in vacuum). A mole of photons (6.022×10^{23} photons) is known as an *einstein*. Thus, for an einstein of radiation, the energy is

$$E = N_0 h\nu = \frac{N_0 hc}{\lambda} \tag{7.74}$$

Intensities of light are measured as the number of photons (or of einsteins, or of energy in J) crossing a 1-m^2 cross section in each second of time. Alternative measures of intensity are often used in different applications.

Light that is transmitted through the sample or is scattered by it does not become absorbed and therefore will not produce any photochemistry. For a photochemical reaction

$$B \xrightarrow{h\nu} P$$

the velocity or photochemical rate is given by

$$v = -\frac{d[B]}{dt} = \frac{\phi I_{abs} \, 1000}{l} \tag{7.75}$$

where I_{abs} is the light absorbed by the sample in einsteins $cm^{-2} s^{-1}$, l is the path length in cm and 1000 converts cm^3 to liters, and ϕ is the quantum yield. The units of v are thus $M s^{-1}$. The photochemical quantum yield is defined as the number of molecules reacted per photon absorbed; it is obviously unitless:

$$\phi = \frac{\text{number of molecules reacted}}{\text{number of photons absorbed}} \tag{7.76}$$

Some representative values are given in table 7.6.

According to the *Einstein law of photochemistry*, in a primary photochemical process, each molecule is excited by the absorption of one photon. However, the excited molecule can then undergo many different reactions. It can

TABLE 7.6 Some Representative Quantum Yields

Photochemical reaction	Quantum yield, ϕ*
$NH_3(g) \rightarrow \frac{1}{2} N_2(g) + \frac{3}{2} H_2(g)$	0.2
$S_2O_8^{2-}(aq) + H_2O \rightarrow 2SO_4^{2-}(aq) + 2H^+(aq) + \frac{1}{2}O_2(g)$	1
$H_2(g) + Cl_2(g) \rightarrow 2HCl(g)$	10^5
$H_2O + CO_2 \xrightarrow{\text{chloroplasts}} \frac{1}{x}[CH_2O]_x + O_2$ (in photosynthesis)	10^{-1}
Rhodopsin $\rightarrow$ Retinal + Opsin (in mammalian eyes)	1
2Thymine $\rightarrow$ Thymine dimer (in DNA)	10^{-2}
Hemoglobin $\cdot$ CO $\rightarrow$ Hemoglobin + CO(g)	1
Hemoglobin $\cdot$ O$_2$ $\rightarrow$ Hemoglobin + O$_2$(g)	10^{-2}

*Values depend on wavelength and other experimental conditions.

form product; it can catalyze a chain reaction; it can fluoresce; it can thermally de-excite; and so forth. Thus, the quantum yield ϕ can be much smaller than 1 or be much greater than 1 (for chain reactions), as seen in Table 7.6. The value of ϕ must be determined experimentally for each photochemical reaction. Recently, exceptions to the Einstein law have been demonstrated at very high intensities using pulsed lasers. At superhigh intensities, two or three photons can be absorbed simultaneously by the same molecule.

The relation between the rate of a photochemical reaction and the absorbed light [Eq. (7.75)] can be written in a more useful form by introducing the Beer–Lambert law (see chapter 10) for light absorption:

$$I = I_0 10^{-\varepsilon l c} = I_0 10^{-A} \tag{7.77}$$

where I_0 = intensity of incident light

I = intensity of transmitted light

l = path length of light in sample

c = concentration of absorbing molecules, mol L^{-1}

ε = molar absorptivity, or molar extinction coefficient, L mol^{-1} cm^{-1}

A = absorbance; it is defined by Eq. (7.77)

The intensity of light absorbed is $(I_0 - I)$; therefore,

$$-\frac{d[B]}{dt} = \phi(I_0 - I)\frac{1000}{l} = \phi I_0 (1 - 10^{-A})\frac{1000}{l} \tag{7.78}$$

The exponential term can be rewritten using a series expansion, as in

$$10^{-A} = 1 - 2.303A + \frac{(2.303A)^2}{2!} \cdots$$

For *dilute solutions*, where $A \ll 1$, the terms involving higher powers are negligible in comparison with the first-order term, and we can substitute

$$1 - 10^{-A} \cong 2.303A$$

into Eq. (7.78) to obtain, for dilute solutions,

$$-\frac{d[B]}{dt} = \frac{(2.303)(1000)\phi I_0 A}{l} = (2.303)(1000)\phi I_0 \varepsilon[B] \tag{7.79}$$

where we have replaced the absorbance A by $\varepsilon l[B]$. For constant incident light intensity, this equation has the form of the usual first-order rate expression. The concentration and time variables can be separated, and the equation can be integrated to give

$$\ln \frac{[B]}{[B]_0} = -2303\phi I_0 \varepsilon t = -kt \qquad (7.80)$$

Most, but not all, photochemical reactions are first order in reactant concentration at sufficiently low concentration. To test for a first-order photochemical reaction one can illuminate identical dilute samples for different intensities and for different times but with $I_0 \cdot t$ a constant for all the experiments. If the reaction is first order in the absorbing species, the amount of photochemical conversion will be the same for each of the experiments.

EXAMPLE 7.10

In the presence of a reducible dye D, chloroplasts from spinach are able to photocatalyze the oxidation of water to O_2 by the reaction

$$2H_2O + 2D \xrightarrow[\text{chloroplasts}]{h\nu} O_2 + 2DH_2$$

The rate of this process at an incident intensity of 40×10^{-12} einstein cm^{-2} s^{-1} of 650-nm light was 6.5×10^{-12} equivalent (moles of electrons transferred) cm^{-3} s^{-1} in a solution of 1-cm path length. The chloroplasts exhibited an absorbance of $A_{650}^{1\,cm} = 0.140$. Calculate the quantum efficiency for this reaction.

Solution

Because the solution is optically dilute ($A_{650} \ll 1$), we can use Eq. (7.79). In a photocatalytic reaction, the concentration of the absorbing chlorophyll in the chloroplasts does not change with time; that is, A is a constant during the process. Therefore,

$$v = \frac{2.303\phi I_0 A}{l}$$

and

$$\phi = \frac{vl}{2.303\phi I_0 A} = \frac{(6.5 \times 10^{-12} \text{ equivalent } cm^{-3} \text{ s}^{-1})(1 \text{ cm})}{2.303(40 \times 10^{-12} \text{ einstein } cm^{-2} \text{ s}^{-1})(0.140)}$$

$$= 0.50 \frac{\text{equivalent}}{\text{einstein}} = 0.50 \frac{\text{electron}}{\text{photon}}$$

[*Note:* We omitted the factor of 1000 in Eq. (7.79) because concentration units of mol cm^{-3} were given.]

We complete our discussion of photochemical kinetics by describing some biologically important processes.

Vision

Visible and near-UV light absorbed by the eye pigment rhodopsin produces a sequence of photoreactions that is still under active investigation. The normal form of rhodopsin in dark-adapted mammalian rod outer segments has retinal in the 11-*cis* form covalently attached as a chromophore ("color bearer") to the protein opsin. Following illumination, the characteristic absorption spectrum of rhodopsin undergoes a series of changes associated with

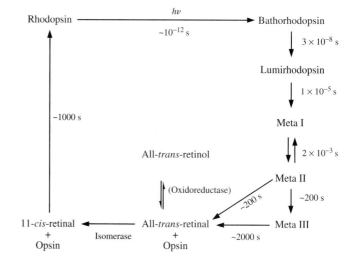

► FIGURE 7.20

Photobleaching and regeneration of rhodopsin. The lifetimes shown are only approximate to within a factor of about 3 in most cases. [For details, see H. Shichi, 1983, *Biochemistry of Vision* (New York: Academic Press.)]

a set of intermediates, as shown in figure 7.20. Some of these steps are very fast at room temperature and have been studied either using flash relaxation methods or by freezing them at low temperature. The primary step in the photo response of rhodopsin occurs with a quantum yield of about unity. In the eye, the light stimulus requires additional processing before it arrives as a signal at the brain. Nevertheless, an incidence of only a few photons per second is sufficient to give a significant visual sensation.

The visual process is a cyclic one, and the rhodopsin is regenerated in the dark by processes that are not well understood. It is likely that the reconstitution of active rhodopsin has a requirement for metabolic energy. There is no evidence that any of the energy of the photon is stored by the rhodopsin photochemistry; light serves simply to trigger or activate an otherwise exergonic ($\Delta G < 0$) reaction. The principal function of the photon is to overcome an activation barrier. This is the most common kind of photochemical process.

Photosynthesis

The process of photosynthesis in plants, algae, and bacteria involves the incorporation of carbon into the various compounds such as carbohydrates, proteins, lipids, nucleic acids, and so on that make up the material substance and the essential functional components of the organism. The source of carbon for plants and most algae is carbon dioxide; in other organisms, small carbon-containing compounds (acetate, succinate, malate, etc.) are required. The representative reaction for higher plants is

$$CO_2 + H_2O \longrightarrow (CH_2O) + O_2$$

where (CH_2O) represents the fixed carbon of carbohydrate, the major metabolic product. This reaction is endothermic by about 485 kJ $(mol\ CO_2)^{-1}$, and the energy required to drive this and the other biosynthetic reactions comes ultimately from sunlight. By contrast with most photochemical reactions and with vision, a portion of the energy of the absorbed photons is retained in photosynthesis in the form of chemical potential of the metabolic products.

For purposes of estimating the efficiency of such processes as photosynthesis, we should compare ΔG for the carbon-fixation process with the free-energy change associated with the absorption of radiation. The value of ΔG for carbon fixation is about 494 kJ $(\text{mol } CO_2)^{-1}$, but the free-energy change for the absorption of radiation is more difficult to estimate. [A detailed analysis for solar irradiation is given by Ross and Calvin (1967).]

The overall process of photosynthesis can be divided into light-dependent and dark reactions. Light absorbed by chlorophyll pigments serves to split water molecules into molecular oxygen and hydrogen atom equivalents. The hydrogen atoms are transferred as electrons and hydrogen ions along a transport chain of cytochromes, quinones, and iron-, manganese-, and copper-containing proteins to nicotinamide adenine dinucleotide phosphate ($NADP^+$), which becomes reduced. During this process, a part of the energy is stored as the high-free-energy product adenosine triphosphate (ATP):

$$2\,H_2O \ + \ 2\,NADP^+ \ \xrightarrow[\substack{\text{chloroplast} \\ \text{membranes}}]{h\nu} \ O_2 \ + \ 2\,NADPH \ + \ 2\,H^+$$

$$2(ADP \ + \ P_i) \qquad\qquad 2\,ATP$$

The NADPH and ATP are then used in a series of dark enzymatic reactions to fix CO_2:

$$CO_2 + 2NADPH + 2ATP + 2H^+ \xrightarrow[\text{enzymes}]{\text{dark}}$$

$$(CH_2O) + 2NADP^+ + 2ADP + 2P_i + H_2O$$

The sum of these reactions is the overall process presented above.

Detailed and extensive studies of the kinetics of photosynthetic light reactions have been made, and the following list summarizes the current state of our knowledge of this subject:

1. The light reactions occur in lipoprotein thylakoid membranes within chloroplasts in higher plants (see chapter 5) or within the cells of blue-green algae or photosynthetic bacteria.

2. Chlorophyll or bacteriochlorophyll are the essential pigments that absorb the light, although carotenoids and other accessory pigments can transfer absorbed photon excitation to the chlorophylls.

3. All wavelengths from the near-UV to the near-infrared can be effective. In particular, wavelengths as great as 700 nm (170 kJ einstein^{-1}) are effective in higher plants and as great as 1000 nm (120 kJ einstein^{-1}) in certain photosynthetic bacteria.

4. The quantum yield varies with growth conditions. Under optimal levels of CO_2 pressure, relative humidity, and soil nutrients, the yield at low light intensities corresponds to 1 mol of CO_2 fixed or O_2 evolved for each 8 to 9 einsteins absorbed. This represents a higher efficiency than it would seem, because each O_2 molecule is produced by the removal of four electrons from two water molecules and it is known that there are two photon acts or light reactions that operate in series for the transfer of each electron. A current view of the mechanism of photosynthetic electron transport is shown in figure 7.21.

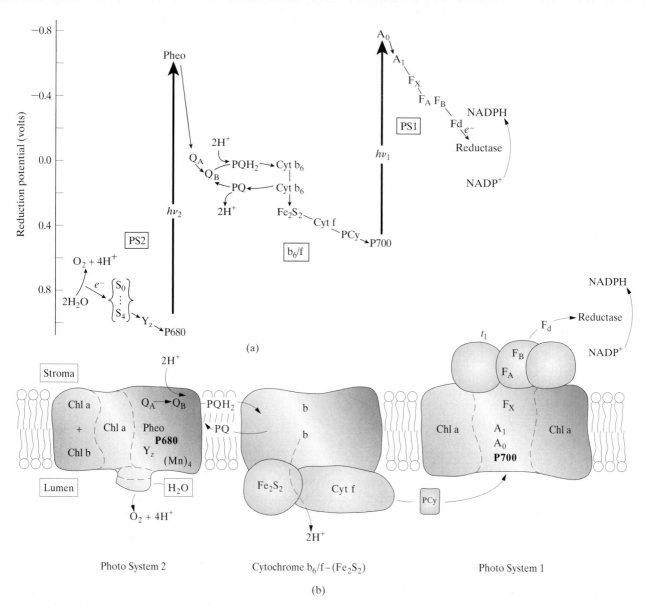

(a)

(b)

Photo System 2   Cytochrome b_6/f – (Fe$_2$S$_2$)   Photo System 1

▲ FIGURE 7.21

Two schematic views of photosynthetic electron transport involving two photosystems (PS1 and PS2) joined by a cytochrome b_6/f complex. (a) The sequence of electron transport from water (left side) to form reductants (NADPH, right side) capable of supporting CO_2 fixation; plotted on the vertical scale is the effective reduction potential of each redox cofactor, in volts. The two light reactions, $h\nu_1$ and $h\nu_2$, shown as heavy vertical arrows, provide the energy input from sunlight that is the source of the chemical potential. (b) A current view of the physical arrangement of the three major membrane associated complexes, their cofactors and their orientations with respect to the thylakoid membrane. (See figures 5.20 and 5.21.) Photosystem 2 (PS2) consists of approximately 25 different proteins, many of which collectively bind more than 200 chlorophyll a and b (Chl a, Chl b) and carotenoid light-harvesting pigments. P680 (Chl a) is the primary PS2 reaction center electron donor and pheophytin a (Pheo) is the primary acceptor. A tetranuclear Mn-containing water oxidation center cycles progressively through S-states (S_0, . . . , S_4) to accumulate oxidizing equivalents for water oxidation and donate electrons through a tyrosine (Y_z) to P680. On the acceptor side, electrons are transferred from Pheo to bound plastoquinones Q_A and Q_B; the latter becomes protonated and shuttles to the b6/f complex, which consists of 7 proteins. Electrons pass through cytochrome b6, a 2Fe–2S (Rieske) center, and cytochrome f to a soluble copper protein, plastocyanin (PCy) located in the lumenal compartment. Upon light activation, P700, the Chl a dimer in the reaction center of PS1, transfers an electron to the primary acceptor A_0 (Chl a) and accepts an electron from reduced PCy to return P700 to its reduced ground state. Electrons are transferred from A_0 to membrane-associated secondary acceptors, A_1 (phylloquinone) and three FeS centers (F_X, F_A, and F_B) to soluble ferredoxin (Fd) in the stromal phase. Reduced Fd donates electrons through a flavoprotein (Fd–NADPH–reductase) to $NADP^+$. Many of the 13 proteins of PS1 in plants bind the 200 Chl a, Chl b, and carotenoid pigments that transfer excitation to the reaction center. Transfer of four electrons to release O_2 and fix CO_2 in the stroma phase occurs by fourfold repetition of the two light reactions. A further consequence of the light reactions is the transfer of protons from the stromal to the lumenal phase. The resulting proton gradient is a major contributor of chemical potential needed for the formation ATP by the thylakoid-associated ATPase (not shown, but see figure 5.21).

380

Summary

Note: We represent concentrations as c_A or [A] and so forth.

Zero-Order Reactions

Rate is constant:

For a product whose concentration is c,

$$\frac{dc}{dt} = k \tag{7.2}$$

$k \equiv$ rate constant; units are concentration time^{-1}—for example, $M\ s^{-1}$.

Concentration is linear in time:

$$c_2 - c_1 = k(t_2 - t_1)$$

Plot of concentration versus time:

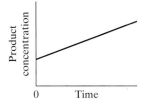

$k =$ slope; concentration at zero time is the y-intercept.

Possible mechanism:

$$A \longrightarrow B + C$$

but concentration of A is held constant—for example, by reacting a saturated solution of A:

$$\frac{dc_B}{dt} = k$$

$$c_B = kt + \text{constant}$$

First-Order Reactions

Rate is proportional to concentration:

For a reactant whose concentration is c,

$$-\frac{dc}{dt} = kc \tag{7.5}$$

$k \equiv$ rate constant; units are time^{-1}—for example, s^{-1}.

Logarithm of reactant concentration is linear in time:

$$\ln c = -kt + \text{constant}$$

$$\ln \frac{c_2}{c_1} = -k(t_2 - t_2)$$

$\ln c \equiv$ natural logarithm of concentration   **(7.10)**

Plot logarithm of reactant concentration versus time:

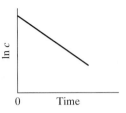

k = slope; logarithm of reactant concentration at zero time is the y-intercept. Any concentration units can be used for the plot.

Reactant concentration is exponential in time:

$$\frac{c}{c_0} = e^{-kt} \quad \text{or} \quad \frac{c}{c_0} = 10^{-0.434kt} \tag{7.11}$$

c = concentration at time t
c_0 = concentration at zero time (note that only the ratio of concentrations is important)

Half-life:

$$\frac{c}{c_0} = 2^{-t/t_{1/2}} \tag{7.13}$$

$t_{1/2}$ ≡ half-life, the time necessary for concentration to become half of its original value
   = $0.693/k$

Relaxation time:

$$\frac{c}{c_0} = e^{-t/\tau} \tag{7.14}$$

τ ≡ relaxation time, the time necessary for concentration to become $1/e (= 0.368)$ of its original value

Possible unimolecular mechanism:

$$A \longrightarrow B$$

$$-\frac{dc_A}{dt} = kc_A \quad \text{and} \quad \frac{dc_B}{dt} = kc_A$$

$$\frac{c_A}{(c_A)_0} = e^{-kt} \qquad \frac{c_B}{(c_A)_0} = 1 - e^{-kt}$$

c_A, c_B = concentrations of A, B at any time t
$(c_A)_0$ = concentration of A at zero time

Second-Order Reactions

Rate is proportional to concentration squared or to the product of two different concentrations:

$$\frac{-dc}{dt} = kc^2 \text{ (class I)} \quad \text{or} \quad \frac{-dc}{dt} = kc_Ac_B \text{ (class II)}$$

k ≡ rate constant; units are in concentration^{-1} time^{-1}, for example, $M^{-1}\,s^{-1}$

Reciprocal of reactant concentration is linear in time (class I):

$$\frac{1}{c} = kt + \text{constant}$$

Plot of reciprocal of reactant concentration versus time (class I):

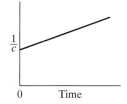

k = slope; reciprocal of concentration at zero time is the y-intercept.

Possible bimolecular mechanism (class I):

$$A + A \longrightarrow B$$

$$-\frac{dc_A}{dt} = kc_A^2 \quad \text{and} \quad \frac{dc_B}{dt} = \tfrac{1}{2}kc_A^2$$

$$\frac{1}{c_A} - \frac{1}{(c_A)_0} = kt \qquad c_B = \frac{(c_A)_0}{2} - \frac{c_A}{2}$$

c_A, c_B = concentrations of A, B at any time t
$(c_A)_0$ = concentration of A at zero time

Possible bimolecular mechanism (class II):

$$A + B \longrightarrow C$$

$$-\frac{dc_A}{dt} = -\frac{dc_B}{dt} = kc_A c_B$$

$$\ln \frac{(c_A)(c_B)_0}{(c_A)_0(c_B)} = [(c_A)_0 - (c_B)_0]kt \qquad \text{(7.23)}$$

c_A, c_B = concentrations of A, B at any time t
$(c_A)_0$ = concentration of A at zero time
$(c_B)_0$ = concentration of B at zero time

If $(c_A)_0 = (c_B)_0$, then $c_A = c_B$ at all times, and

$$\frac{1}{c_A} - \frac{1}{(c_A)_0} = \frac{1}{c_B} - \frac{1}{(c_B)_0} = kt \quad \text{as in Class I}$$

Temperature Dependence
Arrhenius equation:

$$k = Ae^{-E_a/RT} \qquad \text{(7.41)}$$

$k \equiv$ rate constant
$A \equiv$ pre-exponential factor; units are same as rate constant
$E_a =$ activation energy, units are J mol^{-1}
$R =$ gas constant $= 8.314$ J K^{-1} mol^{-1}
$T =$ absolute temperature

Plot of logarithm of k versus reciprocal of absolute temperature:

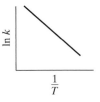

$E_a = -\text{R} \times$ slope
$\ln A = y$ intercept, but extrapolation is often impractical; therefore,
$\ln A = \ln k - (1/T)(\text{slope})$ where $\ln k$ and $(1/T)$ are the coordinates of
  any point on the line

Activation energy:

$$E_a = \frac{RT_2 T_1}{T_2 - T_1} \ln \frac{k_2}{k_1}$$

$E_a =$ activation energy, which is assumed independent of temperature
$k_2, k_1 =$ rate constants at T_2, T_1
$R =$ gas constant $= 8.314$ J K^{-1} mol^{-1}

Eyring equation:

$$k = \frac{k_\text{B} T}{h} e^{\Delta S^{\ddagger}/R} e^{-\Delta H^{\ddagger}/RT} \tag{7.52}$$

$k =$ rate constant; its units of time must be in seconds; its units
  of concentration specify the standard state for $\Delta H^{\ddagger}$, $\Delta S^{\ddagger}$
$\Delta H^{\ddagger} =$ enthalpy of activation $=$ enthalpy difference per mole between
  activated complex and reactants, each in their standard states
$\Delta S^{\ddagger} =$ entropy of activation
$T =$ absolute temperature
$k_\text{B} =$ Boltzmann constant $= R/N_0$; its units are consistent with
  Planck's constant, h

$\dfrac{k_\text{B}}{h} = \dfrac{R}{N_0 h} =$ gas constant(Avogadro's number)$^{-1}$(Planck's constant)$^{-1}$

$$= \frac{8.3144 \text{ J K}^{-1} \text{ mol}^{-1}}{(6.0220 \times 10^{23} \text{ molecules mol}^{-1})(6.6262 \times 10^{-34} \text{ J s})}$$

$$= 2.0837 \times 10^{10} \text{ (s K)}^{-1}$$

Relation of $\Delta H^{\ddagger}$, $\Delta S^{\ddagger}$ to E_a:

$$\Delta H^{\ddagger} = E_a - RT \approx E_a \tag{7.53}$$

$$\Delta S^{\ddagger} \approx R\left(\ln \frac{Ah}{k_\text{B} T} \right) \tag{7.54}$$

A = Arrhenius pre-exponential factor, its units of time must be in s^{-1}; its concentration units specify the standard states of $\Delta H^{\ddagger}$, $\Delta S^{\ddagger}$

Free energy of activation:

$$\Delta G^{\ddagger} = \Delta H^{\ddagger} - T\Delta S^{\ddagger} \tag{7.51}$$

$$k = \frac{RT}{N_0 h} e^{-\Delta G^{\ddagger}/RT}$$

$$\Delta G^{\ddagger} = -RT\left(\ln \frac{k}{T} - \ln \frac{R}{N_0 h} \right)$$

$$= -RT \ln K^{\ddagger} \tag{7.49}$$

$\Delta G^{\ddagger}$ = free energy of activation = free energy difference between transition state and reactants each in their standard states; the standard states are specified by the concentration units used in the rate constant k; if mol L^{-1} is used, the standard state is the infinitely dilute solution state on the molarity scale

$K^{\ddagger}$ = equilibrium constant for the formation of the transition state

Electron Transfer Reactions: Marcus Theory

$$\Delta G^{\ddagger} = \frac{\lambda}{4}\left(1 + \frac{\Delta G^{\circ}}{\lambda} \right)^2 \tag{7.55}$$

$\Delta G^{\ddagger}$ = free energy of activation
λ = the reorganization energy of the solvent, reactants, and products
ΔG° = free energy of the reaction

$$\lambda_{\text{solv}} = (\Delta e)^2 \left(\frac{1}{2a_1} + \frac{1}{2a_2} - \frac{1}{R} \right)\left(\frac{1}{n^2} - \frac{1}{\varepsilon} \right) \tag{7.56}$$

λ_{solv} = the reorganization energy of the solvent
Δe = charge transferred
a_1, a_2 = radii of the reactants
R = center-to-center distance between the reactants
n = refractive index of solvent
ε = dielectric constant of solvent

Relaxation Kinetics

$$A + B \underset{k_{-1}}{\overset{k_1}{\rightleftharpoons}} P$$

$$\Delta[P] = \Delta[P]_0 e^{-(t/\tau)}$$

$\Delta[P] = \Delta[A] = \Delta[B]$ = small displacement from equilibrium concentration
$\Delta[P]_0$ = displacement at zero time
$1/\tau = k_{-1} + k_1([\overline{A}] + [\overline{B}])$
$[\overline{A}], [\overline{B}]$ = equilibrium concentrations of A and B after the perturbation

See table 7.4 for other examples.

Diffusion-Controlled Reactions

Diffusion-controlled reaction M + N → products:

$$A_{diffusion} = \frac{4\pi(r_{MN})(D_M + D_N)N_0}{1000} \tag{7.72}$$

$A_{diffusion}$ = pre-exponential factor, in L mol^{-1} s^{-1}
r_{MN} = encounter distance, cm
D = diffusion constant, cm^2 s^{-1}
N_0 = Avogadro's number = 6.02×10^{23} mol^{-1}

Absorption of Light

Beer–Lambert law:

$$A = \log \frac{I_0}{I} = \varepsilon c l \tag{7.77}$$

A = absorbance (or optical density)
I_0 = incident intensity $\left.\vphantom{\begin{array}{c}a\\b\end{array}}\right\}$ same units for both
I = transmitted intensity
ε = molar absorptivity, M^{-1} cm^{-1}
c = concentration, M
l = path length, cm

Photochemistry

Planck equation:

$$\varepsilon = h\nu = \frac{hc}{\lambda} \tag{7.73}$$

ε = photon energy, J photon^{-1}
h = Planck's constant = 6.6262×10^{-34} J s
ν = frequency of radiation, s^{-1}
c = velocity of light = 2.9979×10^8 m s^{-1}
λ = wavelength of radiation, m

$$E = N_0 h\nu = \frac{N_0 hc}{\lambda} \tag{7.74}$$

E = energy per einstein
N_0 = Avogadro's number = 6.022×10^{23} mol^{-1}
    = 6.022×10^{23} photons einstein^{-1}

Rate of a photochemical reaction:

$$v = \frac{\phi I_{abs} 1000}{l} \tag{7.75}$$

v = velocity of reaction, M s^{-1}
ϕ = quantum yield, moles reacting per einstein absorbed
I_{abs} = flux of light absorbed, einstein cm^{-2} s^{-1}
l = path length in cm

Dilute solution photochemistry:

$$v = \frac{(2.303)(1000)\phi I_0 A}{l} \qquad \text{(where } A \ll 1\text{)}$$

v = velocity of reaction, $M\,s^{-1}$
ϕ = quantum yield, moles reacting per einstein absorbed
I_0 = incident intensity, einstein $cm^{-2}\,s^{-1}$
A = absorbance
l = path length in cm

Mathematics Needed for Chapter 7

A differential equation is an equation containing derivatives. In this chapter, we apply only differential equations that can be solved by separating the variables and integrating each side of the equation separately. Consider a differential equation of the form

$$\frac{dx}{dt} = f(x)f(t)$$

where $f(x)$ is a function of x only and $f(t)$ is a function of t only. We separate the variables x and t,

$$\frac{dx}{f(x)} = f(t)\,dt$$

and integrate each side of the equation

$$\int \frac{dx}{f(x)} = \int f(t)\,dt$$

The integrals we use are

$$\int x^n\,dx = \frac{x^{n+1}}{n+1} + C \qquad \int_{x_1}^{x_2} x^n\,dx = \frac{x_2^{n+1} - x_1^{n+1}}{n+1} \qquad (n \neq -1)$$

$$\int \frac{dx}{x} = \ln x + C \qquad \int_{x_1}^{x_2} \frac{dx}{x} = \ln \frac{x_2}{x_1}$$

$$\int e^{ax}\,dx = \frac{e^{ax}}{a} + C \qquad \int_{x_1}^{x_2} e^{ax}\,dx = \frac{e^{ax_2} - e^{ax_1}}{a}$$

In addition,

$$e^a = 10^{0.4343a}$$

$$10^a = e^{2.303a}$$

$$\ln a = 2.303 \log a$$

References

Kinetics

Connors, K. A. 1990. *Chemical Kinetics*. New York: VCH Publishers.

Espenson, J. H. 1995. *Chemical Kinetics and Reaction Mechanisms*, 2d ed. New York: McGraw-Hill.

Hammes, G. G., ed. 1974. *Investigation of Rates and Mechanisms of Reactions*. Part II: *Investigation of Elementary Reaction Steps in Solution and Very Fast Reactions*. New York: Wiley.

Laidler, K. J. 1987. *Chemical Kinetics*, 3d ed. New York: Harper & Row.

Moore, J. W., and R. G. Pearson. 1981. *Kinetics and Mechanism*, 3d ed. New York: Wiley.

Steinfeld, J. I., J. S. Francisco, and W. L. Hase. 1998. *Chemical Kinetics and Dynamics*, 2d ed. Upper Saddle River, NJ: Prentice Hall.

Suggested Reading

Anderson, J. G., D. W. Toohey, and W. H. Brune. 1991. Free Radicals Within the Antarctic Vortex: The Role of CFCs in Antarctic Ozone Loss. *Science* 251:39–46.

Clayton, R. K. 1980. *Photosynthesis: Physical Mechanisms and Chemical Patterns.* New York: Cambridge University Press.

Doolittle, R. F. 1981. Fibrinogen and Fibrin. *Sci. Am.* (December): 126–35.

Eigen, M., and L. de Maeyer. 1974. Theoretical Basis of Relaxation Spectrometry, in *Investigation of Rates and Mechanisms of Reactions,* 3d ed. Vol. 6, Part II. ed. G. G. Hammes, 63–146. New York: Wiley-Interscience.

Elliott, S., and F. S. Rowland. 1987. Chlorofluorocarbons and Stratospheric Ozone. *J. Chem. Ed.* 64:387–91.

Fasella, P., and G. G. Hammes. 1967. A Temperature Jump Study of Aspartate Aminotransferase. *Biochemistry* 6:1798–1804.

Gruebele, M., and A. Zewail. 1990. Ultrafast Reaction Dynamics. *Physics Today* 43:25–33.

Hammes, G. G., and J. L. Haslam. 1968. A Kinetic Investigation of the Interaction of α-Methylaspartic Acid with Aspartate Aminotransferase. *Biochemistry* 7:1519–25.

Jackson, C. M., and Y. Nemerson. 1980. Blood Coagulation. *Annu. Rev. Biochem.* 49:765–811.

Johnston, H. S. 1975. Ground-Level Effects of Supersonic Transports in the Stratosphere. *Acc. Chem. Res.* 8:289–94.

Lee, Y. T. 1987. Molecular Beam Studies of Elementary Chemical Processes. *Science* 236:793–99.

Marcus, R. A. 1993. Electron Transfer Reactions in Chemistry—Theory and Experiment (Nobel Lecture). *Angewandte Chemie—International Edition in English* 32:1111–21. The article is reprinted in *Pure and Applied Chemistry* 69:13–29 (1997).

Molina, M., and F. S. Rowland. 1974. Stratospheric Sink for Chlorofluoromethanes: Chlorine Atom-Catalyzed Destruction of Ozone. *Nature* 249:810–12.

Newman, E. A., and P. H. Hartline. 1982. The Infrared "Vision" of Snakes. *Sci. Am.* (March): 116–27.

Pörschke, D., O. C. Uhlenbeck, and F. H. Martin. 1973. Thermodynamics and Kinetics of the Helix-Coil Transition of Oligomers Containing GC Base Pairs. *Biopolymers* 12:1313–35.

Ross, R. T., and M. Calvin. 1967. Thermodynamics of Light Emission and Free Energy Storage in Photosynthesis. *Biophys. J.* 7:595–614.

Taylor, R. E. 2000. Fifty Years of Radiocarbon Dating. *Amer. Scientist* 88:60–67.

Turner, D. H., G. W. Flynn, S. K. Lundberg, L. D. Faller, and N. Sutin. 1972. Dimerization of Proflavin by the Laser Raman Temperature Jump Method. *Nature* 239:215–17.

Wan, C., T. Fiebig, S. O. Kelley, C. R. Treadway, J. K. Barton, and A. H. Zewail. 1999. Femtosecond Dynamics of DNA-Mediated Electron Transfer. *Proc. Natl. Acad. Sci. USA* 96:1187–92.

Zewail, A. H., and R. B. Bernstein. 1988. Real-Time Laser Femtosecond Chemistry. *Chem. Engr. News* 66(45): 29–43.

Zimmerle, C. T., and C. Frieden. 1989. Analysis of Progress Curves by Simulations Generated by Numerical Integration. *Biochemical Journal* 258:381–87. This paper describes computer programs that can be used to obtain differential equations for proposed kinetic mechanisms and to obtain their numerical solution. The software can be downloaded from www.kintek-corp.com.

Problems

1. Iodine reacts with a ketone in aqueous solution to give an iodoketone. The stoichiometric equation is

$$I_2 + \text{ketone} \longrightarrow \text{iodoketone} + H^+ + I^-$$

The rate of the reaction can be measured by measuring the disappearance of I_2 with time. Some data for initial rates and initial concentrations follow:

$-d[I_2]/dt$ (mol L^{-1} s^{-1})	$[I_2]$ (mol L^{-1})	[ketone] (mol L^{-1})	$[H^+]$ (mol L^{-1})
7×10^{-5}	5×10^{-4}	0.2	1.0×10^{-2}
7×10^{-5}	3×10^{-4}	0.2	1.0×10^{-2}
1.7×10^{-4}	5×10^{-4}	0.5	1.0×10^{-2}
5.4×10^{-4}	5×10^{-4}	0.5	3.2×10^{-2}

(a) Find the order of the reaction with respect to I_2, ketone, and H^+.

(b) Write a differential equation expressing your findings in part (a) and calculate the average rate constant.

(c) How long will it take to synthesize 10^{-4} mol L^{-1} of the iodoketone starting with 0.5 mol L^{-1} of ketone and 10^{-3} mol L^{-1} of I_2, if the H^+ concentration is held constant at 10^{-1} mol L^{-1}? Will the reaction go faster if we double the concentration of ketone? of iodine? of H^+? How long will it take to synthesize 10^{-1} mol L^{-1} of the iodoketone if all conditions are the same as above?

(d) Propose a mechanism consistent with the experimental results.

2. The saponification (hydrolysis) of ethyl acetate occurs according to the stoichiometric relation $CH_3COOC_2H_5$ + $OH^- \rightarrow CH_3COO^- + C_2H_5OH$. The reaction can be followed by monitoring the disappearance of OH^-. The following experimental results were obtained at 25°C:

Initial concentration of OH^- (M)	Initial concentration of $CH_3COOC_2H_5$ (M)	Half-life $t_{1/2}$ (s)
0.0050	0.0050	2000
0.0100	0.0100	1000

There is no significant dependence on the concentrations of products of the reaction.

(a) What is the overall kinetic order of the reaction?

(b) Calculate a value for the rate constant, including appropriate units.

(c) How long would it take for the concentration of OH^- to reach 0.0025 M for each experiment?

(d) Based on your answer to part (a), what are possible rate laws for this reaction?

(e) Carefully describe a single experiment that would enable you to decide which of the possibilities of part (d) is most nearly correct.

3. The kinetics of the reaction

$$I^- + OCl^- \rightleftharpoons OI^- + Cl^-$$

was studied in basic aqueous media by Chia and Connick [*J. Phys. Chem.* 63:1518 (1959)]. The initial rate of I^- disappearance is given below for mixtures of various initial compositions, at 25°C. (None of the solutions initially contained OI^- or Cl^-.)

Initial composition of I^- (M)	Initial composition of OCl^- (M)	Initial composition of OH^- (M)	Initial rate $\left(\dfrac{\text{mol } I^-}{L \text{ s}}\right)$
2×10^{-3}	1.5×10^{-3}	1.00	1.8×10^{-4}
4×10^{-3}	1.5×10^{-3}	1.00	3.6×10^{-4}
2×10^{-3}	3×10^{-3}	2.00	1.8×10^{-4}
4×10^{-3}	3×10^{-3}	1.00	7.2×10^{-4}

(a) The rate law can be expressed in the form

$$-\frac{d[I^-]}{dt} = k[I^-]^a[OCl^-]^b[OH^-]^c$$

Find the values of a, b, and c.

(b) Calculate the value of k, including its units.

(c) Show whether this rate law is consistent with the mechanism

$$OCl^- + H_2O \xrightleftharpoons{K_1} HOCl + OH^- \qquad \text{(fast, equilibrium)}$$
$$I^- + HOCl \xrightarrow{k_2} HOI + Cl^- \qquad \text{(slow)}$$
$$HOI + OH^- \xrightleftharpoons{K_3} H_2O + OI^- \qquad \text{(fast, equilibrium)}$$

4. The mechanism for a set of reactions is

$$A \xrightarrow{k_1} B$$
$$B + C \xrightarrow{k_2} D$$

(a) Write a differential equation for the disappearance of A.

(b) Write a differential equation for $d[B]/dt$.

(c) Write a differential equation for the appearance of D.

(d) If $[A]_0$ is the concentration of A at zero time, write an equation that gives $[A]$ at any later time.

5. The stoichiometric equation for a reaction is

$$A + B \longrightarrow C + D$$

The initial rate of formation of C is measured with the following results:

Initial concentration of A (M)	Initial concentration of B (M)	Initial rate (M s^{-1})
1.0	1.0	1.0×10^{-3}
2.0	1.0	4.0×10^{-3}
1.0	2.0	1.0×10^{-3}

(a) What is the order of the reaction with respect to A?

(b) What is the order of the reaction with respect to B?

(c) Use your conclusions in parts (a) and (b) to write a differential equation for the appearance of C.

(d) What is the rate constant k for the reaction? Do not omit the units of k.

(e) Give a possible mechanism for the reaction and discuss in words, or give equations, to show how the mechanism is consistent with the experiment.

6. A reaction is zero order in substance S. Starting with the differential rate law, derive an expression for $t_{1/2}$ in terms of the starting concentration $[S]_0$ and the zero-order rate constant k.

7. Write an expression for the rate of appearance of D from each of the following mechanisms:

(a) $$A \underset{k_2}{\overset{k_1}{\rightleftharpoons}} B$$

$$B + C \overset{k_3}{\longrightarrow} D \qquad \text{(assuming steady state of B)}$$

(b) $$A + B \overset{K}{\rightleftharpoons} AB \qquad \text{(fast to equilibrium)}$$

$$AB + C \overset{k}{\longrightarrow} D$$

8. The following data were obtained for the concentration vs. time for a certain chemical reaction. Values were measured at 1.00 s intervals, beginning at 0.00 and ending at 20.00 s. Concentrations in mM are:

10.00, 6.91, 4.98, 4.32, 3.55, 3.21, 2.61

2.50, 2.22, 1.91, 1.80, 1.65, 1.52, 1.36

1.42, 1.23, 1.20, 1.13, 1.09, 1.00, 0.92

(a) Plot concentration (c) vs. time, ln c vs. time, and $1/c$ vs. time.

(b) Decide whether the data best fit zero-order, first-order, or second-order kinetics. Calculate the rate constant (with units) for the reaction and write the simplest mechanism you can for the reaction.

(c) Describe an experimental method which you think might be used to measure the concentrations that are changing so rapidly. The reaction is nearly over in 20 s.

9. The kinetics of double-strand formation for a DNA oligonucleotide containing a G·T base pair was measured by temperature-jump kinetics. The reaction is:

$$2CGTGAATTCGCG \underset{k_{-1}}{\overset{k_1}{\rightleftharpoons}} DUPLEX$$

The following data were obtained:

Temperature (°C)	k_1 ($10^5\ M^{-1}\ s^{-1}$)	k_{-1} (s^{-1})
31.8	0.8	1.00
36.8	2.3	3.20
41.8	3.5	15.4
46.7	6.0	87.0

(a) Determine E_a, $\Delta H^{\ddagger}$, and $\Delta S^{\ddagger}$ for the forward and reverse reactions; assume that the values are independent of temperature.

(b) Determine the standard enthalpy ΔH^0 and standard entropy ΔS^0 changes for the reaction (assumed independent of temperature) and calculate the equilibrium constant for duplex formation at 37°C.

(c) Discuss qualitatively what you would expect to happen to $\Delta H^{\ddagger}$ and the standard enthalpy change for the reaction if the duplex contained one more Watson–Crick base pair. Would each increase, decrease, or remain the same?

10. Equal volumes of two equimolar solutions of reactants A and B are mixed, and the reaction $A + B \rightarrow C$ occurs. At the end of 1 h, A is 90% reacted. How much of A will be left *unreacted* at the end of 2 h if the reaction is:

(a) First order in A and zero order in B?

(b) First order in A and first order in B?

(c) Zero order in both A and B?

(d) First order in A and one-half order in B?

11. In a second experiment, the same reaction mixture of Problem 10 is diluted by a factor of 2 at the time of mixing. How much of A will be left unreacted after 1 h for each of the assumptions (a) through (d) of problem 10?

12. The reaction of a hydrogen halide A with an olefin B to give product P according to the stoichiometric relation $A + B \rightarrow P$ is proposed to occur by the following mechanism:

$$2A \overset{K_1}{\rightleftharpoons} A_2 \qquad \text{(fast to equilibrium)}$$

$$A + B \overset{K_2}{\rightleftharpoons} C \qquad \text{(fast to equilibrium)}$$

$$A_2 + C \overset{k_3}{\longrightarrow} P + 2A \qquad \text{(slow)}$$

(a) Based on this mechanism, derive an expression for the velocity of the reaction as a function of the concentrations of only reactants and stable product.

(b) In a particular experiment, equal concentrations of A and B are mixed together, and both the initial rate, v_0, and the half-life, $t_{1/2}$, are determined. Based on your answer to part (a), predict for a subsequent experiment:

(i) The effect on v_0 if the initial concentrations of both A and B are doubled.

(ii) The effect on $t_{1/2}$ if the initial concentrations of both A and B are doubled.

(iii) The effect on v_0 if the initial concentration of A is unchanged and the initial concentration of B is increased tenfold.

13. The age of water or wine may be determined by measuring its radioactive tritium (^3_1H) content. Tritium, present in a steady state in nature, is formed primarily by cosmic irradiation of water vapor in the upper atmosphere, and it decays spontaneously by a first-order process with a half-life of 12.5 years. The formation reaction does not occur significantly inside a glass bottle at the surface of Earth. Calculate the age of a suspected vintage wine that is 20% as radioactive as a freshly bottled specimen. Would you recommend to a friend that he consider paying a premium price for the "vintage" wine?

14. A radioactive sample produced 1.00×10^5 disintegrations min^{-1}; 28 days later it produced only 0.25×10^5 disintegrations min^{-1} (1 day = 1440 min).

(a) What is the half-life of the radioisotope?

(b) How many radioactive atoms were there in the sample that had 10^5 disintegrations min^{-1}?

15. Consider the following proposed mechanism for the reaction A → P:

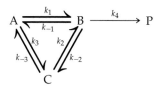

(a) Write a differential equation for the rate of formation of B.

(b) The formation of B from A and C is fast to equilibrium with equilibrium constants $K_1 = k_1/k_{-1}$, $K_2 = k_2/k_{-2}$, and $K_3 = k_3/k_{-3}$. k_4 is much smaller than all other k's. Write a differential equation for the formation of P in terms of concentration of A, equilibrium constants, and k_4.

(c) Write an expression (containing no derivatives) for the concentration of P as a function of time. At zero time, $[A] = [A]_0$ and $[B], [C], [P] = 0$.

16. There is evidence that a critical concentration of a trigger protein is needed for cell division (see *Molecular Biology of the Cell* by Alberts et al., cited in chapter 2). This unstable protein is continually being synthesized and degraded. The rate of protein synthesis controls how long it takes for the trigger protein to build up to the concentration necessary to start DNA synthesis and eventually to cause cell division.

Let's choose a simple mechanism to consider quantitatively. The trigger protein U is being synthesized by a zero-order mechanism with rate constant k_0. It is being degraded by a first-order mechanism with rate constant k_1.

(a) Write a differential equation consistent with the mechanism.

(b) The solution to the correct differential equation is

$$[U] = \frac{k_0}{k_1}(1 - e^{-k_1 t})$$

if [U] is equal to zero at zero time. Show that your equation in part (a) is consistent with this.

(c) If U is being synthesized at a constant rate with $k_0 = 1.00$ nM s^{-1} and its half-life for degradation is 0.500 h, calculate the maximum concentration that U will reach. How long will it take to reach this concentration? Make a plot of [U] vs. time.

(d) If a concentration of U of 1.00 μM is needed to trigger DNA synthesis and cell replication, how long will it take to reach this concentration?

(e) If the rate of synthesis is cut in half ($k_0 = 0.500$ nM s^{-1}), how long will it take for U to reach a concentration of 1.00 μM?

(f) What is the smallest rate of U synthesis k_0 that will allow (slow) cell replication? Assume k_1 remains constant.

17. A simple reaction for a DNA molecule is the exchange of a DNA base proton with a water proton. In the DNA base pair shown below, the imino proton (in boldface) exchanges with water at a rate measurable by NMR.

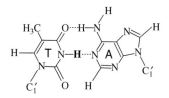

The stoichiometry of the reaction is

$$TH \cdot A + H^*OH^* \longrightarrow TH^* \cdot A + HOH^*$$

A mechanism for its rate of exchange is as follows:

$$(TH \cdot A)_{closed} \underset{k_{cl}}{\overset{k_{op}}{\rightleftharpoons}} (TH \cdots A)_{open}$$

$$(TH \cdots A)_{open} + B + H^*OH^* \overset{k_{tr}}{\rightleftharpoons}$$
$$(TH^* \cdot A)_{closed} + B + HOH^*$$

$(TH \cdot A)_{closed}$ represents a closed base pair. $(TH \cdots A)_{open}$ is the open base pair ready to exchange. B is a base that catalyzes the transfer of H to H*.

(a) Use a steady-state approximation for $[(TH \cdots A)_{open}]$ to obtain an expression for the initial rate of exchange.

$$\frac{d[TH^* \cdot A]}{dt} = k_{ex}[(TH \cdot A)_{closed}]$$

Write k_{ex} as a function of k_{op}, k_{cl}, k_{tr}, and [B].

(b) Show how the rates of exchange vs. concentrations of $[(TH \cdot A)_{closed}]$ and [B] can be used to obtain k_{open} (the rate constant for base-pair opening).

(c) The imino exchange rate for a synthetic polynucleotide, poly dA·dT, was measured as a function of pH at 27°C. Calculate a value for k_{op}, the rate constant for base-pair opening, from the following data:

$[OH^-]$ (μM)	1.26	2.00	3.16	10.0	31.6
k_{ex} (s^{-1})	12.5	18.2	27.2	69.9	140.8

18. A reaction occurs with the following stoichiometry:

$$A + P \longrightarrow AP$$

The concentration of A was measured vs. time after mixing; the data are the following:

[A] (nM)	Time (s)
50.0	0
40.0	100
30.0	229
20.0	411
10.0	721

(a) There is no dependence of rate on P. What is the order of the reaction? Calculate the rate constant and give the units.

(b) Propose a mechanism that is consistent with the kinetics and stoichiometry of the reaction. Write differential equations for your mechanism and relate the measured k to the k's in your mechanism.

(c) The data given above were measured at 0°C. When the reaction was studied at 10°C, the rate constant doubled. Calculate the activation energy in kJ mol^{-1}.

19. The gas-phase decomposition of di-*tert*-butyl peroxide, $(CH_3)_3COOC(CH_3)_3$, is first order in the temperature range 110°C to 280°C, with a first-order rate constant $k = 3.2 \times 10^{16} \exp[-(164 \text{ kJ mol}^{-1})/RT] \text{ s}^{-1}$.

(a) What are the values of $\Delta H^{\ddagger}$ and $\Delta S^{\ddagger}$ for this reaction at 110°C?

(b) Give a brief interpretation of the value of $\Delta S^{\ddagger}$.

(c) At what temperature will the reaction occur ten times faster than at 110°C?

20. Calculate the activation energy that leads to a doubling of the rate of reaction with an increase in temperature from 25°C to 35°C.

21. The mechanism for a reaction is assumed to be

$$A + B \xrightarrow{k} P \qquad (k = 1.0 \times 10^5 \text{ } M^{-1}\text{s}^{-1} \text{ at } 27°C)$$

(a) Calculate the initial rates of formation of P if 0.10 M A is mixed with 0.10 M B at 27°C. State the units.

(b) Calculate the initial rate of formation of P if 1.00×10^{-4} M A is mixed with 1.00×10^{-6} M B at 27°C. State the units.

(c) How long would it take (in seconds) to form 0.050 M product from 0.10 M A and 0.10 M B at 27°C?

(d) At 127°C, the rate constant of the reaction increases by a factor of 10^3. Calculate E_a and $\Delta H^{\ddagger}$ at 27°C.

22. If A and B are mixed together in solution, it is found that the concentration of A decreases with time, but B remains constant. The stoichiometric equation is A → P.

(a) If the initial concentration of A is less than 0.01 M, it is found that the initial rate is $-d[A]/dt = k_1[A][B]$. Write the integrated form of this equation. What is the order of the reaction with respect to A? With respect to B?

(b) If the initial concentration of A is greater than 1 M, it is found that the initial rate is nearly independent of A:

$$-\frac{d[A]}{dt} = k_2[B]$$

Write the integrated form of this equation. What is the order of the reaction with respect to A? With respect to B?

(c) Sketch a plot of the initial rate $-d[A]/dt$ vs. initial concentration [A]. Write one rate equation that is consistent with experiments at both high and low concentrations of A.

23. (a) If we express the rate as the number of molecules reacting per unit time, for a unimolecular reaction the rate depends only on the *number* of reacting molecules present. For a bimolecular reaction, however, the rate depends on their *concentrations*. Explain.

(b) Consider the simple reaction

$$A \underset{k_2}{\overset{k_1}{\rightleftharpoons}} B$$

It is proposed that the forward reaction (k_1) is first order and the reverse reaction (k_2) is zero order. Is this possible? Explain.

24. A reacts to form P. A plot of the reciprocal of the concentration of A vs. time is a straight line. When the initial concentration of A is 1.0×10^{-2} M, its half-life is found to be 20 min.

(a) What is the order of the reaction?

(b) Write a one-line mechanism that is consistent with the kinetics.

(c) What is the value of the rate constant for your mechanism of part (b)?

(d) When the initial concentration of A is 3.0×10^{-3} M, what will be the half-life?

25. A and B react stoichiometrically to form P. If $0.01\ M$ A and $10\ M$ B are mixed, it is found that the log of the concentration of A vs. time is a straight line.

(a) What is the order of the reaction with respect to A?

(b) Write a one-line mechanism that is consistent with the kinetics and stoichiometry.

(c) According to your mechanism, what is the order of the reaction with respect to B?

(d) The half-life for the disappearance of A is 100 s. What would be the predicted half-life if the concentration of B is changed from $10\ M$ to $20\ M$?

(e) From the half-life given in part (d), calculate the rate constant for the reaction of A with B. Specify the units of the rate constant.

26. All living cells have approximately the same ^{14}C to ^{12}C ratio as found in CO_2 in the atmosphere. When an animal or plant or bacterial cell dies, it begins to lose ^{14}C by radioactive decay; thus, over time the ^{14}C to ^{12}C ratio decreases. This ratio can be measured by a mass spectrometer. As a novice archaeologist, you find an old piece of wood and have it analyzed for $^{14}C/^{12}C$. The ratio found is 28% of the present ratio. How long ago was the wood formed?

27. ^{131}I has a radioactive half-life of 7.80 days, but iodine is also removed from the body by excretion at a first-order rate characterized by a "biological half-life" of 26 days.

(a) What will be the "effective half-life" of ^{131}I in the body (the time required for half of the ^{131}I to be removed by the combination of excretion and radioactive decay)?

(b) A laboratory animal injected with ^{131}I should not be reinjected with that isotope for 2 months. Using the result of part (a), explain why this should be so.

28. In a paper by Bada, Protsch, and Schroeder [*Nature* 241:394 (1973)], the rate of isomerization of isoleucine in fossilized bone is used as an indication of the average temperature of the sample since it was deposited. The reaction

$$\text{L-isoleucine} \rightleftharpoons \text{D-alloisoleucine}$$
$$\text{iso} \qquad\qquad\qquad \text{allo}$$

produces a nonbiological amino acid, D-alloisoleucine, that can be measured using an automatic amino acid analyzer. At 20°C, this first-order reaction has a half-life of 125,000 years, and its activation energy is 139.7 kJ mol^{-1}. After a very long time, the ratio allo/iso reaches an equilibrium value of 1.38. You may assume that this equilibrium constant is temperature independent.

(a) For a hippopotamus mandible found near a warm spring in South Africa, the allo/iso ratio was found to be 0.42. Assuming that no allo was present initially, calculate the ratio of the concentration of allo now present to the concentration of allo after a very long time (*Note:* The correct answer is between 0.40 and 0.60.)

(b) Radiocarbon dating, which is temperature independent, indicated an age of 38,600 years for the hippo tooth. Using the result of part (a), estimate the half-life for the process.

(c) Calculate the average temperature of this specimen during its residence in the ground. (The present mean temperature of the spring is 28°C.)

29. In aqueous solution, the reaction of A to form B has the following rate expression:

$$-\frac{d[A]}{dt} = k[A]\{1 + k'[H^+]\}$$

(a) Propose a mechanism that is consistent with this experimental rate. Relate the k's in your mechanism to k and k'.

(b) From the pH dependence of the reaction, k' was found to be $1.0 \times 10^5\ M^{-1}$. In a pH 4.0 buffer, it took 5 min for a $0.30\ M$ solution of A to react to give $0.15\ M$ B. Calculate the value of k and give its units.

30. Transfer RNA can exist in two forms that are in rapid equilibrium with each other; the equilibrium constant at 28°C, $K = [B]^{eq}/[A]^{eq} = 10$. A temperature-jump experiment is done to measure the rates of interconversion. A solution of the tRNA at a concentration of $10\ \mu M$ is quickly (faster than 10 μs) raised in temperature from 25°C to 28°C. An experiment signal is measured with a relaxation time, $\tau = 3$ ms. Assume that the mechanism is:

$$A \underset{k_{-1}}{\overset{k_1}{\rightleftharpoons}} B$$

(a) What are the values of k_1 and k_{-1}, including units?

(b) At what temperature are these values measured?

(c) Would you expect doubling the concentration of tRNA to increase, decrease, or leave unchanged the values of τ, k_1, and k_{-1}?

31. The equilibrium

$$I_2 + I^- \underset{k_{-1}}{\overset{k_1}{\rightleftharpoons}} I_3^-$$

has been studied using a laser-induced temperature-jump technique. Relaxation times were measured at various equilibrium concentrations of the reactants at 25°C:

$[\bar{I}^-]$ (mM)	$[\bar{I}_2]$ (mM)	τ (ns)
0.57	0.36	71
1.58	0.24	50
2.39	0.39	39
2.68	0.16	38
3.45	0.14	32

Source: Turner et al. (1972).

(a) Calculate k_1 and k_{-1} for this system at 25°C.

(b) Compare your results with the value for the equilibrium constant, $K = 720\ M^{-1}$.

(c) Use simple diffusion theory to estimate a value for k_1. For I^- and I_2 the effective radii are 2.16×10^{-8} and 2.52×10^{-8} cm, respectively, and diffusion coefficients are 2.05×10^{-5} and 2.25×10^{-5} cm^2 s^{-1}, respectively. Comparing this result with the experimental value from part (a), what do you conclude about the effectiveness of collisions of I_2 and I^- in leading to reaction?

32. Consider the dimerization of proflavin:

$$2P \underset{k_{-1}}{\overset{k_1}{\rightleftharpoons}} P_2$$

(a) Starting with the usual expression for the chemical relaxation time for a dimerization,

$$\tau = \frac{1}{4k_1[\bar{P}] + k_{-1}}$$

derive the following alternative form:

$$\frac{1}{\tau^2} = k_{-1}^2 + 8k_1 k_{-1}[P]_t$$

where $[P]_t = [\bar{P}] + 2[\bar{P}_2]$ is the total concentration of proflavin in the solution and $[\bar{P}]$ and $[\bar{P}_2]$ are the concentrations of proflavin and its dimer at equilibrium.

(b) A bonus from this analysis is that both k_1 and k_{-1} can be obtained by measuring the relaxation rate as

a function of $[P]_t$. Plot the following data and determine values for k_1 and k_{-1}. The temperature is 25°C.

τ (s)	$[P]_t$ (M)
3.2×10^{-7}	0.5×10^{-3}
1.8×10^{-7}	2.0×10^{-3}
1.4×10^{-7}	3.5×10^{-3}
1.2×10^{-7}	5.0×10^{-3}

(c) What is ΔG^0 at 25°C for the dimerization of proflavin?

33. The ionization constant for NH_4^+ is $K = 5.8 \times 10^{-10}$ at 25°C:

$$NH_4^+ \overset{K}{\rightleftharpoons} NH_3 + H^+$$

Table 7.5 gives

$$H^+ + NH_3 \overset{k_1}{\longrightarrow} NH_4^+ \qquad (k_1 = 4.3 \times 10^{10}\ M^{-1}\,s^{-1})$$

(a) A temperature jump from 20°C to 25°C is made on a 0.1 M NH_4Cl solution at pH 6. Calculate the relaxation time for the reaction, assuming that the mechanism is given by the ionization equation shown.

(b) The relaxation time is a function of pH. At what pH will the relaxation time have its largest value?

(c) How would you measure the activation energy for k_1? Would you expect the activation energy to be less than 50 kJ? Greater than 100 kJ? Explain.

(d) According to Debye–Hückel theory, would you expect the relaxation time to increase, decrease, or remain unchanged if 0.1 M NaCl is added? Explain.

34. Gaseous ozone O_3 undergoes decomposition according to the stoichiometric equation

$$2O_3(g) \longrightarrow 3O_2(g)$$

Two alternative mechanisms have been proposed to account for this reaction:

(I) $2O_3 \overset{k}{\longrightarrow} 3O_2$ (bimolecular)

(II) $O_3 \overset{K_1}{\rightleftharpoons} O_2 + O$ (fast, equilibrium)

      $O + O_3 \overset{k_2}{\longrightarrow} 2O_2$ (slow)

(a) Derive rate laws for the formation of O_2 for each of these mechanisms.

(b) Thermodynamic measurements give standard enthalpies of formation for each of the following species at 298 K:

Substance	ΔH^0_{298} (kJ mol^{-1})
$O_2(g)$	0.0
$O_3(g)$	142.3
$O (g)$	249.4

The observed activation enthalpy $\Delta H^{\ddagger}$ for the overall reaction $2O_3 \rightarrow 3O_2$ is 125.5 kJ mol^{-1} of O_3. Sketch a curve of enthalpy (per mole of O_3) vs. reaction coordinate for each of the two proposed mechanisms. Label the curves with numerical values for the ΔH between reactants, products, intermediates, and transition states.

(c) Can you exclude either of these mechanisms on the basis of the thermodynamic and activation enthalpy values of part (b)? Explain your answer.

(d) Devise a kinetic procedure for distinguishing between the two mechanisms. State clearly the nature of the experiments you would perform and what results you would use to make the distinction.

35. In the complete process of "simple" diffusion of a substance S across a membrane from side 1 to side 2, S first "jumps" into the membrane from the solution at side 1, then diffuses across the membrane interior, and finally "jumps" from the membrane at side 2 into the aqueous solution:

$$(S)_1 \underset{k_{-1}}{\overset{k_1}{\rightleftharpoons}} \Big| (S)_{m1} \underset{k_2}{\overset{k_2}{\rightleftharpoons}} (S)_{m2} \underset{k_1}{\overset{k_{-1}}{\rightleftharpoons}} \Big| (S)_2$$

side 1   membrane   side 2

For a symmetric membrane, the rate constants for "jumping" into and out of the membrane are the same at both surfaces, as indicated. At time zero, (S) is added to the solution on side 1. The area of the membrane is A.

(a) Give the rate law for the initial rate of appearance of (S) on side 2, taking k_1 as the rate-limiting step. Express the rate as

$$\frac{1}{A}\frac{dN_2}{dt}$$

where N_2 is the number of moles of (S) on side 2 and A is the area.

(b) Give the rate law for the initial rate of appearance of (S) on side 2, taking k_2 as the rate-limiting step. Again, express the rate as

$$\frac{1}{A}\frac{dN_2}{dt}$$

Your equation should contain the concentration of (S) on side 1 as the only concentration term.

(c) Experimentally, you can determine the initial rate of appearance of S on side 2 as a function of the initial concentration of S added to side 1. Using this type of kinetic data alone, can you deduce which step (k_1 or k_2) is rate limiting? Explain briefly.

(d) Suppose the membrane surfaces bear a net negative charge and that S is positively charged. What effect, if any, would decreasing the electrolyte concentration have on the initial rate of transport of S across the membrane? Briefly explain your answer.

36. Consider the reaction between substances $A^{\oplus}$ and $B^{\oplus}$, both positively charged. According to transition-state theory, $A^{\oplus}$ and $B^{\oplus}$ collide to form a transition-state complex which decomposes to product:

$$A^{\oplus} + B^{\oplus} \rightleftharpoons \underset{|\leftarrow r^{\ddagger} \rightarrow|}{[A^{\oplus}\cdots B^{\oplus}]} \longrightarrow \text{product}$$

The *electrostatic* contribution to the free energy of activation $\Delta G^{\ddagger}$ in a solvent of dielectric constant ε and in extremely dilute solution so that the ionic strength is essentially zero can be estimated by Coulomb's law: Energy $= q_1 q_2 / \varepsilon r$. In the transition-state complex, the charge centers of $A^{\oplus}$ and $B^{\oplus}$ are $r^{\ddagger}$ apart, as indicated above.

(a) On the basis of electrostatic considerations, would you expect this reaction to be more rapid in a solvent of high or low dielectric constant? Explain briefly.

(b) Suggest an experimental approach that would allow determination of the electrostatic contribution to the free energy of activation.

37. Imidazole (Im) can react with H^+ or H_2O to form positively charged imidazole (ImH^+). The reaction mechanisms are

$$Im + H^+ \underset{k_{-1}}{\overset{k_1}{\rightleftharpoons}} ImH^+$$

$$Im + H_2O \underset{k_{-2}}{\overset{k_2}{\rightleftharpoons}} ImH^+ + OH^-$$

The rate constants in aqueous solution are: $k_1 = 1.5 \times 10^{10} \, M^{-1}\,s^{-1}$; $k_{-1} = 1.5 \times 10^3 \, s^{-1}$; $k_2 = 2.5 \times 10^3 \, s^{-1}$; $k_{-2} = 2.5 \times 10^{10} \, M^{-1}\,s^{-1}$.

(a) What is the value of the equilibrium constant for the ionization of imidazole ($ImH^+ \rightleftharpoons Im + H^+$)?

(b) Write the differential equation for the net rate of formation of ImH^+.

(c) If the pH is suddenly changed for a solution of 0.1 M imidazole in water from pH 7 to pH 4, what is the rate-determining step for the appearance of ImH^+ at pH 4?

(d) What is the value of the initial rate of increase of ImH^+ at pH 4?

(e) The rate constants k_1 and k_{-1} both depend on temperature. Would you expect them to decrease or increase with increasing temperature? Which would you expect to change most with temperature and why?

(f) Predict the sign of the heat of ionization for imidazole based on your answer to part (e). The experimental heat of ionization is positive.

38. In medical treatment, drugs are often applied either continuously or in a series of small, closely spaced doses. Assume that oral ingestion and absorption of a particular drug are rapid compared with the rate of utilization and elimination and that (after a single dose) the latter process occurs by a simple exponential decrease. We can represent this by a scheme (not a mechanism)

$$A \xrightarrow{k_1} B \xrightarrow{k_2} C$$

where [A] is the concentration of the drug at its site of introduction, [B] is the concentration of the drug in blood, and [C] represents the elimination of the drug. In this model, the delivery k_1 of the drug is zero order kinetically, and the elimination k_2 is first order.

(a) Write the rate law, in terms of $d[B]/dt$, that is appropriate to this model.

(b) Assuming that the first introduction of the drug occurs at time zero, draw a curve showing how [B] will change with time during continuous, prolonged administration of the drug. [If this is not intuitively obvious to you, proceed first to part (f).]

(c) After some time, the patient is cured, and the use of the drug is stopped. Add to your drawing a sketch of how [B] changes subsequently.

(d) From an experimental record or monitor of [B] vs. time, how would you determine k_2?

(e) How would you determine k_1?

(f) Derive a mathematical expression for [B] as a function of time during drug administration starting with the rate law from part (a). (A simple variable substitution should convert it to a form that was treated explicitly in this chapter.) Demonstrate that the result is consistent with your rise curve.

(g) To what level does [B] rise after a long time of administration?

39. In view of the current shortage of fossil fuels (chiefly petroleum and natural gas), alternative sources of energy are being sought. An imaginative proposal is to use rubber (latex), which is a plant hydrocarbon, polyisoprene. Latex represents about 50% of the carbon fixed in photosynthesis by rubber trees! Presumably this material could be cracked and processed in refineries in a fashion analogous to that used for petroleum. Under good conditions, a rubber tree in Brazil can produce 10 kg of usable latex per year, growing at a density of about 100 trees per acre. The heat of combustion (fuel value) of hydrocarbons is about 644 kJ mol^{-1} of CH_2, or about 46,000 kJ kg^{-1}.

(a) What fraction of a year's solar energy is stored in rubber crop if the incident solar energy is 4 kJ min^{-1} ft^{-2} and the Sun is shining 500 min day^{-1}, on the average? (1 acre = 43,560 ft^2.) [Compare your answer with the value (energy stored)/(solar energy) = 0.3 that can be obtained from photosynthesis in the laboratory under optimal conditions.]

(b) In 2000 the total fossil fuel consumption in the United States was 14×10^{16} kJ yr^{-1}. What fraction of the area of the United States (3,000,000 mi^2) would have to be devoted to growing rubber trees (if they would grow in the climate of the United States, which they will not) to replace our present sources of fossil fuel by latex? (1 mi^2 = 640 acres.)

(c) In laboratory experiments, it is found that photosynthesis occurs optimally with about 3% CO_2 in the (artificial) atmosphere. CO_2 occurs at only 0.03% in the natural atmosphere. Propose a scheme for increasing the energy yield of agricultural crops (not necessarily rubber trees) based on this observation. List some possible additional advantages and disadvantages of your scheme, based on your background in kinetics and thermodynamics.

40. Consider the following reactions involved in the formation and disappearance of ozone in the upper atmosphere.

$$O_2 \xrightarrow{h\nu \text{ (below 242 nm)}} O + O \qquad \textbf{(A)}$$

$$O + O_2 + M \longrightarrow O_3 + M \qquad \textbf{(B)}$$

$$O_3 \xrightarrow{h\nu \text{ (190–300 nm)}} O_2 + O \qquad \textbf{(C)}$$

where the first and third reactions are photochemical and driven by sunlight and the second reaction is termolecular. (The third body, M, may be N_2, O_2, or any other gas molecule in the atmosphere.) The potential danger of reducing the "ozone shield" by supersonic transports (SSTs) comes from the presence of nitrogen oxides (NO, NO_2, etc.) in the engine exhaust. Although the actual mechanism is more complex, two important reactions in the proposed ozone reaction are

$$NO + O_3 \xrightarrow{k_1} NO_2 + O_2 \qquad \text{(fast)} \qquad \textbf{(1)}$$

$$O + NO_2 \xrightarrow{k_2} NO + O_2 \qquad \text{(fast)} \qquad \textbf{(2)}$$

Note that NO is not used up in this pair of reactions, but it continues to decompose ozone in a pseudo-catalytic fashion (until it eventually diffuses out of the ozone layer, which may be a very slow process taking many months).

(a) At a temperature of 217 K, characteristic of the 20-km altitude at which an SST might fly, the value of k_1 is 4.0×10^{-15} cm^3 molecule^{-1} s^{-1}. From the data on exhaust composition, it is estimated that the steady-state concentration of NO would be about 10^{10} molecules cm^{-3}. If NO were suddenly introduced at this level, what would be the half-time (in hours) for the reduction of ozone in the region of the SST flight paths? (*Note:* You do not need to know the ozone concentration to solve this problem.)

(b) The molecule NO_2 is unstable in the presence of sunlight and decomposes photochemically according to the following reaction:

$$NO_2 \xrightarrow{h\nu\,(260-400\text{ nm})} NO + O \qquad \textbf{(3)}$$

The oxygen atoms are highly reactive and disappear by reactions (B) and (2) above. Consider carefully the contribution of reaction (3) to the overall rate of ozone disappearance and describe qualitatively the effect that you would expect if this reaction were added to the situation approximated in part (a).

41. The photosynthesis efficiency of a strain of algae was measured by irradiating it for 100 s with an absorbed intensity of 10 W and an average wavelength of 550 nm. The yield of O_2 was 5.75×10^{-4} mol. Calculate the quantum yield of O_2 formation. (The quantum yield per "equivalent" of photochemistry is four times this value, since four electrons must be removed from two water molecules to produce each O_2.)

42. The human eye, when completely adapted to darkness, is able to perceive a point source of light against a dark background when the rate of incidence of radiation on the retina is greater than 2×10^{-16} W.

(a) Find the minimum rate of incidence of quanta of radiation on the retina necessary to produce vision, assuming a wavelength of 550 nm.

(b) Assuming that all this energy is converted ultimately to heat in visual cells having a total volume of 10^{-2} cm^3 and assuming that there are no losses of heat, by how much would the temperature of these cells rise in 1 s? Assume a heat capacity and density equal to that of water.

(c) Estimate the safe range of visual intensities (those which you would not expect to cause permanent eye damage).

43. The protein rhodopsin is responsible for the absorption of light in photoreceptor cells. Following the absorption of light, the protein goes through the complex series of reactions shown in figure 7.20. Here we will use the symbols R, BR, LR, MI, and so on, to indicate the different forms of the protein, and we will assume that the lifetimes shown in figure 7.21 are the reciprocals of the corresponding rate constants $k_1, k_2, \ldots, k_n$ at 310 K.

(a) Write the rate law for the formation of MI following a very short (10^{-12} s) flash of light that converts all of the R initially present to BR.

(b) If R is initially present at $1 \times 10^{-3}\,M$, what is the initial rate of formation of MI? State any approximations or assumptions that you make.

(c) Experimentally, it is found that the ratio [MII]/[MI] rapidly (within 5×10^{-3} s) reaches a constant value of 1.0 after light absorption, even though [MI] and [MII] individually decrease slowly with time. What is the value for k_{-4}, the rate constant for the reaction MII $\to$ MI?

(d) It has been proposed that the transformation MI $\to$ MII involves large changes in protein structure. The enthalpy change for this step was determined to be ΔH^0_{310} (MI $\to$ MII) = 42 kJ mol^{-1}. Using the result of part (c), calculate ΔS^0_{310} for this transformation and interpret it in terms of a possible structural change.

(e) Write the rate law for the disappearance of MII.

44. An electronically excited molecule A* can either emit fluorescence to return to its ground state, or it can lose its excitation by collision:

$$A^* \xrightarrow{k_r} A + \text{photon} \qquad \left(\begin{array}{c} k_r \text{ and } k_T \text{ are rate} \\ \text{constants with units} \\ \text{of s}^{-1} \end{array} \right)$$
$$A^* \xrightarrow{k_T} A + \text{heat}$$

(a) The fluorescence intensity of A* can be measured as the number of photons emitted per second. Write an equation relating the fluorescence intensity to the concentration of A*.

(b) Derive an equation for the concentration of A* as a function of time, k_r, k_T, and [A*]$_0$ (the concentration of A* at zero time).

45. Sunlight between 290 and 313 nm can produce sunburn (erythema) in 30 min. The intensity of radiation between these wavelengths in summer, at 45° latitude and at sea level, is about 50 W cm^{-2}. Assuming that each incident photon is absorbed and produces a chemical change in one molecule, how many molecules per square centimeter of human skin must be photochemically affected to produce evidence of sunburn?

46. Psoralen is a three-ring heterocyclic molecule found in rotting celery. Psoralen reacts with DNA in the presence of light to form monoadducts, which can react further to cross-link the DNA. This can lead to dermatitis and hyperpigmentation of the skin of celery workers.

A mechanism for the first step of the reaction with DNA is

$$\text{Psoralen} + \text{DNA} \underset{}{\overset{K}{\rightleftharpoons}} \text{Intercalated complex}$$

$$\text{Intercalated complex} + \text{light} \longrightarrow \text{Monoadduct}$$

The equilibrium constant for intercalation into *E. coli* DNA is $K = 250\,M^{-1}$ at 25°C (the concentration of

DNA is moles of base pairs per liter); the quantum yield for the photochemical reaction is $\phi = 0.030$.

(a) Calculate the concentration of complex at equilibrium in a solution of 1.00×10^{-4} M of psoralen and 1.00×10^{-3} M DNA at 25°C in the dark. Also calculate the fraction of psoralen bound to DNA and the fraction free.

(b) The solution in part (a) was irradiated with 365 nm light with an intensity of 1.00×10^{-11} einstein cm^{-2} s^{-1}. The absorbance of the solution in a 1-cm path-length cell was 0.30 at 365 nm. At this wavelength, only the free psoralen and the complex absorb. They both have the same molar extinction coefficient per psoralen. Calculate the initial rate of formation of monoadduct with units of $M\,s^{-1}$. Remember that only the light absorbed by the complex can lead to product.

(c) After half the psoralen has reacted to form monoadduct, the effective concentration of DNA has changed by only 5%. For these conditions calculate the rate of formation of adduct.

CHAPTER 8

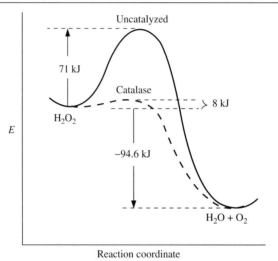

Enzyme Kinetics

Enzymes are the biological catalysts that control all biochemical reactions. It is essential to understand them and to know about their reversible and irreversible inhibitors. Most poisons and many drugs are enzyme inhibitors.

Concepts

An enzyme—a biological catalyst—can increase the rate of a reaction by many orders of magnitude, but it is left unchanged at the end of the reaction. It binds a particular reactant (called a *substrate*) and facilitates its reaction by stabilizing its transition state, leading to a specific product. Many substances present in the surroundings—such as oxygen, acids, bases, and metal ion impurities—can inactivate enzymes. Most enzymes are proteins, but the active sites often contain groups other than amino acids—for example, metalloenzymes, in which zinc, iron, cobalt, manganese, or other transition metals are the catalytic centers. Many enzymes require *cofactors*—small molecules bound to the enzyme—to be effective catalysts (for example, the B family of vitamins are all cofactors). Some enzymes are complexes of protein and RNA in which the RNA also participates in catalysis. One example is the ribosome, a complex assembly of many proteins and three RNAs. The ribosome has a key role in the translation of messenger RNA into proteins.

When Arthur Zaug and Thomas Cech (1986) discovered that RNA by itself could be an enzyme, biochemists' view of biological catalysis was revolutionized. Now we know that enzymes need not be proteins; perhaps other sorts of biological molecules, such as carbohydrates, also have catalytic activity. The catalytic activity of RNA has also provided new ideas about the evolution of enzymes; RNA enzymes—*ribozymes*—may have been primitive enzymes that have now been mainly replaced by the more efficient protein enzymes.

Applications
Catalytic Antibodies and RNA Enzymes—Ribozymes

Antibodies are proteins synthesized by the immune system to protect an organism against foreign proteins or other foreign chemicals. Special cells synthesize antibodies that bind strongly to the foreign molecule. To produce antibodies to any desired chemical, the chemical—called a *hapten*—is first attached covalently to a protein. Used as an antigen, the modified protein induces the cellular production of antibodies. This means that proteins are produced that bind strongly and specifically to different parts of the antigen, including the hapten. Catalysts increase the rate of a reaction by decreasing the free energy of activation; the catalyst stabilizes the transition state. To produce an effective enzyme, the chemical used to induce antibody production can be a transition-state analog of an enzyme substrate. For example, the transition state for hydrolysis of an ester is a tetrahedral species:

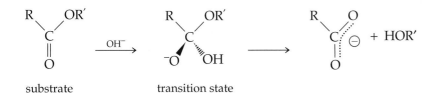

substrate   transition state

An excellent transition-state analog for ester hydrolysis is a phosphate ester:

transition-state analog

Stable phosphonates (shown above) attached to proteins to form antigens produced catalytic antibodies that approach the catalytic efficiency of the protein enzyme chymotrypsin for ester substrates. Scientists have produced catalytic antibodies with enzymatic activity for many different types of substrates. One goal of the research is to produce enzymes that have catalytic activity not found in nature (Schultz and Lerner 1995).

Scientists can produce enzymes that have better catalytic activities than those found in nature. One way to do this is to randomly mutate a naturally occurring enzyme and then to select the mutants with the desired properties. Frances Arnold's group has evolved a variant of a cytochrome P450 enzyme that has 20 times the activity of the natural enzyme and that does not require any cofactors for its activity (Joo, Lin, and Arnold 1999). This P450 class of enzymes can oxidize a wide variety of aromatic organic compounds using NADH and oxygen. The need for the NADH cofactor prevents the use of this enzyme for commercial synthesis of pharmaceuticals and other chemicals. The Arnold group introduced approximately 200,000 random mutants of P450 DNA (prepared by mutagenic polymerase chain reaction, PCR) into bacteria and assayed individual clones for the ability to oxidize naphthalene in the presence of hydrogen peroxide. The clones were assayed by a sensitive fluorescent method that quickly revealed which bacteria contained the most efficient P450 catalysts. Some mutated P450 enzymes also had novel selectivities for producing different oxidized naphthalene products, as revealed by altered fluorescent spectra. The ability to mutate and select enzymes should lead to new catalysts.

Proteins that have new catalytic activities do not overthrow long-held beliefs about enzymes, but RNA molecules that are catalytic clearly do. RNA molecules are transcribed from DNA molecules, but they are often modified—processed—before they are ready to function. The RNA is often cut—to remove an internal piece—and spliced by an enzyme. Cech (Zaug and Cech 1986) was trying to isolate and purify the enzyme that processed a ribosomal RNA from the organism *Tetrahymena thermophila*. He found that the enzymatic activity was not in any of the protein fractions; instead, the activity was always part of the RNA itself. The easiest explanation for this was that the protein enzyme bound tightly to the RNA and could not be removed. Even though all standard methods of removing protein (such as using proteinases to hydrolyze the protein or using hot detergent to denature the protein) did not work, one could still argue that the protein enzyme was either very stable or very active. Any explanation was better than to assume that the enzymatic activity was not a protein. The convincing evidence that the RNA itself was the enzyme came when Cech synthesized the RNA in vitro and showed that it self-processed without any added protein. In 1989 he was awarded the Nobel Prize in Chemistry for his originality and perseverance; the award was shared with Sidney Altman, for his work on ribonuclease P, an RNA-containing protein that processes transfer RNA.

The catalytic unit of the *Tetrahymena* RNA is the part that is removed during processing; it normally excises itself from the pre-ribosomal RNA. Once free, it can act as an enzyme—an RNA enzyme or ribozyme. It has the activity of a ribonuclease (it hydrolyzes RNA) and of an RNA ligase (it splices or ligates RNA). Biochemists have extensively studied its kinetics and have proposed detailed mechanisms (Herschlag and Cech 1990). Other ribozymes have since been found, and researchers expect to find many more.

Enzyme Kinetics

The role of enzymes as very effective catalysts is well illustrated by the decomposition of hydrogen peroxide:

$$2H_2O_2 \longrightarrow 2H_2O + O_2$$

This reaction occurs only very slowly in pure aqueous solution, but its rate is greatly increased by a large variety of catalysts. The increase in rate is typically first order in concentration of catalyst. It may be approximately first order in concentration of H_2O_2 as well. However, for catalysis by the enzyme catalase, the rate becomes zero order in H_2O_2 at higher initial concentrations of H_2O_2. This is a common behavior for enzyme-catalyzed reactions.

Table 8.1 gives approximate values for the kinetics of the rate expression

$$-\frac{d[H_2O_2]}{dt} = k[H_2O_2][\text{catalyst}]$$

For comparison, table 8.1 gives the rates for the extrapolated conditions of 1 M of H_2O_2 and 1 M of catalyst. For catalase, the maximum rate per mole of active sites is given. The rate for the uncatalyzed reaction, as shown in the table, corresponds to a decomposition of only 1% after 11 days at 25°C. Even then, dust particles that are difficult to remove completely from the solutions probably "catalyze" the reaction. Inorganic catalysts, such as iron salts or hydrogen halides, increase the rate of H_2O_2 decomposition by 4 to 5 orders

TABLE 8.1 Catalase and H_2O_2 Decomposition: Comparison of Rates and Activation Energies at 25°C*

Catalyst	Rate, $-d[H_2O_2]/dt$ ($M\,s^{-1}$)	E_a (kJ mol^{-1})
None	10^{-8}	71
HBr	10^{-4}	50
Fe^{2+}/Fe^{3+}	10^{-3}	42
Hematin or hemoglobin	10^{-1}	—
$Fe(OH)_2TETA^+$	10^{3}	29
Catalase	10^{7}	8

*The rate is calculated for 1 M of H_2O_2 and 1 M catalyst (except for catalase). For catalase, the maximum rate is given for 1 M concentration of active sites. This rate is numerically equal to the catalytic constant for catalase. TETA is triethylenetetramine. For a discussion of the mechanism of the reaction and reference to the earlier literature, see J. H. Wang, 1955, *J. Am. Chem. Soc.* 77:4715.

of magnitude per mole of catalyst. The enzyme catalase, which occurs in blood and a variety of tissues (liver, kidney, spleen, etc.), increases the rate by more than 15 powers of 10 over the uncatalyzed rate! To appreciate the speed of this enzymatic reaction, consider that, for the maximum rate, each molecule of catalase can decompose more than 10 million molecules of H_2O_2 per second!

Catalase is a hemoprotein with ferriprotoporphyrin (hematin) as a prosthetic group (the hematin group provides the catalytic function) at the active site. The hematin can be separated from the protein. Alone in solution, hematin exhibits catalytic activity toward H_2O_2 decomposition that is 2 orders of magnitude higher than that of the inorganic catalysts but still more than a millionfold smaller than that of catalase.

The thermodynamics of the decomposition of hydrogen peroxide is also relevant to the function of catalysts. The reaction

$$H_2O_2(aq) \longrightarrow H_2O(l) + \frac{1}{2}O_2(g)$$

is exergonic, $\Delta G^0_{298} = -103.10$ kJ mol^{-1}, with the major contribution to the free-energy change coming from the enthalpy change, $\Delta H^0_{298} = -94.64$ kJ mol^{-1}. Thus, the decomposition of H_2O_2 should go essentially to completion, if there is a suitable reaction path. The slowness of the uncatalyzed reaction must be associated with a large activation barrier. The experimental activation energy is 71 kJ mol^{-1} (table 8.1). Using the observed rate of the uncatalyzed decomposition, $v < 4 \times 10^{-8}$ M s^{-1}, we can calculate an upper limit for the Arrhenius pre-exponential factor of $A < 1 \times 10^5$ s^{-1}, assuming first-order kinetics for the "uncatalyzed" reaction. Although the A factor is significantly less than that observed for other first-order reactions, it is nevertheless clear that the activation energy provides the larger contribution to the reaction barrier. The course of the reaction is illustrated in figure 8.1.

By contrast with the uncatalyzed reaction, the activation energy for the H_2O_2 reaction catalyzed by Fe^{2+}/Fe^{3+} or HBr is only two-thirds as large. In the presence of catalase, the activation energy is only 8 kJ mol^{-1}, and the Arrhenius pre-exponential factor is 1.6×10^8 M^{-1} s^{-1}. It appears that the en-

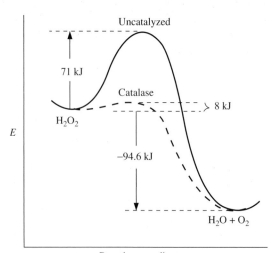

▶ FIGURE 8.1

Role of catalase in lowering the activation energy for the decomposition of H_2O_2.

zyme (and to a lesser extent, the other catalysts) succeeds in speeding the reaction by providing a path with a substantially smaller energy of activation (see figure 8.1). The entropy barrier is also reduced (hence, the large A factor), but its effect on the reaction velocity is less significant at room temperature. Note that the decrease in barrier height for the forward reaction implies a decrease in activation energy for the reverse reaction as well. For this example, the reverse reaction is strongly endergonic; even when the barrier height for the forward reaction is reduced, the reverse reaction is still very slow. However, for most other biochemical reactions that we will consider, the overall reaction has a standard free-energy change closer to zero. In those cases, both the forward and reverse reactions occur at appreciable rates in the presence of the enzyme.

A typical behavior for enzyme-catalyzed reactions with single substrates is that the reaction is first order in substrate at low substrate concentration and then becomes zero order in substrate at high substrate concentration. This means the reaction reaches a maximum velocity with increasing substrate concentration for a constant enzyme concentration. The *catalytic constant*, or *turnover number*, is defined as the maximum rate ($M\,s^{-1}$) divided by the concentration of enzyme active sites (M). Its units are s^{-1}. Concentration of active sites is used rather than concentration of the enzyme so that a fair comparison can be made with enzymes that have more than one active site. Catalase has four active sites per molecule.

It would be misleading to give the impression that catalase activity is entirely typical of enzyme-catalyzed reactions (table 8.2). Catalase is the fastest by several orders of magnitude. A more typical value for the catalytic constant of enzymes is $10^3\,s^{-1}$, and most values fall within a factor of 10 of that value.

TABLE 8.2 Catalytic Constants for Several Enzymes

Enzyme	Substrate	Catalytic constant, k_{cat} (s^{-1})	Reference[*]
Catalase	H_2O_2	9×10^6	a
Acetylcholinesterase	Acetylcholine	1.2×10^4	b
Lactate dehydrogenase (chicken)	Pyruvate	6×10^3	c
Chymotrypsin	Acetyl-L-tyrosine ethyl ester	4.3×10^2	d
Myosin	ATP	3	e
Fumarase	L-Malate / Fumarate	1.1×10^3 / 2.5×10^3	f
Carbonic anhydrase (bovine)	CO_2 / HCO_3^-	8×10^4 / 3×10^4	g

*(a) R. K. Bonnichsen, B. Chance, and H. Theorell, 1947, *Acta Chem. Scand.* 1:685; (b) H. C. Froede and I. B. Wilson, 1971, *The Enzymes* 5:87; (c) J. Everse and N. O. Kaplan, 1973, *Advan. Enzymol.* 37:61; (d) S. Kaufman, H. Neurath, and G. W. Schwert, 1949, *J. Biol. Chem.* 177:793; (e) J. Brahms and C. M. Kay, 1963, *J. Biol. Chem.* 238:198; (f) D. A. Brant, L. B. Barnett, and R. A. Alberty, 1963, *J. Am. Chem. Soc.* 85:2204; (g) H. DeVoe and G. B. Kistiakowsky, 1961, *J. Am. Chem. Soc.* 83:274.

This is not to say that these other enzymes are poor catalysts. The enzyme fumarase catalyzes the hydration of fumarate to L-malate with a catalytic constant of $2.5 \times 10^3 \ s^{-1}$ at 25°C:

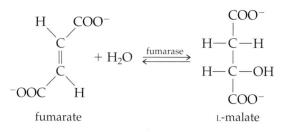

fumarate   L-malate

The rate constants for 1 M fumarate hydrolysis catalyzed by acid (1 M) or base (1 M) are about $10^{-8} \ s^{-1}$. This "ordinary" enzyme still wins by 11 powers of 10.

Michaelis–Menten Kinetics

Because enzymes are such enormously effective catalysts, they are able to exert their influence at exceedingly low concentrations, typically 10^{-10} to $10^{-8} \ M$. At this level, especially if the enzyme is present in a complex cellular soup, it is difficult to make direct measurements of what the enzyme itself is doing. Enzymology therefore began with a study of the kinetics of the disappearance of substrate and formation of products, because their concentrations are typically 10^{-6} to $10^{-3} \ M$. Early observations resulted in the following generalizations about enzyme reactions:

1. The rate of substrate conversion increases linearly with increasing enzyme concentration (figure 8.2).
2. For a fixed enzyme concentration, the rate is linear in substrate concentration [S] at low values of [S] (figure 8.3).
3. The rates of enzyme reactions approach a maximum or saturation rate at high substrate concentration (figure 8.3).

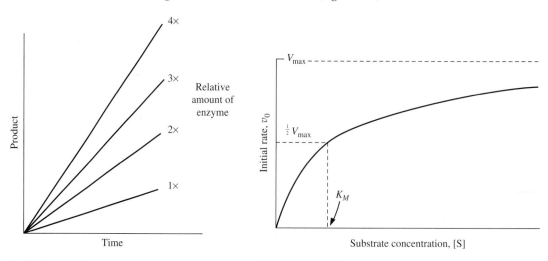

▲ FIGURE 8.2

Increase of product concentration with time for four different amounts of enzyme.

▲ FIGURE 8.3

Initial rate v_0 of an enzyme-catalyzed reaction as a function of substrate concentration [S] for a fixed amount of enzyme.

At low concentrations,

$$v_0 = k[S] \tag{8.1}$$

where v_0 is the initial rate. At high substrate concentrations,

$$v_0 = V_{max} \qquad \text{(maximum rate)} \tag{8.2}$$

This behavior is expected if the enzyme forms a complex with the substrate. At high substrate concentrations, essentially all the enzyme is tied up in the enzyme–substrate complex. Under these conditions, the enzyme is working at full capacity, as measured by its

$$\text{catalytic constant} = k_{cat} = \frac{V_{max}}{[E]_0} \tag{8.3}$$

where $[E]_0$ is the total enzyme-site concentration. At low substrate concentration, the enzyme is not saturated, and turnover is limited partly by the availability of substrate.

Michaelis and Menten presented a kinetic formulation of these ideas, and Briggs and Haldane (1925) made significant improvements. In the simplest picture, the enzyme and substrate reversibly form a complex, followed by dissociation of the complex to form the product and regenerate the free enzyme:

$$E + S \underset{k_{-1}}{\overset{k_1}{\rightleftharpoons}} ES \tag{8.4}$$

$$ES \overset{k_2}{\longrightarrow} E + P \tag{8.5}$$

Because the second step must also be reversible, this mechanism applies strictly only to the initial stages of the reaction before the product concentration has become significant. Under these conditions,

$$v_0 = \left(\frac{d[P]}{dt}\right) = k_2[ES] \tag{8.6}$$

We apply the steady-state approximation:

$$\frac{d[ES]}{dt} = k_1[E][S] - k_{-1}[ES] - k_2[ES] \cong 0$$

Therefore,

$$[ES] = \frac{k_1[E][S]}{k_{-1} + k_2} \tag{8.7}$$

[E] and [S] refer to the concentrations of *free* enzyme and substrate; however, these are difficult to measure, so we write the equation in terms of the measurable *total* enzyme $[E]_0$ and substrate $[S]_0$ concentrations:

$$[E]_0 = [E] + [ES]$$

$$[S]_0 = [S] + [ES] \cong [S] \tag{8.8}$$

We can equate the total substrate concentration to the free substrate concentration because the enzyme–substrate concentration is typically small compared with the substrate concentration. By substituting [E] from Eq. (8.8) into Eq. (8.7) and rearranging terms, we obtain

$$[ES] = \frac{[E]_0}{1 + \dfrac{k_{-1} + k_2}{k_1[S]}} \tag{8.9}$$

Substitution into Eq. (8.6) gives

$$v_0 = \frac{k_2[E]_0}{1 + \dfrac{k_{-1} + k_2}{k_1[S]}} = \frac{V_{max}}{1 + \dfrac{K_M}{[S]}} \tag{8.10}$$

where $K_M = (k_{-1} + k_2)/k_1$ is the Michaelis constant for the enzyme–substrate complex, and V_{max} is $k_2[E]_0$. Equation (8.10) is termed the *Michaelis–Menten equation*. The limiting conditions given by Eqs. (8.1) and (8.2) follow readily from Eq. (8.10). At low substrate concentrations, such that $(K_M/[S]) \gg 1$, Eq. (8.10) gives

$$v_0 = \frac{V_{max}[S]}{K_M} = k[S] \tag{8.1}$$

where $k = V_{max}/K_M$ is a constant at a fixed total enzyme concentration $[E]_0$. At high substrate concentrations such that $(K_M/[S]) \ll 1$, Eq. (8.10) is reduced to

$$v_0 = V_{max} \qquad \text{(maximum rate)} \tag{8.2}$$

$$k_{cat} = \frac{V_{max}}{[E]_0} = k_2 \tag{8.3}$$

When the substrate concentration [S] is equal to K_M, Eq. (8.10) gives

$$v_0 = \frac{1}{2}V_{max}$$

Thus, the Michaelis constant is equal to the substrate concentration that produces half the maximum rate for the enzyme. This is shown graphically in figure 8.3. You should get an intuitive feeling for the magnitude of K_M. A small value of K_M means that the enzyme binds the substrate tightly and small concentrations of substrate are enough to saturate the enzyme and to reach the maximum catalytic efficiency of the enzyme.

In general, enzyme-catalyzed reactions are reversible, and we must consider contributions from the reverse reaction once a significant amount of product is present. The use of initial rates avoids this complication. If the reverse reaction is not important (for example, if the position of equilibrium lies strongly toward products), it is possible to use an integrated form of the Michaelis–Menten equation that gives the substrate concentration as a function of time. Explicitly writing v_0 in Eq. (8.10) as $-d[S]/dt$, we obtain

$$-\frac{d[S]}{dt} = \frac{V_{max}}{1 + \dfrac{K_M}{[S]}} \tag{8.10}$$

By separating the variables in this equation, we obtain

$$\left(\frac{K_m}{[S]} + 1\right) d[S] = -V_{max}\, dt$$

which can be integrated, assuming that $[S] = [S]_0$ when $t = 0$,

$$K_M \ln \frac{[S]}{[S]_0} + [S] - [S]_0 = -V_{max}t$$

Without knowing K_M, there is no simple way to plot data so as to obtain a straight-line relation between a function of $[S]$ and time. With a computer, however, obtaining the best values of K_M and V_{max} for a given data set becomes a straightforward operation.

Very few enzymes act by a simple mechanism with a single enzyme-substrate complex. Even in the more complex situations, however, the kinetic equation often has the form given by Eq. (8.10). The following exercise illustrates one such case. Equations (8.16) and (8.20) provide additional examples in a later section.

EXERCISE 8.1

The enzyme bovine chymotrypsin (represented by EOH below, where OH signifies the hydroxyl group of the serine residue at position 195 of the protein) catalyzes the hydrolysis of peptides. It is one of a group of enzymes called *serine proteases*. A noncovalent complex (EOH·RCONHR′) is formed rapidly between the enzyme (EOH) and the peptide substrate (RCONHR′). This is followed by the acylation of Ser-195 by the bound substrate to give a covalent acyl-enzyme intermediate (EOCOR), and the amine NH$_2$R′ is released. The acyl-enzyme then hydrolyzes to give the enzyme-product complex EOH·RCO$_2$H, which dissociates rapidly to give the enzyme and RCO$_2$H. These steps are as follows:

$$\text{EOH} + \text{RCONHR}' \underset{}{\overset{K_s}{\rightleftharpoons}} \text{EOH}\cdot\text{RCONHR}' \xrightarrow{k_2} \underset{+ \atop \text{NH}_2\text{R}'}{\text{EOCOR}} \overset{+\text{H}_2\text{O}}{\underset{k_3}{\xrightarrow{\hspace{1cm}}}}$$

$$\text{EOH}\cdot\text{RCO}_2\text{H} \underset{\text{fast}}{\xrightarrow{\hspace{1cm}}} \text{EOH} + \text{RCO}_2\text{H}$$

By applying the pre-equilibrium assumption for the K_s step (K_s is the association equilibrium constant) and the steady-state assumption to [EOCOR], show that the overall kinetics of the scheme described above yields the Michaelis–Menten equation with

$$K_M = \frac{1}{K_S}\left(\frac{k_3}{k_2 + k_3}\right)$$

and

$$V_{max} = \frac{k_2 k_3}{k_2 + k_3}[\text{E}]_0$$

Kinetic Data Analysis

For quantitative purposes it is useful to rewrite Eq. (8.10) in a form that suggests a straight-line plot of the data. We can use several approaches; Lineweaver and Burk (1934) proposed the one used most often.

$1/v_0$ versus $1/[S]$: Lineweaver–Burk. Taking the reciprocal of both sides of Eq. (8.10) and rearranging, we obtain

$$\frac{1}{v_0} = \frac{1}{V_{max}} + \frac{K_M}{V_{max}} \cdot \frac{1}{[S]} \tag{8.11}$$

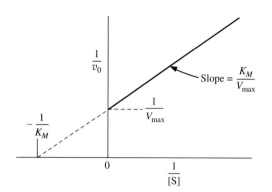

Lineweaver–Burk plot of the reciprocal of initial reaction rate versus reciprocal of initial substrate concentration for a series of experiments at fixed enzyme concentration

Thus, a plot of the reciprocal of initial rate versus the reciprocal of initial substrate concentration for experiments at fixed enzyme concentration should give a straight line. Furthermore, the intercept with the ordinate gives $1/V_{max}$, and the slope is K_M/V_{max} from which these two constants can be determined. Alternatively, K_M can be determined by extrapolation to the abscissa intercept, to give $-1/K_M$, as shown in figure 8.4.

EXERCISE 8.2

Show on a graph such as figure 8.4 that $1/v_0 = 2/V_{max}$ when $[S] = K_M$ and explain how this could be used to determine V_{max} and K_M by a simple procedure.

[S]/v_0 versus [S]: Dixon.   Starting with Eq. (8.11) multiply both sides by [S] to obtain

$$\frac{[S]}{v_0} = \frac{[S]}{V_{max}} + \frac{K_M}{V_{max}} \qquad (8.12)$$

v_0 versus v_0/[S]: Eadie–Hofstee.   Multiply both sides of Eq. (8.11) by $V_{max}v_0$ and rearrange to

$$v_0 = -K_M\frac{v_0}{[S]} + V_{max} \qquad (8.13)$$

Ideally, with no experimental error, each of these equations gives exactly the same desired information from a straight-line plot. In practice, one or the other may be preferable because of the nature of the data involved. Qualitatively, the Eadie–Hofstee plot spreads the values at high substrate concentration (where $v_0 \to V_{max}$), whereas the Lineweaver–Burk plot compresses the points in this region. Analysis using different plots may provide different values for K_M and V_{max} for the same set of experimental data.

The usual method for obtaining values of K_M and V_{max} is to obtain a least-squares fit to the data that produce the linear plots. The values of K_M and V_{max} obtained by least-squares fits to the different plots described by Eqs. (8.11), (8.12), and (8.13) will be different because different sums are minimized to obtain these values. (The equations are given in the section "Mathematics Needed for Chapter 8.") The slope m and intercept b of a line are chosen so as to minimize the sum of the squares of the differences between the experimental y-values and the values calculated from the equation $y = mx + b$. Clearly, our choice of which y and x to plot will affect these values. In the usual least-squares-fit programs (linear regressions) each point is weighted equally, even though different experimental points have different precision. Because of the statistical bias introduced by the different plotting methods, a statistically unbiased method has been developed (Cornish-Bowden and Wharton 1988). Each pair of experimental points of initial rate v_0 and substrate concentration [S] provides a pair of K_M and V_{max} values. If all possible combinations of data points are used to calculate these parameters, many different estimates of the parameters are obtained. The equations are

$$V_{max} = \frac{v_{01}v_{02}([S]_2 - [S]_1)}{[S]_2 v_{01} - [S]_1 v_{02}} \tag{8.14}$$

$$K_M = \frac{[S]_1 [S]_2 (v_{02} - v_{01})}{[S]_2 v_{01} - [S]_1 v_{02}} \tag{8.15}$$

where v_{01} and v_{02} are any two rates corresponding to two substrate concentrations $[S]_1$ and $[S]_2$. The many calculated values of K_M and V_{max} are ordered and the median values found.

The *median values* of K_M and V_{max} are a statistically unbiased representation of the kinetic data. The median value is the middle of an ordered set of numbers; there are as many values above it as below it. Comparison of different plotting methods and the statistically unbiased method is illustrated in example 8.1.

EXAMPLE 8.1

The hydrolysis of carbobenzoxyglycyl-L-tryptophan catalyzed by pancreatic carboxypeptidase occurs according to the reaction

Carbobenzoxyglycyl-L-tryptophan + $H_2O \longrightarrow$
$$\text{carbobenzoxyglycine + L-tryptophan}$$

The following data on the rate of formation of L-tryptophan at 25°C, pH 7.5, were obtained by R. Lumry, E. L. Smith, and R. R. Glantz, 1951, *J. Am. Chem. Soc.* 73:4330:

Substrate concentration (mM)	2.5	5.0	10.0	15.0	20.0
Rate (mM s^{-1})	0.024	0.036	0.053	0.060	0.064

a. Plot these data according to the Lineweaver–Burk method and determine values for K_M and V_{max}.

b. Repeat the analysis using the Eadie–Hofstee method.

c. Find the median of all ten possible values of K_M and V_{max}.

Solution

For the purpose of making these plots, the data are reformulated in the following table:

[S] (mM)	2.5	5.0	10.0	15.0	20.0
$\dfrac{1}{v_0}$ (mM^{-1} s)	41.7	27.8	18.9	16.7	15.6
$\dfrac{1}{[S]}$ (M^{-1})	400	200	100	66.7	50
$10^3 \times \dfrac{v_0}{[S]}$ (s^{-1})	9.6	7.2	5.3	4.0	3.2

a. The Lineweaver–Burk plot is

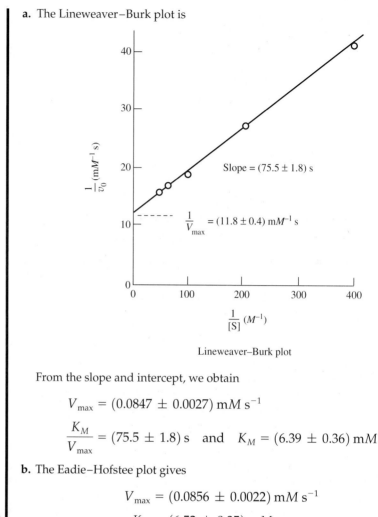

Lineweaver–Burk plot

From the slope and intercept, we obtain

$$V_{max} = (0.0847 \pm 0.0027) \text{ mM s}^{-1}$$

$$\frac{K_M}{V_{max}} = (75.5 \pm 1.8) \text{ s} \quad \text{and} \quad K_M = (6.39 \pm 0.36) \text{ mM}$$

b. The Eadie–Hofstee plot gives

$$V_{max} = (0.0856 \pm 0.0022) \text{ mM s}^{-1}$$

$$K_M = (6.52 \pm 0.35) \text{ mM}$$

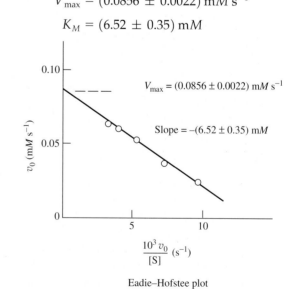

Eadie–Hofstee plot

c. Applying Eqs. (8.14) and (8.15) to the data gives the following ten values for K_M and for V_{max}; each set of numbers has been ordered independently. For five sets of points, there are $4 + 3 + 2 + 1 = 10$ possible pairs; for N sets, there are $N(N - 1)/2$ possibilities.

V_{max} (mM s^{-1})	K_M (mM)
0.0720	5.0000
0.0800	5.0000
0.0808	5.2381
0.0815	5.3846
0.0840	**6.2500**
0.0857	**6.4286**
0.0864	6.7442
0.0887	7.0000
0.0900	7.5000
0.1004	8.9474

The median of each column is the average of the two numbers in the middle (in boldface); for an odd number of values, the median would be simply the center value of the ordered set. The unbiased values are

$$V_{max} = 0.0848 \text{ mM s}^{-1}$$

$$K_M = 6.34 \text{ mM}$$

We note that the values obtained by all three methods agree well. The values are within the standard deviations calculated from the least-squares fits to the plots and the ranges estimated from the columns of numbers above. What is the best method to use? There is no absolute answer. Graphical methods are the most commonly used, and they make it easy to see data points that are obviously in error.

Two Intermediate Complexes

Because most enzymatic reactions are readily reversible, it should not be surprising that there is also evidence for an enzyme–product complex. This is often kinetically distinct from the enzyme–substrate complex. The action of the enzyme fumarase illustrates this nicely:

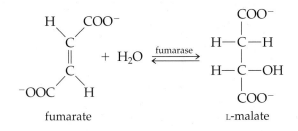

fumarate   L-malate

The enzyme fumarase catalyzes the hydration of fumarate to form malate, and it of course must also catalyze the reverse reaction. At equilibrium, the mixture consists of about 20% fumarate and 80% malate at room temperature.

Alberty and collaborators (see Alberty and Peirce 1957) considered the following mechanism:

$$E + F \underset{k_{-1}}{\overset{k_1}{\rightleftharpoons}} EF \underset{k_{-2}}{\overset{k_2}{\rightleftharpoons}} EM \underset{k_{-3}}{\overset{k_3}{\rightleftharpoons}} E + M$$

There are six kinetic constants in this formalism. As in the case of the simpler mechanism [Eqs. (8.4) and (8.5)], we cannot determine all of them by steady-state experiments. By fairly straightforward treatment similar to that of the earlier section, we can develop the following relations:

Forward reaction, $[M]_0 = 0$

$$v_F = \left(\frac{d[M]}{dt}\right)_0 = k_3[EM]$$

$$= \frac{k_2 k_3 [E]_0 [F]}{(k_2 + k_{-2} + k_3)\left\{\dfrac{k_{-1}k_{-2} + k_{-1}k_3 + k_2 k_3}{k_1(k_2 + k_{-2} + k_3)} + [F]\right\}}$$

which can be written

$$v_F = \frac{V_F[F]}{K_M^F + [F]} \tag{8.16}$$

where

$$V_F = \frac{k_2 k_3}{k_2 + k_{-2} + k_3}[E]_0 \tag{8.17}$$

and

$$K_M^F = \frac{k_{-1}k_{-2} + k_{-1}k_3 + k_2 k_3}{k_1(k_2 + k_{-2} + k_3)} \tag{8.18}$$

Despite the seeming differences, Eq. (8.16) has precisely the same form as Eq. (8.10), and there is no way to detect the presence of the additional intermediate by simple rate measurements in *steady-state experiments*. Methods like those of relaxation kinetics are needed to resolve this question.

Reverse reaction, $[F]_0 = 0$

$$v_R = \left(\frac{d[F]}{dt}\right)_0 = k_{-1}[EF] \tag{8.19}$$

which can be written

$$v_R = \frac{V_R[M]}{K_M^R + [M]} \tag{8.20}$$

where

$$V_R = \frac{k_{-1}k_{-2}}{k_{-1} + k_2 + k_{-2}}[E]_0 \tag{8.21}$$

and

$$K_M^R = \frac{k_{-1}k_{-2} + k_{-1}k_3 + k_2 k_3}{k_{-3}(k_{-1} + k_2 + k_{-2})} \tag{8.22}$$

It is also possible to formulate an expression for the net rate as

$$v = \frac{V_F K_M^R[F] - V_R K_M^F[M]}{K_M^F K_M^R + K_M^R[F] + K_M^F[M]} \tag{8.23}$$

This reduces to the previous equations in the appropriate limits. Note that when equilibrium is reached $v = 0$, and therefore

$$V_F K_M^R [F]^{eq} - V_R K_M^F [M]^{eq} = 0$$

The equilibrium constant K for the reaction is

$$K = \frac{[M]^{eq}}{[F]^{eq}} = \frac{V_F K_M^R}{V_R K_M^F} \tag{8.24}$$

In a process involving a complex sequence of steps, such as occurs in enzymatic reactions, determining the detailed mechanism from steady-state kinetic studies is difficult. The reason is that only one, or a very few, of the steps in the sequence is rate determining. Changes in concentration of substrate or enzyme produce consequences only in terms of the effect on this rate-determining step. Designing steady-state studies so as to extract further information about other steps is difficult but sometimes possible. For example, some of the individual rate constants in the mechanism of the fumarase reaction were estimated from a study of the temperature and pH dependence of the steady-state rates (Brant, Barnett, and Alberty 1963). In other cases, direct measurement of an enzyme complex is possible, and this provides additional information that can be used to establish relations among the rate constants.

Competition and Inhibition
Competition

For any number of enzyme–substrate complexes that exist between substrates and products, the Michaelis–Menten equation still applies:

$$E + S \underset{k_{-1}}{\overset{k_1}{\rightleftharpoons}} ES \overset{k_2}{\longrightarrow} ES' \overset{k_3}{\longrightarrow} ES'' \overset{k_4}{\longrightarrow} E + P$$

$$v_0 = \frac{k_{cat}[E]_0[S]}{K_M + [S]}$$

The parameters k_{cat} and K_M are functions of all the rate constants in the proposed mechanism. However, some generalizations can be made for any mechanism. If there is one slow step (a rate-limiting step) in the reaction path from ES to product, the first-order rate constant for that step is equal to k_{cat}. In the simplest mechanism with only one enzyme–substrate complex, $k_{cat} = k_2$. The Michaelis constant K_M can always be treated as an apparent dissociation constant for all enzyme-bound species. It is equal to the true dissociation constant for dissociation of ES only if the rate constant for its dissociation k_{-1} is much greater than the rate constants for the forward reaction to products; then, $K_M = k_{-1}/k_1$, the equilibrium constant for dissociation of the enzyme–substrate complex.

The ratio k_{cat}/K_M is called the *specificity constant*. It is important because it relates the rate of the reaction to the concentration of *free* enzyme [E] rather than total enzyme $[E]_0$:

$$v_0 = \frac{k_{cat}[E][S]}{K_M}$$

This equation allows us to write the relative rates of reaction for two competing substrates, A and B:

$$\frac{v_A}{v_B} = \frac{(k_{cat}/K_M)_A[A]}{(k_{cat}/K_M)_B[B]} \tag{8.25}$$

Equation (8.25) tells us that the ratio of rates, v_A/v_B, for two substrates is equal to the ratio of their specificity constants times the ratio of concentrations of the substrates. We see that the ratio of k_{cat} to K_M determines the selectivity (or specificity) for competing substrates.

There are nearly always competing substrates for each enzyme in a biological environment. A very important example is DNA polymerase whose biological function is to incorporate the correct nucleotide at each position in the DNA strand being synthesized. The correct nucleotide is the one that is complementary to the template strand. A misincorporation, if not corrected, will lead to a mutation with possible serious results. Enzymatic studies have shown that the selectivity of a DNA polymerase (from the fruit fly *Drosophila*) for equal concentrations of substrates is about 1000 to 1 for the correct substrate (Boosalis, Petruska, and Goodman 1987). Although this is a high selectivity, it is not enough to keep an organism alive; other mechanisms—called *proofreading*—are necessary to produce new DNA with the required fidelity.

Competitive Inhibition

The great specificity and selectivity of enzymatic reactions have been well documented. A consequence of this closeness of fit between enzyme and substrate occurs in the action of inhibitors. Several classes of inhibitors are known, and studies of their behavior made a major early contribution to our knowledge of enzyme mechanisms.

A molecule that resembles the substrate may be able to occupy the catalytic site because of its similarity in structure, but the molecule may be nearly or completely unreactive. By occupying the active site, this molecule acts as a *competitive inhibitor* in preventing normal substrates from being examined and catalyzed. Operationally, competitive inhibitors are those that bind reversibly to the active site. The inhibition can be reversed by (1) diluting the inhibitor or (2) adding a large excess of substrate.

Mechanistically, we can write a step

$$E + I \rightleftharpoons EI \tag{8.26}$$

in addition to the usual Michaelis–Menten formulation. Thus,

$$
\begin{array}{c}
E + S \rightleftharpoons ES \rightleftharpoons E + P \\
+ \\
I \\
\Updownarrow \\
EI
\end{array}
\tag{8.27}
$$

where EI is an inactive form of the enzyme. The kinetic behavior in the presence of the competitive inhibitor is seen in figures 8.5(a) and 8.5(b). Because the equilibria are reversible, the same maximum velocity is reached at sufficiently high substrate concentrations whether the inhibitor is present or not. The Michaelis constant, however, is different in the presence of the inhibitor; that is, the concentration of substrate required to reach one-half V_{max} is greater in the presence of competitive inhibitors than in their absence. It is not difficult to show that the new apparent Michaelis constant, K'_M, is given by

$$K'_M = K_M\left(1 + \frac{[I]}{K_I}\right) \tag{8.28}$$

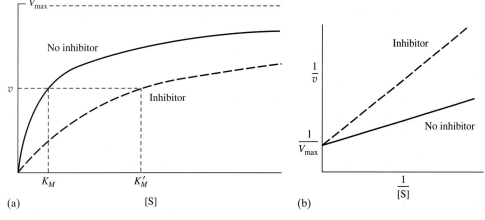

▲ FIGURE 8.5

(a) The effect of a competitive inhibitor on the rate of substrate reaction: an increase of Michaelis constant, but no change in the maximum velocity. (b) In a Lineweaver–Burk plot, the extrapolated maximum rate V_{max} is not changed by a competitive inhibitor.

where [I] is the inhibitor concentration and K_I is the dissociation constant for the enzyme-inhibitor complex. We obtained Eq. (8.28) by writing the total enzyme concentration as

$$[E]_0 = [E] + [ES] + [EI]$$

and using the definition of K_I,

$$K_I = \frac{[E][I]}{[EI]}$$

to obtain

$$[E]_0 = [E]\left(1 + \frac{I}{K_I}\right) + [ES]$$

We solve for [E] and substitute into Eq. (8.7) to obtain an expression for [ES]. It has the same form as Eq. (8.9) except that K_M' replaces K_M.

A classic example of competitive inhibition occurs in the reaction catalyzed by the enzyme succinic dehydrogenase. The normal substrate of this enzyme is succinate, the ionized form of succinic acid, and the enzyme catalyzes its oxidation to fumarate:

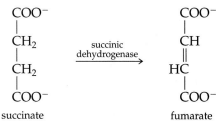

Malonate (from malonic acid), which has the structure

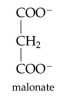

acts as a competitive inhibitor for succinic dehydrogenase. There is a difference of only one methylene group between the structures of succinate and malonate; hence, the two ions resemble each other. In this case, malonate clearly cannot act as an alternative substrate because there is no oxidized form of the molecule analogous to fumarate. The value of K_I for inhibition of yeast succinic dehydrogenase by malonate is $1 \times 10^{-5}\,M$, which means that a concentration of $1 \times 10^{-5}\,M$ malonate decreases the apparent affinity of the enzyme for succinate by a factor of 2.

Inhibition by the product of the reaction is another common example of competitive inhibition. Again, a similarity in structure is implicit in the relation between a substrate and its product. Furthermore, inhibition by the product can serve the useful purpose of turning off (or down) an enzyme's action when it has made sufficient product for the biochemical needs of the cell. It is one of the most immediate forms of regulation or metabolic control.

Noncompetitive Inhibition

Several types of inhibition occur that cannot be overcome by large amounts of substrate. These may occur as a consequence of (1) some permanent (irreversible) modification of the active site; (2) reversible binding of the inhibitor to the enzyme, but not at the active site itself; (3) reversible binding to the enzyme–substrate complex.

The simplest form of noncompetitive inhibition occurs when only the value of V_{max} is affected, but the affinity of the enzyme for substrate, as measured by $1/K_M$, is not affected. The kinetic behavior in this case is illustrated in figure 8.6(a) and 8.6(b). As an example, consider the action of an irreversible modifier, such as the chemical alkylating agent iodoacetamide. This compound reacts with exposed sulfhydryl groups, in particular with cysteine residues of the enzyme protein, to form a covalently modified enzyme:

$$\text{enzyme} - \text{S} - \text{CH}_2\text{CNH}_2$$
$$\overset{\|}{\text{O}}$$

In triose phosphate dehydrogenase and in many other enzymes, the chemically modified enzyme is inactive in its catalytic role. In this case it is clear why V_{max} decreases: The inhibitor effectively inactivates a portion of the enzyme molecules irreversibly. At the same time, the unreacted enzyme molecules are perfectly normal, in the sense that the K_M value is unaffected by the previous addition of the inhibitor.

Other forms of noncompetitive inhibition occur in which both V_{max} and K_M are affected by the inhibitor. A type designated *uncompetitive* results in Lineweaver–Burk plots that are parallel but displaced upward in the presence of inhibitors. In addition, where the enzyme has more than one substrate, the action of an inhibitor must be evaluated with respect to each substrate and to the order of addition of components.

We can interpret the term *inhibitor* in a more general sense. There are many substances that serve to promote a reaction. These may be essential cofactors, such as pyridine nucleotides or ATP, that are really second substrates but operate in cyclic fashion in the overall biochemistry of the cytoplasm. Various ions, such as H^+, OH^-, alkali metals, alkaline earths, and so on can

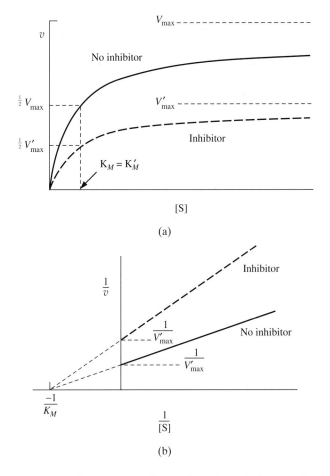

(a) The effect of a noncompetitive inhibitor on the rate of the substrate reaction: a decrease of maximum rate, but no change in the Michaelis constant. (b) In a Lineweaver–Burk plot, the extrapolated Michaelis constant, $1/K_M$, is not changed by a noncompetitive inhibitor.

bind to the enzyme and cause specific and profound changes in the activity of the catalytic site. Transition metals or organic functional prosthetic groups may be involved directly in reactions catalyzed by certain enzymes.

Allosteric Effects

Metabolic enzymes are highly regulated; inhibitors that decrease their rates and effectors that increase their rates often control them. Scientists have proposed many specific mechanisms to explain their behavior, but a general model that applies to any enzyme with multiple subunits has proved very useful. The Monod–Wyman–Changeux (MWC) allosteric theory postulates that the enzyme exists in two distinct forms (allosteric forms) with different binding efficiencies for substrate. *Allosteric effectors* favor the form that binds substrate strongly; *allosteric inhibitors* favor the form that binds substrate weakly. The MWC theory applies to noncatalytic binding proteins, such as hemoglobin (see chapter 5), as well as to enzymes.

The Michaelis–Menten mechanism, with or without competetive or noncompetetive inhibitors, shows a hyperbolic dependence on substrate concentration. The fraction of enzyme that is bound to substrate and the fraction of the maximum rate is

$$\frac{V}{V_{\max}} = \frac{[S]}{K_M + [S]} \qquad \textbf{(8.29)}$$

Figures 8.4 through 8.6 illustrate this behavior. However, many enzymes show a sigmoid (S-shaped) dependence on substrate concentration. The MWC mechanism can produce either type of behavior.

An MWC protein has two or more subunits, each capable of binding a ligand. The binding sites are considered to be identical and independent. The protein can exist in two conformations called *relaxed* and *tense*; the relaxed form binds substrate more tightly (is a better enzyme) than the tense form. These two allosteric forms are in equilibrium; they have different dissociation constants for binding ligands. We illustrate the MWC mechanism for an enzyme that has two subunits. There are three parameters, L, K_R, K_T, that characterize the binding of substrate. The equilibrium constant for the two allosteric forms is L.

$$L = \frac{[T_0]}{[R_0]}$$

where $[T_0]$ is the concentration of the tense form of the enzyme with no substrate bound and $[R_0]$ is the concentration of the relaxed form of the enzyme with no substrate bound. Allosteric inhibitors increase the value of the equilibrium constant L, and allosteric effectors decrease L. The dissociation constant for substrate binding (analogous to the Michaelis constant) by each subunit of the relaxed allosteric form is K_R; the dissociation constant for substrate binding by each subunit of the tense allosteric form is K_T. These constants are assumed to be the same whether or not a substrate is already bound to one of the sites, that is, binding at one of the two sites has no effect on K for the other site. These three parameters, plus the number of subunits in the enzyme (in our example 2), can reproduce a wide variety of curves of enzymatic activity versus substrate concentration, or of extent of binding versus ligand concentration.

To show how the MWC mechanism can produce different-shaped binding curves, we consider an enzyme with two subunits that bind substrate only in the allosteric R-form (K_R is much smaller than K_T):

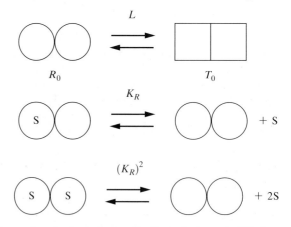

Each of the two subunits has the same dissociation equilibrium constant, and the value does not change if there is a substrate already bound at one site. The sites are identical and independent.

For this example, solving the equilibrium expressions for the concentrations of the free and bound species, we obtain the equation for the fraction of

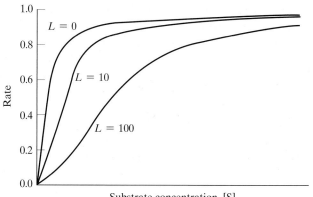

◀ FIGURE 8.7

Plot of relative rates of enzyme-catalyzed activity in the Monod–Wyman–Changeux (MWC) allosteric mechanism. The equilibrium constant for the ratio of the two allosteric forms is $L = [T_0]/[R_0]$. R_0 is the form that binds substrate strongly; T_0 binds substrate weakly. The curves are calculated with Eq. (8.30) for an enzyme with two subunits.

enzymatic sites that have substrate bound, which can be equated to the fraction of the maximum velocity:

$$\frac{V}{V_{max}} = \frac{\alpha(1 + \alpha)}{(1 + \alpha)^2 + L} \tag{8.30}$$

where α is $[S]/K_R$ and L is $[T_0]/[R_0]$.

When L equals zero there is no T-form. This means there is only one form of the enzyme so Eq. (8.30) reduces to the usual Michaelis–Menten hyperbolic shape, with the same form as Eq. (8.29).

$$\frac{V}{V_{max}} = \frac{\alpha}{1 + \alpha} = \frac{[S]}{K_R + [S]} \tag{8.31}$$

As L increases, the binding curve for substrate and the dependence of the rate on substrate concentration become increasingly sigmoid as shown in figure 8.7. The value of L tells us the ratio of the T-form to the R-form of the allosteric enzyme when there is no substrate bound. The R-form of the enzyme binds substrate more tightly than the T-form; in our example, the binding by the T-form is negligible. In the MWC model, the sigmoid shape of the binding curve, or rate curve ($L = [T_0]/[R_0] = 100$ in figure 8.7), is caused by the fact that binding of substrate converts the inactive T-form into the active R-form of the enzyme. Eventually, at high enough substrate concentration, nearly all the enzyme will be in the R-form with bound substrates, and the maximum rate will be obtained. Because the enzyme has two subunits, the concentration of the enzyme species with two substrates bound increases with the square of the substrate concentration. This nonlinear amplification explains why this mechanism requires two or more subunits to give sigmoid curves. The concentration of a three-subunit enzyme with all sites bound increases as the cube of the substrate concentration and produces a sharper sigmoid curve for a constant L. Allosteric effectors lower the value of L (stabilize the R-form), and allosteric inhibitors raise the value of L. In figure 8.7, we see that an effector that changes L from 100 to 0 can change the rate by a factor of 10 at low substrate concentration.

Many metabolic enzymes have multiple catalytic subunits including aldolase (two subunits), pyruvate kinase and glyceraldehyde 3-phosphate dehydrogenase (four subunits), and aspartate transcarbamoylase (six subunits).

The MWC model gives a good qualitative understanding of their regulation, but it cannot explain all the kinetic behavior, and more complex mechanisms have been proposed. The description of other models and derivation of the general form of Eq. (8.30) for any number of subunits are given in Cantor and Schimmel (1980).

Single-Molecule Kinetics

The standard methods for measuring kinetics require millimolar to micromolar concentrations of substrates and nanomolar to picomolar concentrations of enzyme. In a 1-mL container, that means at least 6×10^8 molecules of enzyme and one million times more substrate molecules. It is now possible to measure the rate of a reaction catalyzed by a single enzyme molecule. Can we still apply Michaelis–Menten kinetics to the rate of reaction catalyzed by one molecule? If so, what do the rate constants mean?

The enzyme cholesterol oxidase uses flavin adenine dinucleotide (FAD) in its active site to catalyze the oxidation of cholesterol by O_2; the FAD is reduced to $FADH_2$, and H_2O_2 is produced. The first steps in the mechanism are

$$E \cdot FAD + S \underset{k_{-1}}{\overset{k_1}{\rightleftharpoons}} E \cdot FAD \cdot S \overset{k_2}{\longrightarrow} E \cdot FADH_2 + P$$

where S is cholesterol and P is oxidized cholesterol. In a following reaction, the $FADH_2$ on the enzyme is oxidized by O_2 back to FAD, and H_2O_2 is formed. The enzyme is now ready to catalyze another molecule of cholesterol. The enzyme in the FAD form is fluorescent; the enzyme in the reduced $FADH_2$ form is not. The rate of the reaction can thus be monitored by the fluorescence. Each time the enzyme substrate complex $E \cdot FAD \cdot S$ forms product, P, plus its reduced form, $E \cdot FADH_2$, the fluorescence disappears. The fluorescence reappears when the $E \cdot FADH_2$ is oxidized back to $E \cdot FAD$.

Xie and coworkers have followed the reaction catalyzed by a single molecule of cholesterol oxidase (Lu, Xun, and Xie 1998; Xie and Lu 1999). A dilute solution of the enzyme is placed in an agarose gel so that the enzyme molecules are trapped in the pores of the gel, but the substrates can diffuse freely. The solution is dilute enough that most pores of the gel have only one enzyme trapped. The fluorescence of the FAD in a single enzyme is observed with a microscope. The fluorescence is constant until cholesterol is added. The FAD is then reduced and the fluorescence disappears until O_2 oxidizes the $FADH_2$ on the enzyme and the fluorescence reappears. The fluorescence thus turns on and off with a frequency that depends on the concentrations of O_2 and cholesterol. The concentrations of these two substrates were made large enough that k_2 was rate limiting; the first-order conversion of the enzyme–substrate complex, $E \cdot FAD \cdot S$, to $E \cdot FADH_2$ and P determined the rate. Under these conditions, the fluorescence is on when $E \cdot FAD \cdot S$ is present, and the fluorescence is off otherwise. (The high concentration of S and large k_1 means that $E \cdot FAD$ is negligible.) The length of time the fluorescence is on obviously depends on k_2. To see how to obtain a rate constant from single-molecule kinetics, it is helpful to think about radioactive decay. Consider a single nucleus with a half-life of 10 s. We know that it can decay in 1 s or 100 s, or even shorter or longer. It is only when we observe the lifetimes of many nuclei that we obtain a half-life. For radioactive decay, we must observe different nuclei, but in the single-enzyme experiment we observe the same molecule over and over. To determine a rate constant or a half-life $[k = (\ln 2)/t_{1/2}$ for

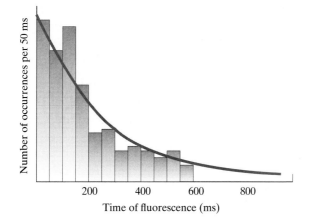

◀ FIGURE 8.8
Plot of the occurrence of different lifetimes for a single molecule of the enzyme substrate complex, $E \cdot FAD \cdot S$, of the enzyme cholesterol oxidase. The substrate S is cholesterol. The single molecule is monitored by its fluorescence when FAD is in its oxidized form; when FAD is reduced to $FADH_2$, there is no fluorescence. The exponential distribution of lifetimes gives a value for the rate constant for formation of product k_2.

the single-molecule enzyme reaction, we measure hundreds of lifetimes of $E \cdot FAD \cdot S$. A plot of the number of lifetimes observed between 0 and 50 ms, between 50 and 100 ms, and so on for an experiment is shown in figure 8.8. The data fit a single exponential with $k_2 = 3.9 \pm 0.5 s^{-1}$, $t_{1/2} = 180 \pm 20$ ms. We emphasize that these measurements are from a single enzyme; the experiment ends when the enzyme is irreversibly inactivated. By changing the conditions, Xie and co-workers were able to measure both k_1 and k_2 and to show that Michaelis–Menten kinetics applies to one molecule as well as many. The rate constants measured for single enzymes are consistent with those measured by standard methods in solution.

Xue and Yeung (1995) have measured single-molecule kinetics of lactate dehydrogenase by having single-enzyme molecules separated in narrow capillaries. The amount of NADH produced from NAD^+ by each enzyme in a 2h period was measured by fluorescence. Each enzyme produces a separate band of NADH; the total amount of NADH produced in each band depends on the integrated kinetics of the reaction. Each enzyme molecule produced a measurably different amount of NADH, even though the enzyme preparations were pure by all standard criteria and the conditions were identical.

Single-molecule kinetic measurements are difficult; they require complex apparatus with very sensitive detectors to obtain data with a large experimental uncertainty. Why go to all the trouble? What is the advantage of studying the properties of one molecule instead of the average properties of many molecules? By studying individual molecules, researchers found that each enzyme is different and that each enzyme has a catalytic efficiency that changes slowly with time. The catalytic constants for different molecules differed as much as a factor of 4. There are many possible explanations, but the most likely is that the enzyme can exist in different conformations that interchange slowly. It may also be that there are slight differences in covalent structure, or in bound ligands, in a population of seemingly identical enzymes. Whatever the explanation, the single-molecule studies may teach us how to make better biological catalysts and allow us to study kinetics in single cells.

Summary
Typical Enzyme Kinetics

Rate of substrate conversion is linear in enzyme concentration:

$$v = k'[E]$$

Rate is linear in substrate concentration at low [S]; first-order reaction:

$$v = k''[S]$$

Rate approaches saturation at high [S]; zero-order reaction:

$$v = V_{max} \qquad \text{(maximum rate)} \tag{8.2}$$

$$\text{catalytic constant} = k_{cat} = \frac{V_{max}}{n[E]_0} \qquad (\text{in } s^{-1}) \tag{8.3}$$

$[E]_0$ = total enzyme concentration, M^{-1}
n = number of active centers molecule^{-1}

Michaelis–Menten Mechanism
Initial rate of forward reaction:

$$E + S \underset{k_{-1}}{\overset{k_1}{\rightleftharpoons}} ES$$

$$ES \xrightarrow{k_2} E + P$$

$$v_0 = \frac{V_{max}}{1 + \dfrac{K_M}{[S]}} \tag{8.10}$$

$V_{max} = k_2[E]_0$ = maximum rate, concentration time^{-1}
$[E]_0 = [E] + [ES]$ = total enzyme concentration
$K_M = \dfrac{k_{-1} + k_2}{k_1}$ = Michaelis constant, M

Lineweaver–Burk:

$$\frac{1}{v_0} = \frac{1}{V_{max}} + \frac{K_M}{V_{max}} \cdot \frac{1}{[S]} \tag{8.11}$$

Eadie–Hofstee:

$$v_0 = -K_M \frac{v_0}{[S]} + V_{max} \tag{8.13}$$

Michaelis–Menten mechanism, including reverse reaction:

$$E + S \underset{k_{-1}}{\overset{k_1}{\rightleftharpoons}} ES \underset{k_{-2}}{\overset{k_2}{\rightleftharpoons}} E + P$$

Rate in either direction:

$$v = \frac{d[P]}{dt} = -\frac{d[S]}{dt} = \frac{V_S K_M^P[S] - V_P K_M^S[P]}{K_M^S K_M^P + K_M^P[S] + K_M^S[P]}$$

V_S = maximum velocity for utilization of S
V_P = maximum velocity for utilization of P
$K_M^S = \dfrac{k_{-1} + k_2}{k_1}$
$K_M^P = \dfrac{k_{-1} + k_2}{k_{-2}}$

The relative rates of reaction for two competing substrates, A and B:

$$\frac{v_A}{v_B} = \frac{\left(\dfrac{k_{cat}}{K_M}\right)_A [A]}{\left(\dfrac{k_{cat}}{K_M}\right)_B [B]}$$

Reversible inhibition:

$$E + S \underset{k_{-1}}{\overset{k_1}{\rightleftharpoons}} ES \underset{k_{-2}}{\overset{k_2}{\rightleftharpoons}} E + P$$

$$+$$

$$I$$

$$\Big\updownarrow K_I$$

$$EI$$

$$v_0 = \frac{V_S}{1 + \dfrac{K'_M}{[S]}}$$

$$K'_M = K_M\left(1 + \frac{[I]}{K_I}\right)$$

$$K_I = \frac{[E][I]}{[EI]}$$

Monod–Wyman–Changeux Mechanism

The MWC equation for an enzyme with two subunits in which only the relaxed form binds substrate:

$$\frac{V}{V_{max}} = \frac{\alpha(1 + \alpha)}{(1 + \alpha)^2 + L} \tag{8.30}$$

where $\alpha = \dfrac{[S]}{K_R}$

$$L = \frac{[T_0]}{[R_0]}$$

Mathematics Needed for Chapter 8

Least-squares programs for fitting the best line to a set of points are available in most programmable calculators. To fit a line to a set of points labeled (x_i, y_i) where each x_i is an abscissa and each y_i is an ordinate, calculate

$$\text{slope} = m = \frac{n \sum_i y_i x_i - \sum_i y_i \sum_i x_i}{D}$$

$$\text{intercept} = b = \frac{\sum_i y_i \sum_i x_i^2 - \sum_i y_i x_i \sum_i x_i}{D}$$

where $D = n \sum_i x_i^2 - \left(\sum_i x_i\right)^2$

The sums are over the total number n of points in the set. The slope and intercept determine the line that minimizes the sum of the squares of the vertical (y_i) distances of each point from the line; this is the best least-squares line. It is always a good idea to plot this line ($y = mx + b$) on the same graph as the points to check that the fit is reasonable.

References

Cantor, C. R., and P. R. Schimmel. 1980. *Biophysical Chemistry,* vol. 3, chap. 17. New York: Freeman.

Dressler, D., and D. H. Potter. 1991. *Discovering Enzymes.* New York: Freeman, Scientific American Library. This beautiful book not only describes the discovery of enzymes and their importance in medicine, physiology, and industry but also relates enzyme mechanism to structure.

Boyer, P. D., ed., *The Enzymes,* 3d ed. New York: Academic Press. This is one volume of a useful many-volume series.

Short papers on many different aspects of enzymology are presented in the continuing two series entitled *Advances in Enzymology* (Wiley) and *Methods in Enzymology* (Academic Press). Volumes 63, 64, and 87 of *Methods in Enzymology* are devoted to kinetics and mechanism. They are:

Purich, D. L., ed. *Enzyme Kinetics and Mechanism.* New York: Academic Press.

Part A (1979) *Initial Rate and Inhibitor Methods* (vol. 63).

Part B (1980) *Isotopic Probes and Complex Enzyme Systems* (vol. 64).

Part C (1982) *Intermediates, Stereochemistry, and Rate Studies* (vol. 87).

Part D (1995) *Developments in Enzyme Dynamics* (vol. 249).

Part E (1999) *Energetics of Enzyme Catalysis* (vol. 308).

The following books treat enzymes and enzyme kinetics in great detail.

Cornish-Bowden, A. 1995. *Fundamentals of Enzyme Kinetics,* Revised ed. London: Portland Press.

Cornish-Bowden, A., and C. W. Wharton. 1988. *Enzyme Kinetics.* Washington, DC: IRL Press.

Fersht, A. 1999. *Structure and Mechanism in Protein Science. A Guide to Enzyme Catalysis and Protein Folding.* This book is an expanded and generalized version of Fersht, A. 1985. *Enzyme Structure and Mechanism,* 2d ed. New York: Freeman.

Gutfreund, H. 1995. *Kinetics for the Life Sciences. Receptors, Transmitters and Catalysts.* Cambridge, England: Cambridge University Press.

See also the biochemistry texts listed in the References for Chapter 1.

Suggested Reading

Alberty, R. A., and W. H. Peirce. 1957. Studies of the Enzyme Fumarase: V. Calculation of Minimum and Maximum Values of Constants for the General Fumarase Mechanism. *J. Am. Chem. Soc.* 79:1526–1530.

Boosalis, M. S., J. Petruska, and M. F. Goodman. 1987. DNA Polymerase Insertion Fidelity: Gel Assay for Site-Specific Kinetics. *J. Biol. Chem.* 262:14,689–14,696.

Brant, D. A., L. B. Barnett, and R. A. Alberty. 1963. The Temperature Dependence of the Steady State Kinetic Parameters of the Fumarase Reaction. *J. Am. Chem. Soc.* 85:2204–2209.

Briggs, G. E., and J. B. S. Haldane. 1925. A Note on the Kinetics of Enzyme Action. *Biochem. J.* 19:338–339.

Herschlag, D., and T. R. Cech. 1990. Catalysis of RNA Cleavage by the *Tetrahymena thermophila* Ribozyme. 1. Kinetic Description of the Reaction of an RNA Substrate Complementary to the Active Site. *Biochemistry* 29:10,159–10,171; 2. Kinetic Description of the Reaction of an RNA Substrate That Forms a Mismatch at the Active Site. *Biochemistry* 29:10,172–10,180.

Joo, H., Z. Lin, and F. H. Arnold. 1999. Laboratory Evolution of Peroxide-Mediated Cytochrome P450 Hydroxylation. *Nature* 399:670–673.

Lineweaver, H., and D. Burk. 1934. The Determination of Enzyme Dissociation Constants. *J. Am. Chem. Soc.* 56:658–666.

Lu, H. P., L. Xun, and X. S. Xie. 1998. Single-Molecule Enzymatic Dynamics. *Science* 282:1877–1882.

Monod, J., J. Wyman, and J.-P. Changeux. 1965. On the Nature of Allosteric Transitions: A Plausible Model. *J. Mol. Biol.* 12:88–96.

Schultz, P. G., and R. A. Lerner. 1995. From Molecular Diversity to Catalysis: Lessons from the Immune System. *Science* 269:1835–1842.

Wang, J. H. 1970. Synthetic Biochemical Models. *Acc. Chem. Res.* 3:90–97.

Xie, X. S., and H. P. Lu. 1999. Single-Molecule Enzymology. *J. Biological Chemistry* 274:15,967–15,970.

Xue, Q., and E. S. Yeung. 1995. Differences in the Chemical Reactivity of Individual Molecules of an Enzyme. *Nature* 373:681–683.

Zaug, A. J., and T. R. Cech. 1986. The Intervening Sequence RNA of *Tetrahymena* Is an Enzyme. *Science* 231:470–475.

Problems

1. The decarboxylation of a β-keto acid catalyzed by a decarboxylation enzyme can be measured by the rate of CO_2 formation. From the initial rates given in the table, determine the Michaelis–Menten constant for the enzyme and the maximum velocity by a graphical method.

Keto acid concentration [M]	Initial velocity (μmol CO_2/2 min)
2.500	0.588
1.000	0.500
0.714	0.417
0.526	0.370
0.250	0.256

2. The hydration of CO_2,

$$CO_2 + H_2O \rightleftharpoons HCO_3^- + H^+$$

is catalyzed by the enzyme carbonic anhydrase. The steady-state kinetics of the forward (hydration) and reverse (dehydration) reactions at pH 7.1, 0.5°C, and 2 mM phosphate buffer was studied using bovine carbonic anhydrase (H. DeVoe and G. B. Kistiakowsky, 1961, *J. Am. Chem. Soc.* 83:274). Some typical results for an enzyme concentration of 2.3 nM are

Hydration		Dehydration	
$\dfrac{1}{v}$ (M^{-1} s)	[CO_2] (mM)	$\dfrac{1}{v}$ (M^{-1} s)	HCO_3^- (mM)
36×10^3	1.25	95×10^3	2
20×10^3	2.5	45×10^3	5
12×10^3	5	29×10^3	10
6×10^3	20	24×10^3	15

Using graph paper, make suitable plots of data and determine the Michaelis constant K_M and the rate constant k_2 for the decomposition of the enzyme–substrate complex to form product for

(a) The hydration reaction.

(b) The dehydration reaction.

(c) From your results, calculate the equilibrium constant for the reaction

$$CO_2 + H_2O \rightleftharpoons HCO_3^- + H^+$$

(*Hint:* Start by writing an equation that defines the equilibrium condition in terms of the Michaelis–Menten formulation of this reaction. Note that the kinetics was measured at pH 7.1.)

3. At pH 7 the measured Michaelis constant and maximum velocity for the enzymatic conversion of fumarate to L-malate,

$$\text{fumarate} + H_2O \longrightarrow \text{L-malate}$$

are 4.0×10^{-6} M and 1.3×10^3 $[E]_0$ s^{-1}, respectively, where $[E]_0$ is the total molar concentration of the enzyme. The Michaelis constant and maximum velocity for the reverse reaction are 1.0×10^{-5} M and $800[E]_0$ s^{-1}, respectively. What is the equilibrium constant for the hydration reaction? (The activity of water is set equal to unity; see chapter 4.)

4. The enzyme pyrophosphatase catalyzes the hydrolysis of inorganic pyrophosphate $[PP_i]$ to orthophosphate $[P_i]$ according to a reaction that can be written

$$PP_i + H_2O \longrightarrow 2P_i$$

The standard free energy change for this reaction is $\Delta G_{310}^{0\prime} = -33.5$ kJ-(mol PP_i)$^{-1}$ at pH 7. The pyrophosphatase from *E. coli* bacteria has a molecular weight of 120,000 daltons and consists of six identical subunits. Purified enzyme has a maximum velocity V_{max} of 2700 units per milligram of enzyme, where a unit of activity is defined as the amount of enzyme that hydrolyzes 10 μmol of PP_i in 15 min at 37°C under standard assay conditions. As the concentration of the substrate PP_i is decreased, the initial velocity decreases to 1350 units per milligram of enzyme when $[PP_i] = 5.0 \times 10^{-6}$ M.

(a) How many moles of substrate are hydrolyzed per second per milligram of enzyme at maximum velocity?

(b) How many moles of active site are there in 1 mg of enzyme? (Assume that each subunit has one active site.)

(c) What is the catalytic constant of each active site of the enzyme?

(d) Estimate the Michaelis constant K_M for this enzyme.

5. Consider the simple Michaelis–Menten mechanism for an enzyme-catalyzed reaction:

$$E + S \underset{k_{-1}}{\overset{k_1}{\rightleftharpoons}} ES \overset{k_2}{\longrightarrow} E + P$$

The following data were obtained:

k_1, k_{-1} very fast

$k_2 = 100$ s^{-1}, $K_M = 1.0 \times 10^{-4}$ M at 280 K

$k_2 = 200$ s^{-1}, $K_M = 1.5 \times 10^{-4}$ M at 300 K

(a)   For $[S] = 0.10\ M$ and $[E]_0 = 1.0 \times 10^{-5}\ M$, calculate the rate of formation of product at 280 K.

(b)   Calculate the activation energy for k_2.

(c)   What is the value of the equilibrium constant at 280 K for the formation of the enzyme–substrate complex ES from E and S?

(d)   What is the sign and magnitude of the standard thermodynamic enthalpy ΔH^0 for the formation of ES from E and S?

6. Assuming that a catalyst is not itself affected by temperature, do you expect the rate of the catalyzed or uncatalyzed reaction to be more sensitive to temperature changes? Explain.

7. The enzyme urease catalyzes the hydrolysis of urea to ammonia and carbon dioxide:

$$H_2O + H_2NCNH_2 \longrightarrow 2\,NH_3 + CO_2$$
$$\underset{O}{\|}$$

The enzyme has a molecular weight of 473,000, and it contains eight independent active sites per molecule.

(a)   Temperature-dependence studies show that the value of the maximum velocity at 25°C is 2.2 times greater than it is at 10°C. The value of K_M is not significantly affected by temperature. Calculate the enthalpy of activation $\Delta H^{\ddagger}$ for this reaction.

(b)   The catalytic constant per active site is about 560 s^{-1} at 25°C. Calculate the value of $\Delta S^{\ddagger}$ for the reaction using the result of part (a). Interpret your answers to parts (a) and (b).

(c)   The hydrolysis of urea also occurs in acidic solutions in the absence of any enzyme, but the rate constant is estimated to be 10^{14} times slower at 25°C. The activation energy for the acid-catalyzed reaction is 105 kJ mol^{-1}. What temperature would be required for the acid-catalyzed reaction to occur as rapidly as the enzyme-catalyzed reaction does at room temperature?

8. Consider the following mechanism for the role of an inhibitor I of an enzyme:

$$E + S \underset{k_{-1}}{\overset{k_1}{\rightleftharpoons}} ES \overset{k_2}{\longrightarrow} E + P$$

$$E + I \underset{k_{-3}}{\overset{k_3}{\rightleftharpoons}} EI$$

The concentrations of [S] and [I] are much larger than the total enzyme concentration $[E]_0$. Derive an expression for the rate of appearance of products.

9. $E + F \underset{k_{-1}}{\overset{k_1}{\rightleftharpoons}} EF \underset{k_{-2}}{\overset{k_2}{\rightleftharpoons}} EM \underset{k_{-3}}{\overset{k_3}{\rightleftharpoons}} E + M$

(a)   What rate constant would you need to measure as a function of temperature to determine the $\Delta H^{\ddagger}$ for formation of EF from E + F? Formation of EM from EF? Formation of EM from E + M?

(b)   What rate constants would you need to measure as a function of temperature to determine ΔH^0 for formation of EF from E + F? Formation of EM from E + M?

(c)   How could you determine ΔH^0 for formation of M from F, using only kinetic experiments?

10. Penicillinase is an enzyme secreted by certain bacteria to inactivate the antibiotic penicillin. A typical form of the enzyme with a molecular weight of 30,000 has a single active site, a catalytic constant of 2000 s^{-1}, and a Michaelis constant, $K_M = 5.0 \times 10^{-5}\ M$. In response to treatment with 5 μmol of penicillin, a 1.00-mL suspension of certain bacterial cells release 0.04 μg of the enzyme.

(a)   Assuming that the enzyme equilibrates quickly with its substrate, what fraction of the enzyme will be complexed with penicillin in the early stages of the reaction?

(b)   How long will it take for half of the penicillin present initially to become inactivated?

(c)   What concentration of penicillin would cause the enzyme to react at half its maximum velocity V_{max}?

(d)   A second suspension of bacterial cells releases 0.06 μg mL^{-1} of the enzyme. Will this affect the answer to parts (b) or (c)? If so, by how much?

(e)   An antibiotically inactive analog of penicillin serves as a competitive inhibitor to penicillinase. If the enzyme has the same affinity for inhibitor as for substrate, what concentration of inhibitor is required to decrease the rate of disappearance of penicillin fivefold at low (subsaturating) penicillin concentration?

11. The hydrolysis of sucrose by the enzyme invertase was followed by measuring the initial rate of change in polarimeter (optical rotation) readings, α, at various initial concentrations of sucrose. The reaction is inhibited reversibly by the addition of urea:

PROBLEM 11 ▶
Data

[Sucrose]$_0$ (mol L^{-1})	0.0292	0.0584	0.0876	0.117	0.175	0.234
Initial rate, $\left(\dfrac{d\alpha}{dt}\right)_0 = v_0$	0.182	0.265	0.311	0.330	0.372	0.371
Initial rate (2 M urea),v_0'	0.083	0.119	0.154	0.167	0.192	0.188

	ΔH (kJ)	E_a (kJ)
Fumarate + H_2O → malate	−15.1	
Fumarate + fumarase → fumarate-fumarase complex, ES	+17.6	
Fumarate-fumarase complex → transition state, [EZ]‡		+25.5
Malate + fumarase → malate-fumarase complex, EP	−5.0	
Malate-fumarase complex → transition state, [EZ]‡		+63.0

◀ PROBLEM 12

Data

(a) Make a plot of the data in the absence of urea and determine the Michaelis constant K_M for this reaction.

(b) Carry out an analysis of the data in the presence of urea and determine whether urea is a competitive or a noncompetitive inhibitor of the enzyme for this reaction. Justify your answer.

12. The reaction of an enzyme E with a substrate S typically passes through at least two intermediates, ES and EP, and three transition states before reaching the dissociated product P:

$$E + S \longrightarrow [ES]^{\ddagger} \longrightarrow ES \longrightarrow [EZ]^{\ddagger} \longrightarrow$$
$$EP \longrightarrow [EP]^{\ddagger} \longrightarrow E + P$$

From the data given above, prepare a diagram of the energy of reaction of fumarate to malate catalyzed by the enzyme fumarase as a function of the reaction coordinate. (See figure 8.1 for an example of a simpler reaction.) Carefully label all enthalpy or energy differences for which you have information.

13. A certain enzyme has an ionizable group at the active site such that the enzyme is active only when this ionizable group is protonated:

$$E + S \rightleftharpoons ES$$
$$\Updownarrow H^+ \qquad\quad \Updownarrow H^+$$
$$H^+E + S \rightleftharpoons H^+ES \xrightarrow{k} products + H^+E$$

Assume that the conversion of H^+ES to products and H^+E is slow so that equilibria are established for the four reversible steps shown.

(a) What is the relation among the equilibrium constants of the four steps forming a cyclic path?

(b) Show that the enzyme obeys Michaelis–Menten kinetics and express K_M and V_{max} in terms of the equilibrium constants of part (a) in addition to k, $[H^+]$, and $[E]_0$.

(c) How does the rate depend on pH at high substrate concentrations?

14. The initial rate of ATP dephosphorylation by the enzyme myosin can be estimated from the amount of phosphate produced in 100 s, $[P_i]_{100}$. Use the follow-

ing data and graphical methods to evaluate K_M and V_{max} for this reaction at 25°C. $[Myosin]_0 = 0.040$ g L^{-1}.

$$ATP \xrightarrow{\text{myosin}} ADP + P_i$$

$[ATP]_0$ (μM)	7.1	11	23	40	77	100
$[P_i]_{100}$ (μM)	2.4	3.5	5.3	6.2	6.7	7.1

Explain any curvature that you observe in the plot of the data. Justify the slope that you choose to characterize the reaction.

15. Since it was first discovered in 1968, the enzyme superoxide dismutase, SOD, has been found to be one of the most widely dispersed in biological organisms. In one form or another, its presence has been demonstrated in mammalian tissues (heart, liver, brain, etc.), blood, invertebrates, plants, algae, and aerobic bacteria. It catalyzes the reaction

$$O_2^- + O_2^- + 2H^+ \xrightarrow{\text{SOD}} O_2 + H_2O_2$$

where O_2^- represents an oxygen molecule with an extra (unpaired) electron. SOD is therefore of great importance in detoxifying tissues from the potentially harmful O_2^-. (Several of the proteins had been known since 1933, but their enzymatic activity went unrecognized for 35 years!)

The enzyme has some unusual properties, as illustrated by the following data (for the velocity of the reaction in terms of O_2 formed) taken from a paper by Bannister et al. (1973, *FEBS Lett.* 32 : 303). The enzyme kinetics is independent of pH in the range 5 to 10.

$[SOD]_0 = 0.4\ \mu M$ buffer pH = 9.1

$\dfrac{1}{v_0}$ (s M^{-1})	$\dfrac{1}{[O_2^-]}$ (M^{-1})
260	13 $\times 10^4$
175	8.5 $\times 10^4$
122	6.0 $\times 10^4$
60	3.0 $\times 10^4$
10	0.50 $\times 10^4$

(a) Use graph paper to plot these data according to the Lineweaver–Burk method.

(b) What values do you obtain for V_{max} and K_M for this reaction? (Do not be surprised if they are somewhat unusual.)

(c) How can you interpret the results of part (b) in terms of a Michaelis–Menten type of mechanism?

(d) What is the kinetic order of the reaction with respect to O_2^-? Explain your answer.

(e) The reaction is independent of pH (in the region investigated) and first order in enzyme concentration. Write the simplest rate law consistent with the experimental observations. What is the value of the rate constant, including appropriate units?

Klug-Roth, Fridovich, and Rabani [*J. Am. Chem. Soc.* 95:2768 (1973)] have proposed a Ping-Pong-like mechanism for this reaction, involving the steps

$$E + O_2^- \xrightarrow{k_1} E^- + O_2$$

$$E^- + O_2^- \xrightarrow[k_2]{2\,H^+} E + H_2O_2$$

where $k_2 = 2k_1$. (The enzyme contains copper that is oxidized in E and reduced in E^-.)

(f) Show that this mechanism is consistent with the rate data of Bannister et al.

(g) What is the ratio $[E]/[E^-]$ under steady-state conditions?

(h) Given the answer in part (e) for the "observed" rate constant, calculate k_1 and k_2.

16. RNA polymerase is thought to bind rapidly to DNA to form an initial complex. This complex then rearranges slowly to form an "open" complex, with the DNA strands now available to act as templates for the synthesis of RNA. A proposed mechanism is

$$A + B \underset{k_{-1}}{\overset{k_1}{\rightleftharpoons}} C \qquad \begin{cases} \text{(very fast, equilibrium)} \\ K_1 = k_1/k_{-1} \end{cases}$$

$$C \underset{k_{-2}}{\overset{k_2}{\rightleftharpoons}} D \qquad \begin{cases} \text{(slow to equilibrium)} \\ K_2 = k_2/k_{-2} \end{cases}$$

At the beginning of the reaction, 1.00 n*M* A and 1.00 n*M* B are present. At final equilibrium, the concentrations are $[\overline{A}] = [\overline{B}] = 0.60$ n*M*, $[\overline{C}] = 0.010$ n*M*, and $[\overline{D}] = 0.39$ n*M*. The bar over a concentration is to remind you that it is an equilibrium value.

(a) Calculate K_1 and K_2 from the equilibrium concentrations.

(b) Write the differential equations for

$$\frac{d[C]}{dt} \quad \text{and} \quad \frac{d[D]}{dt}$$

(c) Sketch curves for [A], [B], [C], and [D] vs. time on the same plot. You have been given their initial concentrations and their final (equilibrium) concen-

trations. The main questions to consider are which concentrations have minima or maxima, roughly where they occur in time, and what are the minimum or maximum concentrations.

(d) The value of k_2 was found to be 0.050 s^{-1}. Calculate the value of k_{-2} in s^{-1}.

17. Sodium succinate S is oxidized in the presence of dissolved oxygen to form sodium fumarate F in the presence of the enzyme succinoxidase E. The following table gives the initial velocity v_0 of this reaction at several succinate concentrations:

[S] (m*M*)	10	2.0	1.0	0.50	0.33
v_0 ($\mu M\,s^{-1}$)	1.17	0.99	0.79	0.62	0.50

(a) Make a suitable plot of the data to test whether they follow the Michaelis–Menten mechanism. Draw the best line or curve through the data.

(b) Calculate values for the maximum velocity V_{max} and the Michaelis constant K_M for this reaction. Show how you obtain them from the plot.

(c) Malonate is a competitive inhibitor of this reaction. In an experiment with initially 1.0 m*M* succinate, the initial velocity is decreased by 50% in the presence of 30 m*M* malonate. Use this result to construct a second curve or line on the plot showing the expected kinetic behavior of this reaction in the presence of 30 m*M* malonate.

18. Alcoholism in humans is a disorder that has serious sociological as well as biochemical consequences. Much research is currently under way to discover the nature of the biochemical consequences, to understand better how this disorder can be treated. You can analyze this problem using a variety of kinetic approaches.

Alcohol taken orally is transferred from the gastrointestinal tract (stomach, etc.) to the bloodstream by a first-order process with $t_{1/2} = 4$ min. (These and subsequent figures are subject to individual variations of ±25%.) The transport by the bloodstream to various aqueous body fluids is very rapid; thus, the ethanol becomes rapidly distributed throughout the approximately 40 L of aqueous fluids of an adult human. These fluids behave roughly like a sponge from which the alcohol must be removed.

The removal occurs in the liver, where the alcohol is oxidized in a process that follows zero-order kinetics. A typical value for this rate of removal is about 10 mL ethanol h^{-1}, or 4×10^{-3} mol (liter of body fluid)$^{-1}$ h^{-1}. Consumption of about 1 mol (46 g, or 60 mL) of ethanol produces a state defined as legally intoxicated. (This amount of alcohol is contained in about

4.5 oz of 80-proof liquor.) At this level, alcohol is rapidly taken up by the body fluids and only slowly removed by the liver.

In the liver, ethanol (C_2H_5OH) is oxidized to acetaldehyde in the presence of the enzyme liver alcohol dehydrogenase (LADH). The overall process may be represented by the equation

$$C_2H_5OH \text{ (stomach)} \xrightarrow[\text{transfer}]{k_1}$$

$$C_2H_5OH \text{ (body fluids)} \xrightarrow[\substack{\text{LADH} \\ \text{liver}}]{k_0} CH_3CHO \downarrow \text{ acetate, etc.}$$

(a) On the basis of the data above, calculate at least three points that will enable you to construct a semi-quantitative ($\pm 10\%$) sketch of the concentration of ethanol in body fluids as a function of time following the consumption of 60 mL of C_2H_5OH at time zero. Pay particular attention to the rising portion, the maximum level reached, and the decaying portion of the curve.

(b) What is the maximum concentration of ethanol attained in the body fluids after consuming 60 mL of C_2H_5OH? Take this to be the threshold level defining legal intoxication and draw a horizontal line at this level. Approximately how many hours are required to reduce the ethanol concentration essentially to zero? This is the recuperation time or hangover period.

(c) On the same graph, add a curve showing the time dependence following initial consumption of 120 mL of ethanol. (Label the two curves "60 mL" and "120 mL," respectively.) The concentration in body fluids remains above the "legally intoxicated" level for about how many hours? The hangover period is how long?

(d) Without constructing further plots, estimate the length of time that a person would remain "legally intoxicated" if initially 180 mL of ethanol is consumed.

(e) If 60 mL of ethanol is consumed initially, what is the maximum amount that this person may drink at subsequent hourly intervals and still avoid exceeding the level of legal intoxication?

(f) It is a popular opinion that a "cocktail party" drinker is less susceptible to intoxication because the consumption of alcohol is stretched over a longer period interspersed with conversation. Using dashes, add a curve to the graph that shows the body-fluid concentration for the social drinker who extends the consumption of 120 mL of ethanol over a period of 1 h in 15-min intervals. Justify your plot and comment on the effect of the social drinker's tactic on the intoxication period.

19. The removal of ethanol in the liver involves its oxidation to acetaldehyde by nicotinamide adenine dinucleotide (NAD^+) catalyzed by the enzyme liver alcohol dehydrogenase (LADH). The overall reaction is

$$C_2H_5OH + NAD^+ \xrightarrow{\text{LADH}} CH_3CHO + NADH + H^+$$

The reaction follows a sequential or ordered mechanism wherein NAD^+ must bind to the enzyme before C_2H_5OH binds to form a ternary complex, and CH_3CHO and H^+ dissociate from the ternary complex before the NADH is released.

(a) Write a detailed set of reactions involving intermediate complexes so as to represent the mechanism of this reaction. Indicate each step as being reversible and designate the rate constants (in order) by k_1, k_2, etc., for forward reaction steps and k_{-1}, k_{-2}, etc., for reverse steps.

(b) Give a plausible explanation for the observation that the removal of ethanol from the body obeys zero-order kinetics.

(c) The rate-limiting step for the overall reaction under physiological conditions and in the presence of an intoxicating level of ethanol is the final dissociation of the LADH·NADH complex. The rate constant for this step has been measured to be 3.1 s^{-1}. Using appropriate data given in problem 18, calculate the amount (in mmol) of the LADH enzyme present in the liver.

(d) Relaxation measurements using the temperature-jump technique were applied to the binding of NAD^+, at two different concentrations, to LADH in the absence of ethanol. The relaxation times measured were

$$\tau = 1.65 \times 10^{-3} \text{ s} \qquad \text{at } [NAD^+] = 1.0 \text{ m}M$$

$$\tau = 7.9 \times 10^{-3} \text{ s} \qquad \text{at } [NAD^+] = 0.1 \text{ m}M$$

where $[LADH] \ll [NAD^+]$. Use these relaxation times to calculate the constants k_1 and k_{-1} for the reactions

$$LADH + NAD^+ \underset{k_{-1}}{\overset{k_1}{\rightleftharpoons}} [LADH \cdot NAD^+]$$

(e) By contrast with ethanol, other simple alcohols, such as methanol (CH_3OH) and ethylene glycol ($HOCH_2CH_2OH$), are highly toxic to humans. Methanol is oxidized to formaldehyde, which reacts irreversibly with proteins (it is a commonly used cross-linking agent, or fixative). Ethylene glycol is not toxic, but its oxidation product oxalic acid is. Each alcohol is about as good as ethanol as a substrate for LADH. A common antidote for methanol or ethylene glycol poisoning is to administer ethanol in large quantities. Propose a rationale for this therapy.

20. Pyrazole has been proposed as a possible nontoxic inhibitor of LADH-catalyzed ethanol oxidation. Its ki-

netics was studied by Li and Theorell [*Acta Chem. Scand.* 23:892 (1969)]. In separate series of experiments, the velocity of the reaction was measured as a function of $[C_2H_5OH]$ or of $[NAD^+]$. Portions of their results are tabulated as follows:

Ethanol as Variable Substrate
[LADH] = 4 μg mL^{-1},
[NAD$^+$] = 350 μM, pH 7.4, 23.5°C

$\dfrac{1}{[C_2H_5OH]} (M^{-1})$	$1/v$ (Relative units)	
	Control (no pyrazole)	$1 \times 10^{-5} M$ pyrazole
0.125×10^3	0.47	0.59
0.5×10^3	0.55	0.97
1.0×10^3	0.66	1.45
1.5×10^3	0.74	1.91

NAD$^+$ as Variable Cofactor
[LADH] = 4 μg mL^{-1},
[C$_2$H$_5$OH] = 5 mM, pH 7.4, 23.5°C

$\dfrac{1}{[NAD^+]} (M^{-1})$	$1/v$ (Relative units)	
	Control (no pyrazole)	$4 \times 10^{-5} M$ pyrazole
0.9×10^4	0.59	1.17
1.8×10^4	0.70	1.41
3.0×10^4	0.77	1.59
6.0×10^4	1.05	2.15

(a) Plot these data as Lineweaver–Burk plots on two separate graphs.

(b) What are the Michaelis constants $K_M(C_2H_5OH)$ and $K_M(NAD^+)$, in the absence of inhibitor?

(c) What type of inhibition is exhibited by pyrazole against C_2H_5OH as a substrate?

(d) What type of inhibition is exhibited against NAD^+ as a cofactor?

(e) With reference to your answer to problem 19(a), suggest a mechanism for the action of this inhibitor.

(f) It has been proposed that the damaging effects of ethanol on the liver result from the pronounced decrease in the ratio $[NAD^+]/[NADH]$ caused by the oxidation of ethanol. (NAD^+ is required for several key dehydrogenation steps in the pathway of glucose synthesis. NADH promotes the accumulation of fat in the liver cells.) Assuming that pyrazole proves to be nontoxic and that it is transported to the liver, evaluate its use as an antidote to liver damage in treating chronic alcoholism.

21. A carboxypeptidase was found to have a Michaelis constant K_M of 2.00 μM and a k_{cat} of 150 s^{-1} for its substrate A.

(a) What is the initial rate of reaction for a substrate concentration of 5.00 μM and an enzyme concentration of 0.01 μM? Give units.

(b) The presence of 5.00 mM of a competitive inhibitor decreased the initial rate above by a factor of 2. What is the dissociation constant for the enzyme-inhibitor complex, K_I, where $K_I = [E][I]/[EI]$?

(c) A competing substrate B is added to the solution in part (a). Its K_M is 10.00 μM and its k_{cat} is 100 s^{-1}. Calculate the relative rates of substrate reaction for equal concentrations of substrates; that is, calculate v_B/v_A.

22. For the enzyme succinoxidase, which adds oxygen to succinate to give fumarate, the following rate data were determined for a solution in which the enzyme concentration is 10.0 μM.

[S] (mM)	0.33	0.50	1.00	2.00	10.00
v (μM s^{-1}) at 20°C	0.50	0.62	0.79	0.99	1.17
v (μM s^{-1}) at 40°C	2.41	2.99	3.81	4.78	5.65

(a) What are the values of the Michaelis constants K_M for this enzyme at the two temperatures?

(b) What are the values of the rate constants k_{cat} for this enzyme at the two temperatures?

(c) What is the activation energy for the reaction?

(d) Would you expect the activation energy to be higher or lower for the same reaction in the absence of enzyme? Explain.

23. A simple noncompetitive inhibitor of acetylcholinesterase binds to the enzyme to affect V_{max} only; it does not affect K_M. For an inhibition constant of $K_I = 2.9 \times 10^{-4} M$, what concentration of inhibitor is needed to give a 90% inhibition of the enzyme?

24. The catalytic activity of a ribozyme (an RNA enzyme) from *Tetrahymena thermophila* is discussed in detail by Herschlag et al. [*Biochemistry* 29:10,159 (1990)]. The ribozyme uses a guanosine cofactor (G) to cleave a ribo-oligonucleotide substrate. A simplified mechanism for the reaction is

$$
\begin{array}{ccc}
S + E & \underset{k_{-1}}{\overset{k_1}{\rightleftarrows}} & ES \\
+ & & + \\
G & & G \\
K_G \updownarrow & & \updownarrow K_G \\
S + GE & \underset{k_{-1}}{\overset{k_1}{\rightleftarrows}} & GES \xrightarrow{k_c} EP \xrightarrow{k_P} E + P
\end{array}
$$

At 50°C, neutral pH, and 10 mM Mg^{2+}, the following data were obtained. Kinetics of binding of substrate S, $E + S \rightarrow ES$, and $GE + S \rightarrow GES$: $k_1 = 9 \times 10^7 \, M^{-1}$ min^{-1}; $k_{-1} = 0.2 \, min^{-1}$. Binding of G is fast to equilibrium with dissociation constant, $K_G = 0.5 \, mM$:

$$K_G = \frac{[E][G]}{[GE]} = \frac{[ES][G]}{[GES]}$$

We note that substrate and cofactor bind independently.

Chemical step for cleavage of substrate: $k_c = 350 \, min^{-1}$

Release of product: $k_P = 0.06 \, min^{-1}$

(a) Derive an expression for the steady state concentration of GES in terms of the total concentration of enzyme, $[E_0]$, $[S]$, $[G]$, k_1, k_{-1}, k_c, and K_G. Remember that $[E_0] = [E] + [ES] + [GE] + [GES]$.

(b) In the steady state, the rate of formation of product can be written as

$$\frac{d[P]}{dt} = k_P[GES]$$

Use your expression from part (a) and the values of rate constants and the equilibrium constant to calculate $d[P]/dt$ as a function of $[S]$ for a constant value of $[E_0] = 100 \, \mu M$ at three different concentrations of G: $[G] = 0.05 \, mM$; $[G] = 0.5 \, mM$; $[G] = 5 \, mM$. Make an approximate plot of the rate vs. $[S]$ for each value of $[G]$; use enough values of $[S]$ to be able to see the trends.

(c) Describe the experiments you would need to do to measure k_1 and k_{-1} under different conditions of temperature and solvent.

25. In a study of the force-generating mechanism in muscle, Goldman et al. [*Nature* 300:701 (1982)] first soaked rabbit skeletal muscle fibers in a solution containing an analog of ATP, P^3-1-(2-nitro)phenylethyladenosine 5′-triphosphate, which has no effect on muscle contraction or relaxation. When the muscle fibers equilibrated with this analog are irradiated with a pulse of intense 347-nm UV light from a frequency-doubled ruby laser, photolysis of the analog occurs and ATP is suddenly generated:

It has been observed that in the presence of excess Mg^{2+}, this generation of ATP leads to relaxation of the muscle fiber according to first-order kinetics. By varying the intensity of the laser beam, different concentrations of ATP can be generated, and the first-order relaxation time t has been measured as a function of ATP concentration:

ATP (μM)	65	107	119	125	168	205	277
τ (ms)	230	59	40	56	42	23	22

(a) From the data given, obtain the best second-order rate constant for the relaxation of the muscle.

(b) Describe briefly why it is important to use a short pulse of light in this experiment.

26. Consider the enzyme-catalyzed hydrolysis of an ester that is competitively inhibited by Mg^{2+}. The following data were obtained: $K_M = 10^{-3} \, M$, $K_I = 10^{-4} \, M$, and $V_{max} = 1 \, M \, s^{-1}$.

(a) Estimate how long it will take for $10^{-5} \, M$ substrate to become $10^{-6} \, M$ in the absence of inhibitor.

(b) Estimate how long it will take 1 M substrate to become $10^{-1} \, M$ in the absence of inhibitor.

(c) Repeat calculations for parts (a) and (b) in the presence of $10^{-4} \, M \, Mg^{2+}$.

27. Consider the stoichiometric reaction $S \rightarrow P$ that is enzyme catalyzed and is found to follow Michaelis–Menten kinetics. Its maximum velocity V_{max} is equal to 100 $M \, s^{-1}$. In the presence of a *competitive* inhibitor (I), the following data are obtained:

Concentration of S (M)	Concentration of I (M)	Velocity ($M \, s^{-1}$)
3×10^{-3}	0	50
6×10^{-3}	1.0×10^{-4}	50

(a) What is the value of the Michaelis constant K_M in the absence of an inhibitor?

(b) What is the value of the dissociation constant for the enzyme-inhibitor complex K_I?

◀ PROBLEM 25

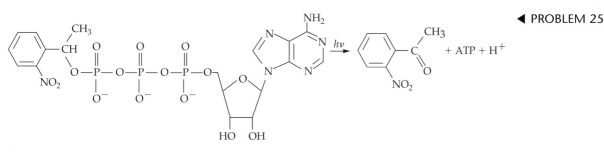

(c) The proposed mechanism for the reaction is

$$E + S \underset{k_{-1}}{\overset{k_1}{\rightleftharpoons}} ES \underset{k_{-2}}{\overset{k_2}{\rightleftharpoons}} EP \underset{k_{-3}}{\overset{k_3}{\rightleftharpoons}} E + P$$

Assume that there is a transition state between each reactant, intermediate, and product. They are, in order, $(ES)^{\ddagger}$, $(EZ)^{\ddagger}$, and $(EP)^{\ddagger}$.

From the following thermodynamic and kinetic data, plot energy vs. reaction coordinate. Label the position of the substrate, each intermediate, and each transition state with its energy value relative to product P, taken as the zero of energy.

Reaction	ΔH^0 (kJ)	$\Delta H^{\ddagger}$ (kJ)
$S \rightarrow P$	-30	—
$S \rightarrow ES$	—	$+12$
$ES \rightarrow EP$	$+4$	$+46$
$EP \rightarrow P$	-25	$+75$

(d) Which kinetic rate constant would you measure as a function of temperature to obtain $\Delta H^{\ddagger}$ for $EP \rightarrow P$? What is the value of $\Delta H^{\ddagger}$ corresponding to k_{-1}?

(e) Which step in the reaction do you think is rate determining? Why?

28. You have measured the following data on the destruction of thiamine by an enzyme, both with and without 0.20×10^{-4} M inhibitor (o-aminobenzyl-4-methylthiazolium chloride).

[Thiamine]	Velocity (M min^{-1})	
(mM)	Uninhibited	Inhibited
0.10	0.746×10^{-6}	0.136×10^{-6}
0.25	0.992×10^{-6}	0.285×10^{-6}
0.50	1.12×10^{-6}	0.618×10^{-6}
1.00	1.26×10^{-6}	0.758×10^{-6}
2.00	1.36×10^{-6}	1.29×10^{-6}

(a) Make suitable plots and use the $V_{\max}$ and K_M values to determine the type of inhibition.

(b) Based on these results, determine the binding constant for the inhibitor.

29. Methylglyoxal, as well as many other 2-oxoaldehydes, reacts readily with gluthathione (γ-L-glutamyl-L-cysteinylglycine, denoted by HSG below) to give a hemimercaptal adduct:

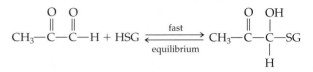

The enzyme glyoxalase I catalyzes the isomerization of the adduct to give S-D-lactoylglutathione:

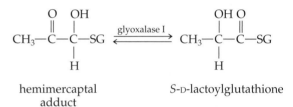

hemimercaptal S-D-lactoylglutathione
adduct

The forward reaction of (2) catalyzed by glyoxalase I purified from human erythrocytes is found to obey Michaelis–Menten kinetics with $K_M^F = 0.057$ mM and $(V_{\max}^F/[E]_0) = 1.17 \, 10^3 \, s^{-1}$ [Sellin et al., 1983, *J. Biol. Chem.* 258:2091]. The reverse reaction of (2) can be demonstrated if free glutathione (HSG) is removed quantitatively from the solution with a thiol reagent such as 5,5'-dithiobis(2-nitrobenzoate), Nbs$_2$, which in turn removes the hemimercaptal adduct according to (1).

The rate v of disappearance of S-D-lactoylglutathione (S) has been measured as a function of [S] upon addition of Nbs$_2$. The double-reciprocal plot shown to the right for data obtained at several Nbs$_2$ concentrations indicates that Nbs$_2$ is also a competitive inhibitor of the enzyme.

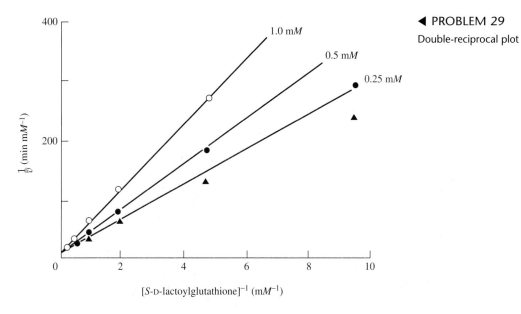

(a) Evaluate K_M^R, V_{max}^R, and K_I from the data.

(b) Calculate the equilibrium constant for reaction (2).

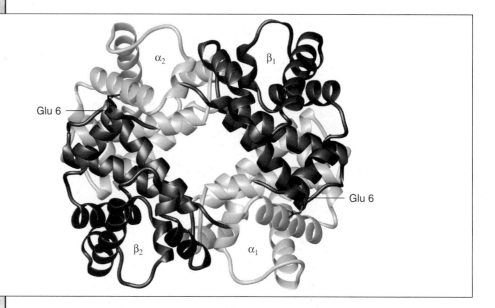

CHAPTER 9

Molecular Structures and Interactions: Theory

Many students of biology seem to fear or hate quantum mechanics; they worry about the mathematics used or they think it is irrelevant to medicine and everyday science. This chapter will show that quantum mechanics is no harder than other topics and that it does apply to the real world.

Concepts

Molecules are made of atoms, which contain charged nuclei and electrons. The balance among repulsion of like-charged particles and attraction of unlike-charged particles leads to the final bonded structure of a molecule. The structure of a molecule explains many of its molecular and macroscopic properties. One goal of chemistry is to predict the behavior of atoms and molecules, using the laws of physics. Since atomic and molecular structures depend on electrostatic interactions that are well understood, it follows that prediction of atomic and molecular properties should be straightforward. However, the universe gets tricky at small scale. Electrons, which determine whether atoms bond to each other, can have only discrete energies, whereas classical physics allows them to have any energy. Thus, *quantum mechanics* is needed to explain the behavior of atoms and molecules. Electrons bound to atoms exist in discrete states; in these states, electrons have precise energies but an imprecise location, characterized by a probability of finding an electron in a given region of space. The definite language of classical physics, with precise energies and location, is exchanged for a language of energy states and probabilities.

Quantum mechanics allows the prediction of atomic properties and how atoms bond to form molecules. Strengths and lengths of bonds are derived from quantum mechanics, as well as the arrangements of bonded atoms that are the key features of molecular structures. Since these calculations reveal electronic energies and regions of electron density, they also suggest the reactivity of molecules. Structure and chemical reactivity go hand in hand. In photosynthesis, both the lowest-energy (ground) state and higher-energy (excited) electronic states play a role in the conversion of light to cellular energy. The energies and properties of these ground and excited states can be understood using quantum mechanics.

In contrast to most reactions that occur on a chemist's benchtop, cellular processes involve many diverse macromolecules, such as proteins and nucleic acids, of molecular weights greater than 10,000 Daltons. A large number of noncovalent interactions among the constituent atoms stabilizes the three-dimensional conformations of these macromolecules. Quantum mechanical calculations are too complex for these large systems to be readily solved. However, the standard interactions of classical physics can approximate the behavior of these large molecules. Classical potential functions, which mimic predictions from quantum mechanics, can approximate the bond vibrations and rotations. Electrostatic and van der Waals potentials similarly approximate noncovalent interactions. Computer simulations, which employ these potential functions in Newton's equations of motion, can probe the physical behavior of a macromolecule. The *molecular dynamics* (change of system with time) or *molecular mechanics* (movement of system to lowest energy) of the macromolecule can then be evaluated.

Applications

Thermodynamics provides the framework to discuss the energies of molecules and their reactions. Thermodynamics tells us whether a reaction can occur, whether a folded conformation for a protein is more stable than an unfolded one, or whether a drug binds to an active site of an enzyme. Thermodynamics is purposely vague, however, at the molecular level. Thermodynamic analysis requires no knowledge of the details of what molecules look like and how they interact. Revealing these details is the goal of molecular biology, biochemistry, and chemistry. The basic rules that determine the detailed interactions of molecules and atoms are derived from quantum mechanics, classical mechanics, and electrostatics.

Proteins and nucleic acids are macromolecules consisting of repeating monomeric building blocks. Proteins are made from 20 amino acids linked together by amide peptide bonds. Nucleic acids such as DNA or RNA are built from nucleotides linked together by phosphodiester bonds. Macromolecules, pieced together with these building blocks, adopt complex three-dimensional structures that serve very specific functions in a cell. The regular double helix of DNA is an elegant means to store genetic information and protect it from the harsh environment of a cell. Hemoglobin is a finely tuned globule that can bind oxygen tightly and then release it in the capillaries. Antibodies are proteins with a common folded structure, yet a small change in amino acid sequence can lead to a very high affinity for a specific chemical compound compared with related compounds. Because there are more than 30,000 different proteins in a human cell and because determining the three-dimensional structure of a protein or nucleic acid is time-consuming (see chapter 11), biochemists want to be able to predict the structure of a protein or nucleic acid from its sequence. Quantum mechanical calculations readily derive the structures and properties of the macromolecular building blocks. Even using the classical approximations described above, it is still not yet possible to predict the structures of proteins. However, this is a vigorous area of current research (see chapter 1).

The structure of a biological molecule provides a powerful starting point for understanding biological function. Certain features of a structure are constant; any change of sequence is then deleterious to function. Changes in other portions of a structure can have little effect on biological activity. Understanding why this occurs may explain genetic diseases, which involve spontaneous changes in the critical part of a macromolecule. By understanding in detail the interactions that allow a protein to form a particular shape and to recognize a specific substrate, we can perhaps explain the origins of these diseases. Linus Pauling, one of the pioneers of quantum mechanical views of chemistry, put forward the first example of such an explanation (figure 9.1). He showed that sickle-cell anemia, a genetic disease, involves a single amino acid change in hemoglobin that causes hemoglobin to aggregate when it releases oxygen in the capillaries. This observation confirmed that biological molecules, despite their complex shapes, are no different from simpler chemical compounds: Changing their compositions can change their chemical properties.

The Process of Vision

Quantum mechanics has played an important role in understanding the process of visual stimulation. In the eyes of mammals and other animals, the visual pigment rhodopsin absorbs visible light and undergoes photochemical changes that result in electrical signals being transmitted by the optic nerves

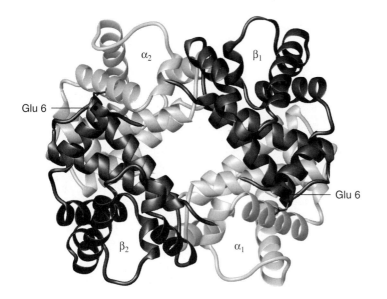

◀ FIGURE 9.1

Three-dimensional structure of hemoglobin. The site of the glutamic acid to valine mutation that is responsible for sickle-cell anemia is highlighted.

to the brain. The rhodopsin molecule is a protein, opsin, that is associated with the chromophore retinal, a polyene closely related to Vitamin A (see figure 9.2). In the dark-adapted retina, the pigment is present in rhodopsin as 11-*cis*-retinal; it is connected at one end by a Schiff's base linkage to a lysine amino acid side chain of the opsin protein. Upon illumination, one of the earliest events of the visual process is the conversion of the 11-*cis*-retinal to all-*trans*-retinal. Each photon absorbed produces a photoexcited rhodopsin that causes the activation of a phosphodiesterase. This enzyme hydrolyzes hundreds of guanosine 3′–5′ cyclic phosphate molecules (cyclic GMP). Cyclic GMP keeps sodium channels open across membranes; when the GMP is hydrolyzed, however, the sodium channels close and the passage of millions of sodium ions is blocked. Thus, the absorption of each photon is amplified many times to produce a nerve impulse (Stryer 1988; Hoff et al. 1997; Kochendoerfer et al. 1999).

The *cis* → *trans* isomerization caused by photoexcitation involves a significant change in the geometry of the chromophore at its attachment site to the protein; nevertheless, it occurs exceedingly rapidly (less than 6 ps) and without dissociation of the link between retinal and opsin. At the same time, events are triggered that result in an electrical signal, associated with the translocation of protons across the membrane in which the rhodopsin molecule is intrinsically embedded. Quantum mechanics can contribute in several ways to understanding how this process may occur. Consider, for example, the light absorption process itself. Visual pigments occur in a variety of forms: Rod cells contain rhodopsin that absorbs maximally at about 500 nm, whereas cone pigments may have absorption maxima at longer wavelengths, to 580 nm. All of these contain 11-*cis*-retinal, which absorbs light maximally only at 380 nm in the ultraviolet (UV) region as the isolated pigment. Even as the Schiff's base, it absorbs only at 450 nm in ordinary solution. Clearly, interaction with the protein opsin produces not only the red shift in absorption that makes rhodopsin exquisitely sensitive to visible light but also the variety

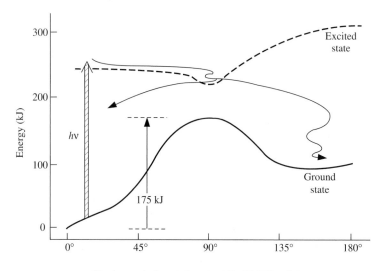

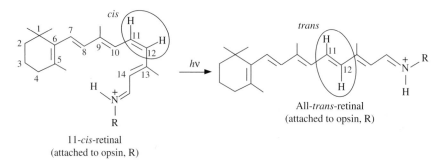

▲ FIGURE 9.2

Light-driven isomerization of 11-*cis*-retinal attached to opsin (R) by a Schiff's base linkage. Theoretical calculations of the energy of both the ground state and an excited electronic state suggest a possible path whereby absorption of a photon will lead to conversion from the *cis* to the *trans* configuration about 60% of the time. The angle of rotation about the 11,12-double bond is plotted as the horizontal axis.

Following the absorption of a photon of visible light (heavy vertical arrow at left of diagram) by the chromophore in the 11-*cis* configuration, the calculations indicate that the molecule follows a trajectory such as shown by the paths (solid curves in upper part of diagram) leading first to the minimum of the excited-state energy surface at a torsion angle of about 90°. After a few oscillations in this geometry, it continues to lose energy by returning to the ground-state energy surface. The branch in the trajectory at this point indicates that it may continue on to the all-*trans* configuration (180° torsion angle) or it may return to the original 11-*cis* form (0° torsion angle). The branching ratio is determined by quantum mechanical considerations. There is experimental evidence to support the conclusion that the entire process depicted occurs within a few picoseconds (see figure 7.20). (Reproduced, with permission, from R. B. Birge, *Annual Review of Biophysics and Bioengineering* 10:315–354, © 1981 by Annual Reviews.)

of absorptions in different cone cells, which is the origin of color vision. Theoretical analysis (Nakanishi et al. 1980) indicates that the origin of these spectroscopic red shifts is electrostatic in nature. The Schiff's base nitrogen atom is protonated in rhodopsin, which places a positive charge on that region of the molecule. The compensating negative charge, which usually comes from a counterion in solution, appears to be associated with a particular negatively charged amino acid (for example, aspartate or glutamate) in the immediate vicinity of the retinal chromophore. Calculations show that the placement of a charge near the 11,12-double bond should produce the kind of red shift that is observed.

Calorimetric studies of rhodopsin photochemistry show that the photo-isomerized product, bathorhodopsin (all *trans*), has 145 kJ mol^{-1} more energy than the starting rhodopsin (11-*cis*). By comparison, there is only a few kilo-joules per molecule difference for *cis/trans* isomers of analogous polyenes in solution. Where does this extra "stored" energy reside in bathorhodopsin? One proposal (Birge 1981) suggests that it resides in increased separation of charge between the protonated Schiff's base nitrogen and negatively charged amino acid side chains of the opsin. The *cis* → *trans* isomerization produces a change in location of the Schiff's base relative to the opsin matrix to which it is bound. If this region of rhodopsin does not incorporate solvent water, then an increased charge separation of only a few angstroms in the low-dielectric medium could result in a substantial "storage" of energy. In visual transduc-tion, the process does not stop there, of course. Other processes follow in which this stored energy is converted, ultimately to the translocation of a proton across the 40- to 50-Å thickness of the embedding membrane.

There is actually an energy barrier that lies at least 175 kJ mol^{-1} above the dark-adapted (11-*cis*) state. It is most important that this barrier is there; otherwise, our eyes would have to deal with an enormous "background" signal generated by thermal energy in the environment. You would see this background even with your eyes closed. The high barrier effectively prevents appreciable triggering by thermal excitation at room temperature. Photons of visible light have energies corresponding to 200 kJ mol^{-1} or more. Thus, the photon energy is more than sufficient to allow the molecule to surmount the barrier between the 11-*cis* and the all-*trans* forms (see figure 9.2). The "noise" level is sufficiently low that light fluxes of a few photons per second per visual cell can be discriminated from the background. Furthermore, the quantum yield of this photoisomerization is quite high. Over one-third of the photons absorbed by rhodopsin lead to bathorhodopsin. A sketch of po-tential energy surfaces for the ground electronic state and the first-excited electronic state of rhodopsin as a function of the angle of rotation of the chro-mophore about the 11,12-double bond is shown in figure 9.2. Also shown are some calculated "trajectories" that attempt to describe typical routes that might be taken during the photoisomerization (or the return to the original 11-*cis* form following those events that failed to lead to isomerization). Stud-ies such as these are examples of the kinds of contributions that quantum mechanics can make to understanding important biological phenomena.

Origins of Quantum Theory

At the end of the nineteenth century, physicists were proud of their achieve-ments. Mechanics, as illustrated by Newton's laws, related forces to poten-tials and was able to predict accurately the trajectories of particles. The major principles of thermodynamics had been outlined. Finally, Maxwell had dis-covered the central laws of electromagnetism, which apparently confirmed the wave nature of light. These achievements, which form the core of what is now called classical physics, led Lord Kelvin to declare that future dis-coveries in physics would be found in the decimal places: Twentieth-century physics would involve filling in the remaining details of the existing theories. But within ten years, the twin revolutions of quantum theory and relativity would be in full force and revolutionize not only physics but also chemistry and biology.

We can summarize the central principles of classical physics that apply to our discussion of atomic and molecular structure as follows: (1) A trajectory can be predicted for a particle, with precise positions and momenta specified at each time point. (2) Wave energy depends on the amplitude of the wave (squared), and waves can be excited with any desired energy dependent on the applied forces. The translational, vibrational, or rotational motions of the system can have any value, given a sufficient force on the system. This summarizes our commonsense view of the macroscopic world: A baseball, given an infinitely strong pitcher, can be thrown at any velocity. A series of critical observations and interpretations overturned this commonsense view and led to the development of quantum theory. The story is illustrative how a few nagging flaws in an otherwise widely accepted theory can ultimately lead to its downfall and to a scientific revolution.

Blackbody Radiation

Several observations demonstrated that systems could only take up (or emit) energy in discrete amounts. The first, made by Planck, correctly provided a theoretical explanation for blackbody radiation. Blackbody radiation is the light given off by an object when it is heated; examples are the metal filament of a lamp or the glow of molten metal. As the temperature is raised, more light is emitted at shorter wavelengths (higher frequencies). Frequency (ν) and wavelength (λ) are related to the speed of light (c) by:

$$\lambda\nu = c \tag{9.1}$$

The blackbody is an ideal version of a heated object, in which the light reflects within a cavity and is emitted from a pinhole. This allows the emitted light to come into thermal equilibrium with the walls of the container, because it has bounced off the walls many time before being emitted from the pinhole. Classical physics could not explain the temperature dependence of blackbody radiation (figure 9.3). The best attempt, by Rayleigh, predicted that the emitted energy by the blackbody depended on $1/\lambda^4$. This successfully predicted the falloff in emitted radation at a longer wavelength. Unfortunately (for Rayleigh and classical physics), it does not predict a maximum but instead predicts increasing emitted intensities at shorter and shorter wavelengths. This failure of classical physics was called the "ultraviolet catastrophe" because it predicted significant radiation in the short wavelength, UV portion of the electromagnetic spectrum. According to this theory objects should glow even at room temperature.

Planck found a theoretical explanation for blackbody radiation that successfully predicted the behavior at both high and low wavelengths. Planck treated the walls of the blackbody as oscillators, which could emit radiation. Rayleigh treated blackbody radiation in a similar manner, but Planck then made the theoretical leap that the permitted energies of these oscillators were *quantized*. This directly conflicted with one of the principles of classical physics discussed earlier. Planck was able to predict the blackbody radiation if the available energy of the oscillator was set to $E = nh\nu$, where n is an integer $= 0, 1, 2$ (etc.) and h is a constant, called Planck's constant, with a value of

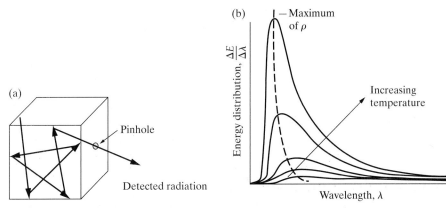

▲ FIGURE 9.3

(a) Schematic of a blackbody radiation experiment. A cavity with a small pinhole is heated. The light exits the pinhole only after reflecting off the internal walls multiple times. (b) Experimental results of a blackbody radiation experiment. The energy-density (energy/volume) distribution as a function of wavelength is shown, as is the behavior predicted by Rayleigh using classical physics.

6.626×10^{-34} J s. This mathematically expressed the proposed quantization. The second major breakaway from classical physics is the proportionality of energy and frequency; in classical physics, the energy of an oscillation is proportional to the square of the amplitude of the oscillation, not frequency. The proposed distribution formula for the energy density (change in energy ΔE for a given incremental change in wavelength $\Delta \lambda$) is

$$\frac{\Delta E}{\Delta \lambda} = \frac{8\pi hc}{\lambda^5} \left(\frac{1}{e^{\frac{hc}{\lambda kT}} - 1} \right) \qquad \textbf{(9.2)}$$

where c is the speed of light and k is the Boltzmann constant. The formula has the correct "limiting behavior"; in other words, it correctly predicts the blackbody radiation at both long and short wavelengths. In fact, at long wavelengths, the Planck expression reduces to Rayleigh's originally predicted formula.

The concept of quantized energy provided a theoretical explanation for the line spectra of atoms and simple molecules. When an electric discharge is sent through a gas of atoms, it emits light only at discrete frequencies. Classical physics could not explain the preference of one frequency over another. The modern view of the atom, as we will see, resolves this dilemma. Electrons exist in discrete states with discrete energies; energy is emitted only when transitions occur from one discrete state to another. The frequency of light emitted is proportional to the energy difference between the states:

$$\Delta E = h\nu \qquad \textbf{(9.3)}$$

Equation (9.3) plays a central role in spectroscopy. Bohr, in his early model of the hydrogen atom, successfully predicted the spectral frequencies observed in the spectrum of hydrogen. We address this point again later when we derive a quantum mechanical treatment of the hydrogen atom. The discrete line spectra of the elements are graphic illustrations of the quantized nature of energy.

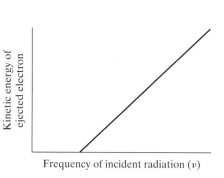

The photoelectric effect. Plot of the kinetic energy of ejected electrons as a function of the frequency of incident radiation. No electrons are emitted until the incident radiation exceeds a threshold frequency; above the threshold frequency, the kinetic energy of the ejected electrons increases linearly with the frequency of incident radiation.

Photoelectric Effect

When light strikes a clean metal surface, electrons may be ejected. This is the *photoelectric effect* (figure 9.4). Three observations concerning the photoelectric effect are unexplained by classical physics. First, the kinetic energy of the ejected electrons is independent of the intensity of the light. Second, electrons are only ejected if the frequency of light exceeds a certain threshold; below that threshold frequency, no electrons are ejected. Finally, if the threshold frequency is exceeded, the kinetic energy of the ejected electrons is proportional to the frequency of the incident light.

From classical physics we learned that light is a wave and that the energy of the wave is proportional to its intensity (amplitude^2). Thus, electrons with higher kinetic energy should be ejected by more intense light. Einstein instead proposed that light behaved like small packets of energy, now called *photons*, with energy $E = h\nu$. He then formulated a simple equation using conservation of energy:

$$\frac{1}{2}mv^2 = h\nu - \Phi \qquad (9.4)$$

where $\frac{1}{2}mv^2$ is the kinetic energy of the ejected photoelectron, $h\nu$ is the energy of the impinging photon, and Φ is the minimum energy to remove an electron from the metal surface (it is sometimes called the *work function*). The kinetic energy cannot be less than zero, so we predict that

$$\Phi = h\nu_0$$

where ν_0 is the minimum threshold frequency needed to eject an electron. In one amazing formula, Einstein explained the photoelectric effect. The essence of his treatment is that light behaves like a particle with energy $E = h\nu$; with the assumption of quantized units of energy, the theory is quite simple. Once a photon has greater energy than the work function, which is a property of the metal, electrons are ejected. The energy of the ejected electron is the difference between the photon's initial energy and the work function of the metal. That explanations of blackbody radiation, spectral lines, and photoelectric effect, which are very disparate phenomena, all yield the same value for Planck's constant underscores the fundamental nature of this constant.

Electrons as Waves

In the 1920s it became clear that for many purposes an electron or any other particle is better understood as a wave phenomenon. One of the most direct experimental verifications of this wavelike character was the observation of electron diffraction by Davisson and Germer in 1927. They projected a fine beam of electrons onto a single crystal of nickel metal and observed a diffraction pattern of concentric rings around the transmitted beam. Such diffraction patterns were well known from studies of light, sound, X rays, ripples on the surface of a liquid, and other phenomena involving waves. From their measurements, Davisson and Germer calculated a wavelength for the electron; the wavelength λ was inversely proportional to the momentum (mv) as predicted two years earlier by de Broglie. The *de Broglie relation* is stated as

$$\lambda = \frac{h}{mv} \tag{9.5}$$

where h is Planck's constant, (6.626×10^{-34} J s), m is the rest mass of the electron (9.11×10^{-31} kg), and v is its velocity (m s^{-1}). The wavelength thus has units of meters. Since the velocity is determined by the electric potential used to accelerate the electron, it is easy to change v by changing the electric potential and to observe the related change in the spacing of the diffraction pattern. Subsequent experiments showed that other particles, such as protons and neutrons, also exhibit wavelike behavior and fit the de Broglie relation when the appropriate rest masses are used. Because these other particles are much heavier than the electron, their characteristic wavelengths are much shorter for the same velocity. The nuclei of atoms are thus particles with wavelengths that are typically very much shorter than those of lighter and faster electrons.

Heisenberg Uncertainty Principle

The "wave-particle duality" is not obvious from our usual macroscopic observations. Most macroscopic particles have significant masses such that their de Broglie wavelength is infinitesimally small. A consequence of the dual nature of light, and a more general result of quantum theory, is the *Heisenberg uncertainty principle:* It is impossible to specify simultaneously the position and momentum of a particle to arbitrary precision. If we know the position of a particle, there is uncertainty in its momentum and vice versa. The mathematical expression of the uncertainty principle is

$$\Delta p \Delta x \geq \frac{1}{2}\left(\frac{h}{2\pi}\right) \tag{9.6}$$

where Δp and Δx are the uncertainties in momentum and position, respectively. The uncertainty principle can be understood qualitatively in terms of the wave nature of particles. According to the de Broglie relation, momentum is inversely proportional to wavelength. It is difficult to think about the "position" of a wave, but we can convert position into amplitude. The amplitude is related to the probability of finding a particle in a given position. To define sharply a local amplitude, a large number of waves of different wavelength need to be added. The amplitudes of these waves add constructively at one specific point and cancel elsewhere. This creates a particle with a highly defined position, but at a cost in terms of momentum. Since the momentum is inversely proportional to wavelength and we have used a large number of waves of different wavelength to define the position, the momentum is very poorly defined. In the limit of precise definition of position (a peak of infinitely narrow width), the momentum is completely undefined. The uncertainty principle has many profound implications for understanding molecular systems, and we address them in future sections.

Quantum Mechanical Calculations

In principle, we can calculate any property of any molecule using quantum mechanics. In practice, the ability to obtain useful results from these a priori calculations decreases rapidly as the number of electrons increases. Calculations on a nucleic acid base pair containing 30 nuclei and 136 electrons are possible but extremely difficult to do accurately. That is why we use the more empirical, classical methods mentioned earlier to investigate the conformations of proteins or nucleic acids. However, we can directly apply quantum mechanical methods to parts of a large molecule, such as the retinal in rhodopsin, the metal-containing catalytic site of an enzyme, or the special chlorophyll in the reaction center of the photosynthetic apparatus.

Biologists should know at least a minimum of quantum mechanics so they can feel confident in using the many spectroscopic methods (see chapter 10). Spectra do not make sense without quantum mechanics. It is also important to know the vocabulary of bonding, orbitals, electron distribution, and charge densities to understand reaction mechanisms and molecular interactions. These concepts are based on relatively simple quantum mechanical descriptions of molecules. The goal of molecular biologists is to explain biological functions in terms of molecular interactions. The goal of quantum biologists, or submolecular biologists, is to explain molecular interactions in terms of electron distribution and motion. In this chapter, we present the fundamental aspects of quantum mechanics that provide a basis for this powerful method.

Wave Mechanics and Wavefunctions

Once the wave nature of matter was recognized, it became possible for Heisenberg and Schrödinger, independently, to formulate mathematical descriptions of electron wave motion; both treatments were later shown to be equivalent. The Schrödinger treatment, which is the more familiar of the two, assigns an amplitude ψ to the electron wave; ψ is known as the *wavefunction* of the system. For classical wave motion, the wavefunction is simply the displacement of the system from its equilibrium position—for example, the height of waves on the surface of a lake. For electron wavefunctions, there is no simple interpretation of the wavefunction itself, but ψ^2 characterizes the distribution of the electron in space.

In general, the wavefunction of a system is a function of both the space coordinates and time. However, we consider only time-independent wavefunctions in this section. Once we know the wavefunction for a molecule, we can calculate all the properties of the molecule. This is a fundamental feature of quantum mechanics and explains the chemist's desire to determine wavefunctions at all costs. To illustrate this, consider a possible molecule, HeH^{2+} (figure 9.5). This molecule has one electron, so its electronic wavefunction is a function of the x-, y-, z-coordinates of the electron:

$$\psi = \psi(x, y, z)$$

The method of obtaining wavefunctions is illustrated in several of the following sections. For the moment, we consider the He and H nuclei fixed, and we are interested in the electron distribution around the nuclei. We see that for each position of the two nuclei there is a set of time-independent states of the electron with each state characterized by a definite energy E_n and wavefunction ψ_n. The lowest energy and the electron distribution corresponding to it is the

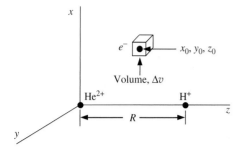

Diatomic molecular HeH^{2+} with a given internuclear distance R. The probability of finding the electron in a volume Δv around point x_0, y_0, z_0 is just the square of the electronic wavefunction multiplied by the volume. HeH^{2+} is a thermodynamically stable molecular ion with an equilibrium internuclear distance of 0.774 Å and a bond dissociation energy of 155 kJ.

ground state. There are excited states corresponding to higher energies and to different electron distributions.

Because there is only one electron in this molecule, it is fairly easy to obtain $\psi_n(x, y, z)$ for many internuclear distances; that is, we obtain ψ_n as a function of possible bond lengths. What properties can we now calculate?

The *electron density* or *distribution* is given by $[\psi_n(x, y, z)]^2$. For a chosen internuclear distance and a particular ground or excited state, the probability of finding the electron in a small volume of size Δv around point x_0, y_0, z_0 is

$$[\psi_n(x_0, y_0, z_0)]^2 \Delta v$$

The probability of finding the electron in a finite volume of space is obtained by integration over the volume:

$$\int [\psi_n(x, y, z)]^2 \, dv \qquad \text{define space} \qquad \textbf{(9.7)}$$

where the integral sign represents integration over all variables and dv represents all the coordinates of integration. The probability of finding the electron in all of space is equal to 1. This is called the *normalization condition*. It states that the electron must be found somewhere; the probability of finding it somewhere is unity.

$$\int [\psi_n(x, y, z)]^2 \, dv = 1 \qquad \text{any space}$$

The relation of $[\psi_n(x, y, z)]^2$ to probability of finding an electron in a given region of space is a reflection of the uncertainty principle and demonstrates the probabilistic description that quantum mechanics takes of matter.

The bonding geometries and chemical properties of molecules derive from the distribution of the constituent electrons. The structures of biological molecules are often determined by X-ray diffraction. The electron distribution ψ_n^2 directly determines the scattering of X rays by the molecule. The electron distribution of course also determines the chemical reactivity of the molecule, but this is harder to quantify.

The *average position* of the electron can be calculated from the wavefunction. As a diatomic molecule is symmetric around the bond, the average position of the electron will be on the z-axis, the internuclear axis. The average position for the positive charges of the nuclei is $R/3$, where R is the internuclear distance measured from He. The average position of negative charge (the electron) may change markedly upon excitation from the ground to the excited state. This was first predicted and then found to be the case for retinal. Evidence about the excited state of retinal is obtained from spectroscopic measurements, such as the effect of an intense electric field on the absorption of light (Mathies and Stryer 1976). A large shift in charge density was ob-

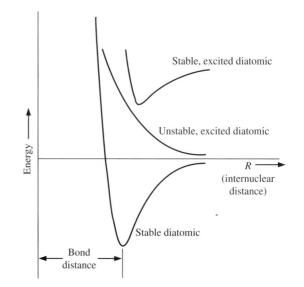

► FIGURE 9.6

Plot of calculated energy against distance R between nuclei in a diatomic molecule.

served for the Schiff base of retinal; the positive charge is moved from the N^+ toward the six-membered ring (see figure 9.2). In this way, the photon energy has been changed to an electrical signal. The hope is that further experimental and theoretical investigations based on quantum mechanics will eventually help us understand the transmission of the signal across the retinal rod membrane and along the optic nerve to the brain.

We can calculate the *energy* of the electron in each electronic state from the corresponding wavefunction. The energy of the electron is of vital importance because it tells us if the molecule is stable and what its stable bond distance is. We can calculate the electronic energy as a function of internuclear distance. This energy decreases (becomes more negative) as the internuclear distance decreases. The repulsion between the positively charged nuclei, which is easily calculated from Coulomb's law for two point charges, increases as the internuclear distance decreases. If there is a minimum in the total (electron plus nuclear) energy at some internuclear distance, a stable molecule can exist (figure 9.6). The energy versus distance can also be calculated for excited electronic states. This tells us when excitation (by light, for example) will cause dissociation of the molecule. If dissociation does not occur, how does the bond distance change on electronic excitation? The position of the minimum in energy for the excited electronic state tells us the stable bond distance in the excited state. We can see how electronic excitation in a molecule such as retinal leads to a change in conformation. The energy as calculated by quantum mechanics is a link to the macroscopic measurements of thermodynamics; for example, we can measure bond dissociation energies and compare them to the quantum mechanical predictions.

Calculation and measurement of energies of molecules in their ground and excited states involves all of spectroscopy (discussed in detail in chapter 10).

We have treated a simple, one-electron, diatomic molecule to make the notation simpler, but the same principles apply to more complicated molecules. For a two-electron molecule, the electronic wavefunction depends on the positions of both electrons:

$$\psi = \psi(x_1, y_1, z_1, x_2, y_2, z_2)$$

where the subscripts refer to electrons 1 and 2. The equations for calculating electron distribution, position, energy, and so on are the same, but we now must integrate over six coordinates instead of three. It is clear why many-electron molecules quickly begin to tax present-day computers. Adenine has 70 electrons, so we have electronic wavefunctions that depend on 210 coordinates. Therefore, at present we need to make approximations to calculate useful energies, bond distances, electron densities, and so on. A useful approximation for many applications is to consider only the π electrons. This approximation is adequate for treating the UV spectrum of adenine, for example.

For the larger molecules encountered by the biochemist or biologist, approximations must be made to tackle the problems at all. Among the most useful approximations are those that correspond to our chemical intuition:

1. Molecules can be thought of as collections of atoms or groups (such as sulfate, ammonium) whose internal structure (nuclei, electrons) need not be known in detail.

2. Electrons may be localized primarily on individual atoms, they may be shared between pairs of atoms, or they may be delocalized over a large portion of the molecule.

3. The motion of electrons is very rapid in comparison with that of the much heavier nuclei. As a consequence, changes in electronic structure or distribution (caused, for example, by photon absorption) occur with the nuclear arrangement essentially frozen.

Schrödinger's Equation

Schrödinger's equation is the differential equation (an equation that involves derivatives) whose solutions give the wavefunctions. From the wavefunctions, measurable properties can be calculated, which are functions of the positions and momenta of the electrons and nuclei in the molecule. Schrödinger (and Heisenberg) deduced the form of *operators*, which operate on a wavefunction to give a measurable property. We use operators constantly in mathematics. For example, $\ln x$, $\sqrt{x}$, dx, and $3x$ are all examples of operators operating on x. The *position operator* in quantum mechanics is simply the coordinate itself. For example, to find the average position of the electron in the molecule HeH^{2+} (figure 9.5), we use the following equation:*

$$\bar{z} = \int \psi_n z \psi_n \, dv$$

Here, z operates on ψ_n simply by multiplying ψ_n. The quantity $\bar{z}$ is known as the *expectation value*, or average value of z. If ψ_n is the wavefunction of an electron, for example, then $\bar{z}$ gives us the *average* position of the electron along the z-axis. Expectation values can be calculated similarly for the momentum, the energy, the dipole moment, and many other useful properties that can be compared directly with experimental physical measurements;

---

*We consider only real wavefunctions here. In general, the equation must be written as $\int \psi_n^* z \psi_n \, dv$, where ψ_n^* is the complex conjugate of ψ_n. This means that wherever the imaginary i appears in ψ_n, it is replaced by $(-i)$ in ψ_n^*.

these are average values of *physical observables* in the language of quantum mechanics. In each case, the evaluation involves integrating the product of operator and wavefunction over all space—at least, everywhere that the product has a significant value. Again we see that the probabilistic interpretation of quantum mechanics extends to other observable parameters beyond position and momentum.

The momentum operator in quantum mechanics is slightly more complicated. In classical mechanics, the momentum is mass times velocity; in the x-direction, it is $p_x = mv_x$. In quantum mechanics, the *momentum operator* in the x-direction is

$$p_x = \frac{\hbar}{i} \frac{d}{dx}$$

$$\hbar \equiv \frac{h}{2\pi} = 1.055 \times 10^{-34} \, \text{J s}$$

$$i \equiv \sqrt{-1}, \text{imaginary} \qquad (i^2 = -1)$$

The quantum mechanical momentum operator operates on ψ and involves taking the derivative of ψ with respect to a space coordinate and multiplying by $\hbar/i$.

$$\bar{p}_x = \frac{\hbar}{i} \int \psi_n \frac{d}{dx}(\psi_n) dv$$

The Schrödinger equation is an equation describing the energy of a system written in terms of its wavefunction ψ. In the operator notation that we have developed and for a single particle of mass m moving in one dimension, this is

$$\text{kinetic energy} + \text{potential energy} = \text{total energy}$$

$$T_x \psi \quad + \quad U(x)\psi \quad = \quad E\psi$$

$$\frac{p_x^2}{2m}\psi \quad + \quad U(x)\psi \quad = \quad E\psi$$

$$-\frac{\hbar^2}{2m}\frac{d^2\psi}{dx^2} \quad + \quad U(x)\psi \quad = \quad E\psi$$

$$\left[-\frac{\hbar^2}{2m}\frac{d^2}{dx^2} \quad + \quad U(x)\right]\psi \quad = \quad E\psi$$

$$\mathcal{H}\psi \quad = \quad E\psi \qquad \text{(9.8)}$$

where the operator $\mathcal{H}$, which operates on the wavefunction to give the energy, is called the *Hamiltonian* and contains the terms that describe the total energy of the system; T_x is the kinetic energy operator; $U(x)$ is the potential energy operator; and E is the total energy, which is a constant for stationary states where time is not involved. Note that

$$T_x = \frac{1}{2}mv_x^2 = \frac{p_x^2}{2m}$$

is the expression for the kinetic energy in classical mechanics for a particle of mass m moving with a velocity v_x.

When E is constant, Eq. (9.8) is the time-independent Schrödinger equation in one dimension for a particle of mass m. It is a differential equation whose solutions give the wavefunctions ψ_n and energies E_n for the particle in its ground and excited states. To emphasize this, we can write the Schrödinger equation as

$$\mathcal{H}\psi_n = E_n\psi_n$$

$$-\frac{\hbar^2}{2m}\frac{d^2\psi_n}{dx^2} + U(x)\psi_n = E_n\psi_n \tag{9.9}$$

where the subscript n is the *quantum number* for the wavefunction, or *eigenfunction*, ψ_n, and the *energy*, or *eigenvalue*, E_n. Before considering applications of the Schrödinger equation to simple problems, let's first note that we are not restricted to working in one dimension or to using an x-, y-, z-coordinate system. For example, the Schrödinger equation written for a particle in three coordinates in the x-, y-, z-system is

$$-\frac{\hbar^2}{2m}\left(\frac{\partial^2\psi}{\partial x^2} + \frac{\partial^2\psi}{\partial y^2} + \frac{\partial^2\psi}{\partial z^2}\right) + U(x, y, z)\psi = E_{n_x, n_y, n_z}\psi \tag{9.10}$$

where $\psi = \psi(x, y, z)$. There are three quantum numbers, n_x, n_y, and n_z, associated with the system. Many problems involving central symmetry, such as a hydrogen atom, are best solved using spherical coordinates, r, θ, and ϕ.

Other coordinate systems are also useful. To write the Schrödinger equation in a form independent of the coordinate system, the Laplace operator is introduced. The *Laplace operator* ∇^2 indicates the operation of taking second derivatives with respect to all coordinates. In Cartesian coordinates, it has the form

$$\nabla^2 = \frac{\partial^2}{\partial x^2} + \frac{\partial^2}{\partial y^2} + \frac{\partial^2}{\partial z^2}$$

Using this operator, we can write the time-independent Schrödinger equation as

$$-\frac{\hbar^2}{2m}\nabla^2\psi + U\psi = E\psi \tag{9.11}$$

or, even more compactly, as

$$\mathcal{H}\psi = E\psi \tag{9.8}$$

where $\mathcal{H}$ is the Hamiltonian operator

$$\mathcal{H} = -\frac{\hbar^2}{2m}\nabla^2 + U \tag{9.12}$$

representing the sum of kinetic and potential energy. For more than one particle in the system, we sum over all particles:

$$\mathcal{H} = -\frac{\hbar^2}{2}\sum_i\frac{\nabla_i^2}{m_i} + U$$

Solving Wave Mechanical Problems

We will now apply the basic elements and procedures of wave mechanics to several simple systems. The examples include a particle in a potential well, an electron in a central force field, a harmonic oscillator, and molecular orbitals constructed as linear combinations of atomic orbitals. We then point out what approximations are needed to approach the real molecular situation. The simple examples are more than mere exercises; insights derived from them will be used to describe and interpret important aspects of biomolecular structure and spectra.

First, let's examine the general strategy used in solving real problems in wave mechanics. We consider only stationary (time-independent) states of molecules so that time will not enter explicitly as a variable. If the system does change with time (by the absorption or emission of radiation), we can still describe the initial and final states using this stationary approximation. The dynamics (time course) of the changes can be treated using wave mechanics as well, but the calculations are mathematically more involved than we need to consider here.

Outline of Wave Mechanical Procedures

I. Definition of the problem

   A. *Write the appropriate potential for the particular problem.* Although the kinetic-energy operator has the same form for all problems, the potential-energy operator distinguishes each problem.

   1. *Particle in a box:* We consider a particle that is contained in a box but is otherwise free. The potential energy is zero inside the box but infinite at the walls and outside the box. For a one-dimensional box,

   $$U(x) = 0 \qquad \text{(for } 0 < x < a)$$

   $$U(x) = \infty \qquad \text{(for } x \le 0 \text{ and } x \ge a)$$

   Think of a bead on a wire capped at each end. If there is no friction between the bead and wire, the bead is free to move along the wire between the caps. This problem relates to electrons confined to molecular orbitals.

   2. *Harmonic oscillator:* A harmonic oscillator represents a particle attached to a spring. The potential energy increases whenever the spring is stretched or compressed. This problem is easy to solve, and it approximately represents the vibrations of nuclei of a molecule:

   $$U(x) = \frac{1}{2}kx^2 \qquad (k = \text{force constant})$$

   3. *Coulomb's law:* This is the dominant potential involving the interactions between nuclei and electrons in molecules; it is the only interaction we consider. Other, much weaker interactions depend on the spin of the electrons and nuclei. For two charges separated by a distance r, the potential energy is

   $$U(r) = \frac{q_1 q_2}{r}$$

   For charges q_1 and q_2 in electrostatic units (esu) and r in centimeters, the potential energy is in ergs. This is a common form used by chemists and biologists. It is consistent with using m in grams and $h = 1.055 \times 10^{-27}$ erg s in the kinetic-energy operator. With SI units, the potential energy is

   $$U(r) = \frac{q_1 q_2}{4\pi\varepsilon_0 r} \qquad \qquad \textbf{(9.13)}$$

Here the charges are in coulombs, r is in meters, ε_0 (the permittivity constant) $= 8.854 \times 10^{-12}\,C^2\,N^{-1}\,m^{-2}$, and the energy is in joules. It is consistent with using m in kilograms and $h = 1.055 \times 10^{-34}\,J\,s$ in the kinetic-energy operator.

We can now write the potential-energy operator for any atom or molecule. For a hydrogen atom, in cgs units,

$$U(r) = -\frac{e^2}{r}$$

where e is the electronic charge in esu and r is the distance between the proton and the electron in cm. For a helium ion He^+,

$$U(r) = -\frac{2e^2}{r}$$

For a helium atom,

$$U(r_1, r_2) = -\frac{2e^2}{r_1} - \frac{2e^2}{r_2} + \frac{e^2}{r_{12}}$$

The coordinates r_1 and r_2 refer to electrons 1 and 2 in the helium atom; r_{12} is the absolute (positive) distance between the electrons. The three terms represent, respectively, the electron–nucleus attraction for each electron and the electron–electron repulsion. For a molecule, we can easily write the potential energy just by adding terms corresponding to each electron–nucleus attraction, each electron–electron repulsion, and each nucleus–nucleus repulsion.

B. *Establish the boundary conditions for the wavefunction.* The wavefunctions are solutions to the Schrödinger equation, but in addition they must satisfy some auxiliary conditions. Examples of these are the following:

1. The wavefunction should be single valued everywhere.
2. The wavefunction of a bound electron, molecule, and so on should vanish at infinite distance and in any region where the potential is plus infinity.
3. The wavefunction should be continuous (no sharp breaks) and have a continuous first derivative, except where the potential energy becomes plus or minus infinity, such as at the walls of the box or at the nucleus of an atom, respectively.

II. Writing the Schrödinger equation for the problem

Write the Schrödinger equation in a suitable coordinate system. Some examples are as follows:

1. Particle in a box (one dimension):

$$-\frac{\hbar^2}{2m}\frac{d^2\psi_n(x)}{dx^2} = E_n\psi_n(x) \qquad (\text{for} \quad 0 < x < a)$$

2. Harmonic oscillator (one dimension) of reduced mass μ:

$$-\frac{\hbar^2}{2\mu}\frac{d^2\psi_v(x)}{dx^2} + \frac{1}{2}kx^2\psi_v(x) = E_v\psi_v(x)$$

3. Hydrogen atom:

$$-\frac{\hbar^2}{2\mu}\nabla^2\psi_{n,l,m}(r,\theta,\phi) - \frac{e^2}{r}\psi_{n,l,m}(r,\theta,\phi) = E_n\psi_{n,l,m}(r,\theta,\phi)$$

In spherical polar coordinates, where r is the radius, θ is a colatitude angle, and ϕ is a longitude, the Laplace operator is given by

$$\nabla^2 = \frac{1}{r^2}\frac{\partial}{\partial r}\left(r^2\frac{\partial}{\partial r}\right) + \frac{1}{r^2\sin^2\theta}\frac{\partial^2}{\partial\phi^2} + \frac{1}{r^2\sin\theta}\frac{\partial}{\partial\theta}\left(\sin\theta\frac{\partial}{\partial\theta}\right)$$

We have introduced a new symbol μ to represent the reduced mass. In the Schrödinger equations for the harmonic oscillator and hydrogen atom, we are interested in relative motions of particles. For example, in the hydrogen atom, we solve for the wavefunction of the electron relative to the nucleus; we ignore the motion of the hydrogen atom through space. However, both the electron and the nucleus move relative to the center of mass of the hydrogen atom. Because of this, we must use the reduced mass of the two particles in the Schrödinger equation. The reduced mass μ for two particles is

$$\frac{1}{\mu} = \frac{1}{m_1} + \frac{1}{m_2} \tag{9.14}$$

Because the proton is nearly 2000 times more massive than the electron, for a hydrogen atom, the reduced mass is nearly equal to the mass of an electron.

III. Solving for the eigenfunctions and for the eigenvalues of energy

Solve the Schrödinger equation, using standard methods of solving differential equations. Obtain the set of wavefunctions ψ_n and eigenvalues E_n that satisfy the Schrödinger equation and the boundary conditions. We simply present some of the solutions that were obtained. It is relatively simple to convince yourself that they are valid; simply substitute the solutions into the Schrödinger equation and show that they satisfy it exactly.

IV. Interpretation of the wavefunctions

The interpretation of the solutions to the Schrödinger equation involves comparison with available experimental data on the system. A great variety of properties can be tested. Some of the most important of these are the following:

1. Energies (eigenvalues)—comparison with values obtained from spectroscopy, ionization potentials, electron affinities, bond dissociation energies, and so on

2. Electron distribution (eigenfunctions)—comparison with atomic and molecular dimensions, dipole moments, probabilities of transition from one state to another, directional character of bonding, and intermolecular forces

3. Systematic changes for different but related systems—comparison of spectra, ionization potentials, bond lengths, bond energies, force constants, dipole moments, and so on, for atoms or molecules that differ from one another in a simple manner

Particle in a Box

The particle-in-a-box problem is the simplest that can be solved using the wave equation. In this problem, the *particle*, which may be an electron, a nucleus, or even a baseball, is considered to move freely ($U = 0$) within a defined region of space but is prohibited ($U = \infty$) from appearing outside that region. The problem can be solved readily in one, two, or three dimensions and for any shape of rectangular box. The solution of the particle-in-a-box Schrödinger equation reveals some of the basic ideas of wave mechanics, which are applicable to more complex problems.

Some biologically relevant applications of this problem include the behavior of π electrons that are delocalized over large portions of molecules, as in linear polyenes (carotenoids, retinal), planar porphyrins (heme, chlorophyll), and large aromatic hydrocarbons; conduction electrons that may move over extensive regions of biopolymers; and the exchange of protons involved in hydrogen bonding between two nucleophilic atoms, as between the oxygens of adjacent water molecules. Even in cases where the real potential well is not infinite and square, the particle-in-a-box calculation is a useful approximation for obtaining the order of magnitude of the energies involved.

The potential energy for a particle in a one-dimensional box,

$$U = 0 \qquad\qquad (0 < x < a)$$

$$U = +\infty \qquad\qquad \begin{cases} (x \le 0) \\ (x \ge a) \end{cases}$$

is shown diagrammatically in figure 9.7. The solution of the Schrödinger equation outside the box is trivial; because the particle cannot exist there, its wavefunction must be zero everywhere outside the box. Within the box, it must satisfy the one-dimensional Schrödinger equation for $U \equiv 0$:

$$-\frac{\hbar^2}{2m}\frac{d^2\psi_n(x)}{dx^2} = E_n\psi_n(x) \tag{9.15}$$

Rather than present the rigorous solution to this second-order differential equation by conventional methods, which may be unfamiliar to you, let's look at the requirements for a wavefunction that solves Eq. (9.15). It must be a function of x such that its second derivative, $d^2\psi/dx^2$, is equal, apart from

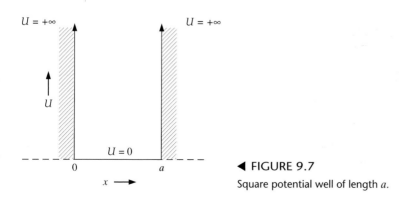

◀ **FIGURE 9.7**

Square potential well of length a.

constant factors, to the original function ψ. Two such functions are $\sin bx$ and $\cos bx$, since

$$\frac{d^2}{dx^2}(\sin bx) = -b^2 \sin bx$$

and

$$\frac{d^2}{dx^2}(\cos bx) = -b^2 \cos bx$$

In fact, a general solution can be written in the form

$$\psi = A \sin bx + B \cos bx$$

We can immediately simplify the solution by applying the continuity condition (I.B.3 in the previous section). We know that the wavefunction must be zero at the boundary of the box (since it is zero everywhere outside); therefore, $\psi(0) = \psi(a) = 0$. When $x = 0$, $\sin bx = 0$, but $\cos bx = 1$. Therefore, we must set $B = 0$ in the equation above to satisfy the boundary condition at $x = 0$. Thus,

$$\psi = A \sin bx$$

To satisfy the boundary condition at $x = a$, $\sin bx$ must $= 0$ when $x = a$. We know that $\sin n\pi = 0$ for any integer n; therefore, the function

$$\psi_n = A \sin \frac{n\pi}{a}x \qquad (n = 1, 2, 3, 4, \ldots)$$

is the only one that satisfies the second part of the boundary conditions as well. Here, a is the length of the box, and n can be any positive integer; it is the quantum number for the problem. Note that the solution for $n = 0$ requires that $\psi_0 = 0$, which is not meaningful because it is associated with zero probability for finding the particle in the box. We obtain A by using the normalization condition:

$$A^2 \int_0^a \sin^2 \frac{n\pi x}{a} \, dx = 1$$

Using a table of integrals, we obtain $A = \sqrt{2/a}$. We now have the wavefunction for a particle in a box in its ground ($n = 1$) and excited ($n = 2, 3, 4, \ldots$) states:

$$\psi_n = \sqrt{\frac{2}{a}} \sin \frac{n\pi x}{a} \qquad (9.16)$$

The solutions to any Schrödinger equation—the eigenfunctions—have the following properties: They are *orthogonal*, and it is convenient to *normalize* them. Two functions are orthogonal when the integral of their product is zero. For a particle in a box, it is easy to show that any two states are orthogonal,

$$\int_0^a \psi_n \psi_m \, dx = 0 \qquad (n \neq m)$$

and that all states are normalized,

$$\int_0^a \psi_n^2 \, dx = 1 \qquad (\text{any } n)$$

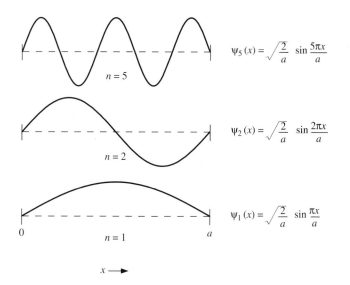

$$\psi_5(x) = \sqrt{\frac{2}{a}} \sin \frac{5\pi x}{a}$$

$$\psi_2(x) = \sqrt{\frac{2}{a}} \sin \frac{2\pi x}{a}$$

$$\psi_1(x) = \sqrt{\frac{2}{a}} \sin \frac{\pi x}{a}$$

◀ FIGURE 9.8

Variation of $\psi_n(x)$ with x for $n = 1$, 2, and 5. The magnitude of $\psi_n(x)$ is plotted versus x. The extreme values of $\psi_n(x)$ are $\pm\sqrt{2/a}$.

The correct wavefunctions for any two different states of a molecule are orthogonal: The orthogonality condition is often used as a criterion to test approximate wavefunctions.

Figure 9.8 shows the behavior of ψ versus x for several values of n. (In each case, the negative of the ψ shown is also a valid solution.) The analogy of a standing wave for the classical vibrating string is worth noting here. The violin string or a rope fixed at both ends can be set into motion as a standing wave with nodes (points where the string has zero displacement) only at the ends or with additional nodes between the ends. These are called the *fundamental mode* ($n = 1$), the *first harmonic* or *overtone* ($n = 2$), the *second overtone* ($n = 3$), and so on. The requirements that the ends are fixed and undergo no displacement are the boundary conditions that limit standing waves to these particular (integral) modes. The properties of the harmonic progression arise naturally from the constraints on the system, just as the quantum numbers arise naturally in wave mechanics from the boundary conditions characteristic of the particular problem. Simply said, quantization arises from the boundary constraints on the system.

The solution of the Schrödinger equation also gives us the energies for the particle in a box. Having determined the eigenfunctions [Eq. (9.16)] that constitute the solutions to the Schrödinger equation for this problem, we proceed to substitute them into Eq. (9.15):

$$-\frac{\hbar^2}{2m} \frac{d^2}{dx^2}\left(A \sin \frac{n\pi x}{a}\right) = E_n\left(A \sin \frac{n\pi x}{a}\right)$$

$$\frac{\hbar^2}{2m} \frac{n^2\pi^2}{a^2}\left(A \sin \frac{n\pi x}{a}\right) = E_n\left(A \sin \frac{n\pi x}{a}\right)$$

from which we can readily extract the eigenvalue solutions

$$E_n = \frac{\hbar^2\pi^2 n^2}{2ma^2} = \frac{h^2 n^2}{8ma^2} \qquad (n = 1, 2, 3, \ldots) \qquad \textbf{(9.17)}$$

Each eigenfunction has an associated energy, or eigenvalue, determined in part by the quantum number n; the energy increases with the square of the

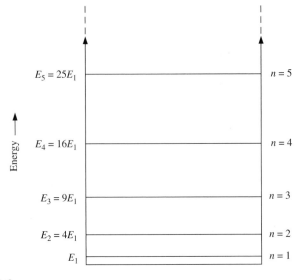

▲ **FIGURE 9.9**

Eigenvalues of energy for a particle in a one-dimensional potential well.

quantum number n, as illustrated in figure 9.9. The lowest energy state, $n = 1$, is called the *ground state*. All higher-energy states ($n = 2, 3, 4, \ldots$) are called *excited states*. Apart from the fundamental constants that enter into Eq. (9.17), the energy depends on m and a. The spacing of the energy levels depends inversely on the mass of the particle and inversely on the square of the length of the box. A light particle such as an electron will have widely spaced energy levels compared with a heavy particle like a nucleus or a baseball. A particle in a small box will have widely spaced eigenvalues compared with the same particle in a larger box.

The particle-in-a-box energies explain why we must be concerned with quantum behavior for electrons in atoms and molecules. The energy-level spacings are greater for light particles (electrons) in small regions of space (atoms or molecules). For a baseball in a room or an electron traversing a vacuum tube, the eigenvalues are so closely spaced that they may be considered to provide a continuum of energy. This is just the approach used by classical physics, and it is entirely appropriate for problems of large dimensions. The classical view of macroscopic particles is a limiting behavior of quantum systems—in this case, in the limit of large mass, large box size or large quantum number. The wave mechanical analysis allows us to determine just when the classical picture fails and gives us a method of dealing with truly submicroscopic problems.

EXAMPLE 9.1

a. Calculate the energies of the two states of lowest energy for an electron in a one-dimensional box of length 2.0 Å.

b. What is the probability that the electron is within 0.5 Å of the center of the box in the lowest energy state?

c. Repeat the calculation of part (b) for the second-lowest energy state.

Solution

a. Using Eq. (9.17), we obtain

$$E_1 = \frac{(6.626 \times 10^{-34} \text{ J s})^2 (1)^2}{8(9.11 \times 10^{-31} \text{ kg})(2.0 \times 10^{-10} \text{ m})^2} = 1.506 \times 10^{-18} \text{ J}$$

$$E_2 = (1.506 \times 10^{-18})(2)^2 = 6.024 \times 10^{-18} \text{ J}$$

b. The probability of finding the electron between $x = \frac{1}{4}a$ and $x = \frac{3}{4}a$ is obtained by evaluating Eq. (9.7) for the appropriate wavefunction (figure 9.8):

$$\int_{a/4}^{3a/4} \psi_1^2(x)\, dx = \frac{2}{a} \int_{a/4}^{3a/4} \sin^2 \frac{\pi x}{a}\, dx$$

$$= \frac{2}{a} \left(\frac{x}{2} - \frac{a}{4\pi} \sin \frac{2\pi x}{a} \right) \Big|_{a/4}^{3a/4} \qquad \text{(see summary at end of chapter)}$$

$$= \frac{2}{a} \left(\frac{3a}{8} - \frac{a}{4\pi} \sin \frac{3\pi}{2} - \frac{a}{8} + \frac{a}{4\pi} \sin \frac{\pi}{2} \right)$$

$$= 2 \left(\frac{1}{4} + \frac{1}{2\pi} \right) = \frac{1}{2} + \frac{1}{\pi} = 0.818$$

Note that the probability of finding a particle in the central half (or any other fraction) of the box is independent of the length of the box.

c. For $n = 2$,

$$\int_{a/4}^{3a/4} \psi_2^2(x)\, dx = \frac{2}{a} \int_{a/4}^{3a/4} \sin^2 \frac{2\pi x}{a}\, dx$$

$$= \frac{2}{a} \left(\frac{x}{2} - \frac{a}{8\pi} \sin \frac{4\pi x}{a} \right) \Big|_{a/4}^{3a/4}$$

$$= \frac{2}{a} \left(\frac{3a}{8} - \frac{a}{8\pi} \sin 3\pi - \frac{a}{8} + \frac{a}{8\pi} \sin \pi \right)$$

$$= 0.500$$

This result is exactly what you expect if you inspect the behavior of $\psi_2(x)$ in figure 9.8. The area under the central half of a plot of $\psi_2(x)$ will be exactly equal to the sum of the area under the two outer quarters. It is also easy to see that the probability that the electron will be in the central portion of the box is greater in the state $n = 1$ than for the state $n = 2$, again confirmed by our calculation.

Example of a particle-in-a-box calculation. The eigenfunctions and eigenvalues for the particle-in-a-box problem are more than mathematical abstractions. To appreciate their physical significance, we choose to look at the family of carotenoids, which are examples of linear polyenes. An example is β-carotene, whose molecular structure is as follows:

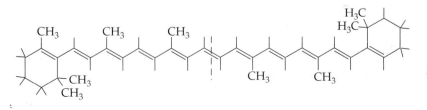

It is a symmetric molecule containing 40 carbon atoms. If the molecule is oxidized, it breaks at the center (dashed line) and forms two molecules of retinal, or vitamin A_1, which is closely related to the pigment of rhodopsin. Many isomers and chemically modified derivatives of β-carotene occur in nature. They play important roles in vision, as sensitizers in photosynthesis, and as protective agents against harmful biological oxidations.

β-Carotene and the other carotenoids possess two classes of bonding electrons. For the present, consider one kind, the sigma (σ) electrons, to be responsible for the single bonds and for half of each double bond. These σ electrons are localized in the small regions of space between adjacent atoms. The other type, the pi (π) electrons, contribute the other half of each double bond. In systems like β-carotene or benzene, where double and single bonds alternate in the classical molecular structure, the π electrons are not strongly confined to regions between adjacent atoms. In fact, they are nearly free to roam the full length of the conjugated system. Thus, they are confined approximately in a one-dimensional potential well that is about as long as the conjugated system and within which the potential energy is virtually constant. The particle in a box provides a reasonable model for predicting the eigenfunctions and eigenvalues for these π electrons. Having adopted this model, we can now proceed to examine its predictions and compare them with physically measurable properties.

β-Carotene has 11 conjugated double bonds, and thus has 22 π electrons. We will assume that the energies and wavefunctions for these 22 electrons can be approximately described by the energies and wavefunctions of a particle in a box. The energy-level pattern shown in figure 9.9 can be extended to any value of n. The electron configuration of the lowest-energy state for the 22 π electrons of β-carotene is obtained by placing the electrons in the levels or orbitals of lowest available energy, such that each orbital has two electrons with opposite spins. This causes 11 orbitals ($n = 1$ to 11) to be completely filled and just uses up the 22 π electrons. If we could somehow measure the energy of the electrons in the $n = 11$ level, we could use Eq. (9.17) to calculate a, the length of the potential well, and compare it with the sum of bond lengths for the conjugated system. There is in fact no good way to measure directly the energy of the bound electrons in a molecule such as β-carotene. We can, however, relate this model to an *energy difference* that can be measured from the absorption spectrum of the molecule.

When light of the appropriate wavelength is incident on a molecule, it may absorb some of the radiation, resulting in the promotion or excitation of electrons to higher-energy orbitals. The process occurs only if the energy of the incident photons corresponds to the difference between two possible states (eigenstates) of the molecule. Because there are many filled and many empty states available, the absorption spectrum is potentially a rich source of information about the location of the energy levels. In practice, the lowest-energy transition, corresponding to the longest-wavelength electronic absorption band, is the easiest to identify and observe. The lowest-energy absorption occurs when an electron in the highest filled energy level is excited to the lowest unfilled energy level. This is shown in figure 9.10 for that segment of the energy-level diagram. Thus, the energy associated with the photons absorbed in the long-wavelength absorption band is just the difference between energy level $n = 11$ and level $n = 12$. As we will see, this is just as useful as knowing E_{11} or E_{12} for the purposes of making an estimate of the length a.

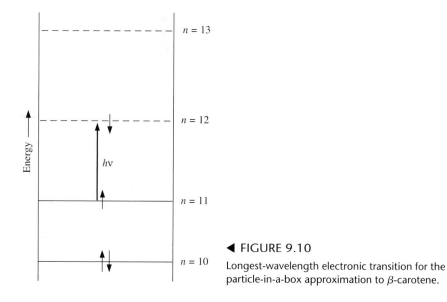

◀ FIGURE 9.10

Longest-wavelength electronic transition for the particle-in-a-box approximation to β-carotene.

Using Eq. (9.17), we can easily derive a general expression for the energy difference between adjacent levels for the particle-in-a-box problem. Let N represent the quantum number of the highest filled level; then $N + 1$ is the quantum number for the lowest unfilled level. The energy difference between these two levels is

$$\Delta E = E_{N+1} - E_N = \frac{h^2}{8ma^2}(N + 1)^2 - \frac{h^2}{8ma^2}N^2$$

$$= \frac{h^2}{8ma^2}(N^2 + 2N + 1 - N^2) = \frac{h^2}{8ma^2}(2N + 1)$$

Because

$$\Delta E = \frac{hc}{\lambda} \quad > \frac{h^2}{8ma}$$

we obtain the result

$$\lambda(N \rightarrow N + 1) = \frac{8mca^2}{h(2N + 1)} \qquad \textbf{(9.18)}$$

For linear polyenes with different extents of conjugation, both the length a and the number of π electrons ($2N$) will be different. The wavelength increases in proportion to the square of the length a and inversely with the number of π electrons. Since the length of the π system increases *linearly* with the number of π bonds, the long-wavelength absorption band increases in wavelength (shifts toward the red) with increasing extent of conjugation.

The long-wavelength absorption of β-carotene is shown in figure 9.11. The structure of the band actually arises from additional vibrational effects and need not concern us here. The long-wavelength maximum is readily identified and occurs at about 480 nm. By substitution into Eq. (9.18) and solving for a, we obtain for $N = 11$,

$$a^2 = \frac{h\lambda(2N + 1)}{8mc} = \frac{(6.626 \times 10^{-34}\,\text{J s})(4.8 \times 10^{-7}\,\text{m})(23)}{8(9.11 \times 10^{-31}\,\text{kg})(3.0 \times 10^8\,\text{m s}^{-1})}$$

$$a = 1.83 \times 10^{-9}\,\text{m} = 18.3\,\text{Å}$$

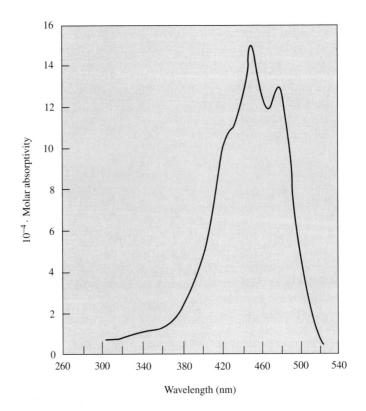

▶ **FIGURE 9.11**

Absorption spectrum of all-*trans* β-carotene.

This is an effective length for the one-dimensional free-electron model of β-carotene.

The actual length of the zigzag chain can also be calculated from the sum of the lengths of the single and double bonds. For such a conjugated system, the length of a single bond is approximately 1.46 Å, and the length of a double bond is about 1.35 Å. There are 11 double bonds and 10 single bonds for the conjugated chain in β-carotene; therefore, the calculated length is 10 × 1.46 Å + 11 × 1.35 Å, or 29 Å. The actual length and effective length are different for several reasons. The actual molecular potential does not have a flat bottom and infinitely steep sides. There are also important electron–electron repulsions. Nevertheless, if we compare the spectra of carotenoids with different lengths of conjugation, there is a very good correlation with the form of Eq. (9.18).

Particles in two- and three-dimensional boxes can be solved just like those in the one-dimensional box. An electron in a two-dimensional box provides a reasonable model for discussing the spectrum of such molecules as benzene, naphthalene, and anthracene. These molecules have conjugated double bonds in cyclic ring structures. A particle in a three-dimensional box accurately describes a dilute gas in a real container. For a box of sides a, b, and c, the wavefunctions are

$$\psi(x, y, z) = \sqrt{\frac{8}{abc}} \sin \frac{n_x \pi x}{a} \sin \frac{n_y \pi y}{b} \sin \frac{n_z \pi z}{c} \qquad (9.19)$$

The energies are

$$E = \frac{h^2}{8m} \left(\frac{n_x^2}{a^2} + \frac{n_y^2}{b^2} + \frac{n_z^2}{c^2} \right) \qquad (9.20)$$

These energies correspond to the translational energies of an ideal gas. They can be used to derive the ideal gas equation and to predict when quantum effects will cause deviations from the classical ideal gas behavior. These quantum effects can become important for He gas at temperatures near absolute zero.

Particles in two- and three-dimensional boxes introduce the concept of *degeneracy*. Whenever several linearly independent wavefunctions have the same energy, the wavefunctions are said to be degenerate. For a particle in a three-dimensional cubical box, different values of n_x, n_y, n_z—and therefore different wavefunctions—can have the same energy. For example, the three different wavefunctions with quantum numbers, $n_x, n_y, n_z = 2, 1, 1,$ or $= 1, 2, 1$ or $= 1, 1, 2$ all have the same energy. The energy level is threefold degenerate. We find that for the hydrogen atom the lowest-energy state ($1s$) is nondegenerate, but the $2s$, $2p_x$, $2p_y$, and $2p_z$ orbitals all have the same energy; there is a fourfold degeneracy.

Using a modified coordinate system, it is possible to describe the motion of an electron on a circular ring as a one-dimensional particle-in-a-box problem. Here, the box has no ends with infinite potential; the potential is zero everywhere on the circumference of the ring. The continuity condition requires that the wavefunctions are continuous with themselves for each 2π rotation about the ring. Some additional states (degeneracies) arise because the particle can move either clockwise or counterclockwise around the ring with wavefunctions having the same energy. The electron-on-a-ring model has been used as a model for the behavior of the π electrons in large aromatic molecules such as porphyrins, heme, and chlorophyll. For more compact molecules such as CH_4 or CCl_4, it is possible to treat the electrons as particles in a spherical (three-dimensional) potential well, but this becomes less useful where the distribution of nuclear charge is highly nonuniform.

The infinite potential well is a convenient abstraction, but in real systems the heights of the barriers are always finite. This leads to the important phenomenon of *tunneling* of the particle into the barrier and even out the other side. Tunneling of electrons and even protons is currently of interest in many biological and biochemical systems. It has been studied with respect to electron transport associated with respiration and photosynthesis and with respect to proton translocation across membranes driven by molecular pumps, such as bacteriorhodopsin.

Tunneling

We discussed a particle in a box in detail because it is the simplest quantum mechanical problem to solve. From now on, we give only the results of solving Schrödinger's equation, and we discuss the applications of the solutions. The box we considered has walls that are infinite in height; thus, the particle cannot escape. Its wavefunction is zero outside the box. What happens for a box with walls of finite height? Classically, the particle can only escape if its energy is equal to or greater than the energy of the walls. However, quantum mechanics shows that the particle can escape even if its energy is less than the repulsion of the walls. The wavefunction has a nonzero value outside the walls. The particle is said to *tunnel* through the barrier. The probability of tunneling depends on the energy and mass of the particle and on the height and width of the barrier. The mass of the particle is important because, as we stated earlier, de Broglie found that the wavelength of a particle is inversely proportional to its momentum, mv. Because the kinetic energy is equal to $\frac{1}{2}mv^2$, we can combine these two equations to obtain

$$\lambda = \frac{h}{(2mE)^{1/2}} \tag{9.21}$$

We see that for a constant energy, the wavelength of the wavefunction of the particle increases as the mass of the particle decreases. The wavelength of a 20 kJ mol^{-1} electron is 27 Å; the wavelength of a proton of the same energy is only 0.63 Å.

Electron tunneling occurs in many oxidation–reduction reactions catalyzed by enzymes. Rapid electron transfer between donor and acceptor sites separated by 10–20 Å in a protein has been established (Axup et al. 1988). Proton tunneling can also occur. Consider the energy barrier that separates reactants from products for a proton transfer from one carbon to another, as shown in figure 9.12. An example is the oxidation of ethanol by the enzyme liver alcohol dehydrogenase to form acetaldehyde. The proton is transferred from the alcohol to the enzyme cofactor nicotinamide adenine dinucleotide, NAD$^+$ (the structure is given in appendix A.9):

$$CH_3CH_2OH + NAD^+ \longrightarrow CH_3HC{=}O + NADH + H^+$$

Classically, the reactants need to have an energy equal to or larger than the activation energy to react, but the proton can sneak through the activation barrier by tunneling instead of going over it. This speeds the reaction. It is not possible to calculate absolute rates of reaction for the classical model or the quantum mechanical model, but it is possible to calculate relative rates for a triton (T = ^{3}H), versus a deuteron (D = ^{2}H), and versus a proton (H = ^{1}H). The only difference in the reactions is the mass of the tunneling particle; all the electronic distributions and the intermolecular interactions are the same. By experimentally measuring the kinetic isotope effects, k_H/k_D, k_H/k_T, and k_D/k_T, Klinman and her co-workers found that proton tunneling occurs in several different alcohol dehydrogenases (Klinman 1989; Kohen and Klinman 1999). The effects are large. For the oxidation of benzyl alcohol to benzaldehyde by yeast alcohol dehydrogenase, a $k_H/k_T = 7.13 \pm 0.07$ was measured;

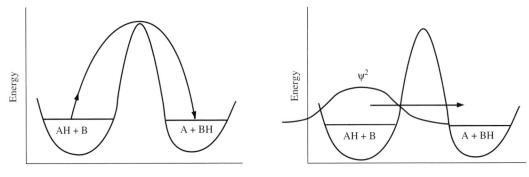

Reaction coordinate
Classical crossing of transition state

Reaction coordinate
Quantum mechanical tunneling

▲ FIGURE 9.12

The transfer of a proton in a reaction can occur by tunneling through the barrier, instead of by going over the barrier. The square of the wavefunction of the proton on the reactant illustrates that there is a small probability that the proton is found on the product instead of the reactant. This speeds the reaction for protons (H) more than for the heavier deuterons (D) or tritons (T).

without tunneling the ratio would be only $k_H/k_T = 5.91 \pm 0.20$. Clearly quantum mechanical tunneling can have a major effect on the rates of reactions involving protons (and by extrapolation an even larger effect on electron transfer). Tunneling becomes negligible for larger particles. A qualitative understanding of the phenomenon is sufficient for our purposes.

Simple Harmonic Oscillator

A simple harmonic oscillator describes the stretching and compression of bonds. The harmonic oscillator is two masses connected by an ideal spring, with force constant (or stiffness) k. From Hooke's law, the force exerted by the spring on the masses is $F = -kx$, where x is the displacement from an equilibrium position (in our molecular case, the equilibrium bond length). The quantum mechanics of a simple harmonic oscillator provides a clearer understanding of the vibration of molecules. Each oscillator has a fundamental vibration frequency, ν_0, which depends on the strength of the bond and the masses attached to the bond. A harmonic potential energy can be written in the form

$$U = \frac{1}{2}kx^2 \tag{9.22}$$

where k is the force constant for the bond and the origin of the coordinate system is chosen at the equilibrium bond length. The functional form is a parabola.

The harmonic oscillator Schrödinger equation is thus

$$-\frac{\hbar^2}{2\mu}\frac{d^2\psi_v}{dx^2} + \frac{1}{2}kx^2\psi_v = E_v\psi_v \tag{9.23}$$

We use the subscript v for the vibrational quantum number. Note that the potential function is very different than the simple square well of the particle in a box. This differential equation can be solved by standard methods. Some of the wavefunctions and their corresponding energies are tabulated in table 9.1.

TABLE 9.1   Some Eigenfunctions and Eigenvalues for a Simple Harmonic Oscillator*

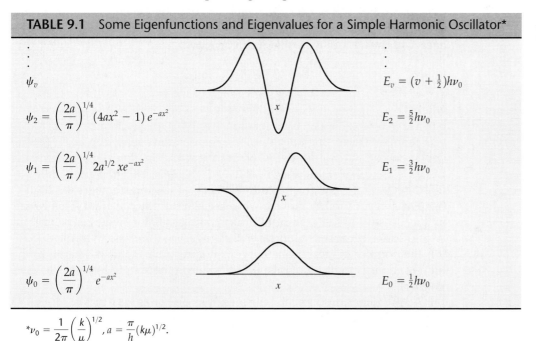

ψ_v		$E_v = (v + \frac{1}{2})h\nu_0$
$\psi_2 = \left(\dfrac{2a}{\pi}\right)^{1/4}(4ax^2 - 1)\,e^{-ax^2}$		$E_2 = \frac{5}{2}h\nu_0$
$\psi_1 = \left(\dfrac{2a}{\pi}\right)^{1/4}2a^{1/2}\,xe^{-ax^2}$		$E_1 = \frac{3}{2}h\nu_0$
$\psi_0 = \left(\dfrac{2a}{\pi}\right)^{1/4}e^{-ax^2}$		$E_0 = \frac{1}{2}h\nu_0$

$*\nu_0 = \dfrac{1}{2\pi}\left(\dfrac{k}{\mu}\right)^{1/2}, a = \dfrac{\pi}{h}(k\mu)^{1/2}.$

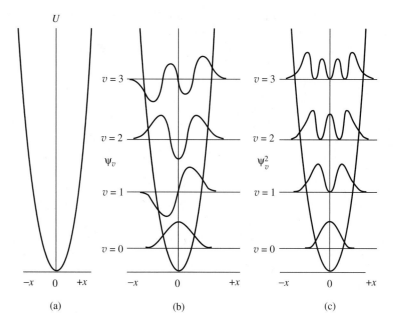

▶ FIGURE 9.13

(a) Potential well for a classical harmonic oscillator. (b) Allowed energy levels and wavefunctions for a quantum mechanical harmonic oscillator. $\psi_v(x)$ is plotted versus x; each different wavefunction is arbitrarily displaced vertically. (c) Probability functions for a quantum mechanical harmonic oscillator. The square of each wavefunction $[\psi_v(x)]^2$ is plotted versus x.

In figure 9.13, the wavefunctions, $\psi_v(x)$ and their squares, $\psi_v^2(x)$, which give the probability of finding the particle at x, are plotted for several quantum numbers.

It is interesting to compare the quantum mechanical results with the classical mechanical results. First, the energy is quantized into energy levels, as is true in quantum mechanics, except for continuum wavefunctions. The energy of the harmonic oscillator can be only

$$E_v = \left(v + \frac{1}{2} \right) h\nu_0 \qquad (9.24)$$

where the quantum number v is a positive integer or zero. As opposed to the particle in a box, the energy levels are evenly spaced with increasing quantum number. According to classical mechanics, the energy can be any value depending on the amplitude. A look at the wavefunction suggests similarities to classical physics. Recall that $[\psi_v(x)]^2$ represents the probability of finding a particle at a given position. As the quantum numbers get higher, the peaks in probability move toward the sides of the potential. This would be the classical turning point of a harmonic oscillator (where a pendulum swings back); because the kinetic energy is zero at the classical turning point, it is the point of highest probability for finding a harmonic oscillator. In the limit of higher quantum numbers, the quantum harmonic oscillator mimics classical behavior.

At low quantum numbers, the quantum system is very different. Classical mechanics allows the oscillator to be at rest and therefore have an energy equal to zero, but according to quantum mechanics, the lowest energy level is $\frac{1}{2}h\nu_0$, called the *zero-point energy*. Both quantum mechanics and classical mechanics give the same fundamental vibration frequency ν_0 of the oscillator, however.

$$\nu_0 = \frac{1}{2\pi}\left(\frac{k}{\mu}\right)^{1/2} \tag{9.25}$$

Here, μ is the reduced mass of the oscillator [Eq. (9.14)], and k is the force constant for the bond. It is also true that when v is large or when k/μ is small, the quantum mechanics and classical mechanics pictures become very similar.

The presence of zero-point energy states that the quantum harmonic oscillator cannot be at rest. This is just a restatement of the Heisenberg uncertainty principle: If the harmonic oscillator is at rest, we know the position (the origin) and momentum (zero) precisely. Zero-point energy ensures that the uncertainty principle is not violated; it requires that molecules and atoms bound in a crystal are always vibrating, even at very low temperature.

The quantum mechanical harmonic oscillator is used in interpreting infrared spectra that correspond to molecular vibrations (see chapter 10). The difference in energy corresponds to a change in vibrational quantum number $\Delta v = 1$:

$$\Delta E = h\nu_0$$

Therefore, light will be absorbed if its frequency ν is equal to a vibrational frequency of a harmonic oscillator. A harmonic oscillator is a good approximation for the vibration of diatomic molecules and the vibration of bonds in polyatomic molecules. Vibrational frequencies are characteristic of a particular type of bond such as $C-H$, $C-C$, $C=C$, and $C-N$, but they can also be used to determine force constants k for the particular vibration. For example, when the force constant for a double-bond vibration decreases (the frequency decreases), it means that the bond has become weaker; the double-bond character has decreased. These differences in bond strength for the same type of bond in different molecules are a useful clue to the distribution of electrons in the molecules. An important application of infrared vibrational spectra is the study of hydrogen bonding. The spectra of $O-H$ and $N-H$ vibrations are very different for the hydrogen-bonded and non-hydrogen-bonded species.

EXAMPLE 9.2

Calculate the fundamental vibrational frequency of carbon monoxide, CO, considering it to behave like a harmonic oscillator with a force constant 1902.5 N m^{-1}. By how much will this frequency change if the molecule contains the isotope ^{13}C instead of the more abundant ^{12}C?

Solution

For the principal isotopic species $^{12}C^{16}O$, the expression for the reduced mass is

$$\mu = \frac{m_C m_O}{m_C + m_O} = \left[\frac{(12.0)(16.0)}{28.0} \times \frac{1}{6.023 \times 10^{26}}\right]\frac{kg}{molecule}$$

$$= 1.1385 \times 10^{-26} \text{ kg molecule}^{-1}$$

According to Eq. (9.25),

$$\nu_0 = \frac{1}{2\pi}\left(\frac{1902.5 \text{ N m}^{-1}}{1.1385 \times 10^{-26} \text{ kg}}\right)^{1/2} = 6.506 \times 10^{13} \text{ s}^{-1}$$

Infrared spectroscopists usually express this in wavenumbers $\bar{\nu}_0$:

$$\bar{\nu}_0 = \frac{\nu_0}{c} \qquad (\text{where } c = \text{velocity of light})$$

$$= \frac{6.506 \times 10^{13} \text{ s}^{-1}}{2.9979 \times 10^{10} \text{ cm s}^{-1}} = 2170 \text{ cm}^{-1}$$

If the molecule contains ^{13}C instead of ^{12}C, the reduced mass will be altered, but not the force constant. (The force constant is the measure of the stiffness or strength of the electronic bond holding the nuclei together.) Thus,

$$\frac{\nu_0(^{13}C^{16}O)}{\nu_0(^{12}C^{16}O)} = \left[\frac{\mu(^{12}C^{16}O)}{\mu(^{13}C^{16}O)}\right]^{1/2}$$

$$\nu_0(^{13}C^{16}O) = \left[\frac{(12.0)(16.0)}{28.0} \times \frac{29.0}{(13.0)(16.0)}\right]^{1/2} (2170 \text{ cm}^{-1})$$

$$= (0.9778)(2170 \text{ cm}^{-1}) = 2122 \text{ cm}^{-1}$$

This difference is readily detected in the infrared spectrum.

Rigid Rotator

Another mode of molecular motion for which quantum mechanics provides important insights is rotation. For gas-phase molecules at low pressure, rotation is relatively free, at least during the time between collisions with other molecules or with the walls. A simple model that reasonably describes such rotational motion is that of the free rigid rotator, a collection of masses (atoms) at fixed distances and relative orientation. This problem can readily be solved using the methods just illustrated for the harmonic oscillator, but using a potential function ($U = 0$) and kinetic energy appropriate to a freely rotating system. The solution is given in standard physical chemistry texts. Despite the fact that the potential is zero, rotational motion is quantized. The boundary condition in this case is that after a single rotation the solution to the wave equation must repeat itself; as stated above, it must be single valued. A set of quantized energy levels is obtained for a linear molecule, with $E_{rot} = J(J + 1)h^2/2I$, where I is the moment of inertia of the molecule and J is a rotational quantum number that takes on integer values $J = 0, 1, 2, 3, \ldots$. The moment of inertia of a diatomic molecule of reduced mass μ is $I = \mu r_0^2$, where r_0 is the equilibrium internuclear distance. The energy levels are not uniformly spaced, by contrast with the harmonic oscillator energy levels.

Furthermore, for all but the lightest molecules, rotation-energy-level spacings are much smaller than those for typical vibrations. The rotational spacings are small compared with thermal energy kT at room temperature, which means that many rotational levels are well populated in gases under these conditions. This means that gas-phase molecules will be distributed over many rotational levels at room temperature. Because of the wide spacing of vibrational levels, most molecules will be in one, or only a few, vibrational levels at room temperature. The spacing of energy levels has important effects on the statistical interpretation of energy and entropy, as described in chapter 11.

The situation for liquids or solids is very different. Here, the model of a freely rotating molecule is not appropriate because, as soon as an individual molecule in a condensed phase begins to rotate, it bumps into its neighbors or experiences a restoring force that changes its direction. Such interrupted or irregular motion, called *libration*, does not produce the sharply defined quantized energy levels characteristic of gas-phase molecules. In liquids and to some extent in solids, the rocking motions or librations that do occur result in a broadening out or smearing of the sharp electronic and vibrational energies. As a consequence, the spectra of molecules in the condensed phase are usually much broader and lack fine resolution compared with gas-phase spectra. There are some exceptions in highly ordered crystals where molecules in a uniform environment may exhibit quite sharp spectra.

Librational motion in pure liquids and solutions is still an important energy reservoir. It reflects the nature of intermolecular forces and local molecular ordering, and it is an important contributor to the motions that lead to chemical reactions.

Hydrogen Atom

The Schrödinger equation for a hydrogen atom can be solved exactly, and the solution provides concepts applicable to all atoms. A hydrogen atom consists of two particles, a nucleus of mass m_1 and charge $+e$ and an electron of mass m_2 and charge $-e$. The force acting between the particles is coulombic, and the potential energy U is equal to $-e^2/r$, where r is the distance between the particles. For SI units, $U = -e^2/4\pi\varepsilon_0 r$. The location of the electron relative to the nucleus (actually relative to the center of mass of the hydrogen atom) can be specified by a wavefunction of three coordinates. Because the potential energy depends only on r (it is spherically symmetric), the wavefunction can be written as a product of two parts; one depends only on r, the other on two angular coordinates. Upon solution of the Schrödinger equation (see the references at the end of the chapter), the wavefunctions and energy levels are obtained.

Three quantum numbers, n, l, and m, characterize the wavefunction; n characterizes the radial distance dependence, and l and m characterize the angular part of the wavefunction (note that there are two angular quantum numbers for the two angular coordinates). These quantum numbers are called the *principal quantum number n*, the *angular momentum quantum number l*, and the *magnetic quantum number m*. Because m is the quantum number for the z-component of angular momentum, l and m are interrelated. The permissible values of n are positive integers $1, 2, 3, \ldots$; the permissible values of l for a given n are zero and positive integers up to $n-1$; the permissible values of m for a given l are integers from $-l$ to $+l$ (consistent with the interrelation of l and m). The symbols s, p, d, and f are assigned to functions with $l = 0, 1, 2,$ and 3, respectively. The wavefunction for a single electron atom, characterized by a unique set of quantum numbers, is called an *orbital*.

The energy E of the hydrogen atom is quantized, as is the energy of the particle in a box, the simple harmonic oscillator, and all other systems. The eigenvalues for E for the hydrogen atom are

$$E_n = -\frac{\mu e^4}{2\hbar^2 n^2} \qquad \textbf{(9.26)}$$

With μ in g, $\hbar$ in erg s, and e in esu, energy is obtained in ergs. Note that the energy of a hydrogen atom depends only on the principal quantum number

n. All three quantum numbers, n, l, and m, characterize the electron distribution, but the energy depends solely on n. Only electrostatic interaction between the nucleus and electron determine energy in this system, which depends only on radial distance from the nucleus.

In SI units, we have for the H atom,

$$E_n = \frac{\mu e^4}{32\pi^2\varepsilon_0^2\hbar^2 n^2} \tag{9.27}$$

Now μ is in kilograms, e in coulombs, and $\hbar$ in J s; E_n is in joules. Because chemists, spectroscopists, and others each seem to use different energy units in discussing the hydrogen atom, it is convenient to write

$$E_n = -\frac{R}{n^2} \tag{9.28}$$

$R = 2.179 \times 10^{-11}$ erg molecule^{-1}
$R = 2.179 \times 10^{-18}$ J molecule^{-1}
$R = 1312$ kJ mol^{-1}
$R = 13.60$ eV molecule^{-1}

The energy is given relative to the separated proton and electron as the zero of energy; the negative value means that the hydrogen atom is stable. The values of the Rydberg constant R given are the ones most used. The first two values correspond to Eqs. (9.26) and (9.27). The value of R in kJ mol^{-1} is useful for comparison with bond energies and heats of chemical reactions. The ionization energy of an atom or molecule is the energy required to remove an electron; it is traditional to quote ionization energies in electron volts (eV). The value of R in electron volts tells us that the ionization energy of the hydrogen atom is $+13.6$ eV.

The Schrödinger equations for other single-electron species (He^+, Li^{2+}, Be^{3+}, etc.) can be solved with the potential energy equal to $-Ze^2/r$, where Z is the atomic number. The hydrogen atom itself corresponds to $Z = 1$. Several of the hydrogen-like wavefunctions are tabulated in table 9.2. The energy of a hydrogen-like orbital is dependent on the quantum number n only, and is

$$E = -\frac{Z^2\mu e^4}{2\hbar^2 n^2} = -\frac{Z^2 R}{n^2} \tag{9.29}$$

The value of R depends slightly on the mass of the nucleus because of the reduced mass μ; however, μ equals the mass of the electron to an accuracy of 0.1%.

Electron Distribution

The electron distribution in molecules is responsible for the enormous range of their properties. The chemical bonds that hold atoms together, the distribution of charge that produces a dipole moment, the forces between molecules that lead to association or to solvation, the interaction with light to produce color or to induce photochemistry—all of these depend critically on how the electrons are distributed in the molecule. Inevitably, this is a matter for quantum mechanical description; however, even the simplest molecules cannot yet be solved exactly by quantum mechanics. An important part of our task now is to develop reasonable models or approximations that will allow us to calculate molecular properties to the desired accuracy.

TABLE 9.2 Hydrogenic Wavefunctions*

n	l	m	
1	0	0	$\psi(1s) = \dfrac{1}{\sqrt{\pi}}\left(\dfrac{Z}{a_0}\right)^{3/2} e^{-Zr/a_0}$
2	0	0	$\psi(2s) = \dfrac{1}{4\sqrt{2\pi}}\left(\dfrac{Z}{a_0}\right)^{3/2}\left(2 - \dfrac{Zr}{a_0}\right) e^{-Zr/2a_0}$
2	1	0	$\psi(2p_z) = \dfrac{1}{4\sqrt{2\pi}}\left(\dfrac{Z}{a_0}\right)^{5/2} (z)\, e^{-Zr/2a_0}$
2	1	± 1	$\begin{cases} \psi(2p_x) = \dfrac{1}{4\sqrt{2\pi}}\left(\dfrac{Z}{a_0}\right)^{5/2} (x)\, e^{-Zr/2a_0} \\[2ex] \psi(2p_y) = \dfrac{1}{4\sqrt{2\pi}}\left(\dfrac{Z}{a_0}\right)^{5/2} (y)\, e^{-Zr/2a_0} \end{cases}$

*$z = r \cos\theta$ $\qquad$ $a_0 =$ Bohr radius $= \dfrac{\hbar^2}{me^2}$

$x = r \sin\theta \cos\phi$

$y = r \sin\theta \sin\phi$ $\qquad\qquad = 0.529$ Å $= 5.29 \times 10^{-2}$ nm

$r^2 = x^2 + y^2 + z^2$ $\qquad$ $Z =$ charge on nucleus

We first look at how electrons in atoms can be described using quantum mechanics. Then we consider molecules as collections of atoms having somewhat modified characteristics. We treat some electrons as localized in chemical bonds, others as delocalized over large regions or, perhaps, the entire molecule. Ultimately, the electron distribution determines what is the most stable configuration or geometry of the molecule. It also determines bond lengths, force constants, and bond angles.

Electron Distribution in a Hydrogen Atom

The square of the wavefunction $\psi(r, \theta, \phi)$ is proportional to the probability of finding the electron at a position r, θ, ϕ, where the origin of the spherical coordinates is at the nucleus (figure 9.14). The hydrogen wavefunctions and their squares represent the hydrogen atom orbitals and their respective

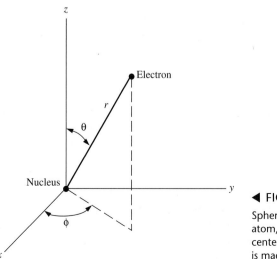

◄ FIGURE 9.14

Spherical coordinate system. For an atom, the origin should be at the center of mass, but negligible error is made by having it at the nucleus.

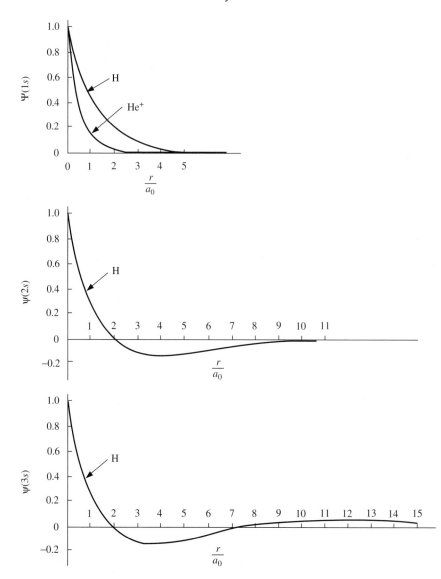

▶ FIGURE 9.15

Plots of hydrogen-like wavefunctions versus distance for spherically symmetric wavefunctions $\psi(ns)$. The distance scale has units of $a_0 = 0.529$ Å. ψ is set equal to 1 at $r = 0$. ψ can be positive or negative; the number of radial nodes ($\psi = 0$), including one at $r/a_0 = \infty$, is equal to the principal quantum number n.

probabilities of finding an electron at a given point. Hydrogen-like orbitals are used to explain orbitals in many-electron atoms and to represent electron distributions in molecules. Figure 9.15 shows the plots of $\psi(1s)$, $\psi(2s)$, and $\psi(3s)$ for a hydrogen atom and $\psi(1s)$ for a helium ion. Each ψ is a maximum at $r = 0$ (the position of the nucleus) and decreases rapidly as the distance from the nucleus increases. The ground state of the hydrogen atom ($n = 1$) has the electron closest to the nucleus. For successive excited states ($n = 2$, 3, . . .), the electron is spread out over more space away from the nucleus. Note how the $\psi(1s)$ for He^+ with a nuclear charge of $+2$ is pulled toward the nucleus relative to $\psi(1s)$ for H with a nuclear charge of $+1$.

A better feeling for the shape of the wavefunctions can be obtained from computer plots of the functions, as illustrated in figure 9.16. The pictures for $\psi(1s)$ and $\psi(2s)$ are just the first two plots in figure 9.15, rotated about the origin. The $\psi(2p_x)$ picture shows a wavefunction that is not symmetric about the origin. $\psi(2p_y)$ and $\psi(2p_z)$ are identical to $\psi(2p_x)$ but are oriented along y and z.

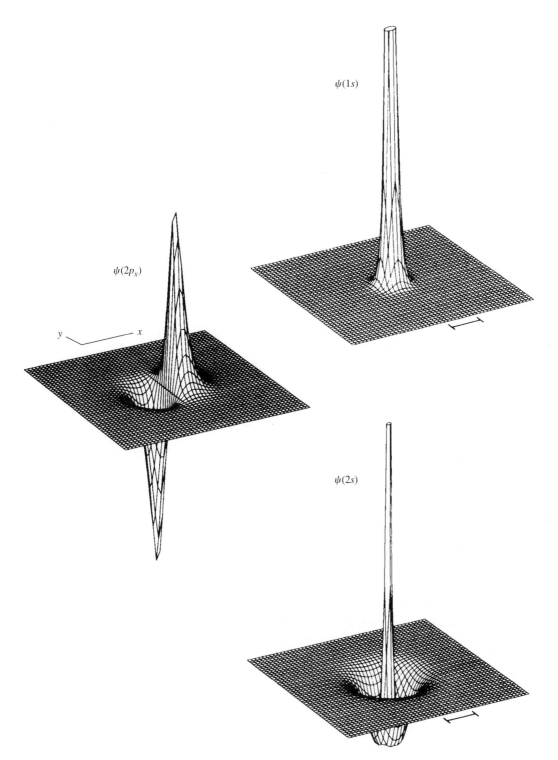

▲ FIGURE 9.16

Computer plots of $\psi(1s)$, $\psi(2s)$, and $\psi(2p_x)$. ψ is plotted in perspective in the *xy*-plane. The marked line in front of the grid is of length $10a_0 = 5.29$ Å. [From A. Streitwieser, Jr., and P. H. Owens, 1973, *Orbital and Electron Density Diagrams* (New York: Macmillan). Copyright 1973 by Macmillan Publishing Company.]

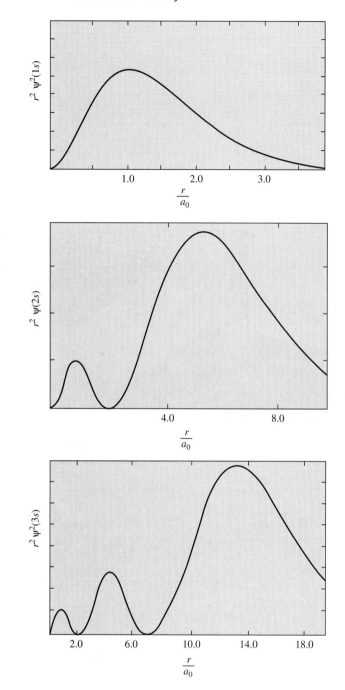

▶ **FIGURE 9.17**

Probability of finding the electron in a hydrogen atom in a spherical shell of radius r around the nucleus; $r^2\psi^2$, which is proportional to this probability, is plotted versus distance in units of $a_0 = 0.529$ Å. Note that the nodes occur at the same locations as in figure 9.15.

The wavefunctions $\psi(2s)$, $\psi(2p_x)$, $\psi(2p_y)$, and $\psi(2p_z)$ all correspond to the same energy for hydrogen atoms; $E_2 = -R/4$.

The probability of finding the electron at a distance r from the nucleus is proportional to the volume of a spherical shell of radius r around the nucleus. The volume of a spherical shell is directly proportional to r^2; therefore, the probability of finding the electron at distance r is proportional to $r^2\psi^2$. This is plotted versus r/a_0 in figure 9.17 for $\psi(1s)$, $\psi(2s)$, and $\psi(3s)$. The probability of finding the electron is a maximum at $r = a_0$ for the ground state,

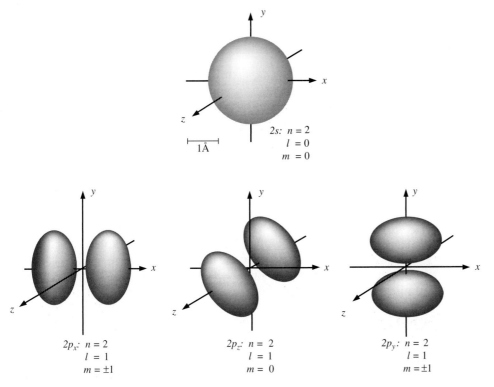

▲ FIGURE 9.18

Pictures that show where there is 90% probability of finding the electron in a hydrogen atom in (2s), (2p_x), (2p_y) orbital. [From G. C. Pimentel and R. D. Spratley, 1971, *Understanding Chemistry* (Oakland, Calif.: Holden-Day), 486.]

$\psi(1s)$, of the hydrogen atom. Bohr had deduced a circular orbit at this distance for the electron in a ground-state hydrogen atom; therefore, this distance is called the *Bohr radius*. Quantum mechanics states that the electron can be anywhere relative to the nucleus, but its most probable value occurs at a_0. For the excited state $\psi(2s)$, the most probable value of r for the electron is at $5.24a_0$.

Still another way of visualizing an electron distribution in space is shown in figure 9.18. Here, the enclosed, shaded areas represent the space where there is 90% probability of finding the electron.

EXAMPLE 9.3

For the 1s orbital of the hydrogen atom, calculate the probability of finding the electron within a distance a_0 (the Bohr radius) from the nucleus.

Solution

Taking $\psi(1s)$ from Table 9.2 and using Eq. (9.7) adapted to a spherical coordinate system, we obtain

$$\int [\psi(1s)]^2 \, dv = 4\pi \int_0^{a_0} \left[\frac{1}{(\pi a_0^3)^{1/2}} e^{-r/a_0} \right]^2 r^2 \, dr$$

$$= 4 \int_0^1 e^{-2\rho} \rho^2 \, d\rho \qquad (\text{where } \rho = r/a_0)$$

$$= -\frac{e^{-2\rho}}{2}(4\rho^2 + 4\rho + 2)\bigg|_0^1 = 1 - 5e^{-2}$$

$$= 0.323$$

Thus, the probability of finding the electron within one Bohr radius of the nucleus of the hydrogen atom in the $1s$ state is 32.3%.

So far we have not mentioned the spin of the electron. Experimental results as well as theoretical considerations indicate that an electron has an intrinsic magnetic moment and behaves as if it were a small magnet; this is exactly what would occur if the electron was spinning, as circulating charge generates a magnetic field. Thus, in addition to the quantum numbers n, l, and m, a fourth quantum number m_s, called the *spin quantum number,* is used to specify the intrinsic magnetic moment. The spin quantum number for an electron can be either $+\frac{1}{2}$ or $-\frac{1}{2}$. We have ignored the spin of the electron up to now because it has an extremely small effect on the energy levels of the hydrogen atom. However, as soon as we consider more than one electron in an atom, the effect of spin becomes important.

Many-Electron Atoms

For atoms with more than one electron, we can easily write the Schrödinger equation, but we cannot solve it exactly. Consider, for example, the helium atom, which has two electrons. The Hamiltonian operator for the two electrons is

$$\mathcal{H} = -\frac{\hbar^2}{2\mu}(\nabla_1^2 + \nabla_2^2) - \left(\frac{2e^2}{r_1} + \frac{2e^2}{r_2} - \frac{e^2}{r_{12}}\right)$$

where μ, the reduced mass, is essentially the mass of an electron and the subscripts 1 and 2 refer to quantities for electrons 1 and 2, respectively. The distance from the nucleus to electron 1 is r_1, the distance to electron 2 is r_2, and r_{12} is the distance between the electrons.

Because of the term containing r_{12}, it is not possible to separate the variables in the Schrödinger equation as we can do for the hydrogen atom. Exact solutions are therefore unavailable. This means we cannot write an exact ψ for helium explicitly as we can for a hydrogen atom. However, we can calculate energy levels, electron distributions, and other electronic properties for helium or any atom to a high degree of accuracy, using mathematical methods to approximate the solution. The calculated values agree with experiments within experimental error. Therefore, we are confident in the predicted values for properties that have not yet been measured.

We discuss many-electron atoms qualitatively; we do not go into details about the approximations used in obtaining wavefunctions and energy levels. We consider a many-electron atom to have hydrogen-like wavefunctions. These orbitals have the shapes shown in figures 9.15 through 9.18, but their energies are very different from those of a hydrogen atom. Because of the repulsion term, the energies of the wavefunctions are no longer a function of only n. The ordering of the energy levels is $1s$, $2s$, $2p_x = 2p_y = 2p_z$, $3s$, $3p$ (three orbitals), $4s$, $3d$ (five orbitals), and so on. The filling of these orbitals with electrons as the atomic number increases is characterized by the *building-up,* or *Aufbau, principle.* Recently, crystallographic methods (chapter 12) have directly revealed the shapes of d-orbitals (figure 9.19) (Zuo et al. 1999).

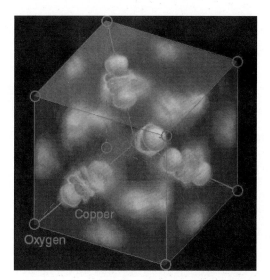

◀ FIGURE 9.19

Direct view of *d*-orbitals in Cu_2O, as observed using high-resolution X-ray diffraction (see chapter 12). (From J. M. Zuo et al. 1999. Reprinted with permission from *Nature*, 1999, Macmillan Magazines, Ltd.)

The *Pauli exclusion principle* states that one, or at most two, electrons can be in each orbital; two electrons in the same orbital must have opposite spins. Pauli deduced that no two electrons in an atom can have the same four quantum numbers: n, l, m, and m_s. Two electrons in the same orbital have identical values of n, l, and m; therefore, they must have antiparallel spins ($m_s = +\frac{1}{2}$ for one and $-\frac{1}{2}$ for the other).

These simple facts are enough to understand the arrangement of elements in the periodic table. Elementary chemistry books provide a more extensive discussion of atomic orbitals and the periodic table. Here, we just introduce the notation for specifying the electron distribution in a many-electron atom. Figure 9.20 illustrates this notation. The prediction of atomic structure, and its correlation with the periodic table, was a great achievement of quantum mechanics. The seemingly mathematical terms of wavefunctions were translated into the chemical principles underlying the periodic table.

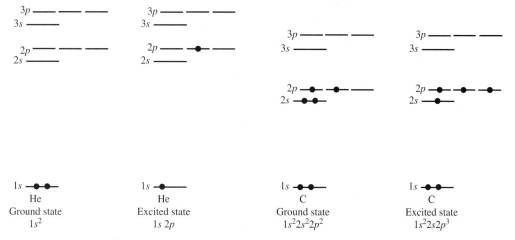

▲ FIGURE 9.20

Energy-ordering and electron orbital notation for many-electron atoms. At most, two electrons are placed in each orbital in accord with the Pauli exclusion principle. The electronic state of the atom is specified by the number of electrons in each orbital. For example, the fluorine ground state is $1s^2 2s^2 2p^5$.

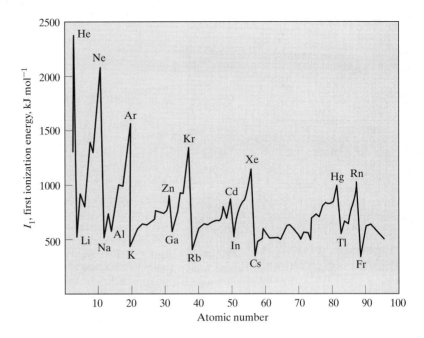

▶ FIGURE 9.21

First ionization energies of the first 80 elements of the periodic table, plotted as a function of atomic number (Z).

Ionization energy and electron affinity are two measurable quantities that directly reflect the orbital energies of many-electron atoms. The first ionization energy (I_1) is the minimum energy required to remove an electron from a many-electron element. The second ionization energy (I_2) is the energy required to remove the second electron. Ionization energies thus probe the energies of the highest-energy electrons. A related quantity is *electron affinity* E_a, which is the energy released upon binding of an electron to a gas-phase atom. Ionization energies are related to the ground-state configuration of an atom and thus show the periodicity of the periodic table (figure 9.21). They generally increase upon going from left to right in a row of the periodic table. This reflects the increased nuclear charge, while additional electrons are placed in orbitals of (roughly) the same principal quantum number n. The additional electrons experience the stronger attraction of the increasingly positively charged nuclei as one goes across the periodic table. In contrast, ionization energies decrease going down a column of the periodic table. Nuclear charge indeed increases, but now the additional electron is added to an orbital of increased n, meaning, on average, that this electron is located farther from the nucleus. The electron is shielded from the increased nuclear charge by the inner orbital electrons. This is called *screening*.

Ionization energies, since they involve the highest-energy electrons, play a critical role in the chemistry of the elements. One parameter, which has significance in biology, is *electronegativity*. One definition of electronegativity, by Mulliken, is

$$\chi = \frac{1}{2}(I_1 + E_a)$$

where I_1 is the first ionization energy and E_a the electron affinity for a given element.

Electronegative elements, such as nitrogen and oxygen, play a central role in biological chemistry. Their electronegativity determines their role as nucleophiles in chemical reactions and gives rise to the hydrogen bond, the critical element of molecular architecture in the cell.

Molecular Orbitals

The electronic wavefunctions of a molecule can be approximated by a set of *molecular orbitals* (MOs) in which one, or at most two, electrons can be placed. The electronic wavefunctions depend on the internuclear distances, but we can consider the nuclei fixed in determining the electronic wavefunctions and orbitals. Each MO can be represented as a *linear combination of atomic orbitals* (LCAOs); the MO is written as a sum of atomic orbitals (AOs) on each nucleus. We know what each of the AOs looks like (figures 9.15 through 9.18); therefore, we know the shapes of the MOs. From a chosen number of AOs, we can write the same number of MOs. Molecular orbitals lead to a detailed understanding of how atoms are bonded together to form molecules.

The first step in using and understanding MOs is to learn the vocabulary. Consider the H_2 molecule with two nuclei labeled A and B. The simplest LCAO we can write is

$$\psi_{H_2} = \frac{1}{\sqrt{2}}[\psi_A(1s) + \psi_B(1s)] \qquad \textbf{(9.30)}$$

This says that an MO of H_2 can be approximated by a $1s$ hydrogen AO on nucleus A plus a $1s$ hydrogen AO on nucleus B. The $1/\sqrt{2}$ is necessary to normalize the H_2 MO. Because we have chosen to use two AOs, we can write two MOs. The other one is

$$\psi_{H_2} = \frac{1}{\sqrt{2}}[\psi_A(1s) - \psi_B(1s)] \qquad \textbf{(9.31)}$$

We can calculate the electron distributions corresponding to these MOs by squaring them, and we can calculate their energy levels by using the Hamiltonian operator for an H_2 molecule. These two MOs are examples of sigma σ and sigma star σ^* MOs, respectively. The distinction depends on the *nodes* of the MOs; a wavefunction is equal to zero at a node.

Equations (9.30) and (9.31) give approximations to the wavefunctions of H_2. To calculate reasonable values for the electronic energies of H_2, much more complicated wavefunctions would be needed. Nevertheless, MOs that are LCAOs are useful concepts to understand electron distributions in molecules qualitatively.

The names of MOs depend on their symmetry properties. A *sigma-bonding orbital* (σ-bonding MO) has cylindrical symmetry around the line joining the nuclei and has no node between the nuclei. It is a bonding orbital because electron density accumulates between the two nuclei. The accumulation of negatively charged electrons between the positively charged nuclei is stabilizing. A *sigma-star antibonding orbital* (σ^*-antibonding MO) has cylindrical symmetry around the line joining the nuclei but has a node between the nuclei; the node tends to drive the two nuclei apart. Figure 9.22 illustrates a σ and a σ^* MO.

The electron density of a σ bond has cylindrical symmetry around the bond, and there is a high probability of finding the electrons between the nuclei. For a σ^* orbital there is still cylindrical symmetry around the bond, but the probability of finding the electrons between the nuclei becomes zero between the nuclei (at the node). The energy calculated for a σ MO is less (more negative) than the sum of the energies for the two AOs, whereas the energy of a σ^* MO is greater than the sum of the two AOs.

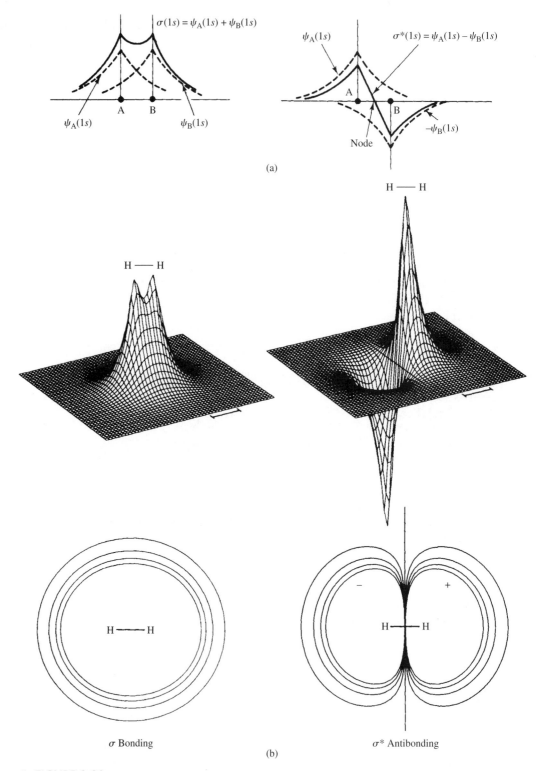

▲ FIGURE 9.22

(a) Sum [$\sigma(1s)$] and difference [$\sigma^*(1s)$] of two $1s$ orbitals. These are the simplest examples of a σ and a σ^* MO. (b) Computer plots of LCAO approximations to the ground state (σ bonding) and first excited state (σ^* antibonding) for the H_2 molecule. In the middle part of the figure, ψ is plotted in perspective in the xy-plane. The marked line in front of the grid is $2a_0 = 1.058$ Å long. The bottom part of the figure shows contour lines on which the value of the wavefunction is constant. The + and − on the σ^* orbital indicate the signs of the wavefunction. [Reprinted with permission from A. Streitwieser, Jr., and P. H. Owens, 1973, *Orbital and Electron Density Diagrams* (New York: Macmillan). Copyright 1973 by Macmillan Publishing Company.]

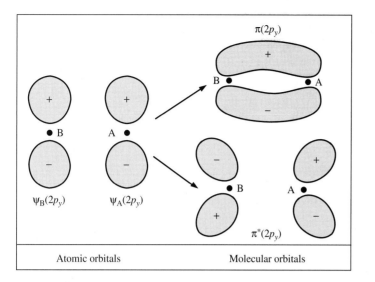

◄ FIGURE 9.23

Combination of two $2p_y$ AOs into a π-bonding and π*-antibonding MO. The drawings represent the angular distributions of the electron densities; the signs + and − specify the signs of the wavefunction.

A π (*pi*)-*bonding molecular orbital* has a node along the line joining the two nuclei. The electron density for a π-bonding MO has a plane of symmetry passing through the bond. The π-bonding orbital can be thought of as the linear combination of two side-by-side p orbitals. The π-bonding orbital leads to net accumulation of electron density between the two nuclei, although this density is not directly concentrated along the axis connecting the nuclei. A π* (*pi star*)-*antibonding molecular orbital* has a node along the line joining the nuclei and also has a nodal plane between the two nuclei. Figure 9.23 illustrates π and π* MOs.

A molecule can be thought of as a collection of MOs. We have discussed σ, σ*, π, and π* MOs; other MOs with different symmetry are sometimes important in metal coordination complexes. The distribution of electrons among these orbitals determines the bonding and stability of the molecule. In general, the proper AOs to be used for the formation of the MOs are those with comparable energies and the same symmetry with respect to the bond axis. Atomic orbitals of greatly different energies are poor choices for constructing MOs.

The principles of MOs are graphically illustrated by homonuclear diatomic molecules; these principles are qualitatively applicable to more complex situations that occur in biological molecules. Let's consider ten MOs that can be formed from combinations of five AOs ($1s$, $2s$, $2p_x$, $2p_y$, $2p_z$) on each of two identical nuclei A and B. These ten MOs can hold up to 20 electrons, so they allow us to discuss the bonding and stability of possible homonuclear diatomic molecules from H_2 to Ne_2. We start adding electrons to the MOs and see what qualitative conclusions we can make about the resulting molecules. Figure 9.24 shows the usual energy-level pattern obtained for the ten MOs. The actual spacing will depend on which nuclei are involved; for some molecules, $\sigma(2p_x)$ may be above the $\pi(2p_y)$, $\pi(2p_z)$ levels. However, this pattern of levels is a useful one to remember for most molecules, not just diatomics. Figure 9.24 is consistent with the known data on the diatomic molecules of the first row of the periodic table. He_2, Be_2, and Ne_2 are unstable; H_2, Li_2, B_2, and F_2 form single bonds; C_2 and O_2 form double bonds; and N_2 forms a triple bond. The number of bonds, sometimes called the *bond order*, is one-half the number of electrons in bonding MOs minus one-half the number of

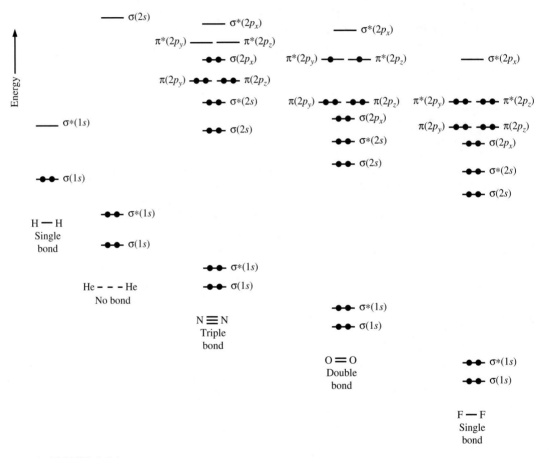

▲ FIGURE 9.24

Energy-level patterns and electron occupation for homonuclear diatomic molecules. (Top) Typical energy levels of MOs formulated as linear combinations of orbitals of the separated atoms. (Bottom) The type of bonding and the relative bond energies can be understood for the first ten elements in the periodic table. The energy of a given type of MO decreases from left to right because increasing nuclear charge results in greater electrostatic interaction with the electrons in the MOs. The molecules are all shown in their ground (lowest-energy) electronic state with no more than two electrons per MO, according to the Pauli principle. Note that O_2 contains two "unpaired" electrons, which are located in two π^* orbitals that have the same energy but different spatial orientations. The unpaired electron spins lead to the *paramagnetism* of O_2 in its interactions with a magnetic field. The other molecules shown are *diamagnetic* in their ground states; all electrons are paired.

electrons in antibonding orbitals. Each pair of filled bonding and antibonding MOs does not contribute to any bonds. For molecules containing C, N, O, F, and higher atomic-number elements, the $\sigma(1s)$ and $\sigma^*(1s)$ MOs involving these atoms are filled and are not involved in bonding. The $1s$ electrons are then called *core electrons* and are not explicitly considered in the MOs.

Figure 9.24 indicates what kind of electronic transitions to expect on excitation by light. For H_2, the longest-wavelength absorption corresponds to a $\sigma \to \sigma^*$ transition. The excited state with one electron in a σ MO and one in a σ^* MO is unstable, as expected. For O_2, a $\pi \to \pi^*$ transition occurs, and the double-bond character decreases on excitation. This results in a weaker bond in the excited molecule.

We must emphasize again that this MO picture is an approximation. However, it can help us remember and systematize the experimental results. There are other reasonable explanations of the bonding in O_2 that we could give. One is based on hybridization of AOs before forming MOs. We will define the various hybrid AOs in the next section, but here we mention that an alternative description of O_2 involves first forming an sp hybrid on each O atom. The $\sigma(2p_x)$ MO in figure 9.24 is replaced by a $\sigma(sp)$, and the $\sigma(2s)$ plus $\sigma^*(2s)$ become two *nonbonding orbitals, n(sp)*, on each O atom. A *nonbonding MO* is an orbital that is located on a single atom only and is thus not involved in bonding.

EXAMPLE 9.4

The molecule C_2 is known in the gas phase. Assuming that the ordering of the MO energy levels is the same as that shown for N_2 in figure 9.24, show the electron occupancy of the ground state of C_2. Do you expect C_2 to be paramagnetic or diamagnetic? If C_2^+ is formed by removing an electron from the highest energy MO of C_2, will the bond energy be smaller or larger than that of C_2?

Solution

The molecule C_2 has two fewer electrons than does N_2. Thus, the electron occupation will be the same as for N_2 except that the $\sigma(2p_x)$ MO will be empty for C_2. All electrons are paired, and the molecule should be diamagnetic, which it is. The bond dissociation energy of C_2 is 600 kJ mol^{-1}, which is reasonable for a double bond between the carbon atoms.

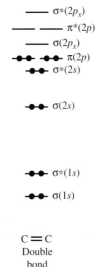

If one electron is removed to form the ion C_2^+, the easiest to remove is one from a $\pi(2p)$ orbital. This is a bonding orbital. Removing an electron from it decreases the bond order from 2 in C_2 to $\frac{3}{2}$ in C_2^+. Thus, the bond should be weaker in C_2^+. A value of 510 kJ mol^{-1} has been determined, which is indeed somewhat weaker than for C_2.

Hybridization

The MOs of a multinuclear molecule can be obtained from the appropriate combinations of AOs of the proper symmetry types. However, it is easier to visualize a complex molecule if the AOs of a given atom are combined into a set of hybrid orbitals first. These hybrid orbitals are then used to form MOs with neighboring atoms. A hybrid AO is just a linear combination of AOs on the same atom. The purpose of forming hybrid orbitals is to rationalize the observed geometry of the molecule. Hybridization can make more orbitals available for bonding (as in CH_4), or it can produce a more favorable bond angle (as in H_2O) than would be the case using unhybridized orbitals.

The three types of hybrid AOs that we consider are sp, sp^2, and sp^3. The names just indicate the number and type of AOs combined to make a hybrid orbital. There are two sp hybrids made from possible combinations of a $2s$ and a $2p$ AO. There are three sp^2 hybrids from possible combinations of a $2s$ and two $2p$ AOs. There are four sp^3 hybrids from possible combinations of a $2s$ and three $2p$ AOs. All four sp^3 hybrids have the same shape, but they are oriented in different directions in space. The three sp^2 hybrids are also identical to each other in shape, as are the two sp hybrids. They are similar in shape to the sp^3 hybrid (figure 9.25), but they are oriented differently in space. Equations for one of each set of hybrid orbitals are

$$\psi(sp) = \frac{1}{\sqrt{2}}\psi_A(2s) + \frac{1}{\sqrt{2}}\psi_A(2p)$$

$$\psi(sp^2) = \frac{1}{\sqrt{3}}\psi_A(2s) + \sqrt{\frac{2}{3}}\psi_A(2p)$$

$$\psi(sp^3) = \frac{1}{\sqrt{4}}\psi_A(2s) + \sqrt{\frac{3}{4}}\psi_A(2p) \tag{9.32}$$

Figure 9.25 can represent any one of these hybrid AOs.

The important difference among the different hybrids is their orientation in space. The two sp hybrids point in opposite directions. The bond

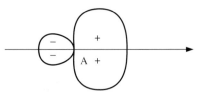

▲ FIGURE 9.25

Angle dependence of an sp^3 hybrid orbital; sp and sp^2 orbitals are similar. The arrow indicates the orientation of the hybrid. The signs $+$ and $-$ indicate the sign of the wavefunctions. A bond can be formed by combining this AO with an AO on another nucleus in the direction of the arrow.

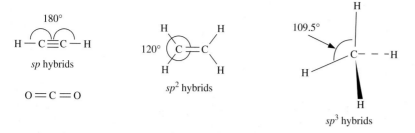

▲ FIGURE 9.26

Examples of molecules whose σ bonds can be represented with hybrid AOs. In acetylene, each C—H bond is a linear combination of an sp hybrid on C plus a $1s$ orbital on H. In CO_2, the sp hybrids on C interact with valence shell orbitals on each O to form σ bonds. In ethylene, the σ bonds involve three sp^2 hybrids on each C. In methane, the four σ bonds are combinations of the four sp^3 hybrids on C plus a $1s$ orbital on each H. The π bonds in acetylene, CO_2, and ethylene are sums of $2p$ orbitals.

angle between two sp hybrids is 180°. The three sp^2 hybrids point toward the corners of an equilateral triangle centered on the nucleus. The bond angle between two sp^2 hybrids is 120°, and all three orbitals lie in the same plane. For the sp and sp^2 hybridized atom, there are two and one remaining atomic p orbitals, respectively. These AOs can interact to form π bonds. The four sp^3 hybrids point toward the corners of a tetrahedron centered on the nucleus. The bond angle between two sp^3 hybrids is 109°28′. Figure 9.26 shows examples of molecules that can be described in terms of hybrid orbitals. Hybrid orbitals are a powerful means to describe the qualitative geometries of organic molecules. For more quantitative evaluation of bond strengths, lengths, and other molecular properties, MO calculations, which require fast computing, are required.

More complex hybrid orbitals can be written for transition metals such as Fe, Cu, Mo, and others that play important biochemical roles. In many such cases, d orbitals will participate in forming hybrids, such as the d^2sp^3 hybrids that result in octahedral coordination of six ligands, along three mutually perpendicular axes. The use of such AOs to account for bonding is known as the *valence-bond* (V-B) approach. It works best where electrons are localized in the region between two participating atoms. For molecules where electrons are delocalized over several nuclei, the MO approach is more satisfactory.

Real molecules, even those with localized electrons, are not always easy to describe using simple atomic or hybrid orbitals. For example, consider the important molecule H_2O, which is a bent molecule with two equal O—H bonds and two pairs of nonbonded electrons associated with the oxygen. Without hybridization, the $2p$ orbitals of O would lead to a 90° bond angle for H_2O because any two $2p$ orbitals chosen to form the two bonds to the two H atoms would be perpendicular to one another. If, on the other hand, sp^3 hybrids are used to form the bonds, then a 109.5° angle is expected. The lone-pair electrons would then occupy the two sp^3 hybrids that are not involved in bonding to H. The measured bond angle for H_2O is 104.5°, indicating that the true V-B picture lies somewhere in-between. The standard explanation in introductory chemistry texts is that lone-pair repulsion decreases the bond angle. Although accurate description of real molecules can become quite elaborate, the simple V-B and MO pictures nonetheless provide clear insights into the general aspects of bonding and structure.

Delocalized Orbitals

We have previously thought of MOs as two-centered, with each orbital localized around two neighboring atoms. Although this approximation is often suitable for σ bonds, it may not be useful for conjugated π bonds. We have already seen, in discussing β-carotene, that the π electrons in a conjugated system behave as if they are free to move throughout the conjugated system. This can be now be explained by the MOs of a conjugated system. We first look at the simplest system: 1,3-butadiene ($CH_2\!\!=\!\!CH\!-\!CH\!\!=\!\!CH_2$). Butadiene is a planar molecule, with angles between adjacent σ bonds close to 120°. This immediately suggests that sp^2-type hybrid orbitals of the carbon atoms are involved: Each carbon atom forms three σ bonds with its three neighboring atoms. If the neighboring atom is a hydrogen, the bonding σ orbital is formed from the $1s$ orbital of the hydrogen and the sp^2 orbital of the carbon. If the neighboring atom is a carbon, the bonding σ orbital is formed from two sp^2 orbitals, one from each carbon. These bonds are localized; each includes two nuclei of the molecule (figure 9.27a).

For each carbon atom, the formation of three sp^2 orbitals from a $2s$ and two $2p$ orbitals leaves it with one $2p$ orbital. If we choose the molecular plane as the xy-plane, the $2p_z$ orbitals of the carbon atoms are not involved in the formation of the sp^2 orbitals and the σ bonds. The $2p_z$ orbitals form a set of multicentered π orbitals, as illustrated in figure 9.27b. The four $2p_z$ orbitals from the four carbons can form four MOs with π or π^* symmetry. Two turn out to be bonding, π, and two antibonding, π^*. The two bonding ones are filled. Thus, two electrons are delocalized over all four carbon atoms, and another two are localized between pairs of carbons. Obviously, the presence of delocalized orbitals means that writing the double bonds as fixed is not very accurate. It is better to think of the double-bond character also being partly on the central C — C bond. This agrees with the barrier to rotation about this central bond. It is energetically more costly to rotate about a π bond than a σ bond because rotation destroys the π bond orbital overlap; there is hindered rotation about this central bond. Because the four π electrons occupy the two lowest-energy orbitals, π_1 and π_2, the central bond has less double-bond character than either of the two end C — C bonds. This is consistent with the experimental bond lengths of 1.476 Å for the central bond and 1.337 Å for the end bonds. The experimental C — C — C bond angle is 122.9°, which is close to that required for sp^2 hybrids to form the σ-bond structure. The power of the MO approach is the ability to predict these molecular features.

On the whole, the MO picture is a much better way of describing the properties of a molecule like 1,3-butadiene than is the use of classical resonance structures, such as

$$CH_2\!\!=\!\!CH\!-\!CH\!\!=\!\!CH_2 \longleftrightarrow \ \dot{C}H_2\!-\!CH\!\!=\!\!CH\!-\!CH_2^. \longleftrightarrow$$

$$\ominus\!:\!CH_2\!-\!CH\!\!=\!\!CH\!-\!CH_2\!\oplus$$

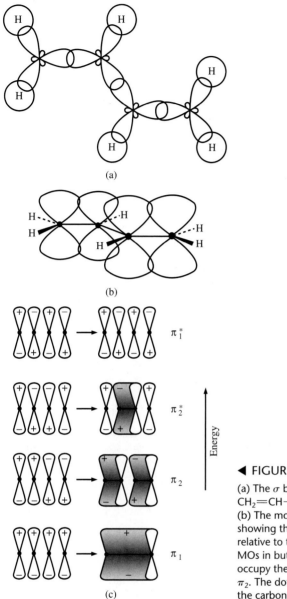

(a)

(b)

(c)

◀ FIGURE 9.27

(a) The σ bonds in butadiene $CH_2{=}CH{-}CH{=}CH_2$ viewed face on. (b) The molecule viewed nearly edge on, showing the arrangement of the p_z orbitals relative to the σ-bond skeleton. (c) The π MOs in butadiene. The four π electrons occupy the two bonding orbitals, π_1 and π_2. The dots (•) represent the positions of the carbon atom nuclei.

There is no experimental evidence for contributions of unpaired electrons or significant charge separation in 1,3-butadiene; as we have seen, the MO description does not require any.

We can label the carbon atoms, a, b, c, and d, starting at the left, and assign each an atomic $2p_z$ orbital, ψ_a, ψ_b, and so on. Four normalized π MOs for butadiene, obtained using methods that are beyond the scope of this book, are

$$\psi(\pi_1) = 0.372\psi_a + 0.602\psi_b + 0.602\psi_c + 0.372\psi_d$$

$$\psi(\pi_2) = 0.602\psi_a + 0.372\psi_b - 0.372\psi_c - 0.602\psi_d$$

$$\psi(\pi_2^*) = 0.602\psi_a - 0.372\psi_b - 0.372\psi_c + 0.602\psi_d$$

$$\psi(\pi_1^*) = 0.372\psi_a - 0.602\psi_b + 0.602\psi_c - 0.372\psi_d$$

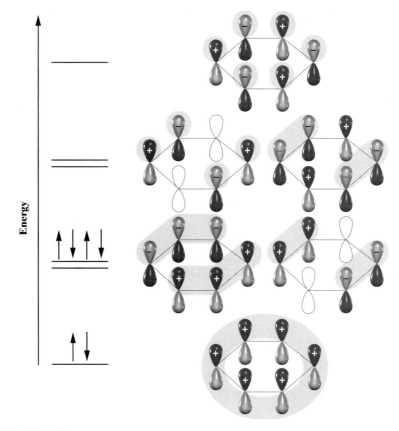

▲ FIGURE 9.28

The combination of six *p* AOs to form six MOs in benzene. The orbitals are ranked according to their relative energy, which increases with the number of nodes. Two electrons fill the lowest-energy orbital and are thus delocalized over all six carbons of the benzene ring. The six electrons from the *p* orbitals fill the three bonding MOs.

The two π-bonding orbitals of 1,3-butadiene are the highest-energy occupied MOs. These highest-energy occupied orbitals and the lowest-energy unoccupied orbitals influence the chemistry of the molecule, since chemical reactions involve changes of the electron density. Chemistry is dominated by the properties of these so-called frontier orbitals. This is a general feature of chemistry and another reason that MO calculations are so important.

An important class of molecules with delocalized orbitals are aromatic ring structures. A simple example is C_6H_6 (benzene). As with 1,3-butadiene, the six carbon atoms each have three sp^2 hybrid orbitals and a remaining $2p_z$ AO. The six $2p_z$ orbitals can form six π MOs, as shown in figure 9.28. The six electrons fill the three bonding orbitals; no antibonding orbitals are filled. The lowest-energy orbital is delocalized over all six carbon atoms in the ring. The delocalization of the electrons gives extra stability to the molecule (recall our discussion of the particle in a box). The bonding structure imposes the flat planar geometry on the aromatic ring. For example, the flat structure of aromatic nucleic acid bases is essential to allow them to stack in the center of the familiar double-helix structure of DNA in a regular fashion. The delocalized bonding electrons in benzene and other aromatic rings give them the properties that are essential to understand their chemistry.

Molecular Structure and Molecular Orbitals

The language of MOs is used to help us systematize and remember facts of chemical structure. It provides a framework that simplifies the logical ordering of the data. The two main ideas we want to emphasize are the following:

1. The electron distribution in a molecule can be described in terms of a set of MOs and corresponding energy levels. In the *ground state* of the molecule, these MOs are filled by adding electrons to the lowest-energy orbitals with at most two electrons in each orbital. In an *excited electronic state,* one or more of the electrons is placed in an orbital of higher energy. The properties of the molecule depend on which MOs are occupied and which excitations of electrons to unfilled MOs can occur.

2. Each MO can be thought of as a sum of hydrogen-like AOs on different nuclei (except for nonbonding MOs, which are localized on a single nucleus).

Geometry and Stereochemistry

Correlating orbitals with bond angles and bond lengths is straightforward. We have mentioned the correlation of sp^3, sp^2, and sp hybrids with 109.5°, 120°, and 180° bond angles. We also know that two unhybridized p orbitals on the same atom form bonds at 90°. We can now use measured bond angles to help us decide on a consistent set of MOs, and we can use assumed MOs to predict bond angles. Similarly, bond types and bond distances can be correlated with bonding orbitals. A single bond (a σ bond) is longer than a double bond (a σ bond plus a π bond), which is longer than a triple bond (a σ bond plus two π bonds). Rotation about a single bond is easy, whereas rotation about a double bond requires enough energy to break the π bond. Figure 9.29 shows some measured bond angles and bond lengths. The expected trends are seen. The $C\equiv C$ bond is 20% shorter than the $C-C$ bond; the benzene $C-C$ bond is intermediate in length between ethane and ethylene. The $H-C-H$ bonds in ethane are tetrahedral. In ethylene, the

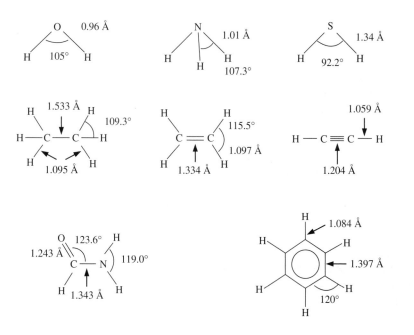

◀ FIGURE 9.29

Molecular geometry. Bond angles are given in degrees, and bond distances are in angstroms (1 Å = 0.1 nm).

H—C—H is less than the 120° trigonal angle expected, but in benzene the C—C—H bond angles are precisely 120°.

The structure of formamide is informative. The H—N—H bond angle of 119° implies that the bonding around the N is trigonal, not tetrahedral pyramidal as in NH_3. The O—C—N bond angle is also trigonal planar, as expected. The C—N bond distance is 1.343 Å, which is shorter than a C—N single bond (1.47 Å), such as is found in methyl amine. These data indicate that formamide is planar and that the C—N bond is partially a double bond. In resonance theory, this is described in terms of migration of the lone pair of electrons on the N atom.

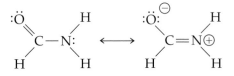

In the MO picture, the π orbitals assumed for formamide are delocalized over three nuclei (C, N, O). The peptide bond is similar to formamide in that it is planar with delocalized π MOs. This has profound consequences for the structures of proteins.

A very important application of MO methods is in the understanding of chemical reactions and reaction mechanisms. The Woodward–Hoffman rules (Hoffman and Woodward 1968) use the symmetry characteristics of MOs to rationalize and predict the stereochemical course of concerted organic reactions. An example is the conversion of cyclobutene to its isomer, butadiene. An isotopically labeled reactant can form either of two products:

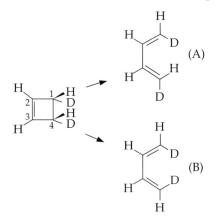

The important MOs are the π and π^* of the C_2=C_3 double bond, the σ and σ^* of the C_1—C_4 single bond of cyclobutene, and the four π, π^* MOs of butadiene. If we can find a path that converts occupied bonding MOs in the reactant to occupied bonding MOs in the product, the reaction should occur easily. Hoffman and Woodward conclude that ground-state cyclobutene should form product (A) and excited-state cyclobutene should form product (B). They thus predict, and find, that thermal reaction gives (A), but photochemical reaction gives (B).

The catalytic efficiency and stereochemical selectivity of enzymes can be related to the ability of the enzyme to constrain reactants and products to particular orientations. This specific orientation will greatly enhance certain reactions and reaction paths over the more random collisions produced by thermal motion. The literature in this field is large, and it often contains MO language (see, for example, Mesecar et al. 1997; Knowles 1989; and Stubbe 1989).

Transition Metal Ligation

The MO picture is very important for describing the interactions between ligands and transition metals, such as Fe in myoglobin, hemoglobin, or cytochromes; Cu in plastocyanin or cytochrome oxidase; Co in vitamin B_{12}; and many others. These metal atoms contain incompletely filled d orbitals in their valence shells: They can accept electrons into these and other low-lying unfilled orbitals from electron-rich ligands, such as Cl^-, OH^-, pyridine, CO, O_2, and the central nitrogens of the heme porphyrin ring. The metal atoms can, at the same time, form favorable interactions by donating some of their valence electrons into favorably oriented, unoccupied orbitals of the ligands. The *ligand field strength* is a measure of the ability of a given ligand to coordinate well with a metal atom. This in turn influences the energy levels of the metal atom and the geometry of the resulting complex.

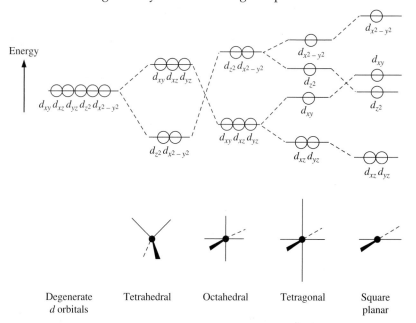

Degenerate *d* orbitals   Tetrahedral   Octahedral   Tetragonal   Square planar

◀ FIGURE 9.30

Ligand-induced splitting of *d* orbitals of transition metal complexes and the relation to geometry. Electrons in orbitals that point in the directions of ligands are increased in energy relative to those that are localized primarily away from the ligand direction. The *d* electrons of the transition metal tend to occupy orbitals of the lowest available energy. There is, however, an additional tendency for electrons to unpair (*Hund's rule*), even if it costs a small amount of energy to promote one of them to a higher orbital to achieve this unpairing. The interplay of these factors is the basis of the rich assortment of spectroscopic, magnetic, and chemical properties exhibited by transition metal complexes.

Figure 9.30 illustrates how the presence and nature of ligands influences the d-orbital energies of metal atoms. In the unliganded atom or ion, the five equivalent d orbitals, designated d_{xy}, d_{xz}, d_{yz}, d_{z^2}, and $d_{x^2-y^2}$, are all degenerate in energy. The presence of ligand interactions breaks this degeneracy in one of several characteristic patterns shown. Which of these is formed by a particular metal–ligand interaction depends on several factors, but ultimately the structure giving the lowest energy will occur.

1. *Nature of the ligand.* Ligand field strength goes, from "weak field" to "strong field," in the order

$$I^- < Br^- < Cl^- < F^- < OH^- < H_2O < NCS^- < NH_3$$

$$< \text{ethylene diamine} < NO_2^- < CN^- \sim CO$$

Strong-field ligands produce larger energy splittings. Ligand field strength is partly a measure of the electrostatic field created by the ligand. Small ions, such as F^-, concentrate the field better than do equivalent large ions, such as I^-. Ligands such as CN^- or CO that have an unfilled antibonding π^* orbital exhibit enhanced ligand field strength because of interactions with d electrons in the metal orbitals that are favorably oriented.

2. *Oxidation state of the metal.* A larger charge on the metal produces a greater electrostatic interaction with the electrons of the ligand.

3. *Number of electrons available to occupy the MOs.*

4. *Geometric placement of the ligands relative to the regions of highest electron density of the orbital.* Overlap occurs best for metal orbitals oriented along axes where ligands are located.

5. *Both σ contributions and π contributions (especially in ligands such as NO_2^-, CN^-, and CO) are important in maximizing the interactions.*

In addition to the geometry of the ligand–metal coordination complex, other properties affected by these factors include the color of the species; the spin state, or paramagnetism; the standard reduction potential; and the chemical reactivity. The familiar differences between the green solutions containing $CuCl_4^{2-}$ and the deep blue of $Cu(NH_3)_4^{2+}$ or the pink of $Co(H_2O)_6^{2+}$ and the deep blue of $CoCl_4^{2-}$ arise from differences in the energy-level splittings caused by the different ligands. The origins of such color differences will be explored further in the next chapter. The spin state, or paramagnetism, of transition metals is controlled by how the electrons in the valence shell are distributed among the available d orbitals. As always, the configuration of lowest overall energy will occur, but several factors contribute. In general, electrons will go into the available orbitals of lowest energy, but Hund's rule states that it costs some energy to pair electrons in the same orbital rather than have them remain unpaired in different orbitals of about the same energy. Thus, the five d electrons of Fe(III) in $Fe(H_2O)_6^{3+}$ or in the heme-containing enzyme horseradish peroxidase are all unpaired and distributed, one in each of the available d orbitals (figure 9.31). This gives the "high-spin" state of maximum paramagnetism. This state is designated as the $S = \frac{5}{2}$ state, where S, the spin quantum number for the ion or molecule as a whole, is the sum of the spin quantum numbers $m_s = \frac{1}{2}$ for the individual electrons. Thus, for $Fe(H_2O)_6^{3+}$, the value $S = 5(\frac{1}{2}) = \frac{5}{2}$ is obtained. By contrast, for $Fe(CN)_6^{3-}$ or the heme in ferricytochrome c, only one of the five d electrons remains un-

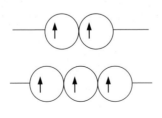

▶ FIGURE 9.31

Electron distribution among d orbitals in octahedral Fe(III) complexes with a weak-field ligand (H_2O) and with a strong-field ligand (CN^-).

$Fe(H_2O)_6^{3+}$ high spin
$S = 5/2$

$Fe(CN)_6^{3-}$ low spin
$S = 1/2$

paired, giving a low-spin state with $S = \frac{1}{2}$. The difference occurs because CN^-, for example, is a strong-field ligand that splits the d orbitals by such a large amount that it costs too much energy to unpair the electrons and promote them from the lower orbital to the upper one. With a weak-field ligand like H_2O, however, the orbital splitting is sufficiently small that a state of maximum unpairing has the lowest overall energy. Many examples of this have been documented, and measurements of the paramagnetic state provide very important information about the structure and chemistry of biochemical substances that contain transition metals.

Charge Distributions and Dipole Moments

An LCAO–MO description of a molecule characterizes the electron distribution in the molecule. When we write, for example, a π MO for formamide (figure 9.29), it has the form

$$\psi(\pi) = C_N\phi_N(2p_z) + C_C\phi_C(2p_z) + C_O\phi_O(2p_z)$$

where N, C, and O refer to the nitrogen, carbon, and oxygen nuclei in formamide; $\phi_N(2p_z)$, and so on are corresponding $2p_z$ AOs on each nucleus; and C_N, C_C, and C_O are numerical coefficients. Each electron that occupies that MO is distributed according to the square of the wavefunction. Therefore, a π electron in that MO will be distributed as follows: C_N^2 probability on N, C_C^2 probability on C, and C_O^2 probability on O. The sum of probabilities is equal to 1. The total electronic charge on each nucleus is just the sum of charges contributed by each occupied MO. In this approximation, the net charge on each nucleus is the nuclear charge minus the electronic charge calculated from the MOs. From the charge distribution, dipole moments can be calculated (discussed later).

Intermolecular and Intramolecular Forces

A central feature of biological chemistry is the presence of macromolecules, which can be defined as molecules with molecular weights greater than several thousand Daltons. These macromolecules are made of repetitive building blocks, such as nucleotides, amino acids, or sugars. The covalent geometry and properties of the building blocks are derived by the principles of quantum mechanics described before; however, the overall molecule is much too large to be treated quantum mechanically. These biological macromolecules have multiple single bonds, which as we see below, have relatively low barriers to rotation. Rotations about single bonds drastically change the orientations of the attached portions of the molecule. The spatial arrangements of a molecule that depend on rotation about single bonds are called *conformations*. For a large molecule, with many freely rotating bonds, there is an extremely large number of possible conformations. However, the thermodynamically favored conformations are often few and very closely related; essentially only one stable conformation exists. We will learn in chapters 10 and 12 the methods that are used to determine the structures of molecules. Here, we seek to understand why only a few of the many possible conformations are thermodynamically preferred. Also, what directs an enzymatic substrate or a drug to a specific portion of a large macromolecule and not elsewhere? The answer lies in the myriad noncovalent interactions that occur among the constituent atoms in the macromolecule. To obtain insight into this problem, we treat the molecule using classical physics; this approximation is sufficient to explain many of the observed interactions. Quantum mechanics has not been abandoned and forgotten. Many features of quantum theory are integrated into the simpler classical treatment, outlined in the following sections.

Bond Stretching and Bond Angle Bending

Bond lengths and bond angles are determined by the interactions of electrons and nuclei as described quantum mechanically, but the forces which change them can be treated by simple physical models. Bond stretching and bond angle bending can be treated as if the atoms were connected by springs. The energy of moving the atoms so as to stretch or compress a bond, or to change a bond angle, depends on the square of the change in bond length, or the square of the change in bond angle (this is the familiar harmonic oscillator potential that was described earlier):

$$U = k_r(r - r_{eq})^2 \qquad (9.33)$$

$$U = k_b(\theta - \theta_{eq})^2 \qquad (9.34)$$

Here $r - r_{eq}$ is the difference between the perturbed bond length and the equilibrium bond angle, and $\theta - \theta_{eq}$ is the difference between the perturbed bond angle and the equilibrium bond angle. The constants k_r and k_b are positive, which means the energy increases when the bonds are perturbed from their equilibrium positions. It takes more energy to change the bond length than it does to change the bond angle by the same percentage. The reason for using the word *perturbed* in the previous sentence is that Eqs. (9.33) and (9.34) apply only to small changes in the bond lengths and bond angles. Changes that would lead to bond breaking have more complicated potential energy curves.

Figures 9.32 and 9.33 show the plots of the bond-stretching and bond-bending potential energies versus the changes in bond length or bond angle for typical bonds. Stretching or compressing a bond requires a large amount of energy; note that changing the bond length of a single bond by 0.1 Å requires about 10 kJ mol^{-1}; a double bond requires about twice as much energy.

Bending a bond requires much less energy than stretching or compressing it. Changing the bond angle 10° requires about 5 kJ mol^{-1} for a typical tetrahedral bond; a trigonal bond, such as the O—C—N bond of an amide

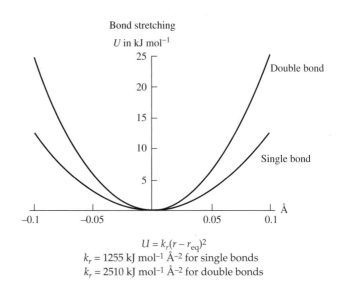

► FIGURE 9.32

Potential energy curves are shown for stretching bonds away from their equilibrium bond lengths. It takes more energy to stretch a double bond than to stretch a single bond. Typical values for bond-stretching force constants are shown.

Bond stretching

U in kJ mol^{-1}

Double bond

Single bond

$$U = k_r(r - r_{eq})^2$$
$k_r = 1255$ kJ mol^{-1} Å^{-2} for single bonds
$k_r = 2510$ kJ mol^{-1} Å^{-2} for double bonds

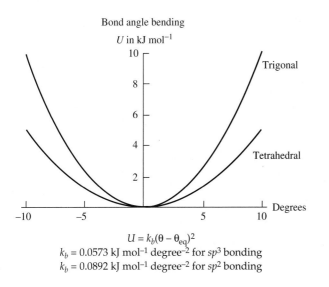

Bond angle bending

U in kJ mol^{-1}

$$U = k_b(\theta - \theta_{eq})^2$$
$k_b = 0.0573$ kJ mol^{-1} degree^{-2} for sp^3 bonding
$k_b = 0.0892$ kJ mol^{-1} degree^{-2} for sp^2 bonding

◀ FIGURE 9.33

Potential energy curves are shown for bending bond angles away from their equilibrium positions. It takes more energy to bend a trigonal bond with sp^2 bonding than to bend a tetrahedral bond with sp^3 bonding. Typical values for bond-bending force constants are given.

requires more energy to bend. A bond angle is defined by two bonds as shown below:

Figure 9.33 shows the energy versus bond angle for tetrahedral and trigonal bonds.

Rotation Around Bonds

As discussed earlier, rotation around single bonds can cause large changes in the conformation of a molecule, and the rotation does not require much energy. For example, at room temperature the benzene ring in phenylalanine rotates rapidly around the bond joining it to the rest of the molecule. This means that the energy barriers to rotation are no larger than the thermal kinetic energy, an RT value of about 2.5 kJ mol^{-1}. The energy necessary to rotate around a C—C single bond will have maxima and minima corresponding to different orientations of the substituents on the carbon atoms. Figure 9.34 illustrates the shape of the energy (called the *torsion energy*) curve as the angle of rotation around a C—C bond changes. For two tetrahedral carbons, there are three minima corresponding to substituents being *trans* ($\phi = 180°$),

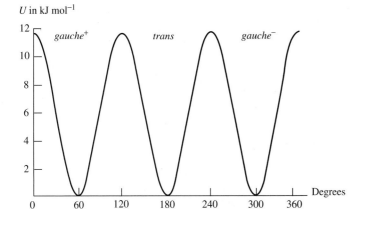

U in kJ mol^{-1}

◀ FIGURE 9.34

The potential energy for rotation around a tetrahedral C—C single bond in ethane. Equation (9.35) is used with the value of $V_3 = 11.6$ kJ mol^{-1}. Note the minima in energy corresponding to *gauche*$^+$, *trans*, and *gauche*$^-$ conformations as shown in figure 9.35.

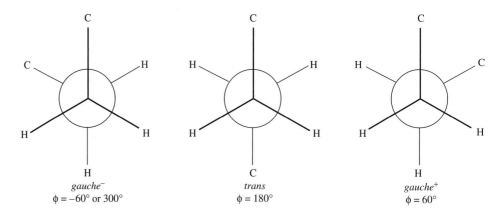

gauche⁻
$\phi = -60°$ or $300°$

trans
$\phi = 180°$

gauche⁺
$\phi = 60°$

▲ FIGURE 9.35

The most stable conformations in an aliphatic C—C chain. In the picture above we are looking along the central C—C bond; these are the atoms labeled B and C in the diagram showing the torsion angle ϕ. The two end carbons are most likely to be found near $\phi = 180°$ or $\pm 60°$.

gauche⁺ ($\phi = +60°$), or *gauche⁻* ($\phi = -60°$). The equation corresponding to this shape is

$$U = \frac{V_3}{2}[1 + \cos(3\phi)] \tag{9.35}$$

The value of V_3 characterizes the barrier to internal rotation; note that, at $\phi = 0°$, the energy $U = V_3$.

A torsion angle is shown below; the angle ϕ (phi) can vary from $0°$ to $360°$ (or from $0°$ to $+180°$ and to $-180°$). Angle ϕ is defined as $0°$ when atoms A, B, C, and D are in the same plane and A and D are on the same side of bond B — C; bonds A — B and C — D are eclipsed:

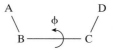

For a tetrahedral C — C bond in an aliphatic chain, . . . C—CH₂—CH₂—C . . . , the end carbon atoms are most stable in gauche or trans conformations as shown in figure 9.35.

It takes a large amount of energy to rotate around a double bond; you have to break the π bond. For a bond with partial double-bond character, such as the C—N bond in formamide, the energy required to rotate around it is in-between that of a double bond and a single bond. There is another difference; both C and N have bonding with trigonal symmetry. Thus, the NH₂ plane will tend to be parallel to the HCO plane:

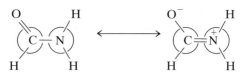

This is characterized by the equation

$$U = \frac{V_2}{2}[1 + \cos(2\phi - 180)] \tag{9.36}$$

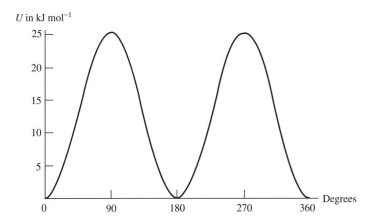

◀ FIGURE 9.36

The potential energy for rotation around a trigonal single bond such as formamide. Equation (9.36) is used with $V_2 = 25$ kJ mol^{-1}.

The potential energy curve is given in figure 9.36. Now there are only two minima—at 0° and 180°—both corresponding to the planar molecule. For a peptide bond with one carbon atom attached to the nitrogen and another to the carbonyl carbon, steric effects make the two minima have different energies. The conformation with the attached carbons on opposite sides of the C—N bond—the *trans* conformation—is lower energy than the *cis* conformation with the carbon atoms on the same side of the C—N bond.

The values for V_3 and V_2 are representative; they depend on the substituents and on the double-bond character. The magnitudes—the barriers to internal rotation—correspond to the maxima in figures 9.35 and 9.36. As a torsion angle is varied, the energy will change because of the barrier to internal rotation as seen in figures 9.35 and 9.36, but interactions from other atoms in the molecule will also contribute to the energy change. These are called *nonbonded,* or *noncovalent, interactions.*

Noncovalent Interactions
Electrostatic Energy and Coulomb's Law

General interactions occur between all molecules and all parts of molecules. *Intra*molecular interactions occur between parts of the same molecule; *inter*molecular interactions occur between two different molecules. The forces can be attractive or repulsive, depending on the properties of the groups and the distances between the interacting groups. Recall from elementary physics that a force can be related to a potential energy of interaction; the force is equal to minus the slope of the potential vs. distance curve.

Like charges repel and unlike charges attract; the electrostatic potential energy of interaction depends on the reciprocal of the distance between charges. This is Coulomb's law. For two charges separated by a distance r, the potential energy is

$$U(r) = \frac{q_1 q_2}{4\pi\varepsilon_0 r} \qquad \textbf{(9.13)}$$

where the charges q_1 and q_2 are in coulombs, r is in meters, ε_0 (the permittivity constant) $= 8.854 \times 10^{-12}\, \mathrm{C^2\, N^{-1}\, m^{-2}}$, and the potential energy is in joules.

Because we most often deal with interactions between electrons, it is convenient to write Coulomb's law in terms of electronic charges and in terms of distances in angstroms, Å (1 Å = 0.1 nm). Angstroms are often used for molecular interactions because bond distances are 1–2 Å. Furthermore, for comparison with thermodynamic energies in kilojoules per molecule, it is convenient to have the potential energy $U(r)$ in kJ mol^{-1}:

$$U(r) = 1389 \frac{q_1 q_2}{r} \qquad \text{(energy in kJ mol}^{-1}) \qquad \textbf{(9.37)}$$

Here, the charges q_1 and q_2 are in multiples of an electronic charge and the distance is in angstroms. For example, for two electrons separated by 1 Å ($q_1 = q_2 = r = 1$), the electrostatic interaction energy is +1389 kJ mol^{-1}. The positive sign means that the electrons repel each other.

If the two charges are separated by a medium, such as a solvent, the potential energy of interaction is decreased. Equations (9.37) must be divided by the dielectric constant (relative permittivity) ε of the medium:

$$U(r) = 1389 \frac{q_1 q_2}{\varepsilon r} \qquad \text{(energy in kJ mol}^{-1})$$

The dielectric constant of a vacuum is defined as 1.000. Dielectric constants of solvents vary from about 2 for nonpolar liquid hydrocarbons to about 80 for liquid water. The large dielectric constant of water is one of its most important properties, because repulsive interactions between like charges are greatly decreased in water. Clearly, the interaction of two charges in a nonpolar membrane is much larger than when they are surrounded by water. Several protein amino acid side chains are charged. When a protein is unfolded, these side chains are exposed in H_2O. Upon folding into a compact conformation, the charged side chain may be buried in the very nonpolar interior of the protein. The interior of protein does not have a dielectric constant of bulk H_2O but is likely much lower. For this reason, buried positively charged side chains are often next to negatively charged side chains.

The atoms in molecules often contain partial charges, since electron density is spread throughout the molecule, as determined by MO calculations. For inter- and intramolecular interactions for most molecules of biological interest (such as proteins and nucleic acids), simplifying approximations must be made. Instead of trying to calculate the interactions of all electrons and all nuclei, the net charge on each atom is considered. An isolated atom is neutral, but electrons are not shared equally between the atoms in a molecule. Some atoms such as oxygen or nitrogen are usually negative; some are positive, such as sodium; and some are near neutral but can be positive or negative, such as carbon and hydrogen. The simplification of using net charges on atoms compared to trying to calculate the interactions of all electrons and nuclei is tremendous because the number of interactions depends on the square of the number of particles. There are 156 electrons plus 39 nuclei in the dipeptide sweetener Aspartame (aspartyl phenylalanine methyl ester) but only 39 atoms. Once there are further constraints placed on the atoms (equilibrium bond lengths and bond angles), a conformational calculation on a dipeptide becomes practical.

EXAMPLE 9.5

Calculate the electrostatic potential energy of interaction for two water molecules arranged (a) favorably for hydrogen bonding (figure 9.37a); (b) with the same distance between oxygen atoms but rotated so that a hydrogen is not between the O's (figure 9.37b).

◀ FIGURE 9.37

Two water molecules separated by 2.500 Å between oxygen atoms. In (a) an attractive hydrogen bond is present; in (b) a repulsive orientation is shown.

Solution

The O—H bond distance in H—O—H is 0.957 Å, and the bond angle is 104.52°. A charge of −0.834 (in units of electronic charges) on the oxygen and +0.417 on each hydrogen atom has been found to give reasonable calculated properties for liquid water. (See W. L. Jorgensen et al., 1983, *J. Chem. Phys.* 79:926 for further details.) We calculate the electrostatic energy for two planar orientations of the water molecules at a typical O···H—O hydrogen-bond distance of 2.500 Å between oxygen atoms. In figure 9.37a, the acceptor water molecule is symmetric with respect to the O···H—O line. The first step is to find xy-coordinates for the atoms so that interatomic distances can be calculated. We place the acceptor oxygen (O1) at the origin of the coordinate system. The x-coordinates of its two hydrogen atoms (H1a, H1b) are at $-0.957 \cos(104.52°/2)$; the y-coordinates are at $\pm 0.957 \sin(104.52°/2)$. The donor oxygen (O2) is placed at $x = 2.500$ and $y = 0.000$, and the bonding hydrogen (H2a) at $x = (2.500 − 0.957)$. The other hydrogen (H2b) is at $x = [2.500 + 0.957 \cos(180° − 104.52°)]$ and $y = −0.957 \sin(180° − 104.52°)$. The values are

	x	y
H1a	−0.586	0.757
H1b	−0.586	−0.757
O1	0.000	0.000
H2a	1.543	0.000
H2b	2.740	−0.926
O2	2.500	0.000

The distances between each atom of the donor water molecule and each atom of the acceptor water molecule are calculated by

$$r_{ij} = \sqrt{(x_j - x_i)^2 + (y_j - y_i)^2}$$

The nine distances (in Å) are

	H1a	H1b	O1
H2a	2.259	2.259	1.543
H2b	3.727	3.330	2.892
O2	3.177	3.177	2.500

The electrostatic energies are calculated using Eq. (9.37):

$$U(\text{H—H terms}) = 1389(0.417)^2$$

$$\times \left[\frac{1}{r(1a - 2a)} + \frac{1}{r(1a - 2b)} + \frac{1}{r(1b - 2a)} + \frac{1}{r(1b - 2b)} \right]$$

$$= +351.1 \text{ kJ mol}^{-1}$$

$$U(\text{O—H terms}) = 1389(0.417)(-0.834)$$

$$\times \left[\frac{1}{r(O1 - 2a)} + \frac{1}{r(O1 - 2b)} + \frac{1}{r(O2 - 1a)} + \frac{1}{r(O2 - 1b)} \right]$$

$$= -784.2 \text{ kJ mol}^{-1}$$

$$U(\text{O—O terms}) = 1389(-0.834)^2 \left[\frac{1}{r(O1 - O2)} \right]$$

$$= +386.5 \text{ kJ mol}^{-1}$$

$$U(\text{total}) = -46.6 \text{ kJ mol}^{-1}$$

There is an attractive force between the water molecules in this orientation. Bringing the two separated water molecules to this distance and this orientation decreases the electrostatic energy by 46.6 kJ mol^{-1}. This is not the hydrogen-bond energy of forming water dimers because there are additional energy contributions. If only electrostatics were acting, the hydrogen atom and oxygen atom would prefer to have zero distance between them; there must also be a repulsive contribution to their interaction. There are also other attractive forces, which we describe shortly.

A similar calculation for the orientation of two water molecules shown in figure 9.37b with the oxygen atoms at the same distance, but with no hydrogen in-between leads to

$$U(\text{H—H terms}) = +253.2 \text{ kJ mol}^{-1}$$

$$U(\text{O—H terms}) = -608.2 \text{ kJ mol}^{-1}$$

$$U(\text{O—O terms}) = +386.5 \text{ kJ mol}^{-1}$$

$$U(\text{total}) = +31.5 \text{ kJ mol}^{-1}$$

This arrangement produces a repulsive force on the water molecules with an unfavorable energy of 31.5 kJ mol^{-1}. Unless other forces are stronger, the water molecules will not dimerize in this orientation.

This calculation illustrates that intermolecular electrostatic energy is important in determining how two molecules will approach each other. The possibility of forming a stable complex and its structure can be investigated by the simple application of Coulomb's law.

To summarize: Using Coulomb's law between net charges on each atom, we can calculate the electrostatic potential energy of interaction between two molecules or between two parts of a single molecule. We sum over all the atoms and their corresponding distances (for an overview of current applications, see Warshel and Papazyan 1998).

Net Atomic Charges and Dipole Moments

The net atomic charges provide a qualitative picture of how molecules interact. Obviously, the negative region of one molecule will be attracted to the

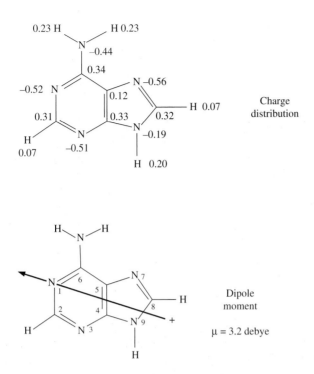

◄ FIGURE 9.38

Approximate quantum mechanical electronic wavefunctions were used to calculate the charge distribution and dipole moment of adenine. 1 debye is 10^{-18} esu cm. The charge distribution is given as the fraction of an electronic charge at each atom relative to the charge (equal to zero) on the isolated atom. (The charge on an electron is -4.803×10^{-10} esu.) (From B. Pullman and A. Pullman, 1969, *Prog. Nucleic Acid Res.* 9:327.)

positive region of another. We gave an example of this in the calculation of the electrostatic contribution to hydrogen bonding (example 9.5). Water is such a good solvent for ionic salts because the negative oxygen atom is strongly attracted to positive ions and the positive hydrogen atoms are strongly attracted to negative ions. The ion–water attractions in solution are stronger than the ion–ion attractions in the crystal.

Net charges also provide insight into chemical properties of a molecule. Figure 9.38 shows net charges on the nucleic acid base adenine, which are calculated using quantum mechanical methods. This gives us some ideas about relative reactivity of various positions on adenine. We cannot make firm predictions because (1) the electron distribution is very approximate, (2) the solvent is important, and (3) calculations on the possible product species would need to be made. However, the charge densities are consistent with our chemical intuition. The nitrogens are all negative, so positive substituents are attracted there. The ring nitrogens are targets for reaction with electrophilic (electron-poor) compounds. Reaction of the adenine base in DNA can disrupt the genetic information stored in the molecule and lead to mutation. In fact, many electrophilic compounds are known carcinogens (cancer-causing compounds). Because the three unsubstituted ring nitrogens are more negative than the amino nitrogen, we could predict that protonation occurs first on the ring. This is found experimentally. Calculation of ionization constants for each nitrogen could be made from the partial charges.

The *dipole moment* $\boldsymbol{\mu}$ is a measure of the polarity of a molecule. If the center of positive charge from all the nuclei is at the same position as the center of negative charge from all the electrons, the molecule has zero dipole moment; it is nonpolar. Otherwise, it is polar and it has a dipole moment. The simplest example is two charges, $+q$ and $-q$, separated by a distance r. In this case, the dipole moment is given by $\boldsymbol{\mu} = q\mathbf{r}$. The dipole moment is a vector because it has a magnitude (as calculated above) and a direction (from the center of positive charge to the center of negative charge). Symmetry is very helpful in calculating dipole moments. Any molecule with a point, or center, of symmetry has zero dipole moment. This includes H_2, N_2, O_2, CH_4, benzene, naphthalene, and so on. For a molecule with a plane of symmetry (such as adenine), the dipole moment, if it exists, must lie in the plane. To calculate the magnitude and direction of the dipole moment for a molecule such as adenine or retinal, we need to know the locations of all negative and positive charges in the molecule. This is done, as described previously, using quantum mechanical methods. The components of the dipole moment vector are

$$\boldsymbol{\mu}_x = \sum_i q_i x_i$$

$$\boldsymbol{\mu}_y = \sum_i q_i y_i$$

$$\boldsymbol{\mu}_z = \sum_i q_i z_i \qquad (9.38)$$

Each q_i is the charge at position x_i, y_i, z_i. For each nucleus, we know its charge and its position, so the sum over nuclei is easy to do. For the electrons, we use the electronic wavefunction.

The electronic wavefunction for all electrons in a molecule tells us the electron density at every point in space. As an approximation, all we need to know is whether a given atom in a molecule has an excess of electronic charge or a deficiency relative to the corresponding neutral atom. Then, by using the coordinates of the nuclei of the atoms, we can estimate the dipole moment. This method is illustrated in figure 9.38.

The magnitude of the dipole moment is

$$|\boldsymbol{\mu}| = \sqrt{\boldsymbol{\mu}_x^2 + \boldsymbol{\mu}_y^2 + \boldsymbol{\mu}_z^2}$$

The magnitude of the dipole moment can be determined from dielectric measurements on solutions containing the molecule. This provides one check on the electron distribution. The calculated dipole moment for adenine is shown in figure 9.38. The dimensions of dipole moments are charge times distance; chemists have traditionally used electrostatic units (esu) times centimeters. The charge on the electron is -4.803×10^{-10} esu. The magnitudes of dipole moments are thus about 10^{-18} esu cm, so a special unit has been defined in honor of Peter Debye, who first explained the difference between polar and nonpolar molecules. One debye (D) equals 10^{-18} esu cm. In SI units, the electronic charge is expressed in coulombs (C) and the distance in meters. Thus, $1 \text{ D} = 3.33564 \times 10^{-30}$ C m.

EXAMPLE 9.6

Estimate the magnitude of the dipole moment of adenine using the calculated charge distribution shown in figure 9.38. The molecule is assumed to be planar, and the atom position coordinates given in the table below are taken from analysis of X-ray diffraction studies of crystals of an adenine derivative.

membering from elementary physics that the force is the negative of the slope of an energy versus distance plot, you can see from figure 9.41 that the repulsive force increases rapidly for distances less than 3 Å. The repulsion can be between two atoms in one molecule (*intra*molecular repulsion) or between two atoms in two different molecules (*inter*molecular repulsion). This universal repulsion of atoms and molecules at short distances is called *van der Waals repulsion* after the Dutch scientist who studied it experimentally. The *van der Waals radius* of an atom is determined from the distance between two atoms where the repulsion energy starts rising rapidly. For two oxygen atoms, the repulsion energy starts rising rapidly at about 3 Å; therefore, the van der Waals radius of an oxygen atom is about 1.5 Å. One way of detecting a hydrogen bond between two atoms is to find them closer than the sum of their van der Waals radii. If two O's, or two N's, or an N and an O are within 2.8 Å of each other, they are nearly always hydrogen-bonded.

London–van der Waals Interaction

The attractive and repulsive energies of interaction can be summed to yield the net interaction, and the net force on the interacting parts. This is illustrated in figure 9.42 where the r^{-6} attraction of figure 9.40 is added to the r^{-12} repulsion of figure 9.41. The curve in figure 9.42 is called a *6-12 potential*. Note the minimum in energy at 3 Å. If there are no other forces acting, the equilibrium distance between the two interacting atoms is 3 Å. If the separation distance starts to increase, an attractive force is felt. If the distance starts to decrease, a repulsion is felt. Clearly the minimum in energy is the preferred position.

The minimum in the energy curve is characterized by the sum of the van der Waals radii of the atoms. This is the closest the atoms approach unless there are other forces, such as electrostatic attraction, acting on them. Covalently bonded oxygens are separated by only about 1.3 Å. Hydrogen-bonded oxygen atoms can be 2.8 Å apart, but unbonded oxygens remain 3.0 Å apart. Two carbon atoms that are not bonded to each other do not approach much closer than 3.4 Å; this is the sum of their van der Waals radii.

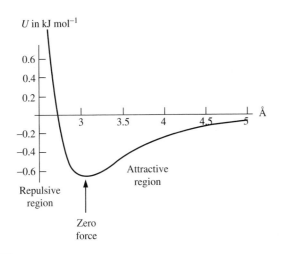

▲ FIGURE 9.42

The London–van der Waals interaction energy of two oxygen atoms in two water molecules. The curve is $U = B/r^{12} - A/r^6$; it is a 6-12 potential.

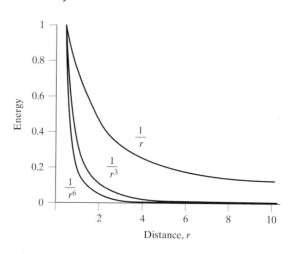

The distance dependence of noncovalent interactions. Relative plots of energy versus distance for charge–charge, dipole–dipole, and London attraction, which show the long-range nature of charge–charge interactions and the short-range nature of London attraction.

As opposed to coulombic or dipole–dipole interaction, the London attraction between two chemical groups is extremely short range. Figure 9.43 shows the distance dependence of the three types of attractive interactions discussed here. London attraction occurs only as two chemical groups approach the sum of their van der Waals radii—in simple parlance, when two chemical groups "contact" each other. As discussed previously, the London attraction increases for chemical groups with high polarizability, or delocalization of electrons. Aromatic bases in particular have strong London attractions for each other, because the high delocalization of electrons in aromatic bases gives them high polarizabilities. This attraction is only apparent when two aromatic rings are in close contact; these are often called *stacking interactions* because the aromatic rings stack like coins in a roll. Stacking of aromatic bases is particularly apparent in the structure of nucleic acid helices; the London dispersion force is one of the dominant stabilizing forces in nucleic acids.

The Lowest-Energy Conformation

We have learned about the different contributions to molecular interactions. They can be divided into two types. Bonded interactions include bond stretching, bond angle bending, and torsion angle rotation. Nonbonded interactions include London–van der Waals and Coulomb interactions. Sometimes all nonbonded interactions are simply called *van der Waals interactions*. The total energy is simply the sum of all of the bonded and nonbonded interactions.

$$U = +k_r(r - r_{eq})^2 \qquad \text{(bond stretching)}$$
$$+k_b(\theta - \theta_{eq})^2 \qquad \text{(bond angle bending)}$$
$$+ (V_n/2)[1 + \cos(n\phi - \phi_0)] \qquad \text{(torsion angle rotation; } n = 2 \text{ or } 3)$$
$$+\frac{B_{ij}}{r_{ij}^{12}} \qquad \text{(van der Waals repulsion; } B_{ij} \text{ is positive)}$$
$$-\frac{A_{ij}}{r_{ij}^{6}} \qquad \text{(London attraction; } A_{ij} \text{ is positive)}$$
$$+C\frac{q_i q_j}{r_{ij}} \qquad \text{(Coulomb interaction; C depends on the units)} \qquad \textbf{(9.42)}$$

These interactions can be used to calculate the low-energy conformations of a molecule. For proteins or nucleic acids, you may know an approximate conformation from experimental data such as nuclear magnetic resonance (NMR) spectra or low-resolution X-ray diffraction data. You vary the torsion angles and calculate the energy, searching for a minimum; this is called *molecular mechanics*. You include nonbonded interactions only between atoms that are separated by more than two bonds. The interactions of atoms that are separated by only one or two bonds are already taken into account in the bond-stretching and bond-angle-bending terms. It is easy to find a minimum energy for a conformation not very different from the starting structure; however, there may be a very different structure with a lower energy. We say that it is easy to find a local minimum, but to find the global minimum for a large molecule is very difficult. Thus, it is extremely difficult to predict the three-dimensional structure of a polypeptide with 100 residues. It is possible to calculate how its structure will change when a mutation changes one amino acid to another or when a substrate is bound to it. Energy minimization is routinely used to improve approximate structures obtained from X-ray diffraction of crystals or NMR data of solutions (for a review, see Karplus and Petsko 1990; Leach 1996).

Energy calculations can predict how a protein or nucleic acid will bind a drug or, in general, how two molecules will interact. The procedure is called *docking*. The interaction energies are calculated as the molecules are brought near each other. The molecules are rotated and moved relative to each other to find a minimum energy orientation. To calculate the interaction energy between two molecules, first calculate the energy of each separately. Then calculate the energy when the two molecules are docked together. The difference is the interaction energy. A likely conformation for the complex of two molecules will minimize the interaction energy.

We have not explicitly discussed two important interactions; hydrogen bonds and hydrophobic effects. A hydrogen bond is a noncovalent bond mediated by a hydrogen atom between two electronegative atoms such as O and N. Its contribution to the energy of a structure can be calculated as a sum of nonbonded interactions, but with parameters different from nonhydrogen-bonded atoms. Experimental values of hydrogen-bond energies are discussed below.

The interactions of biological molecules occur in aqueous solution, so the effect of water must be taken into account. Nonpolar molecules tend to be insoluble in water and thus lower the free energy of a mixture by avoiding the water. This means that the transfer of a nonpolar molecule from a nonpolar environment to an aqueous environment raises the free energy. This *hydrophobic* (literally, "water-fearing") effect helps determine how proteins fold, how they fit into membranes, and how lipids interact in water (see chapter 5). The interactions of water molecules with themselves and with the nonpolar molecules involve the nonbonded interactions discussed previously; they are not new types of interactions. These water effects also involve ordering or disordering of water; large entropy effects are involved.

A complete treatement of the entropy of a system and surroundings is very difficult to implement on a computer. It is thus a complex problem to calculate the overall interaction energy of a large molecule or even its minimum-energy configuration. Nevertheless, efforts have been made to do just this for nucleic acids and proteins with the aid of much calculation time on fast computers. Descriptions of these procedures can be found in Weiner et al.

(1984) and Némethy and Scheraga (1990). An even more difficult calculation is involved in trying to understand the stability of condensed-phase structures that are not covalently linked, such as phospholipid bilayers, micelles, and even the liquid phase itself.

Hydrogen Bonds

A hydrogen atom placed so as to interact simultaneously with two other atoms is said to form a *hydrogen bond.* Hydrogen bonds are extremely important in influencing the structure and chemistry of most biological molecules. The hydrogen bonds of water give rise to unique physical properties that make it ideal as a medium for life processes.

To be capable of forming hydrogen bonds, the H atom in a molecule, HA, must be somewhat acidic. The acidic hydrogen will interact with an electron donor species, B, to form the hydrogen bond:

$$B + HA \longrightarrow B \cdots HA$$

Hydrogen-bond donors include strong acids (HCl), weak acids (H_2O), and even molecules containing acidic C—H bonds ($HCCl_3$). Hydrogens attached to atoms of low electronegativity (CH_4) do not form hydrogen bonds. The best hydrogen-bond acceptors are atoms of the most electronegative elements (F, O, N). The bond energies of some typical hydrogen bonds are given in table 9.4. Others are listed in table 3.1.

Because solvents such as water can serve as both hydrogen-bond donors and acceptors, the enthalpies listed in table 9.4 are strongly solvent dependent. The values given for amide or for the nucleoside base pairs are measured in relatively nonpolar solvents ($HCCl_3$ or CCl_4). Measured in water, the ΔH^0 for the amide hydrogen bond is zero, which means that there is essentially no stabilization of amide intermolecular hydrogen bonding relative to the solvation of individual amides by water. This is not true in the case of nucleoside base pairs, where intramolecular interactions may involve two $(U \cdot A)$ or three $(G \cdot C)$ hydrogen bonds per base pair. In polynucleotides as well as in proteins, there are cooperative interactions involving the formation of many hydrogen bonds per molecule. These are supported, in the case of the polynucleotides, by significant stacking interactions of the aromatic bases in helical structures.

Even in liquid water, extensive networks of clustered molecules interact cooperatively by hydrogen bonding. As a consequence, water has physical properties that are anomalous, considering the related molecules H_2S, H_2Se, and H_2Te formed from atoms in the same column of the periodic table. These anomalous properties of water include a high boiling point, high viscosity, and a high entropy of vaporization. This latter in particular arises from the necessity of breaking up the ordered clusters of molecules in liquid water to form the vapor.

Hydrogen bonds are directional. The maximum stability (lowest energy) occurs when the three atoms A—$H \cdots B$ all lie on a straight line. Bent hydrogen bonds occur, but they usually have decreased stability. The directionality of hydrogen bonds is important for determining the crystal structure of ice. Each water molecule is capable of donating two H atoms to form hydrogen bonds, and each O atom is capable of accepting two hydrogen bonds. To maximize the number of hydrogen bonds in ice, each O atom is surrounded by four other O atoms placed tetrahedrally with respect to the central atom. Between each pair of O atoms lies an H atom involved in hydrogen bond-

TABLE 9.4   Bond Enthalpies for Some Hydrogen Bonds*

	ΔH^0 (kJ mol^{-1})		ΔH^0 (kJ mol^{-1})

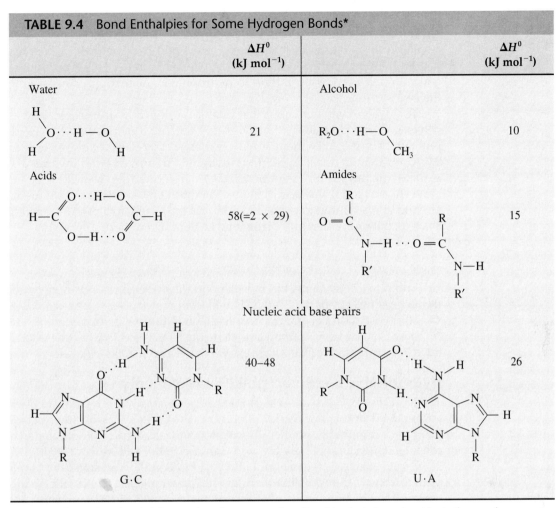

Water — 21

Alcohol — 10

Acids — 58(=2 × 29)

Amides — 15

Nucleic acid base pairs

G·C — 40–48

U·A — 26

*The hydrogen-bond enthalpies are given for water, carboxylic acids, alcohols, and amides in the gas phase. The nucleic acid base pairs were measured in deuteriochloroform at 25°C. In aqueous solution, the net hydrogen-bond strengths would be much weaker because of competition with water.

ing. This is quite compatible with the molecular structure of water, whose H—O—H bond angle of 105° is close to the tetrahedral angle, 109.5°. Thus, each H atom will lie nearly on the line joining the adjacent O atoms. This results in a relatively open, cagelike structure that accounts for the unusual property that the molar volume of ice is about 10% larger than that of liquid water at the freezing point. The stability of the hydrogen-bonded structure prevents the water molecules in ice from forming a more compact structure; when pressure in excess of 2000 atm is applied to ice, however, the ice does undergo a phase transition to a structure with a lower molecular volume.

Molecules that are apolar will not interact strongly with water. Nevertheless, some apolar molecules can form stable crystalline hydrates. These are found to consist of a cagelike shell of ordered water molecules that surrounds the apolar molecule. The stability of such water clusters can be appreciated by noting that the enthalpies of hydration of molecules, such as Ar, Kr, CH_4, C_2H_6, and so on, are all about 60 kJ mol^{-1}. When other, more complex molecules expose an apolar group, such as an aromatic amino acid side chain of a protein, to solvent water, a complete cage of solvent water cannot form. Nevertheless, there may be appreciable stabilization energy that results from smaller, transient water clusters stabilized by their mutual hydrogen bonding at the interface with the apolar residue. This is thought to make an important contribution to hydrophobic interactions, discussed in the next section.

The experimental criterion that establishes that a hydrogen bond exists between A and B (A, B = N or O) is that the A$\cdots$B distance in A—H$\cdots$B is about 2.8 to 3.0 Å. Without a hydrogen bond, that distance is greater than 3.0 Å. A quantum mechanical description of a hydrogen bond involves the electrostatic attraction of the positively charged hydrogen nucleus of the donor to a negatively charged, nonbonding orbital on the acceptor. This attraction of one proton to two nuclei can provide an efficient path for moving charges; the proton can move from one nucleus to another:

$$O—H\cdots O \longrightarrow O\cdots H—O$$

The very high mobility of H^+ in water is explained by this mechanism.

In biological systems, the most important hydrogen bonds occur with the lone-pair electrons of O and N, in whatever molecules these are available. The strongest hydrogen bonds occur with F, but fluorine is not commonly encountered biologically. Carbon and elements in the next row in the periodic table (for example, S or Cl) form very weak hydrogen bonds, if at all.

Hydrophobic and Hydrophilic Environments

The terms *hydrophobic* (water fearing) and *hydrophilic* (water loving) are often encountered in the biochemical literature. The concepts involved are obviously very important, but their description and origins are complicated. Excellent detailed analyses are given by Tanford (1980) and Muller (1990).

Because water is the medium in which biological organisms have developed, the interactions of biomolecules with water are very important. To understand the distinctions between terms like *hydrophobic* and *hydrophilic*, we need to recognize that there is always a competition between the interactions of molecules of two substances—for instance, hexane and water—with one another and their interactions with molecules of the same type. Water molecules are *hydrophilic*, in the sense that there are strong interactions among the water molecules in either pure water or aqueous solutions. Hexane and other nonpolar substances also contain molecules that interact with one another by nonbonding interactions of the *van der Waals* type. It is a mistake to describe such interactions as resulting from "hydrophobic forces" because they occur completely and naturally in the absence of any water whatsoever.

When samples of the two liquids, water and hexane, are brought together, they do not mix. In fact, a small amount of each component is dissolved in the bulk phase of the other at equilibrium, although the two liquids form separate phases with an interface between them. They do not mix because the interactions of water molecules with other water molecules are more favorable (lower ΔG, incorporating contributions both from ΔH and $-T\Delta S$) than are the interactions of water with hexane molecules. A similar situation occurs for the interactions of the hexane molecules. Another way of looking at this is that the transfer of additional water from the liquid water phase to the liquid hexane phase results in an increase in the chemical potential μ. Because this is characteristic of an unfavorable process, it will not happen spontaneously.

Many biomolecules have both hydrophilic and hydrophobic character in the same molecule. Such molecules are said to be *amphiphilic;* examples include phospholipids, many amino acids, nucleotide bases, and even certain proteins and mucopolysaccharides. Detergents are common synthetic examples. In some amphiphiles, one end of the molecule is hydrophilic, and the other end is hydrophobic. Such molecules are good surfactants, and we have seen in chapter 5, some examples of their ability to concentrate at the interface between a lipid or hydrocarbon phase and an aqueous phase. Such molecules also play important roles in forming biological membranes.

An important characteristic of proteins is the way in which the hydrophilic or hydrophobic amino acid side chains are distributed along the polypeptide backbone. The specific sequence of amino acids is of course what distinguishes one protein from another. For water-soluble, globular proteins of known structure, hydrophobic amino acids are typically buried inside the protein, whereas hydrophilic amino acids tend to be at the surface of the protein, where they interact with solvent water. A quantitative description of this phenomenon has been developed in the form of hydropathy plots (Kyte and Doolittle 1982) where a numerical index of the hydrophilic/hydrophobic character of short protein segments is plotted along the chain length and compared with the location of those segments in the native protein structure. A hydropathy plot for lactate dehydrogenase from dogfish is shown in figure 9.44. In other applications, intrinsic membrane proteins contain extensive stretches of 20 or more hydrophobic amino acids in regions where the proteins pass through the interior part of the membrane lipid bilayer. An isolated polar or charged amino acid side chain in such a hydrophobic stretch may be paired with an oppositely charged residue on a neighboring chain, or it may serve as a specific ligand to an associated prosthetic group such as heme or chlorophyll.

It is essential to maintain a balanced view of the interplay of hydrophilic and hydrophobic concepts. One should not picture water as being entirely excluded from a hydrophobic environment, for example. The solubility of water in liquid hydrocarbons is sufficient so that there is a rapid exchange across the interface between the small amount of dissolved water and that in the bulk phase. Most biological membranes are in fact quite freely permeable

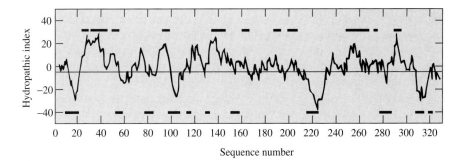

▲ FIGURE 9.44

Hydropathy profile of the enzyme dogfish lactate dehydrogenase. The hydropathy index, which is positive for hydrophobic amino acid side chains and negative for hydrophilic, is averaged over nine adjacent amino acids and plotted at the sequence number of the central amino acid. Bars at the top and bottom of the figure indicate those regions of the protein known to be on the inside and outside, respectively, based on the known X-ray crystallographic structure of the protein. A reasonably good correlation between the hydropathy index and amino acid location is found. (Adapted from Kyte and Doolittle 1982.)

to water, even though the membrane has a hydrocarbon-like interior. The permeability to charged ions and to large hydrophilic solute molecules is very much less. The rapid movement of water across membranes is an important aspect of *homeostasis,* whereby individual cells and biological organisms respond to biochemical stress in a way that does not lead to a serious and potentially disruptive imbalance of forces.

Molecular Dynamics Simulation

A goal of the chemist tackling biological systems is predictive power. We can improve our understanding of the structures and interactions of large molecules by simulating their motions on a computer. The free energy of the system characterizes which reactions occur and which species or conformations are present at equilibrium. To calculate a Gibbs free energy, it is necessary to calculate both the energy (actually enthalpy) and the entropy. We have given formulas for the energy terms [Eq. (9.40)]. The entropy requires knowledge of how many ways the system can change without affecting the energy; the more rearrangements that are possible for the system, the higher is the entropy, and the lower is the free energy. There are two methods for calculating entropy and free energy for a system by simulating the motion of the molecules: *Monte Carlo* and *molecular dynamics.* Both methods calculate the energies for different possible arrangements of the system. The free energy depends on the *average* of $e^{-U/RT}$ averaged over the different arrangements of the system:

$$G = -RT \ln\langle e^{-U/RT}\rangle$$

where $\langle\,\rangle$ represents the average over the possible states of the system. The important concept is that the averaging includes the contribution of the entropy to the free energy. We can understand this by considering two molecules, both of which have the same minimum energy U. For one of them, any change in conformation raises the energy greatly; the molecule is rigid. For the other, changes in conformation do not raise the energy much; it is a flexible molecule. Clearly, the rigid molecule will have a higher average value of $\langle e^{-U/kT}\rangle$ and thus a higher free energy than the flexible molecule. We describe the flexible molecule with the lower free energy as having a higher entropy.

Monte Carlo Method

The Monte Carlo method is named after the famous gambling city in Europe; an equivalent name could be the Las Vegas method. In this method, you start with an arbitrary arrangement of the system and perturb it by slightly moving one molecule or one part of a large molecule. After each move, you calculate the change in energy. If the energy has decreased, you allow the move; if the energy has increased, you compare the increase with the thermal energy kT to decide whether to accept the move or not. You compare the value of $e^{-(\Delta U/kT)}$ with a random number between 0 and 1; here, ΔU is the increase in energy per particle. If $e^{-(\Delta U/kT)}$ is larger than the random number, you allow the move; otherwise, you try another move. The randomness is where the gambling comes in. This procedure makes the system approach a minimum-energy arrangement. Perturbations that lower the energy are always allowed, but perturbations that raise the energy are sometimes allowed. Changes in the system that raise the energy are necessary to prevent it from getting trapped in a local energy minimum. By repeating this procedure many, many times, you will eventually find the lowest *free* energy for the system rather than the lowest energy. The number of perturbations that do not increase the system energy near the minimum is a measure of the entropy. This allows calculation of a free energy.

Molecular Dynamics Method

The molecular dynamics method directly simulates the motions of all molecules in the system. Newton's law is used for each molecule. The force on each atom of a molecule is calculated from the potential energy, using Eq. (9.40); the force depends on the derivatives of the potential energy along different directions in space. The motion of each atom is then calculated from Newton's law:

$$\textbf{Force} = (\text{mass})(\textbf{acceleration})$$

Both **force** and **acceleration** are vectors with a magnitude and a direction. Therefore, they will determine how fast and in what direction each atom in a molecule will move. The velocities and thus average kinetic energy of the molecules in the system determine the temperature, so the motion can be simulated at any temperature. The system is started in an arbitrary arrangement at a temperature near absolute zero; the atoms are nearly stationary. The velocities of the atoms are then allowed to increase so that the average kinetic energy of the system is increased to correspond to the temperature of interest, such as 300 K. Now the motion of all the atoms in the system is simulated at that temperature. The system could be a protein or a nucleic acid fragment surrounded by water molecules. A great deal of computer power is needed to simulate the motion even for a few nanoseconds, because Newton's law must be applied to each atom every one or two femtoseconds to give realistic behavior. However, the simulation allows you to learn how the shape of the molecule fluctuates and how the molecule and the solvent rearrange to facilitate binding of a substrate (Karplus and Petsko 1990). The molecular motions as well as the free energy depend on the potential energy, so it is not surprising that analysis of the motions of all molecules can provide the free energy of the system.

Only differences in free energies can be measured experimentally, so it is important to be able to calculate these differences. For example, the free-energy change of unfolding a protein could be calculated; this would involve changes only in conformation—rotation about single bonds. The binding of a substrate to an enzyme can be calculated, but it is easier to calculate the difference in free energies of binding two substrates, S_1 and S_2. The following thermodynamic cycle is used:

$$
\begin{array}{ccc}
E + S_1 & \xrightarrow{\Delta G_1} & ES_1 \\
\Big\downarrow \Delta G_3 & & \Big\downarrow \Delta G_4 \\
E + S_2 & \xrightarrow{\Delta G_2} & ES_2
\end{array}
$$

The difference in standard binding free energies, $\Delta\Delta G°$, can be obtained experimentally by measuring the equilibrium constants:

$$\Delta\Delta G^0(\text{experimental}) = \Delta G_2^0 - \Delta G_1^0 = -RT \ln \frac{K_2}{K_1}$$

where K_2 and K_1 are the equilibrium binding constants. The difference can be calculated theoretically by simulating the conversion of free substrate S_1 to S_2 (ΔG_3) and the conversion of S_1 to S_2 bound to the enzyme (ΔG_4):

$$\Delta\Delta G^0(\text{calculated}) = \Delta G_4 - \Delta G_3$$

The free energy of conversion is simulated by continuously varying, in the calculation, one atom into another. For example, to calculate the difference in free energy of binding of formate ion versus acetate ion requires the conversion of a methyl group of acetate ion to a hydrogen atom of formate ion. In the calculation, the parameters that specify the interactions of the methyl with its surroundings are changed slightly toward that of a hydrogen atom. After each slight change, enough molecular dynamics is done so that the slight change in free energy can be calculated. Thus, the total free energy change ΔG for the conversion of a methyl to a hydrogen is calculated as the sum of the small changes in free energy caused by each perturbation in parameters. The change in free energy is independent of path; this is true even if a path is chosen that is not experimentally possible. Thus, the free-energy difference of binding formate versus acetate in an enzyme site is calculated by (1) changing formate into acetate surrounded by solvent—this is ΔG_3; (2) changing formate into acetate surrounded by the enzyme-binding site—this is ΔG_4; (3) subtracting (1) from (2). Free energies of chemical reactions can also be calculated by this method. This is done by simulating the conversion of one molecule into another (described by Bash et al. 1987).

These types of calculation have already provided some quantitative predictive power (McCammon 1987; Jorgensen 1989; Kollman and Merz 1990). However, their most important contribution is to provide a qualitative understanding of how reactions and conformations depend on molecular interactions. We can learn to predict qualitatively how a change in a substrate, a protein, or a nucleic acid will affect the reaction.

Outlook

The novel perspectives of quantum mechanics revealed the orbital structures of atoms and molecules. In short, quantum mechanics lets us understand the covalent geometries of molecules. A large number of noncovalent interactions determine the conformations of macromolecules. The sheer number of weak interactions leads to a specific structure of reasonable thermodynamic stability. These weak interactions are easily disrupted individually, yielding conformational flexibility to macromolecules that is essential for function. For example, many enzymes adapt their structure to improve noncovalent contacts with a substrate in an active site.

Our understanding of noncovalent interaction does not yet allow prediction of conformation from a knowledge of a protein's sequence. Nonetheless, there have been great successes in the use of physical principles to understand biological structure. One prominent example is structure-based drug design (Kuntz et al. 1994; Marrone et al. 1997). The human immunodeficiency virus 1 (HIV1) makes the protein protease, whose job is to "chop" a large HIV protein into smaller pieces. The action of this protein is essential for the virus to infect humans. Structural biologists determined the three-dimensional conformation of the HIV-protease, which contained a cleft that is the active site for cleavage of the protein substrate. Using computer mod-

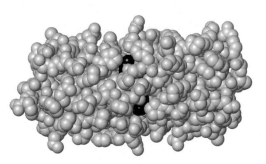

◀ FIGURE 9.45

Structure of HIV-protease with bound inhibitor. On the left, the protein is shown schematically as a ribbon, which represents its peptide backbone. All atoms of the inhibitor are shown in black with their full radii. On the right, both protease (light color) and inhibitor (black) are shown with all their atoms displayed. Note that the protease inhibitor fills a distinct cleft in the protein, which contains the active site for the enzyme. (Coordinates from Lam et al., 1994 *Science* 263:380–384.)

eling and a knowledge of the laws of bonding and noncovalent interactions, scientists were able to design artificial molecules that bind within the active site of the enzyme, preventing the enzyme from performing its task (figure 9.45). These molecules needed to have correct bonding geometry (such that they were chemically feasible), and they needed to form hydrogen bonds, electrostatic and London interactions with the protease, such that the binding affinity would be high. The success of these scientists created a very effective drug against HIV.

Summary
Photoelectric Effect

Einstein equation for kinetic energy of ejected electrons:

$$\frac{1}{2}mv^2 = h\nu - \Phi \tag{9.4}$$

Wave-Particle Duality

De Broglie relation between particle momentum and wavelength:

$$\lambda = \frac{h}{mv} \tag{9.5}$$

Heisenberg Uncertainty Principle

$$\Delta p \Delta x \geq \frac{1}{2}\left(\frac{h}{2\pi}\right) \tag{9.6}$$

Schrödinger's Equation

Schrödinger's time-independent equation:

$$\mathcal{H}\psi = E\psi \tag{9.8}$$

$\mathcal{H}$ = Hamiltonian operator

$$\equiv -\frac{h^2}{8\pi^2} \sum_i \frac{1}{m_i} \cdot \left(\frac{\partial^2}{\partial x_i^2} + \frac{\partial^2}{\partial y_i^2} + \frac{\partial^2}{\partial z_i^2} \right) + U(x_i, y_i, z_i)$$

h = Planck's constant
Σ_i = sum over all particles in system
m_i = mass of each particle i in system
x_i, y_i, z_i = coordinates of each particle i in system
$U(x_i, y_i, z_i)$ = potential-energy operator for system
ψ = eigenfunction or wavefunction, a function of coordinates of all particles in system; there will be an infinite number of wave-functions corresponding to different states of the system and usually each wavefunction will correspond to a different energy of the system
E = energy of the system

The value of any measurable property for a system characterized by wavefunction ψ can be calculated from the appropriate operator:

$$\text{value} = \int \psi^* \text{ operator } \psi \, dv$$

ψ = normalized wavefunction of system
ψ^* = its complex conjugate
$\int \psi^*\psi \, dv = 1$
$\psi^* = \psi$, for real wavefunctions

The probability of finding an electron in a small volume of size Δv around point x_0, y_0, z_0 is $[\psi(x_0, y_0, z_0)]^2 \Delta v$, where $\psi(x_0, y_0, z_0)$ is the wavefunction of the electron evaluated at point x_0, y_0, z_0. To calculate the probability of finding an electron in a finite volume of space, the square of the wavefunction is integrated over the volume.

Some Useful Operators

Energy E:

Operator for Cartesian coordinates for one particle:

$$\mathcal{H} = -\frac{h^2}{8\pi^2 m} \left(\frac{\partial^2}{\partial x^2} + \frac{\partial^2}{\partial y^2} + \frac{\partial^2}{\partial z^2} \right) + U$$

Position x, y, z and momentum p_x, p_y, p_z:

The linear momentum components along x, y, z are p_x, p_y, p_z. The classical definition of momentum is mass · velocity. Units are, for example, g cm s^{-1} or kg m s^{-1}.

Operators in Cartesian coordinates:

$$p_x = \frac{h}{2\pi i} \frac{\partial}{\partial x} \qquad p_y = \frac{h}{2\pi i} \frac{\partial}{\partial y} \qquad p_z = \frac{h}{2\pi i} \frac{\partial}{\partial z}$$

Systems Whose Schrödinger Equation Can Be Solved Exactly

Particle in a box:

Potential energy for a one-dimensional box is $U(x) = 0$ in the box and $U(x) = \infty$ outside the box:

$$-\frac{h^2}{8\pi^2 m}\frac{d^2\psi}{dx^2} = E\psi$$

$$\frac{h^2}{8\pi^2} = \frac{\hbar^2}{2} = 5.561 \times 10^{-69}\ \text{J}^2\,\text{s}^2 \qquad (9.15)$$

m = mass of particle
ψ = wavefunction of particle in box
E = energy of particle in box

Normalized wavefunctions for a particle in a one-dimensional box of length a:

$$\psi_n = \sqrt{\frac{2}{a}}\sin\frac{n\pi x}{a} \qquad (9.16)$$

ψ_n = normalized wavefunction of particle in box in quantum state n
n = quantum number, a positive integer
a = length of box

Energies for a particle in a one-dimensional box of length a are proportional to n^2:

$$E_n = \frac{h^2 n^2}{8ma^2} \qquad (9.17)$$

E_n = energy of particle in quantum state n; each energy corresponds to a unique quantum state; that is, no energy value is degenerate
n = quantum number, a positive integer
$h^2/8$ = $5.4883 \times 10^{-68}\ \text{J}^2\,\text{s}^2$
m = mass of particle
a = length of box

Normalized wavefunctions for a particle in a two-dimensional box:

$$\psi_{n_x n_y} = \frac{2}{\sqrt{ab}}\sin\frac{n_x \pi x}{a}\sin\frac{n_y \pi y}{b}$$

$\psi_{n_x n_y}$ = normalized wavefunction of particle in box of quantum state specified by quantum numbers n_x and n_y
n_x, n_y = quantum numbers, positive integers
a = length of box in x-direction
b = length of box in y-direction

Energies for a particle in a three-dimensional box:

$$E_{n_x n_y n_z} = \frac{h^2}{8m}\left(\frac{n_x^2}{a^2} + \frac{n_y^2}{b^2} + \frac{n_z^2}{c^2}\right) \qquad (9.20)$$

Schrödinger equation for a one-dimensional harmonic oscillator:

$$-\frac{h^2}{8\pi^2 \mu}\frac{d^2\psi}{dx^2} + \frac{k}{2}x^2\psi = E\psi \qquad (9.23)$$

ψ = wavefunction for a linear oscillator whose potential energy is directly proportional to the square of its displacement

E = energy of harmonic oscillator
k = force constant of harmonic oscillator, mass time^{-2}
h = Planck's constant
μ = reduced mass of oscillator (molecule)

Energies for a harmonic oscillator are equally spaced:

$$E_v = \left(v + \frac{1}{2}\right)h\nu_0$$

$$\nu_0 = \frac{1}{2\pi}\sqrt{\frac{k}{\mu}} \qquad (9.24)$$

$E_0 = h\nu_0/2$ = zero-point energy
v = quantum number for harmonic oscillator = 0, or 1, or 2, etc.
h = Planck's constant
ν_0 = fundamental vibration frequency of oscillator
k = force constant of harmonic oscillator, mass time^{-2}
μ = reduced mass of oscillator

Energies for an electron in a hydrogen atom are inversely proportional to n^2:

$$E_n = -\frac{2\pi^2\mu e^4}{(4\pi\varepsilon_0)^2h^2n^2} \qquad \text{(SI units)} \qquad (9.27)$$

$$E_n = -\frac{2\pi^2\mu e^4}{h^2n^2} \qquad \text{(cgs-esu units)} \qquad (9.26)$$

μ = reduced mass of electron and proton = $\dfrac{m_p m_e}{m_p + m_e}$

m_p = mass of proton
m_e = mass of electron
e = elementary charge
h = Planck's constant
ε_0 = permittivity of vacuum
n = principal quantum number

For SI units, $2\pi^2\mu e^4/(4\pi\varepsilon_0)^2h^2 = 2.179 \times 10^{-18}$ J; $m_e = 9.110 \times 10^{-31}$ kg; $e = 1.602 \times 10^{-19}$ C; $h = 6.26 \times 10^{-34}$ J s; $\varepsilon_0 = 8.8542 \times 10^{-12}$ C^2 N^{-1} m^{-2}.
For cgs–esu units, $2\pi^2\mu e^4/h^2 = 2.179 \times 10^{-11}$ erg; $m_e = 9.110 \times 10^{-28}$ g; $e = 4.803 \times 10^{-10}$ esu; $h = 6.626 \times 10^{-27}$ erg s.

$$E_n = \frac{-2.179 \times 10^{-18}}{n^2} \qquad \text{J atom}^{-1}$$

$$E_n = \frac{-313.6}{n^2} \qquad \text{kcal mol}^{-1}$$

$$E_n = \frac{-1312}{n^2} \qquad \text{kJ mol}^{-1}$$

$$E_n = \frac{-13.60}{n^2} \qquad \text{eV atom}^{-1}$$

Coulomb's Law

The energy of interaction between charges:

$$U(r) = \frac{q_1 q_2}{4\pi\varepsilon_0 r} \qquad \text{(energy in joules)} \qquad \textbf{(9.13)}$$

where q_1 and q_2 are in coulombs, r is in meters, and ε_0 (the permittivity constant) $= 8.854 \times 10^{-12}\, C^2\, N^{-1}\, m^{-2}$.

$$U(r) = 1389\frac{q_1 q_2}{r} \qquad \text{(energy in kJ mol}^{-1}\text{)}$$

where q_1 and q_2 are in multiples of an electronic charge and r is in Å.

For charges interacting in a medium of dielectric constant ε, the right side of Eq. (9.13) or (9.37) is divided by ε.

Dipoles and Their Interaction Energy

The components of the dipole moment vector are:

$$\boldsymbol{\mu}_x = \sum_i q_i x_i$$

$$\boldsymbol{\mu}_y = \sum_i q_i y_i$$

$$\boldsymbol{\mu}_z = \sum_i q_i z_i \qquad\qquad \textbf{(9.38)}$$

The interaction between two dipoles is:

$$U(R) = \frac{1389}{R_{12}^3}\frac{\mu_1\mu_2}{}(\cos\theta_{12} - 3\cos\theta_1\cos\theta_2) \qquad \text{(energy in kJ mol}^{-1}\text{)} \textbf{ (9.39)}$$

where μ_1 and μ_2 are in electron Å (1 electron Å = 4.803 debye) and R_{12} is in Å.

θ_1 and θ_2 are the angles each dipole makes with the line joining the two point dipoles; θ_{12} is the angle between the two dipoles. See figure 9.39.

Intramolecular (Within) and Intermolecular (Between) Interactions

$$
\begin{aligned}
U = &+k_r(r - r_{eq})^2 && \text{(bond stretching)}\\
&+k_b(\theta - \theta_{eq})^2 && \text{(bond angle bending)}\\
&+(V_n/2)[1 + \cos(n\phi - \phi_0)] && \text{(torsion angle rotation; } n = 2 \text{ or } 3)\\
&+\frac{B_{ij}}{r_{ij}^{12}} && \text{(van der Waals repulsion;}\\
&&& B_{ij}\text{ is positive)}\\
&-\frac{A_{ij}}{r_{ij}^{6}} && \text{(London attraction; } A_{ij}\text{ is positive)}\\
&+C\frac{q_i q_j}{r_{ij}} && \text{(Coulomb interaction;} \qquad \textbf{(9.42)}\\
&&& \text{C depends on the units)}
\end{aligned}
$$

All terms can contribute to intramolecular interactions; only the last three can contribute to intermolecular interactions.

Mathematics Needed for Chapter 9

In this chapter, for the first time, we use a few trigonometric identities:

$$\cos n\pi = 1 \qquad \text{(for } n = 0 \text{ or an even integer)}$$
$$= -1 \qquad \text{(for } n = \text{an odd integer)}$$
$$\sin n\pi = 0 \qquad \text{(for } n = \text{any integer)}$$
$$\sin(\theta + 2\pi) = \sin\theta$$
$$\cos(\theta + 2\pi) = \cos\theta$$
$$\sin^2\theta + \cos^2\theta = 1 \qquad \text{(for any value of } \theta)$$

We discuss differential equations, but you are not expected to solve any. You should be able to use calculus to verify the solutions, however. Remember that

$$\frac{d}{d\theta}(\sin a\theta) = a\cos a\theta$$

$$\frac{d}{d\theta}(\cos a\theta) = -a\sin a\theta$$

$$\frac{d}{dx}(e^{ax}) = ae^{ax}$$

$$\frac{d}{dx}(e^{ax^2}) = 2axe^{ax^2}$$

Integrals of functions other than powers of x are used in solving some of the problems:

$$\int_0^\infty e^{-a^2x^2}\,dx = \frac{1}{2a}\sqrt{\pi}$$

$$\int \sin^2 ax\,dx = \frac{x}{2} - \frac{\sin 2ax}{4a} + C \qquad (C = \text{a constant})$$

The complex conjugate of a function is obtained by changing i (the imaginary) to $-i$ in the function.

We use vectors to represent dipoles. A vector has a magnitude and a direction; it is represented by a boldface letter such as $\boldsymbol{\mu}$. We can think of a vector as an arrow. The length of the arrow is the absolute magnitude; it points along the direction. The direction of a vector can be characterized quantitatively by specifying its components along the xyz-axes of a coordinate system. The absolute magnitude is

$$|\boldsymbol{\mu}| = \sqrt{\mu_x^2 + \mu_y^2 + \mu_z^2}$$

where μ_x, μ_y, μ_z are the components along x, y, z. The dot product (scalar product) of two vectors, $\mathbf{a}\cdot\mathbf{b}$, produces a scalar (a number). Its numerical value is obtained by either of two equivalent formulas. The components can be used:

$$\mathbf{a} \cdot \mathbf{b} = a_x b_x + a_y b_y + a_z b_z$$

or the angle between the two vectors can be used:

$$\mathbf{a} \cdot \mathbf{b} = |a|\,|b|\cos\theta$$

References

The fundamentals of forces and energies are presented in

Halliday, D., R. Resnick, and J. Walker. 1993. *Fundamentals of Physics*, 4th ed. New York: Wiley.

Good discussions of atomic and molecular orbitals can be found in the following introductory chemistry books and organic chemistry books:

Dickerson, R. E., H. B. Gray, M. Y. Darensbourg, and D. J. Darensbourg. 1984. *Chemical Principles*, 4th ed. Menlo Park, CA: Benjamin/Cummings.

Leach, A. R. 2001. *Molecular Modeling. Principles and Applications*, 2d ed. Harlow, UK: Prentice Hall.

Mahan, B. H., and R. J. Myers. 1987. *University Chemistry*, 4th ed. Menlo Park, CA: Benjamin/Cummings.

Oxtoby, D. W., H. P. Gillis, and N. H. Nachtrieb. 1999. *Principles of Modern Chemistry*, 4th ed. Fort Worth, TX: Saunders.

Streitwieser, A., Jr., C. H. Heathcock, and E. M. Kosower. 1992. *Introduction to Organic Chemistry*, 4th ed. Upper Saddle River, NJ: Prentice Hall.

Vollhardt, K. P. C., and N. E. Schore. 1999. *Organic Chemistry Structure and Function*, 3d ed. New York: Freeman.

Books that provide more depth for the topics discussed here include the following:

DeKock, R. L., and H. B. Gray. 1989. *Chemical Structure and Bonding*, 2d ed. Mill Valley, CA: University Science Books.

McQuarrie, D. A. 1983. *Quantum Chemistry*. Mill Valley, CA: University Science Books.

McQuarrie, D. A., and J. D. Simon. 1997. *Physical Chemistry: A Molecular Approach*. Mill Valley, CA: University Science Books.

Ratner, M. A., and G. C. Schatz. 2001. *Introduction to Quantum Mechanics in Chemistry*. Upper Saddle River, NJ: Prentice Hall.

Tanford, C. 1980. *The Hydrophobic Effect*, 2d ed. New York: Wiley.

Suggested Reading

Axup, A. W., M. Albin, S. L. Mayo, R. J. Crutchley, and H. B. Gray. 1988. Distance Dependence of Photoinduced Long-Range Electron Transfer in Zinc/Ruthenium-Modified Myoglobins, *J. Am. Chem. Soc.* 110:435–439.

Bash, P. A., U. C. Singh, R. Langridge, and P. A. Kollman. 1987. Free Energy Calculations by Computer Simulation. *Science* 236:564–568.

Baylor, D. 1996. How Photons Start Vision. *Proc. Natl Acad. Sci. U.S.A*, 93:560–565.

Birge, R. R. 1981. Photophysics of Light Transduction in Rhodopsin and Bacteriorhodopsin. *Annu. Rev. Biophys. Bioeng.* 10:315–354.

Hoff, W. D., K.-H. Jung and J. L. Spudich. 1997. Molecular Mechanism of Photosignalling by Archael Sensory Rhodopsins. *Ann. Rev. Biophys. Biomol. Struct.* 26:223–258.

Hoffmann, R., and R. B. Woodward. 1968. The Conservation of Orbital Symmetry. *Acc. Chem. Res.* 1:17–22.

Honig, B. 1999. Protein Folding: From the Levinthal Paradox to Structure Prediction. *J. Mol. Biol.* 293:283–293.

Jorgensen, W. L. 1989. Free Energy Calculations: A Breakthrough for Modeling Organic Chemistry in Solution. *Acc. Chem. Res.* 22:184–189.

Karplus, M., and G. A. Petsko. 1990. Molecular Dynamics Simulation in Biology. *Nature* 347:631–638.

Klinman, J. P. 1989. Quantum Mechanical Effects in Enzyme-Catalyzed Hydrogen Transfer Reactions. *Trends Biochem. Sci.* 14:368–373.

Knowles, J. R. 1989. The Mechanism of Biotin-Dependent Enzymes. *Annu. Rev. Biochem.* 58:195–221.

Kochendoerfer, G. G., S. W. Lin, T. P. Sakmar, and R. A. Mathies. 1999. How Color Visual Pigments Are Tuned. *Trends Biochem. Sci.* 24:300–305.

Kohen, A., and J. P. Klinman. 1999. Hydrogen Tunneling in Biology. *Chem Biol.* 6:191–198.

Kollman, P. A., and K. M. Merz, Jr. 1990. Computer Modeling of the Interactions of Complex Molecules. *Acc. Chem. Res.* 23:246–252.

Kuntz, I. D., E. C. Meng, and B. K. Schoichet. 1994. Structure-Based Molecular Design. *Acc. Chem. Res.* 27:117–123.

Kyte, J., and R. F. Doolittle. 1982. A Simple Method for Displaying the Hydropathic Character of a Protein. *J. Mol. Biol.* 170:723–764.

Marrone, T. J., J. M. Briggs, and J. A. McCammon. 1997. Structure-Based Drug Design: Computational Advances. *Ann. Rev. Pharmacol. Toxicol* 37:71–90.

Mathies, R., and L. Stryer. 1976. Retinal Has a Highly Dipolar Vertically Excited Singlet State. *Proc. Natl. Acad. Sci. USA* 73:2169–2173.

McCammon, J. A. 1987. Computer-Aided Molecular Design. *Science* 238:486–491.

Mesecar, A. D., B. L. Stoddard, and D. E. Koshland, Jr. 1997. Orbital Steering in the Catalytic Power of Enzymes: Small Structural Changes with Large Catalytic Consequences. *Science* 277:202–206.

Muller, N. 1990. Search for a Realistic View of Hydrophobic Effects. *Acc. Chem. Res.* 23:23–28.

Nakanishi, K., V. Balogh-Nair, M. Arnaboldi, K. Tsujimoto, and B. Honig. 1980. An External Point-Charge Model for Bacteriorhodopsin to Account for Its Purple Color. *J. Am. Chem. Soc.* 102:7945–7947.

Némethy, G., and H. A. Scheraga. 1990. Theoretical Studies of Protein Conformation by Means of Energy Computations. *FASEB J.* 4:3189–3197.

Richards, W. G., ed. 1989. *Computer-Aided Molecular Design.* New York: VCH Publishers.

Schramm, V. L. 1998. Enzymatic Transition States and Transition State Analog Design. *Ann. Rev. Biochem.* 67:693–720.

Stryer, L. 1988. Molecular Basis of Visual Excitation. *Cold Spring Harbor Symp. Quant. Biol.* 52:283–294.

Stubbe, J. A. 1989. Protein Radical Involvement in Biological Catalysis. *Annu. Rev. Biochem.* 58:257–285.

Warshel, A., and A. Papazyan. 1998. Electrostatic Effects in Macromolecules: Fundamental Concepts and Practical Modeling. *Curr. Opin. Struct. Biol.* 8:211–217.

Weiner, S. J., P. J. Kollman, D. A. Case, U. C. Singh, C. Ghio, G. Alagona, S. Profeta, Jr., and P. Weiner. 1984. A New Force Field for Simulations of Proteins and Nucleic Acids. *J. Am. Chem. Soc.* 106:765–784.

Zuo, J. M., M. Kim, M. O'Keefe, and J. C. H. Spence. 1999. Direct Observation of *d*-Orbital Holes and Cu–Cu Bonding in CU_2O. *Nature* 401:49–52.

Problems

1. (a) The energy required to remove an electron from K metal (called the *work function*) is 2.2eV (where 1 eV = 1.60×10^{-19} J), whereas that of Ni is 5.0 eV. A beam of light impinges on a clean surface of the two metals. Calculate the threshold frequencies and wavelengths of light required to emit photoelectrons from the two metals.

   (b) Will violet light of wavelength 400 nm cause ejection of electrons in K? In Ni?

   (c) Calculate the maximum kinetic energy of electrons emitted in part (b).

2. Biological specimens are often examined by microscopy. However, microscopy with visible light is limited to viewing details of a specimen on the order of the wavelength of light (400 nm). The macromolecules that make up the cell are much smaller, often about 2–10 nm in size, and the interatomic bonds that make up the molecular structure are about 0.2 nm in length. One method that can be used to reveal the atomic-level details of biological molecules is electron microscopy (see chapter 11), in which a beam of electrons is focused onto a biological sample. Modern electron microscopes can emit a beam of electrons with a velocity of 1.5×10^8 m s^{-1}.

   (a) What is the wavelength of an electron particle in this beam? Is the wavelength short enough to reveal molecular details at the atomic level?

   (b) The wavelength of the particle determines the resolution of the microscopy that can be performed. Assume that you desire a minimum uncertainty in the position of the electron of 1.0 Å, 10^{-10} m. Using the uncertainty principle, what is the maximum uncertainty that is acceptable in the momentum of the particle?

3. When spacing between translational energy levels is small compared with random thermal kinetic energies, classical mechanics is a good approximation for quantum mechanics. Thermal kinetic energy is of the order of magnitude of kT, where k is Boltzmann's constant. Consider which particles in which boxes can be treated classically in their translational motion. For each system listed, calculate the energy of transition from the ground state to the first excited state and calculate the ratio of this transition energy to kT. State whether the system can be treated classically or not.

   (a) A helium atom in a 1-μm cube at 25°C and at 1 K.

   (b) A protein with a molecular weight of 25,000 in a 100-Å cube at 25°C and at 1 K.

   (c) A helium atom in a 2-Å cube at 25°C and at 1 K.

4. Consider a one-dimensional box of length l. For parts (a), (b), and (c), calculate the energy of the system in the ground state and the wavelength (in Å) of the first transition (longest wavelength).

   (a) $l = 10$ Å, and the box contains one electron.

   (b) $l = 20$ Å, and the box contains one electron.

   (c) $l = 10$ Å, and the box contains two electrons (assume no interaction between the electrons).

   (d) Write the complete Hamiltonian for the system described in part (c).

(e) What will happen to the answers in part (c) if the potential of interaction between the electrons is included in this Hamiltonian?

5. Consider a particle in a one-dimensional box.

   (a) For a box of length 1 nm, what is the probability of finding the particle within 0.01 nm of the center of the box for the lowest-energy level?

   (b) Answer part (a) for the first excited state.

   (c) The longest-wavelength transition for a particle in a box [not the box in part (a)] is 200 nm. What is the wavelength if the mass of the particle is doubled? What is the wavelength if the charge of the particle is doubled? What is the wavelength if the length of the box is doubled?

6. The exact ground-state energy for a particle in a box is $0.125h^2/ma^2$; the exact wavefunction is $(\sqrt{2}/a)\sin(n\pi x/a)$. Consider an approximate wavefunction $= x(a - x)$.

   (a) Show that the approximate wavefunction fits the boundary conditions for a particle in a box.

   (b) Given this approximate wavefunction, $\psi(x) = x(a - x)$, the approximate energy E' of the system can be calculated from

   $$E' = \frac{\int_0^a \psi(x)\mathcal{H}\psi(x)\,dx}{\int_0^a \psi^2(x)\,dx}$$

   This is called the *variational method*. Calculate the approximate energy E'.

   (c) What is the percent error for the approximate energy?

   (d) According to the variational theorem, the more closely the wavefunction used approximates the correct wavefunction, the lower will be the energy calculated according to part (b). The converse is also true. Propose another approximate wavefunction that could be used to calculate an approximate energy. How can you find out whether it is a better approximation?

7. The π electrons of metal porphyrins, such as the iron-heme of hemoglobin or the magnesium-porphyrin of chlorophyll, can be visualized using a simple model of free electrons in a two-dimensional box.

   (a) Obtain the energy levels of a free electron in a two-dimensional square box of length a.

   (b) Sketch an energy-level diagram for this problem. Set $E_0 = h^2/8ma^2$ and label the energy of each level in units of E_0.

   (c) For a porphyrin-like hemin that contains 26 π electrons, indicate the electron population of the filled

or partly filled π orbitals of the ground state of the molecule. (Note that orbitals that are degenerate in energy, $E_{12} = E_{21}$, can still hold two electrons each.)

   (d) The porphyrin structure measures about 1 nm on a side ($a = 1$ nm). Calculate the longest-wavelength absorption band position for this molecule. (Experimentally these bands occur at about 600 nm.)

8. Nitrogenase is an enzyme that converts N_2 to NH_3. Extraction procedures have shown that nitrogenase contains an iron–sulfur cluster, Fe_4S_4, where the atoms of each type are arranged alternately at the corners of a cube. We wish to attempt to model the bonding in this iron–sulfur cube with the wavefunctions derived for an electron in a cubical (three-dimensional) box whose edges have length a:

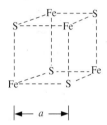

   (a) Write the Hamiltonian for the electron in the box.

   (b) Write general expressions for the energies (eigenvalues) and wavefunctions (eigenfunctions) for the electron in a three-dimensional cubical box as a function of the quantum numbers.

   (c) Construct an energy-level diagram showing the relative ordering of the lowest 11 "molecular orbitals" of this cube. Label the levels in terms of their quantum numbers. Be alert for degeneracies.

   (d) Assuming that the total number of valence electrons available is 20, show the electron configuration on the plot in part (c).

   (e) If $a = 3$ Å, calculate the transition energy for excitation of an electron from the highest filled orbital to the lowest unfilled orbital. At what wavelength in nm should this transition occur?

9. If G is a measurable parameter and g is the quantum mechanical operator corresponding to this parameter, the expected value of G, $\langle G \rangle$, can be calculated from

   $$\langle G \rangle = \frac{\int \psi^* g\psi\,dv}{\int \psi^*\psi\,dv}$$

   where the integration extends over all space important to the problem and ψ^* is the complex conjugate of ψ (if ψ has no imaginary part, $\psi^* = \psi$). Calculate the expected position $\langle x \rangle$, and momentum p for a harmonic oscillator in its ground state.

10. For a diatomic species A—B, the vibration of the molecule can be treated approximately by considering the molecule as a harmonic oscillator.

(a) Calculate the reduced mass for HF.

(b) Take k = 970 m^{-1}; plot the potential energy versus internuclear distance near the equilibrium distance of 0.92 Å.

(c) Sketch in the same diagram horizontal lines representing the allowed vibrational energy levels.

11. The C—C stretching vibration of ethylene can be treated as a harmonic oscillator.

(a) Calculate the ratio of the fundamental frequency for ethylene to that for completely deuterated ethylene.

(b) Putting different substituents on the ethylene can make the C—C bond longer or shorter. For a shorter C—C bond, will the vibration frequency increase or decrease relative to ethylene? State why you think so.

(c) If the fundamental vibration frequency for the ethylene double bond is 2000 cm^{-1}, what is the wavelength in nm for the first harmonic vibration frequency?

12. The fundamental vibration frequency of gaseous $^{14}N^{16}O$ is 1904 cm^{-1}.

(a) Calculate the force constant, using the simple harmonic oscillator model.

(b) Calculate the fundamental vibration frequency of gaseous $^{15}N^{16}O$.

(c) When $^{14}N^{16}O$ is bound to hemoglobin A, an absorption band at 1615 cm^{-1} is observed and is believed to correspond to the bound NO species. Assuming that, when an NO binds to hemoglobin, the oxygen is so anchored that its effective mass becomes infinite, estimate the vibration frequency of bound $^{14}N^{16}O$ from the vibration frequency of gaseous NO.

(d) Using another model—that binding of NO to hemoglobin does not change the reduced mass of the NO vibrator—calculate the force constant of bound NO.

(e) Estimate the vibration of $^{15}N^{16}O$ bound to hemoglobin using, respectively, the assumptions made in parts (c) and (d).

13. The ionization potential of an electron in an atom or molecule is the energy required to remove the electron. Calculate the ionization potential for a 1s electron in He$^+$. Give your answer in units of eV and kJ mol^{-1}.

14. A physicist wishes to determine whether the covalently bonded species (HeH)$^{2-}$ could be expected to exist in an ionized gas.

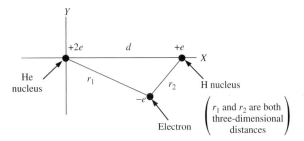

(a) Assuming that the nuclei remain separated by a fixed distance d (the nuclei do not move) and that the electron is free to move in three dimensions, write the Hamiltonian operator for this molecule.

(b) Write out fully the equation that must be solved exactly to determine the various wavefunctions and allowed energies of (HeH)$^{2-}$.

(c) Suppose you have obtained all the exact solutions to the equation in part (b). How can you tell whether the covalently bonded species (HeH)$^{2-}$ is energetically stable?

15. Consider the diatomic molecules NO, NO$^-$, and NO$^+$.

(a) Give MO diagrams for each of these molecules similar to those in figure 9.24.

(b) Rank these species in increasing order of bond energies and in increasing order of bond lengths.

16. The Schrödinger equation can give you the electronic energy levels and wavefunctions for the hydrogen atom.

(a) The electron distribution is characterized by the wavefunctions. Write the equation whose solution would allow you to calculate the radius of the sphere around the hydrogen nucleus within which the electron will be found 90% of the time. What is the radius of the sphere around the hydrogen atom within which the electron will be found 100% of the time?

(b) Calculate the longest wavelength absorption (in nm) for an electronic transition in a hydrogen atom. In what region of the electromagnetic spectrum does this occur?

(c) Write the Schrödinger equation whose solution will give you the translational energy levels and wavefunctions for a hydrogen atom in a cubic box with sides of length of 1 cm. Define all symbols in your equation.

(d) Calculate the longest wavelength absorption (in m) for a translational energy transition of a hydrogen atom in a cubic box of length 1 cm. In what region of the electromagnetic spectrum does this occur?

17. Sketch the spatial distribution of the electron density for all of the valence orbitals of one of the carbon atoms in each of the following molecules. Identify each of the orbitals.

(a) Acetylene

(b) Ethylene

(c) Ethane

18. In a protein, a basic structural unit is the amide residue:

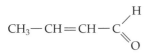

where C_α represents a carbon at the α position to the carbonyl group. A key clue to the solution of the three-dimensional structures of proteins was the realization that the six atoms shown lie in a plane. By constructing the molecular orbitals around the atoms O, C, and N, explain why the planar structure is expected. What are the expected bond angles between adjacent bonds?

19. The molecule crotonaldehyde is an unsaturated aldehyde with the structural formula

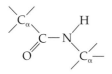

(a) Draw a picture showing the σ-bond structure for crotonaldehyde.

(b) Crotonaldehyde has four π electrons associated with the two double bonds. Draw diagrams that illustrate the four π-type MOs for crotonaldehyde, showing the AOs from which they were constructed, their phase (sign) relations, and with which atoms they are associated. Rank the MOs in energy in terms of bonding vs. antibonding character.

(c) Describe the geometric configuration that you expect for crotonaldehyde. Which groups, if any, will undergo free rotation with respect to the rest of the molecule?

(d) Aldehydes such as crotonaldehyde react with primary aliphatic amines such as n-butyl amine to form substituted imines, or Schiff bases:

R — C⟨H / O + H₂NR' ⟶ RCH=NR' + H₂O
                    Schiff base

What geometry do you expect for the atoms associated with the imine function in a Schiff base?

(e) Schiff bases readily undergo protonation by H^+ in acidic solution. What atom is protonated, and what geometric structure do you expect for the product? Is the resulting bond σ type or π type? What type of orbital (p, sp, sp^2, sp^3, ...) of the protonated atom is involved?

20. The nucleic acid–base adenine is a planar aromatic molecule containing 70 electrons, of which 20 are core electrons. The σ bonding around each aromatic C and N can be represented by sp^2 hybrid orbitals:

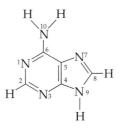

(a) How many π electrons are there in adenine? State which atoms contribute one electron to the π MOs and which atoms contribute two electrons.

(b) How many π or π^* MOs are there in adenine if we consider linear combinations of $2s$ and $2p$ AOs only?

(c) How many MOs of other types (σ, σ^*, n) derived from $2s$ and $2p$ AOs are there in adenine?

(d) Which atoms, if any, have nonbonding orbitals localized on them?

21. An approximate MO calculation gives the following π-electron charges on an amide group:

	Ground-state charges	Excited-state charges
O	−0.33e	+0.08e
C	+0.18e	−0.16e
N	+0.15e	+0.08e

$e = 4.8 \times 10^{-10}$ esu

Use the bond lengths given in figure 9.29 and 120° for the O—C—N bond angle.

(a) Calculate the components of the π dipole moment parallel and perpendicular to the C—N bond for the ground and excited state. Use debyes for units.

(b) Calculate the magnitudes of the π dipole moment for the ground state, for the excited state, and for the change in dipole moment.

(c) Plot the magnitudes and the directions to scale.

22. Consider the MOs of the planar peptide bond. The hypothesis is that the O and the central C and N atoms are sp^2 hybridized:

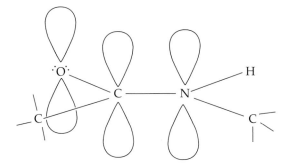

(a) Based on intuition derived from the particle-in-a-box wavefunctions, write the three π-type MOs for the "peptide bond" (ψ_1, ψ_2, ψ_3) in terms of the three equivalent p orbitals (ϕ_O, ϕ_C, ϕ_N).

(b) Normalize ψ_1, ψ_2, and ψ_3. To be properly normalized, the LCAO molecular orbitals

$$\psi_i = C_{i1}\phi_1 + C_{i2}\phi_2 + \cdots + C_{ij}\phi_j + \cdots$$

must satisfy three conditions: (1) normalization, $\sum_j C_{ij}^2 = 1$; (2) orthogonality, $\int \phi_r \phi_s\, d\tau = 0$, for $r \neq s$; (3) unit AO contribution, $\sum_i C_{ij}^2 = 1$. Considerable advantage can be taken of symmetry inherent in the problem; for example, in this case $C_{13} = \pm C_{11} = \pm C_{31} = \pm C_{33}$; $C_{12} = \pm C_{32}$; $C_{21} = \pm C_{23}$. You may wish to practice by testing the normalization conditions above for the MO expressions presented in connection with figure 9.27.

(c) Rank these MOs in order of their energies. Determine how many electrons each atom (O, C, N) contributes to the π system. Give the electron configuration in the π MOs of the peptide bond.

(d) What is the π-electron density for each of the *filled* MOs as a function of ϕ_O, ϕ_C, ϕ_N? State and justify any approximations that you make.

(e) Verify that the integrals of these densities over all space give the total number of π electrons in each orbital.

(f) Integration of the total π-electron density in a region around each atom yields the π-electron density of the atom. Calculate separately the π-electron charge on the O, C, and N atoms.

(g) What is the net charge on these atoms? (Remember that the molecule is neutral.) Calculate the peptide dipole moment given the coordinates $(x{:}y)$ for the three atoms (in Å): C(0:0), N(1.34:0), and O(-0.62, -1.07).

23. (a) Coulomb's law is an important element in determining molecular structure. To understand this better, calculate the interaction energy in kJ mol^{-1} for an electron and a proton at a distance of the Bohr radius of 0.529 Å. Calculate the Coulombic interaction energy between atoms in kJ mol^{-1} for an O$\cdots$H hydrogen-bond distance of 1.5 Å. The partial charge on O is -0.834; the partial charge on H is $+0.417$ in units of electronic charges.

(b) For a molecule to be in a conformation that maximizes attractive forces and minimizes repulsive forces, which bonding interaction is most likely to change: bond stretching, bond angle bending, or torsion angle rotation? Which of these is the most difficult to change? How do you know?

24. (a) A C—C bond has a length of 1.540 Å. For a quadratic potential with a force constant of 1200 kJ mol^{-1} Å^{-2}, how much energy (kJ mol^{-1}) would it take to stretch the bond length to 1.750 Å?

(b) A C—C—C bond angle is nearly tetrahedral with a bond angle of 109.5°. For a quadratic potential with a force constant of 0.0612 kJ mol^{-1} degree^{-2}, how much energy (kJ mol^{-1}) would it take to distort the bond angle to a right angle (90°)?

(c) The potential energy U for rotation around a single bond can be represented by an equation of the form $U = k[1 + \cos(3\phi)]$, where ϕ is the torsion angle. Use this equation to calculate the values of ϕ that give minima and maxima in the potential energy curve vs. ϕ.

(d) All atoms have an attractive interaction with one another that was first explained by Fritz London; it depends on r^{-6}, where r is the distance between the two atoms. Explain what causes this universal attraction.

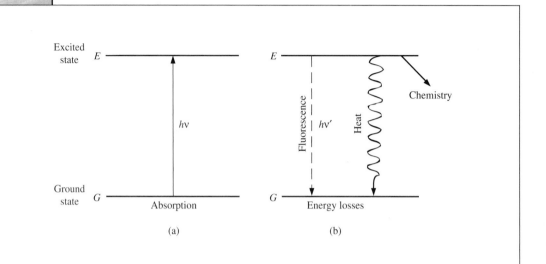

CHAPTER 10

Molecular Structures and Interactions: Spectroscopy

Spectroscopic methods are vital to all measurements. Spectroscopy uses light of any wavelength, from radio waves to far ultraviolet, to "see" molecules. Nuclear magnetic resonance and magnetic resonance imaging are radio-frequency spectroscopies that provide atomic resolution structures of molecules in solution or high-contrast images of soft tissues in patients. Visible and ultraviolet light, including polarized light and fluorescent light, are used to quantitate and characterize all sorts of reactions.

Concepts

Spectroscopy is the study of the interaction of electromagnetic radiation with matter. Figure 10.1 shows the wavelength regions of the electromagnetic spectrum that are useful in spectroscopy. *Absorption* occurs when a molecule undergoes a transition to a higher-energy level. *Emission* occurs when a molecule undergoes a transition from a higher-energy level to a lower one; *fluorescence* and *phosphorescence* are two forms of emission. Spectroscopy at different frequencies monitors different molecular processes, which are rigorously treated by the laws of quantum mechanics. *Ultraviolet* (UV) and *visible spectroscopy* monitors changes between electronic states; *infrared* (IR) *spectroscopy* probes vibrational states. *Circular dichroism* is the differential absorption of right- and left-circularly polarized light. *Nuclear magnetic resonance* (NMR) is the absorption of radiofrequency electromagnetic radiation by nuclei placed in a static magnetic field.

All matter scatters electromagnetic radiation. In *Rayleigh scattering* only the direction of incident light is changed; in *Raman scattering* there is also a change in wavelength. For absorption to occur, there must be an energy-level spacing that corresponds to the wavelength of the incident light, but scattering occurs at all wavelengths. The *refractive index*, the ratio of the speed of light in vacuum to the speed of light in a medium, is a measure of the light scattered in the direction of the incident beam. *Linear birefringence* is the difference in refractive index for light polarized in planes perpendicular to each other. *Circular birefringence* is the difference in refractive index for right- and left-circularly polarized light. *Optical rotation*, which is directly proportional to the circular birefringence, is the rotation of linearly polarized light.

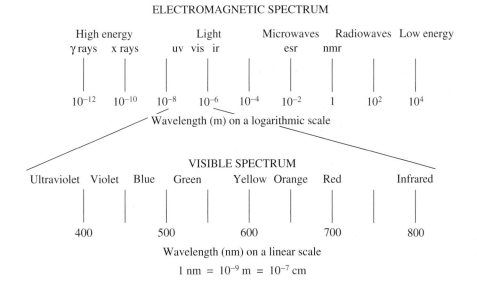

◀ FIGURE 10.1

Wavelength regions of the electromagnetic spectrum.

Applications

Spectroscopy is one of the most powerful physical tools used to characterize biological systems. One application of spectroscopy is quantitative analysis. The identities and concentrations of molecules in any environment are usually determined spectroscopically. The concentrations of proteins and nucleic acids in biochemical experiments are determined by absorbance of UV light. We can detect molecules in biological cells with a fluorescence microscope, or detect CO_2 in the atmosphere by infrared absorption. Fluorescence spectroscopy can even be used to detect individual molecules and to monitor their chemical behavior. Spectroscopic methods are often noninvasive: Light is directed on a sample, and absorbed or emitted radiation is detected. The absorption of light can induce chemical reactions; this is the area of *photochemistry*, of which photosynthesis is an important example.

A second application of spectroscopy is the ability to probe molecular structure. Circular dichroism reveals the helical content of proteins or nucleic acids. Both fluorescence spectroscopy and NMR spectroscopy monitor the relative separations of atoms or chemical groups, which provides three-dimensional conformational information on proteins or nucleic acids. NMR studies of macromolecules can reveal their structures in solutions at atomic resolution.

Spectroscopy can be used to measure changes in biochemical systems, including the binding of biological molecules and chemical reactions. Conformational changes of proteins and nucleic acids can be followed by absorption, fluorescence, or circular dichroism. Spectroscopy thus provides insights into the *dynamics* of chemical systems.

Electromagnetic Spectrum

As discussed in chapter 9, we characterize the electromagnetic spectrum in terms of wavelength, frequency, or energy. Different types of spectroscopists prefer different units. In the visible and near-UV region, wavelengths in nanometers are used: 1 nm, 10^{-9} m $\equiv$ 1 mμm (millimicron) $\equiv$ 10 Å $\equiv 10^{-7}$ cm. In the microwave and radiowave region, frequencies are usually used:

$$\nu = \frac{c}{\lambda} \tag{10.1}$$

where ν = frequency, in Hz (s^{-1})
$\quad c$ = speed of light in a vacuum = 3×10^8 m s^{-1}
$\quad \lambda$ = wavelength in a vacuum, in m

Infrared spectroscopists plot their spectra versus wavenumbers, $\bar{\nu}$:

$$\bar{\nu} = \frac{1}{\lambda} \tag{10.2}$$

where $\bar{\nu}$ = "frequency," in cm^{-1}
$\quad \lambda$ = wavelength in a vacuum, in cm

Wavenumber is the number of wave amplitude maxima per 1-cm length along the wave. *Wavelength* is the distance between successive wave amplitude maxima. Units of energy are popular for spectra in the far-UV, X-ray, and

γ-ray regions. Planck's equation, $\Delta\varepsilon = h\nu$, is used to convert frequencies to energies. For example, 200-nm UV light corresponds to about 6 eV or 598 kJ mol^{-1}. Interaction of radiation with matter may result in the absorption of incident radiation, emission of light as fluorescence or phosphorescence, scattering into new directions, rotation of the plane of polarization, or other changes. Each of these interactions can give useful information about the nature of the sample in which it occurs. Before discussing spectroscopy in detail, let's look at a few familiar examples.

Color and Refractive Index

Color is an obvious property of a substance. The green color of leaves is due to chlorophyll absorption, the orange of a carrot or tomato arises from carotenes, and the red of blood results from hemoglobin. The color is characteristic of the spectrum of light in the visible region transmitted by the substance when white light (or sunlight) shines through it or when light is reflected from it. The quantitative measure of color that we use is the *absorption spectrum*. This is a plot of the variation with wavelength or frequency of the absorption of radiation by the substance. The absorption spectrum is complementary to the transmission of light. Chlorophyll is green because it absorbs strongly in the blue (435 nm) and red (660 nm) regions of the spectrum. Figure 10.2 compares the visible spectra of green chlorophyll, orange β-carotene, and red oxyhemoglobin. Note that the spectra of the three different molecules have both similarities and differences; the spectra are much richer in information than the colors would indicate.

The bent appearance of a pencil partly immersed in a glass of water results from the difference in refractive index of light in air and in water. The refractive index n is simply a measure of the ratio of the velocity of light in vacuum c to the velocity in the medium v. The velocity of light in vacuum is approximately 3×10^8 m s^{-1} (3×10^{10} cm s^{-1}); this corresponds to about 1 ft ns^{-1}:

$$n = \frac{c}{v} \qquad\qquad (10.3)$$

The refractive index of air for visible light is about 1.00028 at 15°C and 1 atm pressure, so the velocity of light in air is very close to that in vacuum. The refractive index at a wavelength of 589 nm for some common substances is given in table 10.1. The reason that we are able to see objects that are not colored is because the changes of refractive index at the boundaries result in bending of the light rays. Techniques such as phase contrast have been used in microscopy to increase the contrast between different regions of biological organelles.* The appearance of detailed internal structure in such organelles occurs because their refractive index differs in the different regions. Because refractive index is an intrinsic property of a material that is related to its composition, a measurement in a refractometer can be used to characterize the substance.

---

*An *organelle* is an organized structure in a cell with a specialized biological function and surrounded by a limiting membrane. Examples are mitochondria, chloroplasts, and nuclei.

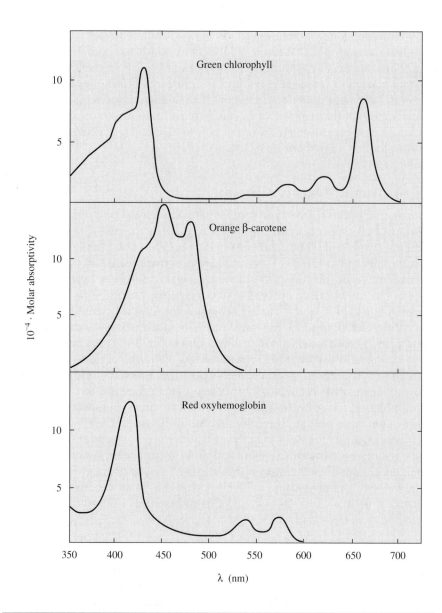

▶ FIGURE 10.2

Comparison of the visible spectra of three molecules.

TABLE 10.1 Refractive Indexes for Some Common Substances at 589 nm

Substance	Conditions		Refractive index, n_{589}
Air	1 atm,	15°C	1.0002765
Water	Liquid,	15°C	1.33341
Ethanol	Liquid,	20°C	1.3611
Carbon tetrachloride	Liquid,	20°C	1.4601
Hexane	Liquid,	20°C	1.37506
Quartz	Fused		1.45843
Protein			1.51–1.54
Sucrose	Crystal,	20°C	1.5376
Lipids			1.40–1.44

Even more information can be gained by examination of the refractive index using polarized light. Plane-polarized light is light in which the electric field of the light oscillates in only one direction; natural light becomes plane polarized upon passage through Polaroid film or other polarizing materials. Objects such as single crystals or biological cells exhibit different refractive indexes, depending on the direction of polarization of the light relative to the object. This anisotropy of refractive index is known as *birefringence* (literally, "double refraction"). It is an indication of ordering of molecules within the sample and can give useful information about intermolecular associations in cell membranes, nerve fibers, muscle, and the like. Crystals of some substances—such as quartz, fluorite (CaF_2), and gypsum ($CaSO_4 \cdot 2\,H_2O$)—exhibit birefringence that is related to the directions of the crystal axes.

Absorption and Emission of Radiation

Light that is incident on a colored sample is partly absorbed by it. This means that at certain wavelengths there is less intensity transmitted. There is no breakdown in the law of conservation of energy. The result of the absorption may appear as *heat* producing a temperature rise in the sample, *luminescence* in which a photon of the same or lower energy is emitted, *chemistry* that incorporates energy into altered bonding structures, or a combination of these. The total energy change of the processes is *always* exactly equal to the energy of the photon that was absorbed. We illustrate the principles of spectroscopy, using absorption of UV or visible light, which results in transitions between electronic states of the molecule; the general ideas are applicable to spectroscopies at different frequencies, although each method has its particular features.

The interactions leading to absorption of light are essentially electromagnetic in origin. The oscillating electromagnetic field associated with the incoming photon generates a force on the charged particles in a molecule. If the electromagnetic force results in a change in the arrangement of the electrons in a molecule, we say that a *transition* to a new electronic state has occurred. The new state will be of higher energy if radiation is absorbed; figure 10.3a, an energy-level diagram, illustrates the process. The absorbed photon results in the excitation of the molecule from its normal, or ground state G to a higher energy, or excited electronic state E. The excited electronic state has a rearranged electron distribution. It can be described approximately as a result of a promotion of one of the ground-state electrons to an orbital of higher energy.

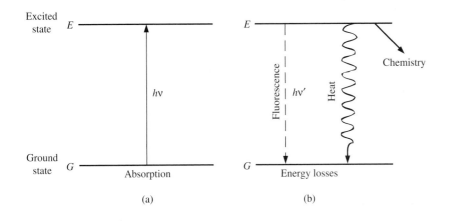

◀ FIGURE 10.3

Energy-level diagram: (a) the absorption of a photon of energy $h\nu$ and (b) three important processes by which the excitation energy is subsequently released or converted.

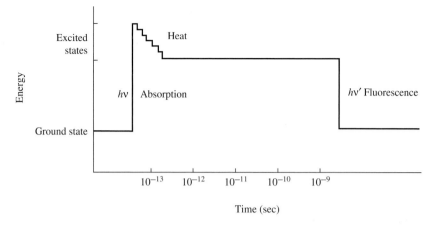

▲ FIGURE 10.4

Energy versus time for a single molecule in the presence of light. The molecule absorbs a photon and is excited to a vibrational level of an excited electronic state. The
excess vibrational energy is rapidly lost as heat, and the molecule reaches the relatively stable lowest vibrational level of the excited electronic state. After a time, the
molecule emits a photon and returns to its ground state. Each molecule has a different excited-state lifetime, but the collection of molecules has a characteristic relaxation time or half-life.

This completes the absorption process, which usually occurs very rapidly
(within 10^{-15} s), but the excited state is usually not stable for very long. Typically, within 10^{-8} s, it has disappeared by fluorescence emission, the dissipa-
tion of heat, or the initiation of photochemistry. Fluorescence and heat
conversion result in the return of the molecule to the ground state, and the
only net effect is the conversion of energy of the absorbed photon into heat
and (perhaps) a photon of lower energy. Figure 10.4 shows the energy of a
molecule versus time, which has absorbed a photon and emitted some of the
energy as fluorescence.

At ordinary intensities, the relaxation of the system back to the ground
state is so rapid that the light produces very little change in the population of
molecules in the ground state. At the very high intensities encountered in
flash-lamp or laser experiments, it is possible to excite a large majority of
molecules out of the ground state simultaneously, but this occurs only under
unusual conditions.

Radiation-Induced Transitions

The dynamics of an atom or molecule placed in a radiation field can be explained using a simple model. Consider any two states of a molecule, such as
the excited (E) state and ground (G) state of figure 10.3. The states will differ
in energy by an amount $E_E - E_G = h\nu$. If the energy separation is large compared with kT, virtually all molecules will be in the ground state in the dark
at ordinary temperatures. In the presence of a radiation field, this population
distribution will be slightly disturbed. We can characterize the radiation
field by a radiation-energy density $\rho(\nu)$, which is the amount of energy near
frequency ν and per unit volume of the sample. The dimensions of $\rho(\nu)$ are

TABLE 10.2 Radiation Densities and Illumination Intensities
Values for Polychromatic Sources Are Given per Nanometer
of Spectral Bandwidth

Source of phenomenon	λ (nm)	Illumination intensity, $I(\lambda)^*$		Radiation density, $\rho(\lambda)^*$	
		$\mu W\ cm^{-2}$	quanta $s^{-1}\ cm^{-2}$	$J\ cm^{-3}$	quanta cm^{-3}
Threshold of completely dark-adapted human eye	507	4×10^{-9}	1×10^4	1.3×10^{-25}	3×10^{-7}
Firefly luminescence	562	5×10^{-2}	1.5×10^{11}	1.7×10^{-18}	5
Full Moon at surface of Earth	500	0.3	1×10^{12}	1×10^{-17}	30
Bright summer sunlight	500	1×10^5	2×10^{17}	3×10^{-12}	1×10^7
Ruby laser, 1 kJ, 1-ms duration	694.3	1×10^{12}	4×10^{24}	3×10^{-5}	1.2×10^{14}

*$I(\lambda)$ and $\rho(\lambda)$ are given for a wavelength range of 1 nm around the λ quoted.

energy volume^{-1}. It is related to the more familiar intensity of illumination of a surface $I(\nu)$, in energy area^{-1} time^{-1}, by the equation

$$I(\nu) = \frac{c}{n}\rho(\nu) \tag{10.4}$$

where c is the velocity of light and n is the refractive index of the medium. Some representative values of approximate illumination intensities and radiation-energy densities are given in table 10.2. Values are also given in terms of quanta or photons, using the Planck equation.

As a consequence of the introduction of the radiation field to an absorbing sample, three processes are induced:

1. *Absorption.* The radiation induces transitions from the ground state G to the excited state E. The rate of this absorption process will depend on the number of molecules in state G, on the radiation density, and on the absorption coefficient of the molecule.

2. *Stimulated emission.* Once molecules are present in the excited state E, the radiation field can stimulate transitions down to the lower state, giving emission of radiation.

3. *Spontaneous emission.* Molecules put into the excited state E by the radiation field are substantially out of equilibrium. Spontaneous relaxation processes restore the system to equilibrium. These are spontaneous in the sense that they do not depend on the radiation density.

Classical Oscillators

The classical theory of light absorption considers matter to consist of an array of charges that can be set into motion by the oscillating electromagnetic field of the light. The electric–dipole oscillators set in motion by the field have certain characteristic, or natural, frequencies ν_i. When the frequency of the radiation is near the oscillation frequency, absorption occurs and the intensity of the radiation decreases on passing through the substance. The refractive index of the substance also undergoes large changes, called *dispersion anomalies*, in the same spectral region. The intensity of the interaction is known as the *oscillator strength f_i*, and it can be thought of as characterizing the number of electrons per molecule that oscillate with the characteristic frequency ν_i.

Although the classical picture has been replaced by the much more informative quantum mechanical description of the absorption of radiation, there are classical holdovers in the current literature. For example, a transition that is fully allowed quantum mechanically is said to have an *oscillator strength* of 1.0. Operationally, the oscillator strength f is related to the intensity of the absorption—that is, to the area under an absorption band plotted versus frequency:

$$f = \frac{2303cm}{\pi N_0 e^2 n} \int \varepsilon(\nu)\, d\nu \qquad (10.5)$$

where ε is the molar absorptivity, c is the velocity of light, m the mass of the electron and e its charge, n the refractive index of the medium, and N_0 is Avogadro's number. The integration is carried out over the frequency range associated with the absorption band. Oscillator strengths can be observed from magnitudes of unity down to very small values ($<10^{-4}$). Measured values greater than unity, as for the blue bands of hemoglobin or chlorophyll, usually indicate the overlap of two or more electronic transitions.

Quantum Mechanical Description

The most complete description of the absorption of radiation by matter is based on time-dependent quantum mechanics. It analyzes what happens to the wavefunctions of molecules in the presence of an external oscillating electromagnetic field, but it requires knowledge of differential equations. Here, we will simply cite some of the useful results obtained from the detailed treatment.

Because the interaction is electromagnetic, the operator that conveniently describes it is the electric dipole-moment operator, $\boldsymbol{\mu} = er$. This operator is just the position operator, $\mathbf{r} = \mathbf{x} + \mathbf{y} + \mathbf{z}$, multiplied by the electronic charge. For example, the *permanent dipole moment* of a molecule in its ground state is obtained by using the ground-state wavefunction of the molecule to calculate the average position of negative charge. When the centers of positive and negative charge coincide, the permanent dipole moment is zero.

A transition from one state to another occurs when the radiation field connects the two states. In wave mechanics, the means for making this connection is described by the *transition dipole moment*

$$\boldsymbol{\mu}_{0A} = \int \psi_0 \boldsymbol{\mu} \psi_A\, dv \qquad (10.6)$$

where ψ_0, ψ_A = wavefunctions for the ground and excited state, respectively

dv = volume element; integration is over all space

The transition dipole moment is nonzero whenever the symmetry of the ground and excited state differ. A hydrogen atom will always have zero permanent dipole moment, but if ψ_0 is a $1s$ orbital and ψ_A is a $2p$ orbital, the transition dipole does not equal zero. Similarly, ethylene has zero permanent dipole moment, but if ψ_0 is a π MO (molecular orbital) and ψ_A is a π^* MO, then $\mu_{\pi\pi^*}$ is not zero. The transition moment has both magnitude and direction; the direction is characterized by the vector components: $\langle\mu_x\rangle_{0A}$, $\langle\mu_y\rangle_{0A}$, and $\langle\mu_z\rangle_{0A}$. Most electronic transitions are polarized, which means that the three vector components are not all equal. In the examples above, the hydrogen-atom transition from s to p_x is polarized along x. The ethylene $\pi \rightarrow \pi^*$ transition is polarized along the $C\!=\!C$ double bond. The magnitude of the transition is characterized by its absolute value squared, which is called the *dipole strength* D_{0A}:

$$D_{0A} = |\mu_{0A}|^2 \tag{10.7}$$

This is a scalar quantity rather than a vector. It has the units of a dipole moment squared. Dipole moments are frequently expressed in *debyes*, D, where $1\,\text{D} = 10^{-18}$ esu cm $= 3.336 \times 10^{-30}$ C m, and the units of the dipole strength are then D^2. The relations among dipole strength, oscillator strength, and spectra are given in table 10.3.

Einstein derived expressions for the rate of absorption, and of stimulated and spontaneous emissions of radiation. The intensity of absorption depends on the difference in population of two states that are connected by a spectroscopic transition. For transitions involving electronic or vibrational energy levels at room temperature, the energy difference involved ($10-400$ kJ mol^{-1}) is much greater than kT (2.5 kJ mol^{-1}); this means that only the ground state is significantly populated at room temperature for these states. In contrast, rotational or nuclear states, which give rise to microwave and NMR signals, have energy separations smaller than kT, such that ground and excited states are both significantly populated. NMR is a significantly less sensitive technique than UV spectroscopy because of the nearly equal populations of the energy levels.

TABLE 10.3 Spectroscopic Relations*

1. Dipole strength $\leftrightarrow$ transition dipole moment:
$$D_{0A} = |\mu_{0A}|^2$$

2. Dipole strength $\leftrightarrow$ absorption spectrum:
$$D_{0A} = \frac{3\hbar \cdot 2303c}{4\pi^2 N_0 n} \int \frac{\varepsilon(\nu)}{\nu}\, d\nu$$
$$= \frac{9.185 \times 10^{-39}}{n} \int \frac{\varepsilon(\nu)}{\nu}\, d\nu \qquad (\text{esu cm})^2$$

3. Oscillator strength $\leftrightarrow$ absorption spectrum:
$$f_{0A} = \frac{2303cm}{\pi N_0 e^2 n} \int \varepsilon(\nu)\, d\nu$$
$$= \frac{1.441 \times 10^{-19}}{n} \int \varepsilon(\nu)\, d\nu \qquad (\text{unitless})$$

---

*Frequency of light, ν, in s^{-1}; N_0 = Avogadro's number; n = refractive index; m, e = mass (g), charge (esu) of electron; $\hbar = h/2\pi$, c = speed of light in cm s^{-1}; $\varepsilon(\nu)$ is the molar absorptivity in M^{-1} cm^{-1}.

The rates of stimulated and spontaneous emission are related by

$$A = \frac{8\pi h\nu^3}{c^3}B \qquad (10.8)$$

where A and B are rate constants for spontaneous and stimulated emission, respectively. This equation shows that spontaneous emission is strongly dependent on the frequency of the transition. For high-energy, high-frequency transitions, spontaneous emission is significant; spontaneous emission plays an important role in electronic transitions. For lower-energy, lower-frequency transitions, such as in NMR spectroscopy, spontaneous emission is insignificant.

Lifetimes and Line Width

We consider the return of excited states to equilibrium after absorption of energy as first-order reactions. The inverse of the rate constant for return to the ground state, which for electronic transitions is dominated by the rate of spontaneous emission, defines the lifetime of the excited state. These lifetimes can vary from 10^{-9} s (ns) for electronic spectra to seconds in NMR spectroscopy. The lifetime has a profound effect on the shape of a spectral peak, which we can understand from the Heisenberg uncertainty principle. Recall in chapter 9 that the Heisenberg uncertainty principle provided a lower limit for the product of uncertainties in measurements of position and momentum. We can prove a similar relation for the product of the time interval during which the energy of a quantum state is observed (Δt) and the uncertainty in its energy (ΔE).

$$\Delta E \Delta t \geq \frac{1}{2}\left(\frac{h}{2\pi}\right) \qquad (10.9)$$

Uncertainties in energy are directly related to a width of a peak, $\Delta\nu$, by $\Delta E = h\Delta\nu$; thus, the greater the lifetime, the smaller is the uncertainty in the energy of a transition, and the narrower is the peak. The affect of the lifetime is called *lifetime broadening,* or *homogeneous broadening,* of a peak (figure 10.5).

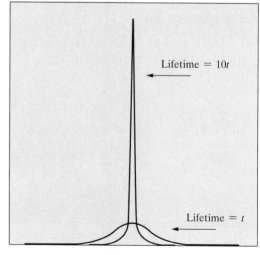

Lifetime = 10*t*

Lifetime = *t*

Frequency

▶ FIGURE 10.5

Schematic of spectral line width versus lifetime of an excited state. For a relatively long excited-state lifetime (10*t*), the energy of the excited state is well defined, and the spectral line is relatively narrow. For a relatively short excited-state lifetime (*t*), the energy of the excited state is less well defined, and the spectral line is relatively broad.

The integrated intensity of a peak is not changed by lifetime broadening, but the increased line width often obscures information in the spectrum. (We will see examples of this later.)

The widths of peaks in spectroscopy are directly influenced by the lifetimes of excited states, which in turn are determined by the rates of spontaneous and stimulated emission. External conditions determine the rates of stimulated emission, and therefore this rate can be modulated. The rate of spontaneous emission depends on only the type of transition and frequency: It determines a limiting "natural lifetime" or "natural line width." Spontaneous emission rates increase with ν^3 [Eq. (10.8)], so corresponding lifetimes will be shorter at higher energies. The result is that line widths of electronic transitions are very large, whereas line widths of vibrational and in particular NMR transitions are exceedingly narrow. The narrow line widths in NMR will be a central element in the power of this spectroscopic method to study biological molecules. External factors can also cause *inhomogeneous broadening* of a spectral line. Different molecules in a bulk sample may experience a different local environment, which can lead to slightly different transition energies. The spread in transition frequencies in the bulk sample broadens a spectral line.

Role of Environment in Electronic Absorption Spectra

The electronic absorption spectrum of a gas is shown schematically in figure 10.6. The electronic spectrum consists of a number of sharp peaks, which represent transitions from the ground electronic and vibrational state, to the first excited electronic state, and a number of excited vibrational states. The energies of the different vibrations are added to the energy of the electronic transition when both the electronic and vibrational quantum numbers change as a result of absorption of a photon. This is called a *vibronic transition*. The energy of the electronic transition itself (without added vibrational energy) is determined from the low-frequency or long-wavelength edge of the

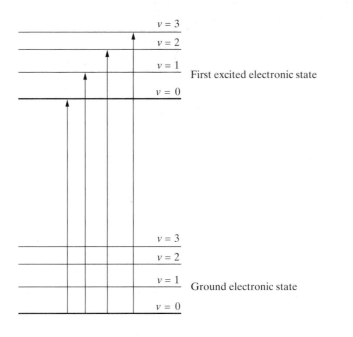

◀ FIGURE 10.6

Schematic of vibronic transitions between the ground and first excited electronic states. Transitions occur mainly between the ground ($v = 0$) vibrational state and many vibrational states in the excited electronic state.

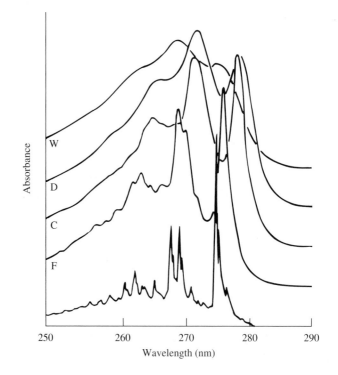

▶ **FIGURE 10.7**

Absorption spectra of anisole vapor and of anisole dissolved in perfluorooctane, F; in cyclohexane, C; in dioxane, D; and in water, W. All at 30°C. The spectra of anisole solutions are displaced vertically to decrease overlap. [From G. L. Tilley, Ph.D. thesis, Purdue University, 1967; cited by M. Laskowski, Jr., 1970, in *Spectroscopic Approaches to Biomolecular Conformation*, ed. D. W. Urry (Chicago: American Medical Association).]

vibronic spectrum. A detailed vibrational analysis may be needed to locate this energy precisely. Because there are many ways in which large molecules can vibrate (many *normal modes* of vibration), the envelope of an electronic transition of a molecule in the gas phase is often very complex. Figure 10.7 illustrates the UV spectrum of the molecule anisole in the gas phase,

and is representative of other aromatic molecules. The spectrum of the vapor (bottom) is typically the "best resolved"; it is similar to our ideal representation, with well-resolved vibronic transitions. The reason that the vibronic structure is so sharp is that most of the molecules in the gas phase are relatively isolated from one another; only a small fraction is in the process of undergoing a collision during the very brief time (about 10^{-15} s) that is required for the photon to be absorbed. Thus, nearly all of the gas-phase molecules are in essentially the same environment and not interacting with one another.

In the liquid state, either pure liquid or in solution, the environment of the molecules is quite different. Close contacts exist between molecules in liquids, and essentially all molecules are in the process of undergoing collisions. The forces between molecules are strong enough to perturb the molecular energy levels significantly and, together with the much greater variety of environments present at the instants of photon absorption, this leads to a broadening and shifting of the spectra. Instead of well-resolved vibronic transitions, we are left with a typical broad absorption spectrum, which represents the full envelope of vibronic transitions. The broadening and spectral shifts induced by solvent depend on the polarity and hydrogen-bonding ability of the solvent, as can be seen in the upper spectra in figure 10.7.

Beer–Lambert Law

Experimentally, absorbance is usually measured in solution. Consider a sample of an absorbing substance (liquid solution, solid, or gas) placed between two parallel windows that will transmit the light, as shown in figure 10.8. The radiation may be in the UV, visible, IR, or other spectral regions. Suppose that light of intensity I_0 is incident from the left, propagates along the x-direction, and exits to the right, with decreased intensity I_t. At any point x within the sample, there is intensity I, which decreases smoothly from left to right. (We ignore for the moment the discontinuous intensity decrease, usually about 10%, which results from reflections at the windows.)

If the sample is homogeneous, the *fractional* decrease in the light intensity is the same across a small interval dx, regardless of the value of x. The fractional decrease depends linearly on the concentration of the absorbing substance. This can be written

$$-\frac{dI}{I} = \alpha c \, dx \tag{10.10}$$

where dI/I is the fractional change in light intensity, c is the concentration of absorber, and α is a constant of proportionality. Equation (10.10) is the differential form of the Beer–Lambert law. It is straightforward to integrate it between limits I_0 at $x = 0$ and I_t at $x = l$, where l is the optical path length. Since neither α nor c depends on x, we can write

$$-\int_{I_0}^{I_t} \frac{dI}{I} = \alpha c \int_0^l dx$$

$$\ln \frac{I_0}{I_t} = \alpha c l \qquad I_t = I_0 e^{-\alpha c l} \tag{10.11}$$

For measurements made with cuvets (optical sample cells) of different path lengths, the transmitted intensity I_t decreases exponentially with increasing path length.

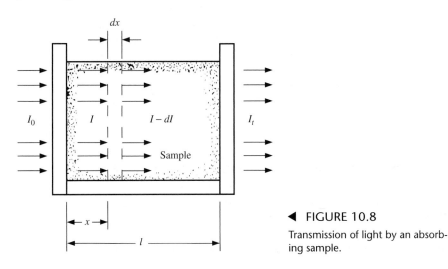

◀ FIGURE 10.8

Transmission of light by an absorbing sample.

Alternatively, for a constant path length, the transmitted intensity decreases exponentially with increasing concentration of an absorbing solute. The *absorbance* is defined for base 10 logarithms rather than natural logarithms,

$$A = \log \frac{I_0}{I_t} = \varepsilon c l \qquad (10.12)$$

where A is the *absorbance* or *optical density* and $\varepsilon = \alpha/2.303$ is the *molar absorptivity* (or molar extinction coefficient), with units M^{-1} cm^{-1}, when the concentration c is in molarity M. Equation (10.12) is the Beer–Lambert law and shows that the absorbance is linearly related to concentration and path length. The relation between absorbance and transmission ($T = I_t/I_0$) is exactly analogous to that between pH and [H^+]:

$$A = -\log T$$

Because absorption depends strongly on wavelength for nearly all compounds, we must specify the wavelength at which measurement is made. This is usually done using a subscript λ, indicating the particular wavelength, as A_λ or ε_λ. For a single substance at a specified wavelength, ε_λ is a constant, characteristic of the absorbing sample, and is *independent* of both c and l. The wavelength dependence of ε_λ (or of A_λ) is known as the *absorption spectrum* of the compound.

To correct for reflection by the cuvet window, as well as for absorption by the solvent, the solution is replaced by solvent in the same cuvet, and a second measurement is made. The transmitted intensity of this second measurement is then used as I_0 in the Beer–Lambert expressions. Alternatively, a double-beam method may be used, where the solution (sample) and solvent (reference) are placed in matched cuvets and their transmissions are measured simultaneously.

Deviations from the Beer–Lambert law can occur for a variety of reasons. Problems arising from insufficiently monochromatic radiation, divergent light beams, or imprecise sample geometry are minimized by the design of most spectrophotometers. Individual samples should be checked to determine whether corrections need to be applied. Beer–Lambert law deviations can arise from inhomogeneous samples, light scattering by the sample, dimerization or other aggregation at high concentrations, or changes in equilibria involving dissociable absorbing solutes. The most common consequence is that the measured absorbance does not increase linearly with increasing concentration or path length. This behavior can be used to obtain useful information about the sample. It is essential to understand the origin of such effects in order for spectrophotometry to be a useful tool for experimental investigations.

Quantitative determinations using Beer's Law

One of the most widely used applications of spectroscopy is for the quantitative determination of the concentration of substances in solution. Noting that both the absorbance A and the molar absorptivity ε depend on wavelength, we now write

$$A_\lambda = \varepsilon_\lambda c l$$

A plot of A_λ versus λ (or A_ν versus ν) is the absorption spectrum of the solution. For a single substance, the absorbance A_λ at any wavelength is directly proportional to concentration c for a fixed path length l. Experimentally, we determine the absorbance by taking the log of the ratio of the incident to the transmitted light intensities.

The absorbance of a solution of more than one independent species is additive. For two components, M and N,

$$A_\lambda = A_\lambda^M + A_\lambda^N = \varepsilon_\lambda^M l[M] + \varepsilon_\lambda^N l[N]$$
$$= (\varepsilon_\lambda^M[M] + \varepsilon_\lambda^N[N])l \qquad \textbf{(10.13)}$$

If measurements are made at two (or more) wavelengths where the ratios of extinction coefficients differ, the resulting equations,

$$A_1 = (\varepsilon_1^M[M] + \varepsilon_1^N[N])l$$
$$A_2 = (\varepsilon_2^M[M] + \varepsilon_2^N[N])l$$

can be solved for the concentrations of the absorbing solutes,

$$[M] = \frac{1}{l} \frac{\varepsilon_2^N A_1 - \varepsilon_1^N A_2}{\varepsilon_1^M \varepsilon_2^N - \varepsilon_2^M \varepsilon_1^N} \qquad same$$
$$[N] = \frac{1}{l} \frac{\varepsilon_1^M A_2 - \varepsilon_2^M A_1}{\varepsilon_1^M \varepsilon_2^N - \varepsilon_2^M \varepsilon_1^N} \qquad \textbf{(10.14)}$$

These relations are widely used in the spectrophotometric analysis of mixtures of absorbing species.

A wavelength at which two or more components have the same extinction coefficient is known as an *isosbestic wavelength* $[\varepsilon_\lambda^M = \varepsilon_\lambda^N = \varepsilon_{iso}]$. At this wavelength, the absorbance can be used to determine the total concentration of the two components:

$$\lambda = \text{isosbestic:} \quad A_{iso} = \varepsilon_{iso}l[M] + \varepsilon_{iso}l[N] = \varepsilon_{iso}l([M] + [N]) \quad \textbf{(10.15)}$$

Measurements at an isosbestic wavelength plus one other wavelength where the extinction coefficients differ for the two components provide a particularly simple solution to the Beer–Lambert law equations:

$$[M] + [N] = \frac{A_{iso}}{\varepsilon_{iso}l}$$

$$\frac{[M]}{[N]} = \frac{\varepsilon_1^N A_{iso} - \varepsilon_{iso}A_1}{\varepsilon_{iso}A_1 - \varepsilon_1^M A_{iso}}$$

EXAMPLE 10.1

Solutions containing the amino acids tryptophan and tyrosine can be analyzed under alkaline conditions (0.1 M KOH) from their different UV spectra. The extinction coefficients under these conditions at 240 and 280 nm are

$$\varepsilon_{240}^{Tyr} = 11{,}300 \ M^{-1} \ cm^{-1} \qquad \varepsilon_{240}^{Trp} = 1960 \ M^{-1} \ cm^{-1}$$
$$\varepsilon_{280}^{Tyr} = 1500 \ M^{-1} \ cm^{-1} \qquad \varepsilon_{280}^{Trp} = 5380 \ M^{-1} \ cm^{-1}$$

A 10-mg sample of the protein glucagon is hydrolyzed to its constituent amino acids and diluted to 100 mL in 0.1 M KOH. The absorbance of this solution (1-cm path) is 0.717 at 240 nm and 0.239 at 280 nm. Estimate the content of tryptophan and tyrosine in μmol (g protein)$^{-1}$.

Solution

Neither of the wavelengths is an isosbestic for these amino acids, so we use Eq. (10.14):

$$[\text{Tyr}] = \frac{(5380)(0.717) - (1960)(0.239)}{(11{,}300)(5380) - (1500)(1960)}$$

$$= \frac{3.39 \times 10^3}{57.9 \times 10^6}$$

$$= 5.85 \times 10^{-5} \, M$$

$$[\text{Trp}] = \frac{(11{,}300)(0.239) - (1500)(0.717)}{57.9 \times 10^6}$$

$$= 2.81 \times 10^{-5} \, M$$

Since 10 mg of protein was hydrolyzed and diluted to 100 mL of solution, the estimated contents are

$$585 \, \mu\text{mol of Tyr (g protein)}^{-1}$$

$$281 \, \mu\text{mol of Trp (g protein)}^{-1}$$

When only two absorbing compounds are present in a solution, one or more isosbestics are frequently encountered if we examine the entire wavelength range of the UV, visible, and IR regions. The isosbestics do not *necessarily* occur, however, for the molar absorptivity of one compound may be less than that of the other in every accessible wavelength region. An easy way to spot isosbestics is to superimpose plots of ε versus λ for the two compounds. Wherever the curves cross, there is an isosbestic. An example is shown in figure 10.9, for cytochrome c.

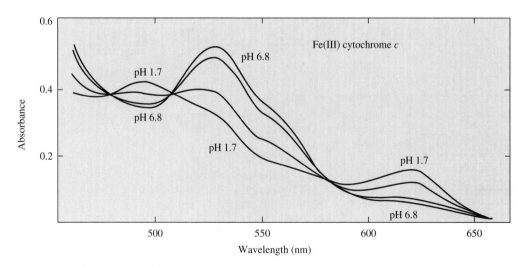

▲ FIGURE 10.9

Absorption spectra of Fe(III) cytochrome c ($5.3 \times 10^{-5} \, M$) at various pH values from 6.8 to 1.7. Isosbestic points occur at 477, 509, and 584 nm. (From E. Yang, unpublished spectra.)

The additivity of absorbance can be generalized to any number of components. Thus, for n components,

$$A = A_1 + A_2 + A_3 + \cdots + A_n = (\varepsilon_1 c_1 + \varepsilon_2 c_2 + \varepsilon_3 c_3 + \cdots + \varepsilon_n c_n)l$$

To determine the n concentrations by absorption methods, we need to make measurements at a minimum of n wavelengths, each characterized by a unique set of ε's. This may be impractical when n is large. Even so, a single component can sometimes be measured accurately in a complex mixture if there is a wavelength where its absorbance is much greater than that of any of the other components. For example, hemoglobin in red blood cells can be measured spectrophotometrically at 541 or 577 nm (figure 10.2), because the other cell constituents (proteins, lipids, carbohydrates, salts, water, and so on) do not absorb significantly at those wavelengths.

Programmable calculators may have a program for solving a set of simultaneous linear equations. Then, n values of the measured absorbances and n^2 values of the ε's can be entered to obtain n concentrations. Digitizing spectrophotometers, which measure absorbances at many wavelengths simultaneously, can calculate concentrations of a mixture directly from the absorbances.

In multicomponent solutions, isosbestic points almost never occur. The probability that three or more compounds have identical molar absorbances at any wavelength is exceedingly small. The probability of two or more isosbestic points for the spectra of the set of compounds is so small as to be completely negligible. This property is useful for diagnostic purposes and results in the following generalization: *The occurrence of two or more isosbestics in the spectra of a series of solutions of the same total concentration demonstrates the presence of two and only two components absorbing in that spectral region.*

It is possible, in principle, for this generalization to be violated, but the chances are vanishingly small. Note that the rule does not necessarily apply to components that do not absorb at all in the wavelength region investigated. It also does not apply if two chemically distinct components have identical absorption spectra (such as ADP and ATP). In this case, the entire spectrum is a set of isosbestics for these two components alone.

Two common examples of the usefulness of isosbestics are the study of equilibria involving absorbing compounds and investigations of reactions involving absorbing reactants and products. The presence of isosbestics is used as evidence that there are no intermediate species of significant concentration between the reactants and products.

EXERCISE 10.1

Consider the reaction of an absorbing reactant M to give an absorbing product P by the reaction

$$M \xrightarrow{k_1} N \xrightarrow{k_2} P \qquad (k_1 \sim k_2)$$

Show that no isosbestic points would be expected for this system, even if the intermediate N has no measurable absorbance in the same spectral region.

Proteins and Nucleic Acids: Ultraviolet Absorption Spectra

Most proteins and all nucleic acids are colorless in the visible region of the spectrum; however, they do exhibit absorption in the near-UV region. These spectroscopic absorptions contain information relevant to the composition of these complex molecules, as well as about the conformation or three-dimensional structure of the molecules in solution. Figure 10.10 shows the UV absorption of serum albumin, a representative protein. There is a distinctive absorption maximum at around 280 nm and a much stronger maximum at 190 to 210 nm. The 280-nm band is assigned to $\pi \rightarrow \pi^*$ transitions in the aromatic amino acids, the 200-nm band to $\pi \rightarrow \pi^*$ transitions in the amide group. These features are sufficiently characteristic that they are often used in the preliminary identification of proteins in biological materials. The spectra of nucleic acids in the same region (figure 10.11) show different characteristics. An absorption maximum occurs at 260 nm, and a shoulder is seen near 200 nm on a background rising to shorter wavelengths. These features are again assigned to $\pi \rightarrow \pi^*$ transitions and are characteristic of all nucleic acids. The observations are sufficiently general that we can use the ratio of absorbances A_{260}/A_{280} to determine quantitatively the ratio of nu-

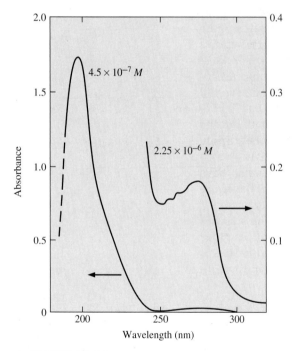

▲ FIGURE 10.10

UV-absorption spectrum of bovine serum albumin. Solution in 10^{-3} M phosphate buffer pH 7.0, 1.0-cm path. The wavelength region above 240 nm was measured at a higher concentration and on an expanded absorbance scale (right) so that the weaker absorption bands in that region would be visible. (From E. Yang, unpublished results.)

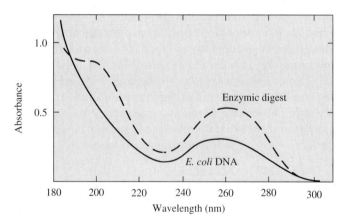

▲ FIGURE 10.11

UV-absorption spectrum of DNA from *E. coli* in the native form at 25°C (solid curve) and as an enzymic digest of nucleotides (dashed curve). (From D. Voet, W. B. Gratzer, R. A. Cox, and P. Doty, 1963, *Biopolymers* 1:193. Reprinted with permission of John Wiley and Sons, Inc.)

cleic acid to protein in a mixture of the two. This requires careful calibration, however, because proteins differ significantly in their content of aromatic amino acids.

   Proteins are natural polyamino acids; that is, they are polymers in which the 20 natural amino acids are connected by amide linkages to form a polypeptide. Nucleic acids are also linear polymers, but the monomer unit is the *nucleotide,* consisting of an aromatic heterocyclic base attached to a sugar phosphate. The polynucleotide chain involves covalent links between the sugars (ribose for RNA, deoxyribose for DNA) and the phosphate linking groups. The structures of amino acids and nucleotides are given in the appendix. To understand the UV absorption of proteins or nucleic acids, we need to examine the various contributions to the spectra. We consider the following important factors:

1. Absorption spectra of individual monomers
2. Contribution of the polymer backbone
3. Secondary and tertiary structure, including helix formation
4. Hydrogen bonding and solvent effects

Amino Acid Spectra

The UV spectra of various amino acids are summarized in figures 10.12 and 10.13. Several distinctive categories of spectra appear. The aromatic amino acids tryptophan, tyrosine, and phenylalanine are the only ones absorbing significantly at wavelengths longer than 230 nm. In particular, at the "characteristic" protein wavelength of 280 nm, absorbance reflects the presence only of tryptophan and tyrosine in a protein and cannot be used for quantitative purposes without supplementary analysis. For example, the absorbance in a 1-cm cell at 280 nm for 1% solutions of proteins in water varies from 3.1 for NAD nucleosidase from pig to 27 for egg-white lysozyme. Between 200 and 230 nm, the aromatic amino acids (especially tryptophan) have absorption and so do histidine, cysteine, and methionine. Between 185 and 200 nm, only phenylalanine and tyrosine have distinct maxima, but the curves of all other amino acids rise sharply to shorter wavelengths, as illustrated by alanine in figure 10.13. Extensive tables of amino acid absorption data are given in the *Handbook of Biochemistry and Molecular Biology* (3d ed., 1976, CRC Press, Cleveland, Ohio).

Polypeptide Spectra

The contribution of the amide linkages to the absorption spectra can be seen by comparing the spectrum of lysine hydrochloride in figure 10.13 with that of poly-L-lysine hydrochloride in the random-coil form (figure 10.14). The broad absorption centered at 192 nm ($\varepsilon_{192} = 7100\ M^{-1}\ cm^{-1}$) is characteristic of the amide linkage in poly-L-lysine and increases the absorbance in this region by eightfold over that of the free amino acid. All proteins have contributions to the absorption spectra in the region around 190 nm (180 to 200 nm) from the polypeptide backbone; however, they are accompanied by absorption contributions from certain of the side chains, especially the aromatic ones.

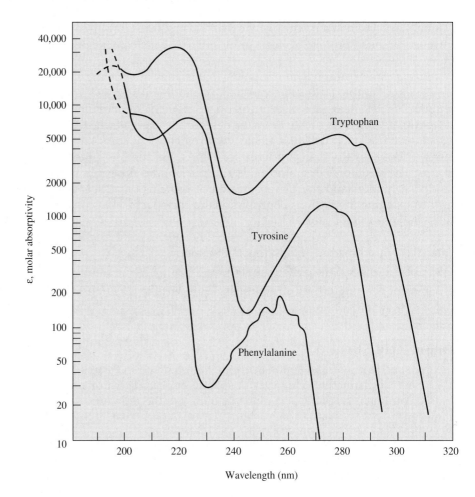

▶ FIGURE 10.12

Absorption spectra of the aromatic amino acids (tryptophan, tyrosine, and phenylalanine) at pH 6. (From S. Malik, cited by D. B. Wetlaufer, 1962, *Adv. Protein Chem.* 17:303.)

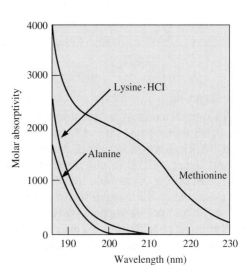

▶ FIGURE 10.13

UV-absorption spectra of three α-amino acids in aqueous solution at pH 5.

550

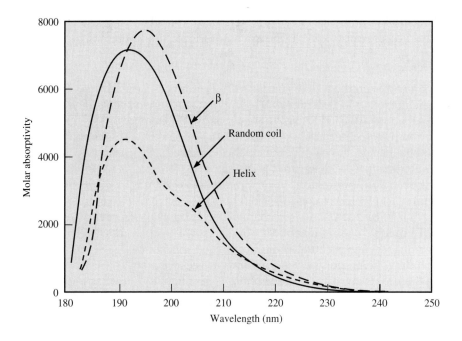

◄ FIGURE 10.14

UV-absorption spectra of poly-L-lysine hydrochloride in aqueous solution; random coil, pH 6.0, 25°C; helix, pH 10.8, 25°C; β form, pH 10.8, 52°C. (From K. Rosenheck and P. Doty, 1961, *Proc. Natl. Acad. Sci. USA* 47:1775.)

Secondary Structure

The secondary structure of a protein describes which residues are in helices or other ordered conformations. The conformation of a protein is sensitively detected by UV spectroscopy. Figure 10.14 shows the effect on the poly-L-lysine spectrum resulting from raising the pH to induce helix formation (by reducing the net positive charge on the lysine side chains) and by raising the temperature, which converts the polypeptide to the β-sheet structure. Proteins are more complicated in their native secondary structure and they usually contain simultaneously several elements of secondary structure, at different locations along the peptide chain. The process of denaturation destroys much of this secondary structure, and changes in the absorption spectra occur as a consequence. The absorption features associated with α helix, β sheet, or random coil are sufficiently distinctive to be used for diagnostic purposes for the native protein (Rosenheck and Doty 1961); circular dichroism spectra (see later in this chapter) can be more informative, however.

Origin of Spectroscopic Changes

Changes in environment of the amino acids contribute to the spectroscopic changes that occur when a protein undergoes a change in secondary structure. Ultimately, most of the effects are electrical in origin because the spectra are sensitive to charge effects on the ground and excited electronic states. Such influences include changes in (1) local charge distribution, (2) the dielectric constant, (3) bonding interactions, such as hydrogen bonds, an (4) dynamic (or resonance) coupling between different parts of the molecule.

The local environment of the amino acids in a native protein depends sensitively on the electric properties of the nearby peptide chain and associated solvent. Careful measurements of the absorption spectrum of a protein near 280 nm show small but distinct differences from the spectrum of the constituent aromatic amino acids. These differences reflect the sum of the effects of the local environments on the aromatic amino acids. Changes of the same magnitude are encountered upon denaturation of the protein.

Nucleic Acids

By contrast with the amino acid units of proteins, the nucleotides that make up the nucleic acid polymers have similar absorption spectra. The aromatic bases that are attached to the ribose- or deoxyribose-phosphates all have absorption maxima near 260 nm (table 10.4). The free base, the nucleoside (the base attached to the sugar), the nucleotide (the base attached to the sugar phosphate), and the denatured polynucleotide all have very similar absorption spectra in this region. For example, the wavelength of maximum absorption is at 260.5 nm for adenine ($\varepsilon = 13.4 \times 10^3$), at 260 nm for adenosine ($\varepsilon = 14.9 \times 10^3$), at 259 nm for adenosine-5'-phosphate ($\varepsilon = 15.4 \times 10^3$), and at 257 nm for the tetranucleotide pApApApA. The latter, however, exhibits a lower absorbance per nucleotide in aqueous solution ($\varepsilon_{259} = 11.3 \times 10^3 \, M^{-1} \, cm^{-1}$). In general, polynucleotides and nucleic acids absorb less per nucleotide than their constituent nucleotides. Also, native double-stranded DNA absorbs less per nucleotide than denatured ("melted") DNA strands. A decrease in absorptivity is called *hypochromicity,* or *hypochromism;* an increase in absorptivity is *hyperchromicity,* or *hyperchromism.* For example, the % hyperchromicity for the melting of a double-stranded DNA is

$$\% \text{ hyperchromicity} = \frac{A \, (\text{melted DNA}) - A \, (\text{native DNA})}{A \, (\text{native DNA})} \times 100$$

Here, each absorbance A is per nucleotide. The percent hypochromicity of a tetranucleotide relative to the mononucleotide is

$$\% \text{ hypochromicity} = \frac{A \, (\text{mononucleotide}) - A \, (\text{tetranucleotide})}{A \, (\text{mononucleotide})} \times 100$$

$$= \frac{15,400 - 11,300}{15,400} \times 100 = 27\%$$

There is a 27% decrease in absorbance of the tetranucleotide relative to the mononucleotide.

TABLE 10.4 Absorption Properties of Nucleotides*

	$\lambda_{max} \, (nm)$	$10^{-3} \varepsilon_{max} \, (M^{-1} cm^{-1})$
Ribonucleotides		
Adenosine-5'-phosphate	259	15.4
Cytidine-5'-phosphate	271	9.1
Guanosine-5'-phosphate	252	13.7
Uridine-5'-phosphate	262	10.0
Deoxyribonucleotides		
Deoxyadenosine-5'-phosphate	258	15.3
Deoxycytidine-5'-phosphate	271	9.3
Deoxyguanosine-5'-phosphate	252	13.7
Thymidine-5'-phosphate	267	10.2

*Wavelengths of maxima and molar absorptivities of nucleotides at pH 7.

The hypochromicity of the polynucleotides or nucleic acids relative to the nucleotides results primarily from interactions between adjacent bases in their stacked arrangement in the helical polymer. The change upon denaturation to single strands or upon hydrolysis to the mononucleotides is easily measured (typically 30 to 40% change) and is often used to follow the kinetics or thermodynamics of the denaturation process. The origin of the hypochromism is electromagnetic in nature. It involves interactions between the electric–dipole transition moments of the individual bases with those of their neighbors. Thus, it depends not only on the intrinsic transition moments of each base, which differ both in magnitude and direction for the chemically distinct bases, but also on the relative orientations of the interacting bases. Stacked bases absorb less per nucleotide than unstacked bases; a stack of two bases has an absorbance slightly less than twice the absorbance of one base; and so on. Thus, stacked base pairs in a double helix absorb less than partially stacked bases in single strands, which absorb less than the mononucleotides.

Rhodopsin: A Chromophoric Protein

Many proteins include groups other than the common amino acids. Often, but not always, these groups are covalently linked to a polypeptide chain. Some examples and the nature of the attached group are glycoproteins (sugars), hemeproteins (iron porphyrins), flavoproteins (flavin), and rhodopsin (retinal; vitamin A). In the last three cases, the group is a *chromophore* and contributes to the absorption spectrum in the visible or near-UV regions. We examine some features of the visual pigment protein rhodopsin, which not only absorbs radiation in the visible region of the spectrum but also undergoes a photochemical transformation that triggers the visual stimulus.

Rhodopsins in nature occur widely in both vertebrates and invertebrates. A form of rhodopsin, called *bacteriorhodopsin,* has been found in the outer membranes of certain halophilic (salt-loving) bacteria. Rhodopsins consist of a colorless protein (or glycoprotein), opsin, to which the chromophore retinal is covalently attached. In mammals the isomer 11-*cis*-retinal is attached to the ε-amino group of a lysine side chain of the opsin by a protonated Schiff's base linkage:

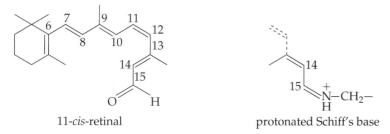

11-*cis*-retinal   protonated Schiff's base

In other organisms the chromophore appears to be the 9-*cis* or 13-*cis* isomer or the molecule 3-dehydroretinal. Rhodopsins have intense broad absorption bands with maxima lying between 440 and 565 nm, depending on the specific chromophore and its environment. Cow rhodopsin absorbs at 500 nm (figure 10.15) and is a bright red orange color; animals that are deficient in pigments (many albinos) show this red-orange in the pupil.

Retinal is an example of the linear polyenes that we examined in chapter 9. The six conjugated double bonds of the aldehyde contain π electrons

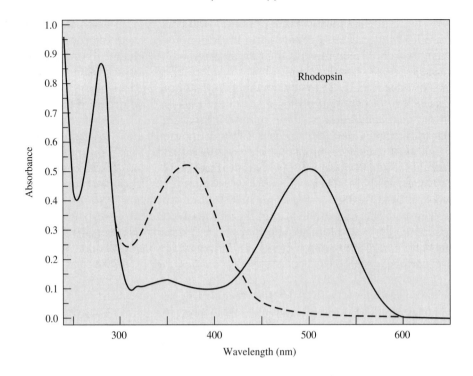

▶ FIGURE 10.15

Absorption spectrum of cow rhodopsin in the UV and visible regions. Purified rhodopsin was measured as isolated from dark-adapted
cow retinas (solid curve) and again
following its conversion by illumination (dashed curve). (From
M. L. Applebury, D. M. Zucker-
man, A. A. Lamola, and T. M. Jovin,
1974, *Biochemistry* 13:3448. Copyright by the American Chemical
Society.)

that can be treated approximately by the particle-in-a-box model. The long-wavelength absorption band (a $\pi \rightarrow \pi^*$ transition) of isolated retinal in a variety of organic solvents occurs between 366 and 377 nm, which is the expected region for polyenes of this length. In fact, the spectrum of rhodopsin after it has been photoconverted (figure 10.15, dashed curves) is very similar to that of retinal in an organic solvent. The origin of the shift of this band to 500-nm or even longer wavelengths in unilluminated rhodopsin has been connected to the presence of a negative charge in the opsin near the C-12, C-13, and C-14 atoms in retinal (see figure 9.2). Derivatives of retinal, with different double bonds reduced, were bound to opsin, and the absorption spectrum was measured. This indicated which part of the π system of electrons was needed to produce the large shift on binding (the opsin shift). Then approximate quantum mechanical calculations were done to show that a single negative charge about 0.3 nm distant from both C-12 and C-14 could account for the observed shift. The same method was applied to bacteriorhodopsin to show that its purple color can be explained by the presence of a negative charge near the β-ionone ring (Nakanishi et al. 1980; Honig et al. 1979).

Fluorescence

Fluorescence has become one of the most widely used form of spectroscopy in biology. Many biological substances emit characteristic fluorescence. Chlorophyll from leaves emits red fluorescence; green fluorescent protein (GFP) from jellyfish fluoresces in the green; many proteins fluoresce in the UV primarily from tryptophan residues; reduced pyridine nucleotides, flavoproteins, and the Y base of transfer RNA also exhibit characteristic fluorescence emission. Interesting and useful information can be obtained from fluorescent species that are added to biochemical molecules or systems. Fluorescent "labels" have been prepared by equilibrium binding or covalent attachment of fluorescent chromophores to enzymes, antibodies, membranes, and so on. Modified fluorescent substrate analogs, such as ε(etheno)-ATP, are useful as probes of enzyme-active sites. In some cases, natural chromophores that are not themselves fluorescent, such as retinal in rhodopsin, can serve as quenchers of the fluorescence of added probe molecules. Fluorescent proteins are widely used as indicators in cellular microscopy (Giulano et al. 1995).

Bioluminescence occurs in a wide variety of biological organisms. Although fireflies are among the most spectacular of these species, by far the greatest number of bioluminescent species occurs in the ocean. The luminescence is produced biochemically and usually does not require prior illumination. It serves a variety of purposes, including visual assistance in minimum-light environments, communication for social or sexual purposes, as camouflage, and for repelling predators. The full role of light emission is only beginning to be appreciated.

In many respects, fluorescence or luminescence spectroscopy is even richer than absorption in the variety of information and the range of sensitivity that it can provide. As with absorption, changes in the shape or position of the fluorescence spectrum reflect the environment in which the fluorescing molecule (*fluorophore*) exists. In addition, the intensity of fluorescence or the fluorescence yield (per photon absorbed) can vary over a large range. In some cases, changes of 100-fold or more can be produced. These changes in fluorescence yield are accompanied by changes in the *fluorescence lifetime* that can be measured directly. Another property that can be measured is the *excitation spectrum*. The fluorescence intensity is measured as the wavelength of the excitation light is varied. Fluorescence cannot occur unless light is first absorbed, so the excitation spectrum is related to the absorption spectrum. As the wavelength of excitation scans the absorption band, the fluorescence intensity will change proportionally to the amount of absorption.

The *polarization* or *depolarization* of the fluorescent light reflects the relative immobilization of the fluorophore, the extent to which the excitation is transferred to other molecules, and relative conformational changes among the different parts of a complex fluorophore. *Quenching* of fluorescence by added molecules provides evidence of molecular interactions, of the accessibility of the surface of organelles, and of the presence of intermediates such as triplet states or free radicals. Fluorescent properties can be used as an accurate meter stick to measure distances at the molecular level.

Simple Theory

Fluorescence is one type of emission of light from an excited electronic state. The origin of fluorescence can be seen in figure 10.3. A molecule absorbs light, which excites it to a higher electronic state. The *Frank–Condon principle* states that this transition is vertical: Nuclei, which are much heavier than electrons, do not change position during an electronic transition. Upon absorption of light, the molecule is in an excited vibrational state within an excited electronic state. The excited electronic state has an electron in a nonbonding or antibonding orbital, in which the bonded atoms are farther apart than in the ground electronic state. Thus, the minimum of the potential well for the excited electronic state is shifted to longer distances (see figure 9.6). The excited molecule can then rapidly lose some of its excited vibrational energy and return to the lowest vibrational state of the excited electronic level; this is called *internal conversion*. This rapid loss of vibrational energy is nonradiative (light is not given off, but instead heat is released); the heat release involves collisions between the molecule and solvent and occurs most efficiently in larger molecules, which have a large number of closely spaced vibrational states. Once in the ground vibrational state of the excited electronic level, the molecule can emit light as fluorescence. This light will be at lower frequency (longer wavelength) than the original absorption transition (shown by the energy-level diagram in figure 10.17).

The shape of a fluorescence-emission spectrum will often resemble that of the absorption, shifted to longer wavelength, because the same electronic and vibrational energy levels are involved in emission and absorption. Spectral shifts in fluorescence can be induced by local environment, as in absorption. However, these shifts are not necessarily the same, because in fluorescence the surrounding "solvent" interacts with an excited electronic state. Fluorescence spectra and maxima of fluorescent emission are often very sensitive to the type of environment, and fluorescence spectra can be used to probe molecular environments. A more sensitive indicator of local environment is the lifetime of the excited state and rate of loss of energy from the excited state by a variety of processes. We now investigate those processes in more detail.

Excited-State Properties

If a sample is illuminated to produce excited electronic states that fluoresce and then the light is extinguished, the fluorescence decays. Kinetically, the process is usually first order and is characterized by a rate law of the form

$$F \propto -\frac{d[M^*]}{dt} = k_d[M^*] \tag{10.16}$$

where F = intensity of fluorescence

$[M^*]$ = concentration of the excited electronic state that undergoes fluorescence

In accordance with this rate law, the decay is commonly exponential with time, characterized by a *fluorescence decay time* or *lifetime* τ given by

$$\tau = \frac{1}{k_d} \tag{10.17}$$

Measurements of the decay of fluorescence of simple molecules in dilute solution after excitation by a flash of light usually give log F versus t plots that are linear. An example is shown for anthracene in figure 10.16.

If fluorescence is the only decay path for the excited state, then the fluorescence decay rate constant (designated k_f) is the reciprocal of the *natural fluorescence lifetime* τ_0. This rate constant k_f is just the Einstein rate constant for spontaneous emission. Thus,

$$k_f = \frac{1}{\tau_0} \tag{10.18}$$

In most cases, however, there are significant nonradiative processes competing for the decay of the excited state. These include thermal deactivation, photochemistry, and the quenching by other molecules, Q. The overall rate

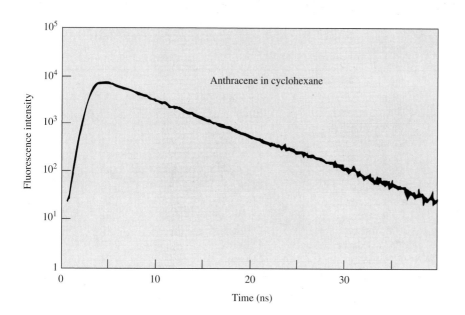

◀ FIGURE 10.16

Fluorescence decay of anthracene in cyclohexane (1.7×10^{-3} *M*), excited by a light flash of 1.4-ns duration (half width) and at approximately 360 nm. Emission measured at 450 nm. Logarithmic decay curve, obtained using single-photon counting method. (From P. R. Hartig, K. Sauer, C. C. Lo, and B. Leskovar, 1976, *Rev. Sci. Instrum.* 47:1122.)

of decay of the excited state is therefore the sum of the rates of all of these processes:

$$-\frac{d[\text{M}^*]}{dt} = k_f[\text{M}^*] + k_t[\text{M}^*] + k_p[\text{M}^*] + k_Q[\text{M}^*][\text{Q}]$$

$$= k_d[\text{M}^*]$$

where k_f, k_t, k_p, and k_Q are the rate constants for fluorescence, thermal deactivation, photochemistry, and quenching, respectively, and

$$k_d = k_f + k_t + k_p + k_Q[\text{Q}]$$

The observed lifetime of fluorescence is then

$$\tau = \frac{1}{k_f + k_t + k_p + k_Q[\text{Q}]} \qquad (10.19)$$

The *quantum yield* of fluorescence, ϕ_f, is the fraction of absorbed photons that lead to fluorescence; it is the number of photons fluoresced divided by the number of photons absorbed. Obviously, the quantum yield is ≤ 1:

$$\phi_f = \frac{\text{number of photons fluoresced}}{\text{number of photons absorbed}} \qquad (10.20)$$

The quantum yield can also be considered as the ratio of the rate of fluorescence to the rate of absorption. But the rate of absorption must equal (in the steady state) the rate of decay of the excited state. Therefore,

$$\phi_f = \frac{k_f[\text{M}^*]}{k_d[\text{M}^*]} = \frac{k_f}{k_d} \qquad (10.21)$$

But, using Eqs. (10.17) and (10.18),

$$\phi_f = \frac{\tau}{\tau_0} \qquad (10.22)$$

TABLE 10.5   Fluorescence Quantum Yields and Radiative Lifetimes

Compound	Medium	τ, ns	ϕ	Reference*
Fluorescein	0.1 M NaOH	4.62	0.93	a
Quinine sulfate	0.5 M H_2SO_4	19.4	0.54	a
9-Aminoacridine	Ethanol	15.15	0.99	a
Phenylalanine	H_2O	6.4	0.004	b
Tyrosine	H_2O	3.2	0.14	b
Tryptophan	H_2O	3.0	0.13	b
Cytidine	H_2O, pH 7	—	0.03	c
Adenylic acid (AMP)	H_2O, pH 1	—	0.004	c
Etheno-AMP	H_2O, pH 6.8	23.8	1.00	d
Chlorophyll a	Diethyl ether	5.0	0.32	e
Chlorophyll b	Diethyl ether	—	0.12	e
Chloroplasts	H_2O	0.35–1.9	0.03–0.08	f
Riboflavin	H_2O, pH 7	4.2	0.26	g
DANSYL sulfonamide[†]	H_2O	3.9	0.55	h
DANSYL sulfonamide + carbonic anhydrase	H_2O	22.1	0.84	h
DANSYL sulfonamide + bovine serum albumin	H_2O	22.0	0.64	h

*(a) W. R. Ware and B. A. Baldwin, 1964, *J. Chem. Phys.* 40:1703; (b) R. F. Chen, 1967, *Anal. Letters* 1:35; (c) S. Udenfriend, 1969, *Fluorescence Assay in Biology and Medicine,* Vol. II (New York: Academic Press); (d) R. D. Spencer et al., 1974, *Eur. J. Biochem.* 45:425; (e) G. Weber and F. W. J. Teale, 1957, *Trans. Faraday Soc.* 53:646; (f) A. Müller, R. Lumry, and M. S. Walker, 1969, *Photochem. Photobiol.* 9:113; (g) R. F. Chen, G. G. Vurek, and N. Alexander, 1967, *Science* 156:949; (h) R. F. Chen and J. C. Karnohan, 1967, *J. Biol. Chem.* 242:5813.
[†]DANSYL sulfonamide is 1-dimethylaminonaphthalene-5-sulfonamide.

Equation (10.22) provides a relation between the quantum yield of fluorescence and the fluorescence lifetime. Table 10.5 gives a summary of fluorescence quantum yields for a variety of fluorophores commonly encountered in biological studies.

Fluorescence almost always occurs from the lowest excited (singlet) state of the molecule. We might otherwise expect to see fluorescence from each of the excited states reached by progressively greater frequency (photon energy) of the exciting radiation absorbed by the fluorophore, but this is almost never observed. (The molecule azulene is an example of one of the rare exceptions; azulene fluorescence comes predominantly from the second excited singlet state.) The reason that only the lowest state normally emits radiation is that the processes of internal conversion of the higher states (thermal deactivation from higher electronic states to the lowest excited state) are exceedingly rapid. This is illustrated for bacteriochlorophyll in figure 10.17, where the absorption and fluorescence spectra are plotted in the vertical direction (turned 90° from the usual orientation) to correspond to the energy-level diagram. Internal conversion from the lowest excited state to the ground state also occurs. It is one of the important sources of thermal deactivation, $k_i[M^*]$, that compete with fluorescence. The rate is often slower for this step, however, partly because of the greater energy separation between the ground state and the first excited electronic state compared with the energy differences among the excited states.

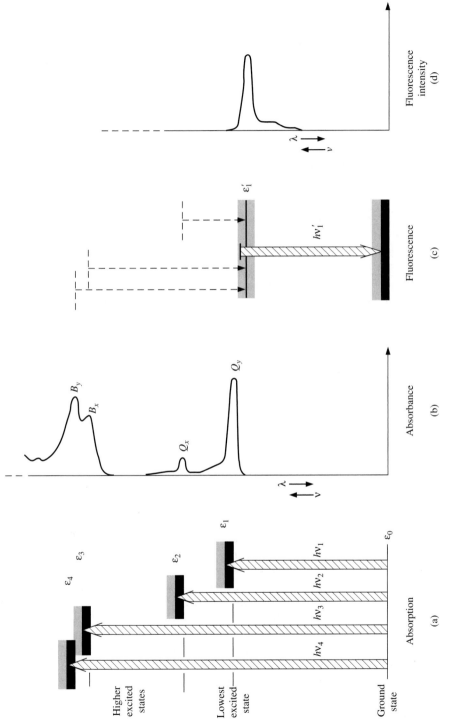

▲ FIGURE 10.17

Absorption and fluorescence of bacteriochlorophyll. (a) Energy-level diagram showing spetral transitions (vertical arrows). The energy levels are broadened (shading) by vibrational sublevels that are not usually resolved in solution spectra. (b) Absorption spectrum corresponding to energy levels of part (a). This spectrum is turned 90° from the usual orientation to show the relation to the energy levels. (c) Radiationless relaxation (dashed arrows) and fluorescence (shaded arrow). (d) Fluorescence emission spectrum corresponding to part (c). Note the red shift of the fluorescence compared with the corresponding Q_y absorption illustrated in parts (a) and (b). [From K. Sauer, 1975, in *Bioenergetics of Photosynthesis*, ed. Govindjee (New York: Academic Press), 115–181.]

Fluorescence Quenching

A decrease in fluorescence intensity or quantum yield occurs by a variety of mechanisms. These include collisional processes with specific quenching molecules, excitation transfer to nonfluorescent species, complex formation or aggregation that forms nonfluorescent species (concentration quenching), and radiative migration leading to self-absorption. There are important biological applications or consequences of each of these quenching mechanisms.

Quenching of fluorescence by added substances or by impurities can occur by a collisional process. Because it is the excited state of the fluorescent molecule that must undergo collisional quenching, the encounters must occur frequently, and the quenching process must be efficient. Molecular oxygen is one of the most widely encountered quenchers because it is a triplet species (two unpaired electrons) in its ground electronic state. Most excited fluorophores emit from a singlet state (no unpaired electrons), and O_2 quenches the fluorescence by means of the reaction

$$M^* \text{ (singlet)} + O_2 \text{ (triplet)} \longrightarrow M^* \text{ (triplet)} + O_2 \text{ (singlet)}$$

Normally, M^* (singlet) is fluorescent and M^* (triplet) is not.

Collisional quenching is a bimolecular process kinetically; however, the excited oxygen molecules quickly return to the ground triplet state upon subsequent collisions or interactions with the solvent. As a consequence, they are not consumed in the process, and collisional quenching obeys pseudo-first-order kinetics. For a generalized quencher molecule Q, the relevant equations are

$$M + h\nu \longrightarrow M^* \qquad \text{(excitation)}$$

$$M^* \xrightarrow{k_f} M + h\nu' \qquad \text{(fluorescence)}$$

$$M^* + Q \xrightarrow{k_Q} M + Q^* \qquad \text{(quenching)}$$

In the absence of quenchers,

$$\phi_f^0 = \frac{k_f}{k_f + k_t + k_p} \tag{10.23}$$

In the presence of a quencher at concentration [Q],

$$\phi_f = \frac{k_f}{k_f + k_t + k_p + k_Q[Q]} \tag{10.24}$$

Therefore, we obtain a result known as the *Stern–Volmer relation:*

$$\frac{\phi_f^0}{\phi_f} = 1 + \frac{k_Q[Q]}{k_f + k_t + k_p} = 1 + K[Q] \tag{10.25}$$

We can also write the equation in terms of the lifetime τ', in the absence of quencher.

$$\frac{\phi_f^0}{\phi_f} - 1 = k_Q \tau'[Q]$$

where

$$\tau' = \frac{1}{k_f + k_t + k_p}$$

Because the intensity of fluorescence F is proportional to the quantum yield ϕ_f, a plot of F^0/F versus [Q] gives a straight line with slope $k_Q\tau'$. Thus, the longer the lifetime of the excited state, the greater is the probability of quenching.

The consequences of concentration quenching can be quite dramatic. The fluorescence of benzene in oxygen-free solutions occurs with a lifetime of 29 ns; in a solution in equilibrium with O_2 at 1 atm pressure, this lifetime is decreased fivefold, to 5.7 ns. Chlorophyll a is strongly fluorescent in dilute solutions ($\phi_f \cong 0.3$), but the fluorescence intensity is quenched essentially to zero with added quinones or carotenoids. (In this case, the mechanism may involve complex formation between chlorophyll and the quencher molecule.) It is clear from these examples that great care must be taken to remove all extraneous quenching species in the determination of the intrinsic properties of fluorescing molecules.

Concentration quenching may occur as a consequence of aggregation, dissociation, or other changes in the fluorophore itself. *Excimers* (excited dimers) may form because of greater interactions of the excited-state species. In each case, a concentration-dependent quantum yield of fluorescence will result. For example, if nonfluorescent excimers form according to

$$M^* + M \xrightarrow{k_e} [M \cdot M]^*$$

then

$$\phi_f = \frac{k_f}{k_f + k_t + k_p + k_e[M]}$$

Chlorophyll a at concentrations of about 10^{-2} M exhibits concentration quenching in most solvents. Depending on the medium, the aggregated chlorophylls may be completely nonfluorescent or they may have a weak but distinctive fluorescence of their own. In photosynthetic membranes, the chlorophyll concentrations locally are typically 0.05–0.1 M, and fluorescence yields are only about one-tenth of the monomer value.

Measurement of changes in fluorescence intensity due to changes in environment of the fluorophore is a major application of fluorescence to biological systems. Proteins contain amino acids that fluoresce; upon change in their local environment—for example, through folding of the protein chain or binding of another molecule—fluorescence intensity can change through quenching mechanisms. Excitation transfer processes, which are discussed next, provide additional paths for fluorescence quenching and more quantitative information about biomolecular structure.

Excitation Transfer

Some of the most valuable applications of fluorescence to biochemical systems involve the transfer of excitation from one chromophore to another. Because this transfer process depends strongly on the distance between the chromophores and on their relative orientations, experiments can be designed to obtain useful information concerning macromolecular geometry. Excitation transfer has been treated theoretically by a number of authors, and several of the most important relations have been verified quantitatively in carefully designed model experiments (Stryer 1978).

We consider the transfer of excitation from an excited donor molecule D^* to an acceptor A, which then fluoresces:

$$D \xrightarrow{h\nu} D^* \qquad \text{(absorption by donor)}$$

$$D^* \xrightarrow{k_f} D + h\nu' \qquad \text{(fluorescence of donor)}$$

$$D^* + A \xrightarrow{k_T} D + A^* \qquad \text{(excitation transfer)}$$

$$D^* \longrightarrow D \qquad \text{(all other de-excitation)}$$

$$A^* \longrightarrow A + h\nu'' \qquad \text{(fluorescence of acceptor)}$$

The transfer of excitation can be measured in three different ways: (1) the decrease in fluorescence quantum yield of the donor due to the presence of the acceptor, (2) the decrease in lifetime of the donor due to the presence of the acceptor, and (3) the increase in fluorescence of the acceptor due to the presence of the donor. The efficiency of excitation transfer is defined as the fraction of D^* that is de-excited by transfer. It is simply related to the rate constants for excitation transfer k_T and for all other processes of de-excitation, including fluorescence k_d:

$$\text{efficiency } (Eff) = \frac{k_T}{k_T + k_d} \tag{10.26}$$

From Eq. (10.19), the quantum yield for fluorescence of the donor alone is

$$\phi_D = \frac{k_f}{k_d} \tag{10.21}$$

and for the donor in the presence of acceptor is

$$\phi_{D+A} = \frac{k_f}{k_d + k_T}$$

Therefore, the efficiency of excitation transfer is directly related to the ratio of quantum yields for the donor in the presence (ϕ_{D+A}) and absence of the acceptor (ϕ_D):

$$Eff = 1 - \frac{\phi_{D+A}}{\phi_D} \tag{10.27}$$

As quantum yields and lifetimes are proportional to each other [Eq. (10.22)], the efficiency is similarly related to the fluorescence lifetimes:

$$Eff = 1 - \frac{\tau_{D+A}}{\tau_D} \tag{10.28}$$

where τ_{D+A} is the fluorescence lifetime in the presence of acceptor and τ_D is the lifetime in the absence of acceptor. The efficiency of excitation transfer depends on the distance between donor and acceptor. In the range from about 1 to 10 nm, *fluorescence resonance energy transfer (FRET)* or *Förster transfer* occurs. For each donor–acceptor pair, the efficiency of transfer depends on r^{-6}, where r is the distance between them. The energy transfer efficiency is

$$Eff = \frac{r_0^6}{r_0^6 + r^6} \tag{10.29}$$

where Eff = efficiency of transfer ($0 \leq Eff \leq 1$)

r_0 = characteristic distance for the donor–acceptor pair; it is the distance for which $Eff = 0.5$

r = distance between donor and acceptor

The value of r_0 depends on the amount of overlap between the fluorescence spectrum of the donor and the absorption spectrum of the acceptor. It also depends on the angular orientation between donor and acceptor, which is often not known. A common assumption is that the orientations are averaged, which may not be true for chromophores with fixed relative geometries:

$$r_0(\text{nm}) = 8.79 \times 10^{-6}(J\kappa^2 n^{-4}\phi_D)^{1/6} \qquad \textbf{(10.30)}$$

where $J = \int \varepsilon_A(\lambda)F_D(\lambda)\lambda^4 \, d\lambda$

$\varepsilon_A(\lambda)$ = absorption spectrum of acceptor

$F_D(\lambda)$ = fluorescence spectrum of donor

n = refractive index of medium

κ^2 = an orientation factor that depends on the angle between transition dipoles [see Eq. (10.6)] of the donor and acceptor; it is zero if the transition dipoles are perpendicular to each other, and four if they are parallel; for random orientations $\langle \kappa^2 \rangle = \frac{2}{3}$

ϕ_D = quantum yield of donor

Molecular Rulers

Extensive experimental tests of FRET have been carried out. Förster (1949) verified the expected concentration dependence in a study of the quenching of fluorescence of trypaflavin in methanol by the dye rhodamine B. In this case, a value $r_0 = 5.8$ nm was obtained, which shows that transfer can occur over distances several times larger than the actual molecular dimensions. The dependence on the inverse sixth power of distance has been tested in an elegant series of experiments by Latt, Cheung, and Blout (1965) and by Stryer and Haugland (1967). In these studies, two chromophores are attached covalently to a rigid molecular framework. Excitation transfer is measured for a series of these synthetic species where the distance between the donor and acceptor molecule is different. Stryer and Haugland examined transfer from a dansyl group at one end to a naphthyl group at the other end of oligoprolines with 1 to 12 monomer units in the rigid chain. The results showed excellent agreement with the r^{-6} dependence from 1.2 to 4.6 nm separation, as shown in figure 10.18.

An important determinant of excitation transfer is the spectral overlap between the donor emission and the acceptor absorption spectrum. This overlap was varied 40-fold by solvent effects for a modified steroid by Haugland, Yguerabide, and Stryer (1969), and the transfer rate varied almost in parallel. These extensive tests have confirmed the validity of the Förster resonance transfer mechanism, at least over distances in the range 1 to 10 nm.

Genetic recombination occurs when two DNA double strands come together and, by a process of cutting and splicing, exchange parts of their sequence; genes are exchanged. The structure of the intermediate is important for understanding the genetic outcome of the process. Fluorescence energy

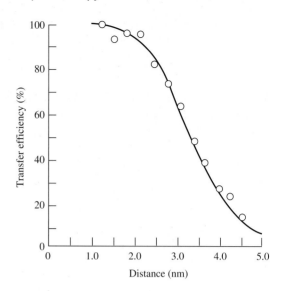

▶ **FIGURE 10.18**

Efficiency of energy transfer as a function of distance in dansyl-(L-prolyl)$_n$-α-naphthyl, $n = 1$ to 12. Energy transfer is 50% efficient at 3.46 nm. Solid line corresponds to r^{-6} dependence. (From Stryer and Haugland 1967, 719.)

transfer was used to determine that the sequences of genes on the two double strands are arranged in antiparallel directions and that the junction forms a right-handed cross (Murchie et al. 1989). A model recombination junction of two DNA fragments of about 30 base pairs each was synthesized. Fluorescein (the donor) was attached to the 5′ end of one strand of one DNA, and rhodamine (the acceptor) was attached to the 5′ end of a strand of the other DNA. The parallel or antiparallel arrangement could thus easily be detected. The amount of fluorescence energy transfer provides the distance between the labeled ends. To decide between a right-handed or left-handed cross, the length of one of the DNA fragments was systematically increased. DNA is a right-handed helix, so lengthening the helix also means that the end rotates around the helix axis. This rotation will move the acceptor closer to or farther from the donor on the other DNA, depending on whether the DNAs form a right- or left-handed cross. Similar experiments have been performed on RNA structures where several helical regions come together. An example is a ribosomal RNA segment that binds to ribosomal protein S15 (Ha et al. 1999). The addition of Mg^{2+} or S15 was seen by FRET to fold the junction. FRET methods are generally applicable to any biochemical structure that can be specifically labeled. (For a general review, see Selvin 2000 and Stryer 1978.)

Fluorescence Polarization

The oscillating electromagnetic field that excites an electronic transition can be linearly polarized. Excitation will preferentially occur for molecules whose transition dipole moments are parallel to the polarized radiation. The polarization of emitted fluorescent radiation depends on the orientation of the emitting transition dipole and any reorientation that has occurred between the time of absorbance and the time of emission. Since molecules are tumbling randomly in solution, measurements of the fluorescence *depolarization* provide insights into molecular motions. The *fluorescence anisotropy r* is defined as

$$r = \frac{I_{\parallel} - I_{\perp}}{I_{\parallel} + 2I_{\perp}}$$

where $I_\parallel$ and $I_\perp$ are the detected fluorescence intensities parallel and perpendicular to the direction of polarization of the exciting radiation. Depolarization occurs due to tumbling in solution. Because small biological macromolecules and other molecules reorient on a 1- to 10-ns timescale and the fluorescence lifetimes are of the same magnitude, depolarization occurs. This is often best measured by *time-resolved fluorimetry.* In such an experiment, a brief pulse of light causes the molecule to be excited, and the subsequent fluorescence anisotropy (with the light turned off) is measured. The decay of the anisotropy provides a direct measure of the rate of reorientation of the molecule. In short, the time resolved experiment provides measurement of *molecular dynamics;* this is an important application of fluorescence spectroscopy.

Depolarization may also occur due to excitation transfer among identical fluorophores. Depolarization by rotational diffusion can be distinguished experimentally because of its sensitivity to temperature and to the viscosity of the medium.

Phosphorescence

An excited singlet state with all electrons paired can become an excited triplet state with two unpaired electrons. The paired electrons (with opposite spins) can occupy two different orbitals or be in the same orbital. The unpaired electrons (with identical spins) must be in separate orbitals as required by the Pauli exclusion principle. The triplet state is usually of lower energy than the singlet state. The unpaired electrons in different orbitals stay farther apart from each other, reducing the electron–electron coulomb repulsion and lowering the energy. Although formation of the triplet is favored by the decrease in energy, the probability of a singlet–triplet transition is usually small. It is a quantum mechanically forbidden transition. The singlet–triplet conversion is catalyzed by certain molecules, such as O_2, which is a triplet in the ground state, and by species with a high atomic number, such as I^-. Once a molecule is in an excited triplet state, it can emit light and return to its ground state. The emission of light from an excited triplet state is *phosphorescence.* Experimentally, phosphorescence is usually distinguished from fluorescence by its lifetime. Phosphorescence lifetimes typically occur in the millisecond and longer range, while fluorescence lifetimes are in the microsecond and shorter range. Phosphorescence occurs at a longer wavelength than do the fluorescence or the absorption, as is illustrated in figure 10.19. The proof of whether luminescence is phosphorescence or fluorescence is to measure the magnetic properties of the excited state. A molecule in a triplet state is paramagnetic and will be attracted by a magnet; its energy is lowered by the magnetic field.

At room temperature, DNA does not phosphoresce; the triplet state is quickly quenched. However, in frozen solution, DNA shows a characteristic phosphorescence due to thymidylic acid. The triplet states of the other nucleotides are either quenched or they transfer excitation to the thymidylic nucleotides, which then emit the phosphorescence.

Single-Molecule Fluorescence Spectroscopy

The intrinsic sensitivity of fluorescence spectroscopy has facilitated its application to individual molecules: This is called *single-molecule fluorescence spectroscopy.* Because fluorescence involves emission of radiation, it can be detected using microscopy methods; the number of photons emitted by a single fluorophore can be reasonably detected by high-sensitivity cameras. The

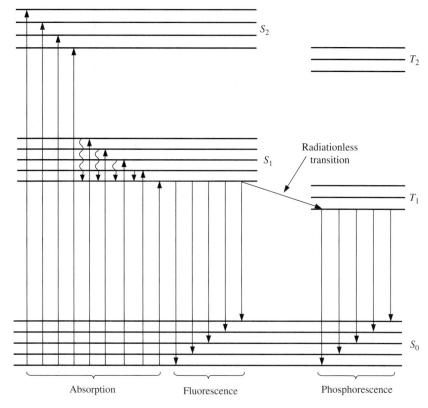

▲ FIGURE 10.19

Energy-level diagram for absorption, fluorescence, and phosphorescence. Vibronic transitions from a ground-state singlet level S_0 with all electrons paired, to an excited-state singlet level S_1 or S_2 with all electrons paired, lead to absorption of light. The transition from an excited singlet to a ground-state singlet leads to fluorescence. A radiationless transition can occur from the excited singlet S_1 to the excited triplet state T_1 with two unpaired electrons. Vibronic transitions from an excited triplet to the ground-state singlet lead to phosphorescence. Other possible transitions are not shown. Direct transitions from the ground-state singlet S_0 to an excited-state triplet T_1 or T_2 lead to weak (forbidden) singlet–triplet absorption bands. Transitions between excited triplet states are allowed (as are transitions between excited singlet states). Measurement depends on the ability to produce sufficient population of the excited state. The longer lifetime of the triplet makes $T_1 \rightarrow T_2$ absorption easier to measure than that of $S_1 \rightarrow S_2$.

spectroscopic studies of single molecules, while technically difficult, involve the same principles as we have described above for measurements of large numbers of fluorescent molecules in solution. The molecule is often immobilized on a glass microscope slide, such that it does not diffuse away from focus during a measurement. Usually, dyes that fluoresce in the visible region of the spectrum are used. Changes in emission wavelength, quenching, and FRET can all be measured using single-molecule approaches. What advantage is there to measuring the fluorescence of a single molecule?

Spectroscopy of single biological molecules can investigate processes that cannot be examined by normal solution experiments. In biological systems, many dynamic processes occur that are obscured by the large numbers of molecules in a usual system. Thus, single-molecule experiments are excellent for extracting molecular behavior from heterogeneous systems. Changes

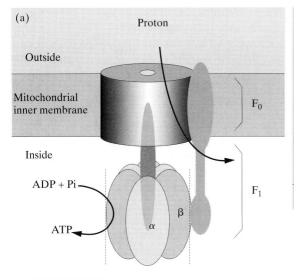

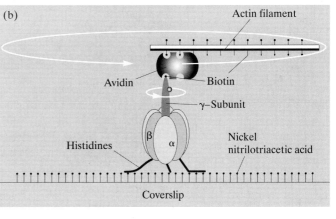

▲ FIGURE 10.20

Single-molecule studies of F_1-ATPase. (a) The F_0F_1–ATP synthesase is shown schematically. The F_0 subunit is embedded in the membrane and is a proton pump. The F_1-ATPase is a rotary motor that synthesizes ATP from ADP and inorganic phosphate. (b) In the experiment, the F_1 subunit was attached to a glass coverslip through nickel–histidine linkages, and a fluorescent actin filament was attached to the gamma subunit. The rotation of a single actin filament could be observed by fluorescence miscroscopy. (Reprinted with permission from Noji et al., 1999, *Science* 283:1689. Copyright 1999.)

in concentration of an ion, binding of another molecule, and so on trigger many biological reactions. Although the biochemist desires to understand on a molecular level the steps that occur in a biological process, it is difficult to experimentally synchronize a large population of molecules to perform on cue. Single-molecule experiments eliminate the need to synchronize molecular events because there is only one molecule.

To illustrate the power of single-molecule analysis, we examine an important biological example. The F_1-ATPase is the molecule that couples proton transport to the synthesis of ATP. Biochemical studies indicated that the F_1-ATPase couples directional proton transfer to ATP synthesis by acting as a rotary motor. A single-molecule measurement graphically confirmed this. As shown in figure 10.20, the F_1-ATPase was attached to a glass surface, through nickel–histidine interactions. To the other side of the motor was attached, through high-affinity noncovalent interaction, a rigid protein fiber (actin) that was tagged with a fluorescent dye. Upon addition of ATP, the rigid, fluorescent fiber was observed to undergo rotary motion in the microscope, like a propeller. The molecule is a trimer, and the rotations were observed to occur in 120° steps. The rotary motion was driven by ATP hydrolysis, and the motor was found to be very efficient. There can be no more graphic elucidation of a biochemical mechanism than to watch it.

Optical Rotatory Dispersion and Circular Dichroism

A property of most biological molecules is molecular asymmetry, or *chirality;* such molecules are not identical to their mirror images. The simplest examples result from the presence of asymmetric carbon atoms in these molecules. A carbon that is tetrahedrally bonded to four different atoms or groups can exist in two different structures that are mirror images of each other. (The rules for characterizing such molecules are described in organic chemistry texts.)

Chiral molecules have distinctive properties that bear emphasis. First, consider some common examples of such "handedness" that are more familiar. Screws or nuts and bolts can be cut with a right-hand or a left-hand thread. A tumbled assortment of left- and right-handed bolts could be sorted by using a test nut. Once the bolts are separated they will have different interactions with nuts of a given handedness. In much the same way, chiral molecules in solution and with random, constantly changing orientations interact differently with light that is polarized in a chiral way (circularly polarized light).

These examples emphasize also that chirality of materials does not necessarily depend on the presence of asymmetrically substituted carbon atoms, although that is commonly the origin for small molecules. In the case of nucleic acids, an important source of chirality is whether the helical polynucleotide winds in a left- or a right-handed sense. Nucleic acids are mainly right-handed, but certain base sequences can form left-handed double helices. Proteins in the α-helical conformation wind in a right-handed sense. The handedness of the helices depends on the stereochemistry of the monomer units: the L (levo, or left) amino acid isomer for proteins and the D (dextro) sugar for nucleic acids. The overall structure of biologically active proteins (enzymes, antibodies) depends on additional determinants of tertiary structure that are not so easily characterized. These are very important, however, for they serve to make the detailed surface structure of an enzyme or antibody nonsuperimposable with its mirror image. As a consequence, the active sites of these molecules are able to distinguish between mirror-image substrate molecules, much as left-handed nuts will interact only with left-handed bolts. Enzymes that are involved with the synthesis of chiral molecules may make a single chiral isomer from a substrate that is not itself asymmetric. It is clear from these few examples that the property of chirality is of fundamental and wide-reaching importance in biology.

Polarized Light

Chiral structures can be distinguished and characterized by polarized light. The optical properties usually measured are *optical rotation*, the rotation of linearly polarized light by the sample, and *circular dichroism* (CD), the difference in absorption of left- and right-circularly polarized light. The spectrum of optical rotation is known as *optical rotatory dispersion* (ORD).

Electromagnetic radiation can be described by an electric-field vector that oscillates with a characteristic frequency in time and space. For *unpolarized light*, the electric vector may oscillate in any direction perpendicular to the direction of propagation. For a large number of photons in an unpolarized beam, all directions are equally represented. The electric vectors can be pictured as radiating spokes on a many-spoked wheel, as shown in figure 10.21 (top). For *plane-polarized light*, the electric vector oscillates in a single plane that includes the propagation direction. A pictorial description of vertically plane-polarized light is shown in figure 10.21 (top and middle).

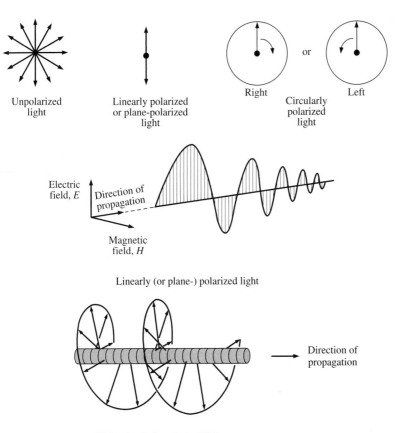

To an observer moving at the photon's velocity, the electric vector appears to be oscillating back and forth along a line. For this reason, plane-polarized light is also referred to as being linearly polarized. *Circularly polarized light* propagates so that the tip of its electric vector sweeps out a helix (figure 10.21, bottom); right-circularly polarized light produces a right-handed helix. To an observer moving with the photons, the electric vector appears to be moving in a circle, like the hands of a clock (figure 10.21, top). The convention is that for left-circularly polarized light the electric vector moves counterclockwise as the light moves away from the observer; for right-circularly polarized light, it moves clockwise. *Elliptically polarized light,* which we do not consider in detail here, propagates such that its electric vector sweeps out an ellipse, as seen by the observer moving with the light. Elliptically polarized light is either right- or left-handed, and it may be thought of as intermediate between circularly and plane-polarized. In fact, optical retardation (birefringent) plates will progressively produce plane, elliptical, circular, elliptical, plane, elliptical, and so on polarizations as the thickness of the retarder is increased.

We can represent the electric vector of light in terms of its components. For light propagating in the z-direction and plane polarized in the x-direction, the x- and y-components of the electric vector for plane-polarized light are:

Plane-polarized light:

$$E_x(z, t) = E_0 \sin 2\pi\nu\left(\frac{z}{c} - t\right) = E_0 \sin \frac{2\pi}{\lambda}(z - ct)$$

$$E_y(z, t) = 0 \tag{10.31}$$

where c is the speed of light in a vacuum, and ν is the frequency and λ the wavelength of light in a vacuum. The x- and y-components of the oscillating electric vector are $E_x(z, t)$ and $E_y(z, t)$; E_0 is the maximum amplitude of the vector. Because the oscillating magnetic field is perpendicular to the electric field, its components are the same as the electric components, *except* that x and y are interchanged. If we specify the electric field, we do not have to consider the magnetic field explicitly because Maxwell's equations characterize the magnetic field. The x- and y-components of the electric vector for circularly polarized light are:

Left-circularly polarized light:

$$E_x(z, t) = E_0 \sin 2\pi\left(\frac{z}{\lambda} - \frac{ct}{\lambda}\right)$$

$$E_y(z, t) = E_0 \sin 2\pi\left(\frac{z}{\lambda} - \frac{ct}{\lambda} + \frac{1}{4}\right) \tag{10.32a}$$

Right-circularly polarized light:

$$E_x(z, t) = E_0 \sin 2\pi\left(\frac{z}{\lambda} - \frac{ct}{\lambda}\right)$$

$$E_y(z, t) = -E_0 \sin 2\pi\left(\frac{z}{\lambda} - \frac{ct}{\lambda} + \frac{1}{4}\right) \tag{10.32b}$$

The y-component of the light is one-fourth of a cycle ahead of (or behind) the x-component. The resultant of the components is an electric field whose magnitude is constant, but which rotates counterclockwise [Eq. (10.32a)] or clockwise [Eq. (10.32b)] as it moves forward in space (along z) and in time. Addition of Eqs. (10.32a) and (10.32b) makes it clear that plane-polarized light is equivalent to the sum of left- and right-circularly polarized light beams propagating in the same direction with the same wavelength and amplitude E_0. Linearly and circularly polarized light can be produced easily from unpolarized light by transmission through appropriate films (such as Polaroid) or crystals (Nicol prism, quarter-wave plate).

We have previously described materials that are (linearly) birefringent toward plane-polarized light. These materials must be geometrically anisotropic; they have different properties along different directions. *Linear birefringence* occurs because the refractive index, and hence the light propagation velocity, is different for polarization planes oriented along different directions of the material. In an absorption band of an anisotropic material, *linear dichroism* occurs; the absorbance is different for different orientations of the plane of polarization of the light. Most crystals are anisotropic and therefore linearly birefringent and linearly dichroic. Polaroid sheets contain oriented polymers that make them linearly dichroic (and linearly birefringent) in the visible range. They absorb light polarized parallel to the oriented polymer and transmit light polarized perpendicular to the polymer; thus a Polaroid sheet is a linear polarizer for incident unpolarized light.

Substances that are optically active (chiral) exhibit *circular birefringence,* where circular birefringence is $n_L - n_R$, the difference between the refractive index for left-circularly polarized and the refractive index for right-circularly polarized light. In other words, the velocities of propagation for left- and right-circularly polarized light are different. There is an important difference between circular birefringence and linear birefringence. By contrast with a linearly birefringent material, which is geometrically anisotropic, with different optical properties viewed from different directions, a homogeneous chiral sample, such as a solution of an optically active solute, is geometrically isotropic; its optical properties are identical viewed from any direction. The circular birefringence results from an intrinsic property of the material that persists even though the orientations of the molecules are random. Chiral molecules also show *circular dichroism* $(A_L - A_R)$, a difference in absorbance for left- and right-circularly polarized light.

A quantum mechanical derivation shows how circular birefringence or dichroism is related to electronic wavefunctions of the ground and excited states of the molecule. An optically active transition must have an electric–dipole transition moment and a magnetic–dipole transition moment that are not perpendicular to each other. This can occur only for molecules that are not superimposable with their mirror images. A qualitative understanding of the "classical" electronic motion in these molecules, during electronic excitation by incident light, is that the electrons do not move in a line or in a circle; instead, they move in a helical path.

Optical Rotation

Optical rotation by chiral samples results from and is a measure of their circular birefringence. The term *optical rotation* comes from the usual experimental measurement procedure. If plane-polarized light is propagated through a transparent chiral sample, the emerging beam will also be plane-polarized, but its plane of polarization (still including the direction of propagation) will be rotated by an angle ϕ with respect to the direction of the polarization of the incident light. If the rotation is clockwise as seen by the observer, ϕ is positive; counterclockwise rotation is assigned a negative value of ϕ (figure 10.22). The origin of optical rotation can be rationalized by considering the incident plane-polarized light to be made up of two opposite circularly polarized but in-phase components, as represented by Eq. (10.32). Because the chiral sample is circularly birefringent $(n_L \neq n_R)$, the two circular-polarized component electric vectors will propagate through the sample with different velocities (c/n), and they will have different wavelengths λ_m in the medium $(\lambda_m = \lambda/n)$. As a consequence, one of them becomes advanced in phase with respect to the other. The resultant light, which remains plane polarized, becomes rotated progressively upon passage through the sample. When it emerges from the other side, it has undergone a rotation. The angle of rotation can be obtained by replacing z in Eq. (10.32b) by $n_R z$ and z in Eq. (10.32a) by $n_L z$ and then adding the x- and y-components. The resultant light is linearly polarized but rotated by an angle ϕ given by

$$\text{rotation (rad cm}^{-1}) = \phi = \frac{\pi}{\lambda}(n_L - n_R) \qquad \textbf{(10.33)}$$

where λ is the wavelength of the light in a vacuum. Note that ϕ is given per centimeter of pathlength in the sample. The actual angle of rotation clearly increases linearly with the path length through the sample. The rotation is measured

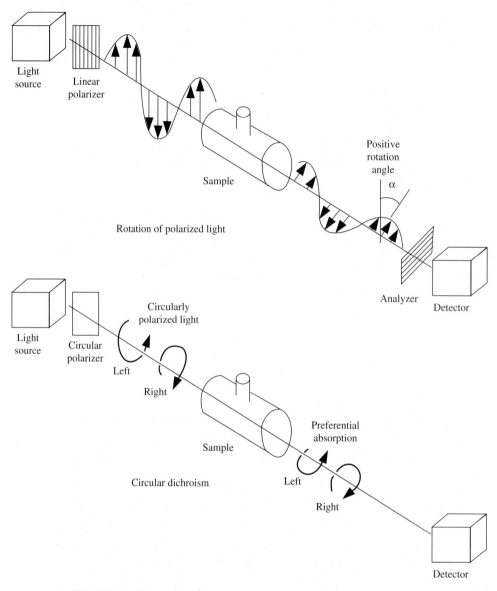

▲ **FIGURE 10.22**

At the top, measurement of the rotation of linearly polarized light. If the sample in the top cell has significant absorbance, the transmitted light will be elliptically polarized. (See figure 10.23.) At the bottom, measurement of circular dichroism, the preferential absorption of circularly polarized light. The detector measures the difference in absorbance of the right- and left-circularly polarized light.

by placing a polarizing element (suitable crystal or Polaroid sheet) called an *analyzer* in the emergent beam. By turning the analyzer so that it is crossed (perpendicular) to the direction of polarization of the emergent beam, the intensity is extinguished, and the angle of rotation can be measured to better than 1 millidegree. To appreciate how sensitively circular birefringence can be measured, consider Eq. (10.33) for a measurement made at $\lambda = 314 \text{ nm}$ ($= \pi \times 10^{-5}$ cm). A rotation of 1 rad cm^{-1} (1 rad = 57.3 deg) corresponds to $n_L - n_R = 10^{-5}$.

Circular Dichroism

Materials that exhibit linear dichroism toward plane-polarized light do so because they have different absorbances as a function of the orientation of the sample with respect to the polarization direction of the incident light. Polaroid sheets are examples of dichroic materials. As with birefringence, linear dichroism results from geometric anisotropy in the sample.

Circular dichroism, analogously, results from a differential absorption of left- and right-circularly polarized light (figure 10.22) by a sample that exhibits molecular asymmetry. A simple expression of the circular dichroism is given by

$$\Delta A = A_L - A_R \qquad \textbf{(10.34)}$$

where A_L and A_R are the absorbances of the sample for pure left- and right-circularly polarized light, respectively. As with circular birefringence, circular dichroism may be either positive or negative. Circular dichroism occurs only in a region of the spectrum where the sample absorbs, whereas circular birefringence occurs in all wavelength regions of an optically active substance. This latter property is of obvious advantage for transparent substances such as sugars, in which the lowest-energy electronic transitions occur in the far UV.

Referring again to figure 10.22, the passage of plane-polarized light E_0 through a circularly dichroic sample produces not only a phase shift due to the circular birefringence (which occurs in the absorption region as well) but also a differential decrease of the amplitudes (the electric vector magnitudes) of the left- and right-circularly polarized components. The emerging beam E is found to be elliptically polarized (figure 10.23) as a consequence. A detailed analysis shows that the ellipticity is given by

$$\theta \, (\text{rad cm}^{-1}) = \frac{2.303(A_L - A_R)}{4l} \qquad \textbf{(10.35)}$$

where l (cm) is the path length through the sample.

Experimental measurements of optical rotation and ellipticity are burdened with a history of measurements in cells of length 10 cm (path length d, in decimeters) and the use of different symbols for the same quantity. The equations summarized in table 10.6 represent a consensus of current usage.

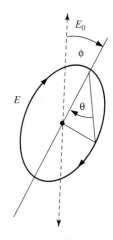

▲ FIGURE 10.23

Elliptically polarized light emerging toward the observer from a circularly dichroic sample. The sign convention is that ϕ is positive for clockwise rotation of the major axis and θ is positive for right-elliptically polarized light, as shown. The rotation angle is sometimes called α; the ellipticity angle is also called ψ.

Circular Dichroism of Nucleic Acids and Proteins

The optical activity of a nucleic acid or protein is the sum of the individual contributions from the monomeric units and the contribution from their interactions in the polymeric arrangement. By comparing the circular dichroism (CD) of the native polymer to that of its monomeric units, their separate contributions can be determined experimentally and subtracted from the optical activity of the intact polymer. The difference then gives the contribution from the interactions in the native polymer conformation. This is illustrated for DNA and RNA in figure 10.24. The dramatic differences between the solid and dashed curves demonstrate that the principal contribution to the CD of the nucleic acids arises from interactions in the polymer.

The synthetic polynucleotide poly(dG-dC)·poly(dG-dC) is a double-stranded helix with a sequence of alternating guanine and cytosine bases on

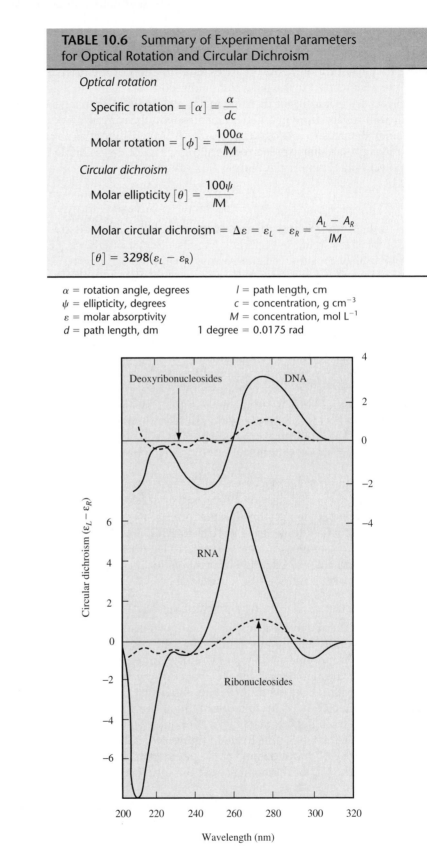

▶ FIGURE 10.24

Circular dichroism of double-stranded DNA and RNA compared with their component mononucleosides. [*M. lysodeikticus* DNA data from F. Allen et al., 1972, *Biopolymers* 11:853. Rice dwarf virus RNA data from T. Samejima et al., 1968, *J. Mol. Biol.* 34:39. Nucleoside spectra calculated from the base composition (72% G + C for the DNA; 44% G + C for the RNA) and CD data of C. R. Cantor et al., 1970, *Biopolymers* 9:1059, 1079.] [From V. A. Bloomfield, D. M. Crothers, and I. Tinoco, Jr., 1974, *Physical Chemistry of Nucleic Acids* (New York: Harper & Row), 134.]

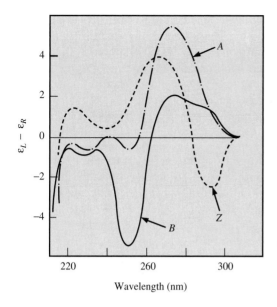

◀ FIGURE 10.25

Circular dichroism of the synthetic polynucleotide poly(dG-dC)·poly(dG-dC) in different conformations. The polynucleotide is a double-stranded helix; each strand has a sequence of alternating deoxyguanylic acid (dG) and deoxycytidilic acid (dC). Different conformations are obtained by changing the solvent. The B form is obtained in 0 to 40% ethanol or 10^{-3} M to 2 M NaCl; it is a right-handed helix with about 10 base pairs per turn of the double helix. The Z form is obtained in 56% ethanol or 3.9 M NaCl; it is a left-handed helix. The A form is obtained in 80% ethanol; it is a right-handed helix with about 11 base pairs per turn. (From F. M. Pohl, 1976, *Nature* 260:365. Copyright © 1976 Macmillan Journals Limited.)

each strand. At high salt concentration, its unusual CD (curve Z in figure 10.25) suggested that the helix could be left-handed. Because this was the first experimental indication of a left-handed DNA helix, there was much skepticism. However, the crystal structure of a double helix of a six-base-pair fragment, $(dC\text{-}dG)_3 \cdot (dC\text{-}dG)_3$, confirmed the presence of a left-handed helix and provided a detailed structure (Wang et al. 1979). CD is a powerful probe of nucleic acid secondary structure.

Several forms of secondary structure are present in native proteins. They include α-helix, parallel and antiparallel β-pleated sheets, and random coil. By measuring homopolypeptides under conditions in which the conformation is uniform throughout the polymer, Gratzer and Cowburn (1969) prepared plots of ORD and CD spectra in the peptide region that show the range of values encountered for polypeptides of different amino acids (figure 10.26). The ranges are relatively small compared with the differences among the CD

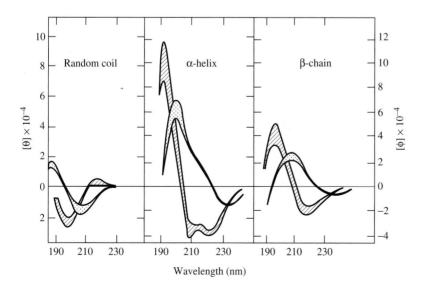

◀ FIGURE 10.26

CD (crosshatched) and ORD (dotted) of homopolypeptides in the random coil, α-helical, and β conformations. The shaded areas indicate the range of values. (From W. B. Gratzer and D. A. Cowburn, 1969, *Nature* 222:426. Copyright © Macmillan Journals Limited.)

spectra of the different conformations. This finding implies that the measured CD of a protein can be used to obtain the amount of each type of secondary structure. Hennessey and Johnson (1981) showed that, after calibration with proteins of known structure, the measured CD from 178 nm to 250 nm yielded correct secondary structures.

Vibrational Spectra, Infrared Absorption, and Raman Scattering

Vibrational spectra arise from transitions between vibrational energy levels. These energy levels correspond to bond stretching, bond bending, and other internal motions of the molecules. Some vibrational frequencies identify particular groups in the molecule, such as a $C-H$ bond, a $C=C$ double bond, or a phenyl group. Other frequencies characterize conformations, such as amide frequencies in proteins or phosphate ester frequencies in nucleic acids. Vibrational spectra occur in the infrared (IR) region, which may be considered to include the region from approximately 10^3 nm to 10^5 nm. The traditional units used by IR spectroscopists are either microns (μm) or wavenumbers (cm^{-1}). In these units, the IR range is 1 to 100 μm, or 10,000 to 100 cm^{-1}.

We learned in chapter 9 that a harmonic oscillator has a vibration frequency equal to

$$\nu_0 = \frac{1}{2\pi}\left(\frac{k}{\mu}\right)^{1/2}$$

where k is the force constant for the oscillator and μ is its reduced mass. This equation applies approximately to vibrations in molecules. It states that stronger bonds with larger force constants have higher vibrational frequencies and that, if the vibration involves a proton attached to a heavy atom, the frequency will decrease by a factor of approximately $\sqrt{2}$ on substitution of hydrogen (mass m_H) by deuterium (m_D):

$$\frac{\nu_0(H)}{\nu_0(D)} = \sqrt{\frac{m_D}{m_H}} = \sqrt{2}$$

Smaller isotope shifts for heavier atoms involved in vibrations are also easily detectable. This is useful in assigning spectra to vibrations in a molecule.

There are two principal ways to measure vibrational spectra: infrared absorption and Raman scattering.

Infrared Absorption

Figure 10.27 illustrates at the left the transitions directly measured in IR absorption. There are selection rules that describe the allowed transitions. A vibration must cause a change of electric–dipole moment of a molecule for there to be absorption of light. Thus, N_2 and O_2 do not absorb in the IR range, but H_2O and CO_2 do. N_2 and O_2 have no electric–dipole moment and a vibration leaves the dipole moment zero. H_2O does have an electric–dipole moment, and stretching or bending vibrations change its magnitude. CO_2 has no electric dipole moment, but an asymmetric stretch of its $C=O$ bonds or an $O=C=O$ bend will break its symmetry and lead to IR absorption. The IR spectrum is measured in a spectrophotometer similar to those used in the visible and UV region. The Beer–Lambert law is generally valid. Infrared is used routinely, as is NMR, in the identification and characterization of or-

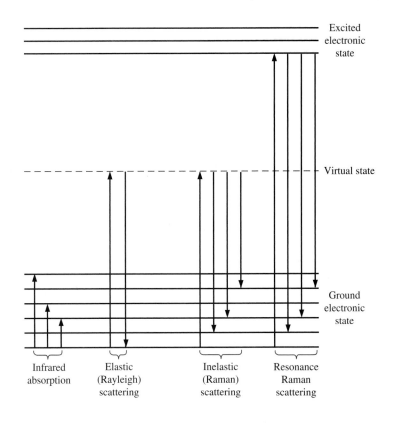

Energy-level diagram for IR absorption, elastic (Rayleigh) scattering, and inelastic (Raman) scattering. Transitions between vibrational energy levels of the ground electronic state absorb IR radiation; not all transitions are allowed. For elastic scattering, the frequency of the incident light is equal to the frequency of the scattered light. For inelastic scattering, the frequency of the scattered light is different from that of the incident light. The virtual state shown can have any energy; the scattering of light occurs in any region of the spectrum.

ganic molecules. One disadvantage for studying biological macromolecules is the strong and broad absorption of liquid H_2O in the IR region. Some very useful data on proteins and nucleic acids have been obtained in H_2O or D_2O solution, but only in a limited wavelength region. Not all the IR region is accessible in aqueous solution.

Raman Scattering

All molecules scatter light. Elastic scattering occurs when a photon interacts with a molecule, but no absorption occurs. The electric field of the light perturbs the electron distribution of the molecule momentarily, but no transition occurs. Because the molecule does not go from one stationary state (ground-state energy level) to another (excited-state energy level), there are no selection rules. Figure 10.27 indicates scattering occurring by a transition to a virtual state (nonstationary state); the molecule immediately returns to the ground electronic state, and the photon is scattered. If there is no change in energy of the scattered photon, the scattering is elastic; it is called *Rayleigh scattering*. The intensity of the scattered light is proportional to the square of the polarizability of the molecule, and if the wavelength is not too close to an absorption band, the scattering intensity is inversely proportional to the fourth power of the wavelength. This wavelength dependence explains blue skies and red sunsets, although for the best red sunsets the scattering particles are dust, not molecules, and the wavelength dependence is more complicated.

The polarizability α is a measure of how easy it is to induce a dipole in a molecule by applying an electric field. The polarizability tells us how easy it is to distort the distribution of electrons of the molecule. The quantitative definition is

$$\mu_{ind} = \alpha E$$

where μ_{ind} is the dipole induced and E is the electric field. The usual units for α are cm^3. The electric field is then in esu-volt cm^{-1}, and μ_{ind} is in esu cm. In SI units, α is in m^3, and

$$\mu_{ind} = 4\pi\varepsilon_0\alpha E$$

with μ_{ind} in coulomb m and E in volt m^{-1}.

In *Raman scattering*, the molecule returns to a different energy level after interaction with the light. In figure 10.27, the final state is an excited vibrational level of the ground electronic state. This is the most common occurrence in Raman scattering. The scattered photon has a longer wavelength (lower energy) than does the incident photon. The scattered photon can have a shorter wavelength if the molecule was originally in an excited vibrational state and returned to the ground vibrational state. In either case, the scattering is inelastic; there has been energy transfer between the molecule and the light. The selection rule for Raman scattering is that the *polarizability* of the molecule must change with the vibration in order to have a transition to a different energy level. Thus, N_2 and O_2 will show vibrational Raman spectra because their vibrations cause a change in polarizability. For CO_2, its asymmetric stretch will cause no change in polarizability, but its symmetric stretch and its bend will. We see that some vibrations can be measured using Raman, but not infrared; some can be measured using infrared, but not Raman; and some can be measured both ways.

In addition to the fundamental difference between Raman and infrared based on selection rules, there is a practical difference. Raman scattering is detected by shining light of one wavelength on the sample and measuring the wavelengths of the scattered light. Most of the light scattered will have the wavelength of the incident light; this is the Rayleigh scattering. There will also be light scattered at slightly different wavelengths; this is the Raman scattering. The difference in frequencies between the incident and scattered light is the frequency of the vibrational transition being measured. Note that using Raman scattering we learn about vibrational transitions (and IR frequencies) using visible or UV light. It does not matter where the solvent absorbs; we can use a wavelength of light away from that absorption band for the Raman studies. This allows great flexibility in the Raman studies, which is particularly useful for biological samples.

Figure 10.28 compares the IR (left) and Raman (right) spectrum of all-*trans*-retinal. Note the large peak at 1008 cm^{-1} in the Raman spectrum that does not occur in the infrared. The strong absorption at 966 cm^{-1} in the infrared has a corresponding weak peak in the Raman; it is assigned to out-of-plane vibrations of *trans*-ethylenic protons at $HC_7 = C_8H$ and $HC_{11} = C_{12}H$ (see figure 9.2). Deuterium substitution and synthesis of demethyl analogs have led to a complete assignment of the vibrational spectra of retinal. Because the Raman spectra of visual receptors and bacteria that use retinal as a key chromophore are being studied in single cells, the changes of conformation of the retinal can be deduced during its biological function (Barry and Mathies 1982).

Resonance Raman spectra occur when the exciting light is in an electronic absorption band of the molecule. There can be gains in sensitivity of a factor

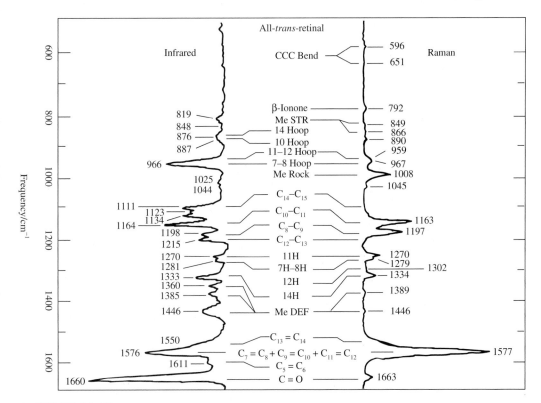

▲ FIGURE 10.28

Raman and IR spectra of all-*trans*-retinal in the region from 600 cm^{-1} to 1600 cm^{-1}. The IR absorbance is versed to facilitate comparison with the Raman. The Raman spectrum was measured in CCl$_4$, using 676.4-nm excitation from a krypton-ion laser. Solvent peaks were subtracted. The IR spectrum was measured with a thin film deposited by evaporation from a pentane solution on a KBr window. All peaks have been assigned to particular vibrations of the molecule. Me STR, Me Rock, and Me DEF refer to the stretching, rocking, and the deformation of methyl groups; Hoop means out-of-plane motion of H atoms. Note that double-bond vibrations occur at higher frequencies than single-bond vibrations. (From Curry, Palings, Broek, Pardoen, Mulder, Lugtenburg, and Mathies, 1984, *J. Phys. Chem* 88:688–702. The figure was kindly supplied by Richard Mathies, University of California, Berkeley.)

of 1000, or more, for vibrations within the absorbing chromophore. This allows one type of molecule to be studied preferentially in a very complicated mixture. Resonance Raman has been applied to heme proteins and to rhodopsin. (For a review, see Spiro 1974.)

Nuclear Magnetic Resonance

Nuclear magnetic resonance (NMR) *spectroscopy* monitors changes in the nuclear spin state. We encountered the idea of spin in our discussion of quantum mechanics. The electron was considered to have a spin of $\frac{1}{2}$; two electrons of different spin can coexist in the same orbital. Spin is an intrinsic property of a particle, which derives from the theory of quantum mechanics that takes into account relativity. Nuclei also have spin, which may be described by quantum number $\frac{1}{2}$, 1, or greater. We will mostly concern ourselves with nuclei of spin $\frac{1}{2}$; these include ^{1}H, ^{13}C ^{15}N and ^{31}P, which are important components of biological molecules.

In the absence of an externally applied magnetic field, different nuclear spin states have the same energy; they are degenerate. In the presence of an

▶ FIGURE 10.29

Schematic of energy levels in NMR. In the absence of an external magnetic field, the nuclear spin states have the same energy (are degenerate). In the presence of an external magnetic field (B_0), the two spin states for a spin $= \frac{1}{2}$ nucleus have different energies.

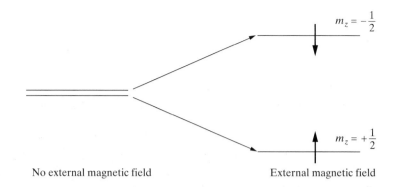

No external magnetic field   External magnetic field

externally applied magnetic field, which can be generated by large superconducting magnets, the spin states have different energy. For a spin $= \frac{1}{2}$ nucleus, there are two possible values for the spin quantum number, $m_z = \pm\frac{1}{2}$, which characterize two different energy levels. A simple classical analog for spin $\frac{1}{2}$ nuclei is the bar magnet, a magnetic dipole, with north and south poles. A dipole aligned with the magnetic field has a lower energy than a dipole aligned against the field (figure 10.29). The spin quantum numbers m_z are related to the magnetic moment along the z-axis, μ_z, by

$$\mu_z = \gamma \frac{h}{2\pi} m_z$$

where γ, the nuclear gyromagnetic ratio, is a constant that is a property of a given nucleus. The energy of a spin state in a magnetic field is given by

$$E = -\mu_z B_0 = -\gamma \frac{h}{2\pi} B_0 m_z$$

where B_0 is the strength of the magnetic field, in tesla (T), directed along the z-axis and μ_z is in J T^{-1}. The energy difference between these two states is

$$\Delta E = \frac{1}{2}\gamma \frac{h}{2\pi} B_0 - \left(-\frac{1}{2}\gamma \frac{h}{2\pi} B_0 \right) = \gamma \frac{h}{2\pi} B_0 \qquad \text{(10.36)}$$

As in any form of spectroscopy, the energy difference between the two states can be related to a frequency for the transition, called the Larmor frequency in NMR, as given by

$$\Delta E = h\nu_L \qquad \text{(10.37)}$$

For a proton in a field B_0 of 11.7 T, available in commercial superconducting magnets, this frequency is 500×10^6 Hz. This frequency corresponds to radio waves in the electromagnetic spectrum. The transitions between nuclear spin states involve very-low-energy transitions, compared with electronic or vibrational transitions. For a large number of nuclear spins, in a bulk sample, the two spin states will be nearly equally populated, since the energy splitting is much smaller than the thermal energy (kT) at room temperature. Because population differences are one determinant of the intensity of a spectral peak, NMR is a much less sensitive technique than, for example, UV spectroscopy. Biomolecular NMR studies are usually done with millimolar molecular concentrations.

The energy splitting between spin states is one determinant of spectral sensitivity in NMR, which is increased by increasing the magnetic field strength. Magnets as large as 18.1 T are available, and higher-field magnets are

TABLE 10.7 Gyromagnetic Ratios, NMR Frequencies (in an 11.7 T field), and Natural Abundances of Selected Nuclei

	$\gamma/10^7 T^{-1} s^{-1}$	ν/MHz	Natural abundance/%
^1H	26.75	500.0	99.985
^2H	4.11	76.8	0.015
^{13}C	6.73	125.8	1.108
^{14}N	1.93	32.4	99.63
^{15}N	−2.71	50.6	0.37
^{17}O	−3.63	67.9	0.037
^{19}F	25.18	470.6	100.0
^{29}Si	−5.32	99.5	4.70
^{31}P	10.84	202.6	100.0

being constructed. At 18.1 T, the *Larmor frequency* (the field strength of a magnet) for protons is 800×10^6 Hz; an NMR instrument with an 18.1 T magnet is often called an "800 MHz NMR." Spectral sensitivity increases with field strength as $B_0^{3/2}$.

The combination of isotope abundance and spectral sensitivity γ determines the spectral sensitivity of a nucleus. As can be seen in table 10.7, ^1H has one of the highest values of γ and a high natural abundance. Other biologically important nuclei that have spin $= \frac{1}{2}$, such as ^{13}C or ^{15}N, have low natural abundance. In modern NMR applications, biological molecules are usually enriched to >95% in these rare isotopes.

The Spectrum

Nuclei absorb energy when the frequency of the electromagnetic radiation matches the Larmor frequency. Spins are excited from the low-energy state to the higher-energy state. The resulting plot of intensity of the absorbed radiation versus frequency gives an NMR spectrum. Modern NMR uses a distinct approach to obtain the NMR spectrum, by first perturbing the system and monitoring the return of the system to equilibrium. Before addressing these methods, it is convenient to introduce a classical picture of NMR that can explain many experiments: the *vector model.* A nucleus with spin in a magnetic field along the z-axis can be considered a spinning top that is tilted away from the z-axis. To conserve angular momentum, spinning tops that are tipped from the z-axis precess around the central axis. In NMR, that precession frequency is the Larmor frequency. If we consider a large number of spins, an excess population of spins will be aligned with the magnetic field; the vector sum of their spin vectors will cancel in the xy-plane and add along the z-axis. The result is a bulk magnetization vector (**M**) along the z-axis that represents the total magnetic moment of the system (figure 10.30). Note that this vector would disappear if the B_0 field was turned off.

If a weak oscillating radiofrequency field B_1 is applied in the xy-plane, the result is a torque on the vector that will tilt the vector toward the xy-plane. This is a standard exercise in introductory physics and involves several mathematical tricks:

1. The frame of reference is said to rotate at the Larmor frequency (the rotating frame).

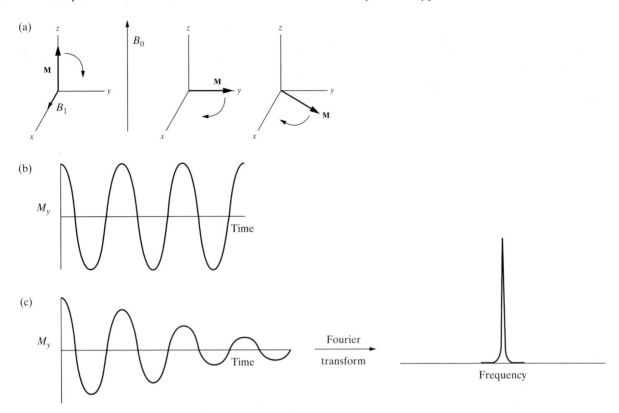

▲ FIGURE 10.30

Vector picture and schematic of a a simple, one-pulse NMR experiment. (a) In the presence of a strong magnetic field, there is a net magnetic vector (**M**) in the z-direction. The weak oscillating field, in this rotating reference frame, is a static vector that is perpendicular to the z-axis. Under the effect of the weak field, the bulk magnetization vector **M** is tilted toward the xy-plane of the rotating frame. If the perturbing field is left on long enough (usually 1–$20\mu s$), the **M** vector rotates into the xy-plane. The perturbing B_1 field is then turned off. (b) To an observer in the laboratory frame, the bulk magnetization vector precesses at the Larmor frequency, which can induce a signal in a wire coil. (c) This oscillating signal, called the free-induction decay, gradually decreases to zero, as the magnetization relaxes back to the $+z$-axis; this system thus returns to equilibrium. Fourier transformation of the free-induction decay yields the frequency spectrum.

2. The oscillating field is at the Larmor frequency; this is a static vector in the rotating frame.

3. The torque on the bulk magnetization vector is the product of **M** and B_1. It operates in a direction at right angles to both B_1 and **M**. This is called a *cross product* in the language of vector mathematics.

Under this torque, the magnetization vector will rotate toward the xy-plane. The electromagnetic field is then turned off when the bulk magnetization vector has been rotated into the xy-plane. This is called a 90° pulse. Since there is no more perturbation on the bulk system, it can return to equilibrium. The vector will precess at the Larmor frequency in the laboratory frame (and be stationary in the rotating frame) and will exponentially return to be aligned along the positive z-axis. The component of the bulk magnetization vector in the y-direction, $\mathbf{M}_y$, is given by

$$\mathbf{M}_y(t) = \mathbf{M} \cos(2\pi\nu t)e^{(-t/T_2)}$$

where T_2 is a time constant for loss of the magnetization signal in the xy-plane. Precession of the $\mathbf{M}$ in the xy-plane induces an oscillating current in the coil surrounding the sample. This oscillating current measures the free-induction decay of the magnetization (shown in figure 10.30). The frequency of oscillation is the Larmor frequency: thus, the spectral information is contained in the free-induction decay. The magnetization $\mathbf{M}(t)$ as a function of time (free-induction decay) and magnetization $\mathbf{M}(\omega)$ as a function of frequency (the spectrum) are related by a mathematical relationship called a *Fourier transformation*. Almost all modern NMR uses the so-called pulse methods just described. The system is perturbed for a short period of time, and then a time dependent signal is measured. Fourier transformation of the time-dependent signal provides the more-familiar frequency dependent spectrum.

$$M(\omega) = \int_{-\infty}^{\infty} M(t)e^{i\omega t}\,dt$$

i = the imaginary = $\sqrt{-1}$
ω = frequency = $2\pi\nu$

The vector model is just a simple way to express the quantum mechanical behavior of the system. The radio-frequency pulse, which rotated the bulk magnetization vector by 90°, drives spins from the ground state to excited states and leads to equal populations in the two states. The free-induction decay is the return of the system to equilibrium in which a large number of spins behave in a coherent fashion. Wavefunctions, like waves, have a phase associated with them, and the wavefunctions can add coherently if their phases are properly related. Expression of the behavior of large number of nuclear spins is well described quantum mechanically, but the vector picture provides a simpler representation. As with all classical views of quantum systems, the vector model has severe limitations.

Interactions in Nuclear Magnetic Resonance

The power of NMR spectroscopy derives from understanding the interactions between nuclei and between nuclei and electrons. These interactions affect either the energy-level splitting (chemical shift or scalar coupling) or the populations of the energy levels (nuclear Overhauser effects).

Chemical Shifts

The magnetic field experienced by a nucleus is slightly different from the applied external field and depends on local environment. The energy-level separation between the two spin states is thus slightly changed from that caused by the applied field, as expressed mathematically by

$$\Delta E = (1 - \sigma)\left(\frac{\gamma B_0}{2\pi}\right)h$$

where σ is the *shielding constant* and characterizes the local environmental effects experienced by a nucleus. The result of this equation is obvious: different protons will have different resonance frequencies, as given by

$$\nu_L = (1 - \sigma)\frac{\gamma B_0}{2\pi}$$

The frequency of a proton peak in NMR is normally expressed with respect to a reference frequency, using a compound with protons that resonate at a high frequency extreme of the spectrum: an example is 2,2-methyl-2-silapentane-5-sulfonate (DSS). The resonance position can be expressed in terms of a frequency difference between the reference peak and the observed peak in a manner that is independent of the magnitude of the applied field. This measure of resonance frequency, called *chemical shift*, is given by

$$\delta = \frac{\nu - \nu_0}{\nu_0} \times 10^6 \tag{10.38}$$

where δ is the chemical shift, ν_0 is the reference frequency, and ν is the observed resonance frequency. Although δ is a unitless number, it is normally expressed in terms of parts per million (ppm), which represents the effect of the local field compared with the overall effect of the static applied field. These small changes in chemical shift due to environment allow the use of NMR for many applications.

The local magnetic fields that give rise to chemical shift are caused by the electrons in the molecule. The applied magnetic field causes the electron distribution around a nucleus to circulate, giving rise to a small magnetic field that opposes the main field. The greater the electron density on the nucleus, the larger is the magnitude of the opposing field, and the greater is the *shielding* of the nucleus. The resonance will appear at relatively low frequency (upfield) and have a small value of chemical shift. Conversely, if there is a low electron density around a nucleus, the shielding will be low, and the resonance will have a relatively high value (downfield) for its chemical shift. The chemical shift of ^1H and other nuclei depend on many factors beyond electron density, including the nature of surrounding substituents. The trends for proton chemical shifts (shown in figure 10.31) roughly follow those expected from the above discussion. Protons attached to carbons with electron withdrawing groups resonate increasingly downfield.

Since electron currents depend on the orbital configuration of the molecule, more complex phenomena occur in molecules with π bonding. For example, the proton of acetylene resonates at lower chemical shift (is more shielded) than predicted by just simple analysis of electron density. An important chemical shift effect in nucleic acids, proteins, and porphyrins is the *ring current effect* (figure 10.32). The π-electron systems of aromatic rings are particularly amenable to charge circulation, which lead to relatively large local-field

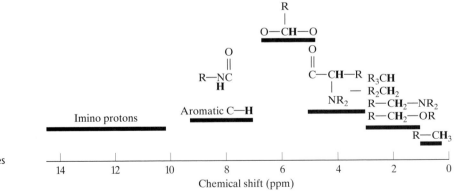

▶ FIGURE 10.31

Approximate chemical shift ranges in ppm relative to a reference for different types of protons

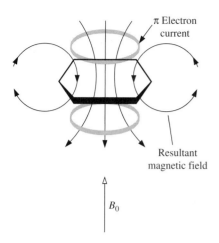

π Electron current

Resultant magnetic field

B_0

◄ FIGURE 10.32

The ring current effect in NMR. When a planar aromatic molecule is placed in a magnetic field B_0, the mobile π electrons can be considered to flow in a current around the periphery of the planar molecule. Classically, a current loop generates a magnetic field as shown in the figure. The magnitude of the field decreases approximately as the inverse cube from the center of the loop. This model is approximate, but it provides a good qualitative understanding of the effect of neighboring aromatic molecules on chemical shifts.

magnitudes. A proton that is in the plane of the aromatic ring experiences a local field aligned with the applied field and thus resonates at relatively high frequency; it has a large chemical shift. The conformation of a biomolecule can place a proton either above or below the plane of an aromatic ring. The local field from the ring current opposes the main field, and the proton resonates at a lower frequency, and experiences an upfield chemical shift in comparison to the proton in the plane of the ring. The sensitivity of chemical shift to local conformation leads to many applications in the study of molecular conformation.

Spin–Spin Coupling, Scalar Coupling, or J-Coupling

Nuclei can interact through electrons in intervening bonds. The spin state (either aligned with or against the magnetic field) of one nucleus affects the spin energies of the neighboring nucleus. This interaction is called *spin–spin coupling, scalar coupling,* or *J-coupling.* The two possible orientations of a spin $= \frac{1}{2}$ nucleus in a magnetic field can split the energy levels of neighboring nuclei. This means that the absorption line of a set of equivalent nuclei is split into a multiplet. The frequency separation between the lines of the multiplet is the spin–spin splitting, J, in hertz. If the spin–spin splitting is less than one-tenth the frequency difference due to the chemical shifts, simple first-order theory, as illustrated in figure 10.33, can be applied. This means that the effects of the chemical shifts and the spin–spin splittings are additive. If the spin–spin splitting between protons A and B (J_{AB}) is comparable to or larger than the difference in their chemical shifts ($\nu_A - \nu_B$), then the spectrum depends on the ratio of J_{AB} to ($\nu_A - \nu_B$), as shown in figure 10.34; the first-order theory used in figure 10.33 is no longer adequate. We can see in figure 10.34 that magnetically equivalent nuclei do not split each other; because their chemical shifts are identical, they have only one line as shown at the bottom of figure 10.34. This means, for example, that methane and ethane each show only one peak in their NMR spectrum. Similarly, the three protons of a methyl group do not split one another, just as the two protons of a methylene do not split one another (figure 10.33). However, chemically equivalent protons are not always magnetically equivalent. If there is not free rotation around C—C bonds, the two protons on a methylene can be in different environments and have different chemical shifts.

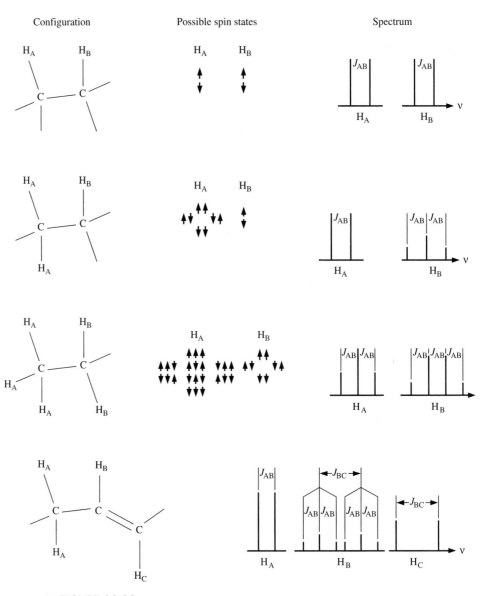

▲ **FIGURE 10.33**

Spin–spin splittings. A group of n equivalent protons will have $(n + 1)$ possible spin states in a magnetic field. This will give rise to a multiplet of $(n + 1)$ lines in adjacent protons. Note that the magnitude of the splitting of A by B equals that of B by A. The central position of the multiplet depends on the chemical shift. The total intensity of each multiplet is a measure of the number of equivalent protons that are split; only relative intensities within each multiplet are indicated in the figure. In the example at the bottom of the figure, proton B is split by two equivalent protons and by a different, single proton. This gives rise to a pair of triplets.

The spin–spin splitting in hertz (unlike the chemical shift in hertz) is independent of the applied magnetic field. The values of J for protons range from 0 to about 20 Hz. If we are measuring proton NMR in a hydrocarbon or carbohydrate, there are no effects from the carbon or oxygen nuclei because they have no magnetic moment (except for negligible amounts of ^{13}C and ^{17}O). Naturally occurring nitrogen (^{14}N) has a spin of 1 and tends to broaden neighboring proton lines rather than split them. Consequently, we often need

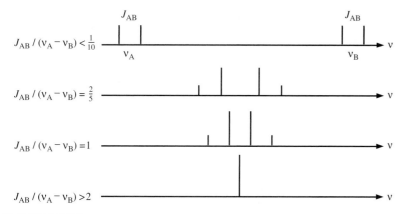

▲ FIGURE 10.34

NMR spectrum of two protons as a function of their ratio of spin–spin splitting (J_{AB}) in hertz to the difference in their chemical shifts ($\nu_A - \nu_B$) in hertz. When the splitting is small compared to the separation of the peaks, we have the simple first-order theory illustrated in figure 10.33. As ($\nu_A - \nu_B$) decreases, the inner lines increase in magnitude and the outer lines decrease. For magnetically equivalent protons ($\nu_A = \nu_B$), only one line (with twice the intensity) is seen. Note that ν_A and ν_B are directly proportional to the magnetic field, but J_{AB} is independent of field. Therefore, a 500-MHz spectrum may show the simple first-order pattern, whereas a 100-MHz spectrum will not. Remember that ($\nu_A - \nu_B$) is five times larger at 500 MHz than at 100 MHz.

to consider only proton–proton splittings. Figure 10.33 gives several examples of spin–spin splittings. Each set of n magnetically equivalent protons, such as the three on a methyl group, split neighboring protons into a multiplet of ($n + 1$) lines. The number of lines and their relative intensities can be easily derived by counting the number of ways of arranging the nuclear spins in a magnetic field. One proton creates a doublet of intensity 1:1. Two equivalent protons create a triplet of intensity 1:2:1. Three equivalent protons create a quadruplet of intensity 1:3:3:1. You should easily be able to obtain the result that n protons create ($n + 1$) lines with relative intensities corresponding to the coefficients of a binomial expansion $(1 + x)^n$.

The magnitude of spin–spin coupling depends on the nuclei involved and the number of bonds separating them. Nuclei with greater gyromagnetic ratios will interact more strongly. The coupling interaction occurs by interaction of the nuclear spins with electron spins in associated orbitals. The interaction is greater when transmitted through fewer bonds. In biomolecular NMR, samples are often prepared with uniform enrichment in the spin = $\frac{1}{2}$ isotopes ^{13}C and ^{15}N. The magnitudes of one-bond coupling constants between these nuclei and their attached protons vary from 20 to 200 Hz. For protons, coupling occurs across two, three, and four bonds; protons separated by four bonds ($H_A - C - C - C - H_B$) often produce negligible spin–spin splitting. In aromatic or conjugated systems, however, a four bond splitting of up to 1 or 2 Hz may be seen.

The values of spin–spin coupling constants contain rich information about molecular configuration and conformation. For protons on adjacent atoms, the spin–spin coupling constant J_{AB} is of order of magnitude 10 Hz, but its exact value is very characteristic of the configuration and conformation. For example, the protons on a carbon–carbon double bond have a $J(cis) \cong 12$ Hz and a $J(trans) \cong 19$ Hz. For protons on a carbon–carbon single

▶ FIGURE 10.35

Schematic of three-bond J-coupling and its dependence on the dihedral angle. (a) The ribose ring of nucleic acids, which can adopt one of two low-energy conformations (right and left). (b) Projections down the C2′–C1′ bond axis in the ribose, showing the change in H1′–H2′ dihedral angle as a function of conformation. When ϕ is about 90°, the scalar coupling is near 0; when ϕ is near 180°, the scalar coupling reaches a maximum of about 9 Hz.

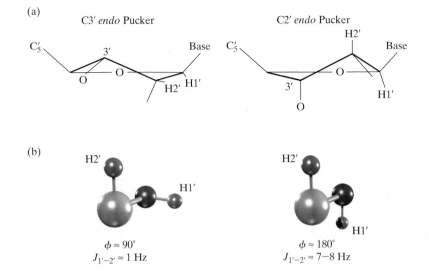

bond, the coupling constant is related to the torsion angle between the protons. The torsion angle is defined as follows:

When this angle is 90°, the spin–spin coupling constant is near zero; when the torsion angle is 0° or 180°, the coupling constant is near 10 Hz (figure 10.35). If there is free rotation about the bond, an average coupling constant is measured. The form of the dependence of coupling constant on conformation is given by the *Karplus equation:*

$$J = A + B \cos \phi + C \cos 2\phi \qquad (10.39)$$

where A, B, and C are parameters that can be calculated from quantum mechanics or can be established from measurements on molecules of known conformation. Their values depend on the other substituents on the carbon atoms.

Relaxation Mechanisms

An NMR experiment involves absorption of energy to change the populations of the two spin states for a spin $= \frac{1}{2}$ system. Once the perturbation is removed, the collection of NMR spins returns to its equilibrium population. The return of a spin population in a magnetic field to its equilibrium population follows first-order kinetics; its relaxation time is defined as the *spin–lattice relaxation time* T_1. One way to measure T_1 is to saturate a signal (NMR peak area is zero), then remove the saturating radiation and measure the peak area with time. The first-order kinetics is exponential:

$$M = M_{eq}(1 - e^{-t/T_1}) \qquad (10.40)$$

The magnetization (measured by peak area) at any time t is M, and M_{eq} is the magnetization at equilibrium.

What does the value of a spin–lattice relaxation time mean? As discussed earlier, only stimulated emission processes lead to relaxation in NMR. Oscillating magnetic fields at the transition frequency can lead to stimulated emission. The magnetic moment of a nucleus interacts with the nuclei around it. The nuclei can be on the same molecule or on different molecules, such as solvent molecules. The important quantity is the magnitude of the magnetic field (caused by surrounding nuclei) at the transition frequency of the nucleus of interest. For a nucleus to contribute effectively to spin–lattice relaxation, it should be nearby and in a molecule that rotates near the transition frequency—of order 500 MHz. The effect of the magnetic field created by a nucleus depends on the inverse sixth power of the distance from the nucleus. The frequencies of the magnetic field created by a nucleus depend on how fast the nucleus reorients—how fast the molecule or group containing the nucleus rotates. The values of T_1's for the nuclei in a molecule thus depend on the size of the molecule (how fast it rotates), its structure (how close neighboring nuclei are), and the solvent.

Measurement of relaxation rates for nuclear spins probes the local environment and dynamics of a molecule. Relaxation measurements can be used to determine whether a biological macromolecule tumbles in solution like a rigid sphere or whether regions of the molecule are moving independently with respect to each other. Measurement of T_1 can provide rotational diffusion coefficients for rigid molecules. Not only can tumbling of a molecule lead to oscillating fields, but rapid local motions of a piece of the molecule can cause relaxation as well. NMR is a powerful method to probe the *internal motions* of biological molecules.

There are other contributions to T_1. If protons whose spin population is saturated exchange with protons with an equilibrium population, this provides a chemical exchange contribution to spin–lattice relaxation T_1. For exchangeable protons, measurement of T_1 by NMR can provide the kinetics of the rate of chemical exchange. This has been used to measure the exchange lifetimes for imino protons on guanine and thymine in nucleic acids in H_2O. Because a G·C base pair must open for the guanine imino proton to exchange with H_2O and an A·T base pair must open for the thymine imino proton to exchange, the exchange rate is in turn a measure of the opening rate of the base pairs. Molecules that bind to DNA, such as ethidium or antibiotics, can stabilize or destabilize the base pairs (Pardi et al. 1983).

Another characteristic relaxation time that is measurable in NMR is the *spin–spin relaxation time* T_2. It characterizes interactions between spins on equivalent nuclei; it does not involve exchange of energy with the environment (the lattice). The spin–spin relaxation time can be measured from the width of the NMR peak

$$\frac{1}{T_2} = \pi \Delta\nu_{1/2} \qquad (10.41)$$

where $\Delta\nu_{1/2}$ is the peak width at half-height. The Heisenberg uncertainty principle [Eq. (10.9)] relates lifetime and line width: The greater the value of T_2, the narrower the line width. Thus, a direct link occurs between efficient relaxation processes and line width. For large molecules, slow tumbling leads to an efficient mode of spin–spin relaxation. Although the mechanisms of relaxation for T_1 and T_2 are similar, the different processes favor fluctuating fields at different frequencies; T_2 relaxation is more efficient at low frequency. Large molecules tumble more slowly and contribute low-frequency fluctuating fields. The result is that the line width of NMR peaks increases with increasing molecular weight. This relaxation problem creates an upper limit to the size of molecules that can be studied by NMR in solution of about 50–100 kDa (see Wider and Wüthrich 1999). Chemical exchange processes can also contribute to NMR line widths and therefore to T_2. In fact, all processes that affect T_1 also affect T_2; T_2 is always smaller than T_1. These two relaxation times can be used to study chemical kinetics and rotational and conformational motion of molecules.

Nuclear Overhauser Effect

We have discussed chemical shifts, spin–spin splittings, and T_1 and T_2; these are all NMR parameters that are used to study conformations and reactions in macromolecules. A very specific NMR method used to determine which nuclei are near each other is the nuclear Overhauser effect (NOE), which is the change of intensity of one NMR peak when another is irradiated. Intense radiation corresponding to the transition frequency of one type of proton is applied to the sample. This saturates the NMR peak; it changes the spin population of those protons so that there are equal numbers of nuclei in the upper and lower energy levels. These nuclei interact with neighboring nuclei and change the spin population of the neighboring nuclei from their equilibrium distribution. The intensity of the NMR peak of each nearby nucleus changes. Because the interaction between magnetic nuclei (the square of the magnetic dipole–dipole interaction) depends on r^{-6} (where r is the distance between the nuclei), the nuclear Overhauser effect is appreciable only for very close neighbors. Normally, the effect is seen only for nuclei closer than about 5 Å (0.5 nm). Although the short range limits its applications, the effect provides great specificity. NOE studies provide quantitative distance information, but they are also used simply to identify protons that are closer than 5 Å. For example, once an amide proton peak is identified in a protein, NOE measurements can be used to assign NMR peaks of protons that are nearby.

Multidimensional NMR Spectroscopy

The NMR spectrum of even a small biological molecule can be very complex. Each proton gives a peak in the spectrum. A protein has 4 to 5 protons per residue, so that a 100 amino acid protein would have 400 to 500 peaks in the spectrum. To use the spectral information, we must assign peaks to individual nuclei in the molecule; this is called *spectral assignment*. A trick of spectroscopy allows the spectral information to be spread from a simple one-dimensional spectrum (the plot of peaks versus frequency) to two-, three-, and four-dimensional NMR data.

In a standard pulse NMR experiment, the spins are perturbed, and the frequencies of the transitions are monitored by the precession of the magnetization vectors in the xy-plane. This leads to the free-induction decay, whose Fourier transform is the spectrum in terms of frequency. In *two-dimensional* (2D) NMR, time-dependent precession of the magnetization vectors is measured during two different time intervals, which gives two frequency axes to the spectrum upon Fourier transformation. During their precession period in the xy-plane, the magnetization vectors precess at their individual frequencies; the spins are said to be *frequency labeled.*

A simple 2D NMR experiment is called a COSY, or correlated spectroscopy. There are two short pulses, separated by a delay period t_1. After the second pulse the free-induction decay is detected during the period t_2. How-

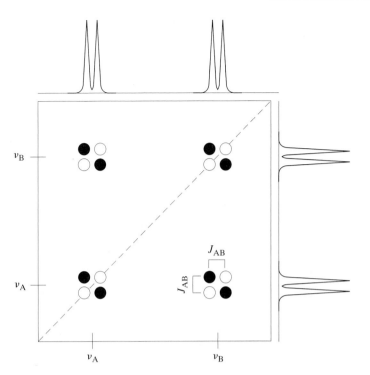

◀ FIGURE 10.36

Schematic of a two-dimensional COSY experiment. The diagonal represents the chemical shifts of the two protons (A and B) that are J-coupled; the one-dimensional spectrum is shown along each axis. Cross peaks occur at the chemical shifts (ν_A, ν_B) of the two protons, indicating that these two protons are J-coupled. The pattern of the cross peak reveals the magnitude of the J-coupling constant.

ever, the signal during t_2 will be a function of t_1. A series of one-dimensional (1D) experiments performed with increasing t_1 values gives a set of spectra that contains peaks whose amplitude oscillates as a function of t_1. Fourier transformation of the amplitudes of these peaks gives frequency information along the second dimension. A closer analysis shows the result of this simple 2D experiment. The spectrum is normally shown as two frequency axes with intensity of the peaks shown as contours. If we take two coupled protons, A and B, that resonate at distinct frequencies and perform the COSY experiment, the result is as shown in figure 10.36. The normal 1D experiment appears along the diagonal of the 2D plot, meaning that spin A precessed at its Larmor frequency during both t_1 and t_2. Because the two spins are coupled, information about spin A is also transferred to spin B. This is achieved by quantum mechanical effects caused by the pulses and precession of the spin system. The practical result is that *cross peaks* are observed in the 2D spectrum that link spin A and spin B. The presence of a cross peak indicates that two spins are coupled to each other.

Each cross-peak in a 2D spectrum corresponds to the interaction between two protons; the type of interaction detected depends on the pulses used. A 2D-COSY has cross peaks for nuclei that interact through spin–spin splitting or J-coupling. Measurable J-coupling occurs for nuclei separated by one bond, two bonds, three bonds, or even four bonds, with generally decreasing coupling constants. A COSY spectrum is the most rigorous method for assigning protons because they must be connected by bonds to produce a cross peak. The pattern of bond connectivities is characteristic of the molecule or group. Consider the COSY connectivity of the amino acids threonine and leucine. Idealized 2D-COSY spectra have the patterns shown in figure 10.37. Once one peak is assigned—such as the methyl resonance with a characteristic chemical shift near 1 ppm—the others follow from the COSY connectiv-

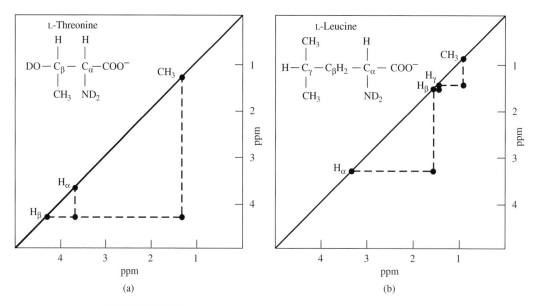

▲ FIGURE 10.37

(a) The proton connectivity seen in a COSY spectrum of L-threonine is illustrated. The spots along the diagonal represent the 1D spectrum. The off-diagonal cross peaks show which protons are correlated by three-bond J-coupling (α-β and β-methyl). (b) The proton COSY connectivity for L-leucine. The off-diagonal cross peaks show three-bond correlations between protons α-β, β-γ, and γ-methyl. The cross peaks on only one side of the diagonal are shown.

ity. In threonine [figure 10.37(a)], the methyl protons are connected only to H_β, and H_β is connected to both H_α and the methyl protons. In leucine [figure 10.37(b)], the COSY connectivity is methyl to H_γ to H_β to H_α. We have assumed in the diagram that the two methyl groups of leucine are magnetically equivalent and that the two methylene protons on the β carbon are equivalent.

The COSY spectrum provides assignments, but it also provides the coupling constants for the correlated protons. Each cross peak in the 2D spectrum (shown simply as a circle in figure 10.37) actually shows the characteristic splittings that provide the J-coupling constants. We remember that the coupling constants are related to torsion angles for rotations around bonds by the Karplus equation [Eq. (10.39)]. We can thus determine torsion angles, or at least ranges of torsion angles, from a COSY spectrum.

Two-dimensional NMR can be used to monitor through-space NOE interactions between protons; this experiment, called a NOESY, is the most powerful means of gaining structural information by solution NMR. In a NOESY experiment, the presence of cross peaks connecting two spins means that two protons have an NOE interaction between them; in other words, they are less than 5–6 Å apart. An example is shown in figure 10.38. The intensity of the cross peak is proportional to $1/r^6$, where r is the distance between the two protons. Since the efficiency of NOE transfer also depends on molecular motions (it is a relaxation process), the cross-peak intensity also depends on the overall tumbling rate and any local motions. The dependence of the NOE on both motion and distance makes precise distance measurments very difficult. Instead, NOE cross peaks are often interpreted as requiring that two protons are within a range ($\pm$1–2Å) of distances. Nonetheless, an NOE can place a powerful restraint on the conformation of a

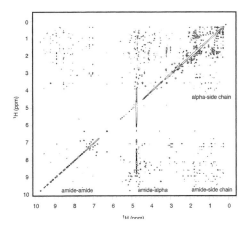

◀ FIGURE 10.38

NOESY spectrum of a protein: Translation initiation factor 1 from *E coli.* The spectrum was acquired at 11.7 T at 25°C; the NOESY mixing time was 150 ms. Various regions of the spectrum are highlighted. (Spectrum courtesy of Dr. R. Gonzalez.)

biological molecule. If an NOE is observed between protons on amino acids that are distant in the sequence, it means that the protein has folded to give a conformation in which these two amino acids are close in space.

NMR spectroscopy can be extended to *three dimensions* and *four dimensions.* All multidimensional NMR experiments use the principles we have described. Each frequency domain has a time delay associated with it. Cross peaks show that two spins interact with each other in a certain way, which is decided by the experiment. The cross peak could represent scalar coupling between spins, as in the COSY experiment, or NOE interactions. The principles hold as well if the two spins are unlike nuclei—for example a ^{13}C and ^{1}H or ^{15}N and ^{1}H. Two-dimensional experiments can readily correlate a ^{15}N with its attached ^{1}H, or ^{13}C with its attached ^{1}H, because these spin pairs are coupled. The chemical shift dispersion of ^{15}N or ^{13}C is much greater than that of ^{1}H, so these heteronuclei can be used to reduce overlap of peaks. An example is shown in figure 10.39, for an RNA molecule. Despite the extensive overlap of ribose protons, the chemical shifts of the different ribose carbons are distinct.

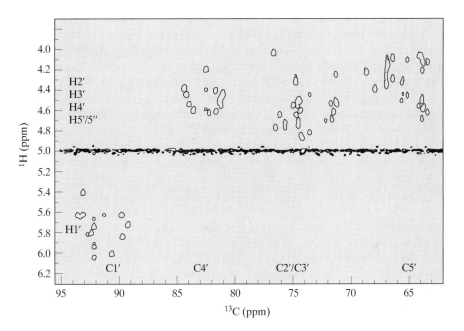

◀ FIGURE 10.39

Two-dimensional proton–carbon correlation experiment for a 29-nucleotide RNA, showing the power of heteronuclear NMR. Cross peaks represent correlations between a proton resonance and its bonded ^{13}C resonance; note that there is no diagonal in this type of experiment because two different sets of chemical shifts are correlated. The ribose proton chemical shifts show significant overlap, and the ribose carbons show significant chemical-shift dispersion: the less crowded spectrum facilitates analysis.

This principle can be used to give increased resolution in a NOESY spectrum; a three-dimensional (3D) NOESY experiment can be performed, where two of the dimensions give the normal proton–proton NOESY we have seen before. The third dimension spreads this information out by a ^{13}C chemical shift. The 3D experiment can be thought of as a book, where the ^{13}C chemical shift (planes of the 3D spectrum) would be an individual page number. NOE cross peaks are observed at a given plane only if the ^{13}C to which a proton is attached resonates at that ^{13}C chemical shift. A large number of so-called heteronuclear NMR experiments, which use combinations of spin $= \frac{1}{2}$ nuclei— ^{1}H, ^{13}C, ^{15}N, and ^{31}P—are used for studies of biomolecules (Pelton and Wemmer 1995).

Determination of Macromolecular Structure by Nuclear Magnetic Resonance

NMR can be used to obtain high-resolution structures of macromolecules in solution (Cavanagh et al. 1996). The accessible size of the molecules keeps increasing as the field strength of the instruments has increased for protons from 500 MHz to 600 MHz to 900 MHz. Isotope labeling plus multidimensional spectra allow assignment and correlation of an increasing number of nuclei. Proteins containing 100 to 200 amino acids, structural elements in nucleic acids, such as double strands, or loops of up to 75 nucleotides can be determined. Complexes of proteins and nucleic acids or drugs bound to proteins or nucleic acids can be studied by isotope labeling only one of the reactants. Special NMR experiments allow the NMR spectra of either the labeled or unlabeled component of a complex to be observed without interference from resonances of the other component. This is called *isotope filtering*. The use of NMR to determine structures in solution and X-ray diffraction to determine structures in crystals (chapter 12) is providing the knowledge necessary for molecular biologists to be able to understand biological and biochemical reactions.

The steps in determining a structure by NMR follow a straightforward logic. First, the NMR spectra must be assigned. This means that each resonance must be assigned to a particular nucleus in the molecule; the proton assignments are the most important and most useful in determination of structure. The assignments are best done through bond connectivities as determined by COSY-type experiments, but NOESY interactions are also very useful. Once the spectra are assigned, torsion angles are estimated from the J-coupling constants (spin–spin splittings), and interproton distances are estimated from the NOESY spectra. The distances and torsion angles characterize the structure. We assume that we know the primary structure—the sequence of the protein or nucleic acid; we want to determine the conformation—the secondary and tertiary structure. This is specified by the torsion angles, but the NMR spectra do not provide all the torsion angles. There is no information from J-coupling constants available for some torsion angles—such as the bonds on either side of the phosphorus atom in a nucleic acid.

The final step in the structure determination is to find the conformation(s) that are consistent with the NMR-derived distances and torsion angles. One way to do this is to add NMR-derived restraints to the potential-energy functions to calculate molecular dynamics and minimum-energy structures as described in chapter 9. For example, if a NOESY cross peak places two protons within 3 ± 1 Å, a potential-energy term is added to the usual bond-stretching terms in Eq. (9.42) that greatly increases the calculated energy of a conformation that has the protons outside this range. Similarly, if a COSY cross peak specifies a torsion angle to be $60° \pm 30°$, a potential-energy term is added that favors conformations with this angle. Molecular dynamics and energy minimization calculations are done that find structures consistent with the NMR data and, of course, with the usual bond lengths and bond angles. Structures with the fewest violations of the NMR restraints are found; these are the structures with the lowest calculated energies. With enough NMR distance and angle restraints, only one structure may be obtained; with fewer restraints, however, a number of structures can be found, all of which are consistent with the data. It may be that the molecule is flexible and that the NMR data indicates this flexibility. Or it may be that not enough NMR data were obtained and that more extensive assignments and correlations will lead to a unique structure.

EXAMPLE 10.2

The ribosome, which catalyzes protein synthesis, is the target of many antibiotics. These small (<1000 Dalton) molecules bind to different parts of the ribosome and inhibit various steps in the process of protein synthesis. Aminoglycoside antibiotics, which include neomycin and gentamicin, bind to ribosomal RNA and cause the ribosome to lose its high fidelity of translation. An RNA oligonucleotide that corresponds to the binding site for aminoglycosides in the ribosome was shown to bind specifically to the drugs. The structure of a complex between the drug gentamicin and the RNA was solved by NMR spectroscopy (figure 10.40). The RNA was labeled with ^{15}N and ^{13}C, and multidimensional NMR experiments were performed (Yoshizawa et al. 1998). A total of 154 dihedral angle restraints were obtained by measurement of spin–spin splittings and 379 distance restraints by NOE measurements. Of critical importance for defining the structure were 47 intermolecular NOE interactions between the drug and the RNA. A set of structures was calculated that agree with the NMR data. A representative structure from this set reveals that the drug binds within a groove of RNA, in a region where the helical

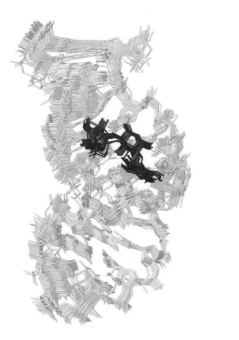
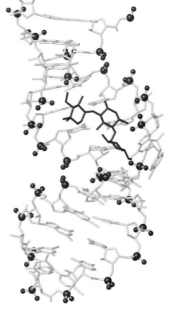

◀ FIGURE 10.40

Structure of a complex between the aminoglycoside antibiotic gentamicin and an RNA oligonucleotide solved using NMR spectroscopy. (a) Final ensemble of structures that agree with the NMR data; these structures are superimposed with respect to each other. (b) A single representative structure is shown. The antibiotic binds in the major groove of the RNA helix.

structure of the RNA is distorted. The drug sits within a specific binding pocket and makes a number of hydrogen bonds and electrostatic contacts with the RNA. The structure explained why resistance to the drug is achieved in bacteria by enzymes that modify the drug; all points of modification are in close contact with the RNA, and their modification would disrupt favorable interactions between the drug and RNA.

Electron Paramagnetic Resonance

Electron paramagnetic resonance (EPR) monitors the change in spin state of electron spin; it can be applied to free radicals or other molecular species that possess unpaired electrons (transition metal complexes, triplet states). The energy of an electron spin in a magnetic field is given by an equation related to Eq. (10.36):

$$E = g_e \mu_B B_0 m_s$$

where $g_e = 2.0023$ for a free electron, μ_B is the Bohr magneton, and m_s is the spin of an electron ($\pm\frac{1}{2}$). The energy difference between the two states is

$$\Delta E = E_{+1/2} - E_{-1/2} = g_e \mu_B B_0$$

When the sample absorbs radiation, the frequency absorbed is given by

$$h\nu = g_e \mu_B B_0$$

Note that the because the spin magnetic moment is much greater for an electron than for a nucleus, the energy splitting between the two states for a given magnetic field is greater for electron than for nuclear spins. Thus, *electron paramagnetic resonance* (EPR) occurs at higher frequencies than nuclear magnetic resonance; most EPR experiments are performed at 0.3–1 T, leading to transition frequences in the microwave region of the electromagnetic spectrum. At 0.3 T, an electron spin transition occurs at about 10 GHz, with a wavelength of 3 cm. Since EPR involves higher-energy transitions, it is a more sensitive spectroscopic technique than NMR. The transition frequency of a given electron is dependent on local covalent geometry (the g_e value can differ), but these variations are much smaller in NMR because of the weak magnetic dipole of the nucleus. Most important in EPR are the splittings that occur due to interactions with nuclei or other unpaired electrons; these are called *hyperfine or fine splittings* and are analogous to the spin–spin coupling in NMR. These interspin interactions provide molecular information regarding local chemical environment, molecular structure, and motional dynamics.

Although its range of applicability is necessarily narrower than that of NMR, EPR can exhibit great sensitivity and is particularly valuable in detecting intermediate or transitory species where electron rearrangements are occurring. It has been applied to heme proteins, ferredoxins, copper–protein complexes, and so on (Swartz et al. 1972). Furthermore, it has been developed as a probe technique. Paramagnetic ions, such as Mn^{2+} or Eu^{3+}, can be used to characterize the nature of ion binding and its role in the catalytic sites of many enzymes. McConnell and co-workers introduced the idea of *spin labeling*, which is the incorporation of a stable free radical spin label (often a nitroxide derivative). Spin labels have been incorporated into the fatty acid chains of lipids (Hubbell and McConnell 1969). When these are introduced into artificial or natural membranes, their EPR spectra give detailed information about the flexibility of the lipid chains and the rotational and translational mobility within the lipid phase. Molecular biology has greatly enhanced the ability to use spin labeling to study biological molecules. Specific amino acids can be mutated to cysteines, which can subsequently react with a spin label; using this strategy, spin labels can be introduced almost anywhere in a protein or nucleic acid. The interaction between two spin labels provides distance relationships between the two unpaired spins. Because of the greater sensitivity of EPR, interactions between unpaired electron spins can give distance information out to 20 Å. Although a single distance measurement does not define a macromolecular structure, changes in a distance upon, for example, ligand binding indicate conformational changes involved in function (see Hubbell et al. 1996 and Ottemann et al. 1999). The coupling between unpaired electrons and magnetic nuclei can be measured by electron-nuclear double resonance (ENDOR) and by pulsed experiments that resolve electron spin echo envelope modulation (ESEEM).

Magnetic Resonance Imaging

NMR is being used to obtain images of the soft tissues of the human body (see Brown and Semelka 1999, Moonenen et al. 1990, Radda 1986). X rays provide good images of bones but are not so good for imaging slight differences in water content of different tissues. Magnetic resonance imaging (MRI) is used most often to image protons (mainly water), but it can also be used to image ^{13}C or phosphorus.

The key to MRI is the use of a magnetic field gradient. The resonance frequency of a nucleus is directly proportional to the external magnetic field [Eq. (10.36)]. Therefore, if an external magnetic field is not uniform in space, the resonance frequency of a nucleus will depend on its position in the field. This is illustrated in figure 10.41. The NMR spectrum of a flask of water placed in a uniform field is compared with the spectrum of a flask in a linear field gradient. In a uniform field, one peak is seen; all protons have the same resonance frequency. In the field gradient, an image of the water in the flask is produced by the spectrum of intensity versus frequency. The protons on the left side of the flask are in a smaller magnetic field and have a lower resonance frequency than

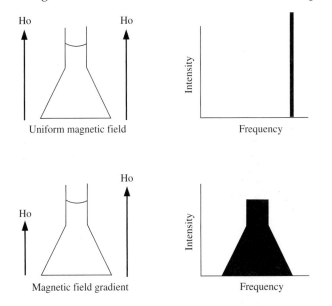

◀ FIGURE 10.41

Magnetic resonance imaging of a flask of water placed in a magnetic field that increases linearly from left to right. A magnetic field gradient means that the resonance frequency of a proton depends on its position in the field. In a magnetic field gradient, a spectrum of intensity of absorption versus frequency represents the number of protons versus position in the field. The spectrum is thus a projection of the image of the water in the flask.

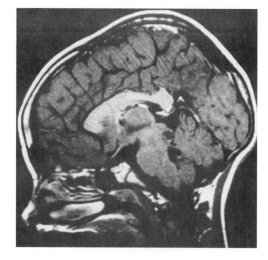

▶ FIGURE 10.42

Magnetic resonance image of a human head. The front of the head is facing left, and a cross section of the brain is obtained without the need to inject dyes or use any other invasive techniques.

those on the right side. The intensity of absorption at each frequency is proportional to the number of protons at each value of the magnetic field. There are more protons in the center of the flask because of the water in the neck, so the center frequencies of the spectrum have the highest intensity. The image obtained is a projection through the flask. To get a complete three-dimensional image, the flask or the field gradient must be rotated to produce several cross sections along different directions. In imaging, the spectrum is produced by the different magnitudes of the magnetic field; effects of chemical shifts of parts per million are completely negligible. An MRI image of a human head is shown in figure 10.42. MRI images can be obtained rapidly, within milliseconds to seconds, such that changes in the image as function of a stimulus can be monitored. This has lead to the exciting field of functional MRI (fMRI), through which regions of the brain that are involved in various functions have been identified (for an example, see Brewer et al. 1998).

Summary

Spectroscopy probes the details of molecular structure. Spectroscopic studies yield general conformational information, as in absorbance or fluorescence spectroscopy, or three-dimensional structure, as with NMR. Dynamic information is revealed by relaxation behavior in fluorescence and NMR spectroscopy. A survey of these and other spectroscopic techniques and their applications is provided in Sauer (1995). Spectroscopy is the most powerful use of quantum mechanics in studies of biological systems. The quantized nature of energy and the uncertainties in measurements of molecular systems are graphically illustrated by the basic ideas of spectroscopy.

Absorption and Emission

Oscillator strength:

$$f = \frac{2303 cm}{\pi N_0 e^2 n} \int \varepsilon(\nu)\, d\nu \qquad\qquad (10.5)$$

c = velocity of light
m = mass of electron
N_0 = Avogadro's number
e = electronic charge
n = refractive index
ε = molar absorptivity
ν = frequency, Hz

Transition dipole moment:

$$\boldsymbol{\mu}_{0A} = \int \psi_0 \boldsymbol{\mu} \psi_A \, dv \qquad (10.6)$$

where $\boldsymbol{\mu} = e \sum_i \mathbf{r}_i$

Dipole strength:

$$D_{0A} = |\boldsymbol{\mu}_{0A}|^2 \qquad (10.7)$$

Transformation relations:

See table 10.3.

Spontaneous and stimulated emission:

$$A = \frac{8\pi h\nu^3 B}{c^3} \qquad (10.8)$$

Heisenberg uncertainty principle for time and energy:

$$\Delta E \Delta t \geq \frac{1}{2}\left(\frac{h}{2\pi}\right) \qquad (10.9)$$

Beer–Lambert law:

$$A = \varepsilon c l \qquad (10.12)$$

$A = \log \dfrac{I_0}{I_t}$ = absorbance
ε = molar absorptivity or molar extinction coefficient, $M^{-1}\,cm^{-1}$
c = concentration, M
l = path length, cm

Experimentally measured fluorescence decay constant:

$$k_d = \frac{1}{\tau} \qquad (10.17)$$

k_d = first-order rate constant for decay of fluorescence; de-excitation can occur by fluorescence, thermal de-excitation, photochemistry, quenching, etc.
τ = experimentally measured lifetime of decay

Fluorescence rate constant:

$$k_f = \frac{1}{\tau_0} \tag{10.18}$$

k_f = first-order rate constant for fluorescence; de-excitation occurs only by fluorescence
τ_0 = natural lifetime of fluorescence

Fluorescence quantum yield:

$$\phi_f = \frac{\text{number of photons fluoresced}}{\text{number of photons absorbed}} = \frac{\tau}{\tau_0} \tag{10.20, 10.22}$$

Excitation Transfer
Energy-transfer efficiency:

$$Eff = 1 - \frac{\tau_{D+A}}{\tau_D} \tag{10.28}$$

τ_D = fluorescence lifetime of donor
τ_{D+A} = fluorescence lifetime of donor in the presence of acceptor

$$Eff = \frac{r_0^6}{r_0^6 + r^6} \tag{10.29}$$

Eff = fractional efficiency of transfer
r_0 = characteristic distance for the donor–acceptor pair
r = distance between donor and acceptor

Optical Rotatory Dispersion and Circular Dichroism

$$\text{Circular birefringence} = n_L - n_R$$

$$\text{Optical rotation} = \phi = \frac{\pi}{\lambda}(n_L - n_R) \text{ rad cm}^{-1} \tag{10.33}$$

n_L and n_R are refractive indices for left- and right-circularly polarized light, respectively, at wavelength λ.

$$\text{Circular dichroism} = \Delta A = A_L - A_R \tag{10.34}$$

$$\text{Ellipticity} = \theta = \frac{2.303(A_L - A_R)}{4l}, \text{ rad cm}^{-1} \tag{10.35}$$

A_L and A_R are absorbances for left- and right-circularly polarized light, respectively; l is the path length in cm.

Experimental parameters:

See table 10.6.

Nuclear Magnetic Resonance
Fourier transform:

$$F(\omega) = \int_{-\infty}^{+\infty} f(t)e^{i\omega t}\, dt$$

Chemical shift:

$$\delta\,(\text{ppm}) = \frac{\nu - \nu_0}{\nu_0} \times 10^6 \qquad\qquad \textbf{(10.38)}$$

Spin–lattice relaxation time T_1:

$$M = M_{\text{eq}}(1 - e^{-t/T_1}) \qquad\qquad \textbf{(10.40)}$$

M = magnetization at time t
M_{eq} = equilibrium magnetization

Spin–spin relaxation time T_2:

$$T_2 = \frac{1}{\pi \Delta \nu_{1/2}} \qquad\qquad \textbf{(10.41)}$$

$\Delta \nu_{1/2}$ = NMR peak width at half-height

References

Brown, M. A., and R. C. Semelka. 1999. *MRI Basic Principles and Applications.* New York: Wiley.

Campbell, I. D., and R. A. Dwek. 1984. *Biological Spectroscopy.* Menlo Park, CA: Benjamin/Cummings.

Cantor, C. R., and P. R. Schimmel. 1980. *Biophysical Chemistry, Part II. Techniques for the Study of Biological Structure and Function.* San Francisco: Freeman.

Carey, P. R. 1982. *Biochemical Applications of Raman and Resonance Raman Spectroscopies.* San Diego: Academic Press.

Cavanagh, J., W. J. Fairbrother, A. G. Palmer III, and N. P. R. Skelton. 1996. *Protein NMR Spectroscopy. Principles and Practice.* San Diego: Academic Press.

Derome, A. E. 1987. *Modern NMR Techniques for Chemistry Research.* Oxford: Pergamon Press.

Gadian, D. G. 1982. *Nuclear Magnetic Resonance and Its Applications to Living Systems.* Oxford: Clarendon Press.

Lakowicz, J. R. 1983. *Principles of Fluorescence Spectroscopy.* New York: Plenum Press.

Morris, P. G. 1986. *Nuclear Magnetic Resonance Imaging in Medicine and Biology.* Oxford: Oxford University Press.

Roberts, G. C. K., ed. 1993. *NMR of Macromolecules, A Practical Approach.* Oxford: IRL Press at Oxford University Press.

Sauer, K., ed. 1995. *Biochemical Spectroscopy.* Vol. 246, *Methods in Enzymology.* San Diego: Academic Press.

Sharma, A., and S. G. Schulman. 1999. *Introduction to Fluorescence Spectroscopy.* New York: Wiley.

Van Holde, K. E., W. C. Johnson, and P. S. Ho. 1998. *Physical Biochemistry.* Upper Saddle River, NJ: Prentice Hall.

Wüthrich, K. 1986. *NMR of Proteins and Nucleic Acids.* New York: Wiley.

Suggested Reading

Absorption

Honig, B., R. Daniel, K. Nakanishi, V. Balogh-Nair, M. A. Gawinowicz, M. Arnaboldi, and M. G. Motto. 1979. An External Point-Charge Model for Wavelength Regulation in Visual Pigments. *J. Am. Chem. Soc.* 101:7084–7086.

Levine, J. S., and E. F. MacNichol, Jr. 1982. Color Vision in Fishes. *Sci. Am.* (February): 140–149.

Nakanishi, K., V. Balogh-Nair, M. Arnaboldi, K. Tsujimoto, and B. Honig. 1980. An External Point Charge Model for Bacteriorhodopsin to Account for Its Purple Color. *J. Am. Chem. Soc.* 102:7947–7949.

Richards-Kortum, R., and E. Sevick-Muraca. 1996. Quantitative Optical Spectroscopy for Tissue Diagnosis. *Ann. Rev. Phys. Chem.* 47:555–606.

Rosenheck, K., and P. Doty. 1961. The Far Ultraviolet Absorption Spectra of Polypeptide and Protein Solutions and Their Dependence on Conformation. *Proc. Natl. Acad. Sci. USA* 47:1775–1785.

Shichi, H. 1983. *Biochemistry of Vision.* New York: Academic Press.

Fluorescence

Guiliano, K. A., P. L. Post, K. M. Hahn, and D. L. Taylor. 1995. Fluorescent Protein Biosensors: Measurement of Molecular Dynamics in Living Cells. *Ann. Rev. Biophys. Biomol. Struct.* 24:405–434.

Ha, T., Zhuang, H. D. Kim, J. W. Orr, J. R. Williamson, and S. Chu. 1999. Ligand-Induced Conformational Changes Observed in Single RNA Molecules. *Proc. Natl Acad. Sci. USA* 96:9077–9082.

Haugland, R. P., J. Yguerabide, and L. Stryer. 1969. Dependence of the Kinetics of Singlet-Singlet Energy Transfer on Spectral Overlap. *Proc. Natl. Acad. Sci. USA* 63:23–30.

Huang, K., R. H. Fairclough, and C. R. Cantor. 1975. Singlet Energy Transfer Studies of the Arrangements of Proteins in the 30S *Escherichia coli* Ribosome. *J. Mol. Biol.* 97:443–470.

Latt, S. A., H. T. Cheung, and E. R. Blout. 1965. Energy Transfer: A System with Relatively Fixed Donor–Acceptor Separation. *J. Am. Chem. Soc.* 87:995–1003.

Mehta, A. D., M. Rief, J. A. Spudich, D. A. Smith, and R. M. Simmons. 1999. Single-Molecule Biomechanics with Optical Methods. *Science* 282:1689–1695.

Murchie, A. I. H., R. M. Clegg, E. von Kitzing, D. R. Duckett, S. Diekmann, and D. M. J. Lilley. 1989. Fluorescence Energy Transfer Shows That the Four-Way DNA Junction Is a Right-Handed Cross of Antiparallel Molecules. *Nature* 341:763–766.

Selvin, P. R. 2000. The Renaissance of Fluorescence Energy Transfer. *Nature Structural Biology* 7:730–734.

Stryer, L. 1978. Fluorescence Energy Transfer as a Spectroscopic Ruler. *Ann. Rev. Biochem.* 47:819–846.

Stryer, L., and R. P. Haugland. 1967. Energy Transfer: A Spectroscopic Ruler. *Proc. Natl. Acad. Sci. USA* 58:719–726.

Weiss, S. 1999. Fluorescence Spectroscopy of Single Biomolecules. *Science* 283:1676–1683.

Yang, M., and D. P. Millar. 1998. Fluorescence Resonance Energy Transfer as a Probe of DNA Structure and Function. *Met. Enzymology* 278:417–444.

Optical Rotatory Dispersion and Circular Dichroism

Gratzer, W. B., and D. A. Cowburn. 1969. Optical Activity of Biopolymers. *Nature* 222:426–431.

Hall, K., P. Cruz, I. Tinoco, Jr., T. M. Jovin, and J. H. van de Sande. 1984. Z-RNA: Evidence for a Left-Handed RNA Double Helix. *Nature* 311:584–586.

Hayward, L. D., and R. N. Totty. 1971. Optical Activity of Symmetric Compounds in Chiral Media: I. Induced Circular Dichroism of Unbound Substrates. *Can. J. Chem.* 49:624–631.

Hennessey, J. P., Jr., and W. C. Johnson, Jr. 1981. Information Content in the Circular Dichroism of Proteins. *Biochemistry* 20:1085–1094.

Johnson, W. C. 1999. Analyzing Protein Circular Dichroism Spectra for Accurate Secondary Structures. *Proteins* 35:307–312.

Wang, A. H.-J., G. J. Quigley, F. J. Kolpak, J. L. Crawford, J. H. van Boom, G. van der Marel, and A. Rich. 1979. Molecular Structure of a Left-Handed Double Helical DNA Fragment at Atomic Resolution. *Nature* 282:680–686.

NMR

Ames, J. B., R. Ishima, T. Tanaka, J. I. Gordon, L. Stryer, and M. Ikura. 1997. Molecular Mechanics of Calcium–Myristoyl Switches. *Nature* 389:198–202.

Bax, A. 1989. Two-Dimensional NMR and Protein Structure. *Annu. Rev. Biochem.* 58:223–256.

Bax, A., and L. Lerner. 1986. Two-Dimensional Nuclear Magnetic Resonance Spectroscopy. *Science* 232:960–967.

Brewer, J. B., Z. Zhao, J. E. Desmond, G. H. Glover, and J. D. Gabrieli. 1998. Making Memories: Brain Activity that Predicts How Well Experience Will Be Remembered. *Science* 281:1185–1187.

Moonenen, C. T. W., P. C. M. van Zijl, J. A. Frank, D. Le Bihan, and E. D. Becker. 1990. Functional Magnetic Resonance Imaging in Medicine and Physiology. *Science* 250:43–61.

Pardi, A., K. M. Morden, D. J. Patel, and I. Tinoco, Jr., 1983. The Kinetics for Exchange of the Imino Protons of the d(C-G-C-G-A-A-T-T-C-G-C-G) Double Helix in Complexes with the Antibiotics Netropsin and/or Actinomycin. *Biochemistry* 22:1107–1113.

Pelton, J. G., and D. E. Wemmer. 1995. Heteronuclear NMR Pulse-Sequences Applied to Biomolecules. *Ann. Rev. Phys. Chem.* 46:139–167.

Radda, G. K. 1986. The Use of NMR Spectroscopy for the Understanding of Disease. *Science* 233:640–645.

Ross, B. D., G. K. Radda, D. G. Gadian, G. Rocker, M. Esiri, and J. Falconer-Smith. 1981. Examination of a Case of Suspected McArdle's Syndrome by ^{31}P Nuclear Magnetic Resonance. *N. Engl. J. Med.* 304:1338–1342.

Schulman, R. G. 1983. NMR Spectroscopy of Living Cells. *Sci. Am.* 248:86–93.

Schulman, R. G., A. M. Blamire, D. L. Rothman, and G. McCarthy. 1993. Nuclear Magnetic Resonance Imaging and Spectroscopy of Human Brain Function. *Proc. Natl. Acad. Sci. USA* 90:3127–3133.

Varani, G., C. Cheong, and I. Tinoco, Jr. 1991. Structure of an Unusually Stable RNA Hairpin. *Biochemistry* 30:3280–3289.

Viani Puglisi, E., and J. D. Puglisi. 1998. Nuclear Magnetic Resonance Spectroscopy of RNA. In *RNA Structure and Function*, ed. R. W. Simons and M. Grunberg-Manago. Cold Spring Harbor, New York: Cold Spring Harbor Laboratory Press.

Wider, G., and K. Wüthrich. 1999. NMR Spectroscopy of Large Molecules and Multimolecular Assemblies in Solution. *Curr. Opin. Struct. Biol.* 9:594–601.

Wüthrich, K. 1989. The Development of Nuclear Magnetic Resonance Spectroscopy as a Technique for Protein Structure Determination. *Acc. Chem. Res.* 22:36–44.

Yoshizawa, S., D. Fourmy, and J. D. Puglisi. 1998. Structural Origins of Gentamicin Antibiotic Action. *EMBO J.* 17:6437–6448.

Raman and Infrared

Barry, B., and R. Mathies. 1982. Resonance Raman Microscopy of Rod and Cone Photoreceptors. *J. Cell Biol.* 94:479–482.

Curry, B., A. Brock, J. Lugtenburg, and R. Mathies. 1982. Vibrational Analysis of All-*Trans*-Retinal. *J. Am. Chem. Soc.* 104:5274–5286.

Mathies, R. A. 1995. Biomolecular Vibrational Spectroscopy. *Methods Enzymol.* 246:377–389.

Spiro, T. G. 1974. Raman Spectra of Biological Materials. In *Chemical and Biochemical Applications of Lasers*, Vol. 1, 29–70. Edited by C. B. Moore. New York: Academic Press.

Other Spectroscopic Methods

Hubbell, W. L., and H. M. McConnell. 1969. Orientation and Motion of Amphiphilic Spin Labels in Membranes. *Proc. Natl. Acad. Sci. USA* 64:20–27.

Hubbell, W. L., H. S. McHaourab, C. Altenbach, and M. A. Lietzow. 1996. Watching Proteins Move Using Site-Directed Spin Labeling. *Structure* 4:1179–1183.

Ottemann, K. M., W. Xiao, Y.-K. Shin, and D. E. Koshland. 1999. A Piston Model for Transmembrane Signalling of the Aspartate Receptor. *Science* 285:1751–1754.

Swartz, H. M., J. R. Bolton, and D. C. Borg, eds. 1972. *Biological Applications of Electron Spin Resonance*, 213–264. New York: Academic Press.

Problems

1. The longest-wavelength absorption band of chlorophyll *a* peaks in vivo at a wavelength of about 680 nm.

   (a) For photons with a wavelength of 680 nm, calculate the energy in J photon^{-1}, in electron volts, and in kJ einstein^{-1}.

   (b) CO_2 fixation in photosynthesis can be represented by the reaction

   $$CO_2 + H_2O \longrightarrow (CH_2O) + O_2$$

   where (CH_2O) represents one-sixth of a carbohydrate molecule such as glucose. The enthalpy change for this endothermic reaction is $\Delta H = +485$ kJ mol^{-1} of CO_2 fixed. What is the minimum number of einsteins of radiation that need to be absorbed to provide the energy needed to fix 1 mol CO_2 via photosynthesis?

   (c) Experimentally, many values have been determined for the number of photons required to fix one CO_2 molecule. The presently accepted "best values" lie between 8 and 9 photons per CO_2 molecule. What is the photochemical quantum yield, based on your answer to part (b), for this process?

2. Using the data given in table 10.2 for a 1-kJ ruby laser pulse of 1-ms duration, estimate the absorbance of safety goggles needed to protect the operator's eyes against damage from a potential accident. Base your estimate on the information in problem 42(b) of chapter 7.

3. The mechanism of Sun tanning is initiated by the absorption of the 0.2% of solar radiation energy lying in the range 2900–3132 Å. Earth's atmosphere cuts off radiation <2900 Å, and the burning efficiency drops from 100% at 2967 Å to 2% at 3132 Å. The *average* efficiency across the 2900- to 3132-Å band is 50%. The total incident radiation from sunlight at the surface of Earth corresponds to 2100 J cm^{-2} day^{-1}.

   (a) Calculate the energy equivalent (in J cm^{-2} s^{-1}) of the radiation involved in sunburning. (Assume that a typical day length is 12 h.)

   (b) Many commercial ointments used to protect against sunburn contain *o*- or *p*-amino benzoates:

   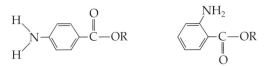

   or titanium dioxide, TiO_2. The latter is a solid powder often used as a "superwhite" pigment for paints. Suggest mechanisms by which these substances might work.

4. The molar absorptivity (molar extinction coefficient) of benzene equals 100 at 260 nm. Assume that this number is independent of solvent.

   (a) What concentration would give an absorbance of 1.0 in a 1-cm cell at 260 nm?

   (b) What concentration would allow only 1% of 260-nm light to be transmitted through a 1-cm cell?

   (c) If the density of liquid benzene is 0.8 g cm^{-3}, what thickness of benzene would give an absorbance of 1.0 at 260 nm?

5. Indicator dyes are often used to measure the hydrogen-ion concentration. From the measured spectra of an indicator in solution, the ratio of acid to base species and the pH can be calculated:

   $$HIn \rightleftharpoons In^- + H^+$$

The molar extinction coefficients are the following:

	ε (300 nm)	ε (400 nm)
HIn	10,000	2000
In$^-$	500	4000

(a) Calculate the ratio of concentrations [In$^-$]/[HIn] for a solution of the indicator that has A (300 nm) = 1.2 in a 1-cm cell and A (400 nm) = 0.7 in a 0.1-cm cell.

(b) What is the pH if the pK of the indicator is 5.0?

6. The content of chlorophylls a and b in leaves or green algae is determined by extracting the pigments into acetone:water (80:20 v/v) solution. The specific absorptivities of the pigments in this solvent system (reported by Mackinney, 1941, *J. Biol. Chem.* 140:315) at 645 and 663 nm are as follows:

Wavelength (nm)	Specific absorptivity (L g^{-1} cm^{-1})	
	Chl a	Chl b
645	16.75	45.60
663	82.04	9.27

Derive expressions, with numerical coefficients evaluated, that give the concentrations in μg mL^{-1} of chlorophyll a, chlorophyll b, and total chlorophylls ($a + b$) for specified values of A_{645} and A_{663} measured using a 1-cm path length.

7. The enzyme chymotrypsin is a protease that is competitively inhibited by a red dye. A solution containing 1.00×10^{-4} M total protein and 2.00×10^{-3} M total dye has an absorbance in a 1-cm cell at 280 nm of $A = 1.4$:

$$\text{protein} + \text{dye} \rightleftharpoons \text{complex}$$

The molar absorptivities at 280 nm are (units of M^{-1} cm^{-1}): ε (protein) = 12,000; ε (dye) = 0; ε (complex) = 15,000.

(a) Calculate the concentration of the complex in solution.

(b) Calculate the equilibrium constant (K with units of M^{-1}).

(c) When the dye binds to the enzyme, the α-helix content of the protein changes. Describe how you would use measurements of circular dichroism to determine the change in α-helix content. What would you measure? How would you relate it to α-helix content?

8. An indicator is a dye whose spectrum changes with pH. Consider the following data for the absorption spectrum of an indicator (pK = 4) in its ionized and nonionized form.

$$\text{InH}^+ \rightleftharpoons \text{In} + \text{H}^+$$

λ (nm)	Molar absorptivity ε (M^{-1} cm^{-1})	
	InH$^+$	In
400	10,000	0
420	15,000	2,000
440	8,000	8,000
460	0	12,000
480	0	3,000

The absorbance of the indicator solution is measured in a 1-cm cell and found to be

λ	400	420	440	460	480
A	0.250	0.425	0.400	0.300	0.075

(a) Calculate the pH of the solution.

(b) Calculate the absorbance of the solution at 440 nm and pH 6.37 for the same total indicator concentration.

(c) If you want to measure the quantum yield of fluorescence of In and InH$^+$ independently, what wavelength would you choose for exciting InH$^+$? For exciting In? Why?

9. Radioactivity is a very sensitive method of detection; let's compare it with fluorescence detection.

(a) ^{32}P has a half-life of 14.3 days. How many disintegrations per second will you obtain from 1000 atoms of ^{32}P? What is the maximum number of counts you can get?

(b) Fluorescein has a molar absorption coefficient of 70,000 M^{-1} cm^{-1} at 485 nm and a quantum yield for fluorescence of 0.93. What is its molecular absorption coefficient in cm^2 per molecule?

(c) 1000 molecules of fluorescein are irradiated with an argon-ion laser that has an intensity of 2 mW cm^{-2} at 485 nm. How many fluorescent photons per second will be emitted?

(d) The quantum yield for photodegradation to nonfluorescent products for fluorescein is 10^{-6}. What is the maximum number of fluorescent photons expected from each fluorescein?

(e) Discuss the relative merits of detecting small numbers of molecules by radioactivity and by fluorescence. What other factors besides counting rate may be important?

10. In a fluorescence-lifetime experiment, a very short pulse of light is used to excite a sample. Then the intensity of the fluorescence is recorded as a function of time. Suppose we study a dansyl fluorophore. The intensity of the fluorescent light versus time is as follows:

Fluorescence intensity	Time (ns)
1.1×10^4	0
4.9×10^3	5
2.3×10^3	10
1.3×10^3	15
5.6×10^2	20

(a) What is the observed fluorescence lifetime?

(b) The fluorescence quantum yield under the same conditions was determined to be 0.7. What is the intrinsic rate constant for radiative fluorescence decay?

(c) It has been determined separately that when the dansyl chromophore is 20 Å away from the 11-*cis*-retinal chromophore in rhodopsin, the efficiency of fluorescence energy transfer is 50%. In a second experiment, rhodopsin is covalently labeled with a dansyl chromophore and the observed fluorescence lifetime for the dansyl–rhodopsin complex is 6 ns. What is the distance between the dansyl label and the retinal chromophore in rhodopsin?

11. Tyrosine has a pK_a of 10.0 for the ionization of the phenolic hydrogen. The approximate absorption spectra for the protonated and deprotonated forms of tyrosine follow:

Suppose we obtain an absorption spectrum of a peptide that contains three tyrosine residues and no tryptophan or phenylalanine. The absorption spectrum of this sample in a 1-cm cell is as follows:

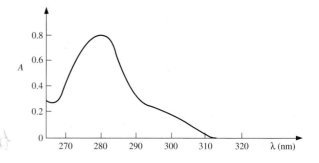

(a) Based on these data, what is the concentration of the peptide? Explain your method clearly.

(b) What is the pH of the solution? Explain your method clearly.

12. Quinine has a fluorescence quantum yield equal to approximately 1. Light at 10^{15} photons s^{-1} is incident on a solution of quinine in a 1-cm cell, and the concentration of quinine is 1.0×10^{-2} M. The wavelength is 300 nm, and the extinction coefficient is 1.0×10^4 at this wavelength.

(a) How many photons are fluoresced per second?

(b) If you place an efficient detector at right angles to the incident light to measure the fluorescence, only a very small fraction of these photons is detected. Why?

(c) If any phosphorescence occurs, is the wavelength of phosphorescence greater or less than that of fluorescence? Why?

(d) A search is being made for mutants of photosynthetic bacteria that lack reaction centers where the first steps in photosynthesis occur. In these mutants, the proteins that convert light energy into chemical energy are missing. The search involves looking for clones that have increased fluorescence. Can you explain why his method can be used?

◀ PROBLEM 11

Approximate absorption spectra

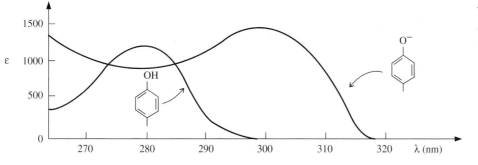

13. Consider a protein that dimerizes:

$$\alpha + \alpha \underset{k_{-1}}{\overset{k_1}{\rightleftharpoons}} \alpha_2$$

Describe how you could measure the equilibrium constant for the dimerization using a spectroscopic method. What would you measure? Write an equation relating the measured properties to the equilibrium constant.

14. The following equilibrium occurs in aqueous solution:

$$A + B \overset{K}{\rightleftharpoons} C$$

The molar extinction coefficients of A and B are ε_A and ε_B. Describe how you could use measurements of absorbance to obtain the equilibrium constant for the reaction. Tell what you would measure and how K would be calculated from the measurements.

15. A mixture of *M. lysodeikticus* DNA and rice dwarf virus RNA gave the following circular dichroism data in a 1-cm cell:

λ (nm)	$A_L - A_R$
300	-0.250×10^{-4}
280	1.16×10^{-4}
260	3.12×10^{-4}
240	-0.24×10^{-4}

The pure components gave the following results for solutions containing 1.00×10^{-4} M concentrations of nucleotides per liter:

λ (nm)	$A_L - A_R$ (DNA)	$A_L - A_R$ (RNA)
300	0.10×10^{-4}	-0.50×10^{-4}
280	3.00×10^{-4}	1.60×10^{-4}
260	0.00×10^{-4}	6.00×10^{-4}
240	-2.20×10^{-4}	0.00×10^{-4}

(a) What is $\varepsilon_L - \varepsilon_R$ for the RNA at 300 nm?

(b) What are the concentrations of DNA and RNA in the mixture?

(c) If you hydrolyze the mixture to its component nucleotides, will the circular dichroism at 260 nm increase or decrease?

(d) Will the absorbance at 260 nm increase or decrease?

(e) Will the optical rotatory dispersion increase or decrease?

16. For the following questions, a single sentence or phrase should be sufficient as an answer.

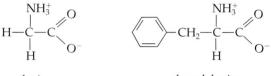

glycine   phenylalanine

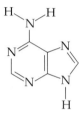

adenine

(a) Which, if any, of these molecules would you expect to rotate the plane of linearly polarized light? Why?

(b) Which, if any, of these molecules would you expect to be circularly dichroic? Why?

(c) Which, if any, of these molecules could in principle emit fluorescence at wavelengths greater than 250 nm when appropriately excited? Why?

(d) If a molecule emitted light on excitation, what experiment would you do to tell if the process was fluorescence or phosphorescence?

17. A fluorescent dansyl group has a lifetime of 21.0 ns when attached to a protein. When a naphthyl group is attached to the amino group of the terminal amino acid in the protein, the dansyl lifetime becomes 15.0 ns. Use the data in figure 10.18 to calculate the distance between the dansyl and the naphthyl group.

18. Cytidylic acid (molecular weight = 323.2) in water at pH 7 has the following optical properties at 280 nm: $\varepsilon = 8000$ M^{-1} cm^{-1}; $\varepsilon_L - \varepsilon_R = 3$ M^{-1} cm^{-1}; $[\phi] = 7500$ deg M^{-1} cm^{-1}.

(a) Calculate the specific rotation $[\alpha]$ and the molar ellipticity $[\theta]$.

(b) For 10^{-4} M cytidylic acid in a 1-cm cell, calculate the absorbance A, the circular dichroism $(A_L - A_R)$, and the rotation angle in degrees at 280 nm.

(c) Calculate the rotation in rad cm^{-1}, the circular birefringence $(n_L - n_R)$, and the ellipticity in rad cm^{-1} for the solution in part (b).

19. The fluorescence lifetime of benzene in cyclohexane is 29 ns when air is completely removed. In the presence of 0.0072 M dissolved O_2, the measured fluorescence lifetime is 5.7 ns because of quenching. Calculate the rate constant k_Q for the quenching reaction. If the relative fluorescence intensity is 100 for pure benzene, what is the relative fluorescence intensity of benzene containing 0.0072 M dissolved O_2?

20. When a double-stranded oligonucleotide (duplex D) in solution is heated, it "melts" to two single strands, S and R. The absorbance at 260 nm increases as the equilibrium between duplex and single strands is shifted toward the single strands.

$$D \rightleftharpoons S + R$$

$$K = \frac{[S][R]}{[D]}$$

A duplex whose concentration is 100 μM is melted in 1 M NaCl, pH 7. Below 10°C, the solution contains only duplex; above 60°C, the solution contains only single strands. Assume that the molar extinction coefficients of the duplex and its single strands are independent of temperature.

(a) Use the data below to calculate the equilibrium constant as a function of temperature for the duplex-to-single-strands equilibrium.

Temperature (°C)	Absorbance
10	0.790
15	0.800
20	0.813
25	0.822
30	0.852
35	0.890
40	0.940
45	0.971
50	0.990
55	1.003
60	1.114

(b) Calculate the melting temperature of the duplex. The melting temperature is defined as the temperature at which the duplex is half-melted.

(c) From the temperature dependence of the equilibrium constant, calculate ΔG^0, ΔH^0, and ΔS^0, all at 25°C. Assume that ΔH^0 and ΔS^0 are independent of temperature.

21. When DNA or RNA is treated with chloracetaldehyde, a fluorescent derivative of adenine (εA) is formed. The following data were measured for the dinucleotide phosphate (ApεA) shown below.

◄ PROBLEM 21

Data

	$T = 2°C$	$T = 80°C$
Molar extinction coefficients, $\varepsilon(260\ nm)$	$3.0 \times 10^4\ M^{-1}\ cm^{-1}$	$3.3 \times 10^4\ M^{-1}\ cm^{-1}$
Molar circular dichroism, $\varepsilon_L - \varepsilon_R(240\ nm)$	$32\ M^{-1}\ cm^{-1}$	$2.5\ M^{-1}\ cm^{-1}$
Fluorescence quantum yield	0.10	0.06
Fluorescence lifetime	8.0 ns	—

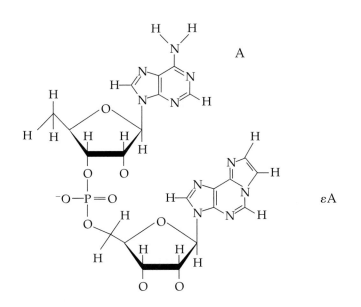

(a) Calculate the absorbance (*A*) of a 0.10 *M* solution of ApεA at 260 nm and 2°C in a 1-cm cell. Would it be experimentally possible to measure the temperature dependence of the absorbance of this concentration in a 1-cm cell? Explain.

(b) Calculate the circular dichroism ($A_L - A_R$) of a 0.10 *M* solution of ApεA at 240 nm and 80°C in a 1-cm cell. Would it be experimentally possible to measure the temperature dependence of the circular dichroism of this concentration in a 1-cm cell? Explain.

(c) Would it be experimentally possible to measure the temperature dependence of the fluorescence at 390 nm for this molecule in a 1-cm cell? Explain.

(d) The change in optical properties of this molecule with temperature is due to a stacking–unstacking equilibrium:

$$S \underset{}{\overset{K}{\rightleftharpoons}} U$$

Under appropriate conditions any optical property could be used to measure *K*. If you had instruments that could measure ε, $\varepsilon_L - \varepsilon_R$, or fluorescence with equal accuracy, that could you use to measure *K*? Why?

If P_s = property of stacked molecule, independent of *T*, if P_u = property of unstacked molecule, independent of *T*, and if P(*T*) = property of mixture of stacked and unstacked species as a function of *T*,

(e) Write an equation for *K*, the equilibrium constant, in terms of the fraction of molecules unstacked.

(f) Write an equation for *K*(*T*), the equilibrium constant at any *T*, in terms of P_s, P_u, and P(*T*).

22. The proton magnetic resonance of selectively deuterated methyl ethyl ether was measured in a 100-MHz spectrometer to give the following spectrum:

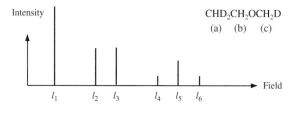

(a) Assign each of the six lines to protons (a), (b), or (c).

(b) If the separation between lines l_1 and l_5 is 200 Hz and is due to a chemical shift, what will be the separation in a 220-MHz spectrometer?

(c) If the spin–spin splitting of proton (a) is 5 Hz, what is the spin–spin splitting of proton (b)? What will be the spin–spin splitting in a 220-MHz spectrometer?

(d) Sketch the proton magnetic resonance spectrum for $CD_3CH_2OCD_3$.

23. Deduce the structure of the following compounds from their NMR spectra. Designate which protons go with which NMR lines. Chemical shifts are given in ppm relative to water.

(a) C_2H_4O

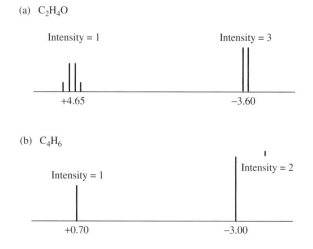

(b) C_4H_6

24. Two possible isomers of C_4H_9Cl are 1-chlorobutane and 2-chlorobutane. Their proton NMR spectra are given in ppm relative to tetramethylsilane. The relative intensities are in parentheses above the peaks.

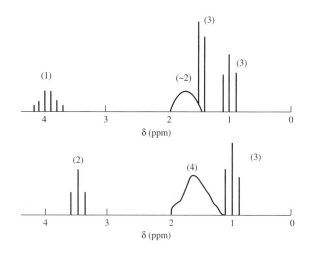

(a) Decide which spectrum goes with which molecule; then assign the protons to their NMR peaks.

(b) Sketch the proton NMR spectrum for another isomer of C_4H_9Cl: trimethylchloromethane.

25. Sketch the proton NMR spectrum of the sugar portion of a deoxynucleoside in D_2O solution with the following parameters. Assume all two-bond spin–

spin splittings are 10 Hz, all three-bond splittings are 4 Hz, and four or more bond splittings are zero.

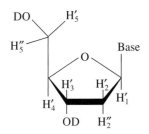

Proton	δ (ppm)
H1'	6.0
H2'	2.0
H2"	2.5
H3'	4.5
H4'	4.0
H5'	3.8
H5"	3.8

(a) A 500-MHz spectrometer is used. Plot a line spectrum vs. frequency in Hz, showing the position of each multiplet. For each multiplet, show the splitting pattern on an expanded frequency scale with correct first-order intensities.

(b) What would change in the spectrum in a 600-MHz spectrometer? Give values of any parameters that would change.

26. Double-stranded nucleic acids can adopt different double helical structures, usually referred to as A-form and B-form. Among the differences are the conformations of the sugars; one conformation is called 2'-endo (the 2'-carbon is above the approximate plane of the four other atoms of the ribose ring); the other is 3'-endo (the 3'-carbon is above the plane). A ribose in an RNA chain and tables for chemical shifts and spin–spin splittings are shown below.

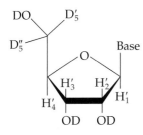

Values of Chemical Shifts

Proton	δ (ppm)
H1'	6.0
H2'	5.0
H3'	4.5
H4'	4.0

Proton ^{1}H—^{1}H Splittings (in Hz)

	A-form	B-form
$J_{1'2'}$	0	8
$J_{2'3'}$	5	5
$J_{3'4'}$	9	0

(a) Sketch the proton NMR spectrum of an A-form RNA; label all splittings. For each proton, state whether it is a singlet, doublet, triplet, and so on in the spectrum.

(b) Sketch the proton NMR spectrum of a B-form RNA; label all splittings. For each proton state whether it is a singlet, doublet, triplet, and so on in the spectrum. B-form RNA has not been observed yet.

(c) A Karplus-type equation relates the ^{1}H—^{1}H splittings to a torsion angle ϕ, which is a quantitative measure of the sugar conformation:

$$J_{1'2'} = 0.9 + 7.7 \cos^2 \phi$$

Calculate the values of ϕ predicted for B-RNA.

27. Consider the side chain of the amino acid arginine that can be modeled in D_2O as:

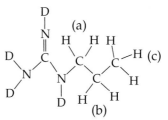

(a) Which type of proton would resonate farthest downfield (large ppm) in the proton NMR spectrum? Which would resonate farthest upfield? Explain or provide evidence for your answer.

(b) Draw a line NMR spectrum with the lengths of the lines to scale. Label the peaks and indicate upfield (low ppm) and downfield (high ppm).

(c) Draw the first-order splitting pattern for each peak, assuming $J_{ab} = 10$ Hz, $J_{bc} = 3$ Hz, and $J_{ac} = 1$ Hz. Show the relative magnitudes and splittings for each line in the pattern for peak a, peak b, and peak c.

28. Consider the two molecules shown below. The α- and β-carbons are tetrahedral, and there is free rotation around the bonds connected to them. Sketch the proton NMR spectrum of each of the molecules dissolved in D_2O. The important elements of your spectrum are the number of peaks, their relative magnitudes, and the splittings. Consider only proton–proton splittings for protons separated by three or fewer bonds. Label the splittings as J_{ab} and so on. The chemical shifts can be estimated from data in the text.

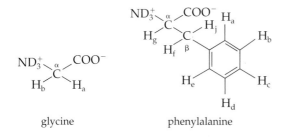

glycine   phenylalanine

29. (a) NMR is very useful in determining molecular structure, but it can also be used in imaging. Describe how an NMR spectrum of intensity vs. frequency can be used to obtain an image. What is the main experimental difference between NMR and MRI (magnetic resonance imaging)?

    (b) A sphere of paraffin (C_nH_{2n+2}) is placed in a magnetic imaging instrument. Plot the image of intensity vs. frequency that you would obtain.

30. The nuclear Overhauser effect (NOE) is a very powerful technique to investigate protein structure because different structures (α-helices, β-sheets) have characteristically different NOE patterns.

    (a) Describe how an NOE is measured experimentally.

    (b) NOEs are used to determine molecular structure. Describe how the measured NOE depends on molecular structure. Be quantitative.

    (c) Typical proton–proton distances (in Å) for the proton on the α carbon (αH) and the amide proton (NH) in α-helices and β-sheets are summarized as follows:

	α-helix	β-sheet (anti-parallel)	β-sheet (parallel)
NH—NH	2.8	4.3	4.2
αH—NH	3.5	2.2	2.2
αH—NH$_{i+3}$	3.4	>5	>5

(The subscript $i + 3$ refers to an amide proton three residues away from the α proton.) In studying the small model oligopeptide: KAALAKAALAKAALAK (K is lysine, A is alanine, L is leucine), the following NOEs are observed:

  i. Each amide proton has a very intense NOE to the next amide proton; for instance, each lysine amide (NH) has a strong NOE to an alanine amide (NH) proton.

  ii. Each αH proton has an NOE of moderate intensity to the next amide proton; for instance, each lysine α proton has an NOE to an alanine amide proton.

  iii. Each αH proton has an NOE of moderate intensity to an amide proton three residues away; for instance, each lysine αH proton has an NOE to the leucine.

What is the structure of the peptide? Is it α-helix, parallel β-sheet, or antiparallel β-sheet? Explain your answer in terms of the measured NOEs.

31. Leucine and isoleucine are amino acids that are isomers of each other; their structures are given in the appendix. The side chain of leucine has two H_β protons, one H_γ proton, and two δ methyls. The side chain of isoleucine has one H_β proton, two H_γ protons, a γ methyl, and a δ methyl. Each amino acid of course also has one H_α proton. Incorporated into a protein, they give characteristic patterns of cross peaks in a COSY experiment.

    (a) Sketch the COSY spectrum for leucine, assuming that each proton has a measurably different chemical shift (except the three protons on a methyl group). The order of chemical shifts can be taken as H_α, H_β, H'_β, H_γ, δ methyl, and δ methyl' in order of decreasing chemical shift. Figure 10.37b shows the COSY pattern for leucine, assuming that the two β protons are not resolved and that the two δ methyls are not resolved.

    (b) Sketch the COSY spectrum for isoleucine, assuming that each proton has a measurably different chemical shift (except the three protons on a methyl group). The order of chemical shifts can be taken as H_α, H_β, H_γ, H'_γ, γ methyl, and δ methyl in order of decreasing chemical shift.

    (c) Describe how the different patterns of cross peaks allow you to distinguish between leucine and isoleucine. If the two β protons of leucine, the two δ methyls of leucine, and the two γ protons of isoleucine are not resolved, can the two amino acids still be distinguished? Explain.

32. The NMR spectrum from a single methyl group (on T) in a DNA oligomer

    d(CGCGTATACGCG)

    is shown as a function of temperature. In general terms, explain what may be occurring to give rise to these spectra.

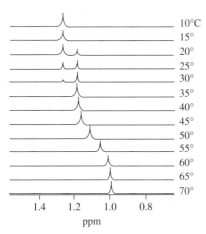

33. The NH resonances in the NMR spectrum at 25°C of a five amino acid (containing three NH groups) peptide YPIDV = Tyr Pro Ile Asp Val are shown below. P = Pro = Proline is unusual among amino acids in that the *cis* and *trans* forms of the peptide bond are more nearly equal in energy than for other amino acids.

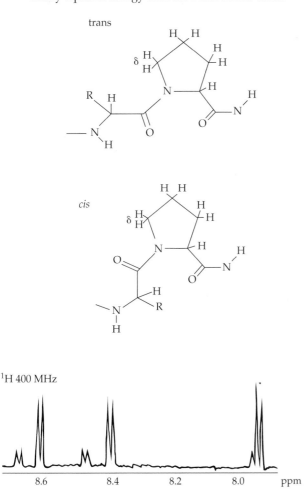

(a) Use the data to estimate the difference in standard free energy between the *cis* and *trans* forms of the peptide at 25°C.

(b) Explain what NMR experiment you would use to distinguish the resonances of the *cis* and *trans* forms. (Explain the experiment and what you would expect to see.)

(c) From these data, what can you say about the rate of interconversion between the *cis* and *trans* isomers? (Explain.)

(d) From the structure of proline shown, predict what the splitting pattern for the δH resonances would be. (Explain, stating any assumptions.)

34. Cytosine and uracil are two nucleic acid bases; their covalent structures are given in the appendix. An RNA structure is studied by NMR, in which a U and C can base pair to each other through hydrogen-bonding interactions. The chemical shifts of protons on the cytidine change drastically as the pH is varied from 5.5 to 7.5. In particular, NH_2 protons of cytidine show two resonances at 7.1 and 6.5 p.p.m at pH 7.5 and a single resonance (of twice the intensity) at 8.1 ppm. at pH 5.5.

(a) Explain the pH dependence of the NMR spectrum for this RNA. Why might chemical shifts of protons change as a function of pH?

(b) Propose base pairs that form between the C and U at pH 5.5 and pH 7.5. Each base pair must have at least two hydrogen bonds. Recall from chapter 9 that a hydrogen bond occurs when NH---O or NH---N interactions occur. How do your geometries explain the observed behavior for the cytosine NH_2 protons at the two solution pHs?

(c) Predict proton–proton NOEs that would be observed in the base pairs proposed in part (b).

(d) A COSY-type experiment is performed on a ^{15}N-labeled version of the RNA to observe scalar couplings between ^{15}N atoms. At high pH, scalar coupling is observed from the N3 position of the uracil to the N3 position of the cytidine. Thus, scalar coupling is observed *across the base pair*. At low pH, no such scalar coupling is observed. Explain how scalar coupling (transmitted through covalent bonds as we've explained) can be observed across a base pair. At low pH, no such cross-base-pair correlation is observed. Is this consistent with the base-pairing geometries you proposed in part (b). Why or why not? (See A. J. Dingley and S. Grzesiek, 1998, *J. Am. Chem. Soc.* 120:8293–8297.)

35. Nuclei that give NMR spectra behave as magnetic dipoles, analogous to the electric dipoles we discussed in chapter 9. The orientation of one dipole affects the energy of the neighboring dipole; this leads to splitting of the energy levels and peaks, exactly

similar to that observed for scalar coupling. The energy of interaction (dipolar coupling constant) between the two nuclei is proportional to

$$\frac{3\cos^2\theta - 1}{r^3}$$

where r is the distance between the two nuclei and θ is the angle between a vector that connects the two nuclei and the direction of the external magnetic (B_0) field.

(a) For a molecule in solution, dipolar splittings are almost always small or nonexistent. Can you explain this?

(b) Solid-state NMR is a powerful method for examining many biological and nonbiological systems. If a solid powder of the same molecule as in part (a) is

examined by NMR, the proton resonances are extremely broad with many splittings; the width of the peaks can exceed 1000 Hz. Can you explain this observation in terms of dipolar couplings?

(c) The dependence of dipolar couplings on the angle between the dipolar vectors and the external field suggests a solution to the large dipolar couplings in solid sample. At what angle of sample orientation would the dipolar couplings for the sample approach zero?

(d) If dipolar couplings could be measured in solution, they would provide advantages compared to NOE or scalar coupling measurements for molecular structure determination. Can you list at least two advantages? (For a recent reference, see N. Tjandra and A. Bax, 1997, *Science* 278:1111–1114.)

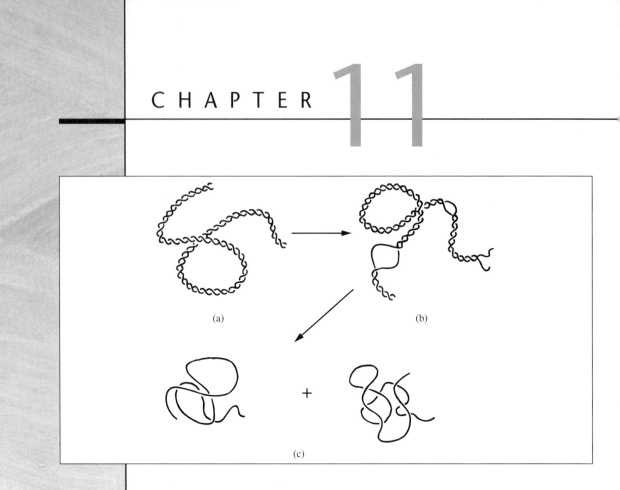

(a)

(b)

+

(c)

Molecular Distributions and Statistical Thermodynamics

Statistical methods and, in particular, statistical thermodynamics relate the average actions of individual molecules to measurable macroscopic properties. We review and summarize topics discussed in earlier chapters, including binding of ligands to multiple sites, random walks, and Boltzmann's statistical interpretation of entropy.

Concepts

All molecules have a range of conformations produced by vibrations of all the bonds and the torsional rotations around the single bonds. Macromolecules can have a very wide range of conformations because they contain so many bonds. DNA from the bacterial virus T4 is a double helix with a molecular weight of about 10^8; in solution this threadlike molecule can range from a tight coil to an extended form. Some properties of the molecule, such as how fast it can diffuse or sediment in solution, depend on the average dimension of the molecule. Other properties depend on the range of possible conformations of the molecule. For example, the probability that the molecule will be broken if the solution is stirred or poured depends on the fraction of molecules in an extended conformation, as well as the average dimension. Statistical methods can provide the distribution of molecular conformations and thus allow calculation of measurable properties. The concept of a random walk is very useful in understanding the possible conformations of a flexible macromolecule.

The binding of small molecules to many sites on a macromolecule and the disruption of hydrogen bonds that lead to helix-coil transitions of polypeptides and polynucleotides can be understood in terms of statistical distributions of bonds formed or broken.

Thermodynamic properties, such as energy, entropy, and free energy, are directly related to the average energy of all molecules in a system and to the distribution of molecules among the possible energies. The heat and work done when there is a change in the system can be calculated from the changes in the energy levels of the system and the changes in distribution of the molecules among the energy levels. These concepts connect molecular interactions and molecular properties to the thermodynamic laws we explored at the beginning of this book.

Applications

If DNA in solution is heated, at a certain temperature the double helix begins to "melt." Base pairing is disrupted, regions of single-stranded loops form, and, if the temperature is high enough, the two strands of the double helix dissociate completely. A schematic illustration of the melting phenomenon is shown in figure 11.1. The transition from the helix to the dissociated strands occurs in a temperature range of only a few degrees; hence, the term *melting* is used to describe this transition. Statistical analysis provides an explanation for the sharp transition. Furthermore, some of the thermodynamic parameters for the transition can be evaluated by a statistical analysis. These parameters are helpful in understanding the replication of the DNA, because the replication of DNA molecules involves the dissociation of the two parent strands and the formation of new double strands for each of the two daughter DNA molecules.

Either inside or outside the cell, the DNA interacts with a variety of small and not so small molecules. Among the small molecules, the antibiotic actinomycin binds to many sites of the DNA and can block transcription of the DNA into RNA. Many heterocyclic dye molecules and polycyclic aromatic

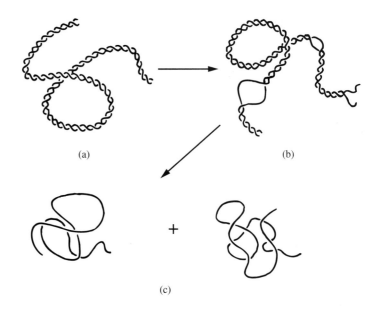

▶ **FIGURE 11.1**

Schematic illustration of the thermal "melting" of a double-stranded DNA (a) to its complementary single strands (c). At an intermediate temperature, single-stranded loops and/or ends can form, as illustrated in (b). See the text for a more detailed discussion.

hydrocarbons, such as 9-amino-acridine and benzo[a]pyrene, can bind to the DNA and cause mutations. Such molecules are often carcinogenic. The binding of small molecules to DNA can be understood by statistical analyses of the binding experiments. Among the larger molecules that interact with the DNA, some bind to only a few specific sites; others bind with much less discrimination. The RNA polymerase of the host bacterium of the T4 bacterial virus, for example, initiates RNA synthesis (transcription) from only a few sites on the DNA. The product of the T4 gene 32 (the gene 32 protein), which is involved in DNA replication and recombination, is much less site specific. It does, however, bind preferentially to single-stranded DNA in a highly cooperative manner. If gene 32 protein is mixed with an excess of single-stranded DNA, some DNA molecules will be completely covered with the protein while other DNA molecules remain bare. Statistical analysis again provides insight into such different patterns of interactions.

Binding of Small Molecules by a Polymer

For the interaction between two simple species, A and B, we found in chapter 4 that the system at equilibrium can be described by an equilibrium constant K and that the standard free-energy change ΔG^0 is related to the equilibrium constant K by the equation $\Delta G^0 = -RT \ln K$. As an example, for the combination of A and B to form AB,

$$A + B \rightleftharpoons AB$$

the equilibrium constant and the standard free-energy change are simply

$$K = \frac{[AB]}{[A][B]} \tag{11.1}$$

and

$$\Delta G^0 = -RT \ln K \tag{11.2}$$

In many cases of biological interest, one of the two interacting species is a complex macromolecule, and the other is a relatively small molecule. Usually, the macromolecule has more than one site where it can bind the small molecule. As a result, many different molecular complexes can form, depending on which sites of the macromolecule are occupied by the small molecules. For example, a hemoglobin molecule has four sites for oxygen. A high-molecular-weight DNA has numerous sites for the binding of actinomycin. In such cases, an expression such as Eq. (11.1) is not very descriptive in comparison with the richness of detail of the different species at the molecular level. We could write an equilibrium constant for each possible species, but the numbers often are very large and such cases can be treated more easily by the use of statistical methods. In the following sections, we explore a number of examples.

Identical-and-Independent-Sites Model

Suppose that a polymer molecule P has a number of identical and independent sites for binding a smaller molecule A. If the sites are *identical* (each has the same affinity for A) and *independent* (the occupation of one site has no effect on the binding to another site), a very simple description of this system can be obtained. Examples for which the identical-and-independent-sites model is applicable are the binding of DNA polymerase molecules to a single-stranded DNA, the binding of the drug ethidium to a DNA double helix, and the binding of substrates to certain enzymes or proteins containing several identical subunits per molecule.

As a simple example, let's consider a polymer P with four identical and independent sites. This polymer molecule might be a tetrameric protein, with each of the four subunits having one site for the binding of a substrate A. Suppose that the four sites are distinguishable and can be labeled 1, 2, 3, and 4. Since the sites are identical, the equilibrium constants of the following reactions are identical:

$$P + A \rightleftharpoons PA^{(1)} \qquad K = \frac{[PA^{(1)}]}{[P][A]}$$

$$P + A \rightleftharpoons PA^{(2)} \qquad K = \frac{[PA^{(2)}]}{[P][A]}$$

$$P + A \rightleftharpoons PA^{(3)} \qquad K = \frac{[PA^{(3)}]}{[P][A]}$$

$$P + A \rightleftharpoons PA^{(4)} \qquad K = \frac{[PA^{(4)}]}{[P][A]}$$

The superscript (1, 2, 3, or 4) in each reaction specifies which site of the polymer is involved.

The total concentration of species with one A molecule per P molecule, [PA], is

$$[PA] = [PA^{(1)}] + [PA^{(2)}] + [PA^{(3)}] + [PA^{(4)}]$$

$$= 4K[P][A]$$

In other words, for the reaction

$$P + A \overset{K_1}{\rightleftharpoons} PA \text{ (any site)}$$

$$K_1 = \frac{[PA]}{[P][A]} = 4K \qquad\qquad \textbf{(11.3)}$$

The factor 4 in Eq. (11.3) is a result of the fact that there are four ways for the formation of a PA (by binding of an A to any one of the four sites of P), but only one way for any PA to dissociate into P and A.

For the reaction

$$PA \text{ (any site)} + A \overset{K_2}{\rightleftharpoons} PA_2 \text{ (any two sites)}$$

$$K_2 = \frac{[PA_2]}{[PA][A]} = \left(\frac{4-1}{2}\right)K = \frac{3}{2}K \qquad (11.4)$$

There are $(4-1)$, or 3, ways of picking one of the three remaining sites in PA for the formation of a PA_2, and there are two ways for the dissociation of a PA_2 into PA and A. Similarly, the equilibrium constants K_3 and K_4 are

$$K_3 = \left(\frac{4-2}{3}\right)K = \frac{2}{3}K \qquad (11.5)$$

and

$$K_4 = \left(\frac{4-3}{4}\right)K = \frac{1}{4}K \qquad (11.6)$$

The concentrations of the various species are

$$[PA] = 4K[P][A] = 4[P](K[A])$$

$$[PA_2] = \frac{3}{2}K[PA][A] = \left(\frac{4 \cdot 3}{2}\right)K^2[P][A]^2 = 6[P](K[A])^2$$

$$[PA_3] = \frac{2}{3}K[PA_2][A] = \left(\frac{4 \cdot 3 \cdot 2}{3 \cdot 2}\right)K^3[P][A]^3 = 4[P](K[A])^3$$

$$[PA_4] = \frac{1}{4}K[PA_3][A] = K^4[P][A]^4 = [P](K[A])^4 \qquad (11.7)$$

Let $K[A] = S$. The total concentration of the polymer is

$$[P]_T = [P] + [PA] + [PA_2] + [PA_3] + [PA_4]$$
$$= [P](1 + 4S + 6S^2 + 4S^3 + S^4)$$
$$= [P](1 + S)^4 \qquad (11.8)$$

The total concentration of A molecules that are bound to the polymer molecules is

$$[A]_{bound} = [PA] + 2[PA_2] + 3[PA_3] + 4[PA_4]$$
$$= [P][4S + 12S^2 + 12S^3 + 4S^4)$$
$$= 4[P]S(1 + 3S + 3S^2 + S^3)$$
$$= 4[P]S(1 + S)^3 \qquad (11.9)$$

The number of bound A molecules per polymer molecule is ν, as defined in Eq. (5.14):

$$\nu \equiv \frac{[A]_{bound}}{[P]_{total}} = \frac{4[P]S(1+S)^3}{[P](1+S)^4} = \frac{4S}{1+S} \qquad (11.10)$$

One can see that, in general, if there are N identical and independent sites,

$$\nu = \frac{NS}{1 + S}$$

$$= \frac{NK[A]}{1 + K[A]} \tag{11.11}$$

From Eq. (11.11), it follows that

$$\nu(1 + K[A]) = NK[A]$$

$$\nu = (N - \nu)K[A]$$

and

$$\frac{\nu}{[A]} = K(N - \nu) \tag{11.12a}$$

or

$$\frac{\nu}{[A]} = KN - K\nu \tag{11.12b}$$

This is the *Scatchard equation*. When N is greater than 2, the parameter ν (the number of bound A molecules per polymer molecule) is usually used to express the extent of binding. If $\nu/[A]$ is plotted versus ν, according to Eq. (11.12a) or (11.12b), a straight line results. The slope of this line is $-K$, where K is the equilibrium constant for binding to one site. The intercept of the line with the ν-axis is N, where N is the total number of sites per polymer molecule. Such a plot is called a *Scatchard plot*. The Scatchard equation and Scatchard plot were first discussed in chapter 5; see, for example, figures 5.5–5.7.

Langmuir Adsorption Isotherm

A binding curve (amount of material bound versus concentration) at a constant temperature is called an *isotherm*. In the independent-sites model, because the occupation of one site has no effect on binding to other sites, the spatial distribution of the sites has no effect on the final binding equation (11.12).

Equation (11.12) can also be changed into the following form:

$$\frac{\nu/N}{1 - \nu/N} = K[A]$$

Because ν is the number of sites occupied per polymer molecule and N is the total number of sites per polymer molecule, ν/N is simply the fraction of sites occupied, which is designated as f. We obtain, then,

$$\frac{f}{1 - f} = K[A] \tag{11.13}$$

Equation (11.13) was first derived by Langmuir in 1916 for the adsorption of a gas on a solid surface.

Nearest-Neighbor Interactions and Statistical Weights

In the previous sections, the many binding sites on a macromolecule are assumed to be independent or noninteracting. Frequently, however, the binding of a molecule to one site affects the binding of molecules to other sites. The interactions between the sites greatly increase the complexity of the problem, and analytical solutions of the binding isotherms can be obtained only for special cases. However, a general method for simplifying the problem involves the use of statistical weights. The statistical weight ω_i of a species is defined as its concentration relative to a reference concentration:

$$\omega_i \equiv \frac{c_i}{c_{\text{ref}}}$$

Statistical weights can simplify the algebra in problems that contain many species. In equations involving ratios of concentrations, statistical weights can replace concentrations. For example, ν can be written in terms of statistical weights instead of concentrations. Because the statistical weight of each species is the relative concentration of that species, the mole fraction of any species X_i is the statistical weight of the species ω_i divided by the sum of the statistical weights of all species:

$$X_i = \frac{\omega_i}{\sum_i \omega_i} \tag{11.14}$$

Since $\sum_i \omega_i$ appears in many calculations, we define a function called the *sum over states:*

$$Z \equiv \sum_i \omega_i \tag{11.15}$$

Now, we consider a special case with N identical but interacting sites arranged in a linear array. In other words, the identical sites form a one-dimensional lattice. First used by Ising, this model is called the *Ising model.* To simplify matters, we first consider only nearest-neighbor interactions. Longer-range interactions will be discussed in a later section.

The N identical sites in a linear array can be expressed by an N-digit number of 0's and 1's such as 00111000 . . . 1011100011, in which the symbol 0 represents a free site and the symbol 1 represents an occupied site. The number written represents a linear polymer with N identical sites, such that the first two sites are unoccupied, the next three sites are occupied, and so on. Let K be the equilibrium constant for the binding of an A molecule to a site with no occupied nearest neighbors and let τK be the equilibrium constant for the binding of an A molecule to a site with one adjacent site occupied. The parameter τ accounts for the interaction between the two adjacent occupied sites. If τ is less than 1, the binding is *anticooperative* (it is more difficult to bind next to an occupied site); if τ is greater than 1, the binding is *cooperative* (it is easier to bind next to an occupied site); if τ is equal to 1, there is no interaction between sites.

With these definitions, the equilibrium constant for any reaction can be written quickly. A few examples follow for a simple case with $N = 3$:

Reaction	Equilibrium constant
$000 + A \rightleftharpoons 001$	K
$001 + A \rightleftharpoons 101$	K
$001 + A \rightleftharpoons 011$	τK
$101 \rightleftharpoons 110$	τ
$101 + A \rightleftharpoons 111$	$\tau^2 K$

The ratio of concentrations of [001] to [000], the statistical weight of [001], is

$$\omega_i = \frac{[001]}{[000]} = K[A]$$

Similarly, the statistical weights of [101], [011], and [111] are $(K[A])^2$, $\tau(K[A])^2$, and $\tau^2(K[A])^3$, respectively. The free species [000] has a statistical weight of one by convention. Defining the product $K[A]$ as S, the statistical weight of each state for the case $N = 3$ is given in table 11.1.

Note that the statistical weight of each state can be written as $\tau^i S^j$, where j is the number of 1's (occupied sites) in the state and i is the number of 1's following 1 (number of nearest-neighbor interactions).

For our present case, summing the statistical weights listed in table 11.1 gives

$$Z = 1 + 3S + S^2 + 2\tau S^2 + \tau^2 S^3 \qquad (11.16)$$

The average number of bound A per P is

$$\begin{aligned}
\nu &= \frac{1[PA] + 2[PA_2] + 3[PA_3]}{[P] + [PA] + [PA_2] + [PA_3]} \\
&= \frac{(3S) + 2(S^2 + 2\tau S^2) + 3(\tau^2 S^3)}{1 + 3S + S^2 + 2\tau S^2 + \tau^2 S^3} \\
&= \frac{3S + 2(S^2 + 2\tau S^2) + 3\tau^2 S^3}{Z}
\end{aligned} \qquad (11.17)$$

TABLE 11.1   Statistical Weights of Species Resulting from Binding to a Trimer

Number of sites occupied	Species	Statistical weight
0	000	1
1	001	S
	010	S
	100	S
2	101	S^2
	011	τS^2
	110	τS^2
3	111	$\tau^2 S^3$

Equation (11.17) can be written in a simple form. Because

$$Z = 1 + 3S + S^2 + 2\tau S^2 + \tau^2 S^3$$

$$\left(\frac{\partial Z}{\partial S}\right)_\tau = 3 + 2S + 4\tau S + 3\tau^2 S^2$$

and

$$S\left(\frac{\partial Z}{\partial S}\right)_\tau = 3S + 2(S^2 + 2\tau S^2) + 3\tau^2 S^3 \tag{11.18}$$

the right side of Eq. (11.18) is identical to the numerator in Eq. (11.17). Thus,

$$\nu = S\frac{(\partial Z/\partial S)_\tau}{Z}$$

$$= \left[\frac{\partial Z/Z}{\partial S/S}\right]_\tau$$

$$= \left(\frac{\partial \ln Z}{\partial \ln S}\right)_\tau \tag{11.19}$$

Eq. (11.19) tells us that we can calculate the number of sites bound ν from the sum of states Z. The derivative of Z with respect to S ($= K[A]$) for any amount of cooperativity τ leads to an equation for ν as a function of S. This equation is the binding curve; it characterizes the number of sites bound versus concentration of free ligand [A]. Although Eq. (11.19) is derived for a special case, it applies to any system in which there is binding to N identical sites with interaction only between adjacent sites.

EXERCISE 11.1

Show that for the case of four identical and independent sites discussed previously, Eq. (11.19) gives the same result as Eq. (11.10).

Cooperative Binding, Anticooperative Binding, and Excluded-Site Binding

It is informative to examine the shape of a plot of the fraction of binding sites occupied, f ($f \equiv \nu/N$) as a function of $S = K[A]$. With $N = 3$, f is plotted against S for the case $\tau = 1$ and $\tau = 10^3$ in figure 11.2. The case $\tau = 1$ is of course equivalent to the case when the sites are independent. For such a case, f increases gradually with S and approaches 1 asymptotically. The case $\tau = 10^3$, or in general for $\tau > 1$, is termed *cooperative binding* because the binding of A to one site makes it easier to bind another A to an adjacent site. Note that the shape of the f versus S plot is sigmoidal (S-shaped). As S increases, f increases slowly at first, showing an upward curvature, and then increases more rapidly. At still higher values of f, a decrease occurs in the slope of the f versus S plot with increasing S, showing a downward curvature. Such a sigmoidal plot of f versus S is indicative of cooperative binding.

For the independent-sites model, the shape of the f versus S curve is not affected by the number of sites N per polymer molecule because N is absent in the binding equation (11.13). For cooperative binding, the shape of the f versus S curve is dependent on N. As N increases, the number of terms in the

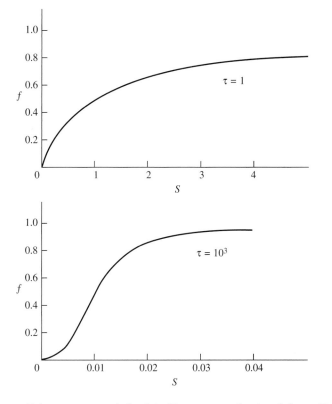

◀ FIGURE 11.2

Fraction of sites occupied f as a function of the substrate concentration times the binding constant for a linear array with three binding sites ($N = 3$). The curves shown are for the cases of no interaction between occupied sites ($\tau = 1$) and for positive interaction between adjacent sites with a value of $\tau = 10^3$.

sum of states Z increases, and the binding curve obtained from Eq. (11.19) changes. For the same value of the interaction parameter τ, the difference in S for a given change in f is less for larger N. Thus, the abruptness of the co-operative binding is even more extreme. If N is very large, it can be shown that f changes from essentially zero to essentially 1 in the range $(1 - 2/\sqrt{\tau})$ $< \tau S < (1 + 2/\sqrt{\tau})$. Thus, a sharp transition occurs at $\tau S = 1$ for large values of τ. At low concentrations of [A], $\tau S \equiv \tau K[A]$ is less than 1, and very little binding occurs; however, as the concentration of [A] is increased so that τS becomes slightly greater than 1, the cooperative binding is suddenly trig-gered, and nearly all sites are filled.

Another important difference between independent-sites binding (non-cooperative binding) and cooperative binding is the relative amounts of the various species. For $N = 3$, let's examine the situation when $\nu = \frac{3}{2}$ (when half of the sites are occupied). The mole fractions of the species with 0, 1, 2, and 3 sites occupied are from table 11.1: $1/Z$, $3S/Z$, $(S^2 + 2\tau S^2)/Z$, and $\tau^2 S^3/Z$, respectively. These quantities are calculated for the case $\tau = 1$ ($\nu = \frac{3}{2}$ at $S = 1$) and $\tau = 10^3$ ($\nu = \frac{3}{2}$ at $S = 9.75 \times 10^{-3}$) and are plotted in figure 11.3.

The difference between the two distributions is evident. In the case $\tau = 1$, the most abundant species center around $\nu = \frac{3}{2}$. In the case $\tau = 10^3$, however, the predominant species are the one with all sites free (47 mol %) and the one with all sites occupied (43.6 mol %). Such an uneven distribution is a con-sequence of the cooperative nature of binding. Because the occupation of one site makes the binding to adjacent sites more favorable, the bound A molecules tend to cluster. When τ is very large, only the species with all sites occupied and the species with all sites free are of importance. This is the *all-or-none limit*.

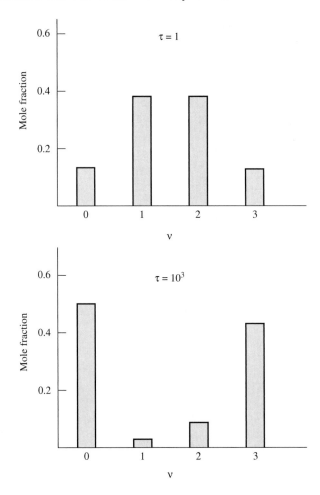

▶ FIGURE 11.3

Distribution of species at half saturation for $N = 3$. Independent sites ($\tau = 1$); cooperative binding ($\tau = 10^3$). Note that the cooperative binding leads to an all-or-none-type binding.

In the all-or-none limit, for our case of $N = 3$, we can simply consider the reaction as

$$P + 3A \longrightarrow PA_3$$

because neither of the other species, PA and PA_2, is present in significant amounts. The equilibrium constant for the reaction is

$$K' = \frac{[PA_3]}{[P][A]^3}$$

and

$$\nu = \frac{3[PA_3]}{[P] + [PA_3]} = \frac{3K'[A]^3}{1 + K'[A]^3} \tag{11.20}$$

We use K' in these equations to remind us that this is not the equilibrium constant for binding at a single site.

Because $\nu/3 = f$, the fraction of sites occupied, simple algebra gives

$$\frac{f}{1 - f} = K'[A]^3 \tag{11.21}$$

and

$$\frac{d \log[f/(1 - f)]}{d \log[A]} = 3$$

In general, if there are N sites per polymer molecule, the all-or-none model (maximum cooperativity) yields

$$\frac{d \log[f/(1 - f)]}{d \log[A]} = N \qquad \textbf{(11.22)}$$

For no cooperativity ($\tau = 1$), the binding curve is independent of N and the left-hand side of Eq. (11.22) is equal to 1. An experimental measure of cooperativity in binding is obtained by a plot of $\log[f/(1 - f)]$ versus $\log[A]$. This plot is a *Hill plot* [see figure (5.10)], and the slope of the plot is n, the *Hill constant*. If the slope n is 1, the sites are independent. If the slope is greater than 1, there is some cooperativity. If the slope is equal to N, the binding can be approximated by the all-or-none model.

Also of interest is the case $\tau < 1$, when binding to one site reduces the affinity of adjacent sites (*anticooperative* or *interfering binding*). In the limit $\tau = 0$, the occupation of one site excludes binding to adjacent sites, and the case is referred to as the *excluded-site model*.

N Identical Sites in a Linear Array with Nearest-Neighbor Interactions

In the preceding section, we discussed quantitatively the binding of small molecules to a polymer molecule with three identical sites in a linear array. A number of important concepts came out of the discussion. These include cooperative and anticooperative binding, the all-or-none limit, and the excluded-site model. The transition from a polymer with $N = 3$ to one with a large number of monomer units requires no new physical concepts, but the mathematics is more involved. This is described in detail in the references given at the end of the chapter. Instead of repeating the lengthy derivations here, we examine some illustrative examples.

EXAMPLE 11.1

Adenosine molecules can form hydrogen-bonded complexes with the bases of polyuridylic acid. In figure 11.4, the fraction of sites occupied in polyuridylic acid is shown as a function of the concentration of free adenosine. The shape of the curve clearly indicates cooperative binding. The interaction parameter τ is calculated to be $\sim 10^4$. The cooperativity is interpreted as a result of the favorable "stacking" interactions between adjacent adenosine molecules.

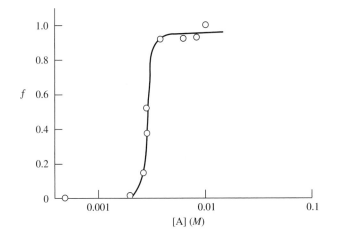

◀ FIGURE 11.4

Binding of adenosine A to polyuridylic acid. The fraction of sites occupied f is plotted against the concentration of free adenosine [A]. (Data from W. M. Huang and P. O. P. Ts'o, 1966, *J. Mol. Biol.* 16:523.)

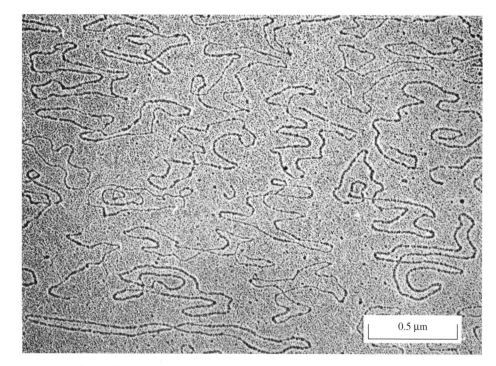

▲ FIGURE 11.5

Electron micrograph showing loops of single-stranded circular DNA molecules. Look carefully and note that some segments of the DNA appear dark and thick, whereas others are light and thin. The thick segments show where the DNA gene 32 protein forms complexes with the DNA. The protein distribution is far from uniform along the DNA. You can find some DNA molecules that are completely thick and others that consist mostly of thin segments. (From H. Delius, N. J. Mantell, and B. Alberts, 1972, *J. Mol. Biol. 67.*)

EXAMPLE 11.2

The bacteriophage T4 gene 32 protein is involved in both DNA recombination and DNA replication. If the amount of the protein is insufficient to saturate the amount of the single-stranded DNA present, the distribution of the protein molecules among the DNA molecules is far from random. Some DNA molecules are nearly saturated with the protein, whereas others are essentially free of bound protein molecules, as shown in the electron micrograph in figure 11.5.

Identical Sites in Nonlinear Arrays with Nearest-Neighbor Interactions

We have described binding to linear arrays such as nucleic acids, but more complex arrays can be treated by similar methods. Here we give only one example of the differences that occur with different arrays.

EXAMPLE 11.3

Glyceraldehyde-3 phosphate dehydrogenase is an enzyme consisting of four identical subunits. It first binds NAD^+ and then catalyzes the oxidation and phosphorylation of glyceraldehyde-3-phosphate to 1,3-diphosphoglycerate. The binding of

NAD^+ to the enzyme isolated from muscle cells was investigated by A. Conway and D. E. Koshland, Jr. (1968, *Biochemistry* 7:411) and was found to occur in four stages, with equilibrium binding constants K_1 10^{11} M^{-1}, K_2 10^9 M^{-1}, K_3 3.3×10^6 M^{-1}, and K_4 3.9×10^4 M^{-1}. Because the binding of successive NAD^+ molecules to a single tetrameric enzyme becomes progressively weaker, this is an example of anticooperativity. (In contrast, the enzyme derived from yeast exhibits positive cooperativity in binding NAD^+.)

To model this binding pattern, it is reasonable to assume that the binding of the first ligand (NAD^+) results in a structural change that weakens the binding to the neighboring subunits. The predicted pattern of binding will be different, depending on whether the subunit interactions occur as (1) a linear array, where each site in the center can have two interactions and each site at each end can have only one; (2) a square array, where each site can have two interactions; or (3) a tetrahedral array, where each site can have three interactions:

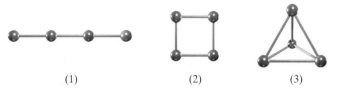

(1)   (2)   (3)

Negative binding interactions are assumed to occur whenever a neighboring subunit has a binding site that is already occupied by a ligand. Because the four experimental binding constants decrease progressively by factors of the order of 10^{-2}, this is consistent with the pattern expected for a tetrahedral array. Let's test this hypothesis.

In the following table representing the site occupation of the tetrahedral model, we use 0's to indicate free sites and 1's to indicate bound sites. We assume that when two ligands are bound, each of the six pairwise interactions is identical and characterized by the parameter τ.

Reaction	Equilibrium Constant
$\begin{array}{c} 0 \\ 0 \\ 0 \quad 0 \end{array} + A = \begin{array}{c} 0 \\ 0 \\ 1 \quad 0 \end{array}$ (4 possible species)	K
$\begin{array}{c} 0 \\ 0 \\ 1 \quad 0 \end{array} + A = \begin{array}{c} 0 \\ 1 \\ 1 \quad 0 \end{array}$ (6 possible species)	τK
$\begin{array}{c} 0 \\ 1 \\ 1 \quad 0 \end{array} + A = \begin{array}{c} 1 \\ 1 \\ 1 \quad 0 \end{array}$ (4 possible species)	$\tau^3 K$
$\begin{array}{c} 1 \\ 1 \\ 1 \quad 0 \end{array} + A = \begin{array}{c} 1 \\ 1 \\ 1 - 1 \end{array}$ (1 species)	$\tau^6 K$

The table makes it evident that the second ligand to bind involves one interaction, the third ligand involves two interactions, and the fourth ligand involves three interactions with subunits having ligands already bound. The fact that each experimentally measured equilibrium constant K_i differs by a factor τ from the previous one is the property that we are seeking to match qualitatively to our theoretical caculations. To test this quantitatively, we need to consider the number of binding arrangements that is possible for a given number of ligands bound and to write the experimental K_i's in terms of the parameters τ and K given in the table.

a. Use reasoning, like that which led to Eq. (11.16), to show that for the tetrahedral model of glyceraldehyde-3-phosphate dehydrogenase the sum of states is

$$Z = 1 + 4S + 6\tau S^2 + 4\tau^3 S^3 + \tau^6 S^4$$

where $S = K[A]$.

b. Now calculate expressions for K_1, K_2, K_3, and K_4 in terms of τ and S. For example,

$$K_1 = \frac{[EA]}{[E][A]} = \frac{4S}{1} \times \frac{1}{[A]} = 4K$$

$$K_2 = \frac{[EA_2]}{[EA][A]} = \frac{6\tau S^2}{4S} \times \frac{1}{[A]} = \frac{3\tau S}{2[A]} = \frac{3}{2}\tau K$$

Next we need to establish the numerical values for τ and K that can be used to best simulate the experimental data. Because the values of K_3 and K_4 were determined with greater precision, we begin with them.

c. Show that your results for K_3 and K_4 in part (b) are consistent with the relation

$$\frac{K_4}{K_3} = \frac{3}{8}\tau = \frac{3.9 \times 10^4 \ M^{-1}}{3.3 \times 10^6 \ M^{-1}}$$

Therefore, $\tau = 3.2 \times 10^{-2}$ and $K = 5.0 \times 10^9$.

Calculate values for K_1 and K_2 and compare them with the experimental values to evaluate the suitability of the tetrahedral, pairwise-interaction model. Remember that only the order of magnitude of these binding constants was determined experimentally. (Answer: $K_2 = 2.4 \times 10^8 \ M^{-1}$ and $K_1 = 2.0 \times 10^{10} \ M^{-1}$.)

Finally, construct a table of interactions for the four-site linear model and the square array model. Based on the behavior of your expressions for the ligand binding equilibrium constants, what is your conclusion about the suitability of the alternative to the tetrahedral model? (For the linear array, $Z = 1 + 4S + 3(\tau + 1)S^2 + 2(\tau^2 + \tau)S^3 + \tau^3 S^4$; for the square array, $Z = 1 + 4S + 2(2\tau + 1)S^2 + 4\tau^2 S^3 + \tau^4 S^4$.)

The Random Walk

In the preceding sections, we discussed the binding of small molecules to a polymer with many sites. We now discuss an important problem in statistics: the random-walk problem or the problem of random flight. The simplest example of such a problem involves someone walking along a sidewalk but unable to decide at each step whether to go forward or back. The direction of each step is random. This is a *one-dimensional random walk*. The *two-dimensional random walk* is analogous to a dazed football player who may take steps randomly in any direction on the field. The problem is also easily generalized to three (or more) dimensions.

Although the formulation of the random-walk problem may seem quite abstract and have little to do with molecules and physical chemistry, we soon see that this is not so. A number of problems of interest, such as molecular diffusion and the average dimension of a long polymer molecule, can be treated as random-walk problems in three dimensions.

To introduce an analysis of the random-walk problem, it is best to start with the one-dimensional version. If we use the symbol 1 for a step forward and the symbol 0 for a step back, the record of a walk of N steps is represented by an N-digit number, such as 11000101011 . . . 0011. Let p be the probability of a step forward and q be the probability of a step back. When we said that the direction of each step was "random," we implied that $p = q$. In general, though, p does not have to be the same as q.

Before going further, let's elaborate on what we mean by "the probability of a step forward being p." Imagine that N similar fleas (N is a very large number) are hopping randomly forward or backward. These N fleas form an ensemble. After each has taken one hop, m are found to have moved forward, and the rest, $N - m$, to have moved backward. The fraction of fleas that moved forward, m/N, is the probability p. Similarly, $(N - m)/N = q$. Obviously, $p + q = 1$, a consequence of the fact that we allow only two possibilities; either a hop forward or a hop back. This imaginary experiment can be performed in a different way. Instead of having N fleas take one hop each, we can allow one flea to take N hops. We expect that, if there are no time-dependent changes, the fraction of steps forward is p. *It is a basic postulate of statistical mechanics that the time average of a property of a system is the same as the ensemble average at any instant.* This means that watching one molecule over a long period of time and averaging its property will give the same result as an average at one time over many molecules. Until recently it was not practical to observe a single molecule to test the basic postulate of statistical mechanics. Now we can follow the diffusion of a single molecule as discussed in chapter 6, and we can measure the kinetics of a single enzyme (see chapter 8). In general, the basic postulate is correct—the time average of a single molecule equals the ensemble average of many molecules. But there do seem to be exceptions—for example, when individual enzymes have different kinetic properties.

Now we come back to the one-dimensional random-walk problem. We need to be able to calculate the probability of taking any number of steps forward and any number backward. Then we will be able to find how far we an expect to move from the starting point after N random steps. The answer is that after N random steps we will just as likely have moved forward as backward. However, this does not mean that we end up at the starting point after N random steps. The square root of the mean-square displacement (a measure of how far we are from the start) is equal to the square root of the number of steps times the length of each step. We derive these results in detail.

For a given sequence of N steps represented by 11000101011 . . . 0011, the probability for this sequence to occur is $ppqqqpqpqpp \ldots qqpp = p^m q^{N-m}$, where m is the number of 1's. There are many ways that we can take N steps with m of them forward. For example, if $N = 4$ and $m = 2$, the possible ways are 0011, 0110, 1100, 1010, 0101, and 1001. Several additional examples are presented in table 11.2.

TABLE 11.2 Some Random-Walk Examples

Total number of steps	Record of random walk	Probability	Sum of probabilities of all possibilities
1	0	q	$q + p$
	1	p	
2	00	q^2	$q^2 + 2qp + p^2$
	01, 10	$2qp$	$= (q + p)^2$
	11	p^2	
3	000	q^3	$q^3 + 3q^2p + 3qp^2 + p^3$
	001, 010, 100	$3q^2p$	$= (q + p)^3$
	011, 110, 101	$3qp^2$	
	111	p^3	
N	0000 . . .	q^N	$(q + p)^N$
	. . .	. . .	
	. . .	. . .	
	1111 . . .	p^N	

In table 11.2, we see that the sum of the probabilities of all possible outcomes of an N-step walk can be represented in the form $(q + p)^N$. Because $q + p = 1$, this sum is equal to 1, but we want to know how the components of the sum depend on q, p, and N. The sum can be expanded by the binomial theorem:

$$(q + p)^N = q^N + Nq^{N-1}p + \frac{N(N - 1)}{2!}q^{N-2}p^2 + \cdots$$

$$+ \frac{N!}{m!(N - m)!}q^{N-m}p^m + \cdots + p^N \qquad (11.23)$$

Remember that

$$N! = N(N - 1)(N - 2)\cdots(3)(2)(1)$$

Each term in the expansion gives the probability of the corresponding outcome. The term q^N, for example, is the probability that all N steps are backward. The term

$$W(m) = \frac{N!}{m!(N - m)!}q^{N-m}p^m \qquad (11.24)$$

gives the probability that m steps are forward and $N - m$ steps are backward. The coefficient

$$\frac{N!}{m!(N - m)!} \qquad (11.24a)$$

is the number of ways one can take N steps with m of them forward. Because $q + p = 1$, the numerical value of $(q + p)^N$ is of course always 1.

Calculation of Some Mean Values for the Random-Walk Problem

The important equation is Eq. (11.24). Knowing the probability of a random walk of N steps with m steps forward, we can now calculate averages of interest, such as the average number of steps forward,

$$\langle m \rangle = \sum_{m=0}^{N} mW(m) \qquad (11.25a)$$

and the mean of the square of the number of steps forward,

$$\langle m^2 \rangle = \sum_{m=0}^{N} m^2 W(m)$$

(11.25b)

Mean displacement

First, let's calculate the average of the displacement. If the pace of each step is l, the net displacement for m steps forward is $[m - (N - m)]l$, or $(2m - N)l$. So if we determine $\langle m \rangle$, the average value of m, the *mean displacement* forward is $(2\langle m \rangle - N)l$.

Substituting Eq. (11.24) into Eq. (11.25a), we obtain

$$\langle m \rangle = 0 \cdot q^N + 1 \cdot Nq^{N-1}p + 2\frac{N(N - 1)}{2!}q^{N-2}p^2 + \cdots$$

$$+ m\frac{N!}{m!(N - m)!}q^{N-m}p^m + \cdots + Np^N$$

(11.26)

We can use this equation to calculate the average number of steps forward, but we note that its terms look like those in Eq. (11.23), except that each term is multiplied by the exponent of p in the term. This suggests that the expression for $\langle m \rangle$ is related to the derivative of Eq. (11.23) and can be greatly simplified. We denote Eq. (11.23) by Z because it is a sum of statistical weights of "species" with m forward steps:

$$Z = (q + p)^N = q^N + Nq^{N-1}p + \frac{N(N - 1)}{2!}q^{N-2}p^2 + \cdots$$

$$+ \frac{N!}{m!(N - m)!}q^{N-m}p^m + \cdots + p^N$$

(11.23)

$$\frac{\partial Z}{\partial p} = Nq^{N-1} + 2\frac{N(N - 1)}{2!}q^{N-2}p + \cdots$$

$$+ m\frac{N!}{m!(N - m)!}q^{N-m}p^{m-1} + \cdots + Np^{N-1}$$

$$p\left(\frac{\partial Z}{\partial p}\right) = Nq^{N-1}p + 2\frac{N(N - 1)}{2!}q^{N-2}p^2 + \cdots$$

$$+ m\frac{N!}{m!(N - m)!}q^{N-m}p^m + \cdots + Np^N$$

(11.27)

Comparing Eqs. (11.27) and (11.26), we see that

$$\langle m \rangle = p\left(\frac{\partial Z}{\partial p}\right)$$

(11.28)

but

$$Z = (q + p)^N$$

Therefore,

$$\left(\frac{\partial Z}{\partial p}\right) = N(q + p)^{N-1}$$

and

$$\langle m \rangle = p\left(\frac{\partial Z}{\partial p}\right) = Np(q + p)^{N-1} = Np$$

(11.29)

The last equality comes from the fact that $(q + p) = 1$; this leads to the simple and intuitively reasonable result. The average number of steps forward is the probability of a forward step times the total number of steps. For the random (unbiased) walk, $p = \frac{1}{2}$, and

$$\langle m \rangle = \frac{N}{2} \tag{11.30}$$

and the net displacement forward is

$$(2\langle m \rangle - N)l = 0 \tag{11.31}$$

This seems completely reasonable. If the probability of going forward is the same as that of going backward, on the average half of the total number of steps taken will be forward. After taking N steps, sometimes a person might be a few paces in front of the origin, sometimes a few paces behind the origin. The average, or mean, displacement comes out to be zero.

Mean-square displacement

After taking a larger and larger number of steps N at random, it becomes increasingly unlikely that we will end up exactly where we started. The *mean-square displacement* is a measure of how far from the origin we can expect to be, on the average. The importance of the difference between the average distance from the origin and the average net displacement forward is that in calculating the average distance we do not distinguish whether we finally end up in front of or behind the origin. For example, if after N steps person A ends up three steps forward of the origin and person B ends up three steps behind the origin, the average of their net displacements is zero; but the average of their distances from the origin is clearly not zero; it is three steps.

One way to characterize the average distance from the origin is to calculate the mean-square displacement. In a single journey of N steps, if m steps are forward, the net displacement forward is

$$d = (2m - N)l \tag{11.32}$$

and the square of the displacement is

$$d^2 = (2m - N)^2 l^2 \tag{11.33}$$

Whether $m > N/2$ (a net displacement forward) or $m < N/2$ (a net displacement backward), d^2 is positive.

The mean value of $(2m - N)^2$ can be calculated as follows, where $\langle \, \rangle$ designates the average of that quantity:

$$\langle (2m - N)^2 \rangle = 4\langle m^2 \rangle - 4\langle m \rangle N + N^2 \tag{11.34}$$

But $\langle m \rangle = N/2$ for a random walk; thus,

$$\langle (2m - N)^2 \rangle = 4\langle m^2 \rangle - 2N^2 + N^2 = 4\langle m^2 \rangle - N^2 \tag{11.35}$$

To calculate $\langle m^2 \rangle$, we start with its definition given in Eq. (11.25b),

$$\langle m^2 \rangle = \sum_{m=0}^{N} m^2 W(m)$$

$$= 0 \cdot q^N + 1 \cdot N q^{N-1} p + 2^2 \frac{N(N-1)}{2!} q^{N-2} p^2 + \cdots$$

$$+ m^2 \frac{N!}{m!(N-m)!} q^{N-m} p^m + \cdots + N^2 p^N \tag{11.36}$$

If we differentiate Eq. (11.27) with respect to p, we obtain

$$\frac{\partial}{\partial p}\left[p\left(\frac{\partial Z}{\partial p}\right)\right] = Nq^{N-1} + 2^2\frac{N(N-1)}{2!}q^{N-2}p^2 + \cdots$$

$$+ m^2\frac{N!}{m!(N-m)!}q^{N-m}p^{m-1} + \cdots + N^2p^{N-1} \quad \textbf{(11.37)}$$

Comparing Eqs. (11.36) and (11.37), we note that multiplying the right side of (11.37) by p gives the right side of (11.36). Therefore,

$$\langle m^2 \rangle = p\left\{\frac{\partial}{\partial p}\left[p\left(\frac{\partial Z}{\partial p}\right)\right]\right\} \quad \textbf{(11.38)}$$

From Eq. (11.29),

$$p\left(\frac{\partial Z}{\partial p}\right) = Np(q + p)^{N-1}$$

Therefore,

$$\frac{\partial}{\partial p}\left[p\left(\frac{\partial Z}{\partial p}\right)\right] = N\left[p\frac{\partial(q + p)^{N-1}}{\partial p} + (q + p)^{N-1}\right]$$

$$= N[p(N - 1)(q + p)^{N-2} + (q + p)^{N-1}]$$

$$= N[p(N - 1) + 1]$$

and

$$p\frac{\partial}{\partial p}\left[p\left(\frac{\partial Z}{\partial p}\right)\right] = Np[p(N - 1) + 1] \quad \textbf{(11.39)}$$

For $p = \frac{1}{2}$,

$$\langle m \rangle = p\frac{\partial}{\partial p}\left[p\left(\frac{\partial Z}{\partial p}\right)\right] = \frac{N}{2}\left(\frac{N - 1}{2} + 1\right) = \frac{N(N + 1)}{4} \quad \textbf{(11.40)}$$

but

$$\langle(2m - N)^2\rangle = 4\langle m^2 \rangle - N^2 \quad \textbf{(11.35)}$$

and, after all these complicated manipulations, we obtain the strikingly simple result that

$$\langle(2m - N)^2\rangle = N(N + 1) - N^2$$

$$= N \quad \textbf{(11.41)}$$

Therefore, the mean-square displacement $\langle d^2 \rangle$ is

$$\langle d^2 \rangle = Nl^2 \quad \textbf{(11.42)}$$

The root-mean-square displacement $\langle d^2 \rangle^{1/2}$ is

$$\langle d^2 \rangle^{1/2} = N^{1/2}l \quad \textbf{(11.43)}$$

Equation (11.43) says that after N random steps, on the average, one is $\sqrt{N}$ paces from where one started. Although we have derived this equation for a one-dimensional random walk, it can be shown that it holds for a random walk in two- or three-dimensional space as well. We presented the derivations for the random-walk displacements in detail to illustrate how average properties are obtained from statistical weights and sums over states.

Diffusion

Molecules, either in the gas phase or the liquid phase, are always undergoing many collisions. Let's assume that the average distance traveled by a molecule between two successive collisions (the mean free path) is l. Whenever a molecule has a collision, its direction changes, depending on the direction of the molecule hitting it. The total number N of collisions for a given molecule is proportional to the time t. There is a close relation between the diffusional process and the random walk. The fact that the individual "steps" in the diffusion process are not all of the same length turns out to be unimportant as long as the measurements are made over distances that are large compared with the mean free path. For a given spatial distribution of molecules (concentration gradient) at time zero, the spatial distribution of the molecules at time t can be obtained by considering each molecule as a random walker. The average value of the square of the displacement for the diffusional process based on Eq. (11.42) is expected to be proportional to N and therefore to the elapsed time. The proportionality constant is a measure of how fast the molecule diffuses and is directly related to the diffusion coefficient D, as was seen in chapter 6.

Using the random-walk model, Einstein was able to derive the diffusion equation from the molecular point of view.

Average Dimension of a Linear Polymer

A flexible polymer can assume many conformations which differ little in energy. Consider a polypeptide,

for example. Along the backbone of the polymer, rotation around the amide bonds $N\text{---}C$ has a high-energy barrier because of the partial double-bond character of these bonds (see chapter 9, problem 18). Rotation around the other bonds is relatively free, however, resulting in many polymer conformations of similar stability. Some of these conformations may be highly extended, whereas others may be much more compact. Thus, in discussing the dimensions of such a polymer, it is useful to specify *average* dimensions. There is an ensemble of polymers with a range of conformations at any time; each polymer will change its conformation over a period of time to sample the range of conformations in the ensemble.

A quantity frequently used to express the average dimension of a flexible polymer is the root-mean-square end-to-end distance, $\langle h^2 \rangle^{1/2}$. h is the straight-line distance between the ends of the molecule; the root-mean-square end-to-end distance is the square root of the average of its square. If a polymer molecule is highly extended, its end-to-end distance is much greater than for the same molecule in a compactly coiled form. A simple and useful model for the evaluation of the average end-to-end distance is the *random-coil*, or *Gaussian, model*. In this model, the linear polymer is considered to be made of N segments each of length l, linked by $(N-1)$ universal joints (unrestricted bending) so that the angle between any pair of adjacent segments can take any value with equal probability.

The reader may recognize immediately that the random-coil model is exactly the same as the random-walk problem we have discussed. From Eq. (11.42), the mean-square end-to-end distance is $\langle h^2 \rangle = \langle d^2 \rangle = Nl^2$. For a random-coil polymer, $\langle h^2 \rangle$ is proportional to N and therefore proportional to the molecular weight of the polymer. We can also write

$$\langle h^2 \rangle = Nl^2 = Nl(l) = Ll \qquad (11.44)$$

where $L \equiv Nl$ is the *contour length* of the molecule. The contour length of a polymer molecule is the length measured along the links of the polymer.

Real polymer molecules are of course made of monomers linked by chemical bonds rather than segments linked by universal joints. The relation between bond lengths and lengths of statistical segments is not simple. In general, the length l of a statistical segment must be determined experimentally.

Let's consider polyethylene,

For simplicity, we consider only the carbon backbone chain,

with a bond angle θ between two adjacent bonds and bond length b for each C—C bond. If θ could assume any value, you could imagine a universal joint at each carbon atom and the mean-square end-to-end distance $\langle h^2 \rangle = Nl^2 = Nb^2$, where N is the number of carbon atoms per polymer chain. We know, however, that the bond angle θ cannot assume any value. Rather, the most stable conformation for

bonds has an angle θ close to the tetrahedral angle of $109°$. The restriction in θ makes the molecule a little less flexible. If the random-coil model is used for polyethylene, we expect that $l > b$; the effective distance l between carbon atoms will be greater than the bond length b. In fact, it can be shown that $l = \sqrt{2}b$ for polyethylene.

For a polymer such as a double-stranded DNA, the effective segment length l of an equivalent random coil cannot be related directly to the bond lengths. Hydrodynamic measurements give $l \approx 10^3$ Å for DNA. Thus, for a flexible molecule such as polyethylene, the segment length is of the order of the bond length $[l = \sqrt{2}b \cong 2.2$ Å$]$; for a stiff molecule such as a double-stranded DNA, the effective length is several orders of magnitude larger than a bond length.

A quantity closely related to mean-square end-to-end distance is the mean-square radius, which is defined by

$$\langle R^2 \rangle = \frac{\Sigma_i m_i r_i^2}{\Sigma_i m_i} \qquad (11.45)$$

for any collection of mass elements, where m_i is the mass of the ith element and r_i is the distance of this element from the center of mass of the collection.

It can be shown that for an open-ended random coil,

$$\langle R^2 \rangle = \frac{Nl^2}{6} \tag{11.46a}$$

and for a circular random coil (formed by linking the two ends of the open-ended chain),

$$\langle R^2 \rangle = \frac{Nl^2}{12} \tag{11.46b}$$

It was first shown by Debye that $\langle R^2 \rangle$ can be measured experimentally by light scattering.

Note that the random-coil model predicts that $\langle h^2 \rangle^{1/2}$ or $\langle R^2 \rangle^{1/2}$ is proportional to $\sqrt{N}$, or the square root of the molecular weight. This relation can be tested experimentally. If it is found to be true, the polymer is said to behave as a random coil, or a Gaussian chain.

For real polymers, in general, interactions occur between adjacent segments, and they have preferred orientations instead of equally probable orientations, as required by the random-coil model. At a given temperature, the random-coil model is expected to be applicable only in solvents ("θ solvents") in which the interactions between polymer segments are balanced by solvent–polymer interactions. Otherwise, the interactions between the polymer segments will cause deviations from the random-coil model.

Helix–Coil Transitions
Helix–Coil Transition in a Polypeptide

In a polypeptide chain, the partial double-bond character of the N⋯C and C⋯O bonds for each amide residue in the chain requires the group of six atoms

to lie in a plane. It was first suggested by Pauling and his co-workers in 1951 that the chain of planar groups can be rotated around the α-carbons (C atoms between the CO group of one amino acid and the NH group of the next) into a helix such that a hydrogen bond is formed between each C=O group and the fourth following NH group. The locations of the hydrogen and oxygen atoms involved in hydrogen bonding are illustrated in figure 11.6 (a projection of the resulting helix, called an α-helix, is shown in figure 3.5). The polypeptide backbone of the α-helix follows closely along the contour described by a helical spring made from a single strand of wire. The locations and amounts of α-helices in a protein is deduced by X-ray diffraction and NMR studies.

For a number of synthetic polypeptides, the entire chain of the molecule can assume the α-helix structure, depending on the solvent, temperature, pH, and so on. In the nonhydrogen-bonded form, the polypeptide chain is

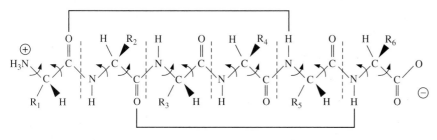

▲ FIGURE 11.6

A polypeptide chain showing the hydrogen bonding between the C=O and the NH groups in an α-helix. Residues are numbered starting from the amino terminal; the residues are separated by dotted lines. Arrows show the two bonds (on either side of each α-carbon) around which rotation can occur in the coil form.

flexible and can assume many conformations, as discussed previously. Such a chain is frequently said to be in the "coiled" form. The transition from the coiled form to the helical form (or vice versa) is referred to as a *coil–helix transition.* For polypeptides of high molecular weight, if one of the parameters (such as temperature) is changed, the transition from the coiled form to the α-helix form (or vice versa) generally occurs within a narrow range.

We can understand the basic features of the helix–coil transition using an approach that is similar to our discussion of the binding of small molecules by a polymer. The conformation of the chain can be specified by the states of the carbonyl oxygen atoms. We use the symbol h (for helix) to represent a carbonyl oxygen if it is hydrogen-bonded and the symbol c (for coil) if it is not. Whenever an oxygen is in a bonded state, it should be understood that it is hydrogen-bonded to the NH group in the fourth residue ahead in the chain (see figure 11.6). We number starting at the amino end of the polypeptide; thus, the CO in the residue 1 can bond to the NH in residue 5, the CO in residue i can bond to the NH in residue $i + 4$, and so forth. The last four residues in a chain by definition cannot be designated as h; their carbonyl groups cannot bond to an appropriate NH. It is further assumed that the sense of the helix is unique (either left-handed or right-handed, but not both). In this way, the conformation of a chain of N residues is represented by N letters, with the last four letters always c. For example, the conformation shown in figure 11.6 is designated hhccccc.

Now consider the transition from a state such as ccccccccc . . . to a state cchccccc . . . , which represents the transition of the third residue from the nonbonded state to the α-helix conformation. In the nonbonded state, there are two covalent bonds in each residue around which rotation may occur. They are the bonds on each side of the α-carbon labeled with arrows in figure 11.6. In a helical form, however, the chain assumes a much more rigid conformation. Because the oxygen atom of the third residue is hydrogen-bonded to the NH group of the seventh residue, the transition of the third residue from the coil state to the helix state means that the fourth, fifth, and sixth residues must be ordered into the helical conformation (even though their oxygen atoms are not hydrogen bonded and they are designated c). In other words, the formation of the first helical element involves the ordering of three residues; the orientation around six bonds is fixed. The transition of the state cchccccc . . . to the state cchhccccc . . . , however, requires the ordering of only one more residue, the seventh, because the fourth, fifth, and sixth are already ordered. The orientation around only two more bonds is fixed. Intuitively, then, we expect that the formation of the first helical element is more difficult than the formation of the next adjacent helical element.

Based on this discussion, we assign statistical weights according to the following rules:

1. For each c, the statistical weight is taken as unity.
2. For each h after an h, the statistical weight is s.
3. For each h after a c, the statistical weight is σs.

Note that s is the equilibrium constant for the reaction

$$\text{cchccccc}\ldots \longrightarrow \text{cchhcccc}\ldots$$

for the *addition* of a helical element to the end of a helix; σs is the equilibrium constant for the reaction

$$\text{ccccccc}\ldots \longrightarrow \text{cchccccc}\ldots$$

for the *initiation* of a helical element. We expect σ to be less than 1 because initiation of an α-helix is more difficult than propagation of the helix.

You have probably recognized the similarity between this treatment of the helix–coil transition and the previously discussed binding problem with nearest-neighbor interactions. The similarity can be made more apparent if the parameters used in the previous discussion on binding, τ and S, are transformed into σ and s by the relation $\tau = 1/\sigma$ and $S = \sigma s$.

There is one difference between the two cases, however. For binding, we consider only nearest-neighbor interactions. For an α-helix formation, the very nature of the α-helix structure dictates more than nearest-neighbor influence. For example, if the third residue goes into a helical state, the fourth, fifth, and sixth residues are brought into the helix conformation. The consequence of this is that the conformations of three adjacent residues are closely related, and conformations with two h's separated by no more than two c's are expected to be rare. To see the last point more clearly, let's consider a reaction

$$\ldots \text{hhhhccccc}\ldots \longrightarrow \ldots \text{hcchccccc}\ldots$$
$$\quad\text{123456789} \qquad\qquad \text{123456789}$$

as an example. For . . . hhhhccccc . . . , the second to seventh residues are held in helical geometry by the hydrogen bonds from residues 1 to 5, 2 to 6, 3 to 7, and 4 to 8. In hcchccccc, the second to seventh residues are still held in helical geometry by hydrogen bonds from residues 1 to 5 and 4 to 8.

Thus, for this reaction, although two hydrogen bonds are broken, there is no gain in freedom for any of the residues. Therefore, the reaction is not favored. This leads to the fourth rule for the assignment of statistical weights:

4. For two h's separated by no more than two c's, the statistical weight is zero.

To assign a statistical weight of zero is of course equivalent to saying that such conformations are not permitted. The correct statistical weight might be 10^{-7}, not 0, but the rule is a good approximation for our purpose.

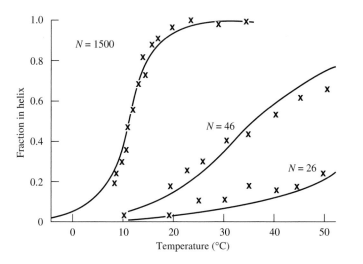

◀ FIGURE 11.7

Temperature dependence of the fraction of monomer units in the helix form for poly-γ-benzyl-L-glutamate (in a 7:3 mixture of dichloracetic acid and 1,2-dichloro-ethane as the solvent) with polymer lengths of 26, 46, and 1500 monomer units. In this solvent the helix is favored at high temperatures. (Data from B. H. Zimm and J. K. Bragg, 1959, *J. Chem. Phys.* 31:526; B. H. Zimm, P. Doty, and K. Iso, 1959, *Proc. Natl. Acad. Sci. USA* 45:1601.)

With these four rules, we can assign the statistical weight for any conformation. The statistical weight for the conformation hhhcccccchccchhcccc is $\sigma^3 s^6$, for example.

Once the rules for the statistical weights have been formulated, the helix–coil transition problem can be treated by mathematical techniques similar to the ones we have discussed for the binding of small molecules to sites with nearest-neighbor interactions. The function Z can be formulated and, in a way similar to Eq. (11.19), $(\partial \ln Z / \partial \ln s)_\sigma$ gives the average number of monomer units in the helical conformation per polymer molecule. We do not give the mathematical details here. Certain features of the coil–helix transition are discussed in the following examples.

EXAMPLE 11.4

Results of the classical treatment of Zimm et al. (1959) for the coil–helix transition of poly-γ-benzyl-L-glutamate are shown in figure 11.7. The curves shown were obtained by the statistical analysis that we outlined. The value of σ used to calculate the curves is 2×10^{-4}. As we have discussed, the parameter s represents the equilibrium constant for the change of a coil element at the end of a helical segment into a helical segment:

$$\ldots \text{hhhhhc} \ldots \underset{}{\overset{s}{\rightleftharpoons}} \ldots \text{hhhhhh} \ldots$$

If the polypeptide molecule is very long, the free-energy difference between a monomer unit in the coil form and in the helix form is zero at the temperature corresponding to the midpoint of the transitions (T_m, the melting temperature) shown in figure 11.7. Thus, s is equal to 1 at T_m. At any other temperature, s can be calculated from

$$\ln \frac{s_2}{s_1} = -\frac{\Delta H^0}{R} \left(\frac{1}{T_2} - \frac{1}{T_1} \right)$$

where ΔH^0 is the enthalpy change for the reaction shown above.

When N is large, the transition is very sharp. The polypeptide changes from 80% in the coil form to 80% in the helix form in a temperature range of $\sim7°$. In this temperature range, the change in s is rather small, from 0.97 to 1.03. The cooperative transition is the result of a small σ. If we consider the formation of a helix from a coil, a small σ means that to start a helical region in the middle of a coil region is difficult, but once a helical region is initiated, it is much easier to extend it. This can be seen from the statistical weights of the species for the following transition:

$$\ldots \text{ccccccc} \ldots \rightleftharpoons \ldots \text{ccchcc} \ldots \rightleftharpoons \ldots \text{cccchhc} \ldots$$

| Statistical weight: | 1 | σs | σs^2 |

For the first reaction, the equilibrium constant is $\sigma s/1$, or σs. For the second reaction, the equilibrium constant is $\sigma s^2/\sigma s$, or s, which is much greater than σs.

Conversely, if we consider the formation of a coil from a helix, to start a coil region in the middle of a helix is difficult, as follows:

$$\ldots \text{ccchhhhhhh} \ldots \rightleftharpoons \ldots \text{ccchhccchh} \ldots \rightleftharpoons \ldots \text{ccchhcccch} \ldots$$

| Statistical weight: | σs^7 | $\sigma^2 s^4$ | $\sigma^2 s^3$ |

The equilibrium constant for the first reaction is $\sigma^2 s^4/\sigma s^7$, or σ/s^3. The equilibrium constant for the second reaction is $\sigma^2 s^3/\sigma^2 s^4$, or $1/s$. Since in the region of interest, $s \approx 1$ and $\sigma \ll 1$, to initiate a coil region in the middle of a helix is more difficult. A consequence of this is that, for short polypeptides, coil regions are expected to be at the ends of the molecules. For long polypeptides, there are so many sites in the middle of the molecules that, in the transition zone, coil regions are expected to be present in the middle of the molecules as well.

The sharpness of the transitions shown in figure 11.7 differs for different N, although all curves can be fitted by the same parameters. For the shorter polypeptides, the transition is broader. This is similar to the binding of small molecules to interacting sites as discussed previously.

EXAMPLE 11.5

The unfolding of ribonuclease A by heating. As shown in figure 11.8, heating ribonuclease A (in 0.04 M glycine buffer, pH 3.15) causes the unfolding of the protein. The unfolding was monitored by the change in light absorption. The transition from the folded form to the unfolded form occurs in a narrow temperature range. The unfolding of a protein is more complex than the helix–coil transition of a simple polypeptide, however. The folded protein has structural features other than the α-helix. In addition, factors such as disulfide bridges, ionic bonds, and interactions between the nonbonded groups can contribute to the stability of the folded form. However, because most of these factors contribute significantly only if the protein is in a correctly folded form, the unfolded to folded, or folded to unfolded, transition is highly cooperative. In fact, the all-or-none model has been used in such studies until very recently, when kinetic studies indicated that there are more than two states involved.

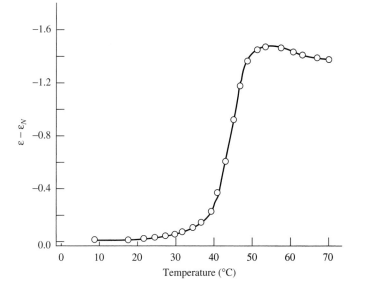

◀ FIGURE 11.8

Temperature dependence of $\varepsilon - \varepsilon_N$ of ribonuclease A. ε is the molar absorptivity of the protein at 287-nm wavelength and temperature T; ε_N is the same quantity when the protein is in its native (folded) form. (Data from J. Brandts and L. Hunt, 1967, *J. Am. Chem. Soc.* 89:4826. Copyright by American Chemical Society.)

Helix–Coil Transition in a Double-Stranded Nucleic Acid

In the preceding discussion on the helix–coil transition in an α-helix, we have seen that the difficulty in initiating the formation of a helical segment ($\sigma \ll 1$) is the essential reason for the cooperative transition. In a double-stranded nucleic acid, a somewhat similar situation exists. Double-stranded DNA is held together by the interactions between complementary bases in two single strands of DNA, as illustrated in figure 3.6. Let's represent a base pair (in the hydrogen-bonded form) by the symbol 1 and a pair of bases in the nonbonded form by the symbol 0. Let s be the equilibrium constant for the formation of an additional bonded pair at the end of a helical segment:

$$11110000\ldots \overset{s}{\rightleftharpoons} 11111000\ldots$$

If we compare the reaction above with the reaction

$$11110000\ldots \overset{\sigma s}{\rightleftharpoons} 11110100\ldots$$

we note that there is an important difference. In the first reaction, the newly formed base pair is directly "stacked" on the base pair at the end of the original helix. For polynucleotides, either DNA or RNA, the stacking of base pairs is thermodynamically favored. In the second reaction, there is little stacking interaction between the newly formed pair and the last pair in the preceding helical segment. If the equilibrium constant is σs for the second reaction, we expect that $\sigma \ll 1$ if stacking interaction is thermodynamically favored. Thus, the helix–coil transition in a double-stranded nucleic acid is similar to that in a polypeptide.

There are several new features in the nucleic acid case, however. First, let's consider the two reactions

$$11110000\ldots \rightleftharpoons 11110100\ldots$$

and

$$11110\underbrace{000\ldots0000}_{j\text{ zeros}}\ldots \rightleftharpoons 11110\underbrace{000\ldots0010}_{j\text{ zeros}}\ldots$$

If j is very large, it means that the pair of bases that are to form the additional base pair are, on the average, far apart in the nonbonded form. From Eq. (11.46), the root-mean-square average distance between this pair of nonbonded bases is proportional to $\sqrt{j}$. It is therefore expected that the larger j is, the more difficult it is to bring the two together to form a base pair. Thus, unlike the case for a polypeptide, σ for a nucleic acid is not a constant but is a function of j. The equilibrium constant for the first reaction is $\sigma_1 s$ and that for the second is $\sigma_j s$. This complicates the situation quite a bit. To compare this case with the binding of small molecules by a polymer, the dependence of σ on j is equivalent to the situation where we have to consider not only the nearest-neighbor interactions ($j = 0$) but also the next-nearest-neighbor interaction ($j = 1$), the next . . . , and so on.

Second, the formation of the very first base pair between two complementary chains means that the two chains, which are free to move about in solution independently of each other, must be brought together. After the formation of the first base pair one of the two chains can still be considered as free to be anywhere, but the second chain is constrained by the pairing to move with the first chain. Thus, for the reaction

$$\underset{\text{all zeros}}{000\ldots000} \longrightarrow \underset{\text{all zeros except one}}{0000100\ldots000}$$

the equilibrium constant is taken as κS, with κ expected to be much less than 1. The parameter κ is dependent on total concentration. The lower the concentration, the harder it is to bring two chains together, and the smaller is κ.

Third, there are two major types of base pairs in a nucleic acid. For a DNA, these are $A \cdot T$ pairs and $G \cdot C$ pairs; for an RNA, they are $A \cdot U$ pairs and $G \cdot C$ pairs. Because the stabilities of the two kinds of base pairs are different, in general, we use the parameters s_A and s_G for these types rather than a single s. Strictly speaking, the stacking interaction between two adjacent pairs is dependent on what kinds of pairs they are, and therefore s_A and s_G are further dependent on at least the nearest-neighbor base sequence.

To include all these considerations in a theoretical analysis of the helix–coil transition is beyond the scope of this text. Nevertheless, from our analyses of the problem, a number of features of the helix–coil transition in a nucleic acid can be understood:

1. When N is large, the transition from the completely helical form to the completely coiled form in a given solvent occurs within a narrow range of temperature. This cooperative process is frequently referred to as the *melting* of a nucleic acid. The temperature corresponding to the midpoint of the transition is designated as T_m, the *melting temperature*. The cooperativity of the transition is primarily a result of the stacking interactions between the adjacent base pairs, which make the initiation of a helical segment, as well as the disruption of a base pair inside a helical segment, difficult.

2. The enthalpy change for the reaction

$$11110000\ldots \overset{s}{\rightleftharpoons} 11111000\ldots$$

is negative; heat is evolved. In most of the solvents, $s_G > s_A$; thus, T_m for a nucleic acid rich in $G + C$ is higher than that for a nucleic acid rich in $A + T$ (or $A + U$). If the base composition of a nucleic acid is intramolecularly heterogeneous—that is, some segments of each molecule are richer in A + T (or A + U) than the other portions—the melting profile is broadened, with the melting of the regions richer in A + T (or A + U) preceding the melting of the (G + C)-rich regions. Because of the cooperative nature of the transition, segments heterogeneous in base composition must be sufficiently long to show independent melting profiles.

3. For a large molecule, the parameter κ contributes negligibly to ln Z. Because κ is the only concentration-dependent term in Z, T_m for a large molecule is expected to be independent of the concentration. For short helices, this is not true. T_m is lower at lower concentration. Similarly, at the same total concentration of nucleotides, T_m for a high-molecular-weight nucleic acid is higher than that of a low-molecular-weight nucleic acid of the same base composition. This dependence is evident only in the molecular-weight range where κ contributes significantly to ln Z.

4. As discussed for the helix–coil transition in a polypeptide, if N is small, the coiled regions should be at the ends of the molecules. This is frequently referred to as "melting from the ends."

EXAMPLE 11.6

The transition temperatures for the helix–coil transitions of DNAs correlate well with the base composition of the DNA. Plots of a large number of such measurements for two sets of experimental conditions are shown in figure 11.9. Analogously, for a particular DNA molecule, the range of temperatures (transition width) over which melting occurs is related to the uniformity of base composition along major segments of the chain. Synthetic DNAs with regular repeating sequences will have sharp melting transitions.

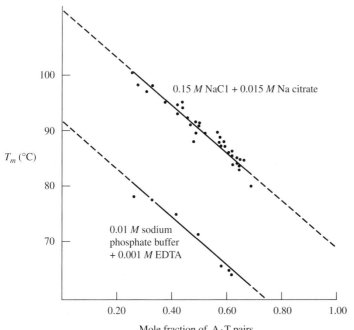

◀ FIGURE 11.9

Melting temperatures of double-stranded DNAs are plotted against the mole fraction of A·T pairs in the DNAs. In a given ionic medium, T_m decreases with increasing A·T content, indicating that $s_A < s_G$. Note also that for a given base composition, T_m is lower in a medium with a lower Na$^+$ concentration. This is because the repulsion between the negatively charged phosphate groups on the two complementary strands is decreased by the positively charged counterions (Na$^+$ in this case). Increasing the Na$^+$ concentration increases the stability of the double helix and therefore increases its T_m. (Data from J. Marmur and P. Doty, 1962, *J. Mol. Biol.* 5:109.)

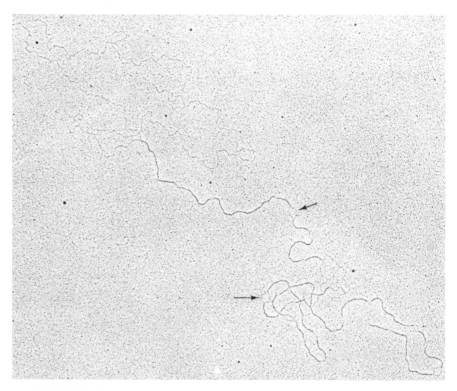

▲ FIGURE 11.10

Electron micrograph showing DNA that was partially denatured by exposure to a high pH (11.2) for 10 min, then processed for viewing. Denatured regions (such as marked by arrows) are shown as thinner and kinkier lines in the micrograph. (Courtesy of R. Inman.)

EXAMPLE 11.7

The DNA of coliphage 186 has a molecular weight of 20×10^6. The base composition of the DNA is not homogeneous along the molecule, and segments rich in $A \cdot T$ pairs can be denatured before the denaturation of segments rich in $G \cdot C$ pairs. The consequences of such inhomogeneous denaturation are seen in figure 11.10. One can see short regions where the double-stranded DNA has opened into two single strands.

EXAMPLE 11.8

The effect on T_m of shearing T2 DNA to reduce its length is seen in figure 11.11. Shorter fragments have lower T_m values and broader melting transitions.

EXAMPLE 11.9

The difference in T_m for short helices of identical base composition but different sequence shows that s is dependent not only on whether a base pair is $A \cdot U$ or $G \cdot C$ but also on the *sequence* of the base pairs (figure 11.12). The nearest-neighbor approximation to the sequence dependence of formation of DNA double strands was discussed in chapter 3. For large double-stranded DNAs and RNAs, the parameter s_A is therefore the *average s* of an $A \cdot T$ pair. Similarly, s_G is the *average s* of a $G \cdot C$ pair.

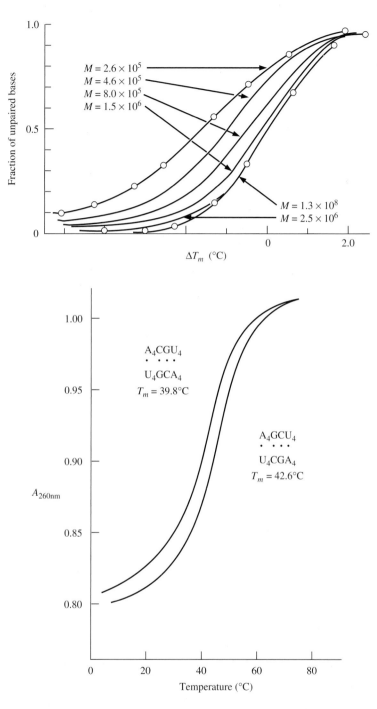

◀ FIGURE 11.11

Fraction of unpaired bases plotted against ΔT_m, the difference between the melting temperature of a DNA of molecular weight M and that of a very large DNA. For $M > 2.5 \times 10^6$, ΔT_m is very small. For DNAs with molecular weights less than 2.5×10^6, ΔT_m is clearly dependent on the size of the DNA. (Data from D. M. Crothers, N. R. Kallenbach, and B. H. Zimm, 1965, *J. Mol. Biol.* 11:802.)

◀ FIGURE 11.12

Helix–coil transition curves for two short double-stranded RNA helices. A_{260nm} is the absorbance at a wavelength of 260 nm, which was measured to monitor the helix–coil transition. The two helices have identical base composition but different sequences and the same concentration in media of identical salt concentrations. (Data from Professor Olke Uhlenbeck, University of Colorado, Boulder.)

Statistical Thermodynamics

In previous sections, we have illustrated the usefulness of statistical concepts in treating problems of biological interest. We stated at the beginning of this chapter that statistical methods also provide a bridge linking the thermodynamic properties of a macroscopic system and the molecular parameters of its microscopic constituents. In the sections that follow, we discuss certain aspects of statistical thermodynamics, which treats an equilibrium system by statistical methods.

To illustrate the basic concepts in simple terms, our *quantitative* discussions are limited to a system of noninteracting particles (that is, an ideal gas). The purpose of these discussions is to provide a molecular basis for the thermodynamic laws. We show how the macroscopic thermodynamic properties—energy E, reversible heat q_{rev}, reversible work w_{rev}, and entropy S—are related to the energy levels of the molecules. In the section "Statistical Mechanical Entropy," some qualitative examples are provided for more complex systems.

Statistical Mechanical Internal Energy

In chapter 2 we discussed the internal energy for a macroscopic system and the relation between the energy E, the heat absorbed by the system q, and the work done on the system w. A *macroscopic system*, such as 1 L of air, 1 drop of water, or 1 carat of diamond, contains many *microscopic* particles. One liter of air at room temperature and atmospheric pressure contains approximately 5×10^{21} molecules of O_2, approximately four times as many molecules of N_2, and many other molecules. A tiny drop of water 1 μm (10^{-4} cm) in diameter contains about 2×10^{10} water molecules, and 1 carat (200 mg) of diamond contains about 10^{22} carbon atoms. In chapter 9 we discussed the *energy levels* for a single molecule. For very simple systems (such as ideal gases), the energy levels can be calculated from quantum mechanics. For more complex systems, the energy levels can be obtained in principle, although the actual computation can be very difficult and, in many cases, not yet possible.

For an ideal gas, the energy levels of a molecule can be assigned with good approximation to translational, rotational, vibrational, and electronic motion. The spacing between energy levels is smallest for translational energy levels and largest for electronic levels. The values of the energy levels can be calculated from quantum mechanics. We derived in chapter 9 that for a particle of mass m in a box of dimensions a, b, and c, its energy levels are

$$E(n_x, n_y, n_z) = \frac{h^2}{8m} \left(\frac{n_x^2}{a^2} + \frac{n_y^2}{b^2} + \frac{n_z^2}{c^2} \right) \tag{9.20}$$

where the quantum numbers n_x, n_y, and n_z are integers. This equation gives the translational energy levels for a molecule contained in the box. For a monatomic ideal gas, this equation characterizes the only significant energy levels (except at high temperatures) that contribute to its thermodynamic properties. At very high temperatures (greater than 1000 K), the electronic energy levels can also become significant. For diatomic and polyatomic molecules in an ideal gas, rotational energy levels and vibrational energy levels are also important for thermodynamic properties.

Each monatomic gas molecule in a container will have an energy E_i given by Eq. (9.20). We can describe the energy distribution of the monatomic gas by specifying the number of molecules $N_1, N_2, \ldots, N_i$ in each possible energy level $E_1, E_2, \ldots, E_i$. Because the molecules of an ideal gas do not interact with one another, the total energy E of the system is simply the sum of the energies of the molecules:

$$E = N_1E_1 + N_2E_2 + N_3E_3 + \cdots + N_iE_i + \cdots$$

$$= \sum_i N_iE_i \tag{11.47}$$

The set of numbers $N_1, N_2, \ldots, N_i$ are called *occupation numbers*. Equation (11.47) tells us that the macroscopic energy of an ideal gas can be calculated from the energy levels of each molecule and the number of molecules in each energy level.

The energy of a system with interacting molecules (such as a real gas, a liquid, or solid) can still be written in the same form as Eq. (11.47). However, now the energies E_i are the energy levels of the entire system, and the N_i's are replaced by the probabilities of finding the system in each energy level. Although it is nearly impossible to do so, in principle the energy levels can be obtained by solving Schrödinger's equation for the whole system. The energy levels E_i obtained will depend on the contents of the system and the volume of the system, but not on the temperature. The temperature controls the N_i. The N_i characterize the distribution of the molecules among the energy levels for an ideal gas. For a system with interacting molecules, the N_i characterize the distribution of the whole system among its energy levels. To summarize our conclusions: Changing the temperature of a system changes the distribution of the system among its energy levels; it does not change the energy levels themselves. Changing the volume of a system changes its energy levels.

Work

Consider a gas in a container with a movable wall (a piston). Let the *external force* in the direction of movement of the piston be F_x. If we move the piston by a distance dx in the direction of the external force, the work done on the gas by the surroundings is

$$dw = F_x\, dx \tag{11.48}$$

The total force against the piston exerted by all molecules is

$$\sum_i N_i F_{ix} \tag{11.49}$$

where F_{ix} is the force in this direction exerted by a molecule in the ith energy level.

If the change dx is carried out *reversibly*, the external force is balanced by the forces exerted by the molecules:

$$F_x = \sum_i N_i F_{ix} \tag{11.50}$$

The work done by a reversible process is

$$dw_{\text{rev}} = F_x\, dx = \sum_i N_i F_{ix}\, dx \tag{11.51}$$

We can equate a force to the negative derivative of an energy. For any molecule of the ideal gas, the force F_{ix} is related to the change in its energy E_i due to the change of the dimension of the container by da. This is expressed by the equation

$$F_{ix} = -\frac{dE_i}{da} \tag{11.52}$$

Substituting Eq. (11.52) into Eq. (11.51), we obtain

$$dw_{\text{rev}} = \sum_i N_i \left(-\frac{dE_i}{da} \right) dx$$

but because $dx = -da$ (a positive dx decreases the size of the container)

$$dw_{\text{rev}} = \sum_i N_i \, dE_i \qquad (11.53)$$

Equation (11.53) states that the work done on the system for a reversible process is related to the change in the energy levels due to the change in the dimension of the container. This is true for any system; it is not limited to an ideal gas. For an ideal gas, changing the dimensions of the container affects only the translational energy. For real gases, liquids, and solids, changing the size of the system affects the molecular interactions as well.

As an example, the translational energy for a molecule in a one-dimensional box of length a is

$$E_i = \frac{h^2 n_x^2}{8ma^2} \qquad (9.17)$$

The change in energy with change in dimension is

$$\frac{dE_i}{da} = \frac{-h^2 n_x^2}{4ma^3}$$

The energy levels are raised with decreasing dimension of the box. The force exerted by each molecule in energy level i is

$$F_{ix} = \frac{-dE_i}{da} = \frac{h^2 n_x^2}{4ma^3} \qquad (11.54)$$

We can obtain the same expression for the force by a different method. As derived first in chapter 6, for a molecule with average velocity u_x, the momentum change per collision with a wall is $2mu_x$. The number of collisions between the particle and one of the walls is $u_x/2a$ per unit time. From Newton's law, the total momentum change with the wall per unit time is equal to the force exerted by the particle on the wall:

$$F_x = 2mu_x \frac{u_x}{2a} = \frac{mu_x^2}{a} \qquad (11.55)$$

However, the translational kinetic energy of the molecule is

$$\frac{1}{2}mu_x^2 = E_i = \frac{h^2 n_x^2}{8ma^2}$$

Thus,

$$\frac{mu_x^2}{a} = \frac{h^2 n_x^2}{4ma^3}$$

Substituting this expression into Eq. (11.55) gives Eq. (11.54).

Heat

Starting with Eq. (11.47), we can write the differential expression for the change of E with change of E_i and N_i.

$$dE = \sum_i N_i \, dE_i + \sum_i E_i \, dN_i \qquad (11.56)$$

Combining Eqs. (11.53) and (11.56), we obtain

$$dE = dw_{\text{rev}} + \sum_i E_i \, dN_i \qquad (11.57)$$

However, from the first law of thermodynamics, Eq. (2.12),

$$dE - dw_{rev} = dq_{rev}$$

Therefore,

$$dq_{rev} = \sum_i E_i \, dN_i \qquad \textbf{(11.58)}$$

Equation (11.58) states that the heat absorbed by a system undergoing a reversible change is related to changes in the number of molecules in the various energy levels. This immediately suggests that dq_{rev} is the part of the total energy that is related to the *distribution* of the molecules among the various energy levels. As heat is absorbed, the number of molecules in the higher-energy levels increases relative to the number in the lower. At 0 K, all the molecules are in the lowest-energy level. As the temperature is raised, the population of the higher-energy levels increases as heat is absorbed.

Most Probable (Boltzmann) Distribution

We have shown how the thermodynamic energy, the reversible heat, and the reversible work are related to the energy levels, E_i, and the number of molecules, N_i, in each energy level. In principle we can calculate the energy levels from quantum mechanics, although this is very difficult except for the simplest systems. However, to obtain thermodynamic properties we must also be able to calculate the N_i, and how the N_i change with temperature.

Equation (11.47) was written for a system of ideal gas molecules, where it is clear that because the molecules do not interact, the energy of the system is the sum of the energies of the individual molecules. The same kind of equation applies to a system containing interacting molecules, but we have to be more careful how we define E_i. It is useful to think of a large number of identical systems in thermal equilibrium; all have the same T. This is called an *ensemble* of systems. The systems of the ensemble all have the same composition, the same volume, and the same temperature. Although all systems have the same average energy (equal to the total energy of the ensemble divided by the total number of systems), each system may have a different energy level, E_i. Now E_i is the energy of a system of interacting molecules. The energy of the ensemble is given by

$$E = \sum_i N_i E_i \qquad \textbf{(11.47)}$$

but E_i is the energy of each system in the ensemble and N_i is the number of systems in energy level E_i. The average energy $\langle E \rangle$ (which is the thermodynamic energy) is

$$\langle E \rangle = \frac{\sum_i N_i E_i}{N}$$

$$N = \sum_i N_i \qquad \textbf{(11.59)}$$

There are many combinations of N_i that give the same $\langle E \rangle$, but we will find that the most probable distribution of N_i is so much more probable than all the others that we need to consider only this (most probable) distribution. For example, in an ensemble of ten systems, it is possible that nine systems have zero energy (the lowest-energy level) and one system has ten times the average energy. This corresponds to $N_i = 1$ for $E_i = 10\langle E\rangle$ and $N_i = 9$ for $E_i = 0$. This is very improbable. A much more probable distribution of energy is that a wide range of E_i values is represented among the systems of the ensemble. As the number of systems increases in the ensemble, all improbable distributions become negligible. We chose the ensemble as a thought experiment to allow us to consider the energies of interacting molecules in a system. Therefore, in our thought experiment, we can have the number of systems become infinite. This means that the most probable distribution of energies among the systems is the only one we need to consider.

Let's consider a simple ensemble where it is easy to calculate all the possible distributions of N_i. We need to find all the distributions that give an arbitrary value of average energy ε_0. To simplify the example, we choose a system in which the energy spacing is constant and equal to ε_0. Figure 11.13 shows the possible distributions of systems among energy levels in an ensemble containing one system ($N = 1$), two systems ($N = 2$), and three systems ($N = 3$). For these small ensembles, we can count all possible distributions consistent with our choice of average energy $\langle E\rangle = \varepsilon_0$. The first point to notice is that some distributions are more probable than others. For two systems in

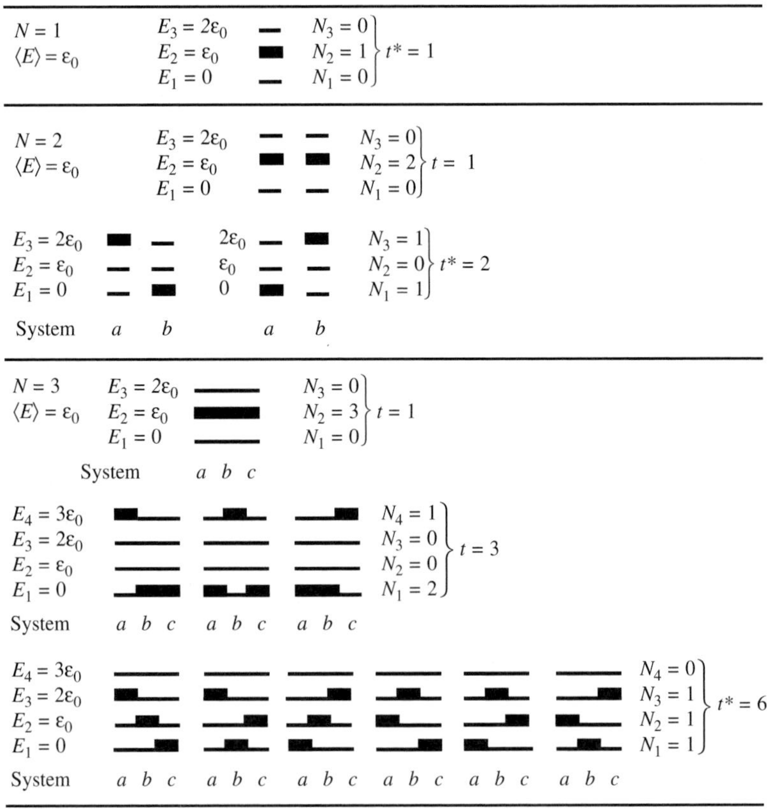

▲ FIGURE 11.13

Possible distributions of systems in an ensemble among energy levels. For simplicity the energy spacing for each system is made constant and equal to ε_0. All possible distributions are shown for three different numbers ($N = 1, 2, 3$) of systems in the ensemble. The average energy of all the systems is chosen to be ε_0. For each distribution, the number of ways (t) of forming the distribution is shown. The most probable distribution is labeled t^*. The dark boxes depict which systems (a, b, c, and so on) are in which energy level (0, ε_0, $2\varepsilon_0$, $3\varepsilon_0$, and so on). Note that for each distribution $\Sigma_i N_i = N$, the total number of systems in the ensemble, and $\langle E\rangle = \varepsilon_0$.

the ensemble, the distribution with one system in the ground level ($E = 0$) and one system in the level $E = 2\varepsilon_0$ is twice as probable as the distribution with both systems in the level $E = \varepsilon_0$. There are two ways of assigning two systems to two different levels; there is only one way of assigning two systems to one level. Similarly, for three systems in the ensemble, the distribution with one system in each of the three lowest energy levels is the most probable; there are six ways of forming it. As the number of systems in the ensemble increases, the most probable distribution (labeled t^* in figure 11.13) has increasingly many more ways of being formed than any other distribution. The most probable distribution is the only one we need consider.

For any distribution the number of ways of assigning systems to energy levels is

$$t = \frac{N!}{N_1! N_2! N_3! \dots} = \frac{N!}{\prod_i N_i!} \qquad \textbf{(11.60)}$$

where $\prod$ means the product over all i. This is the number of ways that N distinguishable objects can be arranged into boxes with N_1 in one box, N_2 in another, and so on. As seen in figure 11.13, the possible values of N_i depend on the value of the average energy chosen, and clearly the sum of N_i must equal N.

EXERCISE 11.2

Find all the possible distributions of N_i for four systems in an ensemble ($N = 4$) for the energy-level pattern in figure 11.13 and $\langle E \rangle = \varepsilon_0$. There are five distributions. Use Eq. (11.60) to calculate the number of ways of forming each possible distribution to make sure that it agrees with your direct counting. There are two equally most probable distributions: $N_1 = 1$, $N_2 = 2$, $N_3 = 1$, $N_4 = 0$, $N_5 = 0$ and $N_1 = 2$, $N_2 = 1$, $N_3 = 0$, $N_4 = 1$, $N_5 = 0$.

The most probable distribution is the one with the largest number of ways of forming it (designated t^* in figure 11.13). As N gets large, it becomes tedious to find all possible distributions and then to calculate all the t's to find the maximum. The maximum value of t in Eq. (11.60) can be found straightforwardly by setting its derivative with respect to all N_i's equal to zero. We actually find the maximum in $\ln t$. This tells us that the maximum value of t occurs for the distributions in which all $N_i = 1$; thus, $t = N!$. However, this is not the distribution we need because it does not fit the necessary constraints that the average energy $\langle E \rangle$ and the total number of systems N must be kept constant [Eq. (11.59)]. To find the maximum in t when these constraints are added requires more complicated mathematics. The result is that the most probable distribution is

$$\frac{N_i}{N} = \frac{e^{-\beta E_i}}{\sum_i e^{-\beta E_i}}$$

where β is a constant that depends on the value of the average energy of the system $\langle E \rangle$. Boltzmann derived this equation and showed that β was proportional to the reciprocal of the absolute temperature. The Boltzmann distribution (or most probable distribution) is written

$$\frac{N_i}{N} = \frac{e^{-E_i/kT}}{\sum_i e^{-E_i/kT}} \qquad \textbf{(11.61)}$$

where k is the Boltzmann constant. This equation is among the most important in statistical thermodynamics; it tells us the probability (N_i/N) of finding a system with a particular value of energy E_i. It was obtained before quantum mechanics, but it applies to quantum mechanical systems with minor modification. We can use Eq. (11.61) as it stands, if we let the sum over i refer to *quantum states* instead of quantum energy levels. An equivalent way of writing it is by explicitly including the *degeneracy* of each energy level. *The degeneracy g_i is the number of quantum states having the same energy.* The Boltzmann distribution is now

$$\frac{N_i}{N} = \frac{g_i e^{-E_i/kT}}{\sum_i g_i e^{-E_i/kT}} \qquad \textbf{(11.62)}$$

where the sum is over energy levels. This equation is also written as

$$\frac{N_i}{N} = \frac{g_i e^{-E_i/kT}}{Z} \qquad \textbf{(11.62a)}$$

where Z is the *partition function*. At the beginning of this chapter, we introduced the partition function as the sum over states:

$$Z = \sum_i g_i e^{-E_i/kT} \qquad \textbf{(11.63)}$$

When we apply Eqs. (11.61) and (11.62a) to ideal gas molecules, the E_i's are the energies of individual gas molecules, and N_i is the number of molecules with each energy E_i. Equations (11.61) and (11.62a) also apply to solids and solutions in which there is strong interaction among molecules. Now, E_i is the energy of the whole system, and (N_i/N) is the probability that the system will have that energy.

Equations (11.61) and (11.62) can be used in a very wide range of applications. From the energy-level distribution (obtained quantum mechanically) and the temperature, we can calculate the average energy (the thermodynamic energy) from Eq. (11.59):

$$\langle E \rangle = \frac{\sum_i N_i E_i}{N} = \frac{\sum_i g_i E_i e^{-E_i/kT}}{\sum_i g_i e^{-E_i/kT}} \qquad \textbf{(11.64)}$$

The ratio of the number of molecules in two energy levels is

$$\frac{N_j}{N_i} = \frac{g_j e^{-E_j/kT}}{g_i e^{-E_i/kT}} = \frac{g_j e^{-(E_j - E_i)/kT}}{g_i} \qquad \textbf{(11.65)}$$

We see that at absolute zero temperature (0 K) only the ground level is populated; all molecules (or all systems) are in the lowest-energy level because the exponential factor in Eq. (11.65) is zero. As the temperature is raised, higher-energy levels are populated, and as the temperature approaches infinity, the distribution becomes independent of the energies—the exponential factors are all equal to 1.

Equation (11.65) can be used to measure temperature. The ratio of the number of particles in two electronic energy levels can be measured from the intensities of emitted light at two different wavelengths or frequencies. The frequencies of the emitted light gives the energies of the two energy levels. Thus, the temperature is the only unknown in Eq. (11.65). The temperatures of stars have been determined with this method.

The significance of the magnitude of kT is apparent from Eqs. (11.61) through (11.65). If the energy spacing $(E_j - E_i)$ is *small* compared to kT, many energy levels will be populated. This applies to translational energy and rotational energy levels at room temperature. If the energy-level spacing is *large* compared to kT, only the lowest-energy level is significantly populated. This applies to electronic energy levels. Vibrational energy levels constitute an intermediate case, where several of the lowest-energy levels are significantly populated at room temperature. At room temperature ($T = 300$ K), kT is 4.14×10^{-21} J. It is more useful to remember the value per mole: $RT \cong 2.5$ kJ mol$^{-1} \cong 0.6$ kcal mol^{-1}.

Quantum Mechanical Distributions

The discussion of the Boltzmann distribution in the preceding section used the formula [Eq. (11.60)] for distributing distinguishable objects among energy levels. However, for quantum systems (all real systems), many objects are indistinguishable, and there are limitations on how many objects can be placed in each energy level.

For molecules, atoms, or subatomic particles, nature has set certain rules governing their distributions into the various quantum states. Particles with half-integral spin, such as electrons, protons, and ^{3}He nuclei, are characterized by the Pauli exclusion principle; that is, no two particles can be in the same quantum state. Fermi–Dirac statistics apply to these particles. Particles with integral spin, such as photons, deuterons, and ^{4}He nuclei, are not characterized by the Pauli exclusion principle. Bose–Einstein statistics represent their distributions. Boltzmann statistics developed before the quantum mechanical rules were realized. No natural particles obey Boltzmann statistics strictly. However, except at temperatures close to absolute zero and a few other special cases, all the distributions become indistinguishable. This means that the most probable Boltzmann distribution presented earlier is nearly always valid.

Statistical Mechanical Entropy

We have used the qualitative concept that entropy is a measure of the disorder of a system (chapter 3). It is then reasonable that entropy is related to t, the number of ways of distributing the molecules in a system among their energy levels (or the number of ways of distributing the systems in an ensemble among their energy levels). Boltzmann showed that the entropy is related to the most probable distribution t^*. We discussed earlier that as long as the number of particles in the system is more than a few hundred, the most probable distribution is the only one we need consider. All other distributions are negligible. Boltzmann deduced that entropy is proportional to the logarithm of the most probable distribution t^*.

$$S = k \ln t^* \qquad \textbf{(11.66)}$$

Equation (11.66) is a very important result; it provides a molecular interpretation of entropy. Let's see how this definition is consistent with the laws of thermodynamics.

As we have discussed, at ordinary temperatures t^* is very large. When the absolute temperature T approaches zero, however, the energy approaches a minimum, and all particles will be in the lowest-energy levels available to them. Thus, t^* is either unity or a very small number compared with the values of t^* at ordinary temperatures. In other words, as T approaches zero, $\ln t^*$ is either rigorously zero or vanishingly small compared with $\ln t^*$ at ordinary temperatures. From the third law of thermodynamics, S is also zero as T approaches zero for a perfectly ordered crystalline substance. From the point of view of statistical mechanics, the third law of thermodynamics is a natural consequence of the occupation of the lowest available energy levels by the particles as T approaches zero.

One statement of the second law of thermodynamics is that for an isolated system, the equilibrium state is the one for which the entropy is a maximum. From the statistical mechanical point of view, the equilibrium state of an isolated system is one that represents the most probable distribution and has the maximum randomness.

Examples of Entropy and Probability

The relation

$$S = k \ln t^* \tag{11.66}$$

is not only important in providing a microscopic interpretation of entropy, but it also contains useful qualitative insights for many problems. In this section we consider a number of quantitative examples.

EXAMPLE 11.10

Consider two chambers, each of volume V, connected through a stopcock as shown in figure 11.14. Initially, the left chamber contains n mol of an ideal gas at pressure P_0 and temperature T, the stopcock is closed, and the right chamber has been evacuated. The stopcock is opened, and the gas expands adiabatically into a vacuum. You may readily show from the discussions in chapters 2 and 3 that for this process

$$q = 0 \quad \text{and} \quad w = 0$$

and, therefore, $\Delta E = 0$, $\Delta T = 0$, and $P_{\text{final}} = P_0/2$. To calculate ΔS from

$$\Delta S = \frac{q_{\text{rev}}}{T}$$

we must carry out a reversible process that arrives at the same final conditions: temperature T and pressure $P_0/2$. This can be done by an isothermal, reversible expansion:

$$\Delta E = 0$$

$$q_{\text{rev}} = w_{\text{rev}} = nRT \ln \frac{2V}{V} = nRT \ln 2$$

▶ FIGURE 11.14

Gas, initially all contained in a volume V at pressure P_0, expands adiabatically to volume $2V$ and pressure $P_0/2$ spontaneously, with a consequent increase in entropy.

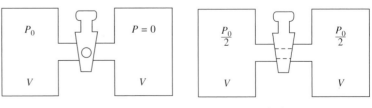

Initial state   Final state

and

$$\Delta S = nR \ln 2$$

We will now see that Eq. (11.66) gives the same result. For the process under consideration,

$$\Delta S = S_2 - S_1 = k \ln t_2^* - k \ln t_1^*$$

$$= k \ln \frac{t_2^*}{t_1^*}$$

where the subscripts 1 and 2 refer to the initial and final states, respectively.

The ratio of the number of ways that the molecules can be arranged, t_2^*/t_1^*, is equivalent to the ratio of the probabilities. Because there are more ways to distribute the gas molecules in a volume $2V$ than in a volume V, the final state is more probable than the initial state. The ratio can be evaluated as follows. Imagine that we place the molecules one by one randomly into either of the two sides. The probability that the first molecule will be in the left side is $\frac{1}{2}$. The probability that both the first and the second molecules will be in the left side is $(\frac{1}{2})(\frac{1}{2}) = (\frac{1}{2})^2$. The probability that all nN_0 molecules are in the left side is $(\frac{1}{2})^{nN_0}$. This corresponds to t_1^*. The probability that all nN_0 molecules are in either side is obviously equal to 1; this corresponds to t_2^*. Thus, $t_2^*/t_1^* = 2^{nN_0}$. Therefore,

$$\Delta S = k \ln 2^{nN_0} = nN_0 k \ln 2 = nR \ln 2$$

The problem can also be looked at in a different way. Imagine that the volume V is divided into y boxes of equal volume. Initially, the nN_0 molecules are to be distributed among the y boxes. Because the molecules of an ideal gas occupy an insignificant volume themselves, we assume that there is no limit to how many molecules can occupy each box. The number of ways of placing one molecule in y boxes is y. The number of ways t_1^* of placing nN_0 molecules in y boxes is $t_1^* = y^{nN_0}$. For the final state, there are $2y$ boxes; therefore, $t_2^* = (2y)^{nN_0}$. It follows then that

$$\Delta S = k \ln \frac{t_2^*}{t_1^*} = k \ln 2^{nN_0} = nR \ln 2$$

EXAMPLE 11.11

Suppose we have n_D mol of an ideal gas D and n_E mol of an ideal gas E, at the same temperature and pressure and separated initially by a partition, as shown in figure 11.15. If the barrier is withdrawn, mixing of the two gases occurs spontaneously. Let the volumes occupied initially by D and E be V_D and V_E, respectively. Because the gases are at the same temperature and pressure,

$$\frac{V_D}{V_E} = \frac{n_D}{n_E}$$

To calculate ΔS for this process, we can divide V_D into y_D and V_E into y_E boxes of equal size. The number of ways of placing $n_D N_0$ molecules of D in y_D boxes is

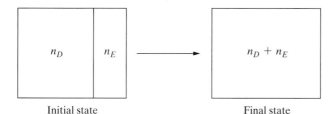

Initial state   Final state

◀ FIGURE 11.15

Two different gases, initially separated by a barrier, will mix spontaneously, and the entropy of the system will increase when the barrier is removed.

$y_D^{n_D N_0}$, and the number of ways of placing $n_E N_0$ molecules of E in y_E boxes is $y_E^{n_E N_0}$. Initially, with the barrier present, the total number of ways t_1^* is therefore

$$t_1^* = y_D^{n_D N_0} \cdot y_E^{n_E N_0}$$

If the barrier is removed, there are $y_D + y_E$ boxes for the gases, and

$$t_2^* = (y_D + y_E)^{n_D N_0}(y_D + y_E)^{n_E N_0}$$

Thus,

$$\frac{t_1^*}{t_2^*} = \left(\frac{y_D}{y_D + y_E}\right)^{n_D N_0}\left(\frac{y_E}{y_D + y_E}\right)^{n_E N_0}$$

However,

$$\frac{y_D}{y_E} = \frac{V_D}{V_E} = \frac{n_D}{n_E}$$

so

$$\frac{y_D}{y_D + y_E} = \frac{n_D}{n_D + n_E} = X_D$$

and

$$\frac{y_E}{y_D + y_E} = \frac{n_E}{n_D + n_E} = X_E$$

where X_D and X_E are the mole fractions of D and E, respectively, in the mixture. Thus,

$$\frac{t_2^*}{t_1^*} = X_D^{-n_D N_0} \cdot X_E^{-n_E N_0}$$

and

$$\Delta S = k \ln \frac{t_2^*}{t_1^*}$$

$$= -(n_D k N_0 \ln X_D + n_E k N_0 \ln X_E)$$

$$= -(n_D R \ln X_D + n_E R \ln X_E)$$

This equation can be generalized to give the ideal entropy of mixing for any number of components:

$$\Delta S = -R \sum_i n_i \ln X_i \tag{11.67}$$

This is the same result presented in chapter 3.

EXAMPLE 11.12

Certain DNA molecules, such as those from phage λ, contain single-stranded ends of complementary base sequences that enable the molecules to circularize, as illustrated diagrammatically in figure 11.16.

Let the length of the double-stranded portion of the DNA be L. Because the ends are very short, L is essentially the length of the whole molecule. For a flexible molecule of linear DNA, we have seen that the mean-square end-to-end distance is

$$\langle h^2 \rangle = Ll \tag{11.44}$$

where l is the effective segment length.

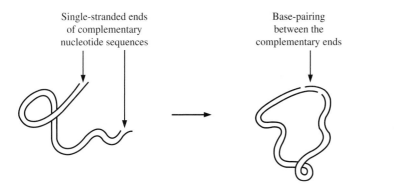

◀ FIGURE 11.16

Cyclization of a DNA molecule with "cohesive" ends. Cyclization involves a loss of entropy proportional to the logarithm of the length of the DNA.

Let's choose one end of the linear DNA molecule as the origin. The other end of the DNA will be found on the average in a sphere of volume $V = (4\pi/3)\langle h^2 \rangle^{3/2}$, whose center is the origin.

In the circular form, the two ends are constrained to be in a much smaller volume v_i. Therefore, the circularization of each DNA molecule is associated with an unfavorable (negative) entropy term:

$$\Delta S_c = k \ln \frac{v_i}{V}$$

$$= k \ln v_i - k \ln \frac{4\pi}{3} - \frac{3}{2}k \ln \langle h^2 \rangle$$

$$= \left(k \ln v_i - k \ln \frac{4\pi}{3} - \frac{3k}{2} \ln l \right) - \frac{3k}{2} \ln L$$

For a group of DNA molecules with the same cohesive ends but of different lengths, the sum of the terms in the parentheses is a constant, and ΔS_c is more negative for larger L. Thus, the longer such a DNA molecule is, the less favorable is ring formation. This has been observed experimentally, as shown in figure 11.17. The curve shown is the function predicted by the ΔS_c term. Note that at the low end of the molecular-weight scale, the discrepancy between the experimental results and the theoretical prediction is large. For such short DNA molecules, the equation $\langle h^2 \rangle = Ll$ (derived from the random-walk formula) is no longer applicable because the stiffness of the double-stranded DNA makes it difficult for short segments to circularize.

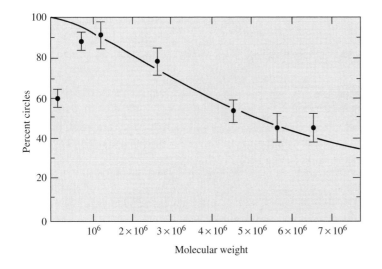

◀ FIGURE 11.17

Percentage of DNA molecules in the circular form as a function of the molecular weight of DNA. The DNA samples were obtained by treating an E. coli F-factor DNA with a restriction enzyme Eco RI, which introduces two staggered breaks at sites with base sequence

-G↓AATTC-
-CTTAA↑G-

yielding DNA molecules with cohesive ends of sequences AATT-. (Data from J. E. Mertz and R. W. Davis, 1972, Proc. Natl. Acad. Sci USA 69:3370.)

Partition Function: Applications

In our treatment leading to Eq. (11.62), we have chosen an isolated system. The partition function

$$Z = \sum_i g_i e^{-E_i/kT} \tag{11.63}$$

sums over all the energy states that a molecule can occupy and is therefore called the *molecular partition function.*

For simple molecules, the various energy levels can be obtained by quantum mechanical calculations or by spectroscopic measurements, and the partition functions can be obtained. If the partition function of a system is known, all the thermodynamic properties of the system can be calculated. For a system of N noninteracting particles, the pressure P, internal energy E, and entropy S are related to the partition function Z by the relations

$$P = NkT \left(\frac{\partial \ln Z}{\partial V} \right)_T \tag{11.68}$$

$$E = NkT^2 \left(\frac{\partial \ln Z}{\partial T} \right)_V \tag{11.69}$$

$$S = kN \ln \frac{Z}{N} + \frac{E}{T} + kN \tag{11.70}$$

EXAMPLE 11.13

We will derive Eq. (11.69). The average energy of one molecule is given by

$$\langle E \rangle = \frac{\sum_i g_i E_i e^{-E_i/kT}}{Z} \tag{11.63}$$

$$Z = \sum_i g_i e^{-E_i/kT} \tag{11.64}$$

The derivative of the partition function Z with respect to T keeping the energy levels constant is specified by keeping V constant:

$$\left(\frac{\partial Z}{\partial T} \right)_V = \frac{1}{kT^2} \sum_i g_i E_i e^{-E_i/kT}$$

We see that $\langle E \rangle$ can be written in terms of this derivative:

$$\langle E \rangle = \frac{kT^2}{Z} \left(\frac{\partial Z}{\partial T} \right)_V$$

But $\dfrac{dZ}{Z} = d \ln Z$, so

$$\langle E \rangle = kT^2 \left(\frac{\partial \ln Z}{\partial T} \right)_V$$

This is the average energy of one molecule; for N molecules, we multiply by N to obtain Eq. (11.69). For derivations of Eqs. (11.68) and (11.70), a standard physical chemistry text should be consulted.

Another important property of the partition function can be seen better if we elect to sum over all quantum states rather than energy levels; that is, the g_i degenerate states are summed up individually:

$$Z = \sum_{\substack{\text{all} \\ \text{quantum} \\ \text{states}}} e^{-E/kT} \tag{11.71}$$

To a high degree of approximation, the energy of a molecule in a particular state is the simple sum of various types of energy, such as translational energy E_{tr}, rotational energy E_{rot}, vibrational energy E_{vib}, electronic energy E_{el}, and so on. If

$$E = E_{tr} + E_{rot} + E_{vib} + E_{el} + \cdots$$

it follows immediately that

$$Z = \left(\sum e^{-E_{tr}/kT} \right)\left(\sum e^{-E_{rot}/kT} \right)\left(\sum e^{-E_{vib}/kT} \right)\left(\sum e^{-E_{el}/kT} \right)$$
$$= Z_{tr}Z_{rot}Z_{vib}Z_{el}\cdots \tag{11.72}$$

In other words, if the energy can be expressed as a sum of terms, the partition function can be partitioned into corresponding terms, the product of which gives the total partition function.

Systems other than an isolated system can also be treated by statistical mechanics. Because systems of chemical or biological interest are seldom isolated systems, it is more useful to obtain the partition functions for closed systems or open systems. It is beyond the scope of this book to treat such problems, however.

For a system consisting of complex molecules, it is not yet possible to obtain the partition function rigorously. Consider the example of a polypeptide molecule. Each amide residue in the chain is made of atoms, and each atom has its translational, and so on, energy levels. Since the atoms interact, the energy levels of one are strongly affected by many others. It is obviously not practical to obtain the complete molecular partition function. On the other hand, if we are interested only in the helix–coil transition, there is no need to know all the details of the energy levels. All we really need are the *relative* contributions to the partition function of a residue in the helix and the coil states of the molecule. These relative contributions are what we have termed *statistical weights* in a number of discussions in the first part of this chapter. In fact, several of the functions that we denoted by Z are examples of partition functions. Therefore, in the first part of this chapter, we have already given a number of examples involving the applications of statistical mechanics to complex systems of biological interest.

Summary
Binding of Small Molecules by a Polymer

Polymer molecule with N identical and independent sites for the binding of A, a small molecule:

$$\frac{\nu}{[A]} = K(N - \nu) \tag{11.12}$$

$$\frac{f}{1 - f} = K[A] \tag{11.13}$$

ν = number of bound A molecules per polymer molecule
f = fraction of sites occupied
N = number of sites per polymer molecule
K = intrinsic binding constant

Polymer with N identical sites in a linear array with nearest-neighbor interactions:

$$\nu = \left(\frac{\partial \ln Z}{\partial \ln S}\right)_\tau \qquad (11.19)$$

ν = number of bound A molecules per polymer molecule
Z = sum of statistical weights of species
$S = K[A]$
τ = cooperativity parameter

Cooperative binding:

If $\tau > 1$, the binding of an A molecule to one site makes it easier to bind another A to an adjacent site. An f versus A plot is sigmoidal in shape. If τ is very large, the binding approaches the all-or-none limit, and the predominant species have $f = 0$ and $f = 1$. In this limit,

$$\frac{d \log[f/(1-f)]}{d \log[A]} = N \qquad (11.22)$$

Anticooperative binding:

If $\tau < 1$, the binding of an A molecule to one site makes it easier to bind another A to an adjacent site. In the limit $\tau = 0$, binding to one site *excludes* binding to adjacent sites.

Random Walk and Related Topics

$$\langle h^2 \rangle = Nl^2$$

$$= Ll \qquad (11.44)$$

$\langle h^2 \rangle$ = mean-square end-to-end distance
N = number of "segments" of a polymer molecule
l = length of a segment;
    l is a measure of the stiffness of the polymer molecule
L = contour length of a polymer

$$\langle R^2 \rangle = \frac{Nl}{6} \qquad \text{(open-ended random coil)} \qquad (11.46a)$$

where $\langle R^2 \rangle$ = mean-square radius.

$$\langle R^2 \rangle = \frac{Nl^2}{12} \qquad \text{(circular random coil)} \qquad (11.46b)$$

Helix–Coil Transitions

Simple polypeptides:

The α-helix is characterized by hydrogen bonds between CO groups and NH groups in the fourth residue ahead in a polypeptide chain. The *initiation* of a helical element in a coiled region is more difficult than the addition of a helical element to the end of a helix. For polypeptides of high molecular weight, helix–coil transitions can be very sharp. Statistical methods have been used successfully to treat the helix–coil transition.

Proteins:

Native proteins are in a highly ordered or folded structure. Unfolding of a protein involves the disruption of hydrogen bonds, disulfide bridges, ionic bonds, and interactions between the nonbonded groups. The transition from a folded to an unfolded form (and vice versa) is usually highly cooperative and approaches the all-or-none limit.

Double-stranded nucleic acids:

Owing to favorable interactions ("stacking interactions") between neighboring base pairs in a double helix, the formation of an additional base pair at the end of a helical region is favored compared with the formation of a base pair in the middle of a coiled region. This leads to a cooperative transition between the helical and coiled forms. The basic features of the transition—such as the sharpness of the transition, the temperature dependence, the effect of the molecular weight of the polynucleotide, and the effect of base composition—can be understood from statistical mechanical considerations of the problem.

Statistical Thermodynamics

The most probable distribution:

The properties of a system at equilibrium is represented, to a high degree of approximation, by the properties of the most probable distribution:

$$\frac{N_i}{N} = \frac{g_i e^{-E_i/kT}}{Z} \tag{11.62a}$$

$$Z = \sum_i g_i e^{-E_i/kT} = \text{partition function} \tag{11.63}$$

$$\frac{N_i}{N_j} = \frac{g_i}{g_j} e^{-(E_i - E_j)/kT} \tag{11.65}$$

N_i = number of molecules in an energy level E_i
N_j = number of molecules in an energy level E_j
N = total number of molecules
g_i, g_j = degeneracy, the number of states with energy E_i, E_j
k = Boltzmann constant
= R/N_0 (R = gas constant and N_0 = Avogadro's number)
Z = molecular partition function, which sums up all the energy states of the molecule; for simple molecules Z can be calculated by quantum mechanics

Entropy:

$$S = k \ln t^* \tag{11.66}$$

Molecular partition function and thermodynamic functions:

All thermodynamic functions, such as P, E, and S, can be obtained from the partition function Z. If the energy of a molecule can be expressed as a sum of terms (translational energy, rotational energy, vibrational energy, electronic energy, etc.), the partition function can be factored (partitioned) into corresponding terms, the product of which gives the total partition function.

Mathematics Needed for Chapter 11

We use permutations and combinations in calculating distributions of systems among energy levels [Eq. (11.60)] and in random walks [Eq. (11.24a)]. The number of *permutations* of n objects is

$$P_n = n! = n(n-1)(n-2)\ldots\cdot 3\cdot 2\cdot 1$$

A simple example is the $3! = 6$ permutations of the three letters, a, b, c.

A combination of objects is a group of objects without respect to order. The three letters (a, b, c) form one combination. The number of *combinations* of n objects taken r at a time is

$$_nC_r = \frac{n!}{r!(n-r)!}$$

A simple example is the number of ways of arranging three pennies into groups of two heads and one tail:

$$_3C_2 = \frac{3!}{2!1!} = 3$$

Stirling's approximation for factorials is useful for large N:

$$\ln N! = N \ln N - N$$

If a group of particles can be arranged in t_1 ways and independently, they can also be arranged in t_2 ways; the total number of arrangements is $t_1 \cdot t_2$. Similarly, if the probability of an event occurring is p_1 and if the probability of an independent event occurring is p_2, then the joint probability of the two independent events occurring is $p_1 \cdot p_2$. For n independent arrangements of n independent probabilities, the expressions are

$$t = \prod_{i=1}^{n} t_i = t_1 \cdot t_2 \cdot t_3 \ldots t_n$$

$$p = \prod_{i=1}^{n} p_i = p_1 \cdot p_2 \cdot p_3 \ldots p_n$$

References

Textbooks on Statistical Mechanics

Chandler, D. 1987. *Introduction to Modern Statistical Mechanics*. New York: Oxford University Press.

Davidson, N. 1962. *Statistical Mechanics*. New York: McGraw-Hill.

Hill, T. L. 1960. *An Introduction to Statistical Thermodynamics*. Reading, MA: Addison-Wesley.

McQuarrie, D. A. 1976. *Statistical Mechanics*. New York: Harper & Row.

Nash, L. K. 1970. *Introduction to Statistical Thermodynamics*. Reading, MA: Addison-Wesley.

For useful books that treat some of the material of this chapter in more detail, see

Cantor, C. R., and P. R. Schimmel. 1980. *Biophysical Chemistry*. Part III: *The Behavior of Biological Macromolecules*. San Francisco: Freeman.

The relation between entropy and probability is discussed entertainingly in

Atkins, P. W. 1984. *The Second Law*. New York: Scientific American Books.

Bent, H. A. 1965. *The Second Law*. New York: Oxford University Press.

Suggested Reading

Ackers, G. K., M. L. Doyle, D. Myers, and M. A. Daugherty. 1992. Molecular Code for Cooperativity in Hemoglobin. *Science* 255: 54–63.

Creighton, T. E. 1983. An Empirical Approach to Protein Conformation Stability and Flexibility. *Biopolymers* 22:49–58.

Crothers, D. M., J. Krak, J. D. Kahn, and S. D. Levene. 1992. DNA Bending, Flexibility, and Helical Repeat by Cyclization Kinetics. *Methods Enzymol.* 212B:3–29.

Levene, S. D., and D. M. Crothers. 1986. Ring Closure Probabilities for DNA Fragments by Monte Carlo Simulation. *J. Mol. Biol.* 189:61–

72; Topological Distributions and the Torsional Rigidity of DNA: A Monte Carlo Study of DNA Circles. *J. Mol. Biol.* 189:73–83.

McGhee, J. D., and P. H. von Hippel. 1974. Theoretical Aspects of DNA–Protein Interactions: Cooperative and Non-Cooperative Binding of Large Ligands to a One-Dimensional Homogenous Lattice. *J. Mol. Biol.* 86:469–480.

Petsko, G. A., and D. Ringe. 1983. Fluctuations in Protein Structure from X-Ray Diffraction. *Annu. Rev. Biophys. Bioeng.* 13:331–371.

Wang, J. C. 1982. DNA Topoisomerases. *Sci. Am.* 247 (July): 94–109.

Problems

1. For a dicarboxylic acid $HO_2C-R-CO_2H$, where the two $-CO_2H$ groups are far apart, using statistical methods show that the ratio of the acid dissociation constants K_1/K_2 is expected to be ~4. (The acid dissociation constants K_1 and K_2 are the equilibrium constants for the reactions

$$HO_2C-R-CO_2H \rightleftharpoons {}^-O_2C-R-CO_2H + H^+$$

and

$${}^-O_2C-R-CO_2H \rightleftharpoons {}^-O_2C-R-CO_2^- + H^+$$

respectively.)

2. A certain macromolecule P has four identical sites in a linear array for the binding of a small molecule A. Prepare a table listing all species with half of the sites occupied (species with two occupied sites and two unoccupied sites) and their statistical weights for the following cases:

   (a) The sites are independent.

   (b) Nearest-neighbor interactions are present.

   (c) Binding to one site excludes the binding to a site immediately adjacent to it.

   (d) Give also, for part (b), an expression for the concentration ratio $[PA_4]/[PA_2]$.

3. A DNA has short single-stranded ends that can join either intramolecularly to form a ring or intermolecularly to form aggregates. Discuss briefly and concisely under what condition you expect ring formation will predominate and under what condition you expect intermolecular aggregation will predominate.

4. An oligopeptide has seven amide linkages (or eight amino acid residues, including the terminal carboxyl group). With the rules for the statistical weights that we have discussed for the formation of an α-helix, obtain the function Z, which is the sum of the statistical

weights of all species. Also obtain an expression for ν, the average number of helical residues per molecule, as a function of σ and s.

5. The H^+ titration curve of 1,2,3,4-tetracarboxyl cyclobutane

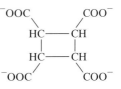

should be interesting. We will try to estimate it by using a simple, nearest-neighbor interaction model with the binding constant for a single COO^-, $K = 10^5$, and $\tau = 10^{-2}$.

   (a) There are 16 possible species involving 0, 1, 2, 3, and $4H^+$'s bound. Give the statistical weight for each in terms of τ and $S = K[H^+]$.

   (b) Write an expression for Z, the sum over states, and use Z to obtain an expression for ν, the number of H^+ bound per molecule.

   (c) Calculate $f = \nu/4$ for pH 4, pH 5, and pH 6 and compare it with the titration curve (f vs. pH) of a single COO^-.

6. Consider the coil–helix transition for a polypeptide containing 50 amides. At a certain temperature, the equilibrium constant for adding an amide to a helical region is $s = 1$; the helix initiation parameter is $\sigma = 10^{-4}$. Statistical weights are relative to the species that is 100% coil.

   (a) Write an expression in terms of σ and s for the statistical weight of the species that has all the possible amides in the α-helix conformation.

   (b) Write an expression in terms of σ and s for the statistical weight of the species that has 20 amides in the α-helix conformation in three separate helical regions.

(c) Which of all the possible species is in the highest concentration? That is, which of all the possible species has the highest statistical weight?

(d) If s were 10, which of all possible species would have the highest concentration?

(e) Roughly sketch a plot of f = fraction of amides in the α-helix conformation vs. temperature. The ΔH^0 for the reaction helix to coil is negative. Label which temperature region corresponds to the helix and which to the coil.

(f) Write an equation for f in terms of σ, s, and Z.

7. An RNA oligonucleotide has the sequence $A_6C_7U_6$. It can form a hairpin loop held together by a maximum of six $A \cdot U$ base pairs:

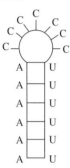

Assume that the helical region can melt only from either end. Use the notation s = equilibrium constant for adding a base pair to a helical region and $\sigma_j s$ = equilibrium constant for initiating the first base pair forming a loop. The subscript $j = 7, 9, 11, 13, 15, 17$ characterizes the number of unpaired bases in the loop.

(a) Calculate the statistical weight of each species that can be present.

(b) Calculate the mole fraction of each species that can be present.

(c) Assume that the molar absorptivity of each species depends on only the number of base pairs formed: ε_0 = absorptivity per mole of mononucleotide for species with no base pairs, ε_1 = absorptivity of all species with one base pair, etc. Write an expression for the absorbance in a 1-cm cell of a solution of $A_6C_7U_6$ in terms of s, σ_j, ε_i, and c = concentration of $A_6C_7U_6$ in moles of nucleotides per liter.

8. The statistical effective segment length l of a DNA molecule is 100 nm. Calculate the mean-square end-to-end distance $\langle h^2 \rangle$ and the contour length L for a

bacterial DNA with 10^7 base pairs. The distance between base pairs is 0.34 nm.

9. Calculate the degeneracy for each of the first five levels for a particle in a three-dimensional cubic box.

10. (a) How many 3-letter "words" can be made from 26 letters? A "word" is any sequence of 3 letters from AAA to ZZZ.

(b) How many different basketball teams of 5 players can be chosen from a group of 100 people?

(c) How many different proteins containing 100 amino acids can be made from the 20 commonly occurring amino acids?

11. Consider two systems: an electron in a cubic box of 1 nm on a side and a He atom in a cubic box of 1 cm on a side. For each system, calculate the ratio of the probabilities (N_2/N_1) of finding the particle in the first two energy levels at (a) 10 K; (b) 1000 K; (c) 10,000 K.

12. What is the change of entropy for the following reactions?

(a) 100 pennies are changed from all heads to all tails.

(b) 100 pennies are changed from all heads to 50 heads plus 50 tails.

(c) 1 mol of heads is mixed with 1 mol of tails to give 2 mol of half heads and half tails.

13. A system of a particle in a magnetic field has a simple energy-level diagram that looks like this:

$$\text{——} \quad \varepsilon_2 = 0, \qquad g_2 = 1$$

$$\text{——} \quad \varepsilon_1 = -D, \qquad g_1 = 2$$

(a) Write an expression for the partition function Z for the system. Your answer should be given in terms of D, T, and universal constants.

(b) Write an expression for the average energy of the system. Your answer should be given in terms of D, T, and universal constants.

(c) Calculate the average energy in the high-temperature limit.

(d) Calculate the average energy in the low-temperature limit.

(e) Write an expression for the ratio of the probabilities of finding the particle in the two states. Your answer should be given in terms of D, T, and universal constants.

(f) Calculate the entropy of the particle in the high-temperature limit.

CHAPTER 12

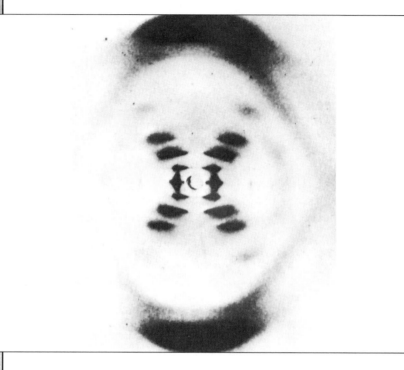

Macromolecular Structure and X-Ray Diffraction

Nearly all atomic resolution structures of proteins and nucleic acids come from X-ray diffraction. Therefore, it is important to understand how this method works and what kind of sample is needed. Light and electron microscopes have been around a long time, but scanning tunneling and scanning force microscopes were invented within the last few years.

Concepts

X rays are electromagnetic radiation of short wavelength and therefore high energy. Visible light has a wavelength range from 4,000 Å to 8,000 Å; X rays used for diffraction studies have wavelengths in the 1-Å range. Like all electromagnetic radiation, X rays are absorbed, scattered, and diffracted by matter. The scattering and diffraction of X rays is caused by interaction with electrons. Electrons in molecules in gases, liquids, or disordered solids scatter X rays. However, electrons in ordered arrays of atoms in crystals scatter X rays only in particular directions; in other directions, the scattering is negligible. *Diffraction* is the scattering of X rays in a few specific directions by crystals. The positions and intensities of the scattered beams produce a diffraction pattern. Diffraction occurs only when the wavelength of the radiation is of the same size as the periodicity in the crystal. Bond lengths in molecules, and X-ray wavelengths, are both in the 1-Å range, so diffraction can occur.

An electromagnetic wave of any wavelength is characterized by an amplitude and a phase. A lens can convert the amplitudes and phases of scattered or transmitted electromagnetic waves from an object into an image of the object. For visible light, excellent lenses exist (including the lenses in our eyes) to produce images. However, at present no lenses exist that can produce images in the 1-Å X-ray wavelength region; we must use some other method. If we could measure the amplitudes and phases of the diffracted X-ray radiation from a crystal, we could calculate an image of the molecules. Detectors of X rays and other electromagnetic radiation measure only the intensity, which is the square of the amplitude. The phase information is lost in this measurement. To obtain information about molecular structure from measured diffraction intensities, it is necessary to obtain the lost phase information by some means. This is the famous *phase problem* in X-ray diffraction. The measured diffraction pattern (diffraction intensities) is sufficient to determine the type and size of the unit cell in a crystal. The *unit cell* is the simplest repeating volume element that produces the crystal. To learn about the contents of the unit cell—the molecular structure—phase information is needed. One way to obtain this phase information is to introduce into the crystal a few heavy atoms that do not change the unit cell. This is the *isomorphous replacement method* for determining macromolecular structures.

There is a wavelength associated with any particle. Therefore, diffraction of electrons and neutrons can also be used to determine molecular structure. Electrons scatter from both electrons and nuclei; neutrons scatter only from nuclei. Magnetic lenses exist that allow focusing of electrons, thus providing images with a resolution in the 1-Å range.

Applications

X-ray diffraction provides structures of molecules with atomic resolution. Positions of atoms can be determined to 0.001 Å for small molecules and 0.1 Å for macromolecules. The most difficult part of the project is to obtain crystals that have enough order to diffract the X rays. The structures of macromolecules determined by X-ray diffraction have led to understanding of the mechanism of their biological function. The binding of oxygen by hemoglobin, and the perturbation of binding caused by a change in an amino acid that occurs in some genetic diseases, was an early example. The structure of a flu virus revealed binding sites for a drug that could inhibit replication of the virus. Rational drug design involves determining the structure of a target enzyme or enzyme–substrate complex, then designing a molecule that will bind to the active site and inhibit the enzyme. Target enzymes are those that exist in disease-causing bacteria or viruses, for example. Some cancers are caused by a single mutation in the base sequence of a DNA, which leads to a single amino acid mutation in a protein. A normal protein is thus turned into a malignant protein. One example is a membrane protein that controls cell division. The structures of the normal and malignant proteins—products of the *ras* oncogene—have been determined. The one amino acid change that distinguishes between normal and malignant is involved in GTP hydrolysis.

The structure of the first nucleic acid determined—transfer RNA (tRNA)—was vital to understanding how protein synthesis occurs. It placed the anticodon, which reads the message of the messenger RNA (mRNA), about 60 Å away from the amino acid coded for by the mRNA. There are 64 codons (four bases read three at a time = 4^3 = 64); all but three code for amino acids. The structure of the complex between a tRNA and the enzyme that specifically attaches the correct amino acid to it helps explain how this important recognition step is accomplished.

Visible Images

If macroscopic objects can be seen by their scattering of light, it seems that we should also be able to see molecules—and the detailed atomic structures within the molecules—by light scattering. There are three important properties that affect our ability to see an object, large or small: contrast, sensitivity, and resolution. *Contrast* is dependent on the amount of light scattered by the object and the amount of light scattered by other things around it. We have difficulty seeing a white fox in the snow because the object and the background scatter about the same amount of light. *Sensitivity* is dependent on the absolute amount of light scattered by the object. A dark-adapted human eye can detect a pulse of visible light containing about 50 photons, and devices of even higher sensitivity are available. *Resolution* is a measure of the spatial separation of two sources. For small enough separation, the two no longer appear separated. The average human eye, for example, has an angular resolution of about 1 minute (there are 60 min in 1 degree). Suppose we write the number 11 on a wall, with the 1's about 1 mm apart. At a distance of about 3.5 m from the wall, it will be difficult for us to tell whether the number is 11 or a fuzzy 1. At this distance, the two 1's subtend an angle of 1×10^{-3} m/3.5 m $\cong 3 \times 10^{-4}$ rad $\cong 1$ min at the eye.

Because of diffraction, even with the help of optically perfect lenses, the minimal resolvable resolution cannot be much less than the wavelength of light. Because the wavelength of light is several thousand angstroms and atomic separation in a molecule is of the order of several angstroms, to see molecular details, other types of radiation of much shorter wavelengths must be employed.

X Rays

In 1895 W. C. Roentgen accidentally discovered that when a beam of fast-moving electrons strikes a solid surface, new radiation is emitted. The new rays, which Roentgen called *X rays,* caused certain minerals to fluoresce, exposed covered photographic plates, and were transmitted through matter.

One of Roentgen's first experiments showed that different kinds of matter are transparent to X rays to different degrees. The bones and flesh of the experimenter's hand could clearly be distinguished when the hand was placed in front of a screen made of a material that fluoresced when X-rayed. Three months after Roentgen's discovery, X rays were put to use in a surgical ward. Eighteen years after the discovery, the first crystal structure (that of NaCl) was solved by X-ray diffraction. Today, X-ray diffraction is one of the most powerful techniques in studying biological structures.

Emission of X Rays

When a target is bombarded by electrons accelerated by an electric field, X rays are emitted. When electrons strike the target atoms, some are deflected by the field of the atoms with a loss of energy. This energy loss is accompanied by the emission of photons with energies corresponding to the energy differences of the incoming and the scattered electrons. Because the maximum energy of the photons emitted cannot exceed the energy of the incoming electrons, there is a sharp, short-wavelength limit of the spectrum at the accelerating voltage. At longer wavelength, spikes occur in the emission spectrum due to discrete energy levels of the orbital electrons of the target atoms. If the energy of an incoming electron is sufficiently high, it can eject an electron from an inner orbital. A photon of a characteristic wavelength is emitted when an electron in an outer orbital falls into the inner orbital to fill the vacancy left by the ejected electron. Thus, X rays are photons (electromagnetic radiation) with wavelengths in the range from tenths of angstroms to a few angstroms. X-ray spectroscopists often characterize the X rays by their energy in kiloelectron volts (keV) instead of their wavelength. Applying the formula $\Delta E = hc/\lambda$, we find that

$$\text{energy (keV)} = \frac{12.39}{\lambda\,(\text{Å})}$$

Image Formation

Because the wavelengths of X rays are of the same order of magnitude as interatomic distances in molecules, we should be able to construct an X-ray camera to provide a detailed picture of a molecule. Let's analyze first how an optical image is formed. When an object is placed in a beam of light, light is scattered by the object in all directions. Because of the wave nature of light, the scattered light in any particular direction is characterized by an amplitude term and a phase term. For example, in Eq. (10.31), E_0 represents the amplitude and $\sin 2\pi(z - ct)/\lambda$ is the phase. The pattern of the scattered waves is called the *diffraction pattern* of the object. In the presence of a lens, part of the diffraction pattern is intercepted by the lens and is refocused to give an image in the image plane. Unfortunately, at present X-ray lenses that are made of concentric rings of thin films (called Fresnel zone plates) have a resolution of only about 500 Å. Another possibility is to make a holographic image of a molecule. Here, the scattered X rays from the object are combined with a reference beam of X rays on a photographic film. A magnified image is obtained by projecting a beam of visible light through the developed photographic film (Boyer et al. 1996). Very intense sources of X rays are needed, and other technological problems need to be solved. Nevertheless, it should be clear that all the information that can be deduced from the image of an object must be present in its diffraction pattern, because a lens or hologram cannot provide any information not present in the diffraction pattern. The determination of molecular structures by X-ray diffraction therefore depends on two aspects: obtaining the diffraction patterns and determining the structures.

Scattering of X Rays

X-ray photons are scattered by electrons in matter. The scattering can be *elastic*, in which the wavelengths of the incident and the scattered radiation are the same, or *inelastic*, in which the wavelengths differ. If inner orbital electrons are ejected by the incident radiation and secondary radiation is emitted when electrons in outer orbitals fall back to fill the vacancies, the emitted secondary radiation is referred to as *fluorescence scattering*. In the following discussions, we consider only elastic scattering. Elastic scattering gives rise to diffraction patterns; inelastic scattered light does not.

As illustrated in figure 12.1(a), light incident on a single electron at a point is scattered. For light polarized perpendicular to the plane of scattering (the plane of the figure), the light is scattered with equal intensity in all directions in the plane. The scattered intensity I_{sca} decreases with the inverse square of the distance from the scattering point; it is proportional to the incident intensity I_0. For one electron at a point

$$I_{sca} = \frac{CI_0}{r^2} \tag{12.1}$$

the distance from the scattering point to the detector is r; C is a proportionality factor that depends on the charge and mass of the electron.

In figure 12.1(b), light is scattered from two electrons. Each electron scatters the light independently in all directions, but the scattered intensity is not

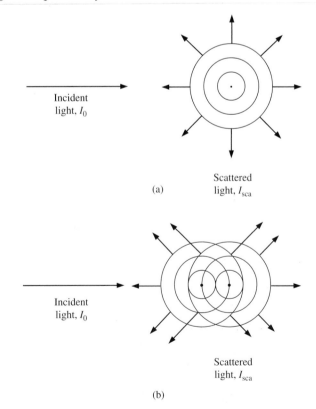

(a)

(b)

▶ FIGURE 12.1

(a) The scattering from a single electron for light polarized perpendicular to the scattering plane is independent of angle in the scattering plane. The scattering intensity decreases inversely proportional to the square of the distance from the scattering electron. (b) For scattering from two electrons, the angle dependence of the scattering amplitude depends on the interference between the two scattered waves.

the sum of the intensities from each electron because of interference between the scattered waves. We must first sum the amplitudes of the waves scattered by each electron; the square of this sum is the scattered intensity. The scattered intensity will still have an r^{-2} dependence, but its angle dependence depends on the distance between the electrons and on their orientation relative to the direction of incident light. Thus, from the scattering pattern, we learn about the structure of the scattering object.

We can calculate the scattered intensity from two electrons at any angle in the plane by simple geometric construction. We need to find the amplitude of the scattered wave at the detector, which is at a large distance r, from the two electrons. The amplitude is the sum of the amplitudes contributed by the two electrons. The amplitude of the wave scattered from each electron will be different because of the difference in distance traveled by each wave. From Eq. (10.31), we know that the electric field (amplitude) of a light beam, or X-ray beam, of wavelength λ traveling along direction z and polarized along x can be written as

$$E = E_0 \sin \frac{2\pi}{\lambda}(z - ct) \qquad (10.31)$$

The intensity E^2 at the detector depends on the path difference for the two scattered beams. We can thus simplify our notation by setting $E_0 = 1$ and $t = 0$. The amplitude of the wave scattered from electron 1 is chosen to be proportional to $\sin 2\pi z/\lambda$; then the amplitude scattered from electron 2 is proportional to $\sin 2\pi(z + \Delta)/\lambda$, where Δ is the path difference for the two rays between source and detector. Figure 12.2 shows how this path difference de-

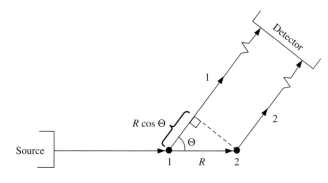

Path difference $\Delta = R \cos \Theta - R$

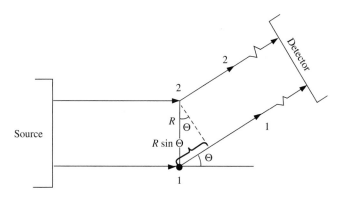

Path difference $\Delta = R \sin \Theta$

◀ FIGURE 12.2

Path difference for two beams scattered from two electrons separated by distance R. The angle between the incident beam and the scattered beam is Θ. At the top, the two scatterers are on a line with the incident beam; at the bottom, the scatterers are perpendicular to the incident beam direction. Trigonometry shows that the path difference Δ is related to the scattering angle Θ and the distance between scattering points R.

pends on the distance between the two electrons R, the orientation of the two electrons relative to the incident beam, and the scattering angle Θ. The phase difference between the two beams is $2\pi\Delta/\lambda$.

The two scattered rays add at the detector to give a wave of amplitude A and phase difference $2\pi\delta/\lambda$:

$$A \sin \frac{2\pi}{\lambda}(z + \delta) = \sin \frac{2\pi z}{\lambda} + \sin \frac{2\pi}{\lambda}(z + \Delta) \tag{12.2}$$

Equation (12.2) states that a sum of sine waves is equal to a new sine wave whose amplitude A and phase δ depend on the amplitudes and phases of its components. However, the intensity, which is what the detector measures, is equal to A^2; therefore, to calculate a scattering pattern, we need only to find the amplitude of the combined beams A. We use a trigonometric identity to rewrite Eq. (12.2):

$$\sin (\alpha + \beta) = \cos \beta \sin \alpha + \sin \beta \cos \alpha$$

$$A \cos \frac{2\pi\delta}{\lambda} \sin \frac{2\pi z}{\lambda} + A \sin \frac{2\pi\delta}{\lambda} \cos \frac{2\pi z}{\lambda}$$

$$= \left(1 + \cos \frac{2\pi\Delta}{\lambda}\right) \sin \frac{2\pi z}{\lambda} + \sin \frac{2\pi\Delta}{\lambda} \cos \frac{2\pi z}{\lambda} \tag{12.3}$$

Equation (12.2) and thus Eq. (12.3) must be true for all values of z because z is a variable that can take on any value. This means that the terms multiplying $\sin(2\pi z/\lambda)$ in Eq. (12.3) are equal to each other, and the terms multiplying $\cos(2\pi z/\lambda)$ in Eq. (12.3) are equal to each other.

$$A \cos \frac{2\pi\delta}{\lambda} = 1 + \cos \frac{2\pi\Delta}{\lambda}$$

and

$$A \sin \frac{2\pi\delta}{\lambda} = \sin \frac{2\pi\Delta}{\lambda}$$

By squaring the two equations and adding them, we can obtain A^2:

$$A^2 \cos^2 \frac{2\pi\delta}{\lambda} + A^2 \sin^2 \frac{2\pi\delta}{\lambda} = \left(1 + \cos \frac{2\pi\Delta}{\lambda}\right)^2 + \sin^2 \frac{2\pi\Delta}{\lambda}$$

But

$$\sin^2 \alpha + \cos^2 \alpha = 1$$

Therefore,

$$A^2 = 2\left(1 + \cos \frac{2\pi\Delta}{\lambda}\right) \tag{12.4}$$

We have found that the sum of two waves, each of which has amplitude equal to 1, gives a wave with an intensity (the square of the amplitude) depending on the phase difference between the waves. Therefore, the intensity scattered by two electrons is not twice Eq. (12.1); it is

$$I_{sca} = 2\frac{CI_0}{r^2}\left(1 + \cos \frac{2\pi\Delta}{\lambda}\right) \tag{12.5}$$

where $2\pi\Delta/\lambda$ is the phase difference α, for the wave scattered from the two electrons. Note that the phase difference depends on the path difference Δ, divided by the wavelength. The path difference can be calculated geometrically as illustrated in figure 12.2.

Let's consider the general behavior of Eq. (12.5) because it illustrates how interference between scattered waves changes the intensity. If $\Delta = 0, \lambda,$ $2\lambda, \ldots$, the phase difference $\alpha = 0, 2\pi, 4\pi, \ldots$, and $\cos \alpha = 1$. Thus, the intensity of scattering from two electrons is four times the scattering from one electron. The scattering amplitude is proportional to the number of electrons; the maximum scattering intensity is proportional to the square of the number of electrons. This is called *total constructive interference*, or *total reinforcement*. If $\Delta = \lambda/2, 3\lambda/2, 5\lambda/2, \ldots, \alpha = \pi, 3\pi, 5\pi, \ldots$, and $\cos \alpha = -1$. Thus, the intensity of scattering from two electrons is zero. This is *total destructive interference*. It is clear that it is the ratio of path difference to wavelength, and thus the phase difference, that determines the intensity. Figure 12.3 illustrates this summing of waves.

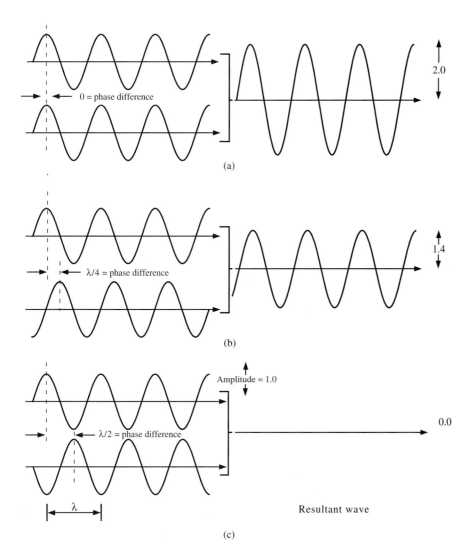

(a)

(b)

(c)

Resultant wave

◀ **FIGURE 12.3**

Summing of two waves: (a) waves completely in phase; (b) waves out of phase by one-quarter wavelength (partial reinforcement occurs); (c) waves out of phase by one-half wavelength (total destructive interference occurs). [From J. P. Glusker and K. N. Trueblood, 1972, *Crystal Structure Analysis: A Primer* (New York: Oxford University Press), p. 19, fig. 5.]

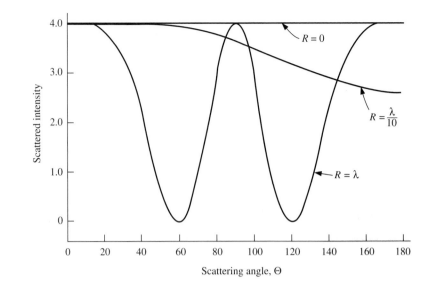

► FIGURE 12.4

Scattered intensity as a function of scattering angle Θ for two-point charges collinear with the incident beam (figure 12.2, top). Equation (12.5) is used to do the calculation. For $R = 0$, the scattered intensity is independent of angle and is four times that of a single charge. As R increases relative to λ, more structure appears in the scattering pattern. The patterns are symmetric around $0°$, so only half the pattern is shown. Note that at $0°$ there is always complete constructive interference.

Figure 12.4 shows the scattering as a function of angle for radiation incident along the line between the two points, as shown in figure 12.2 (top). For this geometry, the path difference is $\Delta = R \cos \Theta - R$; R is the distance between the points, and Θ is the angle between the incident beam and the scattered beam. Equation (12.5) was used to calculate the scattering pattern (angle dependence of the scattering) for these different ratios of R/λ. The patterns are symmetric on either side of zero degrees, so only half the pattern is given. For $R = 0$, or very small relative to the wavelength, the phase difference is negligible, and the scattering is maximum and independent of angle. For $R = \lambda$, the scattering is maximum at $0°$, $90°$, and $180°$; it is zero at $60°$ and $120°$. For $R = \lambda/10$, a small angle dependence is seen. It should be evident that analysis of the scattering patterns of X rays with wavelengths near 1 Å (0.1 nm) can provide accurate measurements of interatomic distances.

Electrons are not points, but electrons in a molecule can be characterized by an electron density at each point in the molecule. The electron density is the number of electrons (or fraction of an electron) in a small volume element around the point. The amplitude of scattered light is proportional to the electron density at each point. To determine molecular structure, we need to know the positions of the nuclei in the molecule. Each nucleus is surrounded by a cloud of electrons, so we must know how this cloud will contribute to the scattering from a molecule.

The scattering from an atom is characterized by the atomic scattering factor f_0, which is the ratio of the amplitude scattered by the atom to the amplitude scattered by a point electron. This is illustrated in figure 12.5 for a three atoms; the relative amplitude of scattering f_0 is plotted versus $(\sin \theta)/\lambda$; θ is half the scattering angle Θ. A spherical electron-density distribution always gives this type of angle dependence. At zero scattering, angle f_0 is equal to the number of electrons in the atom because there is always total constructive interference in the forward direction. It is clear why hydrogen atoms are difficult to locate accurately; their maximum scattering intensity is $\frac{1}{36}$ the scattering intensity of carbon atoms and $\frac{1}{225}$ the scattering intensity of phosphorus atoms. Similarly, a heavy atom such as mercury (atomic number = 80) will have $(\frac{80}{6})^2 = 178$ times the scattering intensity of carbon.

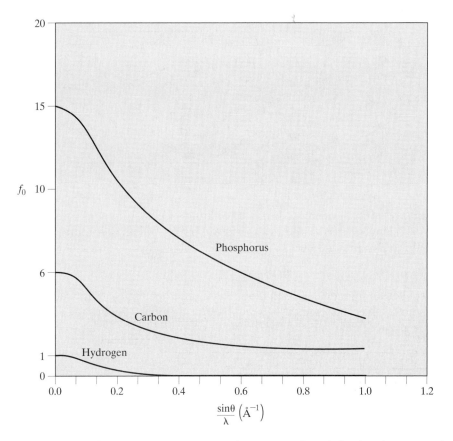

Dependence of atomic scattering f_0 on $(\sin\theta)/\lambda$. The angle between the incident and scattered rays is $\Theta = 2\theta$. The curves shown are for hydrogen (atomic number 1), carbon (atomic number 6), and phosphorus (atomic number 15). Note that when θ approaches zero, the value of f_0 approaches the number of electrons per atom.

In a crystal, the atomic scattering factor is reduced further because of thermally induced vibrations of the atom. This temperature effect can be corrected by multiplying f_0 by a temperature-dependent factor. The symbol f is used for the corrected scattering factor.

Diffraction of X Rays by a Crystal

In the preceding section, we showed how the scattered intensity depends on the distance between two points. We can use exactly the same reasoning to calculate the scattered intensity from a crystal. A crystal is a periodic array of identical scattering elements that differ only in their position within the crystal.

An X-ray diffraction experiment is done by placing a crystal, with dimensions of a millimeter or less, in the X-ray beam. The scattered intensity is measured as a function of angle from the incident beam. The positions of scattered intensity depend on the orientation of the crystal lattice relative to the direction of the incident X-ray beam. By rotating the crystal and measuring the angles and intensities of the scattered radiation, a diffraction pattern is obtained.

Let's consider first the diffraction of X rays by a linear array of identical *point* scatterers that are equally spaced, with a distance a between each adjacent pair. As illustrated in figure 12.6, we let α_0 be the angle the parallel incident rays make with the array. The incident rays are scattered in all directions. In a given direction α, the difference in path of rays scattered by two adjacent scatterers is

$$\text{path difference} = a(\cos\alpha - \cos\alpha_0)$$

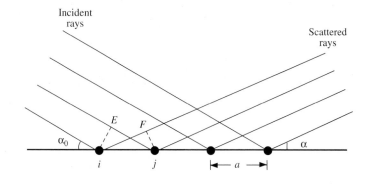

Scattering of incident rays by a row of point scatterers of spacing a. The dashed line iE is perpendicular to the incident beam; the dashed line jF is perpendicular to the scattered beam. The path difference of rays scattered by any two adjacent scatterers i and j is thus $(iF - jE)$, which is equal to $a \cos \alpha - a \cos \alpha_0$, or $a(\cos \alpha - \cos \alpha_0)$.

as illustrated in figure 12.6. If this path difference is an integral multiple of the wavelength, then *all* the rays scattered by the points will reinforce. This condition is expressed by

$$a(\cos \alpha - \cos \alpha_0) = h\lambda \tag{12.6}$$

where h is an integer. This equation tells us at what angles, α, diffraction occurs for a given spacing, a, between scattering points in the crystal and a given angle of incidence, α_0, of the X-ray beam. For a fixed value of a and a fixed incident angle α_0, there are several values of α at which constructive interference occurs, corresponding to $h = 0, 1, 2, \ldots$. For all other values of α, rays scattered from any two scatterers will be out of phase. The resultant wave from a large number of scattered rays with random phases has zero amplitude. Because the number of scattering points N is very large, we need consider only the angles of total constructive interference. At these angles, the scattered intensity is proportional to N^2; at other angles, the scattered intensity is negligible. In summary then, when parallel monochromatic X rays strike a row of equidistant point scatterers at incident angle α_0, the scattered radiation will concentrate in a number of cones, each cone with a single α given by Eq. (12.6). A special case with $\alpha_0 = 90°$ is illustrated in figure 12.7.

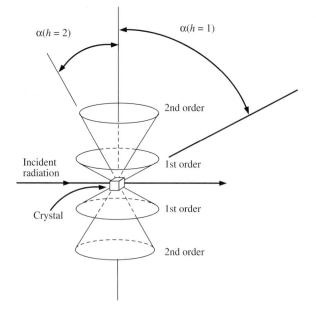

Equidistant point scatterers arranged in the vertical direction at the apices of the cones. With an incident angle $\alpha_0 = 90°$, Eq. (12.6) reduces to $a \cos \alpha = h\lambda$, or $\alpha = \cos^{-1}(h\lambda/a)$. Cones with $h = 1$ (first order) and $h = 2$ (second order) are shown. Note that $\cos \alpha = \cos(-\alpha)$; thus, in this particular case, for each order there are two cones 180° apart. The zero-order diffraction ($h = 0$) is not shown.

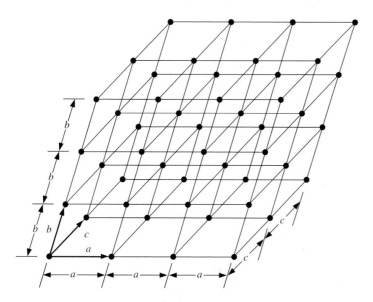

◀ FIGURE 12.8

Lattice of points. The lattice axes are depicted at the lower left corner.

We should note that inelastically scattered rays, because of changes in λ, will be more or less randomly distributed spatially. They contribute to the background intensity but are not important for our discussions.

We can now generalize our treatment to the case of a three-dimensional crystal lattice. Figure 12.8 illustrates a simple lattice formed by three sets of equidistant planes. The points of intersection of these planes form the space lattice. The characteristic interplanar distances are a, b, and c, as indicated. We can also consider this simple lattice as generated by three sets of fundamental translations along three nonplanar axes a, b, and c.

There are only seven fundamental crystal lattices possible. For the most general case, the interaxial angles are unequal, and so are the values of a, b, and c. The lattice is called *triclinic*. If the interaxial angles are all 90° and $a \neq b \neq c$, the lattice is called *orthorhombic*; if the interaxial angles are 90° and $a = b = c$, the lattice is *cubic*. The other four lattices are *monoclinic* (two angles = 90°, $a \neq b \neq c$), *rhombohedral* (all angles equal, but not 90°, $a = b = c$), *tetragonal* (all angles = 90°, $a = b \neq c$), and *hexagonal* (one angle = 120°, two angles = 90°, $a = b \neq c$).

If we have an identical scatterer located at each lattice point in a triclinic lattice, it is easy to generalize Eq. (12.6) to see that total reinforcement occurs if the following *three* equations are satisfied:

$$a(\cos \alpha - \cos \alpha_0) = h\lambda \tag{12.6}$$

$$b(\cos \beta - \cos \beta_0) = k\lambda \tag{12.7}$$

$$c(\cos \gamma - \cos \gamma_0) = l\lambda \tag{12.8}$$

These equations were first derived by Max von Laue and are referred to as *von Laue's equations.* The angles α_0, β_0, and γ_0 and α, β, and γ are the angles that the incident and diffracted beams make with the three axes, respectively.

The diffraction pattern from a crystal will thus consist of spots of scattered intensity whose positions depend on the crystal lattice. The diffraction pattern from a crystalline transfer RNA is shown in figure 12.9.

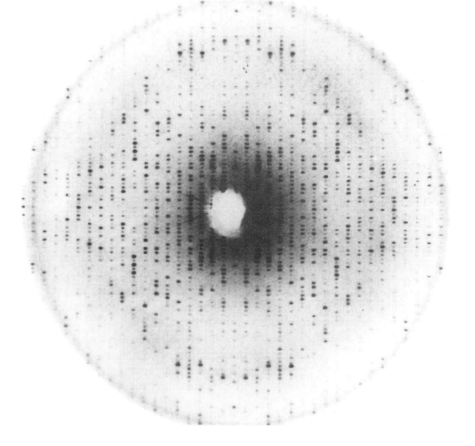

Diffraction pattern for a crystal of yeast phenylalanine transfer RNA. The X-ray beam from a Cu K_α source ($\lambda = 1.54$ Å) is incident perpendicular to the film. A hole is cut in the film to reduce the overexposure caused by the very intense incident beam compared to the diffracted beams. The crystal has been rotated to produce what is technically labeled as a 15° precession photograph. The size of the unit cell is $a = 33$ Å, $b = 56$ Å, $c = 161$ Å. (The photograph was kindly supplied by Prof. Sung-Hou Kim, University of California, Berkeley.)

Measuring the Diffraction Pattern

Diffraction occurs only if all three von Laue's equations [Eqs. (12.6–12.8)] are satisfied. There is a further constraint on the diffraction direction because of geometric constraints on the angles α, β, and γ, caused by the crystal lattice. For example, for an orthorhombic lattice,

$$\cos^2 \alpha + \cos^2 \beta + \cos^2 \gamma = 1$$

Only at very special angles of beam incidence and scattering angles is constructive interference found; four equations must be satisfied.

This discussion shows that to measure the diffraction pattern, a special arrangement must be made. We have several choices:

1. Instead of using a monochromatic incident beam, we an use "white" X rays, which contain a broad distribution of wavelengths. For certain values of λ, the four equations may be satisfied. This option is of historical interest because, when von Laue and his coworkers did their first experiment on X-ray diffraction by a crystal, little was known about X rays and the X-ray source they used gave a broad spectrum of radiation. Now monochromatic radiation is almost always used in X-ray diffraction studies. However, use of a range of X-ray wavelengths for diffraction experiments has been reintroduced to determine structures of macromolecules undergoing rapid changes. The method

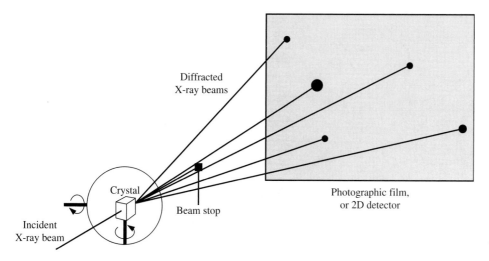

▲ FIGURE 12.10

Experimental arrangement for measuring the X-ray scattering from a crystal. A beam of X rays is incident on a crystal that is mounted so that it can be rotated around perpendicular axes. The intensities of the diffracted X-ray beams are measured with a photographic film or an equivalent imaging plate. A beam stop is used to attenuate the incident X-ray beam.

is called *Laue diffraction*. It uses pulses of X rays from a synchrotron with wavelengths that range, for example, from 0.6 Å to 1.5 Å. Structural information can be obtained for reactions that occur on the nanosecond scale (Stoddard 1998).

2. If monochromatic radiation is used, characteristic diffraction rays can be observed if the orientation of the crystal relative to the incident beam is systematically changed. This can be achieved by rotating or oscillating a crystal about one axis of the crystal at a time or by changing the position of the detector (a photographic film or, most commonly, a solid-state detector called an imaging plate). Figure 12.10 illustrates one experimental arrangement. Sometimes the rotation or the oscillation of the crystal is coupled to the movement of detector so as to obtain a diffraction pattern that is easier to interpret. A particular mode of coupling can be achieved mechanically, or a computer can be programmed to control the movements.

3. If monochromatic radiation is incident upon a powder of little crystals randomly oriented, a fraction of the crystals will have the correct orientation with respect to the incident beam to give characteristic diffractions. Such a picture is known as a *powder pattern*; it consists of a set of concentric rings.

Bragg Reflection of X Rays

So far we have mentioned only that the parameters h, k, and l are integers. The physical meaning of h, k, and l became clear primarily through the work of W. L. Bragg. He showed that the diffraction of an X-ray beam acts as if the X rays are reflected from planes in the crystal. Many groups of parallel planes can be drawn through the lattice points—the positions of the atoms—in a crystal, as shown in figure 12.11. Each group of parallel planes is characterized by three integers h, k, and l; they are the same integers seen in von Laue's

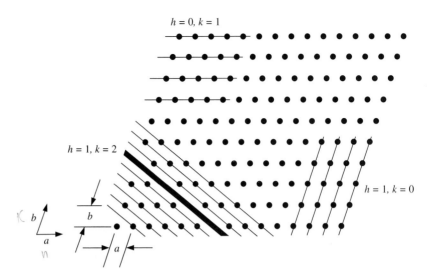

▶ **FIGURE 12.11**

Two-dimensional lattice with some possible lattice lines and their corresponding Miller indices illustrated.

equations [Eqs. (12.6–12.8)]. These integers are known as the *Miller indices* of the planes. They are important because the position of each diffracted beam depends on the group of parallel planes that scatter constructively at this angle. Each spot in figure 12.9 can be labeled by three Miller indices; this is the first step in determining a structure from a diffraction pattern.

Let's see how the Miller indices are obtained. For simplicity, a two-dimensional lattice and several groups of parallel lines going through the lattice points are illustrated in figure 12.11. Each group of parallel lines can be indexed in the following way: Pick any line such as the heavy line in the figure. The intercepts that this line makes with the a-axis and b-axis are $6a$ and $3b$, respectively. The reciprocals of the coefficients are $(\frac{1}{6}, \frac{1}{3})$. Now multiply these reciprocals by the smallest integer that will give integral indices. In this example, the smallest integer is 6, and the indices obtained are (1,2). These indices are called the *Miller indices*. The steps are summarized below:

	Axis	
Steps	a	b
1. Intercepts	6	3
2. Reciprocals	$\frac{1}{6}$	$\frac{1}{3}$
3. Miller indices	1	2

EXERCISE 12.1

Pick a few arbitrary lattice lines parallel to the one we discussed above. Convince yourself that these parallel lines all have the same Miller indices.

The distance d_{hk} between any two adjacent parallel lines is determined by their Miller indices. For example, if the **a**-axis and **b**-axis are perpendicular to each other, it is easy to show that

$$d_{hk} = \frac{1}{\sqrt{\dfrac{h^2}{a^2} + \dfrac{k^2}{b^2}}}$$

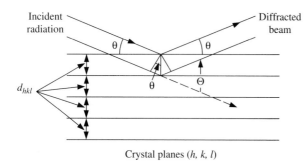

Crystal planes (h, k, l)

Bragg condition: $n\lambda = 2d_{hkl} \sin \theta$

▲ FIGURE 12.12

Diffraction of radiation from a crystal. The parallel lines represent planes of atoms with Miller indices h, k, and l. The Bragg condition for diffraction is for the incident beam and the diffracted beam to make an angle $\theta = \sin^{-1}(n\lambda/2d_{hkl})$ with the planes. Note that the scattering angle (Θ) between incident and scattered beams (defined before in figure 12.2) is 2θ. Simple geometric construction shows that the extra path length traveled by the ray scattered by the top layer relative to the next layer down is $2d_{hkl} \sin \theta$.

If the interaxial angle is different from 90°, the expression for d_{hk} is more cumbersome unless vector–algebra notation is used.

The situation is not much different for a three-dimensional lattice. Any group of parallel planes can be represented by three Miller indices (h, k, l). The interplanar spacing d_{hkl} can be calculated from the indices (for an orthorhombic lattice, $d_{hkl} = 1/[(h/a)^2 + (k/b)^2 + (l/c)^2]^{1/2}$).

For a family of parallel crystal planes (h, k, l) with spacing d_{hkl}, a monochromatic incident beam of wavelength λ will *appear* to be reflected by the planes if the incident angle θ is

$$\theta = \sin^{-1} \frac{n\lambda}{2d_{hkl}} \qquad (12.9)$$

where n is an integer 0, 1, 2, ... To state it in different words, if the incident rays make an angle θ with a family of crystal planes (h, k, l) such that

$$n\lambda = 2d_{hkl} \sin \theta \qquad (12.10)$$

reinforcement of the scattered rays occurs in a direction that also makes an angle θ with the planes. Equation (12.9) or (12.10) is called the *Bragg law*; it is illustrated in figure 12.12. The Bragg law and von Laue's equations can be shown to be equivalent.

We can write the Bragg law in the form

$$2\frac{d_{hkl}}{n} \sin \theta = \lambda \qquad (12.11)$$

The quantity (d_{hkl}/n) can be viewed as the spacing between a family of planes with Miller indices (nh, nk, nl). The higher-order Bragg diffractions can be viewed as coming from these planes.

Intensity of Diffraction

So far we have considered the diffraction of X rays by a lattice of identical *point* scatterers. Von Laue's equations or the equivalent Bragg equation provide the relation between the diffraction pattern and the lattice parameters.

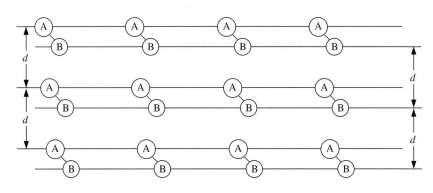

▶ FIGURE 12.13

Three rows of diatomic molecules AB in a crystal. The distance d between adjacent rows of A atoms is necessarily the same as the distance d between rows of B atoms. The distance between adjacent rows of A and B atoms is not the same as d.

For a point scatterer, the intensity of the scattered ray is independent of the scattering angle. If we have a real atom instead of a point scatterer, the intensity of the scattered ray is a function of $(\sin \theta)/\lambda$, as we have discussed already (see figure 12.5). Because constructive interference occurs only at $(\sin \theta)/\lambda = n/2d_{hkl}$ [Eq. (12.11)], the atomic scattering factor for a group of planes of indices (h, k, l) is independent of wavelength. The intensities of rays scattered by different groups of planes differ, of course, because of differences in $(\sin \theta)/\lambda$ values.

The situation is more complex when the lattice contains several kinds of atoms. The lattice still determines the diffraction pattern, but the intensity at each diffraction spot depends on the location of the atoms within the lattice. Analysis of this intensity distribution is what leads to the molecular structure of molecules in a crystal.

Suppose we have a crystal of diatomic molecules AB as illustrated in figure 12.13. Consider the rows of A atoms first. The reinforcement of scattered rays occurs at angles

$$\theta = \sin^{-1} \frac{n\lambda}{2d}$$

where d is the spacing between the rows. For the rows of B atoms, the reinforcement of scattered rays occurs at the same angles

$$\theta = \sin^{-1} \frac{n\lambda}{2d}$$

because the spacing between the rows of B atoms must be the same as the spacing between the rows of A atoms in the crystal. The spacing between a row of A atoms and a row of B atoms is different from d, however. This means that rays scattered by rows of A and rows of B atoms are not in phase at the angle θ. This results in partial interference, and therefore the total intensity at the angle θ is less than the sum of the intensities due to independent scattering by A atoms and by B atoms. At a given Bragg angle, the intensity of rays diffracted by the lattice of A atoms is dependent on f_A, the atomic structure factor for A; the intensity of rays diffracted by the lattice of B atoms is dependent on f_B, the atomic structure factor for B. In addition to its dependence on f_A and f_B, the total intensity of rays diffracted by the two interpenetrating lattices is also dependent on the relative *phases* of rays diffracted by the two lattices. This is exactly analogous to the scattering from two electrons; the scattered intensity depends on the phase difference between the two scattered waves [Eq. (12.5)].

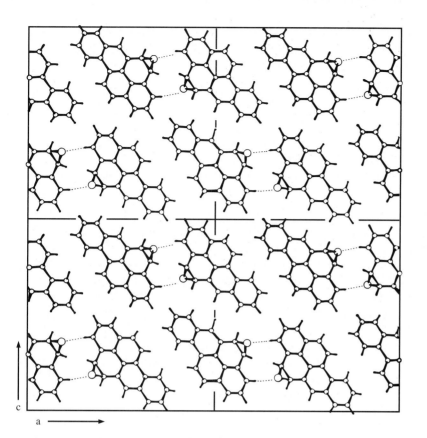

◄ FIGURE 12.14

Crystal structure of benzo[a]-pyrene-4,5-oxide. The molecules form planar layers. Four unit cells are shown; there are four molecules per unit cell. (See Glusker et al. 1976; figure courtesy of Dr. Jenny P. Glusker.)

To generalize from the scattering of two different atoms to the scattering from a molecule it is convenient to use the concept of a unit cell.

Unit Cell

Any crystal can be considered as formed by placing a basic structural unit on every point of a lattice. (The crystal so formed is of course a perfect one. We are not concerned here with crystal imperfections.) An example is shown in figure 12.14.

We can consider the lattice as made of *unit cells.* The translations of the unit cell along the lattice axes generate the lattice. Figure 12.15 illustrates a two-dimensional lattice and several different unit cells that might be chosen.

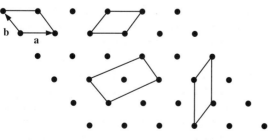

▲ FIGURE 12.15

Two-dimensional lattice of points and some of the unit cells that can be chosen. Each unit cell is defined by two axes **a** and **b,** of length a and b, respectively, which are depicted for the unit cell at the upper left corner. For a three-dimensional lattice, a unit cell is defined by three axes **a, b,** and **c.** Translation of a unit cell along its axes generates the lattice.

For our two-dimensional lattice, a unit cell is specified by two axes **a** and **b**. Let the dimensions of a unit cell be a and b in these directions; then the position of any atom in the unit cell is given by unitless coordinates (X, Y), where X and Y are given in fractions of the unit-cell dimensions. For example, an atom at the center of a unit cell has coordinates $(\frac{1}{2}, \frac{1}{2})$. For a three-dimensional lattice, a unit cell is specified by three axes **a, b,** and **c** and the position of an atom in the unit cell will be given by (X, Y, Z). If we determine the unit-cell size and shape and the coordinates of its atomic contents, we have all the information of the crystal structure.

The first step is to determine the shape of the unit cell (triclinic, cubic, and so on) and the unit-cell dimensions. The number of molecules per unit cell and the symmetry of their arrangement is found next. Figure 12.14 shows the crystal structure of benzo[a]pyrene 4,5-oxide, a metabolite of the carcinogen benzo[a]pyrene (Glusker et al. 1976). The molecule crystallizes in a monoclinic lattice (two angles = 90°, $a \neq b \neq c$) with dimensions $a = 17.341$ Å, $b = 4.095$ Å, and $c = 17.847$ Å. The **b**-axis forms an angle of 90.93° with respect to the perpendicular **a**- and **c**-axes shown in the figure. There are four molecules per unit cell, but only one molecular structure is independent. The coordinates of the other three are related by the specific crystal symmetry (called the *space group*) found for these molecules within the monoclinic lattice. The information about size, shape, and symmetry of the unit cell is obtained from assigning Miller indices to the diffraction pattern, that is, the positions of the diffraction spots and their symmetries—which values of h, k, l have equal intensities or zero intensity—provide the unit-cell properties. Obtaining the molecular structure is more difficult.

Determination of Molecular Structure
Calculation of Diffracted Intensities from Atomic Coordinates: The Structure Factor

Given the atomic coordinates of the molecules in the unit cell, we can calculate the phase differences for the scattered waves from each atom at any angle. The phase differences control the total intensities at each angle. We did this calculation explicitly for two electrons. For two points, the phase difference $\alpha = 2\pi\Delta/\lambda$ depends on the path difference Δ, which depends on the two coordinates and scattering angle as shown in figure 12.2. We now want to generalize this calculation to many atomic scatterers. We need to be able to add the scattered X-ray waves from all the atoms in a unit cell. Each wave is characterized by an amplitude and a phase. The amplitude depends on the atomic number of the scattering atom; the phase depends on its coordinates and the scattering angle. So, from the coordinates of all the atoms in a unit cell, we can calculate the scattered intensities at any angle. However, significant intensities occur only where there is constructive interference of the waves. This means that for a crystal we calculate the intensities only at the Bragg angles specified by the Miller indices h, k, l.

The summation of scattered waves can be done by several mathematical methods. The simplest uses a complex-number notation. Each scattered wave has two parameters: amplitude and phase; it can be represented by a complex number. The waves scattered by a unit cell containing n atoms is the sum of the waves scattered by the atoms. The sum is over all atoms in the unit cell:

$$F(h, k, l) = \sum_{j=1}^{n} f_j e^{i\alpha_j} \tag{12.12}$$

f_j = atomic scattering factor of an atom
α_j = the phase factor for scattering from the atom
$i = \sqrt{-1}$

$F(h, k, l)$ characterizes a scattered wave at the angle specified by Bragg's equation [Eq. (12.10)] and Miller indices h, k, l; it is called the *structure factor* of the planes h, k, l. We can rewrite the structure factor explicitly in terms of the coordinates of the scattering centers at X_j, Y_j, Z_j in the unit cell. The atomic coordinates (X_j, Y_j, Z_j) are unitless; they are expressed in fractions of the unit-cell dimensions, as we have defined them previously.

$$F(h, k, l) = \sum_{j=1}^{n} f_j e^{+2\pi i(hX_j + kY_j + lZ_j)} \tag{12.13}$$

The scattered intensity $I(h, k, l)$ is the square of the structure factor. To square a complex number, we multiply the number by the same number with i changed to minus i (the complex conjugate). We use the mathematical identity

$$e^{ix} \equiv \cos x + i \sin x \tag{12.14}$$

to write Eq. 12.13 as

$$F(h, k, l) = A(h, k, l) + iB(h, k, l) \tag{12.15}$$

with

$$A(h, k, l) = \sum_{j=1}^{n} f_j \cos[2\pi(hX_j + kY_j + lZ_j)] \tag{12.16a}$$

$$B(h, k, l) = \sum_{j=1}^{n} f_j \sin[2\pi(hX_j + kY_j + lZ_j)] \tag{12.16b}$$

Then, because $(A + iB)^2$ is equal to $(A^2 + B^2)$, the scattered intensity is obtained:

$$I(h, k, l) = F(h, k, l)^2 = A(h, k, l)^2 + B(h, k, l)^2 \tag{12.17}$$

The intensity $I(h, k, l)$ is what the detectors measure at each diffraction spot characterized by Miller indices (h, k, l). However, the phase $\alpha(h, k, l)$ of $F(h, k, l)$—which we emphasize *cannot* be measured—can be calculated from Eqs. (12.16a, b) and the trigonometric identity $\tan \alpha = \sin \alpha / \cos \alpha$:

$$\alpha(h, k, l) = \tan^{-1} \frac{B(h, k, l)}{A(h, k, l)} \tag{12.18}$$

These equations tell us how to calculate the complete diffraction pattern —the intensity and phase of every scattered beam—from the coordinates of the atoms in the unit cell. For every group of parallel planes of Miller indices (h, k, l) there is scattered radiation at angle θ given by the Bragg equation ($n\lambda = 2\, d_{hkl}\sin\theta$). From the X-, Y-, Z-coordinates of the atoms in the unit cell and using Eqs. (12.17) and (12.18), we can calculate the diffraction pattern.

Of course, what we really want to be able to do is to go from the measured diffraction pattern to the coordinates of the atoms. We learn about this in the next section. However, once coordinates are obtained, the way we check them and improve them is to calculate the diffraction patterns they predict. The coordinates are improved by minimizing the difference between the measured intensities and the calculated intensities from Eq. (12.17).

Calculation of Atomic Coordinates from Diffracted Intensities

We have described how to calculate a diffraction pattern from a crystal structure; now it is time to learn how to calculate a structure from a diffraction pattern. The problem would be trivial if we could measure the phases of the scattered X-ray beams. Let's see why this is so.

The structure factor $F\,(h, k, l)$ is a function of the coordinates (X, Y, Z) of the contents of a unit cell [Eq. (12.13)]. There is a standard mathematical method called *Fourier transformation* that solves Eq. (12.13) for the positions of the scattering centers. We used a Fourier transform to calculate an NMR spectrum as a function of frequency from a time-dependent free-induction decay (FID) in chapter 10. Here, we use a Fourier transform to calculate the electron density, a function of X, Y, Z from $F\,(h, k, l)$, a function of h, k, l.

$$\rho\,(X, Y, Z) = \frac{1}{V}\sum_{h}\sum_{k}\sum_{l} F\,(h, k, l)\, e^{-2\pi i(hX + kY + lZ)} \qquad \textbf{(12.19a)}$$

Here, $\rho\,(X, Y, Z)$ is the electron density (electrons per unit volume) at position X, Y, Z in the unit cell. X, Y, Z are in fractions of distances along the unit cell, and V is the volume of the unit cell. Equation (12.19a) tells us exactly how to calculate the electron density at every point in the unit cell. We choose a value of X, Y, Z in the unit cell. Then, for every diffracted beam (which we have already assigned to a set of three Miller indices h, k, l), we calculate

$$F\,(h, k, l)\, e^{-2\pi i(hZ + kY + lZ)}$$

Summing over all diffracted beams (that means summing over all Miller indices h, k, l), we calculate the electron density, $\rho\,(X, Y, Z)$, at X, Y, Z. Repeating this for many points in the unit cell, we obtain the distribution of the electrons. From a high-resolution electron-density map [see figure 12.16(c)], the position of each atom, and thus the molecular structure, is clear. For a lower-resolution map [figure 12.16(b)], there will be more uncertainty in the placement of the atoms, but we can still obtain a structure. It is important to note that the entire diffraction pattern—all values of h, k, l—contributes to the calculated electron density at X, Y, Z.

For every Fourier transform there is an inverse, which means that we can use the inverse of Eq. (12.19a) to calculate the diffraction pattern:

$$F\,(h, k, l) = \int dX \int dY \int dZ\, \rho\,(X, Y, Z)\, e^{+2\pi i(hX + kY + lZ)} \qquad \textbf{(12.19b)}$$

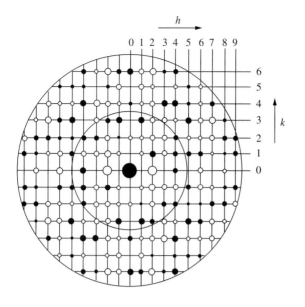

(a)

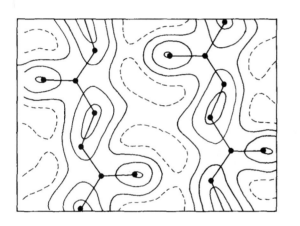

(b)

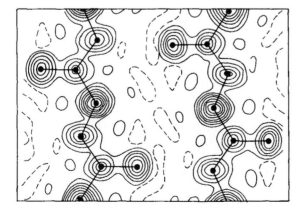

(c)

◄ FIGURE 12.16

An illustration of how X-ray diffraction data are used to generate a structure. In (a), structure-factor data are shown for a two-dimensional projection of a crystal of the polypeptide poly-L-alanine in antiparallel β-sheets. The dark spots represent $F(h, k)$ negative. The size of each spot is proportional to the magnitude of $F(h, k)$. The inner circle includes the data that would be sampled to obtain 2-Å resolution. The corresponding electron-density map is shown in (b). The outer circle in (a) includes the data necessary to obtain 1-Å resolution, as shown in (c). The known position of the nuclei are shown in (b) and (c), superimposed on the electron-density maps. (Adapted from figure 13-27 of C. R. Cantor and P. R. Schimmel, 1980, in *Biophysical Chemistry,* part 2.)

Integrating over the electron density of the unit cell, we can calculate each of the structure factors. Equations (12.19a) and (12.19b) allow us to calculate a structure—an electron density—from a known diffraction pattern and to calculate a diffraction pattern from a known structure.

The Phase Problem

There is a serious problem with using Eq. (12.19a); we cannot measure the structure factor $F(h, k, l)$. The structure factor has an amplitude—the square root of the intensity $I(h, k, l)$—and a phase $\alpha(h, k, l)$:

$$F(h, k, l) = \sqrt{I(h, k, l)}\, e^{i\alpha(h,k,l)} \tag{12.20}$$

We can measure only the amplitude or its square—the intensity of each diffracted beam. The intensity loses the vital phase information, $\alpha(h, k, l)$.

Direct Methods

There are various methods for obtaining the missing phase factors. The method routinely used for small molecules (less than 200 to 300 nonhydrogen atoms) is called the *direct method* (Hauptman 1989). Remember that the intensity of scattering from a hydrogen atom is only $(\frac{1}{6})^2$ of that of a carbon atom, so the scattering from hydrogen atoms can be ignored in the first approximation. In the direct method, you guess at the phases $\alpha(h, k, l)$ that you combine with the measured amplitudes $\sqrt{I(h, k, l)}$ to give the structure factors $F(h, k, l)$, which allows calculation of the electron density. Of course, the guesses of the phases are not random; only certain phases give reasonable results. For example, if a choice of phases gives a calculated negative electron density from Eq. (12.19a), we know it is a poor guess. Herbert Hauptman and Jerome Karle received the Nobel Prize in Chemistry in 1985 (see description by Hendrickson 1986) for devising algorithms to find the phases that, when combined with the measured intensities, will give an electron density that corresponds to a collection of atoms—a molecule. The mathematical problem is equivalent to solving the simultaneous *nonlinear* equations

$$\left|F(h, k, l)\right| = \sqrt{I(h, k, l)} = \left| \sum_{j=1}^{n} f_j e^{+2\pi i(hX_j + kY_j + lZ_j)} \right| \tag{12.21}$$

for the unknown positions of the atoms X_j, Y_j, Z_j. In Eq. (12.21) the vertical lines mean absolute values. The measurable absolute values of the structure factors are related to the absolute values of the sums. All the parameters in Eq. (12.21) are known, except for the three coordinates for each atom.

For 100 atoms in the unit cell ($n = 100$), there are 300 unknowns. The amplitudes $|F(h, k, l)|$ of the thousands of peaks in the diffraction pattern are measured. This means that there may be a few thousand equations specifying the 300 unknowns. The problem is overdetermined; there is enough information in the intensities alone to obtain the atomic coordinates. However, the solution of the simultaneous equations [Eq. (12.21)] is not straightforward, and it becomes increasingly difficult as the size of the molecule increases. Proteins and nucleic acids are at present too large to solve by direct methods.

Isomorphous Replacement

For macromolecules, the most used method for determining phases has been *isomorphous replacement* (*isomorphous* means "same shape"). A few heavy atoms are attached at specific locations to the macromolecule in the crystal, and their positions in the unit cell are determined. The scattering contribution to the intensity from an atom depends on the square of the atomic number, so a few heavy atoms can markedly change the diffraction pattern. Once the heavy atoms are positioned and their contributions to the phases of the structure factors are calculated, the structure factors from the other atoms can be obtained. Now the electron density of the macromolecule can be calculated.

Before describing the concept of the isomorphous replacement method, we introduce a useful function called the *Patterson function*. Patterson defined a function $P(X, Y, Z)$ that can be calculated from the experimental data:

$$P(X, Y, Z) = \frac{1}{V} \sum_h \sum_k \sum_l I(h, k, l) \, e^{-2\pi i(hX + kY + lZ)} \qquad \textbf{(12.22)}$$

The Patterson function, because it contains the square of the structure factor (the scattered intensity), gives information about vectors (specified by components X, Y, Z) between two atoms in the unit cell. This property is useful to us in the method of isomorphous replacement.

Consider a crystal of a protein and a crystal of a heavy-atom derivative of the protein. The two crystals are isomorphous if they have essentially the same unit-cell dimensions and atomic arrangements; the main difference is the substitution of a heavy atom in the derivative with little distortion of the protein structure. The heavy atom can be weakly bound to a side chain, or a heavy metal ion can substitute for a lighter metal ion [Hg(II) for Zn(II)]. From a pair of isomorphous crystals, their *difference* Patterson map usually permits the location of the heavy atom in the unit cell, and therefore the phase of scattering contributions from the heavy atom can be calculated. Suppose that $F_{prot}(h, k, l)$, $F_{deriv}(h, k, l)$, and $f_{atom}(h, k, l)$ are the structure factors for the protein, the derivative, and the heavy atom, respectively, for certain Miller indices (h, k, l). The magnitudes of F_{prot} and F_{deriv} are both known from the intensities, but their phase angles are not known. Both the magnitude and phase angle, α_{atom}, are known for the heavy-atom structure factor, f_{atom}, because we know its position. Each of the three structure factors can be represented as vectors. For two vectors, we know only their magnitudes; for the other—the heavy atom—we know its magnitude and direction.

If the heavy atom scatters much more than the light atom it replaces, then to a good approximation $\mathbf{F}_{deriv}$ should be the vectorial sum of $\mathbf{F}_{prot}$ and $\mathbf{f}_{atom}$:

$$\mathbf{F}_{deriv} = \mathbf{F}_{prot} + \mathbf{f}_{atom} \qquad \textbf{(12.23)}$$

We can therefore obtain the directions of $\mathbf{F}_{deriv}$ and $\mathbf{F}_{prot}$ by the geometric method illustrated in figure 12.17. The vector $\mathbf{AB}$ is drawn with length equal to the magnitude of $\mathbf{f}_{atom}$ and angle α equal to the phase angle of $\mathbf{f}_{atom}$. A circle with radius equal to the magnitude of $\mathbf{F}_{prot}$ and centered at A is first drawn. A vector drawn from any point on this circle to A gives a possible direction (and thus possible phase angle) for $\mathbf{F}_{prot}$. We then draw a second circle with radius equal to the magnitude of $\mathbf{F}_{deriv}$ and centered at B. A vector drawn from any point on this circle to B gives a possible direction (and thus possible phase angle) for $\mathbf{F}_{deriv}$. The points of intersection of the two circles (at C and C' in figure 12.17) gives two solutions for the phase problem. The structure factor for the protein (magnitude and phase) is either the vector from C to A or the one from C' to A. The ambiguity can be resolved from an-

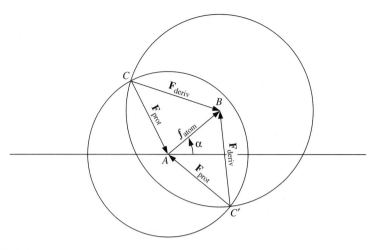

▲ FIGURE 12.17

Phase-angle determination by isomorphous replacement. Once the position of a heavy atom is determined, the phase angle and amplitude of its structure factor can be calculated. This is the vector $\mathbf{f}_{atom}$ going from A to B in the figure; its phase angle is α. The structure factor for the heavy-atom derivative of the protein, $\mathbf{F}_{deriv}$, is the vector sum of the structure factors of the protein and heavy atom, $\mathbf{F}_{prot}$ and $\mathbf{f}_{atom}$. The magnitudes of $\mathbf{F}_{deriv}$ and $\mathbf{F}_{prot}$ are obtained from intensity measurements. The circle centered at A with radius equal to the magnitude of $\mathbf{F}_{prot}$ gives all possible locations for one end of $\mathbf{F}_{prot}$. The circle centered at B with radius equal to the magnitude of $\mathbf{F}_{deriv}$ gives all possible locations for one end of $\mathbf{F}_{deriv}$. The intersection of the two circles at C and C' give two possible solutions to Eq. (12.23). The figure shows two possible vectors for $\mathbf{F}_{deriv}$ (from C to A and from C' to A) and two possible vectors for $\mathbf{F}_{prot}$ (from C to B and from C' to B). Another heavy-atom derivative allows us to choose between the two.

other isomorphous heavy-atom derivative. Thus, at least two crystals with different heavy-atom derivatives are needed, and three or more are routinely used to obtain the structure factors. A Fourier transform of the structure factors gives the electron density [Eq. (12.19a)].

The method of isomorphous replacement, often called *multiple isomorphous replacement* (MIR), has played a crucial role in solving crystal structures of proteins. The method allows us to deduce phase angles and thus structure factors from measured intensities. It is one solution of the phase problem, the problem that must be solved before a structure can be obtained from X-ray diffraction data.

Multiwavelength Anomalous Diffraction

Another method to solve the phase problem is to use *multiwavelength anomalous diffraction* (MAD). The atomic scattering factors shown in figure 12.5 correspond to scattering from atoms when the incident X-ray wavelength is not near any absorption bands of the atoms. However, for a wavelength near an atomic absorption, the scattering becomes anomalous (Hendrickson 1991). The atomic scattering factor adds a phase angle to the diffraction pattern that is wavelength dependent and is specific for each atom. The phase angle added is known for each atom and wavelength, so its effect on the diffraction pattern can be calculated. The change of the diffraction intensities with change in wavelength can be used to obtain the positions of atoms that are scattering anomalously. The method is analogous to using heavy atoms in MIR, but the incident wavelength picks the atoms that act like heavy atoms—the atoms that scatter anomalously. A single crystal can be used to solve the phase problem by applying the MAD method. In MIR we used the protein crystal and crystals of several isomorphous derivatives to obtain the phases (see figure 12.17). In MAD we use one crystal and several wavelengths to obtain the equivalent information.

Atoms with atomic number greater than 20 (Ca to U) have absorption bands and show anomalous scattering in the wavelength range from 0.3 to 3.0 Å. Synchrotrons provide tunable radiation in this wavelength region and can be used to solve a macromolecular structure in days, instead of months, once a suitable crystal is available. One way of introducing a suitable atom for anomalous diffraction into a protein crystal with minimum perturbation is to replace methionine by selenomethionine. Bromine atoms can replace methyl groups in DNA with little disturbance.

Determination of a Crystal Structure

Since the X-ray diffraction method is the most powerful method available to determine structures of large biochemical molecules, such as proteins and nucleic acids, it is important to understand the procedure. The first step is to prepare suitable crystals. Crystals must be large enough to give sufficient scattered intensities, but not so large that the incident X-ray beam is greatly attenuated by absorption. Successful crystal growing depends mainly on educated guesses. Many solvent and temperature conditions are tried until suitable crystals result. The next step, which is usually the easiest, is to obtain a diffraction pattern such as that shown in figure 12.9. From the positions of the spots, the crystal lattice and the dimensions of the unit cell can be determined.

Now the real work starts. Accurate intensity data for many different values of Miller indices h, k, l must be measured. Once the crystal class and unit-cell size are known, it is possible to calculate the crystal orientation and detector position necessary to produce a spot for any value of h, k, l. Because of the reciprocal relation in the Bragg condition [Eq. (12.9)], the spots closest to the incident beam (θ small) correspond to large values of distances (d_{hkl} large). To calculate the electron density with a resolution of 6 Å, for example, all spots corresponding to this distance and larger distances must be measured. For higher resolution, more spots must be measured. Figure 12.16 illustrates this for a two-dimensional projection of the electron density of two strands of poly-L-alanine antiparallel β-sheet. At 2-Å resolution, the strands are barely visible; at 1-Å resolution, the atoms are clearly seen. To obtain 2-Å resolution on a typical protein, about 10,000 values of $I(h, k, l) = F^2(h, k, l)$ must be measured.

The electron-density map is obtained by successive approximations. Heavy-atom derivatives (MIR) or anomalous diffraction (MAD) provide the location of a few atoms in the unit cell. These are used to estimate phases and thus provide approximate values of $F(h, k, l)$. These in turn provide approximate electron-density maps, which are used to obtain locations of more atoms. Least-squares refinement methods are used to obtain the final coordinates. The sum of the squares of the difference between the observed and calculated $|F(h, k, l)|$ values is minimized.

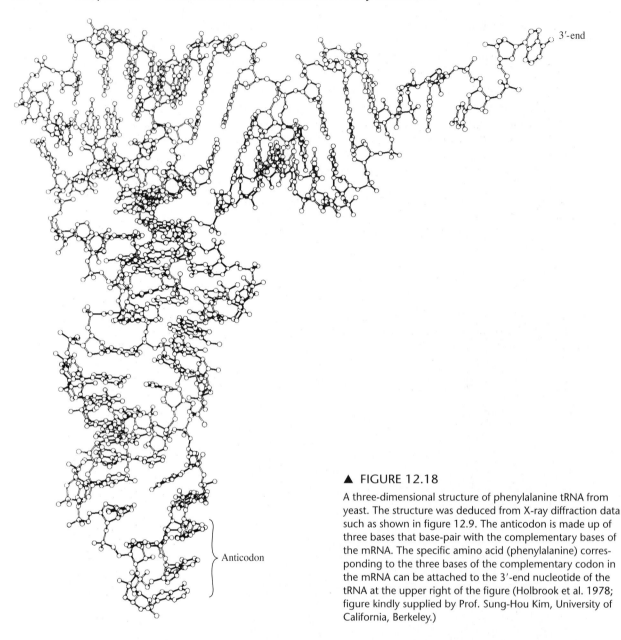

3'-end

Anticodon

▲ FIGURE 12.18

A three-dimensional structure of phenylalanine tRNA from yeast. The structure was deduced from X-ray diffraction data such as shown in figure 12.9. The anticodon is made up of three bases that base-pair with the complementary bases of the mRNA. The specific amino acid (phenylalanine) corresponding to the three bases of the complementary codon in the mRNA can be attached to the 3'-end nucleotide of the tRNA at the upper right of the figure (Holbrook et al. 1978; figure kindly supplied by Prof. Sung-Hou Kim, University of California, Berkeley.)

Figure 12.18 shows a structure for the tRNA molecule from yeast, which adds a phenylalanine molecule to a polypeptide that is being synthesized. The location of the phenylalanine in the polypeptide sequence is determined by the base sequence of the corresponding mRNA. The three anticodon bases of the tRNA form base pairs with the three codon bases of the mRNA. The three-dimensional structure of tRNA is a first step in understanding the mechanism of this complex reaction.

An example of a protein structure determined by X-ray crystallography is a transcription factor that binds to a specific sequence of DNA to control the synthesis of an RNA molecule—the transcription of the DNA sequence into an RNA sequence. The protein is the on–off switch for synthesis of the

RNA. By understanding how transcription factors function, we may be able to control the synthesis of mRNA and thus the proteins they produce. The first step is to learn how the protein binds to the DNA. Nikola Pavletich and Carl Pabo (1991; 1993) chose to study transcription factor IIIA from a frog (*Xenopus*). They used just the DNA-binding domain of the protein, 90 amino acids, and an 11-base-pair fragment of the DNA. Crystals of the polypeptide–oligonucleotide complex were prepared, and diffraction data were obtained. For the native crystal, 34,488 Bragg reflections were measured, but because of crystal symmetry, only 9458 were unique. Heavy-atom derivatives were made by replacing thymine in the DNA with iodouracil; this replaces a methyl group with an iodine atom. Two different heavy-atom derivatives were made and cocrystalized with the polypeptide. This provides values for the phases of the measured diffraction intensities and allows calculation of the structure using Eq. (12.19a). The resulting structure is shown in figure 12.19; it was obtained from a 2.1-Å resolution electron-density map.

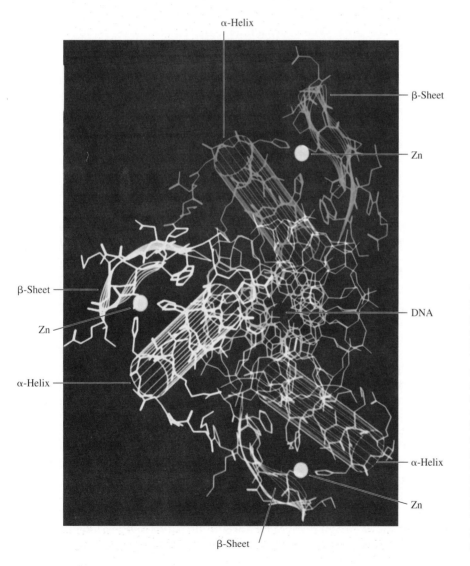

◀ FIGURE 12.19

Part of a transcription factor—a protein that controls the synthesis of RNA on a DNA template—bound to a double-stranded DNA. Three zinc (Zn) fingers binding to one turn of the DNA helix are shown. Each zinc finger contains a zinc ion (seen as a white disc) chelated between imidazole groups from histidines in an α-helix (shown as a cylinder) and cysteines in a β-sheet. The three fingers contact a specific sequence of base pairs in the DNA by forming hydrogen bonds. We are looking down the helix axis of the DNA in this figure. The structure was determined by X-ray crystallography by Pavletich and Pabo (1991).

The structure shows a very common DNA-binding motif called a *zinc finger* (Rhodes and Klug 1993). There are three zinc fingers binding to one turn of the DNA helix. Each zinc finger contains a zinc atom chelated to two cysteines in a β-sheet and two histidines in an α-helix. The three fingers contact a specific sequence of base pairs in the DNA by forming hydrogen bonds between five arginines and one histidine in the protein with six guanines in the nucleic acid. The 90 amino acids and three zinc atoms in the binding site serve to hold these six amino acids in the right place for recognizing the correct DNA sequence.

Scattering of X Rays by Noncrystalline Materials

Because of the long-range three-dimensional order in a single crystal, detailed atomic arrangement can be deduced from X-ray diffraction studies of the crystal. X-ray scattering by noncrystalline materials also yields structural information, although much less detail is obtainable.

For long, threadlike macromolecules such as DNA, it is often possible to obtain fibers of a bundle of molecules with their long axes parallel or nearly parallel to the fiber axes. The orientation of the molecules in a fiber about the fiber axis is random. When such a fiber is placed in an X-ray beam, any long-range regular feature of the fiber will stand out in the diffraction pattern; the disordered features of the fiber will contribute only to the general background scattering. For example, for a DNA fiber with the molecules in the B-form conformation, because the helix axes of the DNA double helices are all parallel to the fiber axis, the strongest long-range regularity is the helical repeat of $p = 34$ Å per helical turn and $h = 3.4$ Å per base pair. These features are reflected in the diffraction pattern shown in figure 12.20. A diffraction pattern like this one is what led James Watson and Francis Crick to discover in 1954 the double-helix structure of DNA.

For solutions of macromolecules, several types of information can be obtained by X-ray scattering measurements. First, as mentioned earlier, at zero scattering angle, constructive interference always occurs for rays scattered by different parts of a scatterer, and the intensity is dependent on the total number of electrons. Therefore, when extrapolated to zero scattering angle and after subtracting scattering due to the solvent, the scattered intensity of a solution of macromolecules is related to the number of electrons per macromolecule. Since we can easily determine the elemental composition of the macromolecule, we can calculate the number of electrons per unit mass and therefore obtain from the corrected zero-angle scattering intensity the molecular weight of the macromolecule. Second, although the macromolecules are randomly oriented in a solution so that little information on the atomic arrangement of the molecule can be deduced from the diffraction pattern, the angle dependence of the scattering intensity does provide us with information on the size of the macromolecule. It can be shown that at very low angles (of the order of milliradians),

$$\ln i(\Theta) = \ln i(0) - \frac{16\pi^4 R^2}{3\lambda^2} \sin^2 \frac{\Theta}{2} \qquad \textbf{(12.24)}$$

where Θ is the angle between the incident and the scattered beams, $i(\Theta)$ is the scattering intensity at Θ, λ is the wavelength of the X rays, and R^2 is the mean-square radius of the macromolecule. Thus, if the experimentally measured $\ln i(\Theta)$ is plotted as a function of $\sin^2(\Theta/2)$, in the low-angle range a straight line results. The intercept gives $\ln i(0)$, from which the molecular

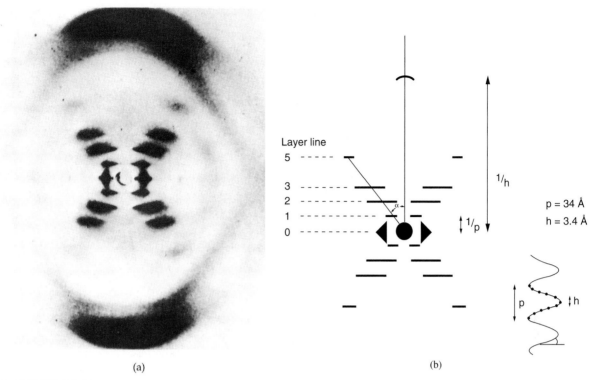

▲ FIGURE 12.20

(a) X-ray diffraction pattern of a NaDNA fiber in the B-helix conformation. Note the character-istic cross and the very strong meridian spots. (b) Line drawings of the pattern above, illustrat-ing the relation between the helix dimensions (lower right) and the layer lines and meridian spots. The DNA helix has repeating distances of $h = 3.4$ Å, the distance between successive base pairs, and pitch, $p = 34$ Å, the distance between successive turns. The angle α character-izes the ratio of the pitch to the radius; $\tan \alpha \cong$ pitch$/2\pi$ radius. [From J. P. Glusker and K. N. Trueblood, 1972, *Crystal Structure Analysis: A Primer* (New York: Oxford University Press), p. 137, fig. 39b.]

weight of the macromolecule can be calculated as we have discussed, and the slope gives $-16\pi^4 R^2/3\lambda^2$, from which we can calculate the mean-square ra-dius. The angle dependence of scattering intensity at larger angles is also de-pendent on the general shape of the macromolecule.

Absorption of X Rays

Because of coherent, incoherent, and fluorescence scattering, the intensity of a beam of X rays is reduced after passing through any material. Similar to Beer–Lambert-law behavior, which we discussed in chapter 10, the absor-bance for monochromatic X rays of a given wavelength is proportional to the thickness of the sample and to the number of electrons per unit volume. For a given material, the absorbance is dependent on the energy (wavelength) of the X rays. Figure 12.21 illustrates this dependence. There are several abrupt rises in the curve going from left to right: three in a region a little over 1000 eV and another near 9000 eV. These absorption edges are due to the ejection of orbital electrons from the L ($2s$, $2p$ electrons) and K ($1s$ electrons) shells, re-spectively, and are labeled in the figure as L and K edges. The energy at which an edge absorption occurs is determined by the energy of the orbital electron, which is different for each element in the periodic table.

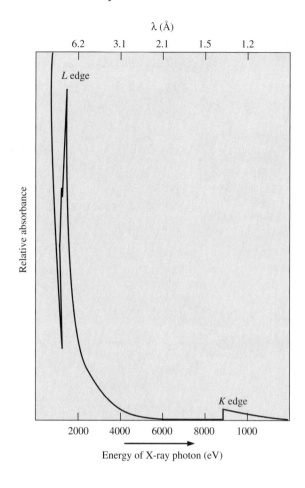

▶ FIGURE 12.21

Dependence of the absorption of X rays by Cu on the energy (wavelength) of the X-ray photon. There are three barely discernible maxima in the L edge.

Extended Fine Structure of Edge Absorption

In figure 12.21, we depict an edge absorption (the K edge, for example) as a sharp rise followed by a smooth gradual drop. Careful experimental measurements show, however, that the drop is not smooth. For a given material, there are characteristic wiggles on the short-wavelength side of the absorption edge, as illustrated in figure 12.22. Qualitatively speaking, such wiggles appear because the probability of absorption of an X-ray photon by an orbital electron depends on the initial and final energy states of the electron. The initial energy state of the electron (in the K shell, for example) is of course quantized (chapter 9). If we have an isolated atom, the final energy state of the ejected electron is not quantized because it can possess different kinetic energies. Indeed, for such an isolated atom, no wiggles would appear in the absorption curve. If the atom that absorbs the X-ray photon is not isolated but is surrounded by neighboring atoms, the ejected electron is back-scattered by the neighboring atoms. This situation is reminiscent of the particle-in-a-box problem (chapter 9). We recall that the energy of a particle in a box is quantized, with the energy levels dependent on the dimensions of the box. Similarly, in the presence of the neighboring atoms, the energy of the electron ejected by an X-ray photon is also quantized, with energy levels dependent on the distances between the atom at which absorption occurs and its neighboring atoms. This quantization of the final energy state of the

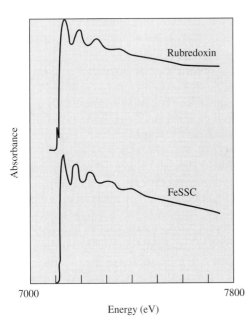

Absorbance

Rubredoxin

FeSSC

7000   7800

Energy (eV)

◀ **FIGURE 12.22**

Extended X-ray absorption fine structure (EXAFS) of the iron–sulfur protein
Peptococcus aerogenes rubredoxin and a
model compound tris (pyrrolidine car-
bodithioate –S, S') iron (III), FeSSC. The
X-ray source was synchrotron radiation
from the Stanford Synchrotron Radiation Laboratory. The protein sample was
a freeze-dried powder. Analysis of the
data leads to a determination of the
iron–sulfur distances in the protein.
(Figure from R. G. Shulman, P. Eisen-
berger, W. E. Blumberg, and N. A.
Stombaugh, 1975, *Proc. Natl. Acad.
Sci. USA* 72:4003.)

ejected electron results in the characteristic wiggles in the fine structure of
the edge absorption curve. By analyzing the edge absorption fine structure,
it is possible to deduce the neighboring atomic arrangement around the absorbing atom.

Because the *K* absorption edge occurs at different wavelengths for different elements, it is possible to select a particular wavelength so that the atomic arrangement around a particular atom can be deduced. For example, if an
enzyme contains a heavy metal atom, the wavelength of the incident X ray
can be chosen to correspond to the absorption edge of the heavy metal.
Analysis of the fine structure of the absorption profile can then yield information on the atomic environment of the heavy metal. It should be noted
that structural analysis by edge absorption fine structure does not require
crystalline material. It is the close neighbors of the absorbing atom that back-
scatter the ejected electron; therefore, long-range order is not necessary for
the phenomenon.

X Rays from Synchrotron Radiation

An intense source of X rays is obtainable from the storage rings of electrons
associated with an electron accelerator such as a synchrotron. In these storage rings, the circulating electrons produce electromagnetic radiation over a
broad range, including the X-ray region. Synchrotron X-ray radiation has
revolutionized the determination of macromolecular structure because of its
tunable wavelength and its high intensity.

The intensity of X rays from a synchrotron storage ring can be tens of
thousands times higher than that from a conventional source. This high intensity permits the use of crystal monochromators that permit the selection
of intense monochromatic radiation of the desired wavelength. The tunable
wavelength is what makes the MAD (multiwavelength anomalous diffraction) method possible. The high intensity of the X rays also means that smaller crystals can be used, and crystals with larger unit cells can be conveniently
studied. Furthermore, Laue diffraction patterns can be obtained in a very

short time using a range of wavelengths, instead of monochromatic radiation. Thus, time-dependent (milliseconds) structural changes can be studied (see "Suggested Reading").

Another important application is to obtain the fine structure of edge absorption discussed in the preceding section. Here, one must be able to tune the wavelength of the X-ray source to cover the regions of interest for a particular element, such as a heavy metal in an enzyme. Especially in biological systems, where the concentrations of such absorbing atoms are very low, the high intensity of X rays from a synchrotron source is crucial.

X-ray lenses are being developed to utilize the higher-resolving power obtainable in principle from the X-ray wavelengths compared to ultraviolet (UV) wavelengths. One strong impetus is the desire to make smaller integrated circuits. At present, visible and UV light is used to project an image of the circuit onto photosensitive plastic, which is then developed to produce the circuit. In principle, 100-Å X rays can give much higher-resolution images, and therefore smaller circuits, than 2000-Å UV light. These X-ray lenses can also provide better resolution images of cells, cell nuclei, and chromosomes. All the spectroscopic techniques based on light, which we described in chapter 10, can be extended into the X-ray region with synchrotron sources.

Electron Diffraction

According to de Broglie's equation ($\lambda = h/p$), a beam of electrons should exhibit wavelike properties. If an electron initially at rest at a point where the electrostatic potential is zero is accelerated by an electric field to a point where the potential is V, its kinetic energy at that point is

$$\frac{p^2}{2m_e} = eV \tag{12.25}$$

where p is the momentum, m_e the rest mass of an electron, and e the charge of an electron. The corresponding wavelength is

$$\lambda = \frac{h}{p} = \frac{h}{\sqrt{2m_e eV}} \tag{12.26}$$

Actually, if the electrons are moving at a velocity not negligible compared with the velocity of light c, a relativistic correction is needed:

$$\lambda = \frac{h}{\sqrt{2m_e eV_{\text{correct}}}} \tag{12.27}$$

where

$$V_{\text{correct}} = V\left(1 + \frac{eV}{2m_e c^2}\right) \tag{12.28}$$

It is easy to show that if the potential is of the order of 10^4 volts, the wavelength is of the same order of magnitude as X rays. Therefore, electron diffraction can also be used for structural determinations of crystals. Electrons are scattered much more strongly by atoms than are X-ray photons. This is a blessing in some ways and a nuisance in others. Because of the strong scattering, a much thinner crystal can be used, and higher diffracted intensity means that less time is needed in data acquisition. Especially for low-energy electrons (V of the order of a few hundred volts or less), scattering is due to the first few layers of the atoms at the surface, and therefore much information can be obtained about the *surface* structures. But also because of the strong scattering, the path for the beam of electrons must be evacuated; otherwise, the beam will be scattered by the gas molecules. This creates problems for many biological samples, which frequently contain water and cannot be kept in a vacuum. Also, effects due to interaction between the diffracted rays with the incident rays, which are not important for X-ray diffraction, are much more pronounced in electron diffraction. They make the interpretation of the electron diffraction patterns more difficult.

Neutron Diffraction

In chapter 6 we gave the translational kinetic energy of a gas as

$$U_{\mathrm{tr}} = \frac{3}{2}RT$$

This equation is not limited to molecular or atomic particles; it applies to neutrons as well. If high-velocity neutrons from an atomic reactor are slowed by collisions with molecules (D_2O is usually used) at around room temperature, it can be readily shown that their energy corresponds to a wavelength of the order of 1 Å. Such thermal neutrons can be used in diffraction studies of crystals and in low-angle-scattering studies of solutions, similarly to the use of X rays.

There are several important differences. X rays are scattered primarily by electrons; neutrons are scattered by nuclei. Because the size of the nucleus is much smaller compared with the wavelength of the thermal neutrons, the neutron scattering factor f_0 of an atom is not much affected by the value of $(\sin\theta)/\lambda$. (Thermal motions of atoms in a crystal, however, do cause the scattering intensity to decrease at larger scattering angles, and it is necessary to correct f_0 for this.)

The scattering factor f_0 of an atom for neutrons is not related to the atomic number or mass in any simple way. The scattering of neutrons by hydrogen is quite high, unlike the scattering of X rays, for which hydrogen is the weakest scatterer. Therefore, the positions of hydrogens can be more readily deduced from the neutron-scattering data of a crystal. The scattering of neutrons by deuterium is quite different from that by hydrogen. Thus, deuterium can be used to replace hydrogen to give isomorphous crystals. For small-angle-scattering studies, neutron diffraction has one important advantage. It can be calculated that the relative scattering factors for H_2O and D_2O (deuterium oxide) are -0.0056 and $+0.064$, respectively. The negative sign means that neutron scattering from H_2O (and from ^1H) is 180° out of phase with the scattering from D_2O and ^2H. Most of the biological substances, such as nucleic acids, proteins, and lipids, have scattering factors between these values. Therefore, by using mixtures of H_2O and D_2O as the solvent, the background scattering of the solvent can be made to be the same as one of the components. Only components that scatter differently will stand out and contribute to the scattering contrast. Thus, if we are studying a protein–nucleic acid complex, we can in essence look at either the protein or the nucleic acid component in the complex by selecting the appropriate H_2O–D_2O mixtures. Similarly, the contrast between membrane and cytoplasm can be varied from positive to negative by increasing the D_2O content of the solvent.

A disadvantage of neutron-scattering studies is that there are very few high-flux neutron sources available. If the flux is low, a larger crystal is needed to give a diffraction pattern in a reasonable time. Therefore, neutron-scattering studies are under way at only a few places.

Electron Microscopy

In the discussion of X-ray diffraction, we explained that the lack of high-resolution X-ray lenses makes it necessary to deduce the structure mathematically from the diffraction pattern. A beam of electrons, however, can be focused by electrostatic or magnetic lenses. Therefore, a beam of electrons, after interacting with a specimen, can be focused to give an image of the specimen on a fluorescent screen or a photographic plate. Because of the short wavelength attainable for a beam of electrons, the resolution of an electron microscope can be made several 100-fold higher than that of a light microscope. As a consequence, electron microscopy has developed into one of the most powerful techniques for studying macromolecules, assemblies of macromolecules, cellular organizations, and so on. One might expect, because the wavelength of electrons can easily be made much less than atomic dimensions, it should be possible to see detailed atomic arrangement within a molecule by electron microscopy. This is difficult to achieve for reasons that will become clearer after brief discussions on resolution, contrast, and specimen damage by electron bombardment.

Resolution, Contrast, and Radiation Damage

For an ideal lens, a point source of light will be imaged into a point in the image plane. In practice, lenses for focusing rays of electrons cannot be constructed to achieve such ideality. A point source will produce instead a circle at the image plane. This lens aberration is called *spherical aberration;* it depends on the lens aperture. The smaller the lens apperture, the less the spherical aberration. However, as we make the lens apperture smaller, diffraction effects become important. A circular apperture produces at the image plane a central spot surrounded by concentric circular maxima of decreasing brightness. The smaller the aperture, or the larger the wavelength, the greater will be the diffraction effect on resolution. At the present time, the point resolution (the smallest separation between two points that can be resolved) of a good electron microscope is about 3 Å. For crystalline materials with periodic spacings, spacing close to 1 Å can be resolved.

At the beginning of this chapter, we mentioned that contrast is an important factor in seeing an object. To see a macromolecular species by electron microscopy, molecules are usually mounted on a supporting film, such as a carbon film (typically about 100 Å thick, although much thinner films can be used). Contrast between a macromolecule and the background is due to a number of phenomena. One important factor is that some of the incident electrons are scattered by the specimen and will not reach the image plane, either because of the beam aperture or because of lens aberration. A strong scatterer will therefore appear as a dark spot in the image. Diffraction of the incident rays by a crystalline material or at the edges of dense particles, and interference between incident and scattered rays, also contribute to contrast. Detailed discussions can be found in books on electron microscopy. Contrast is frequently much more of a problem than resolution in electron microscopy.

Several methods have been developed to enhance the contrast. The macromolecules can be "shadowed" by impinging evaporated metal obliquely onto them. More metal will be deposited on the sides of the macromolecules than on the supporting film. The macromolecules can also be preferentially "stained" with an electron-dense stain (positive staining), or the supporting film can be stained preferentially with an electron-dense stain (negative staining). Contrast is also improved by using a dark-field arrangement: either the incident electron beam is tilted or the aperture is moved off center so that only the scattered electrons will reach the image plane.

We have mentioned the problem of keeping a biological specimen in an evacuated chamber. Another problem of electron microscopy is radiation damage. Energy absorbed by a macromolecule from the incident beam causes heating of the molecule, breakage of chemical bonds, and rearrangement of atoms. Improved detectors allow production of images with smaller amounts of radiation. If the specimen is stained, movement of the stain may also occur. Because the final image of the molecule is the composite picture of the molecule during the period when it is subject to electron bombardment, the image might be quite different from the original structure. Radiation damage is a serious problem when high-resolution experiments are done on individual molecules, but cooling the sample to liquid helium temperature greatly reduces the damage. Cryo-electron microscopy has allowed imaging of macromolecules with 5 Å resolution (Miyazawa et al. 1999).

Transmission and Scanning Electron Microscopes

In a transmission electron microscope, electrons from a source (usually a hot tungsten filament) are first accelerated by an electric field to give the required wavelength. After passing through the specimen, the electrons are detected by a fluorescent screen for viewing, or they are recorded on a photographic plate or a videotape. Lenses are used to guide the paths of the electrons. In a scanning electron microscope, electrons from a source, after acceleration, are first focused by a lens to give a very small spot. The specimen is then scanned by moving the focused spot across it point by point and line by line. This scanning is achieved by the use of scanning coils, in much the same way as in a television camera. A suitable detector analyzes either the transmitted electrons or back-scattered and secondary electrons ejected from the sample by the incident electrons. The signal detected can be displayed on a monitor, point by point and line by line, in synchronization with the movement of the focused spot.

The scanning microscope has several advantages in certain applications. When secondary electrons are detected, the contrast can be made very high for the surface features of a specimen by coating it with a thin layer of gold. Surface topography can be thus revealed. One can also use a detector to measure the energy of the X rays emitted from each point of an uncoated specimen when irradiated by the electrons. Because the energy of the X rays emitted is characteristic of the elements irradiated, a scanning microscope in this mode can be used as a microanalyzer for the elemental composition of the specimen.

Image Enhancement and Reconstruction

If a specimen is composed of a lattice of subunits, the periodicity of the structure will be present in its image recorded on an electron micrograph.

We have already discussed direct studies of periodic structures (crystals) by X-ray diffraction methods. Diffraction techniques can also be applied to study the *image* recorded on a micrograph. Because the image has been magnified many times from the original molecular assembly, its diffraction pattern is studied with visible light. Structural features not seen in the original electron micrograph image can be characterized from the optical diffraction of the image. Undesirable features of the image, such as noise (which has no periodicity), can be filtered out. The filtered diffraction pattern can be reconstructed if necessary to give an improved image by the use of an appropriate lens. Diffraction, filtration, and reconstruction of the image is usually done mathematically. The image is digitized, Fourier transformed into a diffraction pattern, and then analyzed to obtain a high-resolution image.

Because the image recorded on an electron micrograph is a two-dimensional superposition of different layers in a three-dimensional structure, it is possible to deduce the three-dimensional structure from several images obtained with different viewing angles. Such three-dimensional image reconstruction is facilitated if the specimen is composed of symmetrically arranged subunits. The image of a structure composed of helically arranged subunits, for example, contains many different views of the subunit, and therefore our image is sufficient for deducing the three-dimensional structure of a subunit. Three-dimensional image reconstruction has been done in recent years for a number of ordered structures, and resolutions in the range of 5–20 Å have been achieved.

Image reconstruction does not require periodic structures. Tens of thousands of images of an object from different angles can be combined to obtain a three-dimensional picture. Figure 12.23 shows four views of an *E. coli* ribosome at 11 Å resolution reconstructed from over 73,000 images obtained with a cryo-electron microscope. The ribosomes were photographed with and without initiator tRNA (formylmethionine tRNA, fMet-tRNA) to identify the site for binding tRNA.

Scanning Tunneling and Atomic Force Microscopy

A scanning tunneling microscope moves a sharp tip over a sample to produce an image. The electron-tunneling current between the sample and the tip is measured as the tip scans the sample. If two conducting surfaces are separated by an insulator a few angstroms thick, a current tunnels across the intervening insulator when a voltage is applied. The electron current depends exponentially on the distance between the two conductors; it is thus a very sensitive measure of the distance between them. If an atomically sharp conducting tip is scanned over a conducting surface in air or vacuum, individual surface atoms can be seen. Atomic holes in the surface decrease the tunneling current, and atomic bumps increase the current. Experimentally, the tunneling current is kept constant (at 0.1 nA, for example) and the variation of a few angstroms in the height of the tip is measured. Of course, molecules deposited on the surface can also be imaged. The main problem in studying biological molecules by scanning tunneling microscopy (STM) is in the sample preparation. The macromolecules are usually evaporated from a buffer solution onto the conducting gold or graphite support, so it is important to test for drying artifacts or other sample distortions. Images are obtained, but relating the STM image to biomolecular structure is difficult.

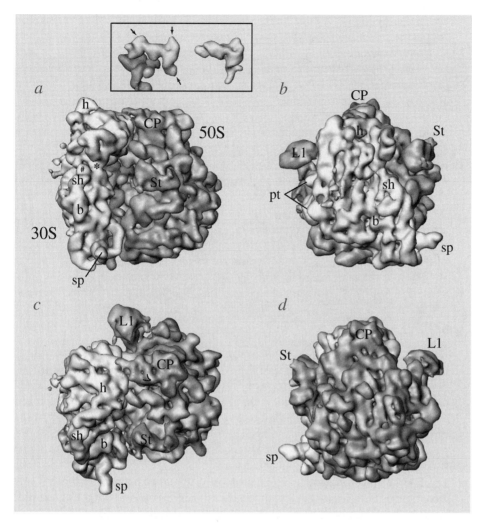

▲ FIGURE 12.23

Four views of a 70S *E. coli* ribosome (Gabashvilli et al. 2000): (a) The 30S subunit is on the left and 50S is on the right. (b) The image in (a) is rotated 90° to show the solvent side of the 30S subunit. (c) The image in (a) is seen from the top. (d) The image in (a) is rotated 90° to show the solvent side of the 50S subunit. A spur on the ribosome labeled sp is a good marker for ribosome orientation. The inset in the figure shows fMet-tRNA as seen in the ribosome compared with a 10-Å resolution calculated from the atomic resolution X-ray structure of the tRNA; they are very similar. (The figure was kindly supplied by Dr. Joachim Frank, HHMI, State University of New York at Albany.)

Atomic force microscopy (AFM) uses the forces between all atoms (the London–van der Waals and electrostatic interactions) to image molecules. A tip on a cantilever arm is scanned over the surface of the sample and the angstrom-size displacement is measured. It is surprising but true that the spring force constants for atomic repulsion—of order 10 Newtons per meter—are larger than that of a cantilever arm made from household aluminum foil 4 mm long and 1 mm wide (as described by Rugar and Hansma 1990). Thus, the slight up-and-down motion of the tip as it is scanned over the sample provides a surface image of the sample. The tremendous advan-

▶ FIGURE 12.24

An AFM picture of a double-stranded DNA plasmid (psK 31). The DNA molecules were deposited on freshly cleaved mica pretreated with magnesium acetate. The DNA on mica was stored over a desiccant then imaged under propanol. The sharp tip on a narrow silicon nitride cantilever arm is scanned over the 400-nm × 400-nm area; about 1 min is required to obtain an image. The apparent width of the DNA is about 3 nm; B-form DNA has a width of about 2.4 nm. (The picture was kindly provided by Prof. Carlos Bustamante, University of California; further description of the results is given in Hansma et al. 1992.)

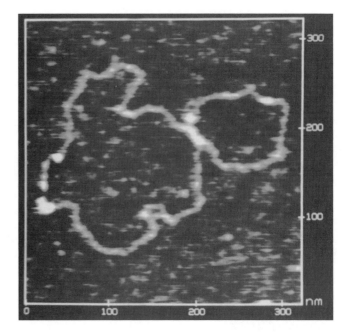

tage for biological samples is that the AFM can be done in aqueous solution. No drying, staining, metal deposition, or other fixing need be done (Radmacher et al. 1992). The polymerization of fibrinogen to form fibrin, the main component of blood clots, has been followed by AFM (Drake et al. 1989). Figure 12.24 shows an AFM image of a double-stranded DNA plasmid (psK 31) on a mica support under propanol. The apparent width of the DNA is 3 nm; B-form DNA has a width of about 2.4 nm.

The atomic force microscope not only visualizes macromolecules but also manipulates them. For example, the tip can be used to cut a DNA molecule at a specific place (Hansma et al. 1992), such as 25% of the distance from one end. By attaching one end of a molecule (a protein or a nucleic acid) to the microscope tip and the other end to the surface under the tip, one can unfold the molecule. The force required to unfold the molecule can be measured as a function of the distance between the ends to obtain the work (force times distance) of unfolding. If the unfolding is done slowly enough to make it reversible, the reversible work is equal to the Gibbs free energy at constant temperature and pressure (chapter 4). Thus, the free energy can be obtained for a transition from a folded native structure to an unfolded denatured conformation (Rief et al. 1997).

Summary
X-ray Diffraction

X rays are photons (electromagnetic radiation) with wavelengths in the range from a few tenths of 1 Å to several angstroms. X rays are usually produced by bombarding a metal target with electrons. Synchrotrons produce radiation covering the entire range from hard X rays (1 Å) to soft X rays (100 Å) to the ultraviolet (1000 Å) and beyond.

Scattering from one point:

$$I_{\text{sca}} = \frac{CI_0}{r^2} \qquad (12.1)$$

I_0 = intensity of incident plane-polarized light
I_{sca} = intensity of scattered light in plane perpendicular
    to plane of polarization
r = distance from scattering point to detector
C = proportionality constant

Scattering from two points:

$$I_{\text{sca}} = \frac{2CI_0}{r^2}\left(1 + \cos\frac{2\pi\Delta}{\lambda}\right) \qquad (12.5)$$

Δ = path difference between the rays scattered from the two points;
    the path difference is the difference in distance traveled from
    source to each scattering point to detector
λ = wavelength of radiation
$2\pi\Delta/\lambda$ = phase difference between the rays scattered from the two points

von Laue's equations:

For a crystal made of identical scatterers at the lattice points, the conditions for diffraction are

$$a(\cos\alpha - \cos\alpha_0) = h\lambda \qquad (12.6)$$

$$b(\cos\beta - \cos\beta_0) = k\lambda \qquad (12.7)$$

$$c(\cos\gamma - \cos\gamma_0) = l\lambda \qquad (12.8)$$

λ = wavelength of X rays
a, b, c = lattice spacings along the axes **a, b, c,** respectively
$\alpha_0, \beta_0, \gamma_0$ = angles the incident light makes with the three axes **a, b,** and **c**
α, β, γ = angles the diffracted beam makes with the three axes **a, b,** and **c**
h, k, l = integers

Miller indices:

Any plane of Miller indices (h, k, l) is parallel to one that makes intercepts a/h, b/k, and c/l with the lattice axes **a, b,** and **c.** For an orthorhombic lattice, the interplanar spacing for planes of Miller indices (h, k, l) is

$$d_{hkl} = \left[\left(\frac{h}{a}\right)^2 + \left(\frac{k}{b}\right)^2 + \left(\frac{l}{c}\right)^2\right]^{-1/2}$$

Bragg equation:

For a family of crystal planes of Miller indices (h, k, l) and an interplanar spacing d_{hkl}, the incident X-ray beam of wavelength λ appears to be reflected by the planes when the incident angle θ with respect to the planes satisfies the equation

$$\theta = \sin^{-1}\frac{n\lambda}{2d_{hkl}} \qquad (n = 0, 1, 2, \ldots) \qquad (12.9)$$

$$n\lambda = 2d_{hkl}\sin\theta \qquad (12.10)$$

The Bragg law is equivalent to von Laue's equations. We can also write

$$2\frac{d_{hkl}}{n}\sin\theta = \lambda \tag{12.11}$$

d_{hkl}/n can be considered to be the interplanar spacing of planes of Miller indices (nh, nk, nl).

Unit cell:

The unit cell is the basic unit of a crystal structure. Translation of the unit cell along the three lattice axes generates the crystal.

Structure factor:

At a Bragg angle $\theta = \sin^{-1}(\lambda/2d_{hkl})$ corresponding to diffraction by planes of indices (h, k, l), the amplitude of the rays scattered by a unit cell can be represented by a complex number:

$$F(h, k, l) = A(h, k, l) + iB(h, k, l) \tag{12.15}$$

$$A(h, k, l) = \sum_{j=1}^{n} f_j \cos[2\pi(hX_j + kY_j + lZ_j)] \tag{12.16a}$$

$$B(h, k, l) = \sum_{j=1}^{n} f_j \sin[2\pi(hX_j + kY_j + lZ_j)] \tag{12.16b}$$

$$f_j = \text{atomic scattering factor of atom } j$$
$$X_j, Y_j, Z_j = \text{unitless coordinates of atom } j, \text{ fractions of}$$
$$\text{the unit-cell dimensions}$$
$$n = \text{total number of atoms in the unit cell}$$

$F(h, k, l)$ is the structure factor. The intensity of the diffracted rays by planes of indices (h, k, l) is proportional to $A^2 + B^2$. The phase of F is

$$\alpha(h, k, l) = \tan^{-1}\frac{B(h, k, l)}{A(h, k, l)} \tag{12.18}$$

Phase problem:

A critical step in obtaining the structure of the crystal from its diffraction pattern is to determine the phases of the structure factors. An important method for phase determination of macromolecules is the isomorphous replacement method.

Equations (12.15), (12.16a), and (12.16b) can be written in an equivalent form:

$$F(h, k, l) = \sum_{j=1}^{n} f_j e^{+2\pi i(hX_j + kY_j + lZ_j)} \qquad (i = \sqrt{-1}) \tag{12.13}$$

The Fourier transform of this equation is the electron density of the unit cell:

$$\rho(X, Y, Z) = \frac{1}{V}\sum_{h}\sum_{k}\sum_{l} F(h, k, l)\, e^{-2\pi i(hX + kY + lZ)} \tag{12.19a}$$

$$\rho(X, Y, Z) = \text{electron density (electrons volume}^{-1}) \text{ at coordinate } X, Y, Z$$
$$\text{of the unit cell}$$
$$V = \text{volume of unit cell}$$

F (h, k, l) is needed to calculate the electron density; F^2 (h, k, l) or $|F$ $(h, k, l)|$ is not sufficient.

Scattering by noncrystalline materials:

Some molecular parameters, such as molecular weight, mean-square radius, and neighboring atomic arrangement around an atom from which an electron has been ejected, can be deduced for noncrystalline materials. Structures of fibers can also be studied by X-ray diffraction.

Neutron Diffraction

The de Broglie wavelength of thermal neutrons is of the order of interatomic distances in crystals. Neutrons are scattered by nuclei; X-ray photons are scattered primarily by electrons.

The neutron-scattering factor of an atom is not related to the atomic number or mass in any simple way. The scattering of neutrons by hydrogen is quite high, making neutron diffraction a valuable tool for determining the positions of the hydrogens in a crystal.

The isotopes hydrogen and deuterium scatter neutrons differently. This fact is useful in structural studies of macromolecules by neutron diffraction.

Electron Microsopy

The de Broglie wavelength of electrons accelerated by a voltage V is

$$\lambda = \frac{h}{\sqrt{2m_e eV_{correct}}} \qquad (12.27)$$

$$V_{correct} = V\left(1 + \frac{eV}{2m_e c^2}\right) \qquad (12.28)$$

c = speed of light in vacuum
m_e = rest mass of an electron
e = charge of an electron
h = Planck's constant

Electrostatic or magnetic lenses can be used to control the paths of a beam of electrons. Electron microscopes with a point resolution of a few angstroms have been constructed. In addition to resolution and contrast, specimen damage by the electron beam is an important consideration in electron microscopy.

Mathematics Needed for Chapter 12

Trigonometric identities:

$$\sin^2\theta + \cos^2\theta = 1$$

$$\sin(\alpha + \beta) = \sin\alpha\cos\beta + \sin\beta\cos\alpha$$

$$\sin n\pi = 0 \qquad (\text{for } n = \text{integer})$$

$$\sin\left(\frac{\pi}{2} + n\pi\right) = 1 \qquad (\text{for } n = \text{even integer})$$

$$\sin\left(\frac{\pi}{2} + n\pi\right) = -1 \qquad (\text{for } n = \text{odd integer})$$

$$\cos n\pi = 1 \qquad \text{(for } n = \text{even integer)}$$

$$\cos n\pi = -1 \qquad \text{(for } n = \text{odd integer)}$$

$$\cos \frac{n\pi}{2} = 0 \qquad \text{(for } n = \text{odd integer)}$$

If $y = \sin \theta$, $\theta = \sin^{-1} y$.

Complex numbers:

A complex number an be written as a sum of a real part and an imaginary part:

$$a + ib$$

a = real part
b = imaginary part
$i = \sqrt{-1}$

The magnitude of a complex number, $a + ib$, is equal to $\sqrt{a^2 + b^2}$.

A complex number can also be written in exponential form:

$$Fe^{i\theta}$$

F = magnitude
θ = phase angle
$e^{i\theta} = \cos \theta + i \sin \theta$

By definition of the magnitude of a complex number, we understand why F is called the magnitude.

$$Fe^{i\theta} = F \cos \theta + iF \sin \theta$$

$$\text{magnitude} = \sqrt{F^2 \cos^2 \theta + F^2 \sin^2 \theta} = \sqrt{F^2} = F$$

References

Glusker, J. P., M. Lewis, and M. Rossi. 1994. *Crystal Structure Analysis for Chemists and Biologists.* New York: VCH Publishers.

Ladd, M. F. C., and R. A. Palmer. 1985. *Structure Determination by X-Ray Crystallography,* 2d ed. New York: Plenum Press.

Marti, O., and M. Amrein, eds. 1993. *STM and SFM in Biology.* San Diego: Academic Press.

McRee, D. E. 1993. *Practical Protein Crystallography.* San Diego: Academic Press.

Slayter, E. M., and H. S. Slayter. 1992. *Light and Electron Microscopy.* Cambridge, England: Cambridge University Press.

Stout, G. H., and L. H. Jensen. 1989. *X-Ray Structure Determination.* New York: John Wiley.

Wiesendanger, R. 1994. *Scanning Probe Microscopy and Spectroscopy.* Cambridge, England: Cambridge University Press.

Suggested Reading

Boyer, K., J. C. Solem, J. W. Longworth, A. Borisov, and C. K. Rhodes. 1996. Biomedical Three-dimensional Holographic Microimaging at Visible, Ultraviolet and X-ray Wavelengths. *Nature Medicine* 8:939–941.

Bugg, C. E., W. C. Carson, and J. A. Montgomery. 1993. Drugs by Design. *Sci. Am.* (December): 92–98.

Chapman, M. S., I. Minor, M. G. Rossmann, G. D. Diana, et al. 1991. Human Rhinovirus-14 Complexed with Antiviral Compound r-61837. *J. Mol. Biol.* 217:455–463.

Dickerson, R. E., and I. Geis. 1983. *Hemoglobin: Structure, Function, Evolution, and Pathology.* Menlo Park, CA: Benjamin/Cummings.

Drake, B., C. B. Prater, A. L. Weisenhorn, S. A. C. Gould, T. R. Albrecht, C. F. Quate, D. S. Cannel, H. G. Hansma, and P. K. Hansma. 1989. Imaging Crystals, Polymers, and Processes in Water with the Atomic Force Microscope. *Science* 243:1586–1589.

Engel, A. 1991. Biological Applications of Probe Microscopes. *Annu. Rev. Biophys. Biophys. Chem.* 20:79–108.

Gabashvilli, I. S., R. K. Agrawal, C. M. T. Spahn, R. A. Grassucci, D. I. Svergun, J. Frank, P. Penczek. 2000. Solution Structure of the *E. coli* 70S Ribosome at 11.5 Å Resolution. *Cell* 100:537–549.

Glusker, J. P., D. E. Zacharias, H. L. Carrell, P. P. Fu, and R. G. Harvey. 1976. Molecular Structure of Benzo[a]pyrene 4,5-Oxide. *Cancer Research* 36:3951–3957.

Hauptman, H. A. 1989. The Phase Problem of X-Ray Crystallography. *Physics Today* (November): 24–29.

Hansma, H. G., J. Vesenka, C. Siegerist, G. Kelderman, H. Morrett, R. L. Sinsheimer, V. Ellings, C. Bustamante, and P. K. Hansma. 1992. Reproducible Imaging and Dissection of Plasmid DNA Under Liquid with the Atomic Force Microscope. *Science* 256:1180–1184.

Hendrickson, W. A. 1986. The 1985 Nobel Prize in Chemistry. *Science* 231:362–364. A description of the direct method for determining phases in X-ray diffraction.

———. 1991. Determination of Macromolecular Structures from Anomalous Diffraction of Synchrotron Radiation. *Science* 254:51–58.

———. 1995. X Rays in Molecular Biophysics. *Physics Today* (November): 42–48.

Holbrook, S. R., J. L. Sussman, R. W. Warrant, and S.-H. Kim. 1978. Crystal Structure of Yeast Phenylalanine Transfer RNA. II. Structural Features and Functional Implications. *J. Molecular Biology* 123:631–660.

Malhotra, A., P. Penczek, R. K. Agrawal, I. S. Gabashvili, R. A. Grassucci, R. Jünemann, N. Burkhardt, K. H. Nierhaus, and J. Frank. 1998. *Escherichia coli* 70 S Ribosome at 15-Å Resolution by Cryo-Electron Microscopy: Localization of fMet-tRNAfMet and Fitting of L1 Protein. *J. Molecular Biology* 280:103–116.

Miyazawa, A., Y. Fujiyoshi, M. Stowell, and N. Unwin. 1999. Nicotinic Acetylcholine Receptor at 4.6 Å Resolution: Transverse Tunnels in the Channel Wall. *J. Molecular Biology* 288:765–786.

Pavletich, N. P., and C. O. Pabo. 1991. Zinc Finger–DNA Recognition: Crystal Structure of a Zif268–DNA Complex at 2.1 Å. *Science* 252:809–817.

Pavletich, N. P., and C. O. Pabo. 1993. Crystal Structure of a Five-Finger GLI–DNA Complex: New Perspectives on Zinc Fingers. *Science* 261:1701–1707.

Perutz, M. 1992. *Protein Structure: New Approaches to Disease and Therapy.* New York: Freeman. The first chapter is entitled "Diffraction Without Tears: A Pictorial Introduction to X-Ray Analysis of Crystal Structures."

Radmacher, M., R. W. Tillman, M. Fritz, and H. E. Gaub. 1992. From Molecules to Cells: Imaging Soft Samples with the Atomic Force Microscope. *Science* 257:1900–1905.

Rhodes, D., and A. Klug. 1993. Zinc Fingers. *Sci. Am.* (February): 56–65.

Rief, M., M. Gautel, F. Oesterhelt, J. M. Fernandez, and H. E. Gaub. 1997. Reversible Unfolding of Individual Titin Immunoglobulin Domains by AFM. *Science* 276:1109–1112.

Rossmann, M. G., and J. E. Johnson. 1989. Icosahedral RNA Virus Structure. *Annu. Rev. Biochem.* 58:533–573.

Rugar, D., and P. Hansma. 1990. Atomic Force Microscopy. *Physics Today* (October): 23–29.

Srajer, V., T. Teng, T. Ursby, C. Pradervand, Z. Ren, S. Adachi, W. Schildkamp, D. Bourgeois, M. Wulff, and K. Moffat. 1996. Photolysis of the Carbon Monoxide Complex of Myoglobin: Nanosecond Time-Resolved Crystallography. *Science* 274:1726–1729.

Stoddard, B. L. 1998. New Results Using Laue Diffraction and Time-Resolved Crystallography. *Curr. Opin. Structural Biology* 8:612–618.

Stoddard, B. L., B. E. Cohen, M. Brubaker, A. D. Mesecar, and D. E. Koshland, Jr. 1998. Millisecond Laue Structures of an Enzyme–Product Complex Using Photocaged Substrate Analogs. *Nature Structural Biology* 5:891–897.

Tong, L., A. M. DeVos, M. V. Milburn, and S.-H. Kim. 1991. Crystal Structures at 2.2 Å Resolution of the Catalytic Domains of Normal Ras Protein and an Oncogenic Mutant Complexed with GDP. *J. Mol. Biol.* 217:503–516.

Problems

1. The energy of an electron upon acceleration by a potential of V volts is eV, where e is the charge. In units of electron volts, the energy is just V electron volts.

   (a) Calculate the energy in joules of an electron accelerated by a potential of 40 kV.

   (b) When the electron strikes a target atom, the maximum energy of the photon emitted corresponds to the conversion of all the energy of the incoming electron to that of the photon. Calculate the wavelength of such photons.

   (c) Show that in general,

   $$\lambda_s = \frac{12{,}390}{V}\,\text{Å}$$

   where λ_s is the shortest wavelength emitted when the electron strikes the target.

2. (a) For Cu metal, the K absorption edge is at $\lambda = 1.380$ Å (figure 12.21). For Ag metal, the absorption edge is at $\lambda = 0.4858$ Å. Calculate the energy required for the ejection of an electron from the K shell of Cu and that from the K shell of Ag.

   (b) The wavelengths of four emission lines from Ag are 1.54433 Å (α_2), 1.54051 Å (α_1), 1.39217 Å (β) (which is actually a doublet at high resolution; the value given is the average of the doublet), and 1.38102 Å (γ). These lines result from the emission of X-ray photons when the electrons in the *L*-shell energy levels fall into the K shell vacated by the ejected electrons. Calculate the L-shell energy levels for Ag corresponding to these emission lines.

3. (a) For the arrangement shown in figure 12.7, calculate α for the first- and second-order diffraction cones if the wavelength of the incident X rays is 1.54 Å and the spacing between the equidistant scattering centers is 4 Å.

(b) A cylindrically shaped photographic film, with its axis coincident with the row of scatterers, is used to record the diffracted rays. It is easy to see that the diffraction cones will intersect the film to give layer lines. Calculate the distance between the layer lines formed by the first and the second diffraction cones. The radius of the cylindrical film is 3 cm, and all other parameters are the same as in part (a).

4. In figure 12.2, the path difference for the scattered rays is calculated for two points parallel and perpendicular to the incident beam. Derive an equation for the general case in which the incident beam makes an angle ϕ with the line joining the two points. Obtain $\Delta = f(R, \Theta, \phi)$.

5. Assume that you can measure scattered intensities to 1%. What is the smallest separation between two points which you could distinguish from a single point using 1.54-Å radiation? That is, what value of R gives a scattering pattern that differs by at least 1% from that of a point scatterer?

6. For $\lambda = 1.54$-Å X rays, obtain the first-order ($n = 1$) Bragg angles for interplanar spacings of 5 Å, 10 Å, and 1000 Å.

7. From Eq. (12.11) it can be seen that for a fixed λ, the interplanar spacing (d_{hkl}/n) is inversely proportional to $\sin \theta$. In other words, scattering corresponding to the smallest spacing occurs at the maximum value of $\sin \theta$ (when $\theta = 90°$). If $\lambda = 1.54$-Å X rays are used, what is the theoretical limit of resolution? (That is, what is the minimum spacing observable?)

8. (a) The lattice of NaCl is cubic with a unit-cell dimension of 5.64 Å. Each unit cell contains $4Na^+$ and $4Cl^-$ ions. The density of NaCl is 2.163 g cm^{-3}. Calculate Avogadro's number, using this information.

(b) The dimensions of the unit cell of crystalline lysozyme are $a = 79.1$ Å, $b = 79.1$ Å, and $c = 37.9$ Å. The interaxial angles are all 90°. Each unit cell contains eight lysozyme molecules. The density of crystalline lysozyme is 1.242 g cm^{-3}, and chemical analysis shows that 64.4% (by weight) is lysozyme, the rest being water and salt. From these data, calculate the molecular weight of lysozyme.

9. (a) Show that for an orthorhombic lattice, the interplanar spacing d_{hkl} between planes of Miller indices (h, k, l) is

$$d_{hkl} = \frac{1}{\sqrt{\dfrac{h^2}{a^2} + \dfrac{k^2}{b^2} + \dfrac{l^2}{c^2}}}$$

where a, b, and c are the lattice spacings.

(b) Show that for a cubic lattice, $d_{100}:d_{110}:d_{111} = 1:0.707:0.578$.

10. (a) A cubic unit cell contains one atom at the origin. Calculate the relative intensities of the diffractions by the 100, 200, and 110 planes.

(b) Do the same calculations for a unit cell containing two identical atoms, one at the origin and one at $X = Y = Z = \frac{1}{2}$.

11. Consider a two-dimensional rectangular lattice with unit cell dimensions 50 Å × 30 Å along the a- and b-axes, respectively.

(a) Sketch an accurate picture of this lattice, giving at least nine lattice sites.

(b) On the drawing in part (a), sketch and label three representative planes having the Miller indices $(h, k) = (1, 0)$. Repeat this operation for the (0, 1) and (1, 1) planes.

(c) Make another drawing of the lattice. Sketch and label (2, 0), (1, 2), and (2, 1) planes.

(d) What are the distances between the planes in the six cases above?

(e) If we use 1.54-Å radiation, what will be the angle of the (1, 0), (0, 1), (1, 1), and (2, 2) reflections in first order?

(f) Suppose the unit cell contains one copy of a protein that has been labeled with two lead atoms at (x, y) positions given by (12 Å, 7 Å) and (37 Å, 18 Å). The atomic scattering factor for Pb is 82. Determine the intensity of the (1, 0) and (2, 0) reflections and the phase of the respective structure factors. Discuss the physical basis behind the difference between these two intensities with reference to the drawings in parts (b) and (c).

12. Calculate the de Broglie wavelengths of electrons accelerated by a voltage of
(a) 10 kV.
(b) 100 kV.

13. Calculate the de Broglie wavelength of thermal neutrons at 300 K.

14. To convince yourself that you understand X-ray diffraction,
(a) Describe a diffraction pattern. How is it experimentally measured?

(b)   How is the electron density determined from a diffraction pattern? Describe what this has to do with phases, amplitudes, and intensities (if anything).

(d)   How are coordinates for the atoms of a molecule obtained?

(d)   What, if anything, does the structure of a molecule determined in a crystal have to do with biology?

15.   Consider the crystal structure of flatein, a two-dimensional protein. The unit cell is a two-dimensional rectangle with cell dimensions of 80 Å × 90 Å along a and b, respectively.

(a)   Sketch the unit cell and draw and label three planes having Miller indices: 0,1; 2,0; 1,0; and 1,1.

(b)   What are the distances between the planes above?

(c)   What is the first-order Bragg diffraction angle using 1.75-Å radiation for a distance between planes of 7.00 Å?

(d)   You use heavy-atom isomorphous substitution to solve the phase problem. There is one protein per unit cell with one heavy atom per protein. The atom is Au, which has an atomic scattering factor of 74. Determine the intensity of the diffraction spot for the 2,0 reflection, and the phase of the structure factor. The location of the Au atom within the unit cell is $x = 20$ Å, along a and $y = 30$ Å along b.

16.   (a)   An X-ray beam of wavelength 1.54 Å shines on a crystal. Two beams scattered from the crystal are exactly in phase at the detector. What are the possible path differences in angstroms for the two scattered beams? One answer is zero. Give three other possible answers.

(b)   The two beams in part (a) are exactly out of phase at the detector. Give three possible values in angstroms for their possible path differences.

(c)   A Bragg reflection occurs in first order at 5.00°. What is the distance in angstroms between the scattering planes causing this reflection?

(d)   A cubic crystal has a unit cell that is 5 Å on a side. Many different scattering planes can be drawn through the crystal; they are labeled by Miller indices h, k, l. What is the largest distance (d_{hkl}) between two scattering planes that can occur in this crystal?

(e)   What is the smallest value of d_{hkl} that can be detected with 1.54-Å X rays?

(f)   Consider a one-dimensional crystal of carbon atoms; the unit cell has a spacing of 3 Å. There are two atoms per unit cell at coordinates $X = 0$ and $X = \frac{2}{3}$.

The scattering intensities will depend on one Miller index, h. What are the possible values of h? Write an expression for the structure factor F_h and calculate the intensity I_h for $h = 1$. The atomic scattering factor for C is $f_c = 6$.

17.   The ovals below represent four proteins that form a complex in solution that was crystallized and studied by X-ray diffraction:

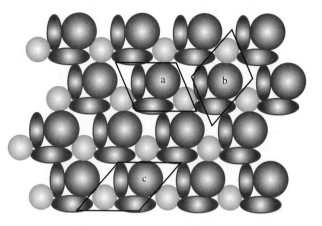

(a)   Three possible unit cells are shown and labeled a, b, c. Define a unit cell and state which, if any, of the parallelograms can represent correct unit cells. Assume that the third axis is chosen correctly for all the unit cells.

(b)   From the crystal structure, can you tell whether the group of proteins in the unit cell corresponds to the complex present in solution?

(c)   Sketch the diffraction pattern that you would expect from a two-dimensional crystal of these proteins. The symmetry of the diffraction pattern is the part to emphasize.

(d)   If the lightly shaded spheres were crystallized separately, how would the spacing of the diffraction pattern change?

(e)   If the wavelength of the X rays were decreased by a factor of 2, how would the spacing of the diffraction pattern change?

(f)   If the lightly shaded spheres were removed but the packing of the other three proteins remained the same so that the unit-cell size and symmetry did not change, how would the spacing of the diffraction pattern change?

APPENDIX

TABLE A.1 Useful Physical Constants

		SI units	cgs-esu units
Gas constant	R	$8.3145 \text{ J K}^{-1} \text{ mol}^{-1}$	$8.3145 \times 10^7 \text{ erg deg}^{-1} \text{ mol}^{-1}$
			$1.987 \text{ cal deg}^{-1} \text{ mol}^{-1}$
			$0.08205 \text{ L atm deg}^{-1} \text{ mol}^{-1}$
Avogadro's number	N_0	$6.0221 \times 10^{23} \text{ mol}^{-1}$	$6.0221 \times 10^{23} \text{ molecules mol}^{-1}$
Boltzmann constant	k	$1.3807 \times 10^{-23} \text{ J K}^{-1}$	$1.3807 \times 10^{-16} \text{ erg deg}^{-1}$
Faraday constant	F	$9.6485 \times 10^4 \text{ C mol}^{-1}$	$9.6485 \times 10^4 \text{ C mol}^{-1}$
Speed of light	c	$2.9979 \times 10^8 \text{ m s}^{-1}$	$2.9979 \times 10^{10} \text{ cm s}^{-1}$
Planck constant	h	$6.6261 \times 10^{-34} \text{ J s}$	$6.6261 \times 10^{-27} \text{ erg s}$
Elementary charge	e	$1.6022 \times 10^{-19} \text{ C}$	$4.8030 \times 10^{-10} \text{ esu}$
Electron mass	m_e	$9.1094 \times 10^{-31} \text{ kg}$	$9.1094 \times 10^{-28} \text{ g}$
Proton mass	m_p	$1.6726 \times 10^{-27} \text{ kg}$	$1.6726 \times 10^{-24} \text{ g}$
Standard gravity	g	9.8066 m s^{-2}	980.66 cm s^{-2}
Permittivity of vacuum	ε_0	$8.8542 \times 10^{-12} \text{ C}^2 \text{ N}^{-1} \text{ m}^{-2}$	

C	=	coulomb	L =	liter
g	=	gram	m =	meter
J	=	joule	N =	newton
K	=	Kelvin	s =	second
kg	=	kilogram		

TABLE A.2 Definition of Prefixes

yotta $\equiv$ Y	means multiply by 10^{24}
zetta $\equiv$ Z	means multiply by 10^{21}
exa $\equiv$ E	means multiply by 10^{18}
peta $\equiv$ P	means multiply by 10^{15}
tera $\equiv$ T	means multiply by 10^{12}
giga $\equiv$ G	means multiply by 10^{9}
mega $\equiv$ M	means multiply by 10^{6}
kilo $\equiv$ k	means multiply by 10^{3}
centi $\equiv$ c	means multiply by 10^{-2}
milli $\equiv$ m	means multiply by 10^{-3}
micro $\equiv$ μ	means multiply by 10^{-6}
nano $\equiv$ n	means multiply by 10^{-9}
pico $\equiv$ p	means multiply by 10^{-12}
femto $\equiv$ f	means multiply by 10^{-15}
atto $\equiv$ a	means multiply by 10^{-18}
zepto $\equiv$ z	means multiply by 10^{-21}
yocto $\equiv$ y	means multiply by 10^{-24}

TABLE A.3 Energy Conversion Factors

To convert from:	To:	Multiply by:
calories	ergs	4.184×10^7
calories	joules	4.184
calories	kilowatt-hours	1.162×10^{-6}
electron volts	kcal mol^{-1}	23.058
electron volts	ergs	1.602×10^{-12}
electron volts	joules	1.602×10^{-19}
electron volts	kilojoules mol^{-1}	96.474
ergs	calories	2.390×10^{-8}
ergs	joules	10^{-7}
ergs molecule^{-1}	joules mol^{-1}	6.022×10^{16}
ergs molecule^{-1}	kcal mol^{-1}	1.439×10^{13}
joules	calories	0.2390
joules	ergs	10^7
joules	kilowatt-hours	2.778×10^{-7}
joules mol^{-1}	ergs molecule^{-1}	1.661×10^{-17}
joules mol^{-1}	kcal mol^{-1}	2.390×10^{-4}
kilowatt-hours	calories	8.606×10^5
kilowatt-hours	joules	3.600×10^6
kcal mol^{-1}	electron volts	0.04337
kcal mol^{-1}	ergs molecule^{-1}	6.949×10^{-14}
kcal mol^{-1}	joules mol^{-1}	4.184×10^3
kilojoules mol^{-1}	electron volts	0.01036
kilojoules mol^{-1}	ergs molecule^{-1}	1.661×10^{-14}
kilojoules mol^{-1}	kcal mol^{-1}	0.2390

TABLE A.4   Miscellaneous Conversions and Abbreviations

Length

1 ångstrom $= 10^{-10}$ m $= 10^{-8}$ cm $=$ 0.1 nm
1 inch $=$ 2.54 cm
1 foot $=$ 30.48 cm
1 mile $=$ 5280 ft $=$ 1609 m

Mass

1 pound $=$ 453.6 g
1 kg $=$ 2.205 pounds

Energy

1 erg $= 1$ g cm^2 s^{-2}
1 joule $= 1$ kg m^2 s^{-2}

Force

1 newton $= 1$ kg m s^{-2} $= 10^5$ dyn
1 dyne $= 1$ g cm s^{-2} $= 10^{-5}$ N

Pressure

1 atmosphere $=$ 760 mmHg (Torr.) $=$ 14.70 lb in^{-2}
    $= 1.013 \times 10^6$ dyn cm^{-2}
    $= 1.013 \times 10^5$ newton m^{-2}
    $= 1.033 \times 10^4$ kg-force m^{-2}
    $= 1.013 \times 10^5$ pascals
1 newton m^{-2} $=$ 1 pascal

Power

1 watt $= 1$ J s^{-1} $=$ 1 VA

Volume

1 gallon $=$ 3.785 L

Viscosity

1 poise (P) $= 1$ g cm^{-1} s^{-1} $= 10^{-1}$ Pa s
1 pascal sec (Pa s) $= 1$ kg m^{-1} s^{-1} $=$ 10 P
1 mPa s $=$ 1 cP

Å	= ångstrom	P	= poise
A	= ampere	rad	= radian
cal	= calorie	V	= volt
hertz (Hz)	= cycles per second	W	= watt
hr	= hour	y	= year

TABLE A.5   Inorganic Compounds*

	$\Delta H_f^0 \equiv \bar{H}^0$ (kJ mol^{-1})	$\bar{S}^0$ (J K^{-1} mol^{-1})	$\Delta G_f^0 \equiv \bar{G}^0$ (kJ mol^{-1})
Ag(s)	0	42.55	0
Ag$^+$(aq)†	105.579	72.68	77.107
AgCl(s)	−127.068	96.2	−109.789
C(g)	716.682	158.096	671.257
C(s, graphite)	0	5.740	0
C(s, diamond)	1.895	2.377	2.900
Ca(s)	0	41.42	0
CaCO$_3$(s, calcite)	−1206.92	92.9	−1128.79
Cl$_2$(g)	0	223.066	0
Cl$^-$(aq)	−167.159	56.5	−131.228
CO(g)	−110.525	197.674	−137.168
CO$_2$(g)	−393.509	213.74	−394.359
CO$_2$(aq)	−413.80	117.6	−385.98
HCO$_3^-$(aq)	−691.99	91.2	−586.77
CO$_3^{2-}$(aq)	−677.14	−56.9	−527.81
Fe(s)	0	27.28	0
Fe$_2$O$_3$(s)	−824.2	87.40	−742.2
H$_2$(g)	0	130.684	0
H$_2$O(g)	−241.818	188.825	−228.572
H$_2$O(l)	−285.830	69.91	−237.129
H$^+$(aq)	0	0	0
OH$^-$(aq)	−229.994	−10.75	−157.244
H$_2$O$_2$(aq)	−191.17	143.9	−134.03
H$_2$S(g)	−20.63	205.79	−33.56
N$_2$(g)	0	191.61	0
NH$_3$(g)	−46.11	192.45	−16.45
NH$_3$(aq)	−80.29	111.3	−26.50
NH$_4^+$(aq)	−132.51	113.4	−79.31
NO(g)	90.25	210.761	86.55
NO$_2$(g)	33.18	240.06	51.31
NO$_3^-$(aq)	−205.0	146.4	−108.74
Na$^+$(aq)	−240.12	59.0	−261.905
NaCl(s)	−411.153	72.13	−384.138
NaCl(aq)	−407.27	115.5	−393.133
NaOH(s)	−425.609	64.455	−379.494
O$_2$(g)	0	205.138	0
O$_3$(g)	142.7	238.93	163.2
S(rhombic)	0	31.80	0
SO$_2$(g)	−296.830	248.22	−300.194
SO$_3$(g)	−395.72	256.76	−371.06

* Standard thermodynamic values at 25°C (298 K) and 1 atm pressure. Values for ions refer to an aqueous solution at unit activity on the molarity scale. Standard enthalpy of formation, ΔH_f^0, third-law entropies, S^0, and standard Gibbs free energy of formation, ΔG_f^0, are given.
† The standard state for all ions and for species labeled (aq) is that of a solute on the molarity scale.
SOURCE: Data from *The NBS Tables of Thermodynamic Properties*, D. D. Wagman et al., eds., *J. Phys. Chem. Ref. Data, 11,* Suppl. 2 (1982).

TABLE A.6   Hydrocarbons*

	$\Delta H_f^0 \equiv \overline{H}^0$ (kJ mol^{-1})	$\overline{S}^0$ (J K^{-1} mol^{-1})	$\Delta G_f^0 \equiv \overline{G}^0$ (kJ mol^{-1})
Acetylene, $C_2H_2(g)$	226.73	200.94	209.20
Benzene, $C_6H_6(g)$	82.93	269.20	129.66
Benzene, $C_6H_6(l)$	49.04	173.26	124.35
n-Butane, $C_4H_{10}(g)$	−126.15	310.12	−17.15
Cyclohexane, $C_6H_{12}(g)$	−123.14	298.24	31.76
Ethane, $C_2H_6(g)$	−84.68	229.60	−32.82
Ethylene, $C_2H_4(g)$	52.26	219.56	68.15
n-Heptane, $C_7H_{16}(g)$	−187.78	427.90	7.99
n-Hexane, $C_6H_{14}(g)$	−167.19	388.40	−0.25
Isobutane, $C_4H_{10}(g)$	−134.52	294.64	−20.88
Methane, $CH_4(g)$	−74.81	186.264	−50.72
Napthalene, $C_{10}H_8(g)$	150.96	335.64	223.59
n-Octane, $C_8H_{18}(g)$	−208.45	466.73	16.40
n-Pentane, $C_5H_{12}(g)$	−146.44	348.95	−8.37
Propane, $C_3H_8(g)$	−103.85	269.91	−23.47
Propylene, $C_3H_6(g)$	20.42	266.94	62.72

* Standard thermodynamic values at 25°C (298 K) and 1 atm pressure. Standard enthalpy of formation, ΔH_f^0, third-law entropies, S^0, and standard Gibbs free energy of formation, ΔG_f^0, are given.
SOURCE: Data from D. R. Stull, E. F. Westrum, Jr., and G. C. Sinke, *The Chemical Thermodynamics of Organic Compounds,* John Wiley, New York (1969).

TABLE A.7 Organic Compounds*

	$\Delta H_f^0 \equiv \bar{H}^0$ (kJ mol^{-1})	$\bar{S}^0$ (J K^{-1} mol^{-1})	$\Delta G_f^0 \equiv \bar{G}^0$ (kJ mol^{-1})	$\Delta G_f^0 \equiv \bar{G}^0$ (1 M activity, aq) (kJ mol^{-1})
Acetaldehyde $CH_3CHO(g)$	−166.36	264.22	−133.30	−139.24
Acetate$^-$(aq)	—	—	—	−372.334
Acetic acid $CH_3CO_2H(l)$	−484.1	159.83	−389.36	−396.60
Acetone $CH_3COCH_3(l)$	−248.1	200.4	−155.39	−161.00
Adenine $C_5H_5N_5(s)$	95.98	151.00	299.49	—
L-Alanine $CH_3CHNH_2COOH(s)$	−562.7	129.20	−370.24	−371.71
L-Alanylglycine $C_5H_{10}N_2O_3(s)$	−826.42	195.05	−532.62	—
L-Aspartate^{+--}(aq)	—	—	—	−698.69
Aspartic acid $C_4H_7NO_4(s)$	−973.37	170.12	−730.23	−719.98
Butyric acid $C_3H_7COOH(s)$	−533.9	226.4	−377.69	—
Citrate^{3-}(aq) $C_6H_5O_7$	—	—	—	−1168.34
Creatine $C_4H_9N_3O_2(s)$	−537.18	189.5	−264.93	—
L-Cysteine $HSCH_2CHNH_2COOH(s)$	−533.9	169.9	−343.97	−340.33
L-Cystine $C_6H_{12}N_2O_4S_2(s)$	−1051.9	280.58	−693.33	−674.29
Ethanol $C_2H_5OH(l)$	−276.98	160.67	−174.14	−180.92
Formaldehyde $CH_2O(g)$	−115.90	218.78	−109.91	−130.5
Formamide $HCONH_2(g)$	−186.2	248.45	−141.04	—
Formic acid $HCOOH(l)$	−424.76	128.95	−361.46	—
Fumarate$^-$(aq)	—	—	—	−604.21
Fumaric acid $trans$-(=CHCOOH)$_2$(s)	−811.07	166.1	−653.67	−646.05
α-D-Galactose $C_6H_{12}O_6(s)$	−1285.37	205.4	−919.43	−924.58
α-D-Glucose $C_6H_{12}O_6(s)$	−1274.4	212.1	−910.52	−917.47
L-Glutamate^{+--}(aq)	—	—	—	−694.00
L-Glutamic acid $C_5H_9NO_4(s)$	−1009.68	118.20	−731.28	−722.70
Glycerol $HOCH_2CHOHCH_2OH(l)$	−668.6	204.47	−477.06	−488.52
Glycine $H_2CNH_2COOH(s)$	−537.2	103.51	−377.69	−379.9
Glycylglycine $C_4H_8N_2O_3(s)$	−745.25	189.95	−490.57	—

TABLE A.7   Organic Compounds* (cont.)

	$\Delta H_f^0 \equiv \overline{H}^0$ (kJ mol^{-1})	$\overline{S}^0$ (J K^{-1} mol^{-1})	$\Delta G_f^0 \equiv \overline{G}^0$ (kJ mol^{-1})	$\Delta G_f^0 \equiv \overline{G}^0$ (1 M activity, aq) (kJ mol^{-1})
Guanine $C_5H_5N_5O(s)$	−183.93	160.2	47.40	—
L-Isoleucine $C_6H_{13}NO_2(s)$	−638.1	207.99	−347.15	—
Lactate$^-$(aq)	—	—	—	−517.812
L-Lactic acid $CH_3CHOHCOOH(s)$	−694.08	142.26	−522.92	—
β-Lactose $C_{12}H_{22}O_{11}(s)$	−2236.72	386.2	−1566.99	−1569.92
L-Leucine $C_6H_{13}NO_2(s)$	−646.8	211.79	−357.06	−353.09
Maleic acid cis-(=CHCOOH)$_2$(s)	−790.61	159.4	−631.20	—
Methanol $CH_3OH(l)$	−238.57	126.8	−166.23	−175.23
L-Methionine $C_5H_{11}NO_2S(s)$	−758.6	231.08	−505.76	—
Oxalic acid (—COOH)$_2$(s)	−829.94	120.08	−701.15	—
Oxaloacetate^{--}(aq) $C_4H_2O_5$	—	—	—	−797.18
L-Phenylalanine $C_9H_{11}NO_2(s)$	−466.9	213.63	−211.59	—
Pyruvate$^-$(aq)	—	—	—	−474.33
Pyruvic acid $CH_3COCOOH(l)$	−584.5	179.5	−463.38	—
L-Serine $HOCH_2CHNH_2COOH(s)$	−726.3	149.16	−509.19	—
Succinate^{2-}(aq)	—	—	—	−690.23
Succinic acid (—CH$_2$COOH)$_2$(s)	−940.90	175.7	−747.43	−746.22
Sucrose $C_{12}H_{22}O_{11}(s)$	−2222.1	360.2	−1544.65	−1551.76
L-Tryptophan $C_{11}H_{12}N_2O(s)$	−415.0	251.04	−119.41	—
L-Tyrosine $C_9H_{11}NO_3(s)$	−671.5	214.01	−385.68	−370.83
Urea $NH_2CONH_2(s)$	−333.17	104.60	−197.15	−203.84
L-Valine $C_5H_{11}NO_2(s)$	−617.98	178.86	−358.99	—

* Standard thermodynamic values at 25°C (298 K) and 1 atm pressure. Values for ions refer to an aqueous solution at unit activity on the molarity scale. Standard enthalpy of formation, ΔH_f^0 third-law entropies, S^0, and standard Gibbs free energy of formation, ΔG_f^0 are given.
Sources: Data from D. R. Stull, E. F. Westrum, Jr., and G. C. Sinke, *The Chemical Thermodynamics of Organic Compounds,* John Wiley, New York (1969) and from J. T. Edsall and J. Wyman, Biophysical Chemistry, Vol. 1, Academic Press, New York (1958).

TABLE A.8   Atomic Weights of the Elements*

Element	Symbol	Atomic number	Atomic weight	Element	Symbol	Atomic number	Atomic weight
Actinium	Ac	89	(227)	Lutetium	Lu	71	174.97
Aluminum	Al	13	26.98	Magnesium	Mg	12	24.312
Americium	Am	95	(243)	Manganese	Mn	25	54.94
Antimony	Sb	51	121.75	Meitnerium	Mt	109	(266)
Argon	Ar	18	39.948	Mendelevium	Md	101	(258)
Arsenic	As	33	74.92	Mercury	Hg	80	200.59
Astatine	At	85	(210)	Molybdenum	Mo	42	95.94
Barium	Ba	56	137.34	Neodymium	Nd	60	144.24
Berkelium	Bk	97	(247)	Neon	Ne	10	20.183
Beryllium	Be	4	9.012	Neptunium	Np	93	237.048
Bismuth	Bi	83	208.98	Nickel	Ni	28	58.71
Bohrium	Bh	107	(264)	Niobium	Nb	41	92.91
Boron	B	5	10.81	Nitrogen	N	7	14.007
Bromine	Br	35	79.909	Nobelium	No	102	(259)
Cadmium	Cd	48	112.40	Osmium	Os	76	190.2
Calcium	Ca	20	40.08	Oxygen	O	8	15.9994
Californium	Cf	98	(251)	Palladium	Pd	46	106.4
Carbon	C	6	12.011	Phosphorus	P	15	30.974
Cerium	Ce	58	140.12	Platinum	Pt	78	195.09
Cesium	Cs	55	132.91	Plutonium	Pu	94	(244)
Chlorine	Cl	17	35.453	Polonium	Po	84	(209)
Chromium	Cr	24	52.00	Potassium	K	19	39.102
Cobalt	Co	27	58.93	Praseodymium	Pr	59	140.91
Copper	Cu	29	63.54	Promethium	Pm	61	(145)
Curium	Cm	96	(247)	Protactinium	Pa	91	231.036
Dubnium	Db	105	(262)	Radium	Ra	88	226.025
Dysprosium	Dy	66	162.50	Radon	Rn	86	(222)
Einsteinium	Es	99	(252)	Rhenium	Re	75	186.23
Erbium	Er	68	167.26	Rhodium	Rh	45	102.91
Europium	Eu	63	151.96	Rubidium	Rb	37	85.47
Fermium	Fm	100	(257)	Ruthenium	Ru	44	101.1
Fluorine	F	9	19.00	Rutherfordium	Rf	104	(261)
Francium	Fr	87	(223)	Samarium	Sm	62	150.35
Gadolinium	Gd	64	157.25	Scandium	Sc	21	44.96
Gallium	Ga	31	69.72	Seaborgium	Sg	106	(266)
Germanium	Ge	32	72.59	Selenium	Se	34	78.96
Gold	Au	79	196.97	Silicon	Si	14	28.09
Hafnium	Hf	72	178.49	Silver	Ag	47	107.870
Hassium	Hs	108	(269)	Sodium	Na	11	22.9898
Helium	He	2	4.003	Strontium	Sr	38	87.62
Holmium	Ho	67	164.93	Sulfur	S	16	32.064
Hydrogen	H	1	1.0080	Tantalum	Ta	73	180.95
Indium	In	49	114.82	Technetium	Tc	43	(98)
Iodine	I	53	126.90	Tellurium	Te	52	127.60
Iridium	Ir	77	192.2	Terbium	Tb	65	158.92
Iron	Fe	26	55.85	Thallium	Tl	81	204.37
Krypton	Kr	36	83.80	Thorium	Th	90	232.04
Lanthanum	La	57	138.91	Thulium	Tm	69	168.93
Lawrencium	Lw	103	(262)	Tin	Sn	50	118.69
Lead	Pb	82	207.19	Titanium	Ti	22	47.90
Lithium	Li	3	6.939	Tungsten	W	74	183.85

TABLE A.8   Atomic weights of the elements* *(cont.)*

Element	Symbol	Atomic number	Atomic weight	Element	Symbol	Atomic number	Atomic weight
Uranium	U	92	238.03	Yttrium	Y	39	88.91
Vanadium	V	23	50.94	Zinc	Zn	30	65.37
Xenon	Xe	54	131.30	Zirconium	Zr	40	91.22
Ytterbium	Yb	70	173.04				

* Based on mass of ^{12}C at 12.000 . . . . The ratio of these weights to those on the older chemical scale (in which oxygen of natural isotopic composition was assigned a mass of 16.000 . . .) is 1.000050. (Values in parentheses represent the most stable known isotopes.) Data for trans-actinide elements taken from D.C. Hoffman and D.M. Lee, J. Chem. Ed. 76, 331–347 (1999).

TABLE A.9   Biochemical Compounds

Amino acids found in proteins:

R **groups:**

Glycine	H—
Alanine	CH_3—
Valine	
Leucine	
Isoleucine	
Phenylalanine	
Tyrosine	
Tryptophan	
Threonine	
Methionine	$CH_3SCH_2CH_2$—
Cysteine	$HSCH_2$—
Proline (amino acid)	

TABLE A.9 Biochemical Compounds *(cont.)*

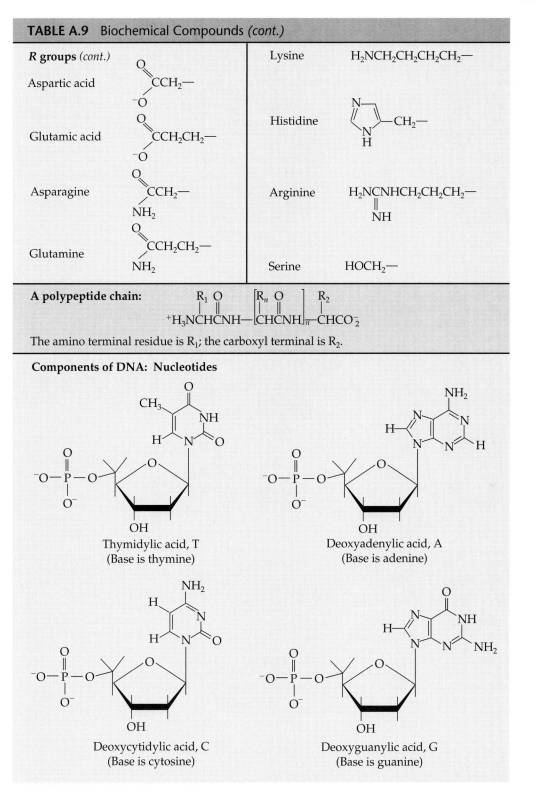

R groups *(cont.)*

Aspartic acid

Glutamic acid

Asparagine

Glutamine

Lysine $H_2NCH_2CH_2CH_2CH_2-$

Histidine

Arginine $H_2NCNHCH_2CH_2CH_2-$
                           $\parallel$
                           NH

Serine $HOCH_2-$

A polypeptide chain:

$$^+H_3NCHCNH-[CHCNH]_n-CHCO_2^-$$

with $R_1\ O$, $[R_n\ O]$, R_2

The amino terminal residue is R_1; the carboxyl terminal is R_2.

Components of DNA: Nucleotides

Thymidylic acid, T
(Base is thymine)

Deoxyadenylic acid, A
(Base is adenine)

Deoxycytidylic acid, C
(Base is cytosine)

Deoxyguanylic acid, G
(Base is guanine)

TABLE A.9   Biochemical Compounds *(cont.)*

Components of RNA: Nucleotides *(cont.)*

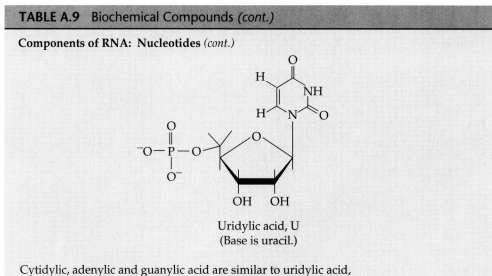

Uridylic acid, U
(Base is uracil.)

Cytidylic, adenylic and guanylic acid are similar to uridylic acid,
but with cytosine, adenine, and guanine replacing uracil.

A polynucleotide chain, a single strand of DNA. (A single strand of RNA would have a
hydroxyl group at each 2′ position.)

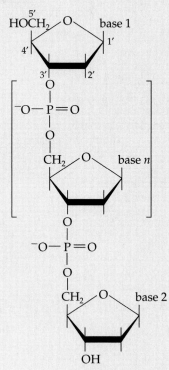

Base 1 is at the 5′ terminal end of the chain; base 2 is at the 3′ terminal end.

TABLE A.9 Biochemical Compounds *(cont.)*

Miscellaneous:

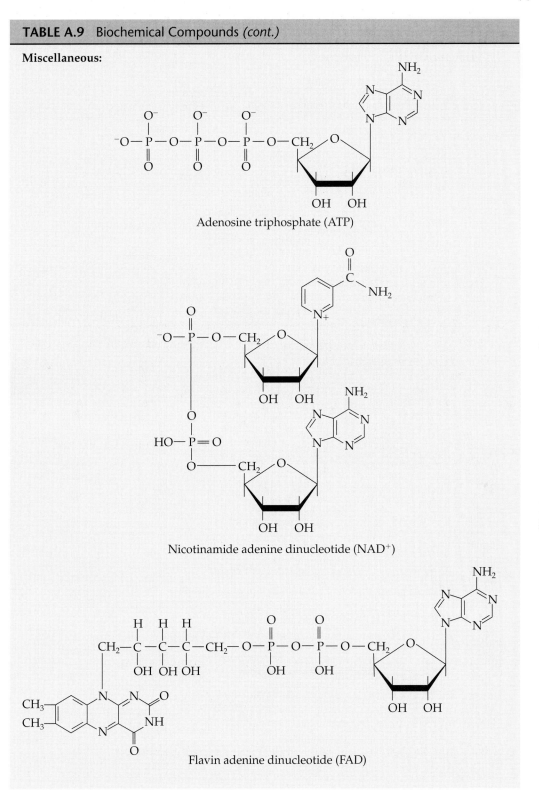

Adenosine triphosphate (ATP)

Nicotinamide adenine dinucleotide (NAD$^+$)

Flavin adenine dinucleotide (FAD)

TABLE A.9 Biochemical Compounds *(cont.)*

Miscellaneous *(cont.)*

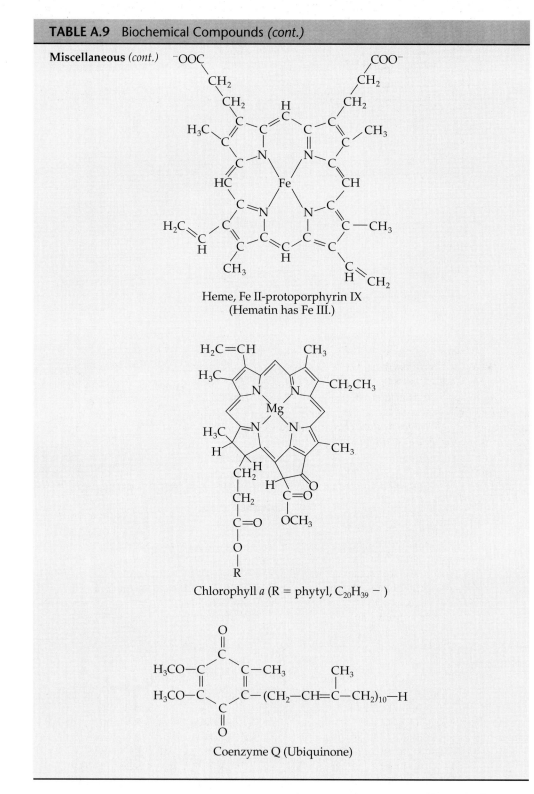

Heme, Fe II-protoporphyrin IX
(Hematin has Fe III.)

Chlorophyll *a* (R = phytyl, $C_{20}H_{39}-$)

Coenzyme Q (Ubiquinone)

ANSWERS TO SELECTED PROBLEMS

Chapter 2

2. (a) 6.8×10^6 L; (c) 1.3 yr. **3.** (d) -203 J **6.** (a) 41.9 kJ; (b) -33.3 kJ **7.** (a) -2.49 kJ
10. (a) $q = 0$, $w < 0$, $\Delta E < 0$, $\Delta H < 0$ **11.** (a) $q_p = \Delta H = 40.66$ kJ **12.** (a) 761 J
15. (b) $q = 0$, $w > 0$, $\Delta E > 0$, $\Delta H > 0$ **17.** 1 degree per hour. About 3 hours. **20.** 42°C
22. (a) $q = -66.6$ kJ mol^{-1}; (b) 2.4% **26.** (a) -66.6 kJ mol^{-1}; (b) -1052 kJ mol^{-1};
(c) -2802 kJ mol^{-1} **29.** (a) -141.8 kJ g^{-1}; (b) -48.3 kJ g^{-1}; (c) on a weight basis, $H_2(g)$
produces 3 times as much heat on burning as does n-octane. **33.** (a) -217.28 kJ mol^{-1};
(b) 140.5 kJ mol^{-1}; (c) -189.3 kJ mol^{-1}; (d) 25.23°C

Chapter 3

1. (b) $\Delta E = -1.247$ kJ; $\Delta S = -0.217$ J K^{-1} **2.** (a) $q_1 = 133$ kJ; $q_3 = -33$ kJ; (b) $q_1 = -133$ kJ;
$q_3 = 33$ kJ; (c) $q_1' = 125$ kJ; $q_3' = -25$ kJ **5.** (a) 100°C; (b) 8.3×10^{-4} mol; (c) 4.83×10^{-7} kJ K^{-1}
8. (a) $\Delta H_{298}^0 = -141{,}780$ kJ; $\Delta S_{298}^0 = -80.94$ kJ K^{-1}; $\Delta G_{298}^0 = -117{,}650$ kJ; (b) $\Delta H_{298}^0 =$
$-48{,}254$ kJ; $\Delta S_{298}^0 = -6.087$ kJ K^{-1}; $\Delta G_{298}^0 = -46.440$ kJ **10.** (a) irreversibly; (c) enthalpy
change **14.** (a) -6.007 kJ mol^{-1}; (b) -6.007 kJ mol^{-1}; (c) -22.2 J K^{-1} mol^{-1}; (d) zero;
(e) -6.007 kJ mol^{-1}; heat is released; (f) -0.17 J mol^{-1}; work is done by the system
15. (a) decrease; (b) remain unchanged; (g) remain unchanged; (h) more negative than
17. (a) If more solvent is bound by the coil than the helix, the entropy and enthalpy can
decrease. Increasing the temperature will not favor the helix-coil transition. (b) Yes,
$\Delta G^0 = -0.26$ kJ (mol amide)$^{-1}$ (c) $T_m = 60$°C (d) No **24.** (a) temperature, internal energy;
(c) increases **27.** 2668 atm

Chapter 4

1. (a) $K = 77$, $\Delta G_{298}^{0\prime} = -10.8$ kJ mol^{-1}; (b) $\Delta G_{298}^{0\prime} = 20.2$ kJ mol^{-1}, $K = 2.9 \times 10^{-4}$
3. (c) 3.1×10^{-5} **5.** (a) 0.22 **8.** (a) -3.4 kJ mol^{-1}; (b) 0; (c) -3.4 kJ mol^{-1}; (d) -6.8 kJ mol^{-1};
(e) 52.9 kJ mol^{-1}; (f) 189 J K^{-1} mol^{-1} **10.** 5×10^{-49} atm. Do not worry. **13.** (b) $[H^+] = 1.0$
$\times 10^{-7} M$, $[OH^-] = 1.0 \times 10^{-7} M$, $[Na^+] = 0.138 M$, $[H_2PO_4^-] = 0.062 M$, $[HPO_4^{2-}] = 0.038 M$,
$[PO_4^{3-} = 1.7 \times 10^{-7} M$, $[H_3PO_4] = 8.7 \times 10^{-7} M$ **15.** (a) -46.25 kJ mol^{-1}; (b) -1.72 kJ mol^{-1};
(c) $[GTP] = 1.9 \times 10^{-8} M$, $[GDP] = 0.055 M$, $[P_i] = 0.065 M$ **18.** (a) $+0.363$ V
20. (c) -183.3 kJ mol^{-1} **21.** (a) 5×10^{-8}; (b) -0.439 V **25.** (a) -1.38 V; (b) 133 kJ mol^{-1};
(c) -13.5 kJ mol^{-1} **27.** (a) -215 kJ mol^{-1} **31.** (a) For the sequence 5′GGGCCC3′,
$\Delta G^0 = -32.0$ kJ mol^{-1}, $\Delta H^0 = -174.2$ kJ mol^{-1}, $\Delta S^0 = -460$ J K^{-1} mol^{-1}. For the sequence
5′GGTTCC3′ + 5′GGAACC3′, $\Delta G^0 = -22.9$ kJ mol^{-1}, $\Delta H^0 = -167.6$ kJ mol^{-1}, $\Delta S^0 =$
-470 J K^{-1} mol^{-1}. (b) For the sequence 5′GGGCCC3′, $T_m = 52$°C. For the sequence
5′GGTTCC3′ + 5′GGAACC3′, $T_m = 30$°C.

Chapter 5

1. (a) 7.99 kg m^{-2}; (b) 1.57×10^5 L; (c) 5.28 kg rain, 12,920 kJ, therefore temperature rises; (d) 15.9 atm   **5.** (a) 45.9 kJ mol^{-1}; (b) 22.07 kJ mol^{-1}; not enough; (c) 2.1   **7.** $n = 5$, $K = 1.1 \times 10^5 \, M^{-1}$ **13.** (a) $-76°C$; (b) 97.2 J K^{-1} mol^{-1}. Hydrogen bonding and association in liquid decreases its entropy. (c) 128.1 kJ mol^{-1}. Pure liquid is standard state. (d) NH_4Cl dissociates completely in NH_3. Boiling point corresponds to a 2 molal solution.
16. (a) true; (c) false; (e) false   **18.** (a) 17.52 Torr; (b) 1015 Torr; (c) 0.999; (d) 1548 Torr
21. (a) -2.79 kJ; (b) -22.2 kJ; (c) -7.0 kJ; (d) Yes, the free energy of transfer of the side chain of valine to the interior is negative; (e) $K(\mathrm{val})/K(\mathrm{gly}) = 16.9$ **25.** 0.84 atm   **26.** (a) -0.062 V; (b) 31.2 kJ mol^{-1} **32.** (c) 271 K

Chapter 6

1. (a) 1.84×10^5 cm s^{-1}; (b) 3.4 kJ mol^{-1}; (c) 2.7×10^{19} molecule cm^{-3}; (d) 1.34×10^{-5} cm; (e) 1.26×10^{10} s^{-1}; (f) 1.70×10^{29} cm^{-3} s^{-1} **2.** (a) 0 K; (b) ∞ K; (c) 320 K
3. (b) 6.7×10^{-8} g s^{-1} **6.** (a) 5.05×10^7 g mol^{-1}; (b) 2.58×10^7 g mol^{-1}
7. $\dfrac{s_2}{s_1} = \left(\dfrac{M_2}{M_1}\right)^{2/3}; \dfrac{D_2}{D_1} = \left(\dfrac{M_1}{M_2}\right)^{1/3}; \dfrac{[\eta_2]}{[\eta_1]} = 1$
10. (a) 271×10^3 g mol^{-1} (c) 17.4×10^3 g mol^{-1} **13.** (a) 7.3×10^{-8} s   **17.** (a) 68,400
18. ring/linear = 0.328   **23.** (a) 62.2 Å; (b) 60.1 Å; (c) Either shape is not spherical, or hydration changes on ligation.   **25.** 2.02 cm^3 g^{-1}

Chapter 7

1. (a) I_2 is zero order; ketone is first order; H^+ is first order; (b) 0.034 M^{-1} s^{-1} **5.** (e) A + A $\to$ A$_2$ (slow), A$_2$ + B $\to$ C + D + A (fast)   **7.** (a) $d(D)/dt = k_1 k_3 (A)(C)/[k_2 + k_3(C)]$
10. (a) 1.0%; (b) 5.3%   **13.** (a) A 29-year-old wine usually commands a premium price.
16. (a) $\dfrac{d[U]}{dt} = k_0 - k_1[U]$; (b) $\dfrac{d[U]}{dt} = k_0 e^{k_1 t}$ from derivative of expression for [U] given in (b); substitute expression for [U] given in (b) into answer for (a) to show derivatives are equal; (c) Max. occurs at $t \to \infty$; $[U]_{max} = k_0/k_1 = 2.60 \, \mu M$; (d) 1260 s; (e) 3817 s; (f) 0.385 nM s^{-1}.
20. 52.9 kJ   **22.** (a) $\ln[(A)/(A)_0] = -k_1(B)t$, first order in (A) and in (B); (b) $(A)_0 - (A) = k_2(B)t$, zero order in (A) and first order in (B).   **24.** (a) second order; (b) A + A $\to$ P; (c) 5 M^{-1} min^{-1}; (d) 67 min   **28.** (c) 26.0°C   **32.** (b) $k_1 = 8.5 \times 10^8 \, M^{-1}$ s^{-1}; $k_{-1} = 2.0 \times 10^{-6}$ s^{-1} **34.** (a) (I) $d(O_2)/dt = 3k(O_3)^2$, (II) $d(O_2)/dt = 2(k_2 k_1/k_{-1})(O_3)^2/(O_2)$
40. (a) 4.8 hr   **42.** (c) 4 mW   **45.** 1.4×10^{17}

Chapter 8

1. $K_M = 0.42 \, M$, V = 0.35 μmol CO$_2$ min^{-1} **2.** (c) $K'_{eq} = 3.75$ at pH 7.1, $K_{eq} = 3.0 \times 10^{-7}$
5. (a) $10^{-3} \, M$ s^{-1}; (b) 24 kJ mol^{-1}; (c) $10^4 \, M^{-1}$; (d) -14 kJ mol^{-1}
8. $\dfrac{d(P)}{dt} = \dfrac{k_2 (E)_0}{1 + \dfrac{K_M}{(S)}\left(1 + \dfrac{(I)}{K_I}\right)}$, where $K_I = \dfrac{k_{-3}}{k_3}$ **11.** (a) 44 mM; (b) noncompetitive
15. (b) $\dfrac{1}{V_{max}}$ and $\dfrac{1}{K_M} = 0$; (c) $k_2 \gg (k_1 + k_{-1})$; (d) first order; (e) $1.25 \times 10^9 \, M^{-1}$ s^{-1}; (g) 2
18. (b) 0.025 M, 6 hr; (c) 0.050 M, 6 hr, 12 hr; (d) 12 hr; (e) 10 mL hr^{-1}
20. (b) $K_M(C_2H_5OH) = 0.44$ mM, $K_M(NAD^+) = 18 \, \mu M$; (c) competitive; (d) noncompetitive   **23.** 2.61 mM **26.** (a) 2.3 ms; (b) 0.9 s; (c) 4.6 ms, 0.9 s

Chapter 9

1. (a) $\lambda_0(K) = 564$ nm; (c) 1.44×10^{-19} J   **4.** (a) $E = 6.02 \times 10^{-20}$ J, $\lambda = 1100$ nm; (c) $E = 12.04 \times 10^{-20}$ J, $\lambda = 1100$ nm
6. (b) $\dfrac{5h^2}{4\pi^2 ma^2}$ **7.** (d) 660 nm   **10.** (c) $E_0 = 4.12 \times 10^{-20}$ J, $E_1 = 12.36 \times 10^{-20}$ J, etc.

12. (a) 1597 N m^{-1}; (b) 1870 cm^{-1}; (c) 1390 cm^{-1}; (d) 1148 N m^{-1};
(e) ν(c) = 1343 cm^{-1}, ν(d) = 1586 cm^{-1}
15. (b) Bond energies: NO$^-$ < NO < NO$^+$
     Bond lengths: NO$^+$ < NO < NO$^-$
21. (a) Ground state: $\mu_\parallel$ = 1.95 D, $\mu_\perp$ = 1.70 D
     Excited state: $\mu_\parallel$ = 0.28 D, $\mu_\perp$ = 0.41 D

Chapter 10

1. (a) 2.91×10^{-19} J photon^{-1} = 1.82 eV = 176 kJ einstein^{-1}; (b) 2.76; (c) 30–35%
4. (a) 0.01 M, (b) 0.02 M; (c) 9.8×10^{-4} cm
6. c(Chl a, μg ml^{-1}) = 12.80 A_{663} − 2.585 A_{645}
  c(Chl b, μg ml^{-1}) = 22.88 A_{645} − 4.67 A_{663}
  c(Chl total, μg ml^{-1}) = 8.13 A_{663} + 20.30 A_{645}
8. (a) 4.0; (b) 0.400 **10.** (a) 6.7 ns; (b) 1.05×10^8 s^{-1} **15.** (a) −0.50 M^{-1} cm^{-1};
(b) (DNA) = 0.109×10^{-4} M, (RNA) = 0.52×10^{-4} M; (c) decrease; (d) increase
18. (a) $[\alpha]$ = 2320 deg dm^{-1} cm^3 g^{-1}; $[\theta]$ = 9894 deg M^{-1} cm^{-1}; (b) A = 0.80,
$A_L - A_R = 3 \times 10^{-4}$, α = 0.0075 deg; (c) $\alpha = 1.31 \times 10^{-4}$ radians cm^{-1}, $n_L - n_R = 1.16 \times 10^{-9}$,
$\psi = 1.73 \times 10^{-4}$ radians cm^{-1}
23. (a) CH$_3$CHO; (b)

30. (c) α-helix **33.** (a) Peak ratio $\cong$ 3/1; $\Delta G^0 = -2.72$ kJ mol^{-1}

Chapter 11

2. (d) $\dfrac{\tau^3 s^2}{3(1 + \tau)}$ **5.** (b) $Z = 1 + 4S + 4\tau S^2 + 2S^2 + 4\tau^2 S^3 + \tau^4 S^4$

7. (c) $A = \left(\dfrac{c}{Z}\right)[\varepsilon_0 + \varepsilon_1 s(\sigma_7 + \sigma_9 + \sigma_{11} + \sigma_{13} + \sigma_{15} + \sigma_{17}) + \varepsilon_2 s^2(\sigma_7 + \sigma_9 + \sigma_{11} + \sigma_{13} + \sigma_{15})$

$+ \varepsilon_3 s^3(\sigma_7 + \sigma_9 + \sigma_{11} + \sigma_{13}) + \varepsilon_4 s^4(\sigma_7 + \sigma_9 + \sigma_{11}) + \varepsilon_5 s^5(\sigma_7 + \sigma_9) + \varepsilon_6 s^6 \sigma_7]$

Z = term in brackets with at ε_i set equal to 1. **10.** (a) 26^3 = 17,576
11. (a) Electron: N_2 / N_1 = 0, He: 3.0

Chapter 12

1. (a) 6.4×10^{-15} J; (b) 0.31 Å **3.** (a) 67.4° and 39.6°, respectively; (b) 2.38 cm **7.** 0.77 Å
8. (a) 6.025×10^{23}; (b) 14,300 **10.** (a) 1:1:1 **13.** 1.45 Å
16. (a) λ, 2λ, 3λ; (b) $\dfrac{\lambda}{2}, \dfrac{3\lambda}{2}, \dfrac{5\lambda}{2}$; (c) 8.83 Å; (d) 5.0 Å; (e) $\dfrac{\lambda}{2}$

INDEX

728

Useful Physical Constants

		SI units	cgs-esu units
Gas constant	R	$8.3145 \text{ J K}^{-1} \text{ mol}^{-1}$	$8.3145 \times 10^7 \text{ erg deg}^{-1} \text{ mol}^{-1}$
			$1.987 \text{ cal deg}^{-1} \text{ mol}^{-1}$
			$0.08205 \text{ L atm deg}^{-1} \text{ mol}^{-1}$
Avogadro's number	N_0	$6.0221 \times 10^{23} \text{ mol}^{-1}$	$6.0221 \times 10^{23} \text{ molecules mol}^{-1}$
Boltzmann constant	k	$1.3807 \times 10^{-23} \text{ J K}^{-1}$	$1.3807 \times 10^{-16} \text{ erg deg}^{-1}$
Faraday constant	F	$9.6485 \times 10^4 \text{ C mol}^{-1}$	$9.6485 \times 10^4 \text{ C mol}^{-1}$
Speed of light	c	$2.9979 \times 10^8 \text{ m s}^{-1}$	$2.9979 \times 10^{10} \text{ cm s}^{-1}$
Planck constant	h	$6.6261 \times 10^{-34} \text{ J s}$	$6.6261 \times 10^{-27} \text{ erg s}$
Elementary charge	e	$1.6022 \times 10^{-19} \text{ C}$	$4.8030 \times 10^{-10} \text{ esu}$
Electron mass	m_e	$9.1094 \times 10^{-31} \text{ kg}$	$9.1094 \times 10^{-28} \text{ g}$
Proton mass	m_p	$1.6726 \times 10^{-27} \text{ kg}$	$1.6726 \times 10^{-24} \text{ g}$
Standard gravity	g	9.8066 m s^{-2}	980.66 cm s^{-2}
Permittivity of vacuum	ε_0	$8.8542 \times 10^{-12} \text{ C}^2 \text{ N}^{-1} \text{ m}^{-2}$	

C	= coulomb	L	= liter
g	= gram	m	= meter
J	= joule	N	= newton
K	= Kelvin	s	= second
kg	= kilogram		